Linear Control Systems Engineering

McGRAW-HILL SERIES IN MECHANICAL ENGINEERING

CONSULTING EDITORS

Jack P. Holman, *Southern Methodist University*
John R. Lloyd, *Michigan State University*

Anderson: Modern Compressible Flow: With Historical Perspective
Arora: Introduction to Optimum Design
Bray and Stanley: Nondestructive Evaluation: A Tool for Design, Manufacturing, and Service
Burton: Introduction to Dynamic Systems Analysis
Culp: Principles of Energy Conversion
Dally: Packaging of Electronic Systems: A Mechanical Engineering Approach
Dieter: Engineering Design: A Materials and Processing Approach
Driels: Linear Control Systems Engineering
Eckert and Drake: Analysis of Heat and Mass Transfer
Edwards and McKee: Fundamentals of Mechanical Component Design
Gebhart: Heat Conduction and Mass Diffusion
Gibson: Principles of Composite Material Mechanics
Hamrock: Fundamentals of Fluid Film Lubrication
Heywood: Internal Combustion Engine Fundamentals
Hinze: Turbulence
Holman: Experimental Methods for Engineers
Howell and Buckius: Fundamentals of Engineering Thermodynamics
Hutton: Applied Mechanical Vibrations
Juvinall: Engineering Considerations of Stress, Strain, and Strength
Kane and Levinson: Dynamics: Theory and Applications
Kays and Crawford: Convective Heat and Mass Transfer
Kelly: Fundamentals of Mechanical Vibrations
Kimbrell: Kinematics Analysis and Synthesis
Krieder and Rabl: Heating and Cooling of Buildings
Martin: Kinematics and Dynamics of Machines
Modest: Radiative Heat Transfer
Norton: Design of Machinery
Phelan: Fundamentals of Mechanical Design
Raven: Automatic Control Engineering
Reddy: An Introduction to the Finite Element Method
Rosenberg and Karnopp: Introduction to Physical Systems Dynamics
Schlichting: Boundary-Layer Theory
Shames: Mechanics of Fluids
Sherman: Viscous Flow
Shigley: Kinematic Analysis of Mechanisms
Shigley and Mischke: Mechanical Engineering Design
Shigley and Uicker: Theory of Machines and Mechanisms
Stiffler: Design with Microprocessors for Mechanical Engineers
Stoecker and Jones: Refrigeration and Air Conditioning
Ullman: The Mechanical Design Process
Vanderplaats: Numerical Optimization: Techniques for Engineering Design, with Applications
White: Viscous Fluid Flow
Zeid: CAD/CAM Theory and Practice

Also available from McGraw-Hill

SCHAUM'S OUTLINE SERIES IN MECHANICAL ENGINEERING

Most outlines include basic theory, definitions and hundreds of example problems solved in step-by-step detail, and supplementary problems with answers.

Related titles on the current list include:

Acoustics
Continuum Mechanics
Elementary Statics & Strength of Materials
Engineering Mechanics
Fluid Dynamics
Fluid Mechanics & Hydraulics
Heat Transfer
Lagrangian Dynamics
Machine Design
Mathematical Handbook of Formulas & Tables
Mechanical Vibrations
Statics & Mechanics of Materials
Strength of Materials
Theoretical Mechanics
Thermodynamics for Engineers
Thermodynamics with Chemical Applications

SCHAUM'S SOLVED PROBLEMS BOOKS

Each title in this series is a complete and expert source of solved problems with solutions worked out in step-by-step detail.

Related titles on the current list include:

3000 Solved Problems in Calculus
2500 Solved Problems in Differential Equations
2500 Solved Problems in Fluid Mechanics & Hydraulics
1000 Solved Problems in Heat Transfer
3000 Solved Problems in Linear Algebra
2000 Solved Problems in Mechanical Engineering Thermodynamics
2000 Solved Problems in Numerical Analysis
700 Solved Problems in Vector Mechanics for Engineers: Dynamics
800 Solved Problems in Vector Mechanics for Engineers: Statics

Available at most college bookstores, or for a complete list of titles and prices, write to: Schaum Division
McGraw-Hill, Inc.
1221 Avenue of the Americas
New York, NY 10020

Linear Control Systems Engineering

MORRIS DRIELS

U. S. Naval Postgraduate School, Monterey, California

McGRAW-HILL, INC.

New York St. Louis San Francisco Auckland Bogotá Caracas Lisbon
London Madrid Mexico City Milan Montreal New Delhi
San Juan Singapore Sydney Tokyo Toronto

LINEAR CONTROL SYSTEMS ENGINEERING

This book is printed on acid-free paper.

2 3 4 5 6 7 8 9 0 DOC DOC 9 0 9 8 7 6 5

ISBN 0-07-017824-0

This book was set in Times Roman by Beacon Graphics, Corporation.
The editors were John J. Corrigan and Eleanor Castellano;
the designer was Joan Greenfield;
the production supervisor was Denise L. Puryear.
R. R. Donnelley & Sons Company was printer and binder.

Cover photo courtesy of NASA.

Library of Congress Cataloging-in-Publication Data

Driels, Morris
Linear control systems engineering / Morris Driels.
p. cm.—(McGraw-Hill series in mechanical engineering)
Includes index.
ISBN 0-07-017824-0
1. Linear control systems. I. Title. II. Series.
TJ220.D75 1996
629.8′32—dc20 94-11331

About the Author

MORRIS DRIELS received his B.S. in Mechanical Engineering from the University of Surrey, and Ph.D. from City University, London. After working in the aerospace industry for some time, he became a Lecturer at Edinburgh University, Scotland. Moving to the United States in 1982, he held positions at the University of Rhode Island and Texas A&M. He is currently Professor in the Mechanical Engineering Department at the Naval Postgraduate School in Monterey, California. Dr. Driels is a Member of ASME, IEEE and a Fellow of The Institution of Mechanical Engineers.

To my wife Jenny, son Christopher, and daughters Joanne and Fiona.

Contents

Preface xvii

MODULE 1 INTRODUCTION TO FEEDBACK CONTROL 1

MODULE 2 TRANSFER FUNCTIONS AND BLOCK DIAGRAM ALGEBRA 22

Transfer Functions 22

Block Diagram Algebra 23

MODULE 3 FIRST-ORDER SYSTEMS 37

Impulse Response 39

Step Response 40

Ramp Response 41

Harmonic Response 41

First-Order Feedback Systems 43

Complex-Plane Representation: Poles and Zeros 45

Poles and Zeros of First-Order Systems 46

Dominant Poles 47

MODULE 4 SECOND-ORDER SYSTEMS 57

Second-Order Electrical System 63

Step Response 64

MODULE 5 SECOND-ORDER SYSTEM TIME-DOMAIN RESPONSE 75

Ramp Response 75

Harmonic Response 76

Relationship between System Poles and Transient Response 78

Time-Domain Performance Specifications 81

MODULE 6 SECOND-ORDER SYSTEMS: DISTURBANCE REJECTION AND RATE FEEDBACK 93

Open- and Closed-Loop Disturbance Rejection 96

Effect of Velocity Feedback 99

MODULE 7 HIGHER-ORDER SYSTEMS 111

Reduction to Lower-Order Systems 111

Third-Order Systems 112

Effect of a Closed-Loop Zero 114

Occurrence of Closed-Loop Zeros 117

MODULE 8 SYSTEM TYPE: STEADY-STATE ERRORS 125

Impulse Input 127

Step Input 128

Ramp Input 129

Acceleration Input 130

Non-Unity-Feedback Control Systems 132

MODULE 9 ROUTH'S METHOD, ROOT LOCUS: MAGNITUDE AND PHASE EQUATIONS 145

Routh's Stability Criterion 145

Root Locus Method: Magnitude and Phase Equations 148

MODULE 10 RULES FOR PLOTTING THE ROOT LOCUS 173

MODULE 11 SYSTEM DESIGN USING THE ROOT LOCUS 199

MultiLoop System 199

System Design in the Complex Plane 202

Performance Requirements as Complex-Plane Constraints 203

Steady-State Error 204

Desirable Areas of Complex Plane for "Good" Response 205

MODULE 12 FREQUENCY RESPONSE AND NYQUIST DIAGRAMS 223
Frequency Response 224
Nyquist Diagrams from Transfer Functions 225

MODULE 13 NYQUIST STABILITY CRITERION 241
Conformal Mapping: Cauchy's Theorem 241
Application to Stability 245
Some Comments on Nyquist Stability 252
Alternative Approach to Nyquist Stability Criterion 254

MODULE 14 NYQUIST ANALYSIS AND RELATIVE STABILITY 272
Conditional Stability 272
Gain and Phase Margins 274

MODULE 15 BODE DIAGRAMS 289
Bode Diagrams of Simple Transfer Functions 289
Bode Diagrams of Compound Transfer Functions 293
Elemental Bode Diagrams 297

MODULE 16 BODE ANALYSIS, STABILITY, AND GAIN AND PHASE MARGINS 319
Conditional Stability 319
Gain and Phase Margins in the Bode Diagram 321
System Type and Steady-State Error from Bode Diagrams 323
Further Discussion of Gain and Phase Margins 326

MODULE 17 TIME RESPONSE FROM FREQUENCY RESPONSE 341
Bode Diagram from the Root Locus 341
Closed-Loop Time Response from Open-Loop Phase Margin 344
Time Response of Higher-Order Systems 346

MODULE 18 FREQUENCY-DOMAIN SPECIFICATIONS AND CLOSED-LOOP FREQUENCY RESPONSE 361
Frequency-Domain Specifications 361
Closed-Loop Frequency Response from Nyquist Diagram 365
Closed-Loop Frequency Response from Bode Diagram 371
Gain for a Desired M_p from the Nyquist Diagram 374
Gain For a Desired M_p from the Nichols Chart 377
Non-Unity-Feedback Gain Systems 377

MODULE 19 PHASE LEAD COMPENSATION 396

Multiple-Design Constraints 396

Transfer Function of Phase Lead Element 399

Phase Lead Compensation Process 402

Comments on the Applicability and Results of Phase Lead Compensation 409

MODULE 20 PHASE LAG AND LEAD-LAG COMPENSATION 431

Transfer Function of Phase Lag Element 431

Phase Lag Compensation Process 433

Comments on Phase Lag Compensation 435

Lead-Lag Compensation 436

Transfer Function of a Lead-Lag Element 438

Lead-Lag Compensation Process 440

MODULE 21 MULTIMODE CONTROLLERS 463

Proportional Control 464

Proportional-Plus-Integral Control 466

Proportional-Plus-Derivative Control 468

Proportional-Plus-Integral-Plus-Derivative Control 471

MODULE 22 STATE-SPACE SYSTEM DESCRIPTIONS 487

State-Space Form Equations from Transfer Functions 492

Transfer Function from State-Space Form 495

Transformation of State Variable and Invariability of System Eigenvectors 496

Canonical Forms and Decoupled Systems 497

Relationship between Eigenvalues and System Poles 500

MODULE 23 STATE-SPACE SYSTEM RESPONSE, CONTROLLABILITY, AND OBSERVABILITY 515

Direct Numerical Solution of the State Equation 515

Solution Using State Transition Matrix 516

Solution Using Laplace Transforms 518

System Stability 518

Controllability and Observability 519

MODULE 24 STATE-SPACE CONTROLLER DESIGN 531

Direct Calculation of Gains by Comparison with Characteristic Equation 532

Pole Placement via Control Canonical Form of State Equations 534

Pole Placement via Ackermann's Formula 539

MODULE 25 STATE-SPACE OBSERVER DESIGN 550

Observer Synthesis 550

Compensator Design 555

CONTROL SYSTEM DESIGN: CASE STUDIES

MODULE 26 WAVE ENERGY ABSORBTION DEVICE 569

Open loop frequency response, bandwidth, selection of feedback gains, closed loop frequency response, Nichols charts

MODULE 27 MISSILE ATTITUDE CONTROLLER 574

Model construction, block diagram representation, multimode controller design, root locus, state-space analysis and controller design, pole placement

MODULE 28 ROBOTIC HAND DESIGN 582

Multi-loop feedback systems, steady state values of force and position, control system synthesis, adaptive control

MODULE 29 PUMPED STORAGE FLOW CONTROL SYSTEM 589

Hydraulic system modeling, characteristic equation, P + I controller, state-space analysis, controllability, Ackermann's method

MODULE 30 SHIP STEERING CONTROL SYSTEM 597

Modeling, root locus, stabilization of unstable systems, performance constraints, iterative root locus, rate feedback

MODULE 31 CRUISE MISSILE ALTITUDE CONTROL SYSTEM 605

Design in frequency domain, signal and noise, design constraint boundaries, open loop design from closed loop requirements, lead-lag controller

MODULE 32 **MACHINE TOOL POWER DRIVE SYSTEM WITH FLEXIBILITY** 613
System modeling, P + D control, poor performance, state space model, pole placement, comparison of performance

APPENDIX 1 **REVIEW OF LAPLACE TRANSFORMS AND THEIR USE IN SOLVING DIFFERENTIAL EQUATIONS** 620
Linear Properties 620
Shifting Theorem 620
Time Differentials 621
Final-Value Theorem 622
Inverse Transforms 622
Solving Linear Differential Equations 622

Index 625

Preface

Although I have been teaching linear control systems engineering to mechanical engineering undergraduate students for the last twenty years or so, I was never motivated to write a book on the subject until I had the opportunity to teach controls together with a first course in dynamics in the same academic quarter. This enabled me to directly compare the structure, style, and student use of the required texts in each subject. Such a comparison proved quite enlightening.

Both books had been around for more than a decade and had been through several editions. The controls book was, at the time, the best seller, as was the dynamics book. Their styles, however, were very different. The dynamics book was written in the following format:

- The subject matter was grouped into discrete amounts of material that could be comfortably covered in one lecture.
- Following this, two or three worked problems showed the student how this material is used to solve engineering problems.
- Finally, several homework problems were provided to enable the student to test his or her knowledge of the material.

The controls book was written in a more traditional style comprising chapters of around fifty pages followed by twenty or so problems.

In teaching both courses, it was apparent that the students made more use of the dynamics book; they were not overwhelmed by the amount of material covered in a class, and the abundance of solved and homework problems ensured a self-assessment of their understanding of the material and gave students a better perspective of the structure of the subject. From a professor's point of view, having the material already divided up into lecture-size pieces made the job of planning the course program much easier.

In this book, I have attempted to use the same philosophy as the dynamics books I have just described. I hope that the modular nature of the material will enable the book to be closely allied to the course of lectures, although there is still sufficient flexibility to allow the instructor the option of including additional topics or skipping over material he or she thinks the student already knows. Based on the student reviews of controls courses I have taught, the consensus on problem solving seems to be (a) there can never be too many solved problems and

(b) more detailed solutions to solved problems are welcome. I have attempted to address both these issues in this book. In particular, detailed solved problems are included at the end of each module so that the student may see the applicability of the material just covered. In order to provide students with an understanding of how control system analysis provides a basic tool in the design of complex engineering systems, I have also added several design case studies after Module 25. In many cases, these examples show alternative methods to achieve the required performance, and provide the student with a perspective of how the various analytical topics presented in the book may be used, combined, and applied to real engineering systems.

Originally, this book was intended primarily for undergraduate mechanical engineering students, although other engineering disciplines should find the material not too far from their own area. In most of these areas, a traditional systems stream would comprise:

- An introductory systems modeling course—sophomore level
- A linear controls systems course—junior level
- An advanced controls course—senior level

This book is aimed at the junior-level course and assumes the student is already familiar with systems modeling. Some material in this area is included in this book, but only to provide a smooth transition into controls rather than to teach techniques for modeling physical systems.

With regard to the issue of provision of software for the book, the objective has been to emphasize the fundamentals of control and not to become focused on computational techniques or tools. The student is encouraged to use whatever software is available to him or her, and where appropriate, examples have been given using FORTRAN, BASIC code, as well as proprietary packages such as MATLAB. For some of the problems in the text, involving laborious, though not difficult, manipulations, commercial packages such as MATLAB or MATRIX_X are highly recommended.

Finally, I must thank a great number of people who have helped me write this book. The staff at McGraw-Hill have made the production process as near enjoyable as any author could reasonably expect. My students and colleagues at Edinburgh University, University of Rhode Island, Texas A&M, and the Naval Postgraduate School all deserve mention for their enthusiasm in reviewing material, discussing problems, and generally supporting my efforts over a considerable number of years. Special mention is due to Alan Linnett, Fotis Papoulias, and Tony Healey. The following reviewers provided many helpful comments and suggestions: Larry Banta, West Virginia University; Neyram Hemati, Drexel University; David Hullendar, University of Texas at Arlington; Leo LaFrance, New Mexico State University; Ronald A. Perez, University of Wisconsin, Milwaukee; and Gary Young, Oklahoma State University. All errors in the text are mine. Finally, the most thanks are due to my wife, Jenny, and children, Joanne, Chris, and Fiona. They supplied endless encouragement, help, motivation, and the ability to view all of life's pleasures and disappointments in the correct perspective.

Morris Driels

Linear Control Systems Engineering

MODULE 1

Introduction to Feedback Control

Feedback control systems are at work all around us. The study of control systems involves not so much the development of new engineering components or machines, but taking combinations of existing hardware to achieve a predetermined goal. A control *system* is the collection of components connected in such a way as to effect *control* over certain aspects of the domain in which the system operates. Control systems operate in almost every aspect of human activity, including walking, talking, and handling objects. In addition, control systems exist that require no human interaction, such as aircraft automatic pilots and automobile cruise control systems.

In dealing with control systems, particularly engineering control systems, we will deal with a variety of components, indicating that the subject is an interdisciplinary one. The control engineer needs a working knowledge of mechanics, electronics, electrical machines, fluid mechanics, thermodynamics, structures, material properties, and so on. Obviously not every control system contains elements from each of the above domains, but most useful control systems contain elements from more than one discipline.

Control system *analysis* involves the uniform treatment of different engineering components. What this means is that we try to represent the system elements in a common format and identify the connections between the elements in a similar way. When we do this, most control systems look the same in schematic form and lend themselves to common methods of analysis. This process usually involves a technique known as *block diagram representation,* discussed in Module 2, where each component is reduced to its basic function with one input variable and one output variable, the relationship between them known as a *transfer function.*

At this stage it is best to focus the discussion so far into a simple example. Suppose we attempt to analyze the mechanism at work when we adjust the water

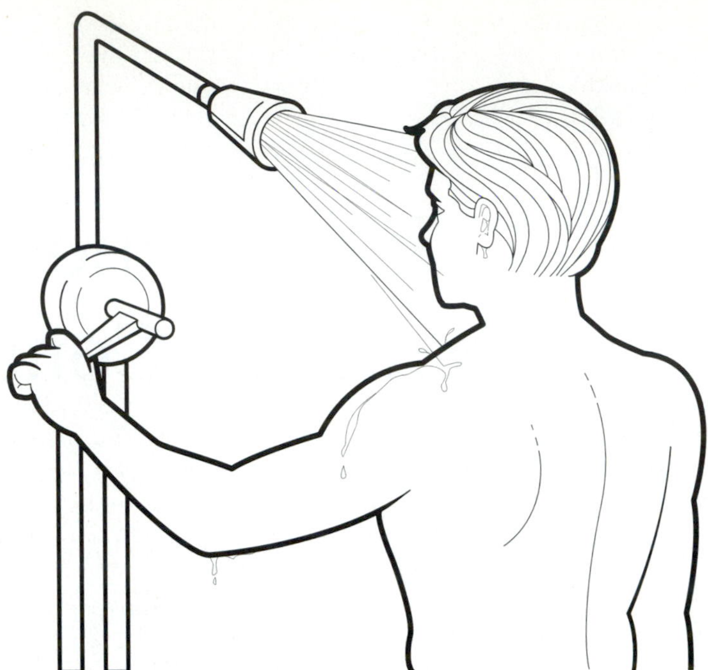

Fig. 1.1 Water temperature control system

temperature while taking a shower. The major components of the system are shown in Fig. 1.1. When we get into the shower, we have some idea of the water temperature we want. This temperature is not known in an absolute sense, such as 82 degrees, but qualitatively, such as cold, warm, or hot. Temperature sensors in our skin effectively measure the water temperature and convey the information to the brain, where it is compared to the water temperature we want. The brain computes the difference in terms of "too hot" or "too cold" and causes the hand muscles to manipulate the hot and cold mixer valve to reduce the temperature if it is too hot or increase the temperature if it is too cold. Once corrective action is taken, the process is repeated until the required water temperature is achieved.

The operation of the system and its major components are shown in Fig. 1.2. The boxes in the diagram represent processes that perform subtasks of the overall

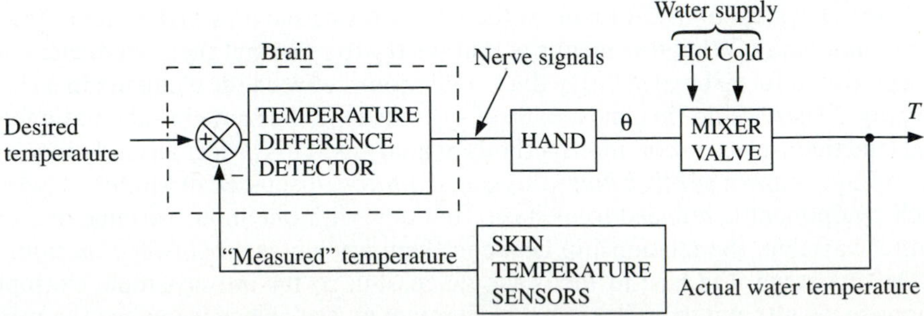

Fig. 1.2 System representation of temperature control

objective, such as measure the water temperature or actuate (move) the mixer valve. Such boxes transfer the input variable to the output variable by means of the transfer function mentioned previously. Some transfer functions are easily calculated, such as the mixer valve. This element has the valve handle angle θ as input variable and water temperature T as output variable. Making assumptions regarding the linearity of the valve might lead to the relationship

$$T = K_T\theta \tag{1.1}$$

where K_T may be defined as the valve temperature constant. Other transfer functions, such as the relationship between the nerve signals passing between the brain and the hand and the rotation of the mixer handle, will be much more difficult to represent in simple mathematical form.

The system described above will now be represented in the somewhat abstract form shown in Fig. 1.3. The purpose of doing this is because most control systems can be represented in the form of Fig. 1.3, and so analytical methods developed for use on this system will be applicable to most control problems without reference to the physical embodiment of the various elements. Shown in Fig. 1.3 is some of the terminology used throughout the book. Generally the plant represents the major component that is being controlled, and its transfer function is usually fixed. The controller is a component that the engineer designs using techniques outlined later in the book, so that the "best" performance may be obtained from the overall system. The feedback path is a critical part of the system and indicates how the output variable of interest from the plant (temperature in this case) is measured and fed back to be compared with the desired value. The magnitude of the error causes changes in the input to the plant, resulting in further changes in the output.

In the simplest controller design, the output is made proportional to the input. In our example, if the water temperature is far too cold, then the mixer handle is turned to maximize the hot-water content downstream of the valve. As the water temperature approaches the required value, smaller changes in the handle position occur. When the water is at the desired temperature, the two inputs to the differencing junction are equal and the output is zero, as is the output of the controller. The plant is therefore unperturbed and the whole system is in equilibrium.

The water temperature control system described above has a human as part of the control loop. For the immediate future, many control systems will continue

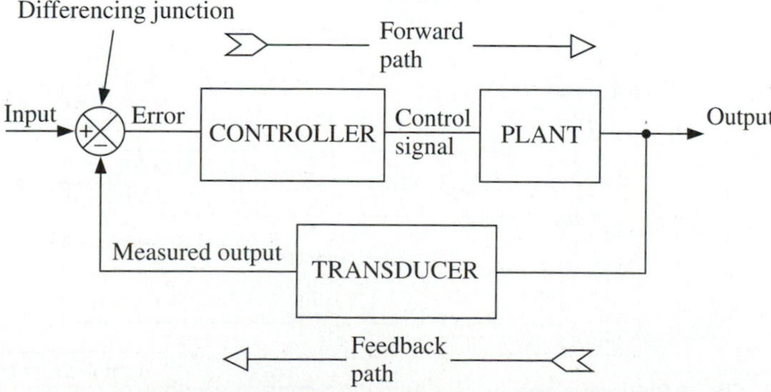

Fig. 1.3 Generalized feedback control system

to have human computational and reasoning capability as part of the system since computers and actuation hardware do not have the performance to provide a practical substitute. Such systems will include driving an automobile in heavy traffic, performing surgery on humans, playing a game of tennis, or performing classical music on a concert piano. There are, however, a vast number of situations where the human has been replaced in the control loop, resulting in an *automatic* control system. In the water temperature control problem, a temperature sensor placed in the mixer valve delivery pipe could generate a voltage proportional to the temperature, which is compared with that set by a potentiometer. The difference, or error, drives a motor that rotates the mixer handle, enabling the human operator to simply dial the temperature required and leave the rest to the control system. Such a system is shown in physical form in Fig. 1.4 and in block diagram form in Fig. 1.5.

So far, we have only discussed closed-loop control systems, but many objectives may be accomplished using open-loop systems. By definition, these are systems that do not measure and feed back the physical variable of interest. To examine the operation of an open-loop system, consider the shower temperature problem again. If we know the input hot- and cold-water temperatures and the mixer valve characteristics, it is possible to set the mixer valve angle θ to give us the exact water temperature we desire. Such a system may be represented in

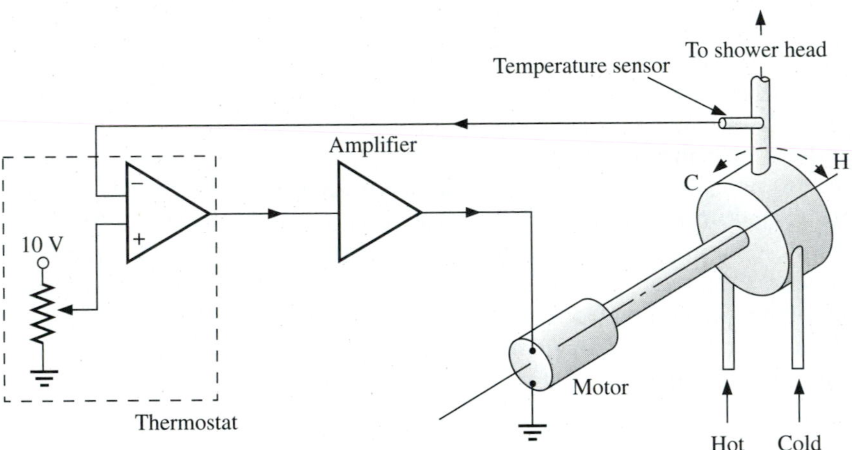

Fig. 1.4 Automatic water temperature control system

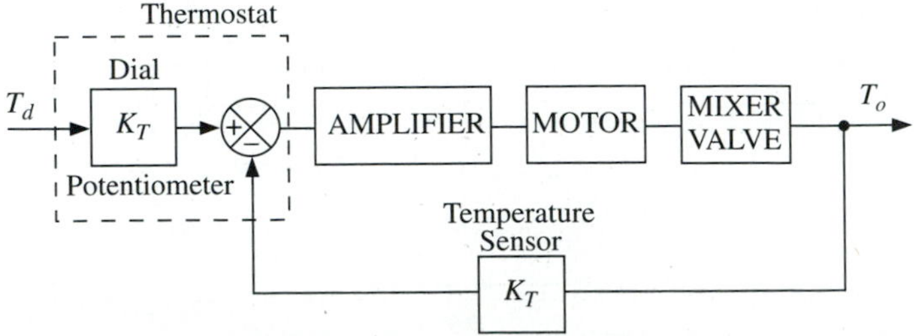

Fig. 1.5 Block diagram of automatic temperature control system

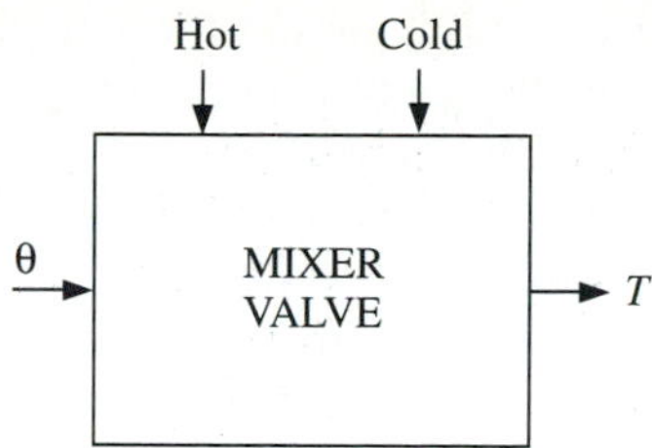

Fig. 1.6 Open-loop water temperature control system

block diagram form in Fig. 1.6. Clearly this is much simpler than a closed-loop control system and consequently is less expensive, since no temperature sensor or comparator is needed, as in the case of the automated system shown in Fig. 1.4. This is the main advantage of open-loop systems compared to closed-loop systems. A major disadvantage of open-loop systems is the need to know an accurate model of the individual components making up the system so that the input may be set appropriately. Consider that the closed-loop system will work whether we know, for example, the transfer function of the mixing valve or not.

Another advantage of closed-loop systems over open-loop systems is their ability to recover from external, unwanted disturbances. A common situation occurs in the shower when someone flushes a toilet, reducing the supply of cold water to the mixing valve, resulting in a rapid increase in water temperature. If the system is open loop, the mixing valve cannot be adjusted in response to the rapid temperature rise; otherwise it becomes a closed-loop system. All we can do is wait until the toilet has filled up, and the initial cold-water flow is restored. This recovery is shown in Fig. 1.7. On the other hand, if the person in the shower operates the mixer valve in a closed-loop mode, less hot and more cold water may be mixed, producing the correct temperature much faster, again as shown in Fig. 1.7. As the toilet fills up, continual adjustment of the mixer valve is required since a decreasing temperature will be sensed, but the user will experience a reasonably constant water temperature. Once the toilet has refilled and the initial water flow conditions have been restored, the mixer valve is in its original position, and no further adjustments are necessary. It may be concluded, therefore, that

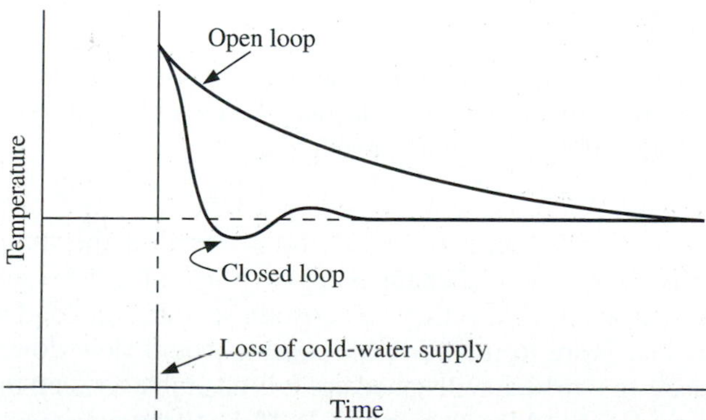

Fig. 1.7 Temperature recovery from external disturbance

although open-loop systems are simpler and less expensive to construct, they require detailed knowledge of each component in order to determine the input value for a required output, and they do not recover from unwanted disturbances as fast as closed-loop systems. We shall see later that closed-loop systems have other significant advantages over open-loop systems.

Many of the automatic control systems that will be studied in this book may be classed as *servomechanisms.* These are simply automatic control systems that have as output some mechanical quantity such as position, force, or temperature. Our closed-loop system using a temperature sensor and motor to adjust the mixer valve is a servomechanism.

It is apparent that the closed-loop water temperature control system works, as will the automated version in Fig. 1.5, yet these systems were constructed using only intuitive knowledge of feedback and actuation. The question arises: Why study the detailed analysis and performance of control systems if they can be constructed, and will apparently work, without technical knowledge of the components in the system? Historically, control system implementation did indeed precede control system analysis by about 100 years. The first sample problem discusses a steam engine speed regulation system invented in the eighteenth century, while control system theory did not develop until the middle part of the twentieth century. The reason to analyze control systems is the desire to determine performance before the system is constructed. We shall see later that it is quite easy for feedback systems to become unstable, and in the design of, say, the control system that regulates the power output of a nuclear power plant by positioning carbon rods in the reactor core, it would be useful to determine whether such unstable behavior is expected before it is tested in practice. Also, it may be important that one control system perform better than another, even though the improvement is small. If one combat aircraft can outmaneuver another, even by a few percent, the consequences are significant.

This concludes the introduction to feedback control. At this point the student should be able to analyze a physical feedback control system so as to draw a block diagram representation of it and identify elements comprising the plant, controller, transducer(s), differencing junction, input, output, and forward and feedback paths. A few further examples of performing this process are given in the sample problems, followed by problems the student may attempt individually.

Sample Problem 1.1

Figure SP1.1.1 shows a speed control system, originally developed by James Watt in 1769, for reciprocating steam engines. Draw a block diagram of the control system and identify the various parts.

Solution

The purpose of the system is to maintain the speed of the steam turbine at a constant value when the load changes. If the inlet valve were set to a constant opening, the turbine would achieve a constant speed if the load were also constant. If the load were to increase, the turbine would slow down, while if the load decreases, the turbine will speed up. It is desirable to keep the turbine running at a constant speed for many reasons. If the turbine drives an alternating-current (AC) generator, the frequency of the supply is directly proportional to speed and must therefore be maintained constant. If the turbine suddenly

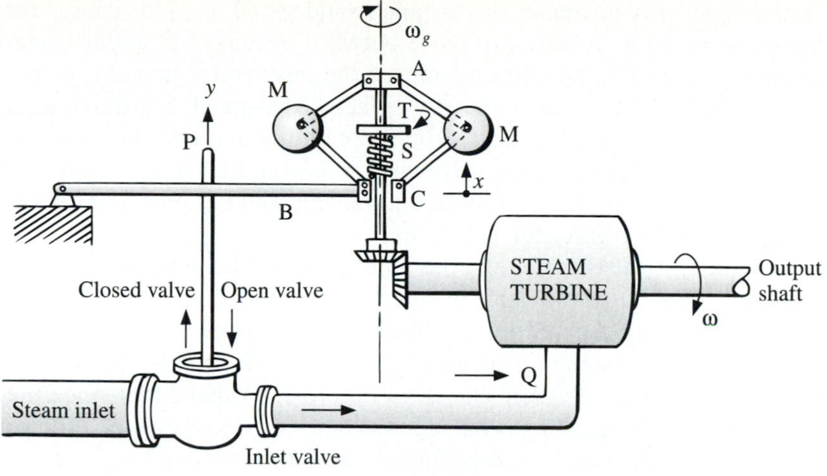

Fig. SP1.1.1

loses load, the speed may increase to the point where the blades break, causing the total failure of the machine.

Speed control is achieved by adjustment of the inlet valve, which alters the steam supply to the turbine. The component that is critical to the operation of the system is Watt's fly-ball governor. This device is attached to the turbine shaft and rotates at a speed proportional to it. Two collars A and C also rotate with the governor shaft. Collar A is fixed to the top while collar C may slide up and down the shaft. Two large masses M are connected by pin-jointed links to A and C. As the governor rotates, centripetal force causes the masses to move outward, making collar C slide up the shaft. Spring S resists this movement. Also attached to collar C is the bar B connected to push rod P, which operates the steam inlet valve. Although collar C rotates with the governor shaft, bar B does not and is attached to C in the manner shown in Fig. SP1.1.2 Suppose the system is in equilibrium at some speed. If the load

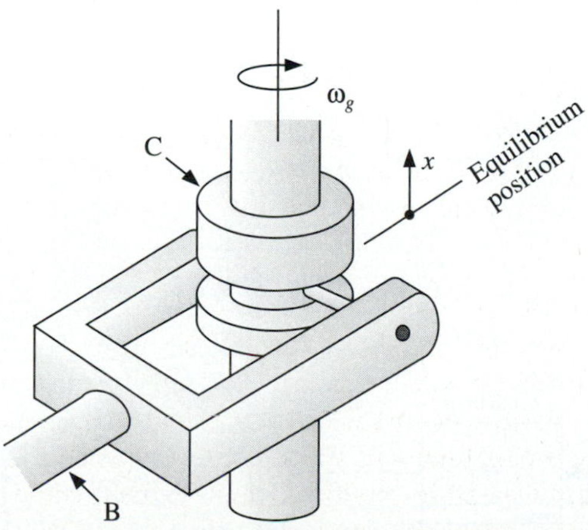

Fig. SP1.1.2

were to suddenly decrease, the turbine would speed up. This causes the rotating masses on the governor to move outward, resulting in collar C sliding up the governor shaft. This in turn moves the push rod P upward, reducing the steam supply to the turbine, thereby reducing the speed. Similarly, a speed reduction would reduce the centripetal force on the masses, allowing the spring to push C down, opening the valve and increasing the steam supply to the turbine. The system is simple, robust, and reliable and has operated satisfactorily in engine speed control for 200 years.

The plant is the turbine, and the input to it may be defined as Q, the steam flow rate. The output variable of interest from the plant is ω, the speed. The plant may be represented by the block diagram shown in Fig. SP1.1.3. The governor shaft rotates at a speed of $\omega_g = N\omega$, where N is the gearbox ratio, resulting in a block diagram shown in Fig. SP1.1.4. Note that since the transfer function of the gearbox is known, it is inserted into the block. The governor performs the function of comparing ω_g with the set speed ω_d and represents the error in terms of a displacement of collar C that will be called x. If we assume that x is proportional to the speed *error*, the block diagram of the governor may be represented by that shown in Fig. SP1.1.5, where K_g is the constant of proportionality with units of millimeter seconds per radian. Note how x is also represented in Fig. SP1.1.2. Assume that the push rod P is connected halfway between C and the pivot. The inlet valve displacement y is given by

$$y = \tfrac{1}{2}x$$

resulting in the block shown in Fig. SP1.1.6. Finally, the valve may be considered as an element with input y and output Q, the steam flow rate. Valves are known to be highly nonlinear in the relationship between stem motion and flow rate, so it would be unwise to make the assumption that Q is proportional to y. This block is shown in Fig. SP1.1.7. This completes the analysis of the individual components, and a complete block diagram of the turbine speed control system is shown in Fig. SP1.1.8. In comparing Fig. SP1.1.8 with Fig. 1.3, several points emerge.

1. Some transfer functions are easy to determine while others are not. Some may be simple constants (the linkage) while others are nonlinear functions

Fig. SP1.1.3

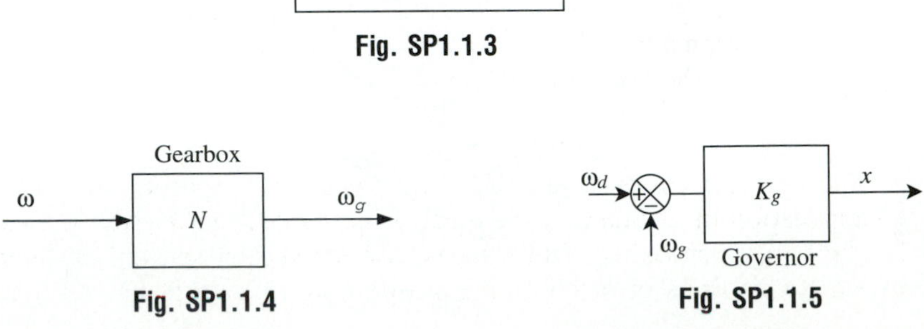

Fig. SP1.1.4

Fig. SP1.1.5

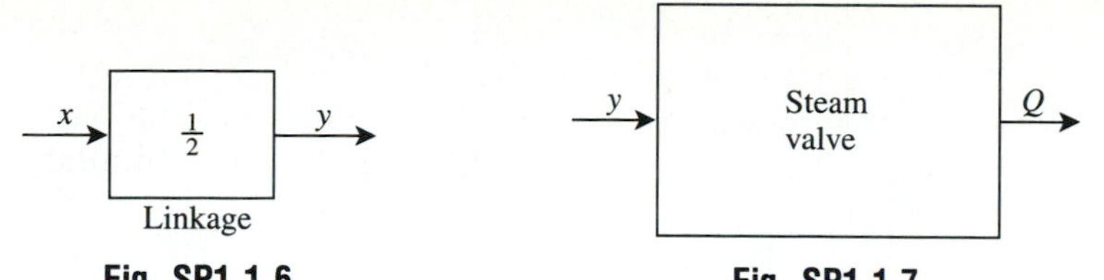

Fig. SP1.1.6

Fig. SP1.1.7

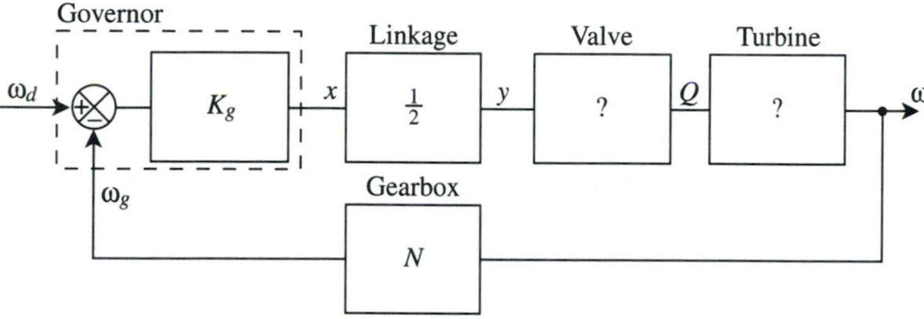

Fig. SP1.1.8

(the valve). Some transfer functions, such as the turbine, may only be identified by experimental means.

2. In many cases there are more "boxes" than there are indicated in Fig. 1.3, and it is difficult to allocate elements to the plant or the controller. In this example, it is not clear if the valve is part of the plant or the controller. It will be shown that this categorization is in fact not important, but deciding if an element is part of the forward path or feedback path may be of more significance.
3. Transfer functions, because they change one variable into another, usually have dimensions. The valve and turbine do have dimensions associated with their transfer functions, but the linkage and gearbox transfer functions are dimensionless.
4. The differencing junction can only subtract two variables of the same dimensions, producing an error variable of the same dimensions. For example, achieved speed can be subtracted from demanded speed to produce a speed error, but speed cannot be subtracted from displacement.

Sample Problem 1.2

Analyze the system shown in Fig. SP1.2.1, which uses a direct-current (DC) motor to position a load according to the angular position set by the input potentiometer. Draw the block diagram and identify the principal elements.

Solution

The arrangement shown in Fig. SP1.2.1 is a very important one in that it finds application in a wide range of electromechanical servosystems. Examples include the positioning of telescopes, camera shutter setting, movement of the individual links of a robot arm, positioning of dive planes on a submarine,

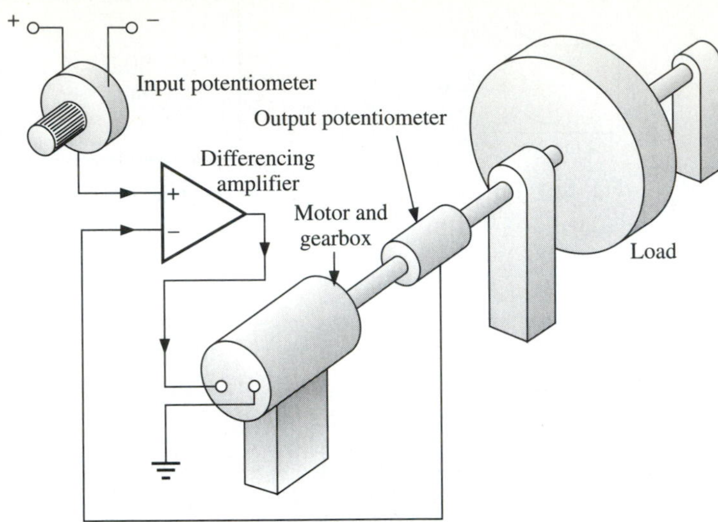

Fig. SP1.2.1

movement of the read/write head on a floppy disk drive, and so on. Although these examples have very different power requirements, they all have the basic form shown in Fig. SP1.2.1 and operate in the same manner.

Students should be encouraged to build their own system, since the cost is minimal and parts are available from most electronics stores. With this in mind, the system studied will be a small one but will have the operational characteristics of one of the most useful control systems available.

In order to see how the system works, the individual elements will be studied first; then the system as a whole may be analyzed. The first component to investigate is a simple potentiometer, shown in schematic form in Fig. SP1.2.2. This device is a single-turn type connected between a positive and negative supply that matches the voltage requirements of the differencing junction. Typical values would be 10–12 V. The potentiometer wiper would then deliver a voltage between these values depending on the position of the potentiometer knob. The potentiometer is therefore characterized by a coefficient, or transfer function, K_p with units of volts per radian. If the potentiometer is capable of a mechanical rotation of 270° and is connected across a ±10-V supply, then

$$K_p = \frac{20 \times 180}{270 \times \pi} = 4.244 \text{ V/rad}$$

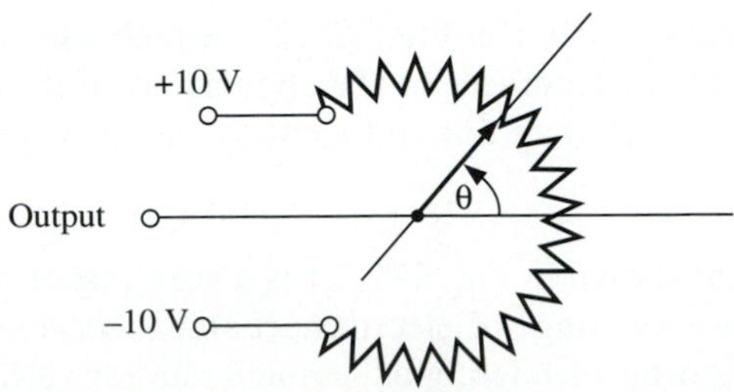

Fig. SP1.2.2

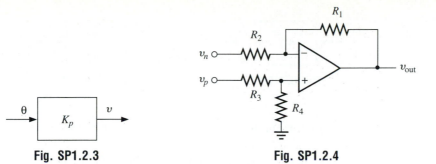

θ
K_p
v

Fig. SP1.2.3

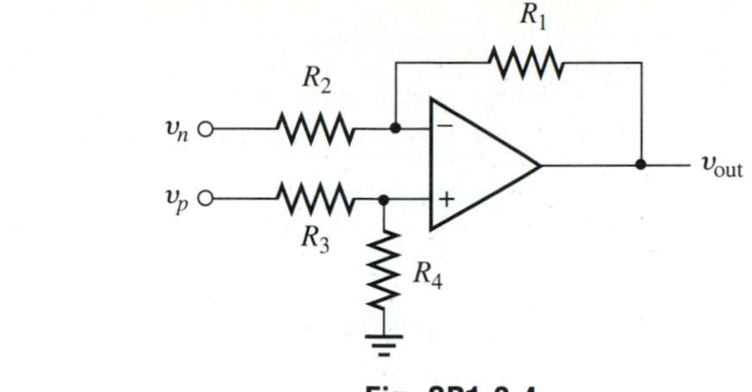

Fig. SP1.2.4

The block diagram for the potentiometer will be as shown in Fig. SP1.2.3. The differencing junction compares the voltage representing the demanded position with the voltage representing the achieved position and subtracts them, providing a measure of position error. This differencing junction may be constructed using an operational amplifier and resistors and takes the form shown in Fig. SP1.2.4. The output voltage v_{out} is given by the following equation:

$$v_{out} = \frac{R_4}{R_3} v_p - \frac{R_1}{R_2} v_n$$

The amplifier may be made to subtract the two voltages and multiply by a gain K_a so that

$$v_{out} = K_a(v_p - v_n)$$

where

$$K_a = \frac{R_4}{R_3} = \frac{R_1}{R_2}$$

The differencing junction will take the schematic form shown in Fig. SP1.2.5. In this example the plant will comprise the motor and load. The input variable to the plant is voltage, while the output variable is angular position of the motor shaft. In applications such as this, there is usually a gearbox between the motor and load shaft because (a) only part of one revolution is required for the rotation of the load and (b) the torque exerted by the motor needs amplification in order to overcome the load inertia. A gearbox achieves both of these objectives. Common experience reveals that when a constant voltage is applied to a DC motor, it produces a constant speed, although, because of the load inertia, it may take some time to achieve it. If this dynamic lag is ignored for now, the block diagram of the plant will be as shown in Fig. SP1.2.6, where v_m is the motor input voltage, ω is the motor and load rotational speed, and K_m is

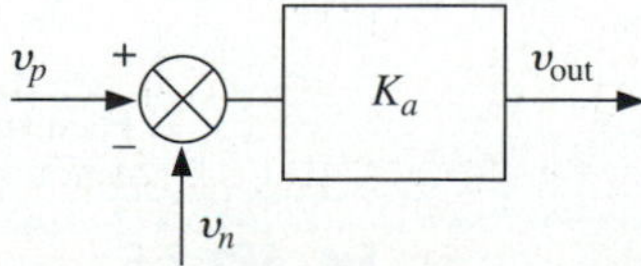

Fig. SP1.2.5

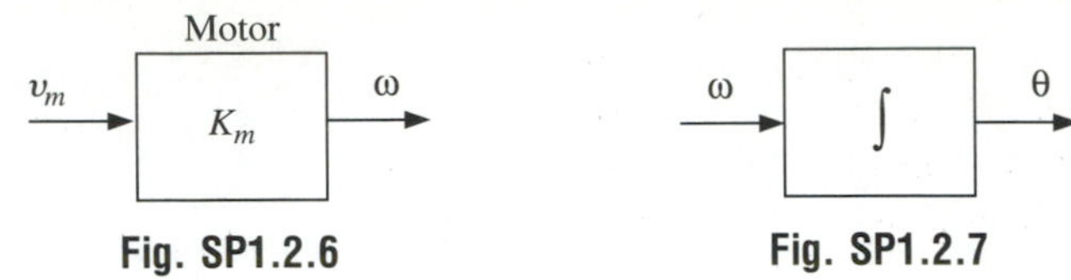

Fig. SP1.2.6

Fig. SP1.2.7

the constant of proportionality between them. The potentiometer mounted on the output shaft measures output position, not output speed; therefore a relationship between them has to be established. Obviously, position is the integral of speed, so this relationship may be expressed in block diagram form, as shown in Fig. SP1.2.7, where the symbol $\int$ denotes integration. The output potentiometer should have identical characteristics to the input potentiometer discussed previously. Now that all elements have been discussed in detail, they are combined in Fig. SP1.2.8 to form the complete system. The operation is fairly straightforward. Suppose the system is in equilibrium at some position when the input potentiometer is turned so that the wiper produces a larger voltage. Initially the output potentiometer signal remains unchanged so the differencing junction produces a positive error voltage. This causes the motor to turn, producing an increasing voltage from the output potentiometer, reducing the error voltage, and bringing the motor and load to rest at the desired position. Notice that connecting the motor leads the wrong way will cause rotation in the opposite direction, reducing the feedback voltage and increasing the error. This leads to motor and load runaway caused by unstable operation of the servomechanism.

For those constructing their own servo, here are a few design guidelines. Choose a motor with output speed less than 20 rpm and small enough to be driven directly by an operational amplifier. A 12-V motor that consumes less than 10 mA is adequate. Determine which way around the motor leads should go before connecting. If the system is unstable, the potentiometers may be damaged if driven hard against their mechanical end stops. Figure SP1.2.9 shows a system constructed according to the preceding design using an operational amplifier circuit driving a small permanent-magnet motor directly.

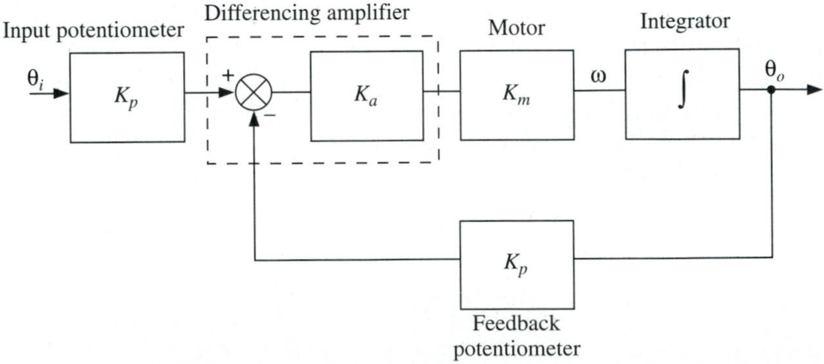

Fig. SP1.2.8

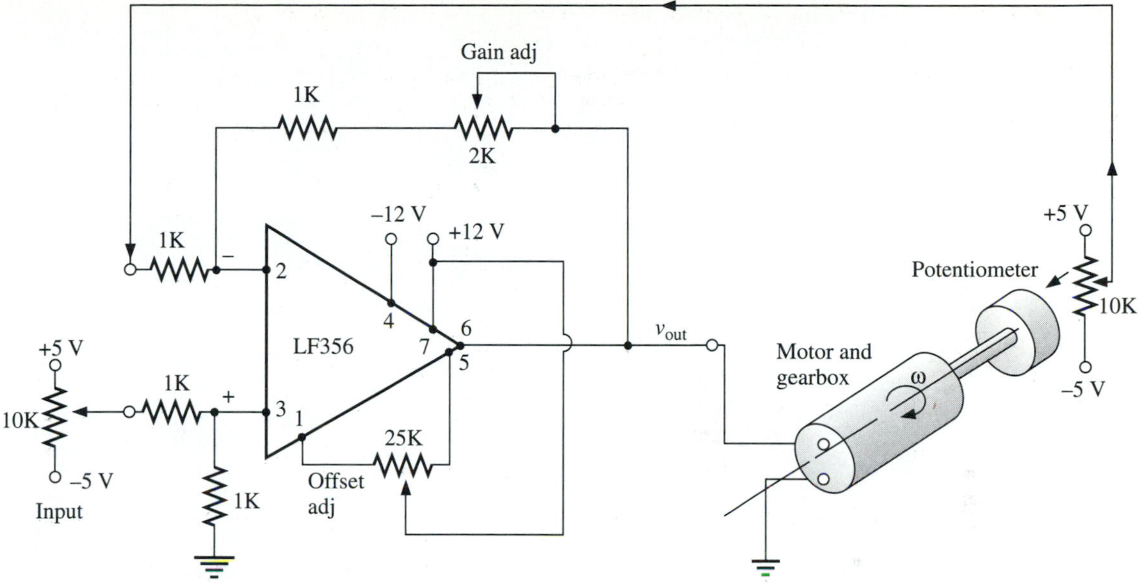

Amplifier gain ≃ 10
Motor and gearbox: Permanent magnet, output speed = 15 rpm for 12 V input, maximum current = 10 MA

Fig. SP1.2.9

Sample Problem 1.3

In order to land the Boeing 747 jet shown in Fig. SP1.3.1, the pilot must set the movable control surfaces at the trailing edge of each wing, known as flaps, to an angle of 25°. To achieve this, a servosystem comprising electrical and hydraulic components is used, as shown in Fig. SP1.3.2. Analyze the operation of the control system, draw the block diagram of each component and the system as a whole, and attempt to write down as many of the component transfer functions as possible.

Solution

An important component used in this servomechanism is the electrohydraulic servovalve and hydraulic actuator, and so this will be analyzed first. This is a common means of controlling the position of large loads, such as aircraft control surfaces, since high-pressure hydraulic fluid moving at high flow rates conveys considerable amounts of power. The actuator may be relatively small; yet it can exert substantial force. Oil at 5000 lbf/in^2 inside a 2-in-diameter cylinder exerts a force of over 15,700 lbf. In addition, the electrohydraulic servovalve may be connected directly to a low-voltage electrical control signal and responds quickly to changes in the input.

A schematic diagram of one form of electrohydraulic servovalve is shown in Fig. SP1.3.3 and is commonly known as a jet pipe valve. High-pressure oil flows into a slender tube, the jet pipe, which discharges equally into two ports A and B, which in turn are connected to each side of the hydraulic actuator.

Fig. SP1.3.1 (courtesy the Boeing Company)

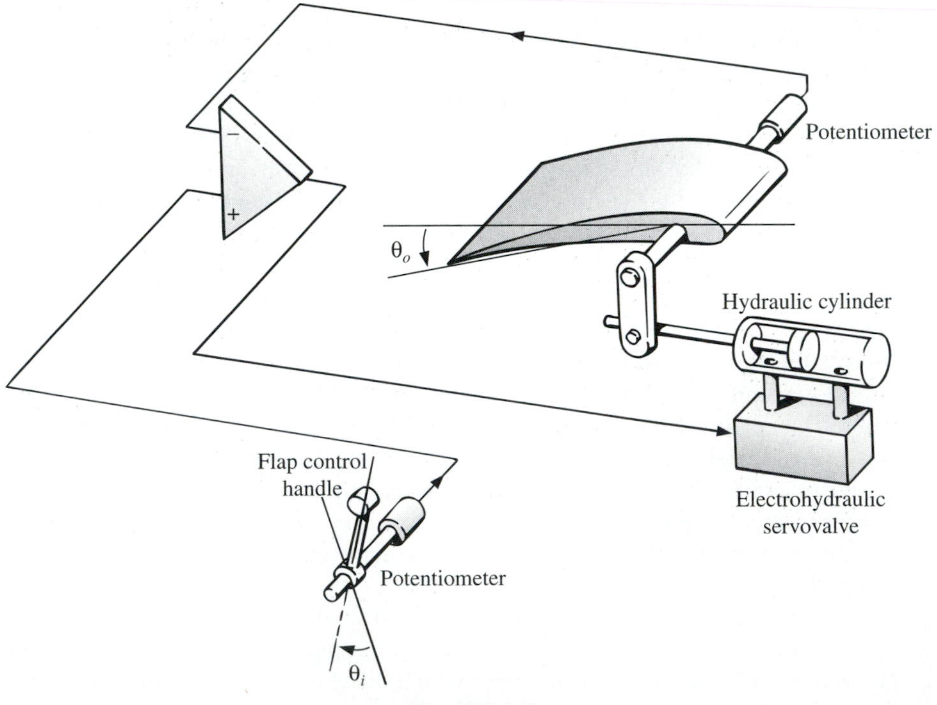

Fig. SP1.3.2

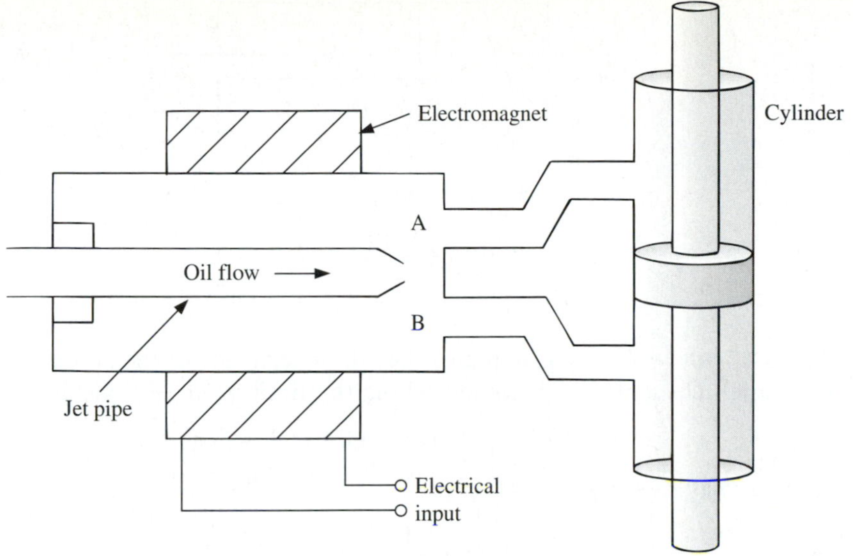

Fig. SP1.3.3

Because of equal flow into A and B, both sides of the cylinder experience the same pressure and the cylinder is therefore balanced. The input electrical signal is used to change the magnetic field established by an electromagnet placed around the chamber containing the jet pipe. Because the jet pipe is manufactured from ferrous materials, the asymmetrical magnetic field causes it to bend slightly, forcing more oil into one of the ports and less into the other, causing the hydraulic actuator to move. The transfer function of the combined valve and actuator is easily derived if assumptions are made regarding the linearity of operation of the various subcomponents. If it is assumed that the input signal produces a proportional change in oil flow rate, the servovalve has the transfer function shown in Fig. SP1.3.4, where K_v has units of cubic inches/ per second per volt. Consideration of the hydraulic cylinder indicates that a flow of oil Q into the cylinder produces a *velocity* of the cylinder rod. Since, however, the primary output variable of interest from the actuator is position, it is necessary to try and develop a transfer function between input oil flow and cylinder rod position. Figure SP1.3.5 shows the cylinder rod movement when a small volume of oil dV flows in, causing the rod to move a distance dx. Equating the increase in volume due to piston movement to the amount of oil entering the cylinder,

$$dV = A\ dx$$

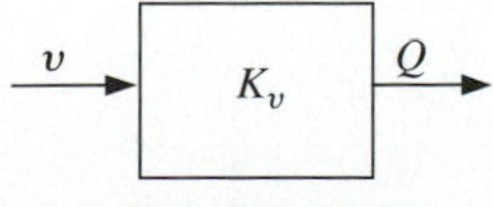

Fig. SP1.3.4

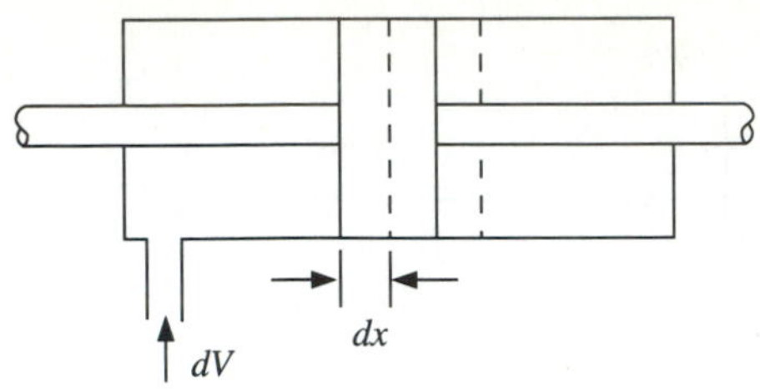

Fig. SP1.3.5

where A is the effective piston area. Dividing each side of the above equation by a small element of time dt and taking the limit as $dt \to 0$ yields

$$Q = A\dot{x}$$

Another way of expressing this is

$$x = \frac{1}{A}\int Q\,dt$$

resulting in the block diagram and transfer function shown in Fig. SP1.3.6. Notice the similarity between this actuation system and the electric drive system shown in Fig. SP1.2.8.

It is now possible to analyze the other elements of the system shown in Fig. SP1.3.2 in a straightforward manner. Assuming there is a linear relationship between the rotation of the flap θ_o and the cylinder displacement x, the block diagram of the linkage and flap is as shown in Fig. SP1.3.7. This rotation is measured with a potentiometer, giving a voltage proportional to θ_o, which is compared to a voltage generated by an identical potentiometer representing the demanded flap angle θ_d set by the pilot in the cockpit. The comparison is made by an operational amplifier–based differencing junction identical to the one described in the previous example.

The complete system block diagram and its operation are easily observed in Fig. SP1.3.8. In the last two examples it will be noticed that in each case an element has occurred where the output variable has been the integral of the input variable, and this operation has been represented symbolically by an integral sign. In analyzing the operation of feedback control systems, it is necessary to represent the transfer function by analytical, rather than symbolic, functions. It is known that variables that are linked by differentiation or integration are usually related by a differential equation, such as

$$Q = A\frac{dx}{dt} = A\dot{x}$$

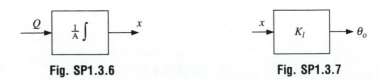

Fig. SP1.3.6

Fig. SP1.3.7

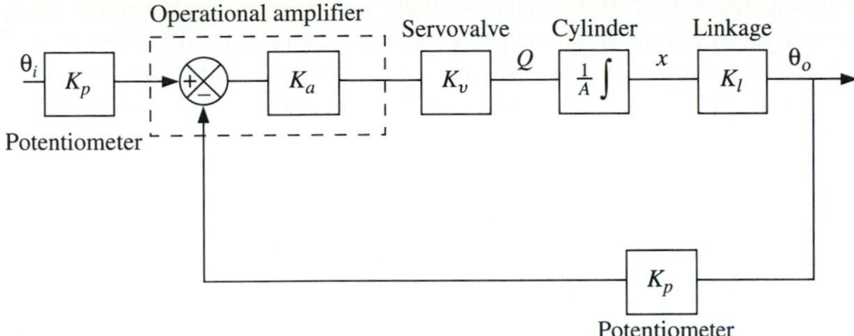

Fig. SP1.3.8

In the study of elements where the input-output variables are related in this fashion, the use of differential equations to relate the variables is one of the fundamental tools of control systems analysis. Invariably, it is necessary to solve for one variable, the output of the system, for example, given the input variable as a function of time. For this reason, the student needs to become proficient at solving differential equations. Although there are many commonly accepted methods for solving differential equations, one method is given emphasis over all others: the Laplace transform method. This method is the most widely used technique in the control systems community and has distinct advantages over other methods when applied to particular problems of interest to control engineers. Our studies continue with a review of Laplace transform methods, their application in solving differential equations, and their use in expressing transfer functions in a concise and meaningful way.

◆ PROBLEMS

1.1 Analyze the process by which a person reaches out and picks up an object resting on a table. Draw a block diagram of the system and identify as many parts of the general feedback control system as possible.

1.2 Redesign the turbine speed control system discussed in Sample Problem 1.1, but replace the fly-ball governor with the tachometer shown in Fig. P1.2. A tachometer consists basically of a small DC motor operated in reverse as a generator, the shaft being rotated continuously, producing a DC voltage proportional to shaft speed.

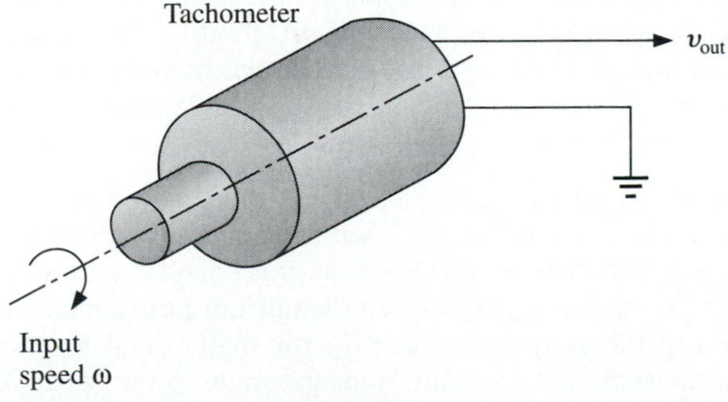

Fig. P1.2

1.3 The system in Fig. P1.3 shows an operator attempting to maintain the speed of a conveyor carrying manufactured parts through a spray booth. When more boxes are placed on the conveyor, it slows down, while fewer boxes allow it to speed up. Draw a block diagram of the system indicating the principal elements and transfer functions where possible.

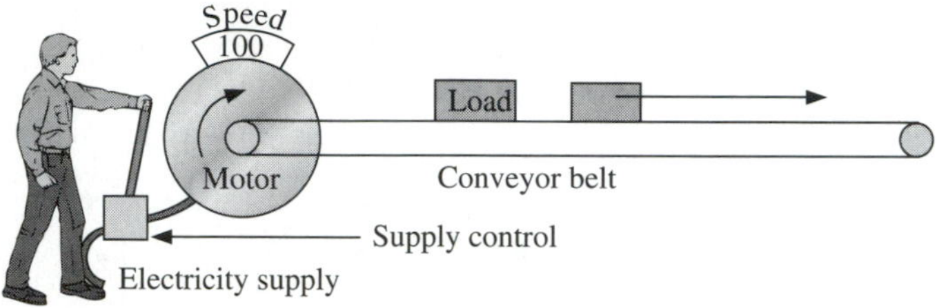

Fig. P1.3

1.4 Shown in Fig. P1.4 is a water-level control system comprising a tank, inlet pipe, slide valve, and float. Details of the operation of the flow valve are also shown in the figure. Draw a block diagram of the feedback control system and identify the main elements, writing down the mathematical transfer functions where appropriate.

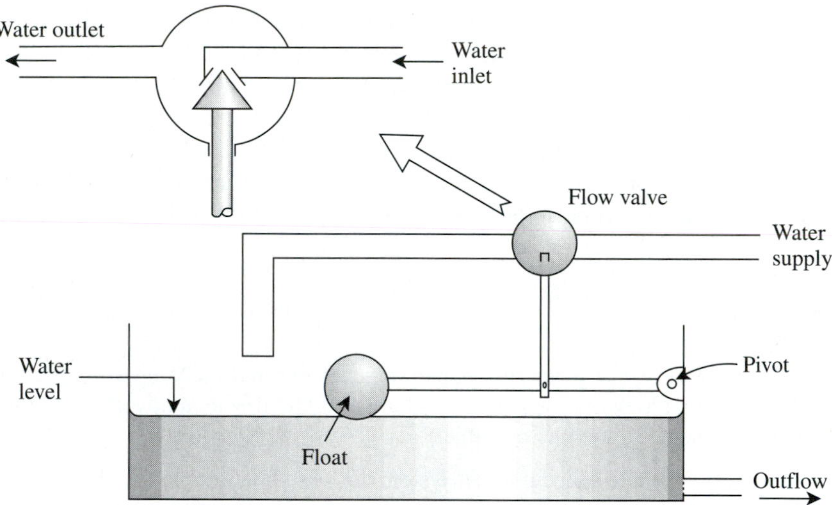

Fig. P1.4

1.5 A method of producing a displacement proportional to an input displacement but with a much larger output force is shown in Fig. P1.5. The input displacement x causes the movement of the spool valve to produce a differential flow to the hydraulic actuator. Draw a block diagram of the system, label the principal parts of the control loop, and identify as many of the transfer functions as possible. State any assumptions made in the analysis.

1.6 Consider Fig. P1.6, which shows a stick balanced on top of a cart that is free to move horizontally. The angle of the stick with the vertical is measured by means of a potentiometer. Using components discussed so far, specify a control system that will move the cart so as to maintain the stick in the upright position when disturbed. Draw the block diagram of the system and identify the major elements. Can the system be used to maintain the stick at a constant, nonzero angle to the vertical?

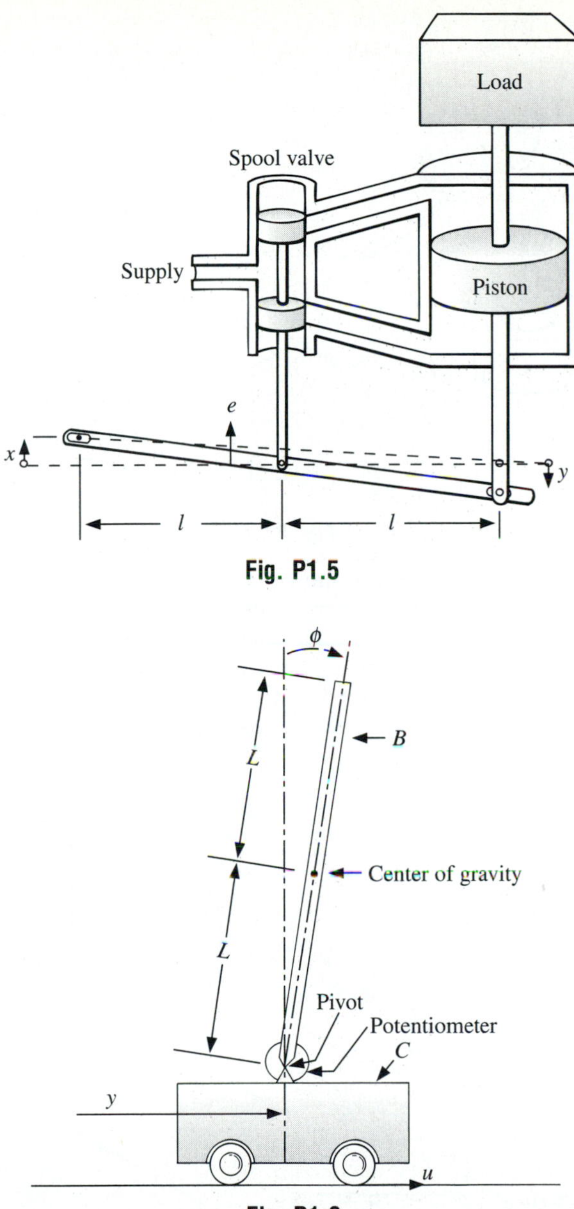

Fig. P1.5

Fig. P1.6

1.7 Figure P1.7 shows a beam with a V groove cut into it and a steel ball that is free to roll in the groove. Along one face of the groove is a carbon track of fixed resistance. A 10-V supply is connected to one end of one track while the other end is grounded. The ball connects the energized track to a wire on the other face of the V, which assumes a voltage proportional to the position of the ball x from the grounded end, the whole system working as a linear potentiometer. The beam is free to rotate about its center and is actuated by a geared motor at one end that moves a lead screw with velocity proportional to input voltage. Design and draw a labeled block diagram of a control system capable of positioning the ball anywhere on the beam.

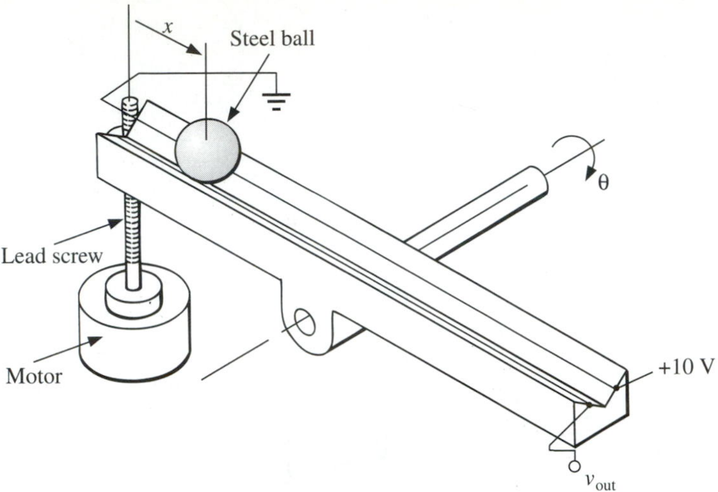

Fig. P1.7

1.8 The automatically guided vehicle shown in Fig. P1.8a follows a white line painted on a factory floor by means of the system shown in Fig. P1.8b. The line is illuminated with a lamp and the reflected light is measured by a photodetector. As the vehicle wanders off track, the photodetector signal decreases due to the decrease in the white-line area beneath it, as shown by the dashed line in Fig. P1.8b. Design a control system to ensure accurate tracking, and draw a block diagram of the principal elements.

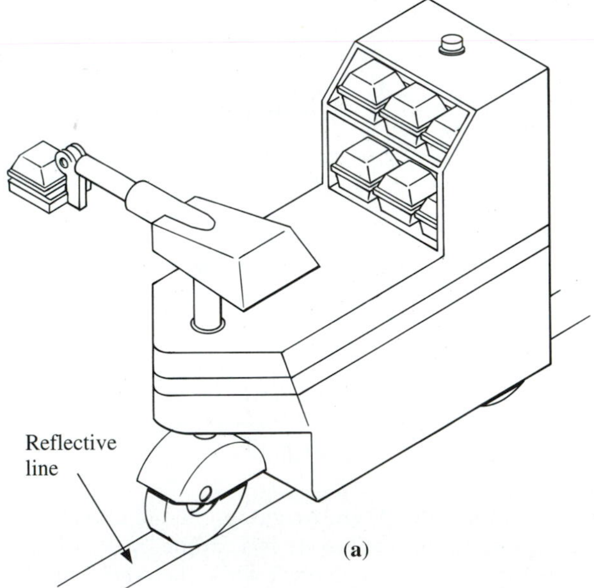

Fig. P1.8a

1.9 Design a control system that will maintain the depth of a submarine at a prespecified value. The dive planes are actuated by means of the arrangement shown in Fig. P1.9. A pressure-sensor signal is available that delivers a voltage proportional to water pressure, with a known calibration coefficient. Design a system that uses this voltage to drive the submarine to a preset depth and describe its operation.

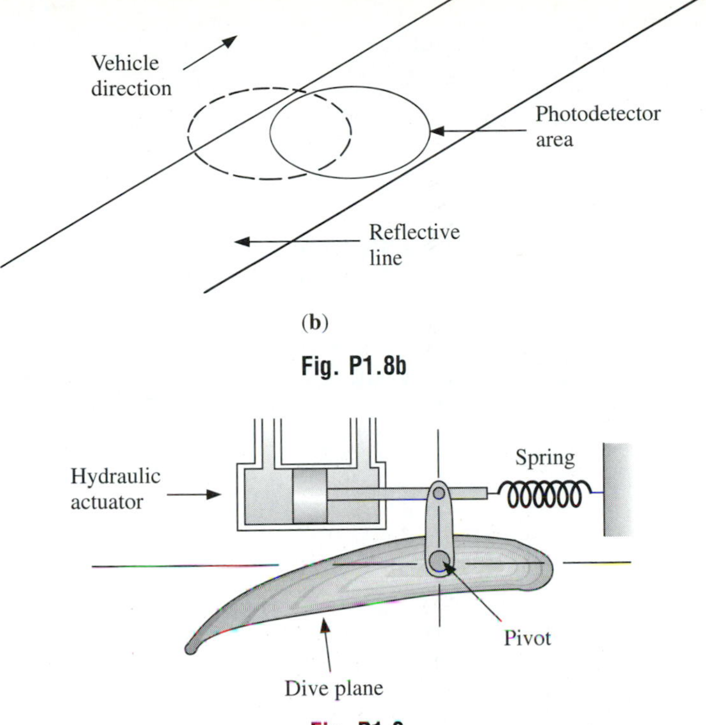

(b)

Fig. P1.8b

Fig. P1.9

1.10 A control system has to be designed to maintain the thickness of a steel plate as it is hot rolled, as shown in Fig. P1.10. The thickness is measured by a roller mounted on a swinging arm connected to and pivoting about a potentiometer and fed back to the control room where it may be compared to the thickness dialed by the operator on an identical potentiometer. The top roller is adjusted in height by means of the hydraulic ram shown in the figure. Complete the design indicating all elements of the control system by a transfer function where possible.

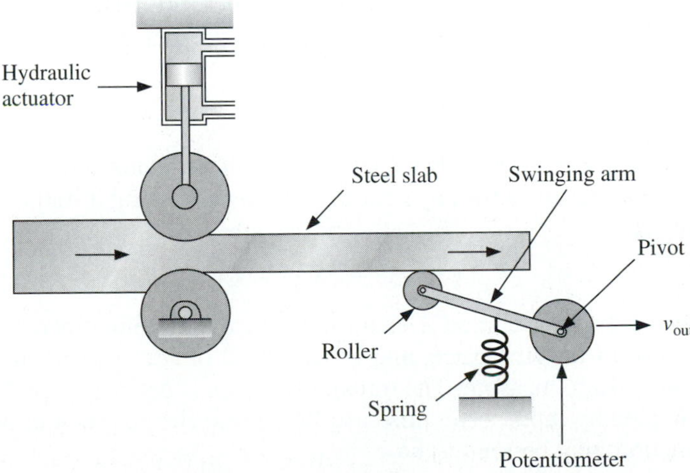

Fig. P1.10

MODULE 2

Transfer Functions and Block Diagram Algebra

◆ TRANSFER FUNCTIONS

The development of transfer functions invariably leads through system differential equations to the use of Laplace transforms. It is assumed that the reader is familiar with using such techniques for representing and solving linear differential equations. However, a brief review is provided in Appendix A.

The concept of a transfer function becomes apparent by regarding a physical system in a particular way and was introduced in the previous module. For example, the dynamic system shown in Fig. 2.1 shows the inner workings of an accelerometer, such that the acceleration of the cart is measured by the movement y of a pointer on a scale. Using Newtonian mechanics, the differential equation of motion relating the cart displacement x, the pointer displacement y, and the various system parameters may be shown to be

$$\ddot{x} = \ddot{y} + c'\dot{y} + k'y \tag{2.1}$$

where $c' = c/m$ and $k' = k/m$. If we let $\ddot{x} = A$, the acceleration of the cart, and take Laplace transforms of the differential equation assuming initial conditions to be zero, we get

$$A = (s^2 + c's + k')Y \tag{2.2}$$

We may regard the acceleration A as an input to the system, since it causes the system to move in the first place, and Y the output of the system, since it is the parameter we wish to measure. The transfer funciton is defined to be the quotient of output divided by input, or in this case Y/A. From the equation of motion, the transfer function may be seen to be

$$\frac{Y}{A} = \frac{1}{s^2 + c's + k'} \tag{2.3}$$

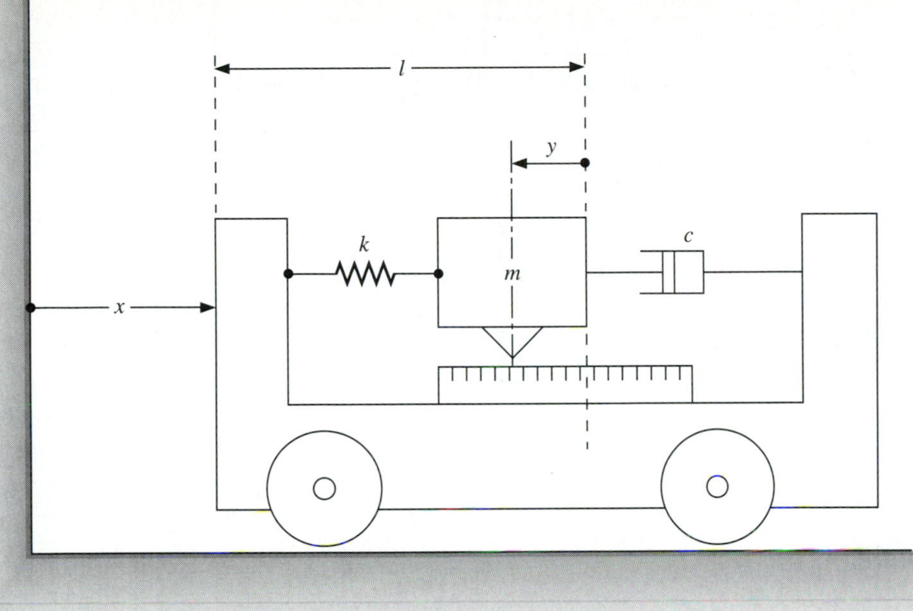

Fig. 2.1 Accelerometer schematic

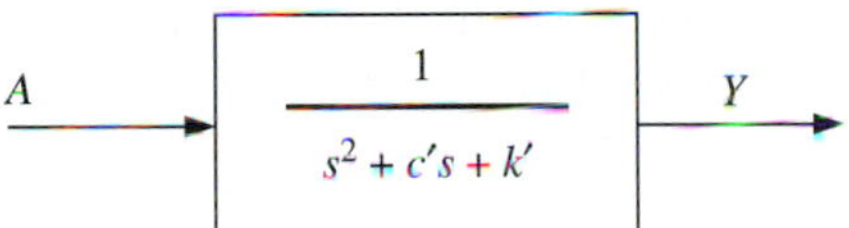

Fig. 2.2 System block diagram

From the transfer function, the resulting acceleration may be determined if the input acceleration is known and its Laplace transform may be found. The system is usually represented as a block, as shown in Fig. 2.2. The transfer function, if it is known analytically, is inserted in the block.

◆ BLOCK DIAGRAM ALGEBRA

In order to analyze complex feedback control systems, it is usual to rearrange the block diagram so as to enable it to be easily analyzed. In doing so, the objective is to produce a block diagram composed of the two basic elements discussed below.

Serial Elements

This case is shown in Fig. 2.3 and consists of one or more individual blocks where each transfer function is known. Since

$$V(s) = U(s)G_1(s) \tag{2.4}$$

$$W(s) = V(s)G_2(s) \tag{2.5}$$

$$X(s) = W(s)G_3(s) \tag{2.6}$$

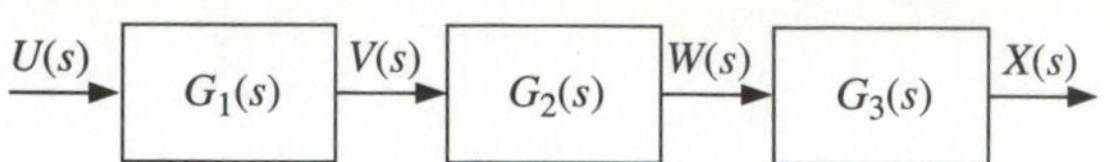

Fig. 2.3 Block diagram of serial elements

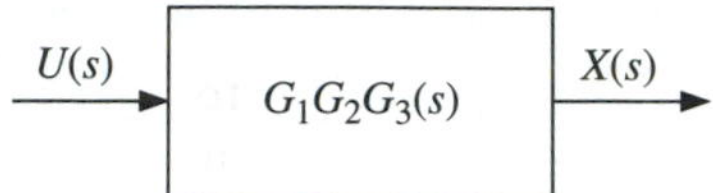

Fig. 2.4 Reduction of serial elements

then

$$W(s) = G_1G_2U(s) \tag{2.7}$$

$$X(s) = G_2G_3V(s) \tag{2.8}$$

$$= G_1G_2G_3U(s) \tag{2.9}$$

resulting in Fig. 2.4.

General Feedback Case

As shown in Fig. 2.5, this system comprises a forward-path transfer function $G(s)$ and a feedback transfer function $H(s)$. For a unity-feedback system, $H(\mathrm{s}) = 1$. From the diagram it is known that

$$E = R - CH(s) \tag{2.10}$$

$$C = EG(s) \tag{2.11}$$

Hence

$$C = EG = G(R - CH) = GR - CGH \tag{2.12}$$

$$C(1 + GH) = GR \tag{2.13}$$

$$\frac{C}{R} = \frac{G}{1 + GH} \tag{2.14}$$

Note that if the feedback loop were positive instead of negative, then

$$\frac{C}{R} = \frac{G}{1 - GH} \tag{2.15}$$

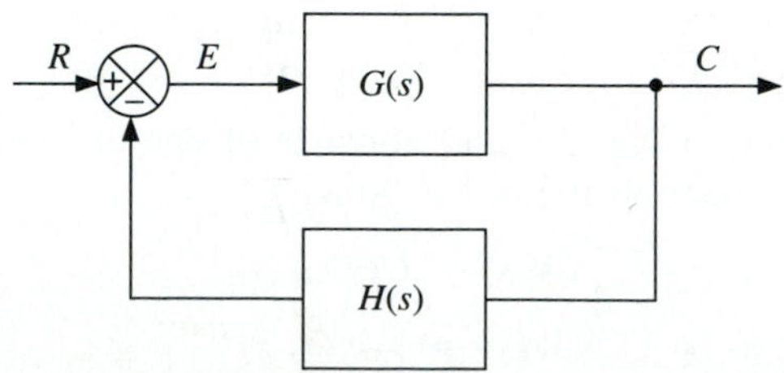

Fig. 2.5 General feedback control system block diagram

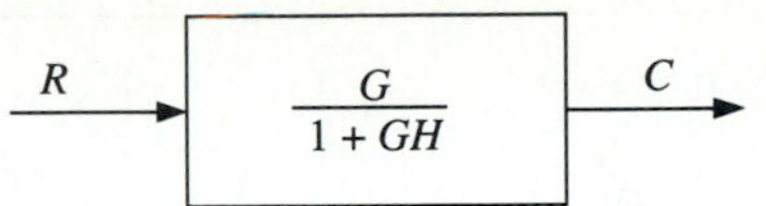

Fig. 2.6 Reduction of feedback loop

Therefore the feedback loop may be reduced to a single element, as shown in Fig. 2.6. The quantity *GH*, which we will encounter frequently in later modules, is defined to be the open-loop transfer function. In order to move individual blocks across summing junctions or pick off points, there are a number of other rules that will be useful, and these are summarized in Fig. 2.7. Most of these are self-explanatory, and their use will be illustrated by one of the sample problems. In general, the simplification of complex block diagrams may be accomplished by systematic modification of the diagram using the rules in Fig. 2.7 applied in the following order:

Step 1: Combine all serial blocks (rule 1).

Step 2: Combine all parallel blocks (rule 2).

Step 3: Close all inner loops (rule 3).

Step 4: Move summing junctions to the left and pick off points to the right (rules 4–7).

Multiple Inputs

In some cases a feedback control system may have several inputs, and it is necessary to determine the output when each input takes a particular form. Providing the system is *linear,* i.e., described by a linear differential equation, this can be achieved by the following:

Step 1: Set all but one input to zero.

Step 2: By rearranging the block diagram if necessary, determine the transfer function from the single nonzero input to the output.

Step 3: Repeat step 2 for all inputs.

Step 4: Add all transfer functions together to obtain the output to all inputs.

Example

Determine the output *C* to the two inputs *R* and *D* for the system shown in Fig. 2.8. Setting *R* equal to zero and rearranging the block diagram produce the system shown in Fig. 2.9. The closed-loop transfer function becomes

$$\frac{C}{D} = \frac{G_2}{1 + G_1G_2H_1} \tag{2.16}$$

or

$$C = \frac{G_2}{1 + G_1G_2H_1}D \tag{2.17}$$

RULE #	PROCESS	ORIGINAL BLOCK DIAGRAM	EQUIVALENT BLOCK DIAGRAM
1	Combining serial blocks	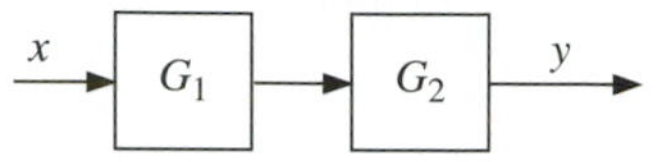	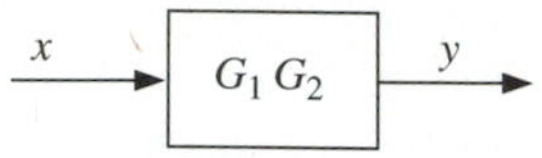
2	Combining parallel blocks	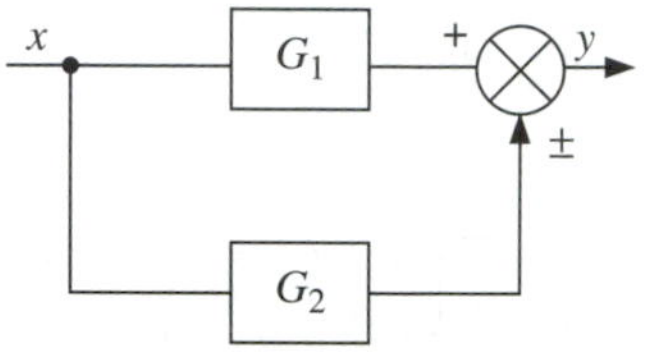	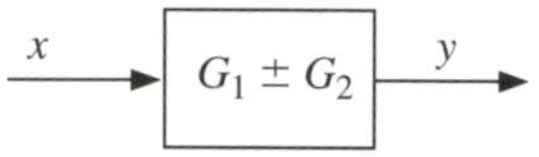
3	Closing a feedback loop	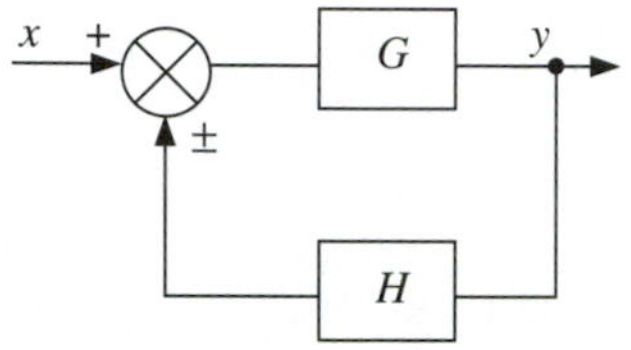	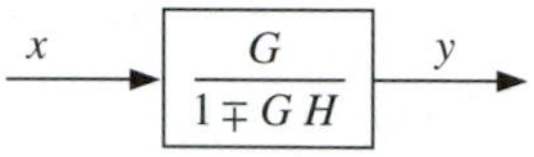
4	Moving a summing junction ahead of a block	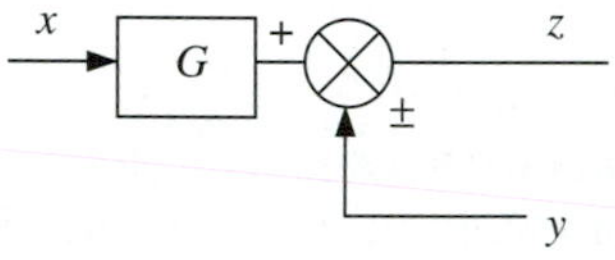	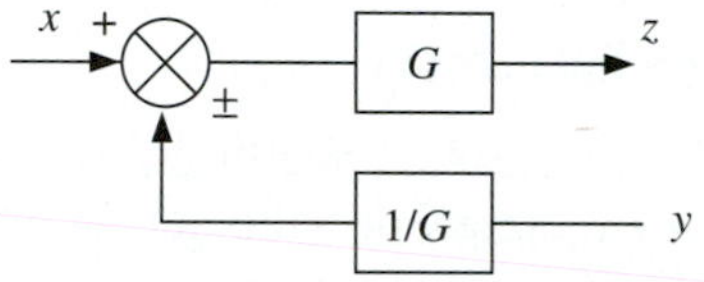
5	Moving a summing junction past a block	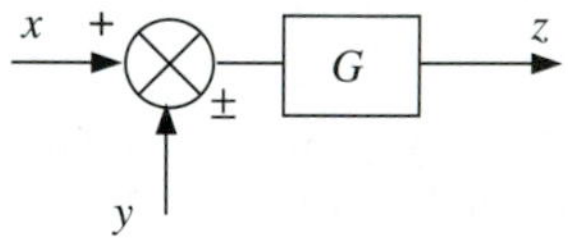	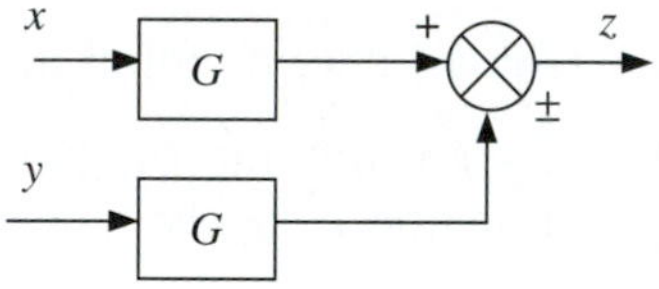
6	Moving a pickoff point ahead of a block	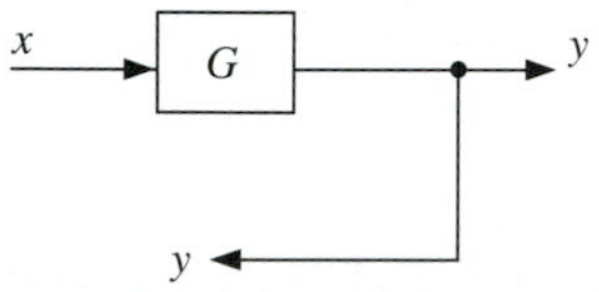	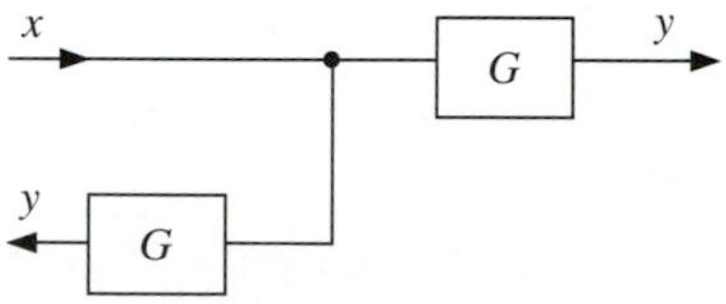
7	Moving a pickoff point past a block	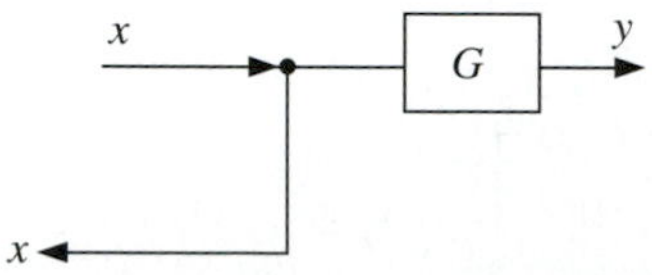	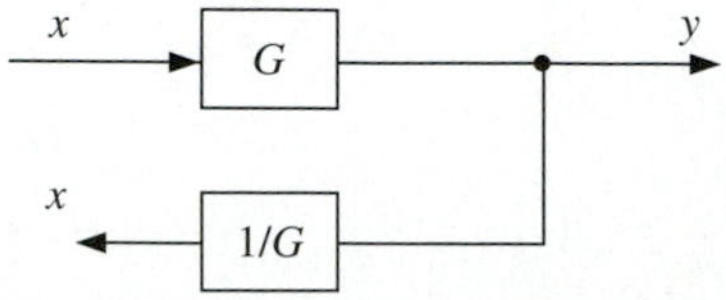

Fig. 2.7 Block manipulation rules

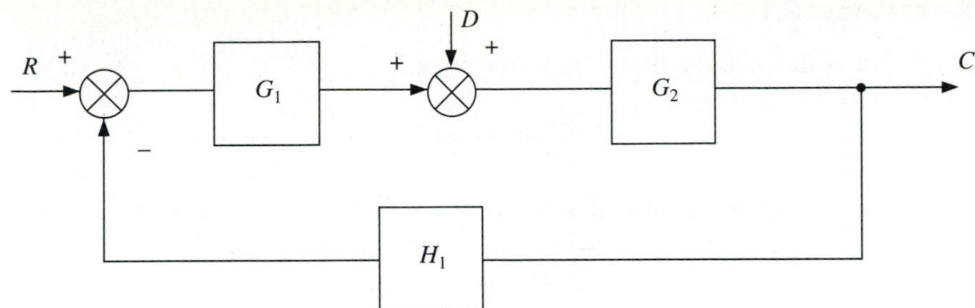

Fig. 2.8 Linear system with two inputs

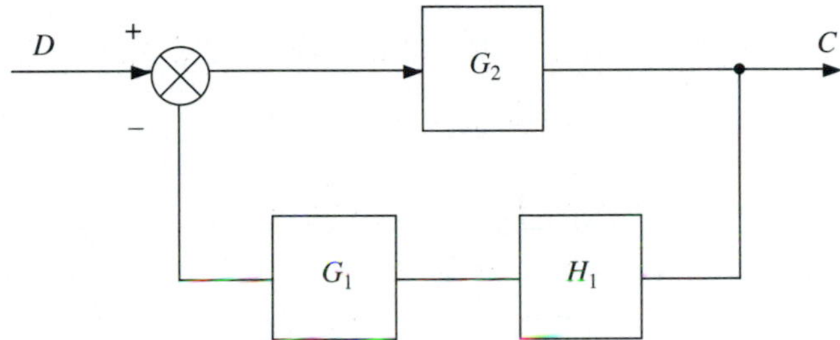

Fig. 2.9 System with one input zero

Now setting the other input D to zero produces the block diagram shown in Fig. 2.10. For this system, the transfer function becomes

$$\frac{C}{R} = \frac{G_1G_2}{1 + G_1G_2H_1} \tag{2.18}$$

or

$$C = \frac{G_1G_2}{1 + G_1G_2H_1}R \tag{2.19}$$

Hence, adding equations 2.17 and 2.19 gives the result

$$C = \frac{G_2}{1 + G_1G_2H_1}D + \frac{G_1G_2}{1 + G_1G_2H_1}R \tag{2.20}$$

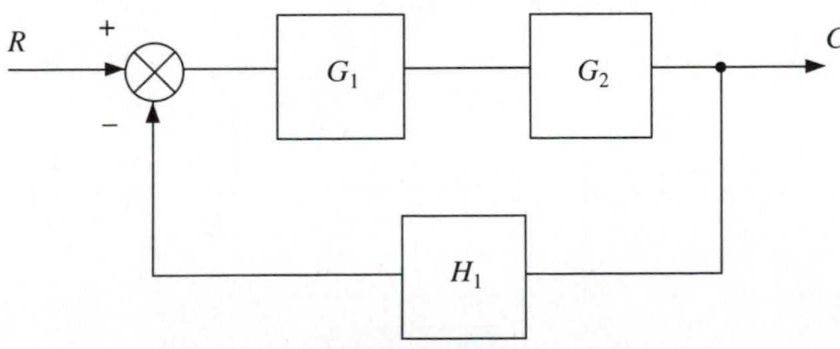

Fig. 2.10 System with other input zero

SAMPLE PROBLEM 2.1

A system is defined by the differential equation

$$\frac{d^2y}{dt^2} + 3\frac{dy}{dt} + 8y = 2x + \frac{dx}{dt}$$

Considering x to be the input and y the output, determine the transfer function of the system, and calculate the output if the input is a unit ramp function. Assume all initial conditions are zero for both input and output.

Solution

Taking the Laplace transform of both sides,

$$s^2Y + 3sY + 8Y = 2X + sX$$

$$Y(s^2 + 3s + 8) = X(s + 2)$$

$$\frac{Y}{X} = \frac{s + 2}{s^2 + 3s + 8}$$

This is the transfer function of the system, which may be represented in block diagram form, as shown in Fig. SP2.1.1. If the input x is a unit ramp, that is, $x(t) = t$, then

$$X(s) = \frac{1}{s^2}$$

Hence

$$Y(s) = \frac{s + 2}{s^2(s^2 + 3s + 8)}$$

$$= \frac{c_1}{s^2} + \frac{c_2}{s} + \frac{c_3s + c_4}{s^2 + 3s + 8}$$

Clearing fractions yields

$$s + 2 = c_1(s^2 + 3s + 8) + c_2s(s^2 + 3s + 8) + s^2(c_3s + c_4)$$

Equating powers of s gives

$$s^3: \quad 0 = c_2 + c_3$$

$$s^2: \quad 0 = c_1 + 3c_2 + c_4$$

$$s^1: \quad 1 = 3c_1 + 8c_2$$

$$s^0: \quad 2 = 8c_1$$

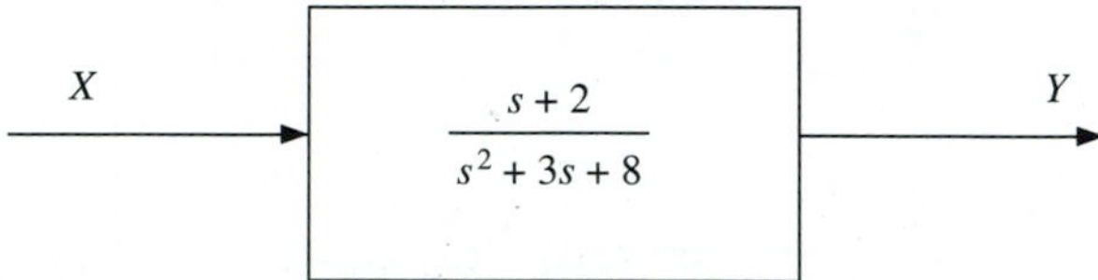

Fig. SP2.1.1

Solving the above produces the partial-fraction coefficients

$$c_1 = \tfrac{1}{4} \qquad c_2 = \tfrac{1}{32} \qquad c_3 = -\tfrac{1}{32} \qquad c_4 = -\tfrac{11}{32}$$

The output is given by the inverse transform

$$y(t) = c_1 t + c_2 + c_3 e^{-3t/2} \cos t\sqrt{\tfrac{23}{4}} + c_4\sqrt{\tfrac{4}{23}}e^{-3t/2} \sin t\sqrt{\tfrac{23}{4}}$$

$$= \frac{t}{4} + \frac{1}{32} - e^{-3t/2}\left[\tfrac{1}{32} \cos t\sqrt{\tfrac{23}{4}} + \tfrac{11}{32}\sqrt{\tfrac{4}{23}} \sin t\sqrt{\tfrac{23}{4}}\right]$$

which is the result required.

Sample Problem 2.2

Figure SP2.2.1 shows a schematic diagram of a permanent-magnet DC motor. Determine the transfer function between the input voltage v_i and the output speed of the motor.

Solution

Direct-current electric motors may be configured in one of two ways: field-controlled or armature-controlled. Figure SP2.2.1 shows an armature-controlled motor that will be studied here. The case of a field-controlled motor will be dealt with in Module 3. In order to determine the dynamic response of

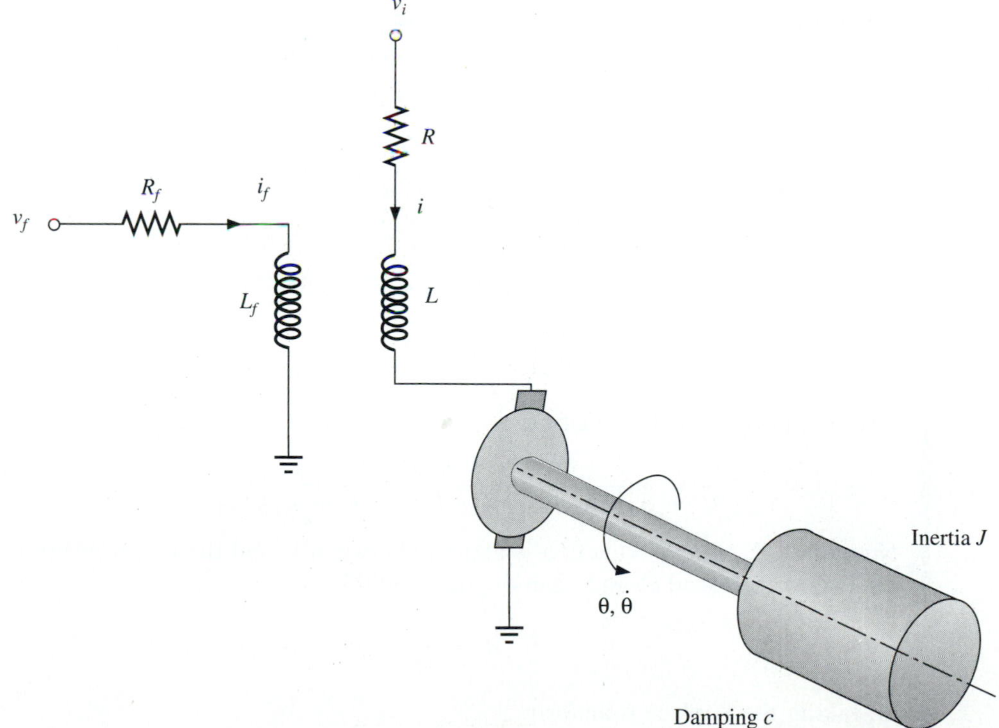

Fig. SP2.2.1

the motor, both the electrical and mechanical responses have to be considered. The motor works in the following manner. In an armature-controlled motor, the field current is constant, and the speed is controlled by the variable armature current, which in turn is controlled by the armature input voltage v_i. When a step change in v_i occurs, for example, when a constant voltage is applied to the motor terminals, a current is not immediately established in the circuit, due to the inductance L of the armature windings. Once a current has begun to flow in the circuit, the motor starts to turn, since the torque developed by the armature is proportional to the current flowing. After rotation of the armature begins, a back electromotive force (EMF) defined by v_b is generated in the armature and opposes the input voltage v_i. The torque causing the motor to rotate has to overcome the inertia of the armature and viscous friction in the bearings, which may be considered proportional to the rotational speed of the motor.

Using the symbols defined in Fig. SP2.2.1, the differential equation relating the electrical variables may be written as

$$L\frac{di}{dt} + Ri = v_i - K_v\frac{d\theta}{dt}$$

where K_v is the back EMF constant in volts per radian per second. The differential equation of the mechanical variables is obtained from Newton's equation for rotational systems:

$$J\frac{d^2\theta}{dt^2} + c\frac{d\theta}{dt} = K_t i$$

where K_t is the motor torque constant in newton meters per ampere. Taking Laplace transforms of both equations and ignoring initial conditions give

$$LsI + RI = V_i - K_v s\Theta$$

$$Js^2\Theta + cs\Theta = K_t I$$

Eliminating I from the above equations results in

$$(Js^2 + cs)(Ls + R)\Theta = K_t V_i - K_t K_v s\Theta$$

$$[(Js^2 + cs)(Ls + R) + K_v K_t s]\Theta = K_t V_i$$

$$\left\{s\left[s^2 + s\left(\frac{R}{L} + \frac{c}{J}\right) + \left(\frac{cR + K_v K_t}{JL}\right)\right]\right\}\Theta = \frac{K_t}{JL}V_i$$

resulting in the transfer function

$$\frac{\Theta}{V_i} = \frac{K_t/JL}{s[s^2 + s(R/L + c/J) + (cR + K_v K_t/JL)]}$$

Note that this transfer function is actually between V_i and the motor position Θ. The motor speed Ω and position are related by

$$\omega = \frac{d\theta}{dt}$$

or, taking the Laplace transform,

$$\Omega = s\Theta$$

Hence the final result is

$$\frac{\Omega}{V_i} = \frac{K_t/JL}{s^2 + s(R/L + c/J) + (cR + K_vK_t/JL)}$$

Again this is a second-order system like the accelerometer and will be studied in more detail in a later module.

Sample Problem 2.3

Simplify the block diagram shown in Fig. SP2.3.1 and determine the closed-loop transfer function C/R.

Solution

Observing Fig. SP2.3.1, it may be seen that there are no serial elements, no parallel blocks, and no simple feedback loop to be rationalized. The simplification begins with step 4, by moving G_1 to the left of summing junction s_2. This rearrangement is shown in Fig. SP2.3.2. Again using step 4, the block G_1 in the feedback loop is moved to the right of the pickoff point p_1, resulting in Fig. SP2.3.3. Now, the part of Fig. SP2.3.3 formed by s_1, s_3, and p_1 may be

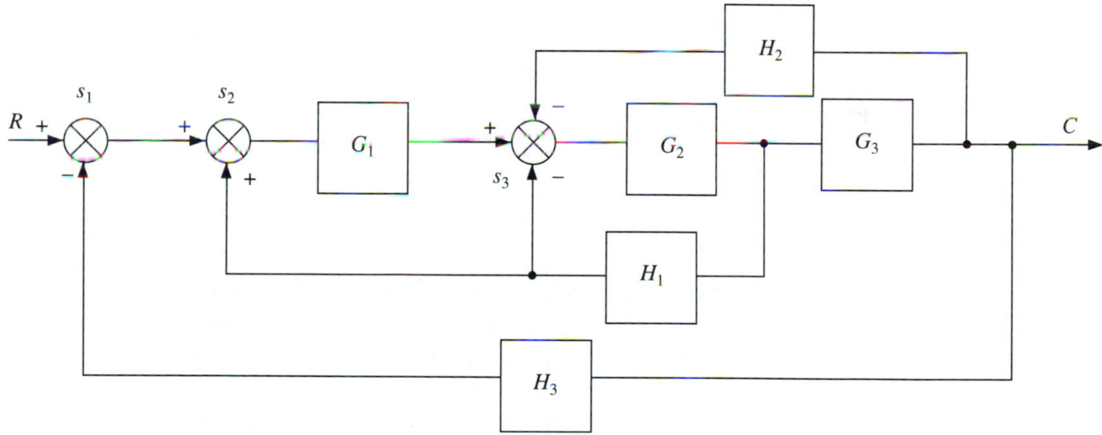

Fig. SP2.3.1

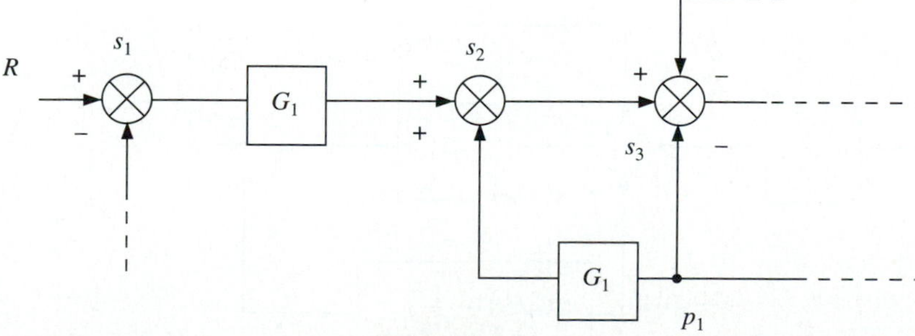

Fig. SP2.3.2

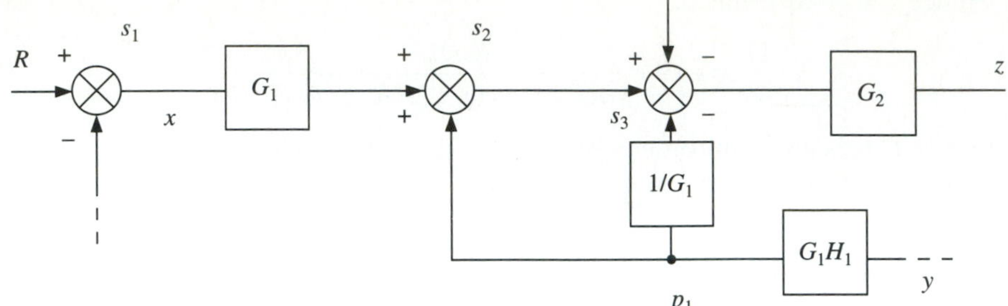

Fig. SP2.3.3

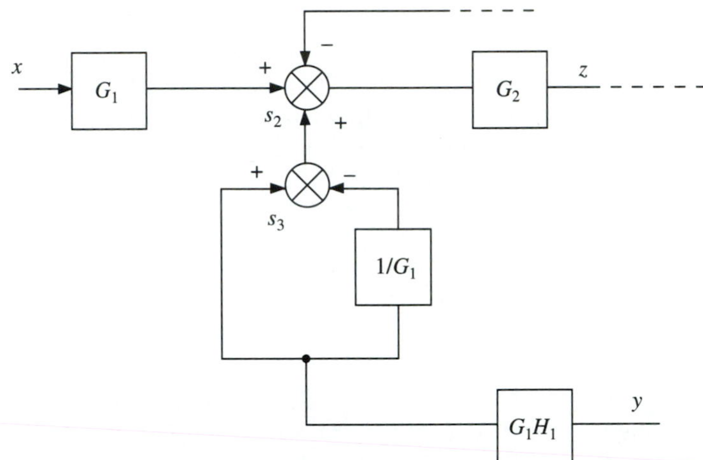

Fig. SP2.3.4

identified as a parallel network and may be reduced to the equivalent parallel network shown in Fig. 2.3.4. Be sure to check that the transfer function between signals z and x and between z and y are the same in Figs. SP2.3.3 and SP2.3.4. The two parallel paths may be combined (using rule 2) to produce Fig. SP2.3.5. The inner loop may now be simplified (using rule 3) to

$$\frac{G}{1 + GH} = \frac{G_2}{1 - G_1G_2H_1(1 - 1/G_1)} = \frac{G_2}{1 + G_2H_1 - G_1G_2H_1}$$

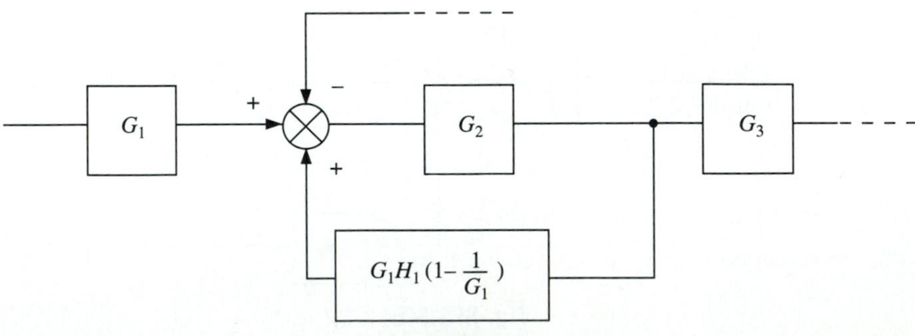

Fig. SP2.3.5

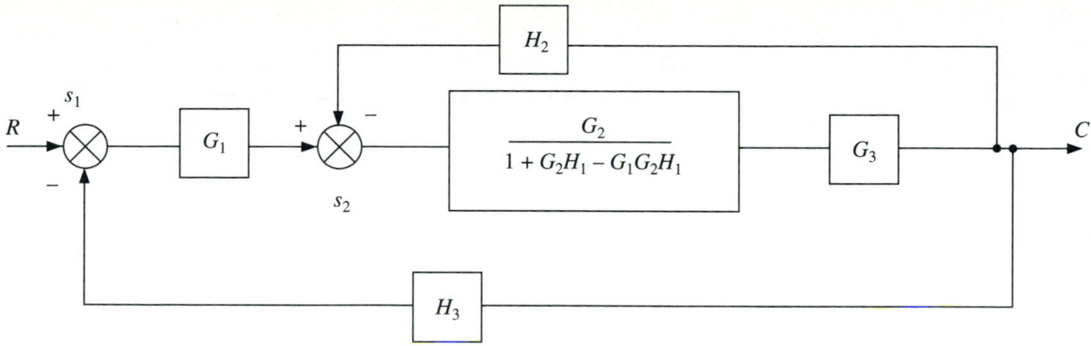

Fig. SP2.3.6

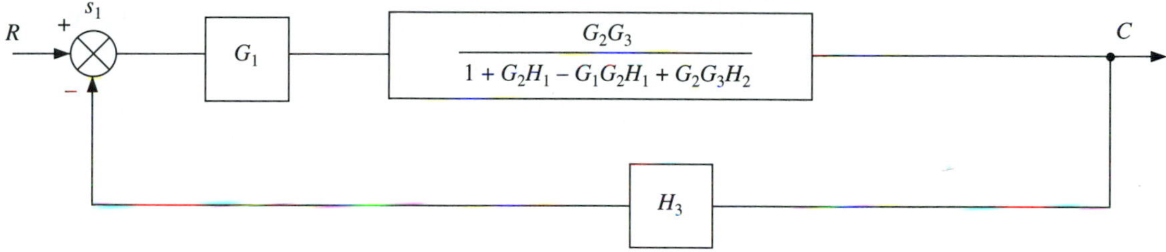

Fig. SP2.3.7

as shown in Fig. SP2.3.6. Now the loop with feedback element H_2 may be closed to yield

$$\frac{G}{1 + GH} = \frac{G_2G_3}{1 + G_2H_1 - G_1G_2H_1 + G_2G_3H_2}$$

as shown in Fig. SP2.3.7. Finally the closed-loop transfer function may be written as

$$\frac{C}{R} = \frac{G_1G_2G_3}{1 + G_2H_1 - G_1G_2H_1 + G_2G_3H_2 + G_1G_2G_3H_3}$$

◆ PROBLEMS

2.1 A system of unknown transfer function is shown in Fig. P2.1. If a unit impulse applied at the input produces at the output a signal described by the time function

$$r(t) = 2e^{-3t}$$

determine the unknown transfer function.

2.2 Find the solution of the differential equation

$$\frac{d^2x}{dt^2} + \frac{dx}{dt} + 8x = \frac{dz}{dt} + 3z$$

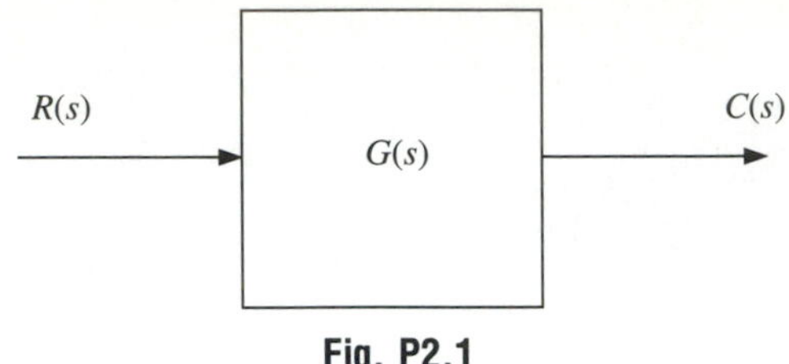

Fig. P2.1

when

$$z(t) = e^{-2t}$$

and all other initial conditions are zero.

2.3 For the system shown in Fig. P2.3, determine the relationship between voltage and current, express this relationship in the form of a transfer function, and determine the current as a function of time when the voltage is a step change from zero to 10 V.

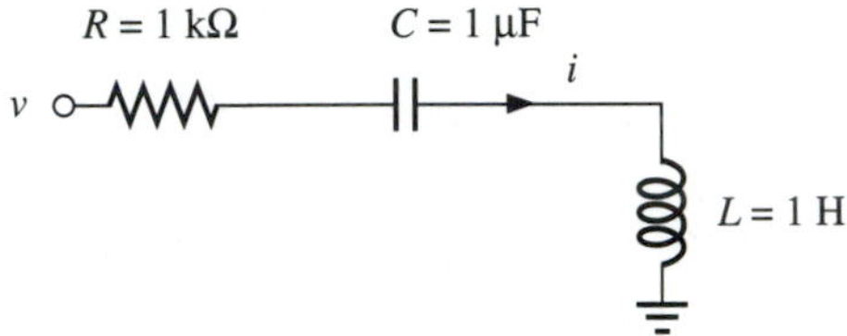

Fig. P2.3

2.4 A simple gearbox is shown in Fig. P2.4, where the input shaft rotates with speed ω_i and has a torque τ_i applied to it while the output shaft rotates with speed ω_o and has torque τ_o. The input gear has N_1 teeth while the output gear has N_2 teeth, where

$$N_1 < N_2$$

Write down the transfer function of the gearbox for **(a)** speed, **(b)** torque, and **(c)** power.

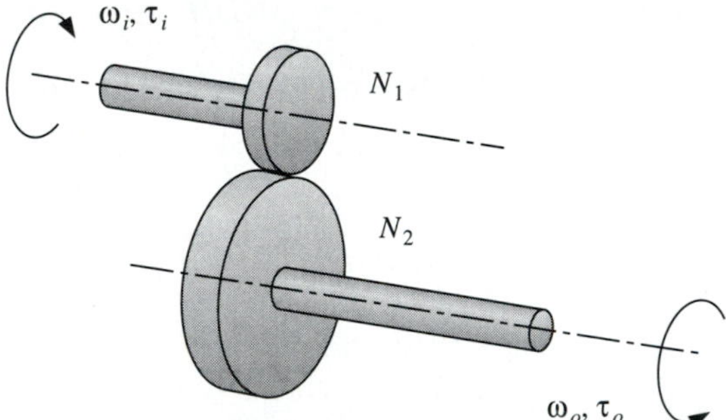

Fig. P2.4

2.5 Figure P2.5 shows the same gearbox in problem 2.4, but this time the output shaft has an inertial load J attached to it. Determine the relationship between the torque applied to the input shaft τ_i and the position of the output shaft θ_i.

2.6 For the mechanical system shown in Fig. 2.6, determine the transfer function between the input displacement x and the output displacement y.

2.7 Find the displacement from the equilibrium position x of the mass shown in Fig. P2.7 when the input displacement y is a step function at $t = 0$ and of magnitude 10 cm.

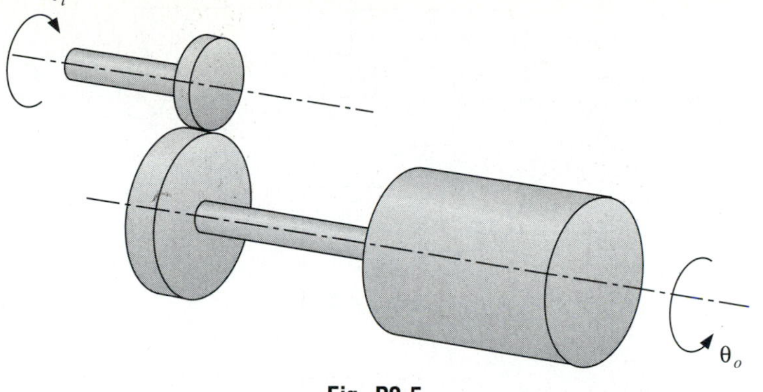

Fig. P2.5

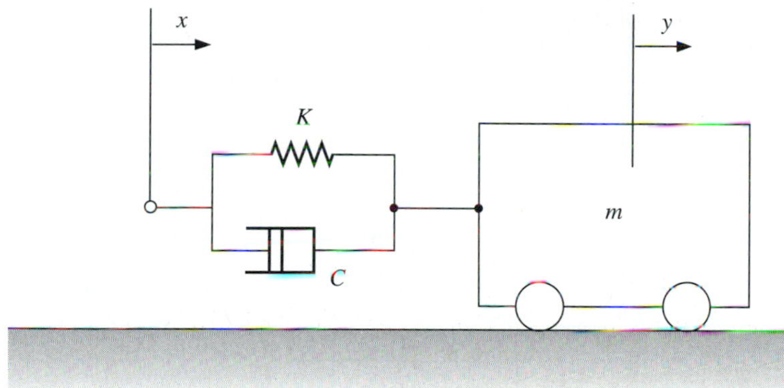

Fig. P2.6

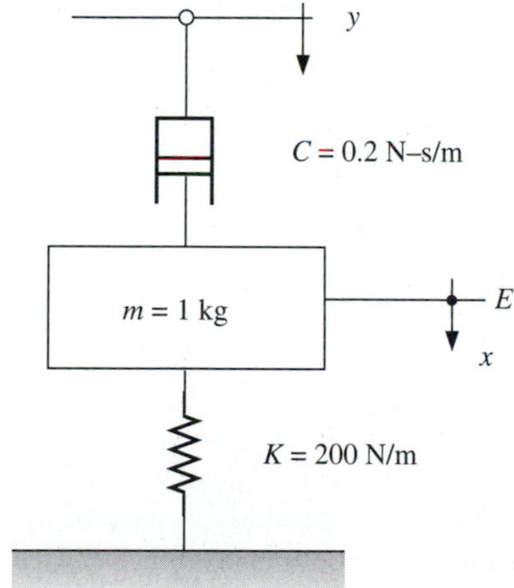

Fig. P2.7

2.8 For the system shown in Fig. P2.8 determine the closed-loop transfer function C/R.

2.9 For the single-input system shown in Fig. P2.9, find the transfer function of output to input C/R.

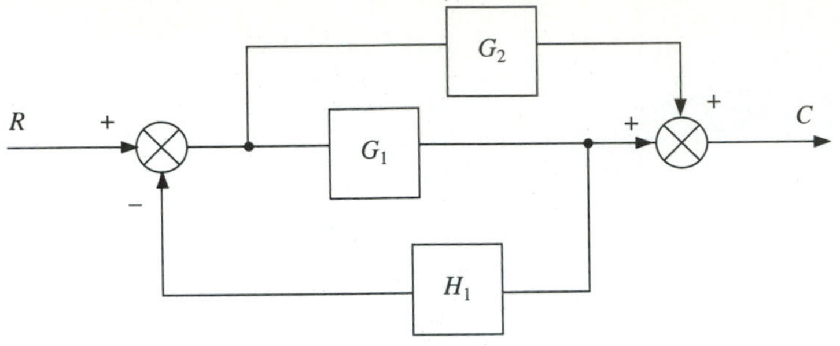

Fig. P2.8

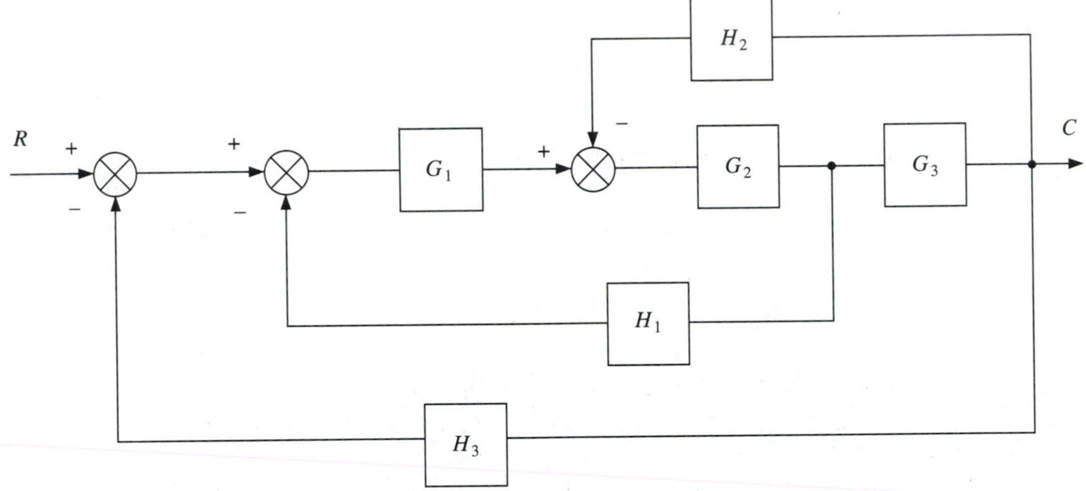

Fig. P2.9

2.10 Determine C as a function of R_1, R_2, and R_3 for the feedback control system shown in Fig. P2.10.

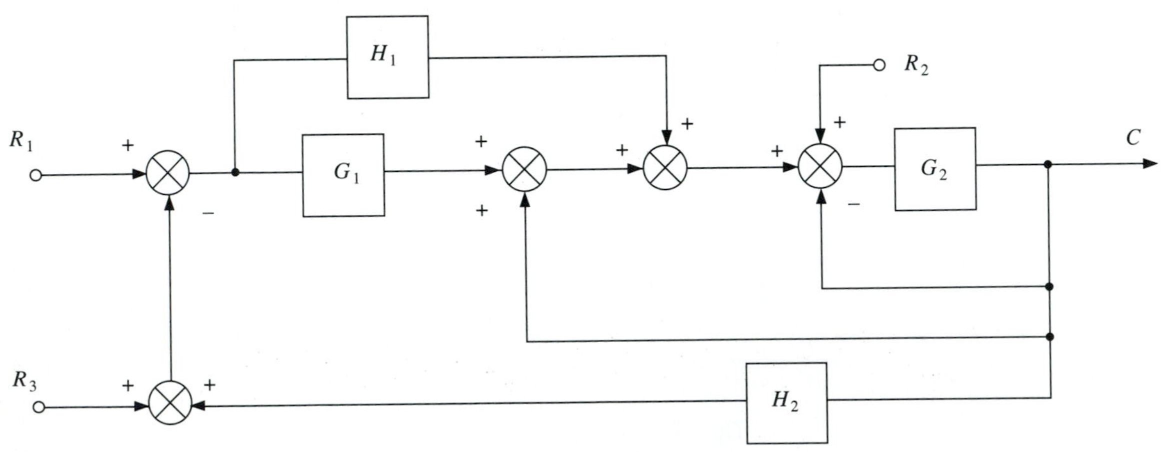

Fig. P2.10

First-Order Systems

One of the simplest systems studied is that represented by a first-order differential equation. Such a system is known as a first-order system. There are many examples of first-order systems; the first one to be studied is mechanical in nature and is shown in Fig. 3.1. The input variable x_i and output variable x_o are related by the equation obtained by equating the forces at A:

$$K(x_i - x_o) = C\dot{x}_o \tag{3.1}$$

or

$$\frac{C}{K}\dot{x}_o + x_o = x_i \tag{3.2}$$

Taking Laplace transforms, assuming initial conditions to be zero,

$$X_o\left(s\frac{C}{K} + 1\right) = X_i \tag{3.3}$$

leading to the transfer function

$$\frac{X_o}{X_i} = \frac{1}{1 + (C/K)s} = \frac{1}{1 + \tau s} \tag{3.4}$$

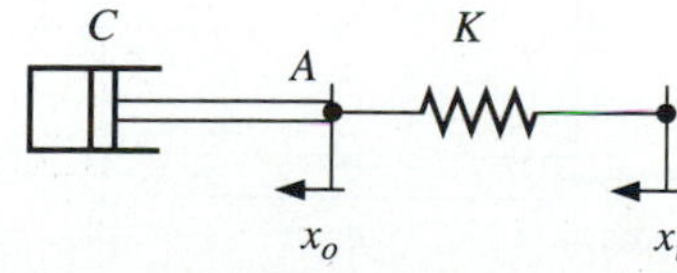

Fig. 3.1 Mechanical first-order system

where the time constant τ is given by

$$\tau = \frac{C}{K} \tag{3.5}$$

and may be shown to have units of seconds. Another first-order system is shown in Fig. 3.2. In this system the input is voltage v_i while the output is voltage v_o, and they are related by applying Kirchhoff's law to the circuit. This gives

$$v_i - v_o = iR \qquad i = C\frac{dv_o}{dt} \tag{3.6}$$

Taking Laplace transforms, assuming zero initial conditions, and eliminating i result in

$$V_i - V_o = RCsV_o \tag{3.7}$$

or

$$V_i = V_o(1 + RCs) \tag{3.8}$$

leading to the transfer function

$$\frac{V_o}{V_i} = \frac{1}{1 + RCs} = \frac{1}{1 + \tau s} \tag{3.9}$$

where the time constant is given by

$$\tau = RC \tag{3.10}$$

A third first-order system is shown in Fig. 3.3, representing a pair of hydraulic tanks connected by a pipe containing a laminar restrictor. This restrictor produces a pressure drop linearly proportional to the flow through it. We know from hydrostatics, however, that pressure and head may be related by the equation

$$p = \rho gh \tag{3.11}$$

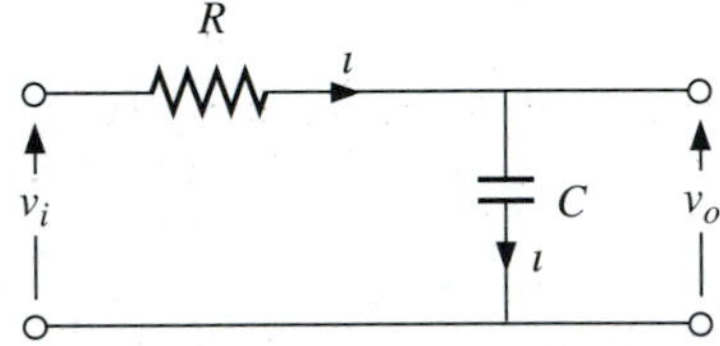

Fig. 3.2 Electrical first-order system

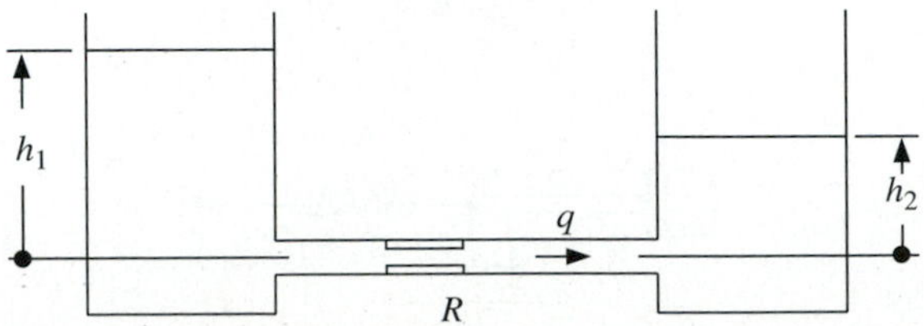

Fig. 3.3 Hydraulic first-order system

The input variable is h_1 while the output variable is h_2. The equation of the laminar restrictor may be written as

$$h_1 - h_2 = qR \tag{3.12}$$

where q is the volumetric flow through the pipe and R is the flow constant (resistance) of the restrictor. In addition, it may be seen from conservation of mass that

$$q = A\frac{dh_2}{dt} \tag{3.13}$$

where A is the area of the second tank. Eliminating the variable q, we get

$$h_1 - h_2 = AR\frac{dh_2}{dt} \tag{3.14}$$

Taking Laplace transforms with zero initial conditions as before,

$$H_1 - H_2 = ARsH_2 \tag{3.15}$$

or

$$H_2(1 + ARs) = H_1 \tag{3.16}$$

leads to the transfer function

$$\frac{H_2}{H_1} = \frac{1}{1 + ARs} = \frac{1}{1 + \tau s} \tag{3.17}$$

Here, the time constant is

$$\tau = AR \tag{3.18}$$

It may be seen from the previous examples that many different systems may be represented in first-order form, and the generalized transfer function between the input $R(s)$ and output $C(s)$ may be related by the equation

$$\frac{C(s)}{R(s)} = \frac{1}{1 + \tau s} \tag{3.19}$$

and shown in block diagram form as in Fig. 3.4. This generalized system will be studied in some detail by determining its response to some common inputs.

◆ IMPULSE RESPONSE

If the input is a unit impulse, then

$$R(s) = 1 \tag{3.20}$$

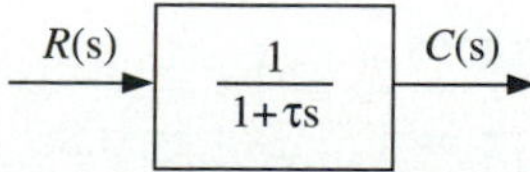

Fig. 3.4 First-order system transfer function

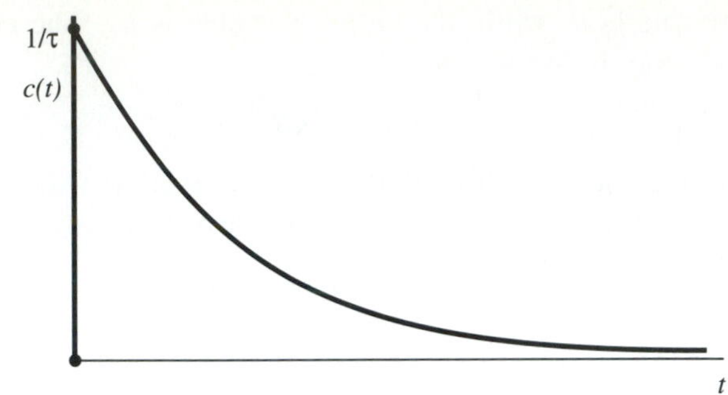

Fig. 3.5 Impulse response

leading to

$$C(s) = \frac{1}{1 + \tau s} \tag{3.21}$$

Taking the inverse Laplace transform gives

$$c(t) = \frac{1}{\tau} e^{-t/\tau} \tag{3.22}$$

Figure 3.5 shows the output of the system. Since we have assumed zero initial conditions, the output must change instantaneously from 0 at $t = (0-)$ to $1/\tau$ at $t = (0+)$.

◆ STEP RESPONSE

In this case, for a unit step,

$$R(s) = \frac{1}{s} \tag{3.23}$$

Hence the output becomes

$$C(s) = \frac{1}{s(1 + \tau s)} = \frac{1}{s} - \frac{\tau}{1 + \tau s} = \frac{1}{s} - \frac{1}{s + 1/\tau} \tag{3.24}$$

giving the time-domain response

$$c(t) = 1 - e^{-t/\tau} \tag{3.25}$$

Again the input and output are plotted in Fig. 3.6. Consider the magnitude of the output when $t = \tau$. This may be calculated from

$$c = 1 - e^{-1} = 0.63 \tag{3.26}$$

When $t = 2\tau$, the output becomes

$$c = 1 - e^{-2} = 0.86 \tag{3.27}$$

First-order systems may be described as reaching 63% of their final value in one time constant and 86% of their final value in 2τ, 95% in 3τ, and 98% in 4τ.

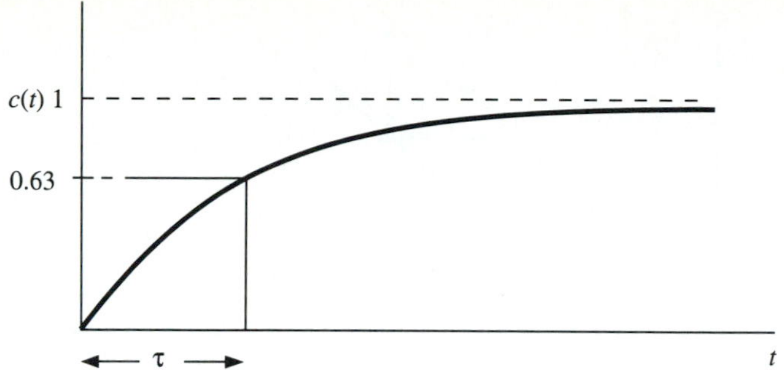

Fig. 3.6 Step response

◆ RAMP RESPONSE

If the input is the unit ramp

$$r(t) = t \tag{3.28}$$

then

$$R(s) = \frac{1}{s^2} \tag{3.29}$$

leading to

$$C(s) = \frac{1}{s^2(1 + \tau s)} \tag{3.30}$$

$$= \frac{1}{s^2} - \frac{\tau}{s} + \frac{\tau^2}{1 + \tau s} \tag{3.31}$$

$$= \frac{1}{s^2} - \frac{\tau}{s} + \frac{\tau}{s + \frac{1}{\tau}} \tag{3.32}$$

The time-domain response becomes

$$c(t) = t - \tau - \tau e^{-t/\tau} = t - \tau(1 - e^{-t/\tau}) \tag{3.33}$$

Figure 3.7 shows the relationship between the input and output for this case. The response consists of a transient part and a steady-state part, and in the steady state the output lags the input by a time equal to the time constant τ. It may be shown that if the input were not the unit ramp but

$$r(t) = At \tag{3.34}$$

then at a fixed time the difference between the input and steady-state output would be $A\tau$.

◆ HARMONIC RESPONSE

This example is somewhat complex to solve in the general case, so to simplify the situation, we will stipulate that only the steady-state solution will be determined.

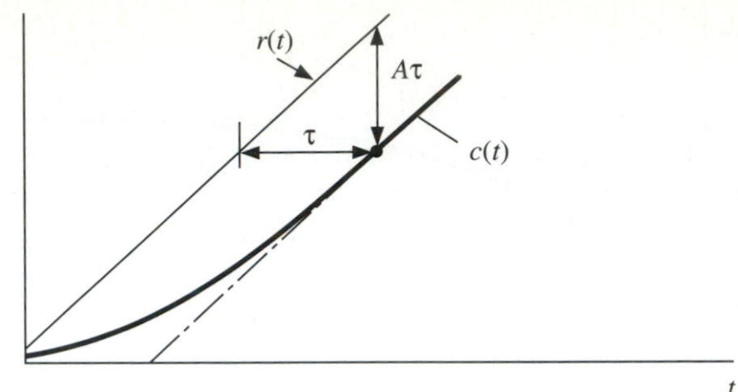

Fig. 3.7 Ramp response

The method of solution will not involve Laplace transforms, since an interesting point will emerge if another approach is adopted. Since it is known that the system under study is linear time invariant, the response to the input

$$r(t) = r_0 e^{j\omega t} \tag{3.35}$$

may be written as

$$c(t) = Ae^{j(\omega t - \phi)} = A\frac{e^{j\omega t}}{e^{j\phi}} = c_0 e^{j\omega t} \tag{3.36}$$

where c_0 is usually a complex number. Hence, substituting into

$$\tau\dot{c} + c = r \tag{3.37}$$

gives

$$j\omega\tau c_0 e^{j\omega t} + c_0 e^{j\omega t} = r_0 e^{j\omega t} \tag{3.38}$$

or

$$c_0(1 + j\omega\tau) = r_0 \tag{3.39}$$

results in

$$\frac{c_0}{r_0} = \frac{1}{1 + j\omega\tau} \tag{3.40}$$

Three important points emerge from the above equation. First, the output will be modified in magnitude from the input, the amount of change being equal to the magnitude of the complex number c_0. Second, the output will be shifted in phase compared to the input, the phase shift corresponding to the argument of the complex number c_0. Third, the above equation could have been obtained simply by substituting $s = j\omega$ directly into the transfer function. It is easily shown that this is always the case when the input is a harmonic signal and is a fact that will be made use of later in the book. From equation 3.40 the magnitude of the output is given by the magnitude of c_0, where

$$A = |c_0| = \left| r_0 \frac{1 - j\omega\tau}{1 + \omega^2\tau^2} \right| \tag{3.41}$$

$$= \frac{r_0}{\sqrt{1 + \omega^2\tau^2}} \tag{3.42}$$

Similarly, the phase of c with respect to r is given by

$$\angle c_0 = \phi = -\tan^{-1} \omega\tau \tag{3.43}$$

It may be seen that as the frequency of the input signal increases from zero to infinity, the amplitude of the output decreases from $c_0 = r_0$ to $c_0 = 0$. The phase meanwhile changes from zero to $-90°$ as the frequency increases. It is customary to plot both magnitude and phase as a function of the frequency, as shown in Fig. 3.8, while Fig. 3.9 shows the time-domain relationship between the input and output at a particular frequency.

◆ FIRST-ORDER FEEDBACK SYSTEMS

Now that the response of first-order systems has been studied in some detail, one of the principal benefits of feedback control may be illustrated. Suppose it is necessary to control a system known to be of first order. The simplest method will be to implement open-loop control as illustrated in Fig. 3.4. If the time constant is assumed to be $\tau = 1$ s, then the response to a unit step will be of the form

$$c(t) = 1 - e^{-t} \tag{3.44}$$

as shown in Fig. 3.6, reaching 63% of the steady-state output in 1 s. Now, suppose the same first-order system is considered to be the plant in a feedback control system with a variable amplifier gain as controller, as shown in Fig. 3.10. This time the relationship between the input and output is given by the closed-loop transfer function

$$\frac{C}{R} = \frac{KG}{1 + KGH} = \frac{K/(1 + \tau s)}{1 + K/(1 + \tau s)} = \frac{K}{1 + K + \tau s} \tag{3.45}$$

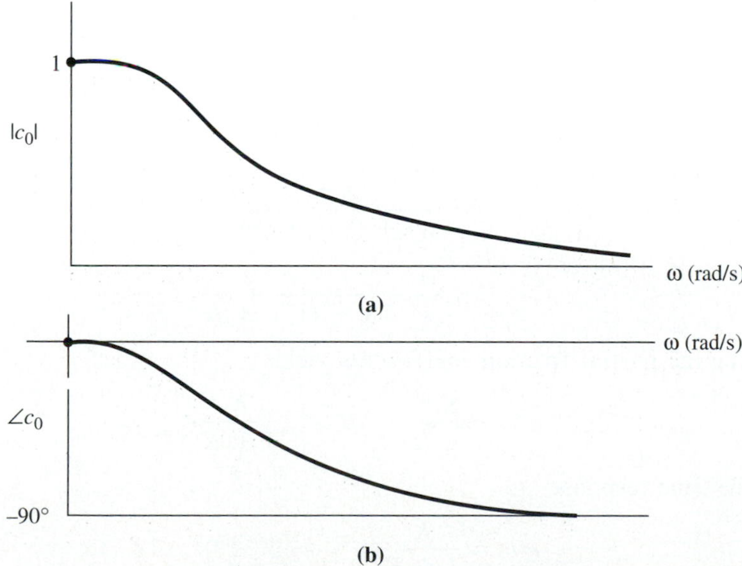

Fig. 3.8 Magnitude and phase of output signal

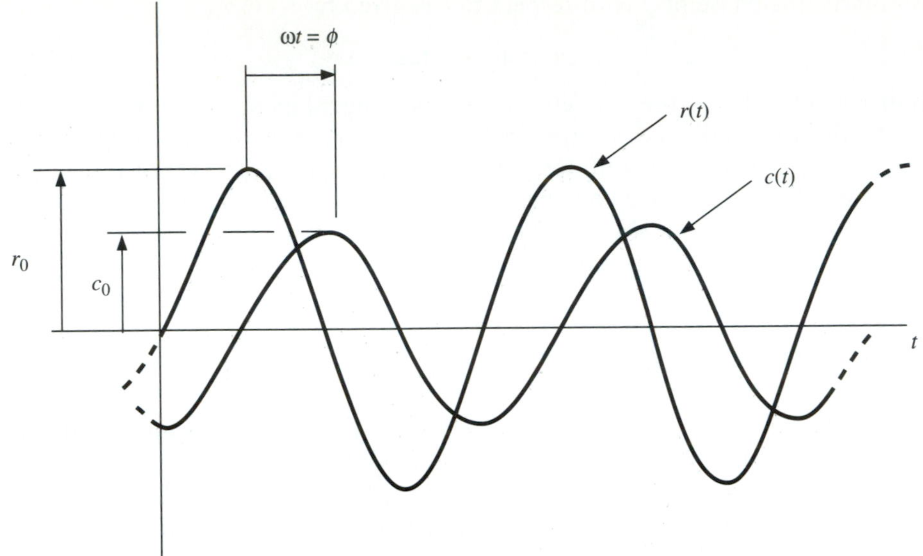

Fig. 3.9 Time-domain relationship between input and output

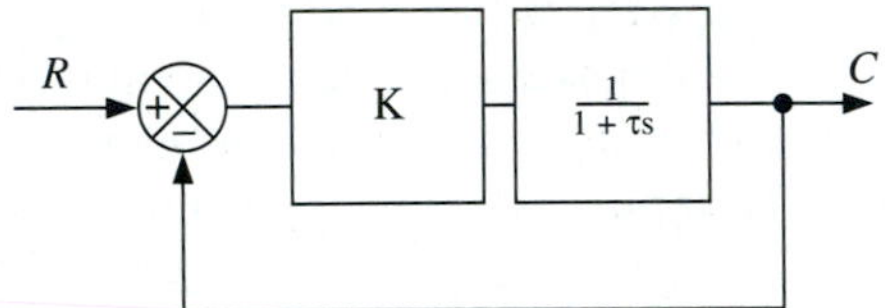

Fig. 3.10 First-order feedback control system

If the input is a unit step, then

$$C(s) = \frac{K}{s(1 + K + \tau s)} \tag{3.46}$$

$$= \frac{K}{s(\tau s + 1 + K)} \tag{3.47}$$

$$= \frac{K/\tau}{s(s + (1 + K)/\tau)} \tag{3.48}$$

$$= \frac{a}{s} + \frac{b}{s + (1 + K)/\tau} \tag{3.49}$$

Evaluating the partial-fraction coefficients yields

$$a = \frac{K}{1 + K} \qquad b = -\frac{K}{1 + K} \tag{3.50}$$

Hence the time response

$$c(t) = \frac{K}{1 + K}(1 - e^{-[(1+K)/\tau]t}) \tag{3.51}$$

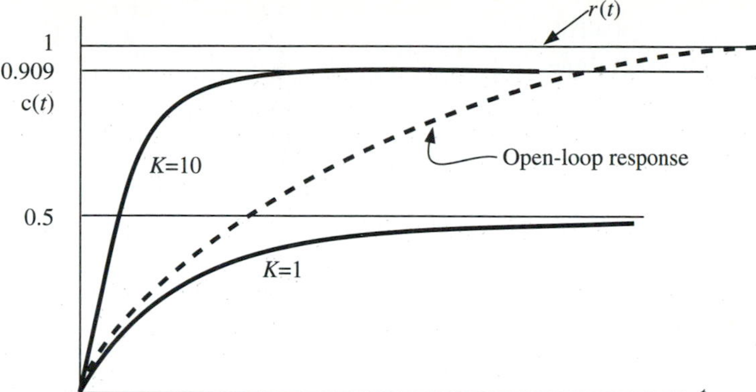

Fig. 3.11 Step response of feedback system

Suppose that $\tau = 1$ and we arbitrarily set $K = 1$; we get

$$c(t) = 0.5(1 - e^{-2t}) \tag{3.52}$$

This output is shown in Fig. 3.11, which also includes the step response for the open-loop system. Note the following important observations:

1. The feedback system has a steady-state error between the input and output.
2. The feedback system response is faster than the open-loop system.

Hence we have improved the system response, which is good; however, the steady-state error, which in this case is considerable, is a serious impediment to the system performance.

Notice now, however, that if the gain K is increased to 10, then the output becomes

$$c(t) = 0.909(1 - e^{-11t}) \tag{3.53}$$

This case, which is also plotted in Fig. 3.11, indicates that the steady-state error is now reduced to less than 10% while the closed-loop time constant is 0.09 s, resulting in over 10 times the response speed of the open-loop system.

This example clearly shows that through the use of feedback *it is possible to change the apparent physical properties of a system in order to improve its dynamic response*. This process may, as in the case of this example, have some detrimental effects such as the appearance of a steady-state error; however, by using a large value of gain, not only does the system response become very fast, but the steady-state error becomes smaller. In practical cases the gain cannot be increased indefinitely without signal saturation occurring, causing the system to become *nonlinear*.

◆ COMPLEX-PLANE REPRESENTATION: POLES AND ZEROS

We have seen that the use of Laplace transforms allows the representation of many different control systems in terms of a notation independent of the physical

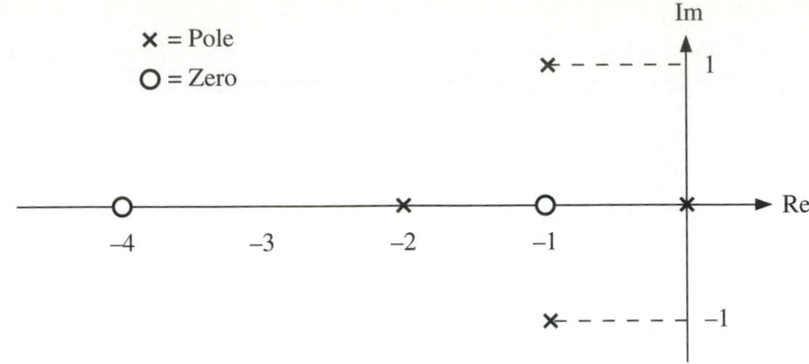

Fig. 3.12 Poles and zeros in the complex plane

origin of the resulting transfer function. The notation of *poles* and *zeros* is normally used, for example, to describe transfer functions in further detail. Suppose $X(s)$ is written in a generalized form,

$$X(s) = \frac{K(s + z_1)(s + z_2)\cdots(s + z_m)}{s^k(s + p_1)(s + p_2)\cdots(s + p_n)} \tag{3.54}$$

The zeros of X are defined as the values of s that make the numerator zero and, for the above equation, take the values $s = -z_1, -z_2, \ldots, -z_m$. The poles of X are the values of s that make the denominator zero and consist of k poles for $s = 0$ and n poles for $s = -p_1, -p_2, \ldots, -p_n$.

In general, the poles and zeros may be complex, as in the case where the numerator or denominator of X contains a quadratic or higher-order factors. It is customary to plot the poles and zeros on a graph representing the complex plane, or s plane. It will be shown in the next section that these locations provide a qualitative insight into the expected time-domain response of the system. Figure 3.12 shows the s plane with the poles and zeros of the transfer function

$$G(s) = \frac{K(s + 1)(s + 4)}{s(s + 2)(s^2 + 2s + 2)} \tag{3.55}$$

noting that the quadratic term in the denominator yields a pair of poles at $s = -1 \pm j$. Any complex poles or zeros will always occur as complex conjugates.

◆ POLES AND ZEROS OF FIRST-ORDER SYSTEMS

Consider the transfer function of the first-order system developed in detail earlier in the module,

$$G(s) = \frac{1}{1 + \tau s} \tag{3.56}$$

It is seen that this system has no zeros and one pole at

$$s = -\frac{1}{\tau} \tag{3.57}$$

Recall that the transient response of the system, for any input, was characterized by the time constant τ, or more specifically $1/\tau$, the *system pole*. This feature is not limited to first-order systems and may be stated in a wider context as follows:

> A system's transient response is determined by the poles of the transfer function, which in turn are obtained by setting the denominator of the transfer function to zero and solving the resulting expression, known as the *characteristic equation*.

The student may observe that it is the poles that determine the character of the system response, since they will appear in the denominator of the output $C(s)$ for *any* input $R(s)$. The role of the zeros is in the determination of the partial-fraction coefficients prior to taking the inverse Laplace transform to obtain the time-domain response. In the general feedback case, the characteristic equation is of the form

$$1 + GH(s) = 0 \tag{3.58}$$

In the example of the transfer function $G(s)$ above, the pole is called an *open-loop pole* since the transfer function from which it is derived is an open-loop transfer function. In the case of the closed-loop first-order system described by equation 4.45,

$$\frac{C}{R} = \frac{K}{1 + K + \tau s} = \frac{K/(1 + K)}{1 + [\tau/(1 + K)]s} = \frac{K/(1 + K)}{1 + \tau' s} \tag{3.59}$$

The pole at

$$s = -\frac{1 + K}{\tau} = -\frac{1}{\tau'} \tag{3.60}$$

is known as a *closed-loop pole* since it was derived from the closed-loop transfer function. The above equation shows that this closed-loop pole is a function of K and the system transient response will therefore be a function of this parameter, confirming our previous observation of the effects of feedback on the first-order system. From equation 3.60, larger values of K result in larger values of s, producing smaller effective time constants τ' and a faster response.

◆ DOMINANT POLES

Consider the case where there are two real-axis poles and the transfer function is given by

$$\frac{C}{R} = \frac{100}{(s + 1)(s + 100)} \tag{3.61}$$

In this case, one pole is located close to the imaginary axis at $s = -1$ while the other is located far from the imaginary axis at $s = -100$, as shown in Fig. 3.13. Assuming that the system is subjected to a unit impulse input, the output may be written as

$$C(s) = \frac{100}{(s + 1)(s + 100)} = \frac{A}{s + 1} + \frac{B}{s + 100} \tag{3.62}$$

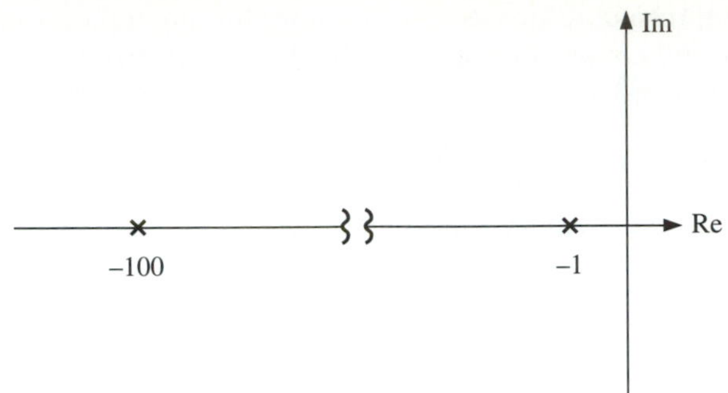

Fig. 3.13 Two-pole system

Calculating $A = 100/99$ and $B = -100/99$, the output may be written as

$$c(t) = 1.01e^{-t} - 1.01e^{-100t} \tag{3.63}$$

$$= 1.01(e^{-t} - e^{-100t}) \tag{3.64}$$

For a given value of t, the term e^{-100t} will be considerably smaller than the term e^{-t}, and for all but $t \to 0$, the contribution from e^{-100t} may be neglected. Observe that this pole originates from the pole at $s = -100$ in the transfer function. This conclusion may be generalized by stating that

> the closed-loop poles close to the imaginary axis *dominate* those farther away, and in many cases the response may be approximated by the response of the closed-loop poles closest to the imaginary axis.

This argument applies equally to complex poles as it does to real poles. For example, the response to any input of a system characterized by two pairs of complex poles at $s = -1 \pm j$ and $s = -100 \pm j$ will be dominated by the poles at $s = -1 \pm j$. This observation is useful in obtaining approximate responses for higher-order systems by ignoring the contributions from those poles farthest away from the imaginary axis.

Sample Problem 3.1

A first-order system is described by the transfer function

$$G(s) = \frac{2}{s + 2}$$

If the open-loop system is subject to the input,

$$r(t) = 5 \sin \omega t$$

Plot the amplitude ratio and phase as a function of frequency, and find the value of ω that makes the phase $-\pi/4$. At this frequency, what is the amplitude of the output?

Solution

We may write

$$G(s) = \frac{1}{1 + s/2}$$

Hence $\tau = 0.5$. Now, since we are dealing with an open-loop system,

$$G(s) = \frac{C}{R}$$

and

$$r(t) = 5 \sin \omega t$$

Then from equation 3.42 it may be seen that

$$|c_0| = \frac{5}{\sqrt{1 + \omega^2/4}}$$

while equation 3.43 gives

$$\phi = -\tan^{-1} \omega/2$$

Figure SP3.1.1 shows the amplitude ratio

$$\frac{c_0}{r_0} = \frac{|c_0|}{5} = \frac{1}{\sqrt{1 + \omega^2/4}}$$

and phase ϕ plotted against frequency. The frequency at which the phase is $-\pi/4$ may be found from the plot or from the phase angle equation

$$\tan \phi = -1 = -\omega/2$$

Hence the required frequency is

$$\omega = 2$$

Note from equation 3.43 that the frequency at which the phase lag is $-\pi/4$ is given by

$$\omega = \frac{1}{\tau}$$

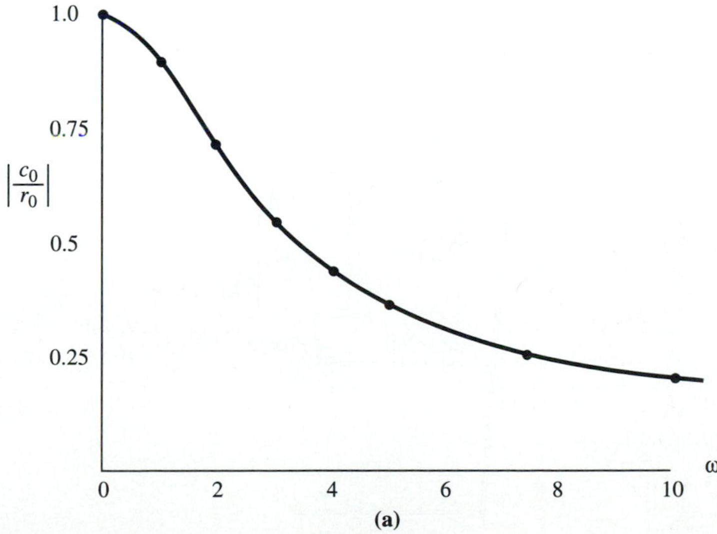

Fig. SP3.1.1a

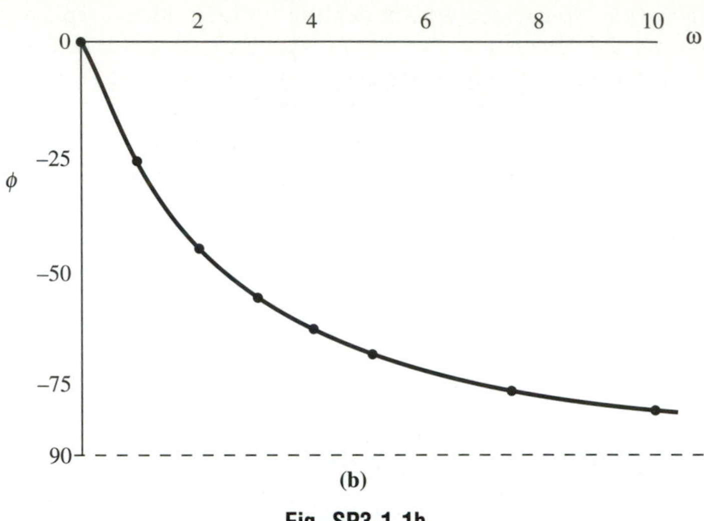

Fig. SP3.1.1b

as shown above. At the frequency $\omega = 2$ rad/s, the amplitude of the output is obtained from

$$|c_0| = \frac{5}{\sqrt{1 + \omega^2/4}} = 3.54$$

which is the required result.

Sample Problem 3.2

Figure SP3.2.1 shows a pneumatic control system used to provide a high-force-level output displacement y from a low-force-level input displacement x. Determine the transfer function of the system and calculate the time constant in terms of the system parameters.

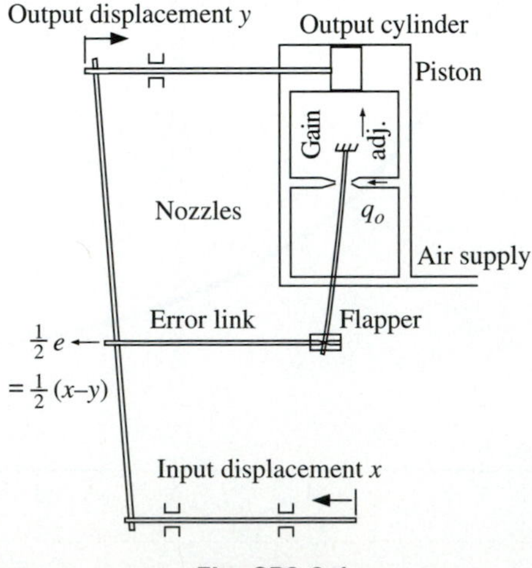

Fig. SP3.2.1

Solution

If the input is a displacement x to the left and the output is a displacement y to the right, then the error link, connected halfway along the connecting link, moves a distance $0.5(x - y) = 0.5e$ in the direction shown. The movement of the flapper at the nozzle is $0.5e/G$, where G is a constant that may be varied by altering the gain. This is achieved by moving the upper end of the flapper beam relative to the nozzles.

If the inertia and friction in the load may be neglected, the pressure on each side of the cylinder will remain constant. If it is assumed that the amount of air passing through the nozzle q_0 varies linearly with flapper displacement and it equals q_i when the flapper is centered between the nozzles, the reduced flow out of the left nozzle when the flapper moves to the left is

$$q_0 = q_i - 0.5k\frac{e}{G}$$

where k is a constant. This causes an additional flow into the left side of the piston,

$$q = q_i - q_0 = 0.5k\frac{e}{G}$$

But since

$$q = A\frac{dy}{dt}$$

where A is the cross-sectional area of the cylinder, then

$$\frac{dy}{dt} = \frac{k}{A}0.5\frac{e}{G} = 0.5K\frac{e}{G}$$

or

$$y = \frac{K}{s}\left(0.5\frac{e}{G}\right)$$

where $K = k/A$. This shows that the piston displacement y is proportional to the time integral of flapper displacement, resulting in the block diagram shown in Fig. SP3.2.2. The relationship between the input x and output y is

$$\dot{y} = 0.5K\frac{e}{G} = \frac{K}{2G}(x - y)$$

which may be written

$$T\dot{y} + y = x$$

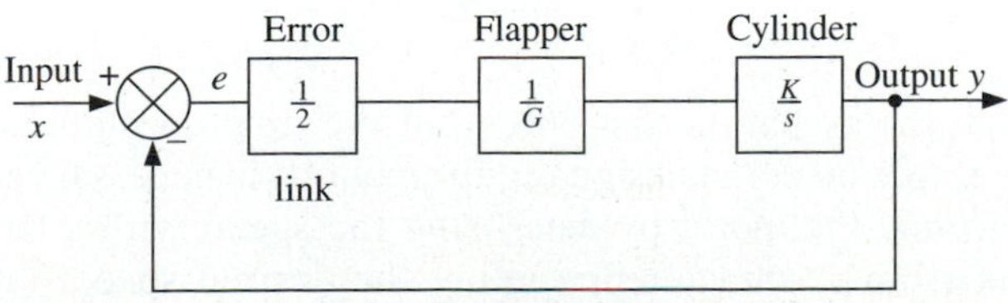

Fig. SP3.2.2

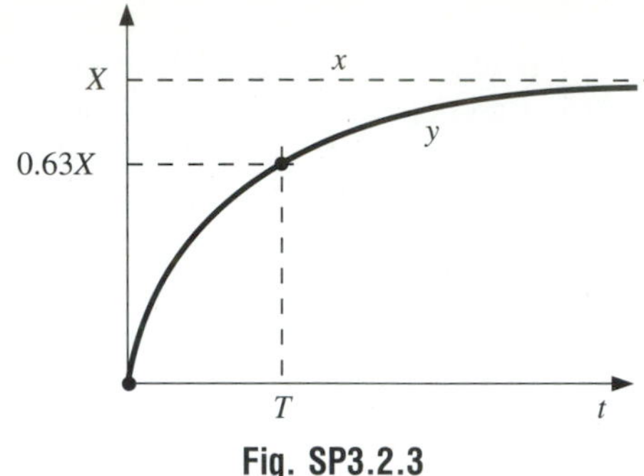

Fig. SP3.2.3

or

$$\frac{y}{x} = \frac{1}{1 + sT}$$

where the time constant $T = 2G/K$. Recall that for first-order systems subject to a step input of amplitude X, the output achieves 63% of the input in one time constant, as shown in Fig. SP3.2.3,

$$y = X(1 - e^{-t/T})$$

A simple experiment may be performed in which the input is moved quickly by some amount, producing an output signal similar to that shown in Fig. SP3.2.3. From this figure the time constant may be measured directly.

Sample Problem 3.3

A small DC motor is used in a speed control system for a computer tape unit. It is required that the time constant of the system be less than 0.1 s. By investigating the possibility of using open- or closed-loop configurations, design a suitable control system given that the transfer function of the motor from input voltage to output speed is given by

$$\frac{\Omega}{V_i} = \frac{2.5}{s + 5}$$

For each case, show the location of the system poles and zeros in the complex, or s, plane.

Solution

The transfer function is written in the form

$$\frac{\Omega}{V_i} = \frac{2.5}{s + 5} = \frac{0.5}{1 + 0.2s}$$

It is seen that the system is first order, but the time constant is $\tau = 0.2$, and does not therefore meet the design requirement. It is necessary to form a feedback loop around the motor by measuring the speed with a tachometer, subtracting that from a voltage representing the desired speed, multiplying this

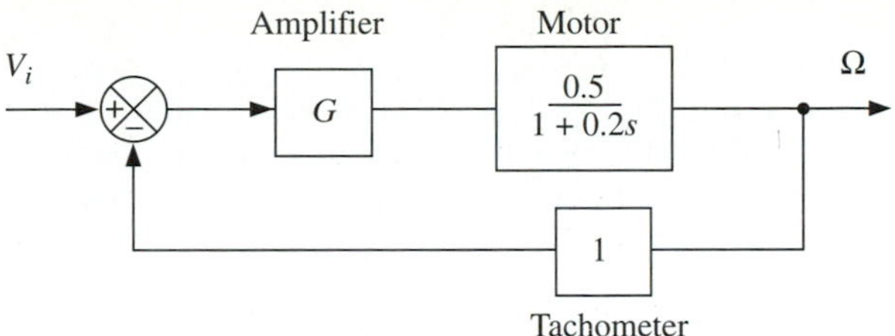

Fig. SP3.3.1

error voltage by a gain G, and using this to drive the motor. Such a closed-loop system is shown in Fig. SP3.3.1, where the tachometer transfer function is assumed to be unity. The closed-loop transfer function may now be calculated as

$$\frac{\Omega}{V_i} = \frac{2.5G}{s + 5 + 2.5G} = \frac{K'}{1 + \tau s}$$

where

$$K' = \frac{2.5G}{5 + 2.5G} \qquad \tau = \frac{1}{5 + 2.5G}$$

In order for the closed-loop system to meet the design requirements of $\tau <$ 0.1 s, it is necessary that

$$G > 2$$

Figure SP3.3.2 shows that both the open- and closed-loop systems are represented by a single pole, but for the feedback configuration the location of the pole is determined by the parameter G. Several locations of the closed-loop pole are indicated. Interestingly, the open- and closed-loop poles are coincident for $G = 0$. Note further that the time-domain constraint $\tau < 0.1$ may be interpreted as an s-plane constraint, namely that the closed-loop pole must lie to the left of a vertical line through the point $s = -10$. This correspondence between the time domain and the s-plane representation of the system is important and will receive further emphasis in later modules.

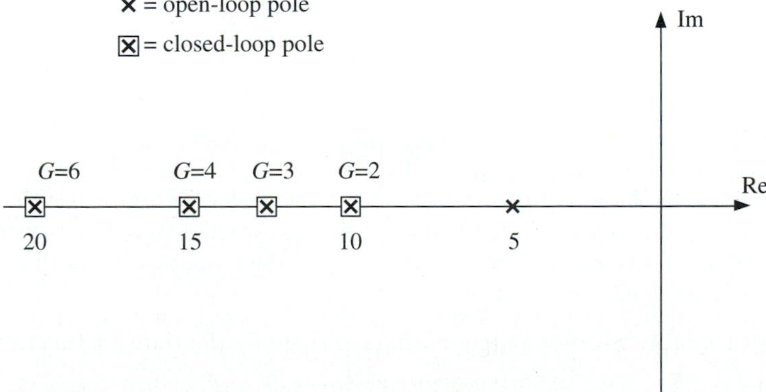

Fig. SP3.3.2

◆ PROBLEMS

3.1 For the system shown in Fig. P3.1, determine the transfer function and time constant relating (**a**) the force F and output displacement y and (**b**) the input displacement x and y when $F = 0$. Plot the system pole on the s plane in each case. (Note: F and x are not applied simultaneously.)

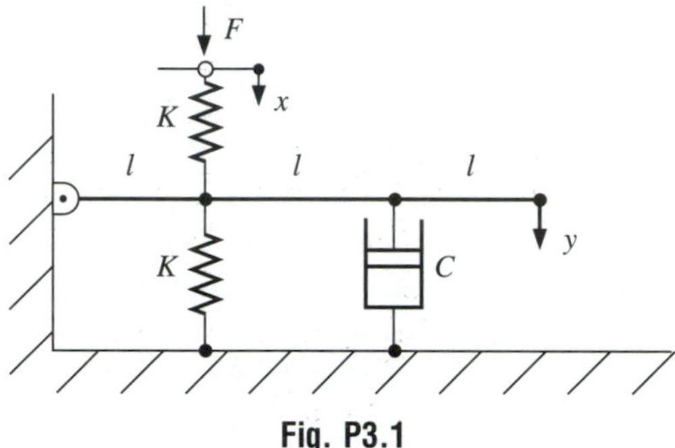

Fig. P3.1

3.2 Determine the output of the open-loop system

$$G(s) = \frac{a}{1 + sT}$$

to the input

$$r(t) = t$$

Sketch both input and output as functions of time, and determine the steady-state error between the input and output. Compare the result with that given by Fig. 3.7.

3.3 The massless bar shown in Fig. P3.3 has been displaced a distance x_0 and is subjected to a unit impulse δ in the direction shown. Find the response of the system for $t > 0$ and sketch the result as a function of time. Confirm the steady-state response using the final-value theorem.

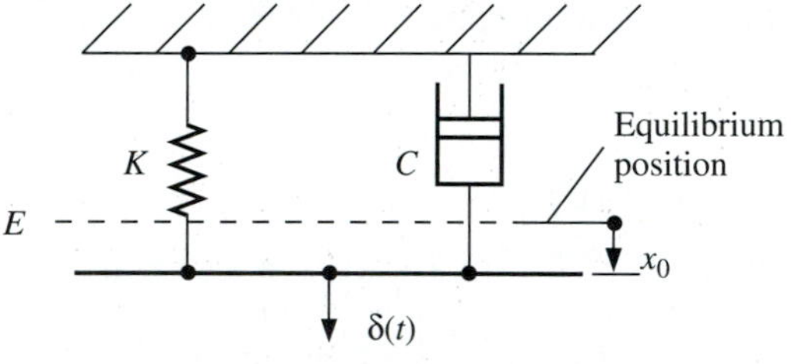

Fig. P3.3

3.4 If the steady-state output is said to have been achieved once it is within 1% of the final value, how long does the transient response last when a first-order system with a time constant of 10 s is given a unit step input? Does the numerical value of the numerator of the transfer function affect the answer?

3.5 An open-loop first-order system is characterized by the transfer function

$$G(s) = \frac{1}{1 + \tau s}$$

where the time constant is $\tau = 5$ s. Calculate the steady-state error when the system input is

$$r(t) = 6t + 1$$

Confirm the result by using the final-value theorem.

3.6 A system described by the differential equation

$$2\dot{y} + 3y = x$$

is subjected to a sinusoidal input x of constant amplitude and variable frequency. At what frequency is the output y equal to 50% of its DC ($\omega \rightarrow 0$) value? What is the phase lag of the output with respect to the input at this frequency?

3.7 One definition of the *bandwidth* of a system is the frequency range over which the amplitude of the output signal is greater than 70% of the input signal amplitude when a system is subjected to a harmonic input. Find a relationship between the bandwidth and time constant of a first-order system. What is the phase angle at the bandwidth frequency?

3.8 Figure P3.8 shows the experimentally obtained voltage output of an unknown system subjected to a step input of +10 V. Determine the transfer function of the system and locate its pole on the complex plane.

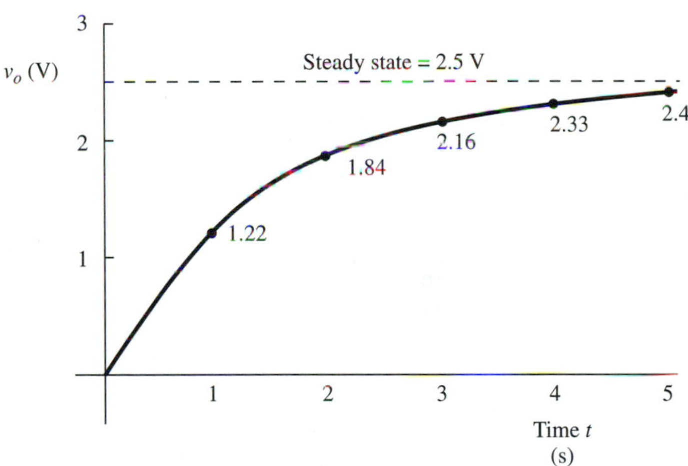

Fig. P3.8

3.9 The circuit shown in Fig. P3.9a is known as a simple *lag* compensation network. Determine the transfer function of this circuit. Figure P3.9b shows two lag circuits connected together. Determine the transfer function for this combination. Why does the transfer function of the combination not equal the product of the two individual lag transfer functions?

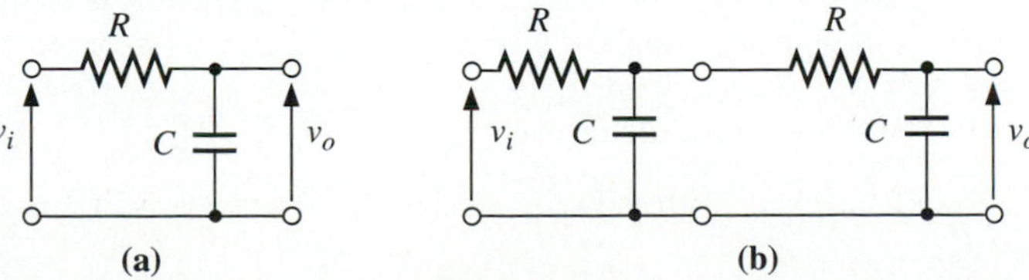

Fig. P3.9

3.10 Two tanks (Fig. P3.10) initially empty and of diameter 10 cm, are connected by a pipe with a laminar restrictor. The second tank discharges to the atmosphere through another laminar restrictor. An inlet pipe supplies $q_i = 5$ liters/min of water into Tank 1. If $R_1 = R_2 = 0.1$ m-min/liter, find a differential equation for the height of water in each tank and calculate the steady-state level.

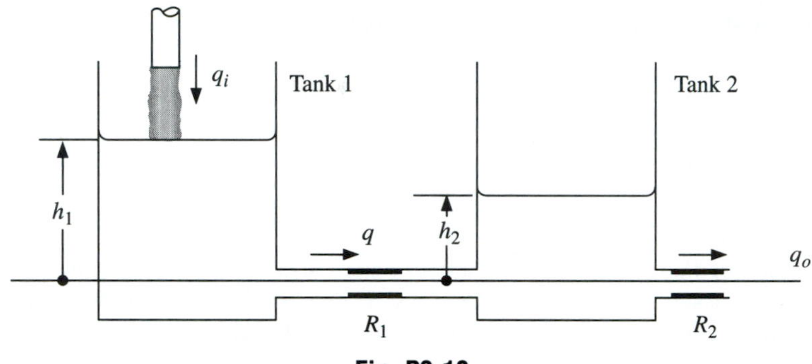

Fig. P3.10

MODULE 4

Second-Order Systems

The next class of systems to be investigated are called second order because they will be described by differential equations containing the second derivative of the output variable. Second-order systems are important because their behavior is very different from first-order systems and may exhibit features such as oscillatory response, or overshoot. In studying such systems, the approach will be as before, namely to investigate in detail a physical system of second order and determine the response of the system to several types of input. This will then be followed by additional studies illustrating the benefits of using feedback control of second-order systems.

Our choice of a system to study is the use of a DC electric motor to position an inertial load through a gearbox. Such a system is very common and illustrates the combination of electrical and mechanical theory using a common mathematical description, the Laplace transform. Although the operation of this system was described in a qualitative manner in Module 1, a more mathematical development will be presented here. The system is shown in schematic form in Fig. 4.1. In operation, the set point is generated by means of a potentiometer. The angular position θ_d of the potentiometer generates a proportional bipolar voltage according to the potentiometer's transfer function K_p, in volts per radian. This voltage is compared to the achieved position of the load as measured by another potentiometer. Matters are simplified considerably if the two potentiometers are identical. The potentiometers and summing junction are shown in block diagram form in Fig. 4.2.

The error voltage is usually passed through an amplifier since the motor may require greater power levels than those obtained directly from the summing junction. The amplifier may be considered a simple gain K_a that is adjustable in most cases. This transfer function is shown in Fig. 4.3.

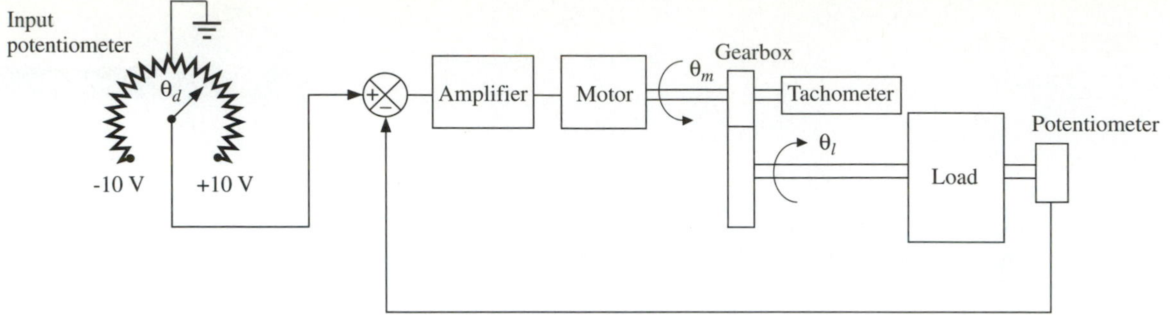

Fig. 4.1 Motor-positioned load

Fig. 4.2 Potentiometers and summing junction

Fig. 4.3 Amplifier

The voltage output from the amplifier drives the motor. It will be remembered that DC motors rotate due to the interaction of two magnetic fields, one in the stator and one in the rotor, and the rotational speed may be varied by controlling the strength of one of these fields. The stationary magnetic field may be generated by passing a current through the field winding or by using permanent magnets for small motors. The rotating field is generated by passing a current through the armature winding. We will consider the motor to be field-controlled, complementing the armature-controlled case dealt with in Module 2. In analyzing the response of the motor to the applied voltage v_a, it is convenient to think of the electrical and mechanical aspects separately. Figure 4.4 shows the circuit representing the field winding as having some resistance R and inductance L.

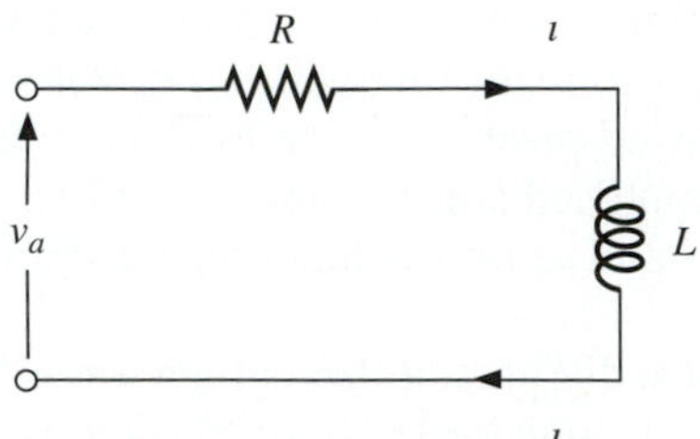

Fig. 4.4 Motor electrical circuit

The relationship between the applied voltage and current may be found from Kirchhoff's method:

$$v_a = \imath R + L\frac{d\imath}{dt} \tag{4.1}$$

Taking Laplace transforms yields

$$V_a = I(R + sL) \tag{4.2}$$

This gives the transfer function

$$\frac{I}{V_a} = \frac{1}{R + sL} \tag{4.3}$$

The block diagram of the electrical portion of the motor is shown in Fig. 4.5. The field current establishes a magnetic flux that interacts with the armature field to produce a mechanical torque on the armature. The torque is given by

$$\tau_m = K_m \imath \tag{4.4}$$

where K_m is the motor torque constant with units of Newton meters per ampere. This element is shown in Fig. 4.6.

This completes the study of the electrical part of the motor, and attention may now be focused on the mechanical aspects, namely what does the torque τ_m do? In order to answer this question, the free-body diagram of the motor, gearbox, and load shown in Fig. 4.7 has to be analyzed. Note from the figure that not all of the torque τ_m is available at the gearbox input shaft, since the armature inertia has to be overcome, as does the viscous friction in the armature bearings. If the gear ratio is N (>1), the system equations may be written using Newton's second law, which may be expressed as

$$\sum \tau = J\dot{\omega} \tag{4.5}$$

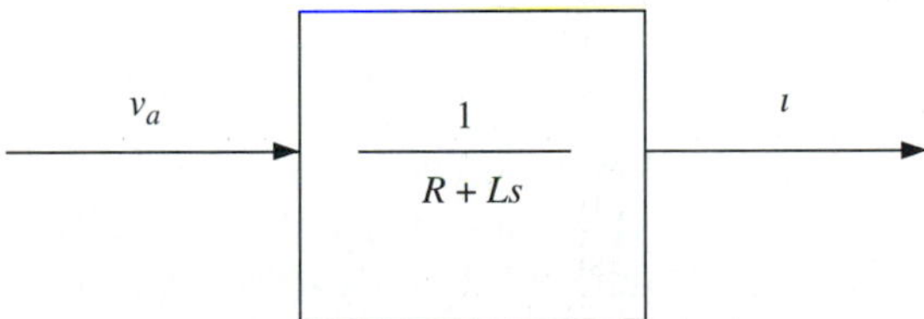

Fig. 4.5 Motor electrical block diagram

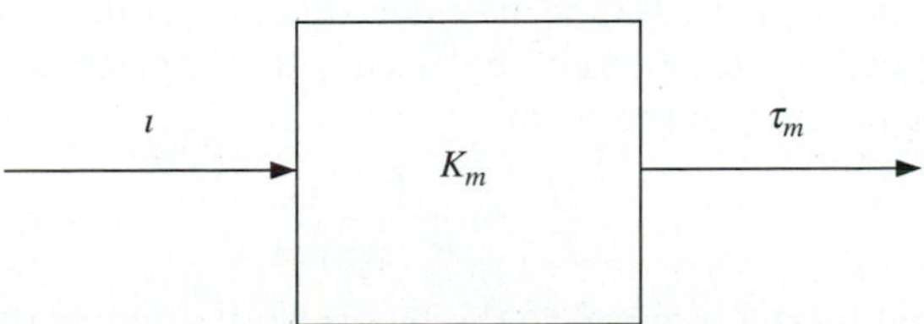

Fig. 4.6 Motor torque constant

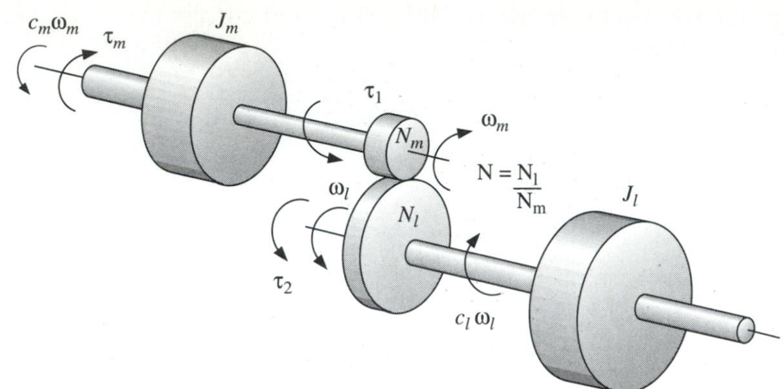

Fig. 4.7 Free-body diagram of motor, gearbox, and load

Applying this to the free-body diagrams of the system yields

$$\tau_m - \tau_1 - c_m\omega_m = J_m\dot{\omega}_m \qquad \text{(for the motor shaft)} \tag{4.6}$$

$$\tau_2 - c_l\omega_l = J_l\dot{\omega}_l \qquad \text{(for the load shaft)} \tag{4.7}$$

$$\omega_l = \omega_m/N \qquad \tau_2 = N\tau_1 \qquad \text{(for the gearbox)} \tag{4.8}$$

Referring all variables to the motor shaft gives

$$\tau_m - c_m\omega_m - \frac{\tau_2}{N} = J_m\dot{\omega}_m \tag{4.9}$$

$$\tau_m - \frac{J_l\dot{\omega}_l}{N} - c_m\omega_m - \frac{c_l\omega_l}{N} = J_m\dot{\omega}_m \tag{4.10}$$

$$\tau_m - \frac{J_l\dot{\omega}_m}{N^2} - c_m\omega_m - \frac{c_l\omega_m}{N^2} = J_m\dot{\omega}_m \tag{4.11}$$

resulting in the final equation

$$\tau_m = \left(J_m + \frac{J_l}{N^2}\right)\dot{\omega}_m + \left(c_m + \frac{c_l}{N^2}\right)\omega_m \tag{4.12}$$

It is seen from the above equation that the result is similar to that expected from a single rotational inertia with damping, except that the effect of the gearbox is to reduce both the load inertia and load shaft damping by a factor of N^2. Although it may seem that the inertia of the armature may be neglected compared to the inertia of the load, it is not uncommon to find gearbox ratios in the order of several thousand. Under these circumstances it is the load inertia rather than the motor inertia that may be neglected! The transfer function from motor torque τ_m to motor angular speed ω_m is obtained by taking the Laplace transform of equation 4.12 and rearranging in the form

$$\frac{\Omega}{\Gamma} = \frac{1}{Js + c} \tag{4.13}$$

where J is the total inertia referred to the motor shaft and c is the total damping, also referred to the motor shaft. The block diagram of the mechanical parts of the motor are shown in Fig. 4.8.

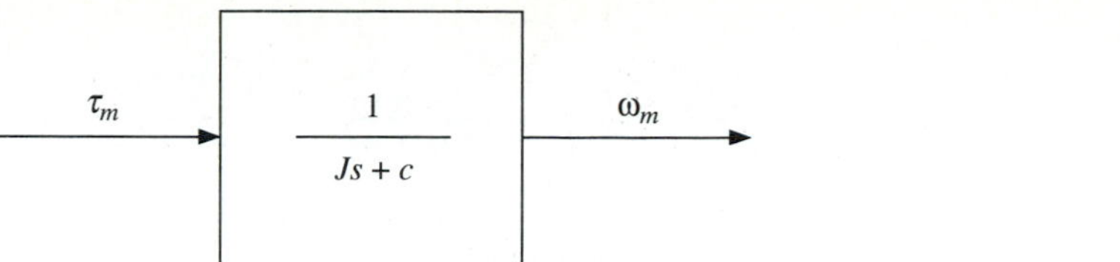

Fig. 4.8 Motor, gearbox, and load transfer function

Referring back to Fig. 4.1, it is noted that the tachometer is attached to the motor shaft, and its transfer function

$$v_t = K_t \omega_m \tag{4.14}$$

may be represented by the block diagram shown in Fig. 4.9. The potentiometer measuring the output position of the load is of course attached to the load shaft, and the block diagram from ω_m to load position θ_l is shown in Fig. 4.10.

Figure 4.11 shows the block diagram of the motor from input voltage v_a to motor speed ω_m and shows that the electrical system and mechanical system may each be represented by a first-order transfer function. It is customary, however, to observe that the time constant of the electrical system is much smaller than the time constant of the mechanical system, i.e.,

$$\frac{L}{R} < \frac{J}{c} \tag{4.15}$$

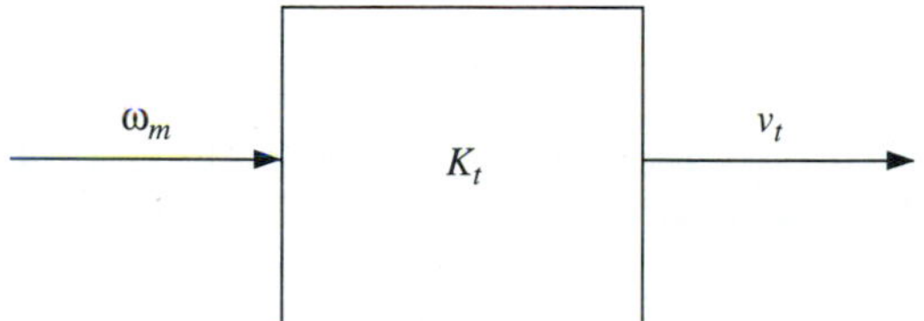

Fig. 4.9 Tachometer

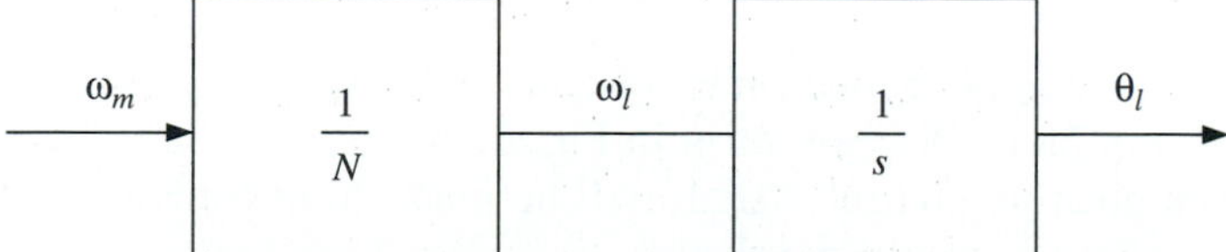

Fig. 4.10 Load position

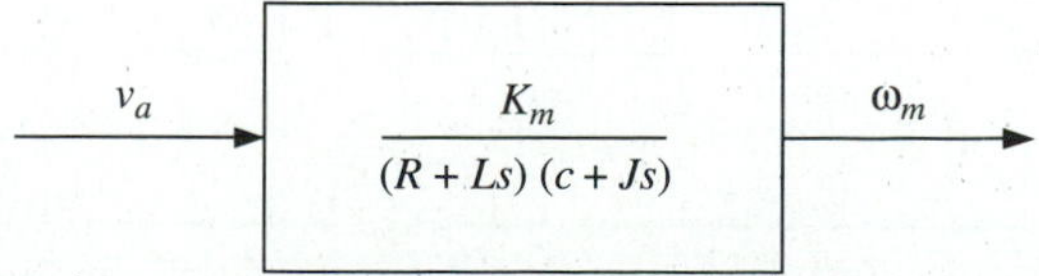

Fig. 4.11 Motor transfer function

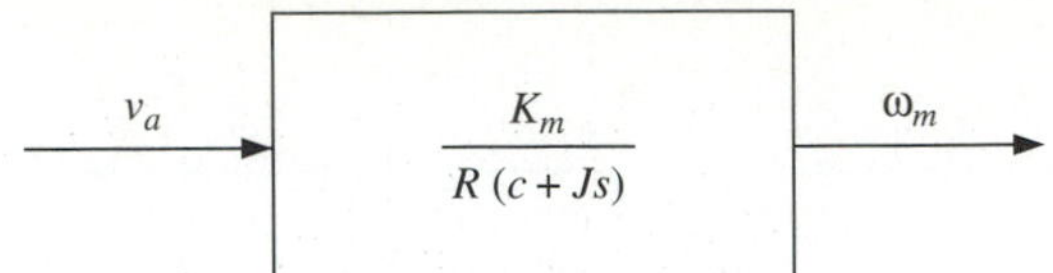

Fig. 4.12 Modified motor transfer function

Physically, this states that the current in the field is established much more quickly than the time needed for the motor to achieve a steady-state speed. Under this assumption, the transfer function of the motor becomes that shown in Fig. 4.12, where the inductance of the field is ignored.

This completes the analysis of the servomechanism, and the block diagram showing all elements is given in Fig. 4.13. By taking the potentiometer block K_p to the right of the summing junction, the closed-loop transfer function may be written as

$$\frac{\theta_l}{\theta_d} = \frac{G}{1 + GH} = \frac{K_p K_a K_m / NR}{Js^2 + cs + K_p K_a K_m / NR} \tag{4.16}$$

From this it may be observed that the differential equation relating the input and output is of degree 2; hence the position control system is of second order. Since second-order systems may be obtained from many engineering disciplines, it is usual to adopt a more generalized notation for describing the system that is independent of the discipline from which the system was obtained. The usual notation takes the form

$$\frac{C}{R} = \frac{\omega_n^2}{s^2 + 2\zeta\omega_n s + \omega_n^2} \tag{4.17}$$

where ζ is the damping ratio and ω_n is the undamped natural frequency. It may be shown that for the system just described

$$\omega_n = \sqrt{\frac{K_p K_a K_m}{NRJ}} \tag{4.18}$$

$$\zeta = \frac{c}{2}\sqrt{\frac{NR}{K_p K_a K_m J}} \tag{4.19}$$

Subsequent analysis of second-order systems will use the generalized notation above, shown in block diagram form in Fig. 4.14, through reference back to the motor-driven position control system will be made from time to time to give a physical basis for the results developed.

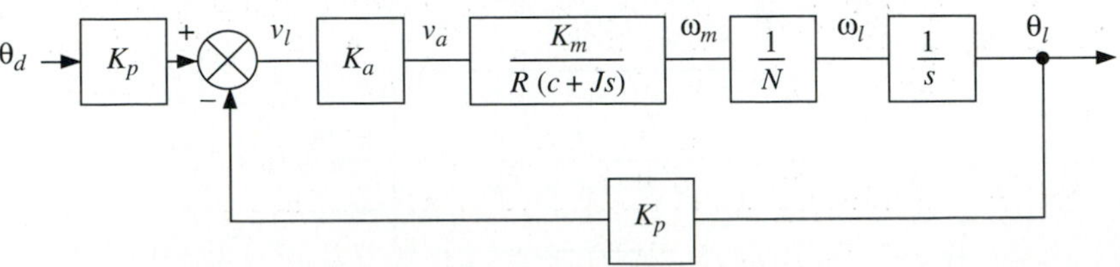

Fig. 4.13 Position control system block diagram

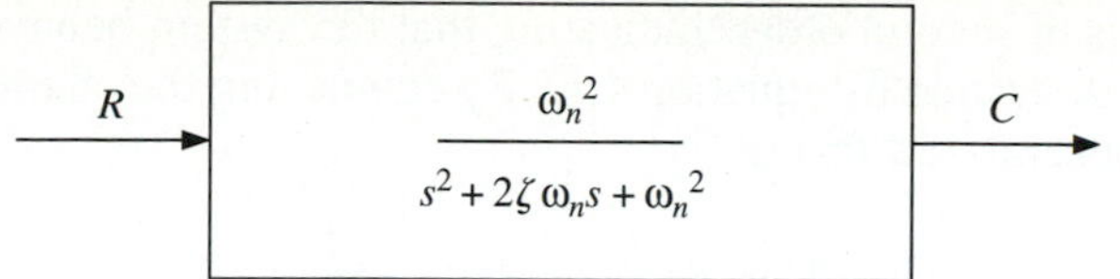

Fig. 4.14 Generalized closed-loop transfer function

◆ SECOND-ORDER ELECTRICAL SYSTEM

As a further example of a system described by a second-order differential equation, we will investigate the behavior of the electrical circuit shown in Fig. 4.15. In this circuit the input variable is considered to be the applied voltage v_i while the output variable is the voltage across the capacitor. As the current becomes established in the circuit, each component generates an impedance that depends upon the physical relationship between current and voltage. Kirchhoff's voltage law requires that the sum of these impedances is equal to the applied voltage, namely,

$$v_i = v_R + v_L + v_C = R\imath + L\frac{d\imath}{dt} + \frac{1}{C}\int_0^t \imath\, dt \tag{4.20}$$

The equation relating current and voltage for the capacitor is

$$v_o = \frac{1}{C}\int_0^t \imath\, dt \tag{4.21}$$

Taking Laplace transforms, assuming zero initial conditions, reduces these equations to

$$\left(Ls + R + \frac{1}{Cs}\right)I = V_i \tag{4.22}$$

and

$$\frac{I}{Cs} = V_o \tag{4.23}$$

Eliminating the variable i yields the transfer function

$$\frac{V_o}{V_i} = \frac{1}{LCs^2 + RCs + 1} \tag{4.24}$$

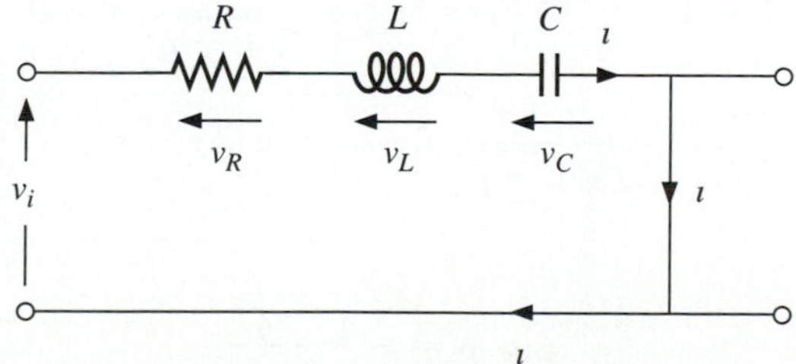

Fig. 4.15 Passive electrical circuit

This equation is of second order, indicating that the system behaves in a similar manner to that described by equation 4.17. By comparing the characteristic equation with the generalized form,

$$2\zeta\omega_n = \frac{R}{L} \qquad \omega_n^2 = \frac{1}{LC} \tag{4.25}$$

This leads to

$$\omega_n = \sqrt{\frac{1}{LC}} \qquad \zeta = \frac{R}{2}\sqrt{\frac{C}{L}} \tag{4.26}$$

◆ STEP RESPONSE

Considerable information may be obtained regarding the nature of response of a second-order system by studying its behavior when subjected to a step input. In such a case, equation 4.17 indicates that the output is given by

$$C(s) = \frac{\omega_n^2}{s(s^2 + 2\zeta\omega_n s + \omega_n^2)} \tag{4.27}$$

$$= \frac{1}{s} - \frac{s + 2\zeta\omega_n}{(s + \zeta\omega_n)^2 + \omega_n^2(1 - \zeta^2)} \tag{4.28}$$

If it is assumed that the system is *underdamped,* that is, $\zeta < 1$, then the damped natural frequency ω_d may be defined as

$$\omega_d = \omega_n\sqrt{1 - \zeta^2} \tag{4.29}$$

Substituting above results in

$$C(s) = \frac{1}{s} - \frac{s + 2\zeta\omega_n}{(s + \zeta\omega_n)^2 + \omega_d^2} \tag{4.30}$$

$$= \frac{1}{s} - \frac{s + \zeta\omega_n}{(s + \zeta\omega_n)^2 + \omega_d^2} - \frac{\zeta\omega_n}{(s + \zeta\omega_n)^2 + \omega_d^2} \tag{4.31}$$

Taking the inverse transform yields

$$c(t) = 1 - e^{-\zeta\omega_n t}\cos\omega_d t - \frac{\zeta\omega_n}{\omega_d}e^{-\zeta\omega_n t}\sin\omega_d t \tag{4.32}$$

$$= 1 - e^{-\zeta\omega_n t}\left(\cos\omega_d t + \frac{\zeta}{\sqrt{1 - \zeta^2}}\sin\omega_d t\right) \tag{4.33}$$

$$= 1 - \frac{e^{-\zeta\omega_n t}}{\sqrt{1 - \zeta^2}}\sin(\omega_d t + \phi) \tag{4.34}$$

where

$$\tan\phi = \frac{\sqrt{1 - \zeta^2}}{\zeta} \tag{4.35}$$

It is left to the student to prove that the response for the overdamped case ($\zeta > 1$) is given by

$$c(t) = 1 - \frac{e^{-\zeta\omega_n t}}{\sqrt{\zeta^2 - 1}} \sinh(\omega_d t + \phi) \tag{4.36}$$

where

$$\omega_d = \omega_n \sqrt{\zeta^2 - 1} \qquad \tan\phi = \frac{\sqrt{\zeta^2 - 1}}{\zeta} \tag{4.37}$$

while the critically damped response ($\zeta = 1$) is

$$c(t) = 1 - e^{-\omega_n t}(1 + \omega_n t) \tag{4.38}$$

Figure 4.16 shows the step response for various values of damping ratio ζ and for the particular case $\omega_n = 1$ rad/s. At this stage, it is important to recognize that when $\zeta < 1$, the figure shows that

1. the system oscillates at frequency ω_d,
2. the response decays due to the exponent $-\zeta\omega_n$, and
3. the overshoot depends on the value of ζ.

In the design of feedback control systems, it is useful to know the relationship between the percentage overshoot a system has and the corresponding value of damping ratio ζ. This relationship is shown in Fig. 4.17. In the following module,

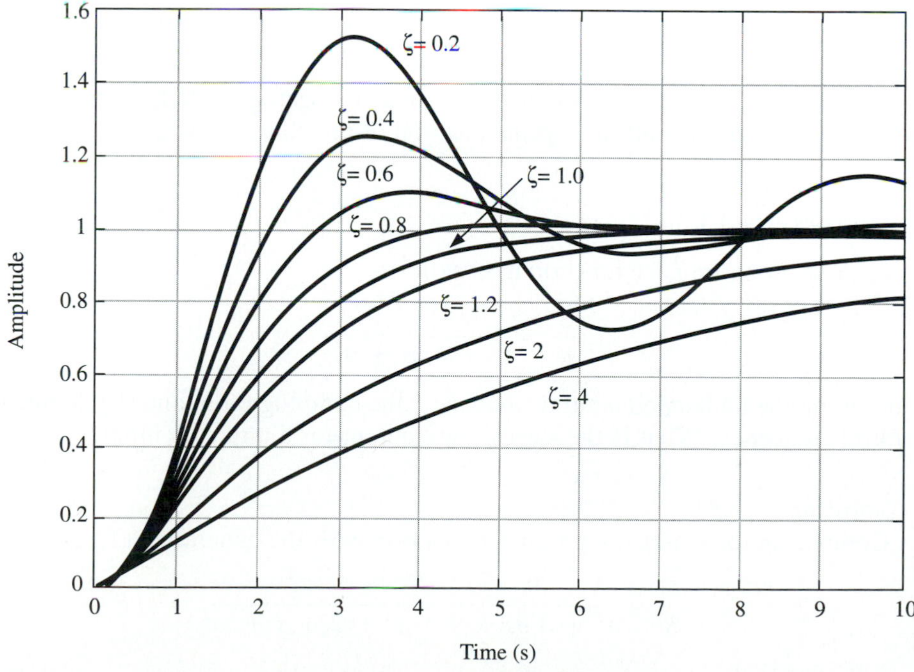

Fig. 4.16 Second-order step response

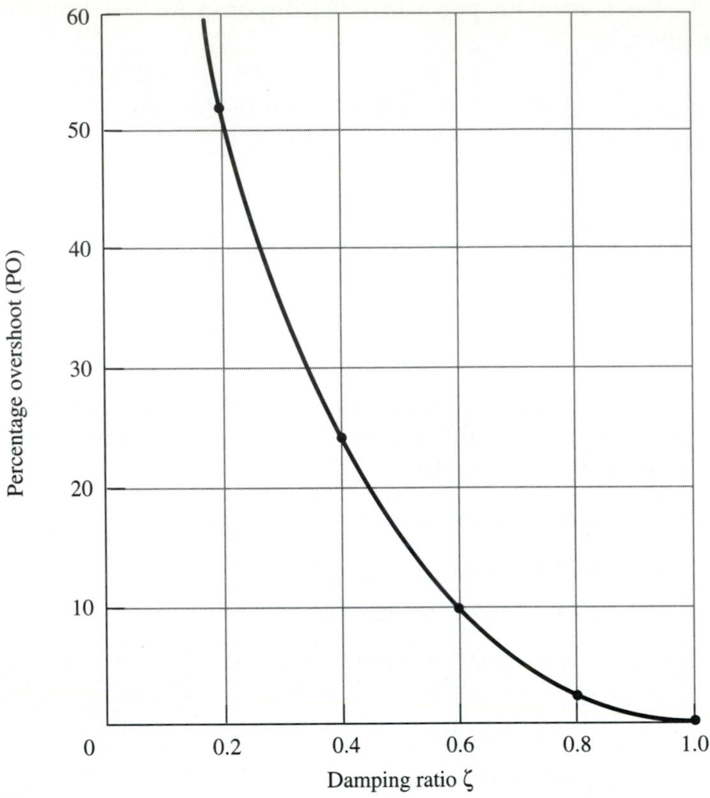

Fig. 4.17 Relationship between overshoot and damping ratio

more detailed system responses will be investigated, together with the relationship between the response and the s-plane location of the closed-loop poles.

Sample Problem 4.1

A closed-loop transfer function has the form

$$\frac{C}{R} = \frac{9}{s^2 + 4.5s + 9}$$

Determine the undamped natural frequency, the damping ratio, and the damped natural frequency. What is the steady-state output for a unit step input?

Solution

Comparing the closed-loop transfer function with the generalized form

$$\frac{C}{R} = \frac{9}{s^2 + 4.5s + 9} = \frac{\omega_n^2}{s^2 + 2\zeta\omega_n s + \omega_n^2}$$

it is seen that

$$\omega_n = 3 \text{ rad/s}$$

and that

$$\zeta = \frac{4.5}{2\omega_n} = 0.75$$

The damped natural frequency is given by

$$\omega_d = \omega_n\sqrt{1 - \zeta^2} = 1.98 \text{ rad/s}$$

If the system is subject to a unit step input, the output may be written as

$$C(s) = \frac{9R(s)}{s^2 + 4.5s + 9} = \frac{\frac{9}{s}}{s^2 + 4.5s + 9}$$

Applying the final-value theorem yields

$$\lim_{t\to} c(t) = \lim_{s\to 0} sC(s) = 1$$

indicating that the steady-state output eventually reaches the input value, resulting in zero error.

Sample Problem 4.2

Figure SP4.2.1 shows a spring-mass damper system subject to an external force p. Show that the transfer function between the input p and the output displacement x is of second order, and develop expressions for the damping ratio and damped natural frequency in terms of the system parameters. Determine the damped natural frequency, the undamped natural frequency, the damping ratio, and the location of the system poles given the following data: spring constant k = 10 N/m, damping constant c = 4 N-s/m, and mass m = 1 kg.

Solution

Considering the mass m displaced a distance x from its equilibrium position, the free-body diagram of the mass will be as shown in Fig. SP4.2.2. Using

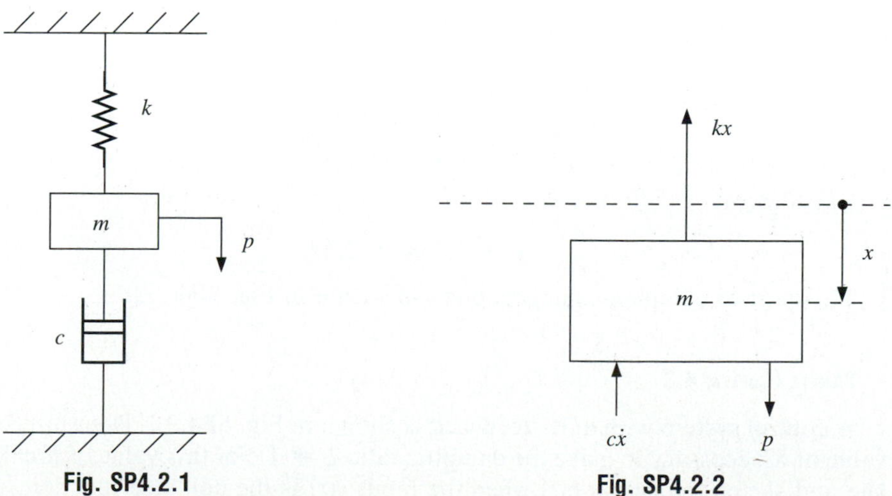

Fig. SP4.2.1

Fig. SP4.2.2

Newton's second law of motion,

$$p - kx - c\dot{x} = m\ddot{x}$$

$$m\ddot{x} + c\dot{x} + kx = p$$

Taking Laplace transforms, assuming zero initial conditions,

$$X(ms^2 + cs + k) = P$$

results in the transfer function

$$\frac{X}{P} = \frac{1/m}{s^2 + (c/m)s + k/m} = \frac{1}{k}\frac{k/m}{s^2 + (c/m)s + k/m}$$

This is a second-order transfer function where

$$\omega_n^2 = \frac{k}{m}$$

and

$$\zeta = \frac{c}{2\omega_n m} = \frac{c}{2\sqrt{km}}$$

The damped natural frequency is given by

$$\omega_d = \omega_n\sqrt{1 - \zeta^2} = \sqrt{\frac{k}{m}}\sqrt{1 - \frac{c^2}{4km}} = \sqrt{\frac{k}{m} - \left(\frac{c}{2m}\right)^2}$$

Using the given data,

$$\omega_n = \sqrt{\frac{k}{m}} = \sqrt{10} = 3.16 \text{ rad/s}$$

$$\zeta = \frac{c}{2\sqrt{km}} = \frac{2}{\sqrt{10}} = 0.63$$

$$\omega_d = \omega_n\sqrt{1 - \zeta^2} = 2.45 \text{ rad/s}$$

The system transfer function indicates that there are no zeros and the two poles are the solution of the characteristic equation

$$s^2 + \frac{c}{m}s + \frac{k}{m} = 0$$

$$s^2 + 4s + 10 = 0$$

which gives

$$s = -2.00 \pm 2.45j$$

This pair of complex-conjugate poles is shown in Fig. SP4.2.3.

Sample Problem 4.3

A control system with unity feedback is shown in Fig. SP4.3.1. Determine the value of K necessary to make the damping ratio $\zeta = 1$. For this value of K calculate and sketch the output $c(t)$ when the input $r(t)$ is the unit step function.

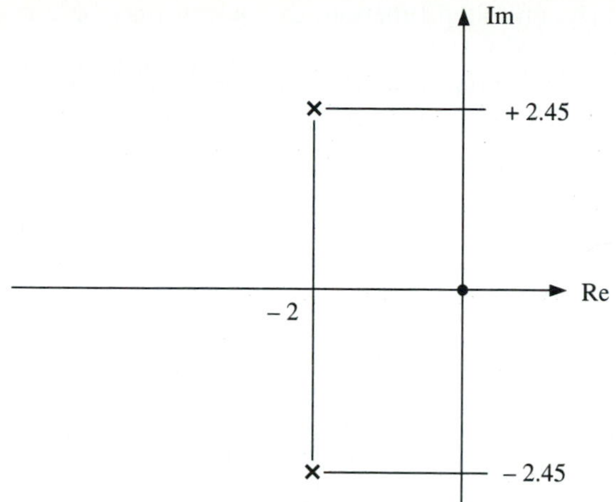

Fig. SP4.2.3

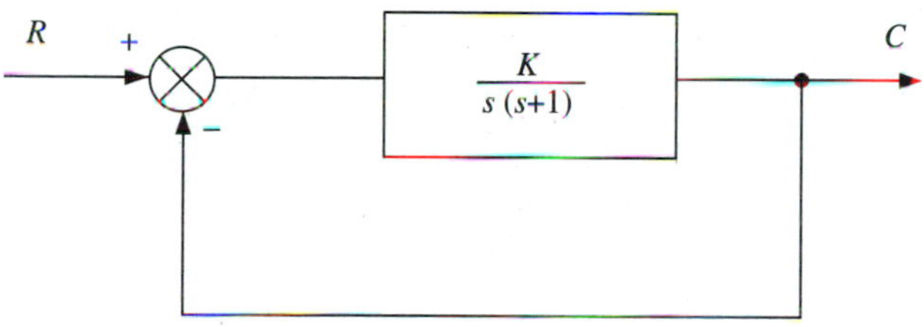

Fig. SP4.3.1

Solution

The closed-loop transfer function is

$$\frac{C}{R} = \frac{K/s(s+1)}{1 + K/s(s+1)} = \frac{K}{s^2 + s + K}$$

Comparing with the generalized second-order system, we get

$$\omega_n = \sqrt{K}$$

and

$$2\zeta\omega_n = 1$$

$$\zeta = \frac{1}{2\zeta\omega_n} = \frac{1}{2\sqrt{K}}$$

If it required that $\zeta = 1$, then from the above

$$K = \tfrac{1}{4}$$

When K takes this value, the transfer function becomes

$$\frac{C}{R} = \frac{\frac{1}{4}}{s^2 + s + \frac{1}{4}} = \frac{\frac{1}{4}}{(s + \frac{1}{2})^2}$$

If the input is the unit step function, the output may be written as

$$C(s) = \frac{\frac{1}{4}}{s(s+\frac{1}{2})^2} = \frac{A}{s} + \frac{Bs + C}{(s+\frac{1}{2})^2}$$

After calculating the partial-fraction coefficients,

$$\begin{aligned} C(s) &= \frac{1}{s} - \frac{s+1}{(s+\frac{1}{2})^2} \\ &= \frac{1}{s} - \frac{s+\frac{1}{2}}{(s+\frac{1}{2})^2} - \frac{\frac{1}{2}}{(s+\frac{1}{2})^2} \\ &= \frac{1}{s} - \frac{1}{(s+\frac{1}{2})} - \frac{\frac{1}{2}}{(s+\frac{1}{2})^2} \end{aligned}$$

The time-domain response is therefore

$$c(t) = 1 - e^{-(1/2)t} - \tfrac{1}{2}te^{-(1/2)t}$$

producing the result

$$c(t) = 1 - e^{-(1/2)t}(1 + \tfrac{1}{2}t)$$

It can be appreciated that the exponential term diminishes faster than the term t increases, resulting in the time response shown in Fig. SP4.3.2.

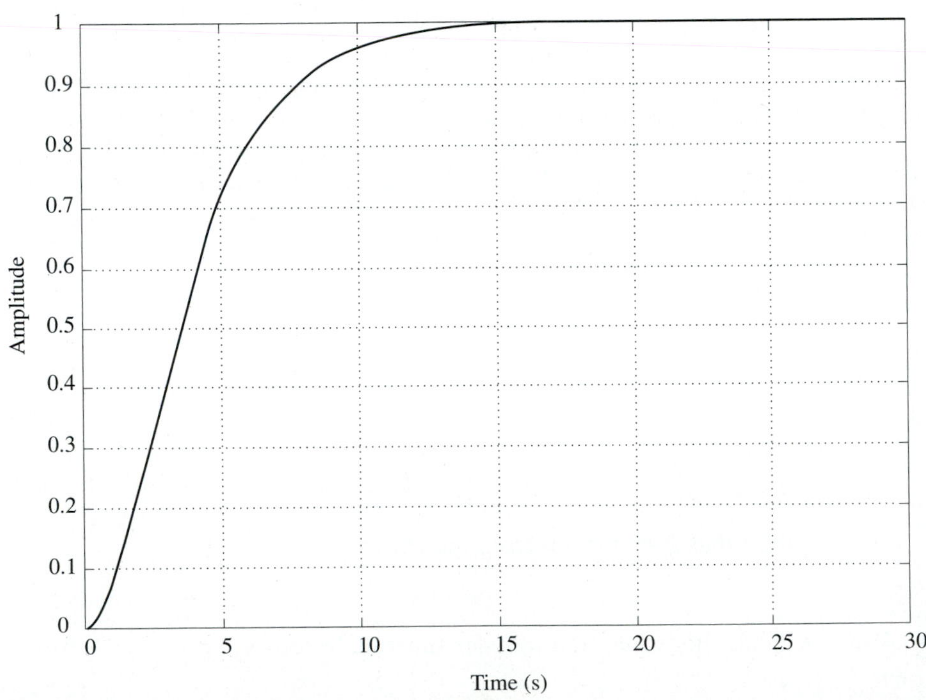

Fig. SP4.3.2

4.1 Figure P4.1 shows a closed-loop feedback system with a second-order plant. Determine the damped natural frequency and damping ratio of the closed-loop response.

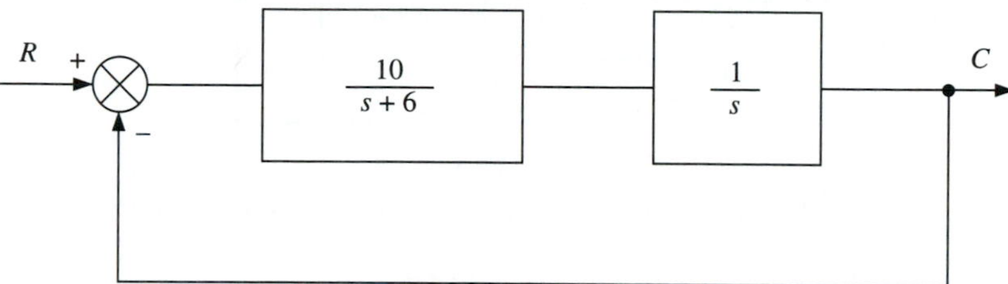

Fig. P4.1

4.2 Determine the displacement transfer function y_o/y_i for the mechanical system shown in Fig. P4.2.

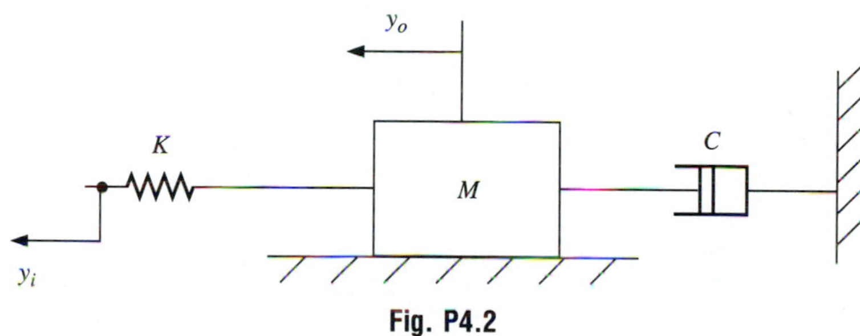

Fig. P4.2

4.3 Figure P4.3 shows a schematic of a vehicle suspension system. Calculate the value of the shock absorber damping constant c such that the vehicle response is critically damped ($\zeta = 1$).

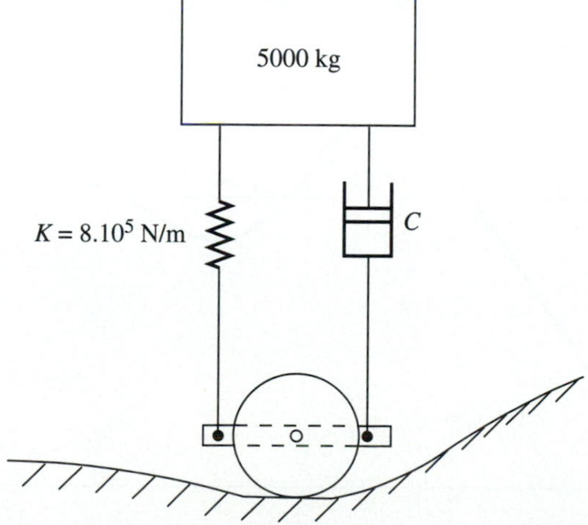

Fig. P4.3

4.4 Calculate the required value of gain K shown in Fig. P4.4 such that the closed-loop response of the system to a step input is limited to no more than 10% overshoot. Plot the closed-loop poles on the complex plane for the same value of gain.

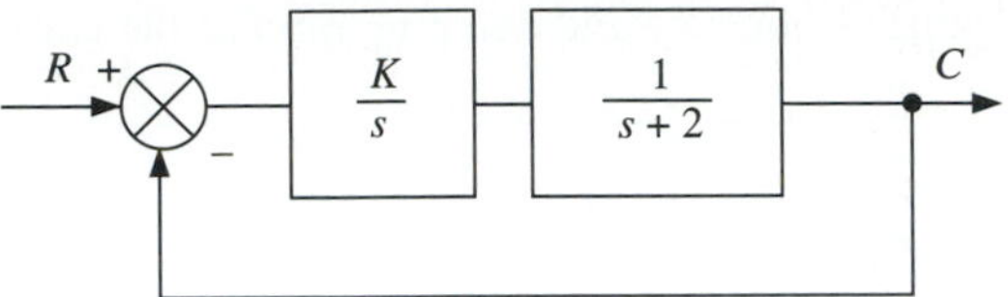

Fig. P4.4

4.5 A unity-feedback control system has the forward-path transfer function

$$G(s) = \frac{K}{s(s + 10)}$$

Find the closed-loop transfer function, and develop expressions for the damping ratio and damped natural frequency in terms of K. Plot the closed-loop poles on the complex plane for $K = 0, 10, 25, 50, 100$. For each value of K calculate the corresponding damping ratio and damped natural frequency. What conclusions can you draw from the plot?

4.6 For the electrical network shown in Fig. P4.6, find a differential equation relating the constant current i_0 in the circuit and the voltage v_b, after the switch is closed. Then determine the steady-state voltage across the capacitor C.

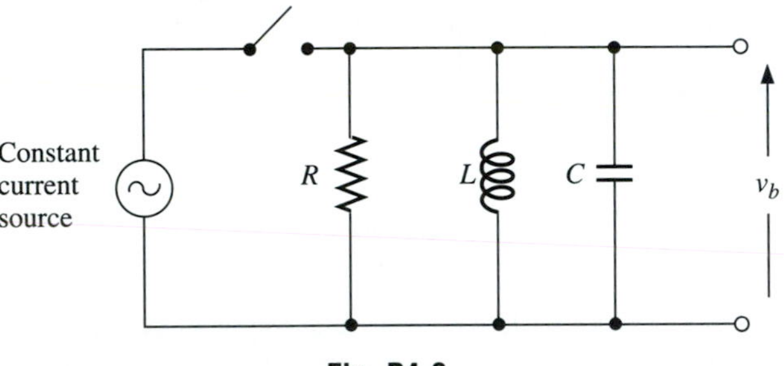

Fig. P4.6

4.7 Prove that for an underdamped second-order system subject to a step input, the percentage overshoot above the steady-state output (as shown in Fig. P4.7) is a function only of the damping ratio ζ.

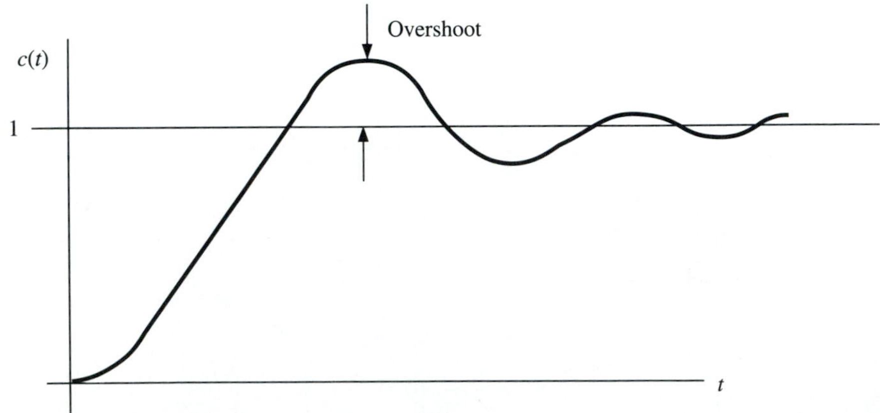

Fig. P4.7

4.8 The system shown in Fig. P4.8 represents a model of a tall building, where the displacement x of the mass is equivalent to the sway of the building. If an earthquake may be represented by an impulse function, calculate and plot the earthquake response of the building. Assume the following data: $m = 2.10^6$ kg, $k = 5.10^4$ N/m, $c = 250$ N-s/m, and impulse value $= 10^4$ N. Estimate the maximum sway of the building.

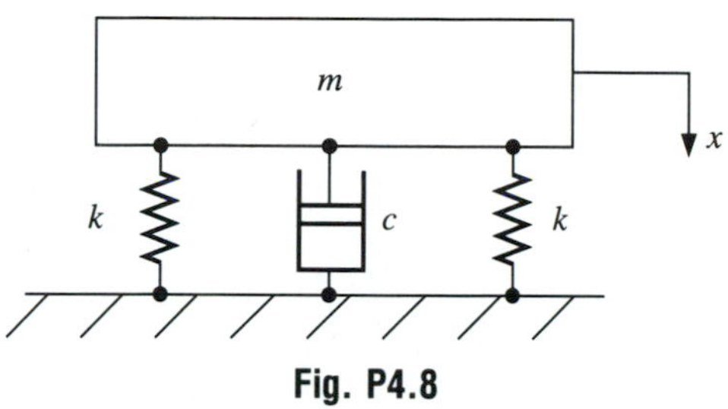

Fig. P4.8

4.9 Figure P4.9 shows a robot arm that rotates in the horizontal plane. The position of the arm is controlled by a feedback control system. The motor inertia is 0.01 kg-m², the arm inertia is 5.0 kg-m², both motor and arm natural damping coefficients are 0.02 N-m/s, $K_T/R = 1$ and the gearbox ratio is 100. What is the percentage change in damping ratio of the closed-loop system when the arm grasps a 20-kg mass 150 cm from the axis of rotation? If the robot is of the direct-drive type (gear ratio = 1), what is the percentage change in the damping ratio?

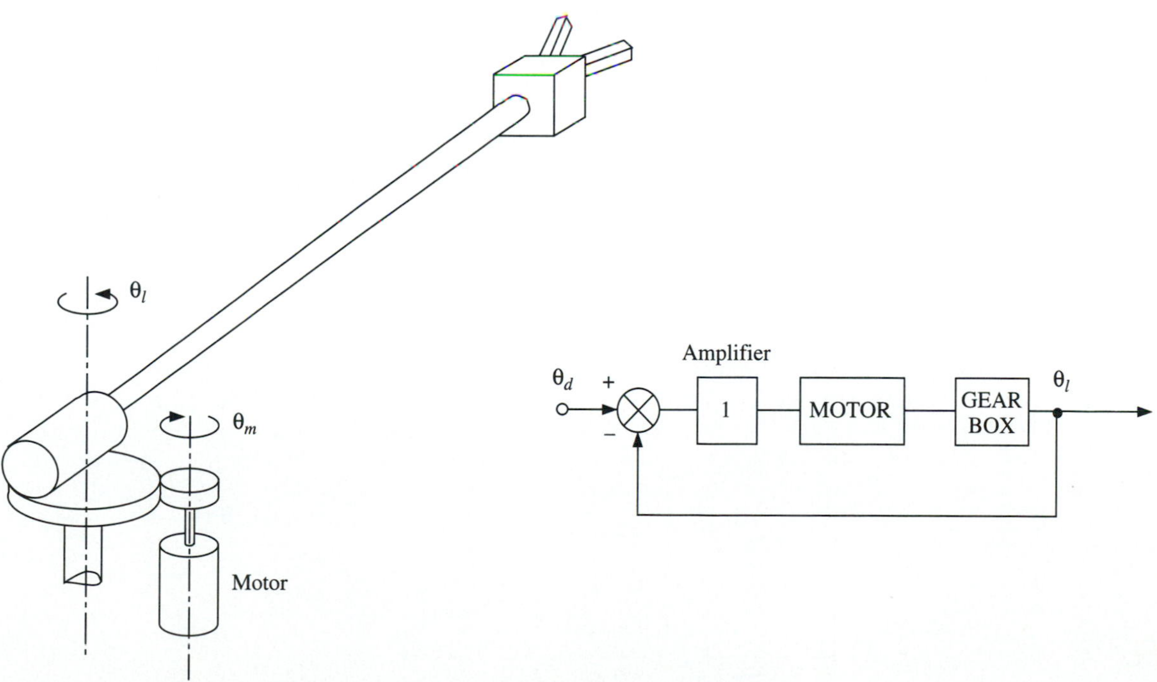

Fig. P4.9

4.10 A feedback control system consists of a DC motor driving a load through a gearbox. The load shaft is connected to a rotational spring, as shown in Fig. P4.10. Determine the closed-loop transfer function for the system, obtain an expression for the damping ratio of the output, and calculate the steady-state output for a step-input voltage θ_d.

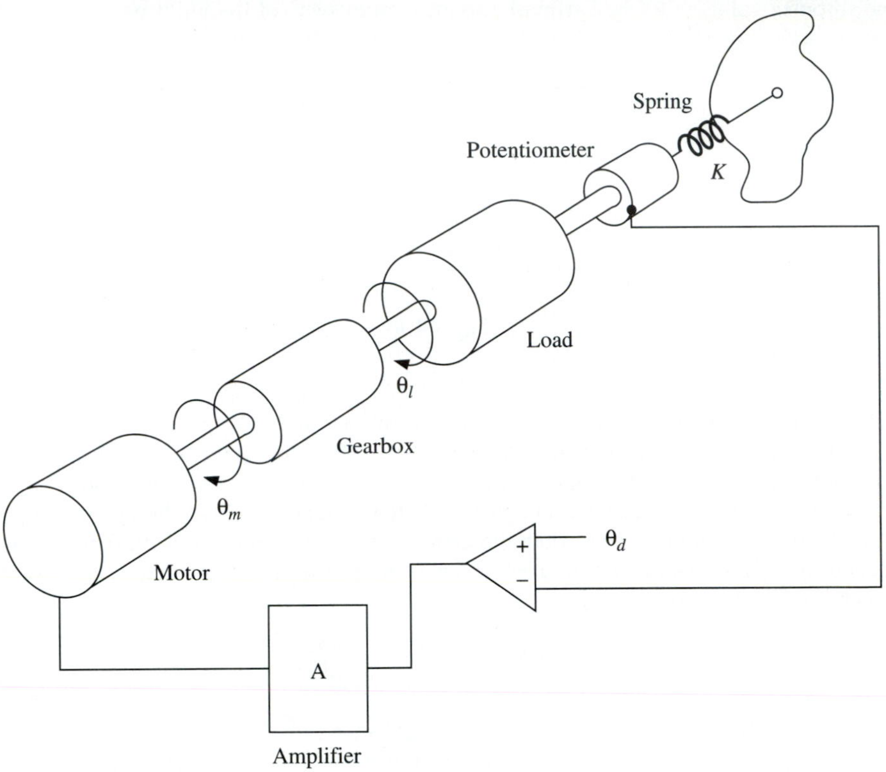

Fig. P4.10

MODULE 5

Second-Order System Time-Domain Response

Recalling that the generalized second-order system may be written as

$$\frac{C}{R} = \frac{\omega_n^2}{s^2 + 2\zeta\omega_n s + \omega_n^2} \tag{5.1}$$

we now look in some detail at the way second-order systems respond to other inputs.

◆ RAMP RESPONSE

If $R(s) = 1/s^2$, then

$$\begin{aligned} C(s) &= \frac{\omega_n^2}{s^2(s^2 + 2\zeta\omega_n s + \omega_n^2)} \\ &= \frac{1 - (2\zeta/\omega_n)s}{s^2} + \frac{(2\zeta/\omega_n)s + (4\zeta^2 - 1)}{s^2 + 2\zeta\omega_n s + \omega_n^2} \\ &= \frac{1}{s^2} - \frac{2\zeta}{\omega_n s} + \frac{(2\zeta/\omega_n)s + (4\zeta^2 - 1)}{(s + \zeta\omega_n)^2 + \omega_n^2(1 - \zeta^2)} \end{aligned} \tag{5.2}$$

Again assuming that the system in underdamped ($\zeta < 1$), the damped natural frequency ω_d will be given by

$$\omega_d^2 = \omega_n^2(1 - \zeta^2) \tag{5.3}$$

The output becomes

$$C(s) = \frac{1}{s^2} - \frac{2\zeta}{\omega_n s} + \frac{2\zeta}{\omega_n}\left(\frac{s + \zeta\omega_n}{(s + \zeta\omega_n)^2 + \omega_d^2} + \frac{\zeta\omega_n - \omega_n/2\zeta}{(s + \zeta\omega_n)^2 + \omega_d^2}\right) \tag{5.4}$$

The inverse transform gives the time response

$$c(t) = t - \frac{2\zeta}{\omega_n} + \frac{e^{-\zeta\omega_n t}}{\omega_n}\left(2\zeta \cos \omega_d t + \frac{2\zeta^2 - 1}{\sqrt{1 - \zeta^2}} \sin \omega_d t\right) \tag{5.5}$$

The three terms in the response may be identified as

1. t, the ideal output, equal to the input;
2. $2\zeta/\omega_n$, a constant steady-state error; and
3. $e^{-\zeta\omega_n t}/\omega_n\,(\ldots)$, a transient error that decays with time.

Figure 5.1 shows the input and output as functions of time. It is left to the student to prove that the critical and overdamped responses are

$$\zeta = 1\colon c(t) = t - \frac{2}{\omega_n} + e^{-\omega_n t}\left(\frac{2}{\omega_n} + t\right) \tag{5.6}$$

$$\zeta > 1\colon c(t) = t - \frac{2\zeta}{\omega_n} + \frac{e^{-\zeta\omega_n t}}{\omega_n}\left(2\zeta \cosh \omega_d t + \frac{2\zeta^2 - 1}{\sqrt{\zeta^2 - 1}} \sinh \omega_d t\right) \tag{5.7}$$

where $\omega_d = \omega_n\sqrt{\zeta^2 - 1}$.

◆ HARMONIC RESPONSE

The analysis begins by writing the transfer function in the form

$$\frac{C}{R}(s) = \frac{1}{s^2/\omega_n^2 + 2\zeta s/\omega_n + 1} \tag{5.8}$$

For a harmonic input, use will be made of the observation made in Module 3, where the output is obtained by substituting $s = j\omega$ into the appropriate transfer

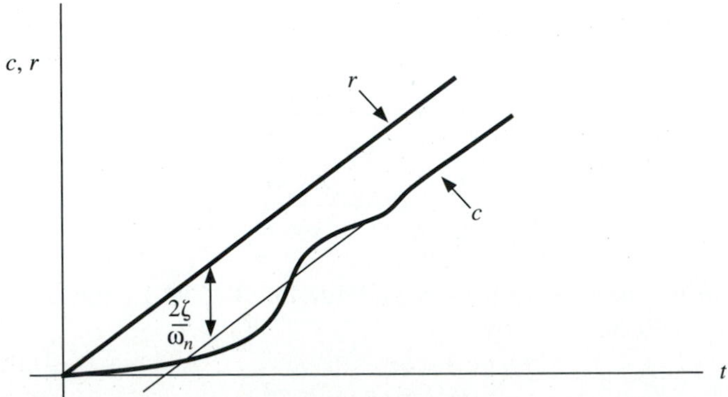

Fig. 5.1 Ramp response

function. It is important to realize that this technique yields only the *steady-state* response and provides no information regarding the transient:

$$\frac{C}{R}(j\omega) = \frac{1}{(1 - \omega^2/\omega_n^2) + (2\zeta\omega/\omega_n)j} \tag{5.9}$$

Rationalizing this complex number yields

$$\frac{C}{R}(j\omega) = \frac{1}{(1 - \omega^2/\omega_n^2)^2 + 4\zeta^2\omega^2/\omega_n^2}\left[\left(1 - \frac{\omega^2}{\omega_n^2}\right) - \frac{2\zeta\omega}{\omega_n}j\right] \tag{5.10}$$

from which the modulus of the output divided by the input is

$$M = \left|\frac{C}{R}\right| = \frac{1}{\sqrt{(1 - \omega^2/\omega_n^2)^2 + 4\zeta^2\omega^2/\omega_n^2}} \tag{5.11}$$

and the phase

$$\phi = \tan^{-1}\left(\frac{-2\zeta\omega/\omega_n}{1 - \omega^2/\omega_n^2}\right) \tag{5.12}$$

These equations for the modulus and phase of the output relative to the input are shown in Figs. 5.2 and 5.3, respectively, as functions of the dimensionless frequency ratio ω/ω_n. The figures show that the peak value of the modulus depends upon the damping ratio ζ, exhibiting a resonance phenomenon for $\zeta < 0.7$ and

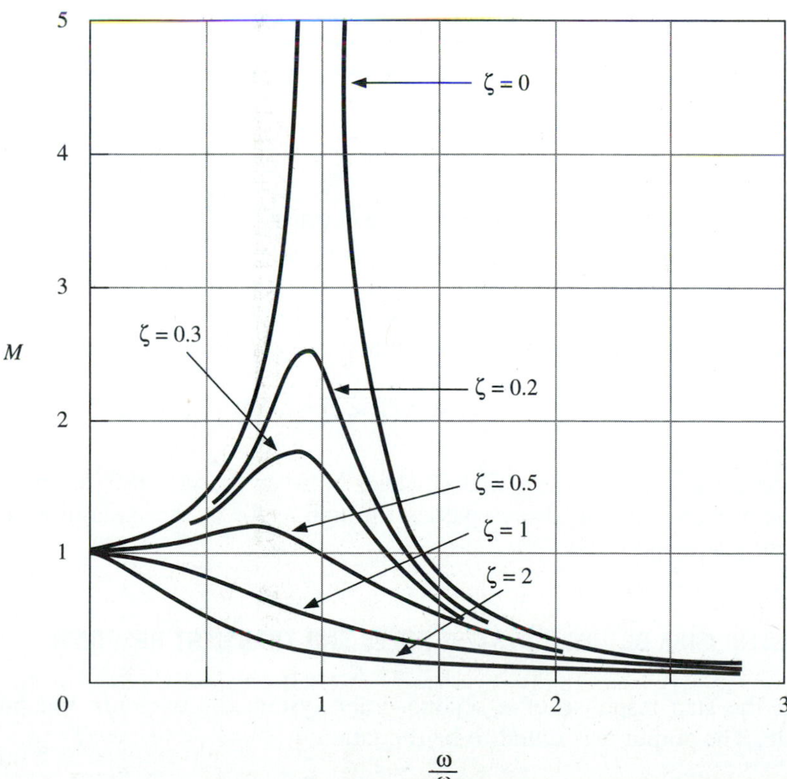

Fig. 5.2 Modulus of output

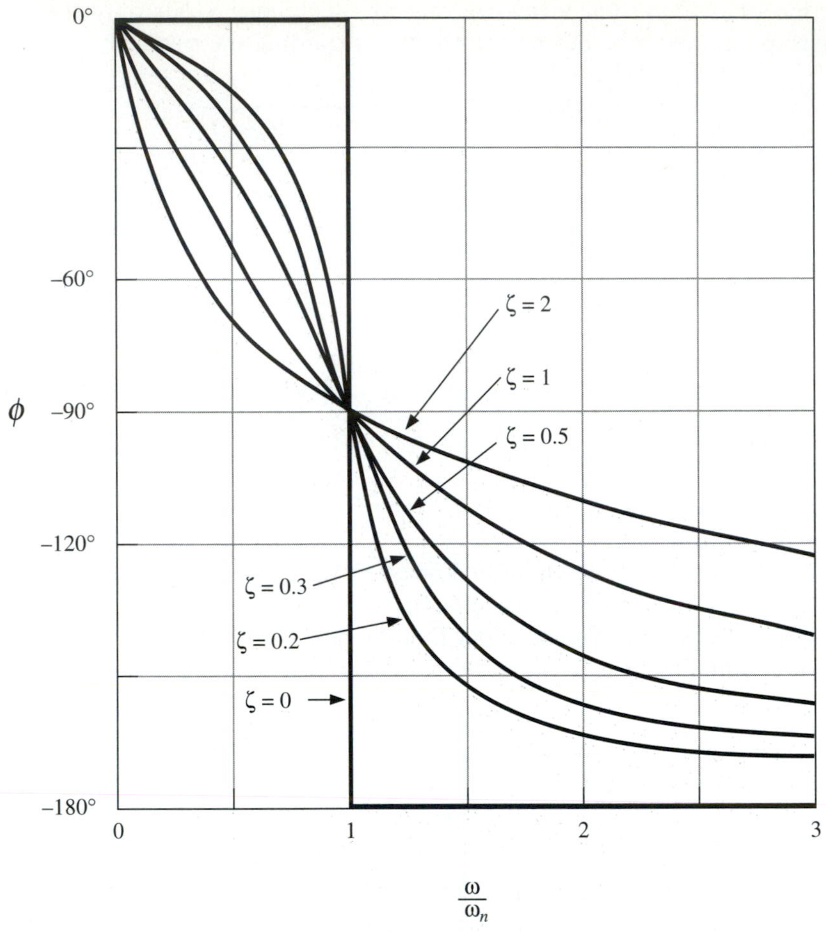

Fig. 5.3 Phase of output

infinite resonance amplitude for the undamped case $\zeta = 0$. It may be shown that the maximum resonance peak occurs at

$$\frac{\omega}{\omega_n} = \sqrt{1 - 2\zeta^2} \tag{5.13}$$

The phase angle curve shows that ϕ varies between 0 and $-180°$ as the forcing frequency increases but always passes through $-90°$ at the undamped natural frequency ω_n.

◆ RELATIONSHIP BETWEEN SYSTEM POLES AND TRANSIENT RESPONSE

Recall the step response of a second-order system discussed in the previous module. The output was found to be (equation 4.30)

$$C(s) = \frac{1}{s} - \frac{s + 2\zeta\omega_n}{(s + \zeta\omega_n)^2 + \omega_d^2} \tag{5.14}$$

Clearly the first term will produce the steady-state response once the inverse Laplace transform is taken. The transient portion of the response is obtained from the second term, namely,

$$C(s)_{\text{tran}} = \frac{s + 2\zeta\omega_n}{(s + \zeta\omega_n)^2 + \omega_d^2} \tag{5.15}$$

The poles of the system are given by

$$(s + \zeta\omega_n)^2 + \omega_d^2 = 0 \tag{5.16}$$

which gives

$$s = -\zeta\omega_n \pm j\omega_d \tag{5.17}$$

These poles are shown on the complex plane in Fig. 5.4. Note the following derived equations:

1. The length of the line joining the origin to the pole is given by

$$l^2 = (\zeta\omega_n)^2 + (\omega_d)^2 \tag{5.18}$$

$$= \omega_n^2 \tag{5.19}$$

2. The angle between the line joining the pole to the origin and the imaginary axis is given by

$$\sin\theta = \frac{\zeta\omega_n}{\omega_n} = \zeta \tag{5.20}$$

Again remembering the time-domain response for a step input to be

$$c(t) = 1 - \frac{e^{-\zeta\omega_n t}}{\sqrt{1 - \zeta^2}} \sin(\omega_d t + \phi) \tag{5.21}$$

the relationship between the location of the system poles on the complex plane and the corresponding transient time response may be summarized as follows:

1. The frequency of oscillation ω_d is given by the imaginary part of the pole.
2. The damping ratio is the sine of the angle between the pole-origin line and the imaginary axis.

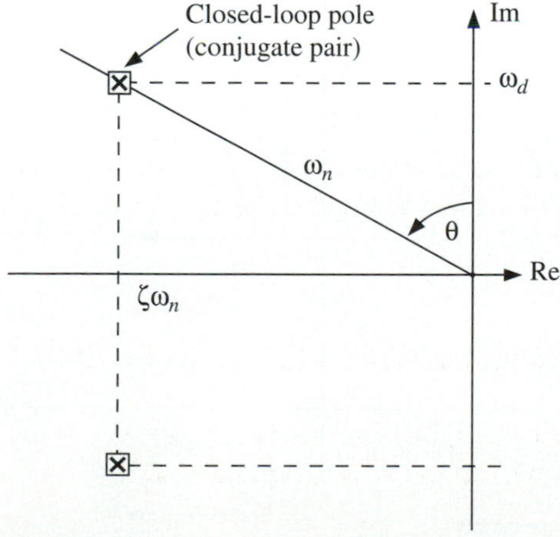

Fig. 5.4 Relationship between poles and time-domain parameters

3. The decay exponent $\zeta\omega_n$ is given by the real part of the pole.
4. The undamped natural frequency ω_n is the distance from the pole to the origin.

These features are also shown in Fig. 5.4. The following important observation should be understood.

> Just as in the case of the first-order system discussed previously, the effect of feedback around a second-order system with fixed parameters has the effect of making the resulting second-order system parameters ω_n and ζ functions of a system gain, such as K_a, as indicated in equations 4.17 and 4.18. Since the system poles are determined uniquely from ζ and ω_n, adjustment of K_a allows some choice in the location of these closed-loop poles and hence some ability to modify the system response to various inputs.

For simple systems, the mathematical relationship between system gain and the resulting closed-loop poles is easy to determine, but for more complicated systems, it is not so. Qualitative methods to deal with such cases are deferred until later in the book when a method, known as the root locus, for handling more involved systems is discussed in detail.

The relationship between the time-domain response of a second-order system and the corresponding location of the poles in the complex plane is most easily understood with the aid of Fig. 5.5. This figure shows the location of some sample poles and a sketch of the corresponding step response, indicating the amount of oscillation, decay, and frequency. Although only the upper part of the complex plane is shown ($j\omega > 0$), all complex poles will have a conjugate below the real axis. The following observations are apparent from Fig. 5.5:

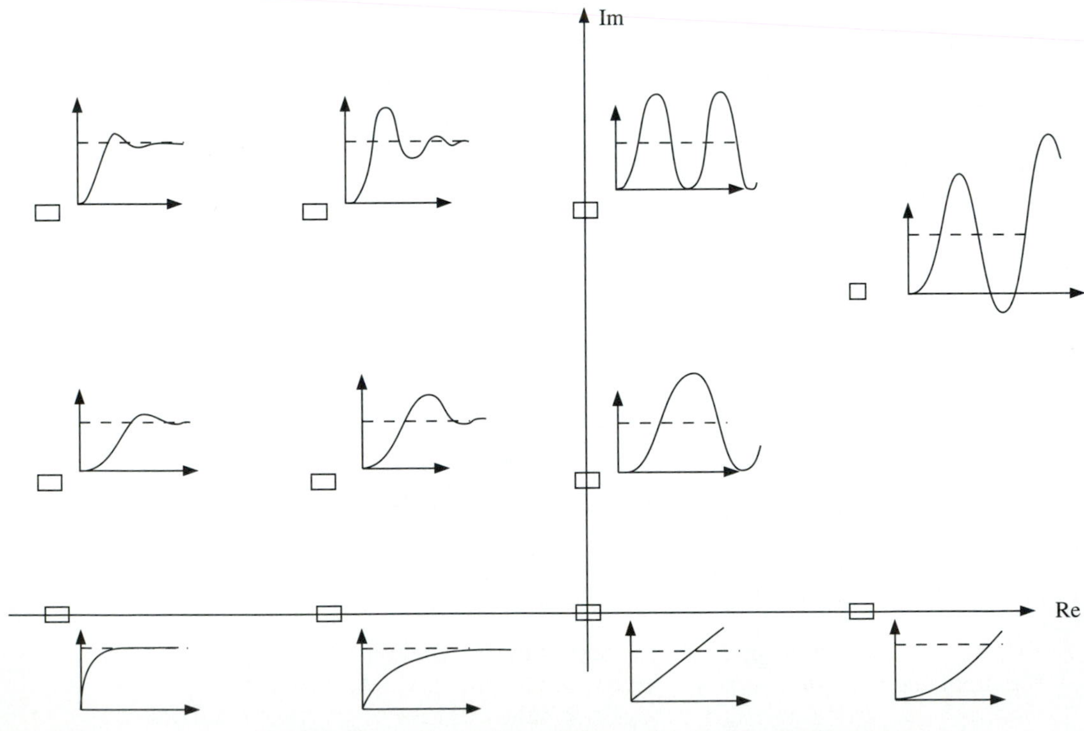

Fig. 5.5 Step response from various poles on the complex plane

1. Lines of constant damping are radial lines centered on the origin.
2. Lines of constant undamped natural frequency are circles centered on the origin.
3. First-order systems, which do not oscillate when subjected to a step input, are represented by a single pole located on the real axis.
4. Overdamped ($\zeta > 1$) second-order systems are represented by two poles located on the real axis. This is consistent with a damping ratio of unity or greater producing a radial line inclined at 90° to the imaginary axis.
5. Poles located on the imaginary axis produce a time-domain response that oscillates with constant amplitude, consistent with a damping ratio of zero.
6. Poles located on the right-hand half of the complex plane correspond to an *unstable* system. For first-order systems this corresponds to a positive exponent in the time response, while for a second-order system it corresponds to an exponentially increasing envelope to the harmonic part of the response.
7. A pole at the origin produces different responses depending on the input. For example, an impulse input will produce a constant, nonzero output, while a step input will produce an output that grows with time.

Figure 5.5 is extremely useful in providing a qualitative approach to estimating the time-domain response of a system based on the location of the system poles, avoiding the necessity to accurately calculate the poles and perform the inverse Laplace transform. It does, however, need to be used with some caution. The responses indicated in Fig. 5.5 apply only for systems represented by the transfer function

$$GH(s) = \frac{\omega_n^2}{s^2 + 2\zeta\omega_n s + \omega_n^2} \tag{5.22}$$

Higher-order systems may be *approximated* to the above transfer function if the poles representing the higher orders are located sufficiently far from the dominant complex poles. The response will be modified from that indicated above if the higher-order poles are close to the dominant poles. Similarly, the existence of a zero in the transfer function may produce unexpected time responses depending on the location of the zero. These two cases will be covered in some detail in a later module.

◆ TIME-DOMAIN PERFORMANCE SPECIFICATIONS

The specification of a required time-domain response is usually given for a second-order system subject to a step input. The resulting output may be used to define quantities that specify the response. A typical underdamped response is shown in Fig. 5.6. The following terms may be defined:

1. *Rise Time*. This is a measure of the speed with which the system responds to the input and is defined to be the time taken for the system to first achieve the final value of the output. The steady-state value for the output is not taken as the input value since for an overdamped system with a steady-state error, the

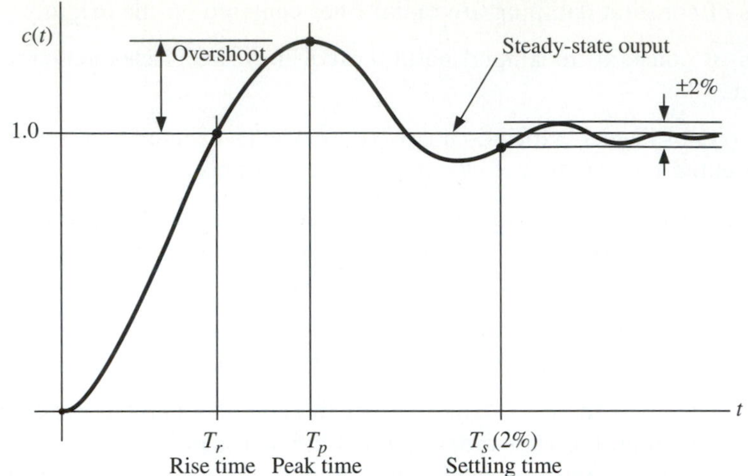

Fig. 5.6 Representative second-order step response

input value is never achieved. It may be shown (see problem 5.7) that the rise time is given by

$$T_r = \frac{1}{\omega_d}\tan^{-1}\left(\frac{-\sqrt{1-\zeta^2}}{\zeta}\right) \qquad (\zeta < 1) \tag{5.23}$$

2. *Peak Time*. The peak time is also a measure of the speed of response and is the time to the maximum of the first overshoot. Since the output oscillates at frequency ω_d, the peak time is one half cycle, i.e.,

$$T_p = \frac{\pi}{\omega_d} = \frac{\pi}{\omega_n\sqrt{1-\zeta^2}} \tag{5.24}$$

3. *Equivalent Time Constant*. Although the concept of a time constant is usually associated with first-order systems, the equivalent time constant of a second-order system may be defined as

$$\tau_e = \frac{1}{\zeta\omega_n} \tag{5.25}$$

   The reason behind this is that $\zeta\omega_n$ is the decay exponent in the transient response, corresponding to the reciprocal of the time constant τ as the decay exponent for a first-order response.

4. *Settling Time*. This is defined to be the elapsed time before the output becomes bounded by two equispaced limits on either side of the steady-state output. Typically these limits may be specified as ±5%, ±2%, or ±1%, with corresponding increases in the settling time. It may be shown that for a second-order system, the settling time may be given in terms of the equivalent time constant

$$T_s = 4.6\tau_e = \frac{4.6}{\zeta\omega_n} \qquad (\pm 1\%) \tag{5.26}$$

$$= 4.0\tau_e = \frac{4.0}{\zeta\omega_n} \qquad (\pm 2\%) \tag{5.27}$$

$$= 3.0\tau_e = \frac{3.0}{\zeta\omega_n} \qquad (\pm 5\%) \tag{5.28}$$

5. *Steady-State Error.* This is a significant performance parameter and is a function of the system transfer function and the type of input. For the system shown, of second order subject to a step input, the steady-state error turns out to be zero. An in-depth study of steady-state error is deferred until a later module.

6. *Percentage Overshoot.* The output of the system is given by equation 4.34 from the previous module:

$$c(t) = 1 - \frac{e^{-\zeta\omega_n t}}{\sqrt{1-\zeta^2}} \sin(\omega_d t + \phi) \tag{5.29}$$

The peak value of output at the time

$$t = T_p \tag{5.30}$$

is given by

$$c(T_p) = 1 + \frac{1}{\sqrt{1-\zeta^2}} e^{-\zeta\omega_n T_p/\sqrt{1-\zeta^2}} \sin\phi = 1 + e^{-\zeta\pi/\sqrt{1-\zeta^2}} \tag{5.31}$$

The percentage overshoot is therefore

$$\text{PO} = 100e^{-\zeta\pi/\sqrt{1-\zeta^2}} \tag{5.32}$$

and results are shown in the graph in Fig. 4.17.

Performance specifications will usually be given in terms of the above parameters, although it may not be possible to satisfy many design requirements with a single value of ζ or ω_n, and some compromises may be required.

Sample Problem 5.1

High-quality audio amplifiers are usually specified in terms of their *bandwidth*, which is defined as the frequency range over which the amplitude of the harmonic output signal varies by no more than ± 3 db[1] from the amplitude of the harmonic input signal. If an amplifier transfer function may be represented by a standard second-order system with an undamped natural frequency of 15 kHz, find the value of damping constant ζ that maximizes the bandwidth, and compare it to the bandwidth for a critically damped response. What is the phase lag at the optimum bandwidth? (*Hint*: if N is a number, the value of N in decibels is given by $N_{\text{db}} = 20 \log_{10} N$.)

Solution

The situation may be visualized as shown in Fig. SP5.1.1, where the amplifier has a voltage signal generator connected to the input. The amplitude of the input signal is held constant while the frequency is varied over an appropriate

[1]Sometimes the bandwidth may be defined as the frequency range only to the -3-db point.

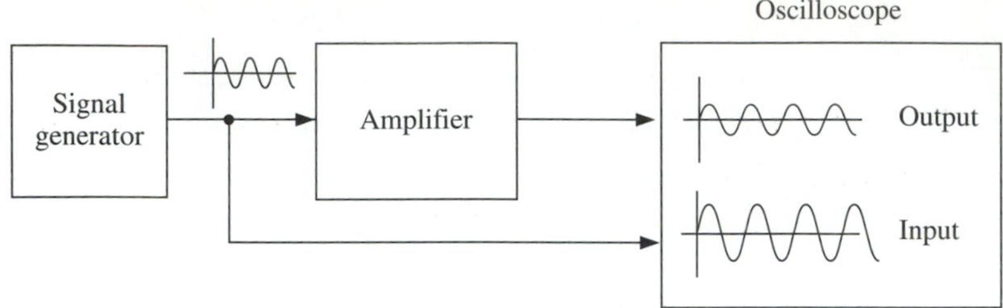

Fig. SP5.1.1

range. In this case, an appropriate range might be 0–20 kHz, which is the audible frequency range for humans. The output voltage is displayed on an oscilloscope, along with the input voltage. The ratio of the output amplitude divided by the input amplitude and the phase lag of the output with respect to the input is recorded for each frequency. Such a test is called a *frequency response test.* If the amplifier may be represented by the standard second-order transfer function

$$\frac{v_o}{v_i} = \frac{\omega_n^2}{s^2 + 2\zeta\omega_n s + \omega_n^2}$$

where the amplifier designer has some control over the damping ratio ζ but not over the natural frequency ω_n, then the frequency response test will yield a result similar to one pair of curves shown in Figs. 5.2 and 5.3.

The problem, therefore, is to determine the value of ζ that gives a modulus response of maximum frequency range lying within a ±3-db boundary. The decibel scale has to be converted to an amplitude ratio. Using the conversion data provided, it is seen that the limiting condition on the modulus of

$$-3 \text{ db} < M < +3 \text{ db}$$

is equivalent to

$$0.707 < M < 1.414$$

The situation is as shown in Fig. SP5.1.2, where it is required to determine the damping ratio that produces a resonance peak of amplitude ratio 1.414. The modulus is given by equation 5.11, which has a maximum value when the denominator is a minimum. Setting

$$\frac{d[(1 - p^2)^2 + 4\zeta^2p^2]}{dp} = 0$$

where $p = \omega/\omega_n$, yields

$$p^2 = 1 - 2\zeta^2$$

which is equation 5.13. Substituting back into 5.11 gives the corresponding peak modulus

$$M_p = \frac{1}{2\zeta\sqrt{1 - \zeta^2}}$$

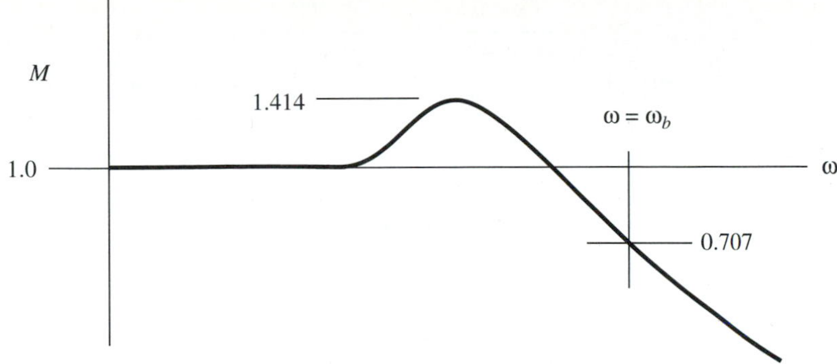

Fig. SP5.1.2

Squaring gives

$$4M_p^2\zeta^2(1 - \zeta^2) = 1$$

For $M_p = 1.414$ the damping ratio is found to be

$$\zeta = 0.38$$

Now that the required damping ratio has been found, equation 5.11 may be used again to determine the frequency at which the modulus falls to 0.707:

$$M = 0.707 = \frac{1}{\sqrt{(1 - p^2)^2 + 0.544p^2}}$$

yielding $p = 1.40$. The bandwith ω_b of the amplifier is then given by

$$\omega_b = 1.40\omega_n = 21 \text{ kHz}$$

For the critically damped case, the modulus equation becomes

$$M = 0.707 = \frac{1}{\sqrt{(1 - p^2)^2 + 4p^2}}$$

which, when solved, gives $p = 0.64$. The bandwidth for the critically damped case is therefore

$$\omega_b = 0.64\omega_n = 9.6 \text{ kHz}$$

indicating that suitable modification of the damping factor may more than double the bandwidth. The phase lag at the optimum bandwidth ($p = 1.40$) is obtained from equation 5.12:

$$\phi = \tan^{-1}\left(\frac{-2\zeta p}{1 - p^2}\right) = -132°$$

Sample Problem 5.2

Figure SP5.2.1 shows a feedback control system with a fixed plant but a controller with the gain and placement of the pole to be specified by the system

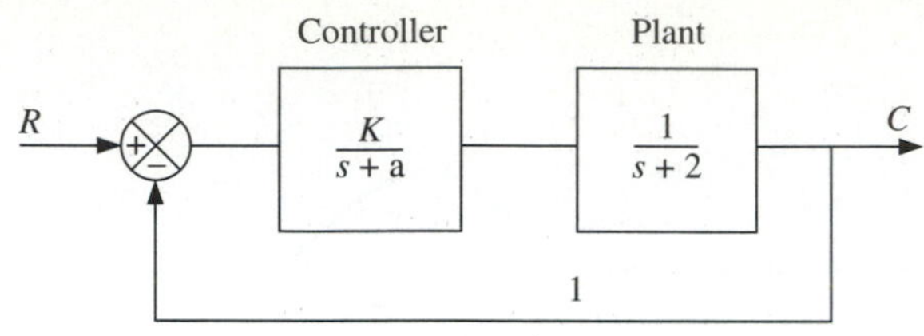

Fig. SP5.2.1

designer. Determine the location of the pole and the value of K that satisfy the following step-response design requirements:

1. a 5% settling time of no more than 1 s and
2. the overshoot not to exceed 10%.

After calculating the appropriate system parameters, determine (a) the damped and undamped natural frequencies, (b) the peak time T_p, and (c) the characteristic equation. Draw the complex plane and indicate the location of the system closed-loop poles and sketch as accurately as possible the unit step response of the closed-loop system.

Solution

From the overshoot requirement, Fig. 4.17 gives

$$\zeta = 0.6$$

The closed-loop transfer function of the system takes the form

$$\frac{C}{R} = \frac{K}{s^2 + s(2 + a) + (2a + K)}$$

Comparing with the standard form of a second-order system leads to

$$\omega_n = \sqrt{(2a + K)} \qquad \zeta = \frac{2 + a}{2\sqrt{(2a + K)}}$$

The 5% settling time is obtained from equation 5.28:

$$T_s = \frac{3}{\zeta\omega_n} = 1\ s$$

yielding the value

$$\omega_n = 5 \text{ rad/s}$$

Substituting ζ and ω_n into the equatons for ω_n and ζ gives

$$25 = 2a + K$$

$$1.44(2a + K) = (2 + a)^2$$

Solving yields the result

$$a = 4 \qquad K = 17$$

The design requirements may therefore be met by setting $K = 17$ and $a = 4$. The characteristic equation is obtained from

$$1 + KGH(s) = 0$$

or

$$s^2 + 6s + 25 = 0$$

which may be obtained directly from the denominator of the closed-loop transfer function derived previously. The roots of this equation prove to be

$$s = -3 \pm 4j$$

Figure SP5.2.2 indicates the location of the closed-loop poles (roots of the characteristic equation) on the complex plane. The damped natural frequency is seen to be

$$\omega_d = 4 \text{ rad/s}$$

while the undamped natural frequency is calculated as the distance from the origin to the closed-loop poles:

$$\omega_n = 5 \text{ rad/s}$$

The time to the first (largest) overshoot is the peak time

$$T_p = \frac{\pi}{\omega_d} = 0.785 \text{ s}$$

Using the data obtained, Fig. SP5.2.3 shows the unit step response of the system.

Note that for a unit step input, the steady-state output is 0.680.

Sample Problem 5.3

Figure SP5.3.1 shows a control system which has a pole on the right-hand half of the complex plane. Discuss the stability of the closed-loop system and

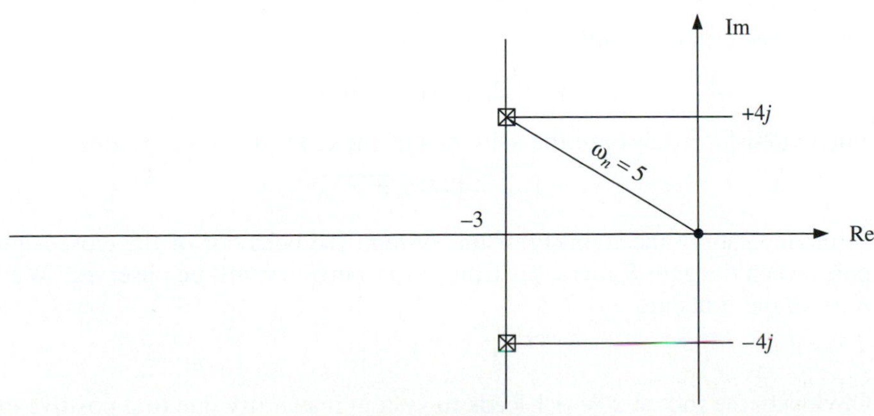

Fig. SP5.2.2

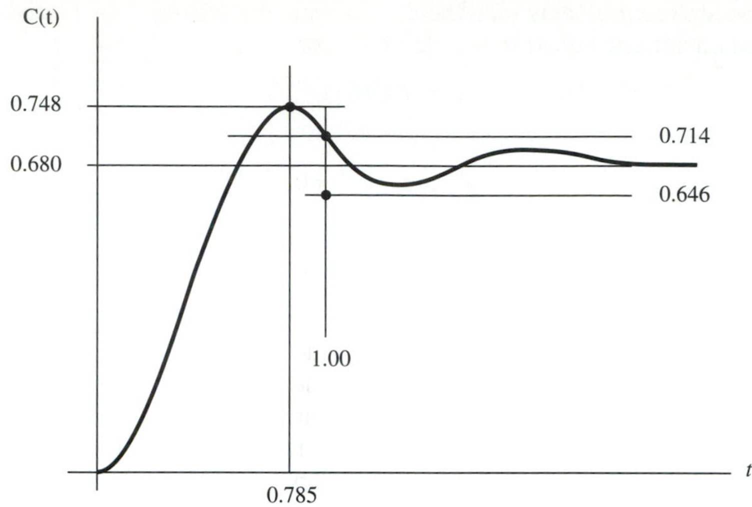

Fig. SP5.2.3

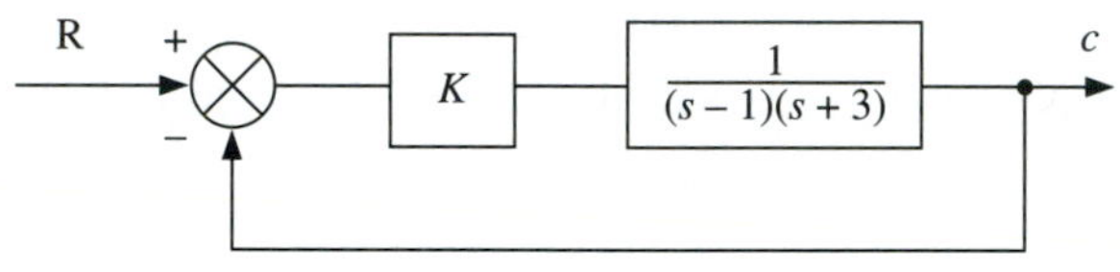

Fig. SP5.3.1

determine the range of K for which the closed-loop poles lie on the left-hand half of the complex plane.

Solution

The closed-loop transfer function of the system may be written as

$$\frac{C}{R} = \frac{K/(s-1)(s+3)}{1 + K/(s-1)(s+3)} = \frac{K}{(s-1)(s+3) + K}$$

The characteristic equation is

$$s^2 + 2s + (K - 3) = 0$$

The closed-loop poles are the solutions of the characteristic equation

$$s = -1 \pm \sqrt{4 - K}$$

In order to study the stability of the system, the behavior of the closed-loop poles when the gain K increases from zero to infinity will be observed. When $K = 0$, the roots are

$$s = -3, +1$$

Obviously the root at $s = +1$ leads to system instability due to a positive exponent in the time-domain response. Note, also, that the closed-loop poles are located at the same locations as the open-loop poles for this value of K. If K is increased a little, to $K = 0.5$, the closed-loop poles are now located at

$$s = -2.87, +0.87$$

while for $K = 1$, the roots are

$$s = -2.73, +0.73$$

and for $K = 2$ they are

$$s = -2.41, +0.41$$

The locations of the closed-loop poles are plotted on the complex plane, as shown in Fig. SP5.3.2 for representative values of K, and indicate that the poles appear to "move" on the complex plane as the gain is continuously varied. The system will be unstable as long as there is at least one root on the right-hand half plane. Instability no longer occurs once the root "moving" from the open-loop pole at $s = +1.0$ reaches the imaginary axis. From the general expression for the pole location above, it is seen that only the positive square root can cause an unstable pole. Therefore this condition disappears when

$$-1 + \sqrt{4 - K} = 0$$

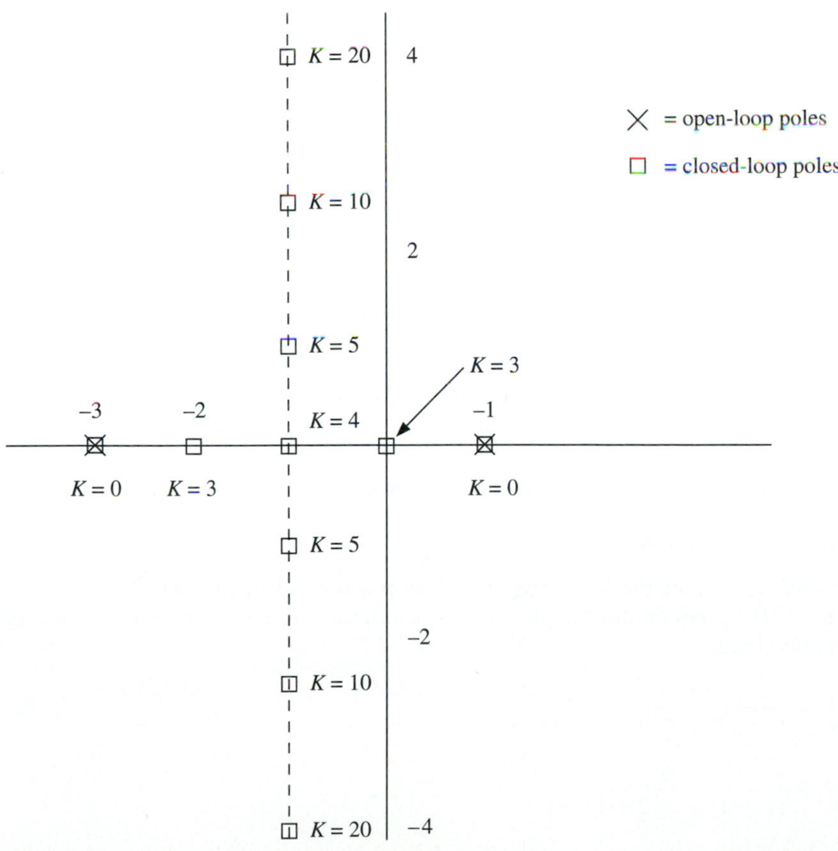

Fig. SP5.3.2

or

$$K = 3$$

Although the system is theoretically stable at this point, the closed-loop pole at the origin may still generate an unfavorable response as discussed earlier. When $K = 4$, we will have two coincident closed-loop poles at $s = -1$. For gain values $K > 4$, the roots become complex and take the form

$$s = -1 \pm j\sqrt{K - 4} \qquad (K > 4)$$

as further indicated in Fig. SP5.3.2. Although the closed-loop poles will have large imaginary components as K increases, leading to smaller damping ratios, the response is still stable, though increasingly oscillatory.

In later modules this observation that the roots of the characteristic equation trace out loci on the complex plane as the gain K varies will be dealt with in more detail. Not surprisingly, the technique is known as the root locus method.

◆ PROBLEMS

5.1 For a standard second-order system subject to a ramp input $r(t) = \Omega t$ show that at a given instant in time the difference between the input and steady-state output is given by

$$r - c_{ss} = \frac{2\zeta\Omega}{\omega_n}$$

5.2 A second-order system is given by the transfer function

$$\frac{C}{R} = \frac{10}{s^2 + 2s + 9}$$

Determine:

a. The damping ratio.
b. The damped and undamped natural frequency.
c. Maximum peak modulus M_p.
d. The frequency at which M_p occurs.
e. The bandwidth, defined as the frequency range over which the modulus does not fall more then 3 db below the low-frequency (DC) value.
f. The steady-state output for a unit step input.
g. The location of the closed-loop poles on the complex plane.
h. The rise time.
i. The 2% settling time.

5.3 Consider each of the following closed-loop transfer functions. By considering the location of the poles on the complex plane, sketch the unit step response, explaining the results obtained:

a. $\dfrac{C}{R} = \dfrac{20}{(s + 2)(s + 10)}$

b. $\dfrac{C}{R} = \dfrac{6}{(s + 1)(s + 2)(s + 3)}$

c. $\dfrac{C}{R} = \dfrac{9}{s^2 + s + 3}$

d. $\dfrac{C}{R} = \dfrac{10}{(s^2 + 4s + 5)(s + 0.5)}$

e. $\dfrac{C}{R} = \dfrac{1}{(s + 5)(s^2 + 2s + 5)}$

f. $\dfrac{C}{R} = \dfrac{100}{(s^2 + 6s + 13)(s^2 + 12s + 40)}$

5.4 Consider the closed-loop system shown in Fig. P5.4. Plot the location of the system closed-loop poles when K takes the values 2, 10, and 20. What are the corresponding damping ratios for each value of gain?

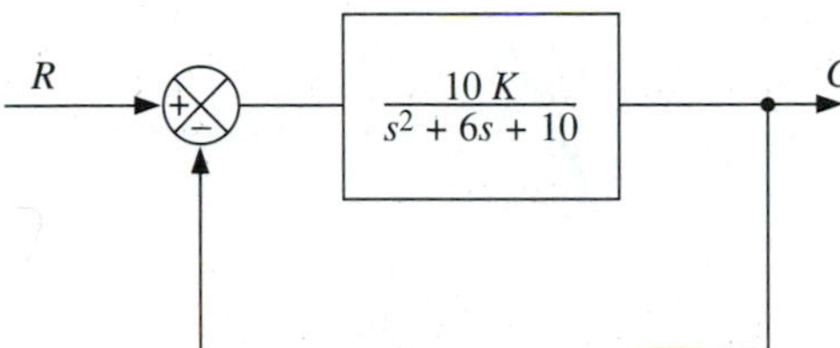

Fig. P5.4

5.5 For the servomechanism shown in Fig. P5.5, determine the values of K and a that satisfy the following closed-loop system design requirements.

a. Maximum of 40% overshoot.

b. Peak time of 4 s.

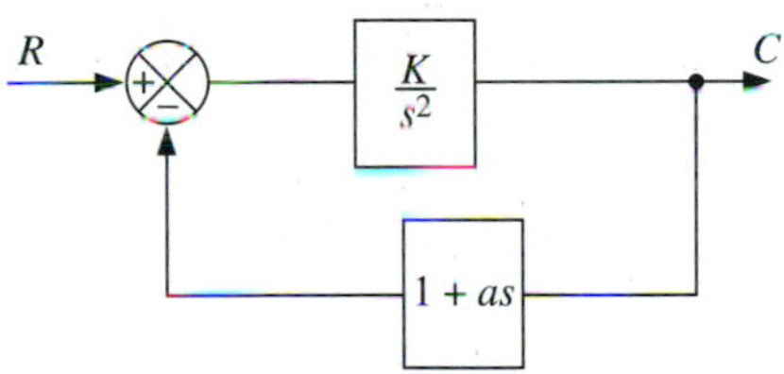

Fig. P5.5

5.6 Determine the magnitude of the first overshoot of an underdamped, second-order system subjected to a unit impuse input. Plot a graph of this magnitude as a function of damping ratio ζ.

5.7 Prove that the rise time T_r of a second-order system with a unit step input is given by

$$T_r = \frac{1}{\omega_d}\tan^{-1}\frac{-\omega_d}{\zeta\omega_n} = \frac{1}{\omega_d}\tan^{-1}\frac{-\sqrt{1-\zeta^2}}{\zeta}$$

Plot the rise time against the damping ratio.

5.8 Consider the response of an underdamped second-order system to a unit impulse. Prove that the ratio of successive peak amplitudes is a constant. Show that this constant, known as the *logarithmic decrement*, is a function of ζ only and is given by

$$\frac{\hat{x}_i}{\hat{x}_{i+1}} = \exp\left(\frac{2\pi\zeta}{\sqrt{1-\zeta^2}}\right)$$

5.9 The open-loop transfer function of a unity-feedback control system is

$$GH(s) = \frac{4K}{(s - 2)(s + 5)}$$

Find the minimum value of K for which the system is stable and the value of K that gives the system a damping ratio of 0.5, and plot the closed-loop poles on the complex plane for $0 < K < \infty$.

5.10 A transient response parameter known as the *delay time* is sometimes used. This is defined as the time taken for the output to achieve 50% of the steady-state value for a unit step input, as shown in Fig. P5.10. For a standard second-order system where $\zeta = 0.6$ and $\omega_n = 2$ rad/s, calculate the delay time.

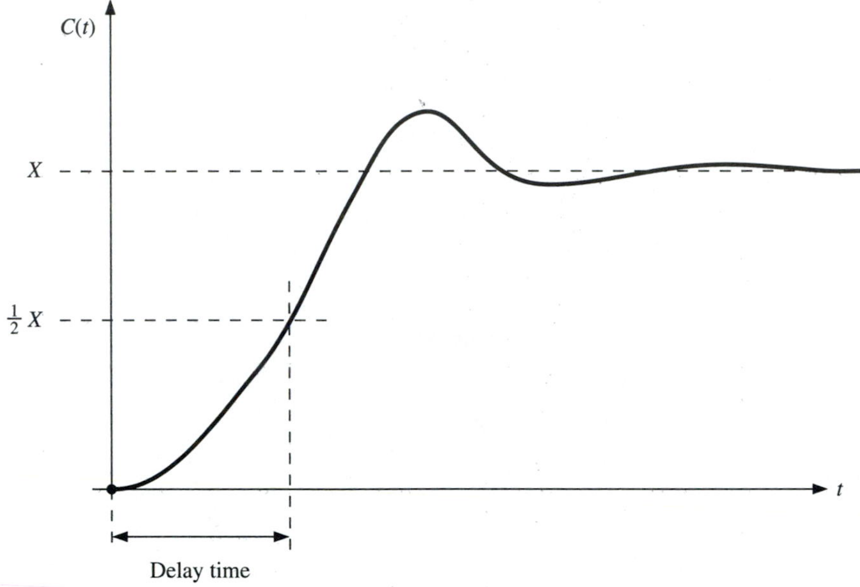

Fig. P5.10

Second-Order Systems: Disturbance Rejection and Rate Feedback

Apart from improving the dynamic response beyond that available with an open-loop system, one of the other benefits of utilizing feedback control is to make a system less responsive to unwanted inputs, known as disturbances. The ability of a system to ignore unwanted inputs is often called its *disturbance rejection* capability.

In order to investigate the ability of a system to reject disturbances and to determine where such disturbances may be represented in block diagrams, a specific example will be considered. Suppose the second-order position control system described in Module 4 is modified so that a disturbing torque is applied to the load shaft, and this torque may be represented by a simple brake mechanism, as shown in Fig. 6.1. For the moment, the nature of the disturbance is not important, but

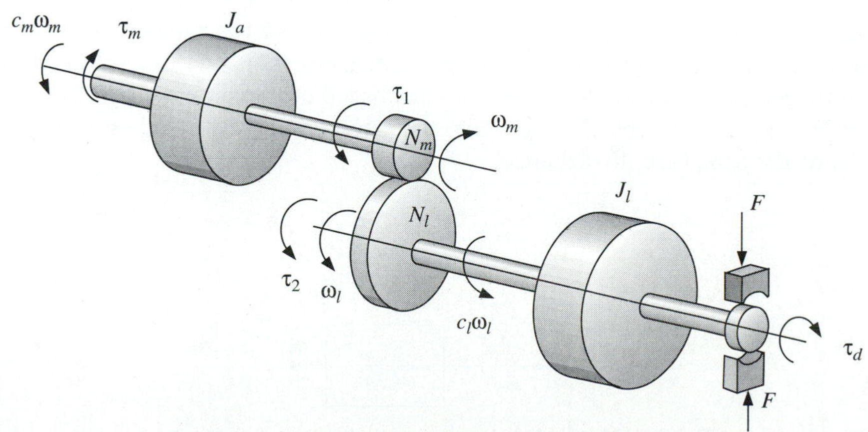

Fig. 6.1 Motor-driven inertial load with brake torque

for simplicity it may be regarded as a constant torque applied about the axis of the load shaft. The analysis of the system will proceed in the same manner as before, except that the equation of the load shaft will be modified due to the disturbing torque τ_d. For completeness, the equations of motion of both motor and load shafts, together with the gearbox equations, will be derived again here. For the motor shaft

$$\tau_m - \tau_1 - c_m\omega_m = J_a\dot{\omega}_m \tag{6.1}$$

For the load shaft with the disturbance

$$\tau_2 - c_l\omega_l - \tau_d = J_l\dot{\omega}_l \tag{6.2}$$

For the gearbox

$$\omega_l = \omega_m/N \qquad \tau_2 = N\tau_1 \tag{6.3}$$

Referring all variables to the motor shaft gives

$$\tau_m - c_m\omega_m - \frac{\tau_2}{N} = J_a\dot{\omega}_m \tag{6.4}$$

$$\tau_m - \frac{J_l\dot{\omega}_l}{N} - c_m\omega_m - \frac{c_l\omega_l}{N} - \frac{\tau_d}{N} = J_a\dot{\omega}_m \tag{6.5}$$

$$\tau_m - \frac{J_l\dot{\omega}_m}{N^2} - c_m\omega_m - \frac{c_2\omega_m}{N^2} - \frac{\tau_d}{N} = J_a\dot{\omega}_m \tag{6.6}$$

resulting in the final equation

$$\tau_m - \frac{\tau_d}{N} = \left(J_a + \frac{J_l}{N^2}\right)\dot{\omega}_m + \left(c_m + \frac{c_l}{N^2}\right)\omega_m \tag{6.7}$$

As before, this equation is represented by a transfer function relating the motor torque τ_m, the disturbance τ_d, and the resulting angular velocity of the motor shaft ω_m, as shown in Fig. 6.2. In general, the effect of the disturbance on the control system has to be determined from the physical model, and the point where the disturbance is represented in the feedback loop will be different for disturbances originating from different sources. In the case discussed above, for example, a disturbance might originate in the electrical section of the motor due to variations in the armature current, previously assumed constant. In this case the analysis of the motor circuit would differ from that based on constant armature current, resulting in a different block diagram to that shown in Fig. 4.5 and indicating the effect of the armature disturbance.

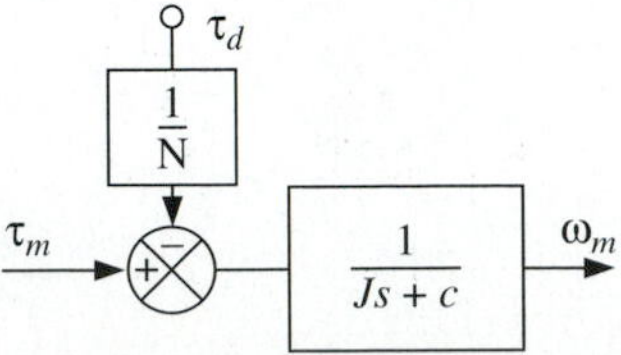

Fig. 6.2 Modified motor dynamics due to a disturbance

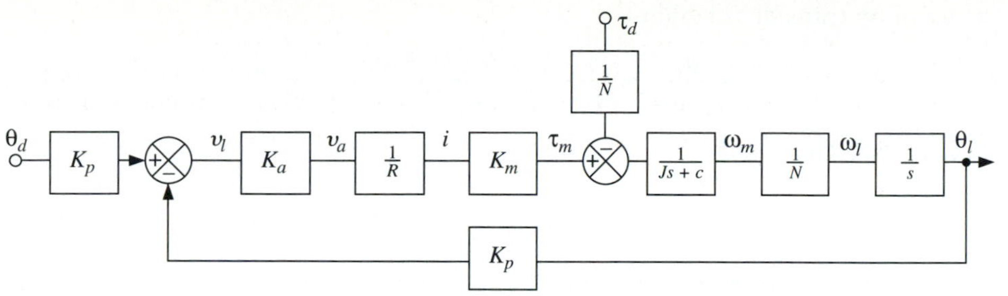

Fig. 6.3 Representation of disturbance on the control system

Whatever the cause of the disturbance, the resulting effect on the control system will be functionally the same as shown in Fig. 6.3, where the system has two inputs and one output. From Fig. 6.3 the output θ_l is given by

$$\theta_l = \frac{1}{Ns(Js + c)}\left[\frac{-\tau_d}{N} + \frac{K_a K_m}{R}(K_p\theta_d - K_p\theta_l)\right] \tag{6.8}$$

Grouping the terms involving θ_l yields

$$\theta_l\left[1 + \frac{K_p K_a K_m}{RNs(Js + c)}\right] = \frac{K_p K_a K_m}{RNs(Js + c)}\theta_d - \frac{1}{N^2 s(Js + c)}\tau_d \tag{6.9}$$

$$\theta_l[RNs(Js + c) + K_p K_a K_m] = K_p K_a K_m \theta_d - \frac{R}{N}\tau_d \tag{6.10}$$

resulting in

$$\theta_l = \left[\frac{K_p K_a K_m}{RNs(Js + c) + K_p K_a K_m}\right]\theta_d - \left[\frac{R/N}{RNs(Js + c) + K_p K_a K_m}\right]\tau_d \tag{6.11}$$

This equation shows how the output position is influenced by both the demanded position θ_d and the disturbing torque τ_d. The first term may be identified as the right-hand side of equation 4.15, indicating that this portion of the output could be obtained by considering the disturbing torque to be zero, and writing down the closed-loop transfer function relating the output position to input position. Consequently, the contribution from the disturbing torque may be obtained from Fig. 6.3 by considering the input position θ_d to be zero and writing the closed-loop transfer function relating θ_l to τ_d. This may be easier to visualize if the forward path and feedback path in Fig. 6.3 are redrawn as shown in Fig. 6.4. The

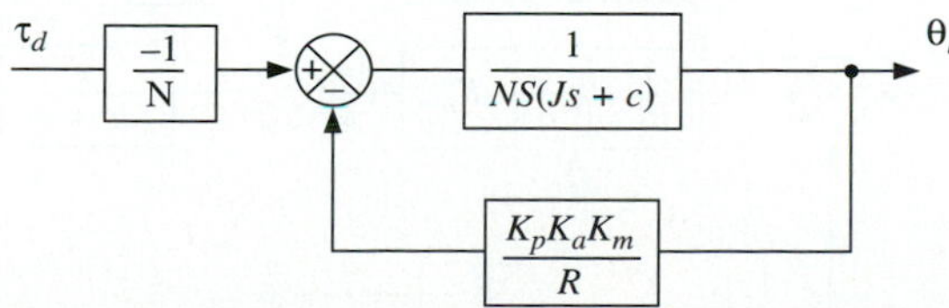

Fig. 6.4 Modified control loop for zero-input θ_2

closed-loop transfer function is

$$\frac{\theta_l}{-\tau_d/N} = \frac{1/Ns(Js + c)}{1 + K_pK_aK_m/RNs(Js + c)} \tag{6.12}$$

which yields the result

$$\frac{\theta_l}{\tau_d} = \frac{-R/N}{RNs(Js + c) + K_pK_aK_m} \tag{6.13}$$

which is seen to be the second term of equation 6.11. The result could also have been obtained using the superposition property of linear systems, which states:

> If a system with multiple inputs is linear (described by linear differential equations with constant coefficients), then the output may be obtained by setting all but one input to zero in turn and calculating the output due to that remaining nonzero input. The total output of the system is then the sum of the individual outputs.

◆ OPEN- AND CLOSED-LOOP DISTURBANCE REJECTION

Although the previous analysis was performed for a disturbance to a second-order system, the ability of a closed-loop system to reject disturbances better than open-loop systems is quite general and is not restricted to systems of a particular order. To illustrate this, a speed control problem will be analyzed using the same motor, load, and disturbance described in the previous section, but configured as shown in Fig. 6.5. In this figure, the disturbance torque is applied, as before, to the output shaft and consequently changes the dynamics of the motor shaft. For the speed control system, the motor speed ω_m is measured by a tachometer mounted on the motor shaft, and the generated voltage is compared to ω_d, a voltage corresponding to the desired motor speed. The tachometer constant is K_v and has units of volts per radian per second, therefore the demanded speed ω_d must be multiplied by the same gain prior to the summing junction. The analysis proceeds as before by determining the closed-loop transfer function:

$$\omega_m = \frac{1}{Js + c}\left[-\frac{\tau_d}{N} + \frac{K_m}{R}(K_v\omega_d - K_v\omega_m)\right] \tag{6.14}$$

This leads to

$$\omega_m\left[1 + \frac{K_mK_v}{R(Js + c)}\right] = \frac{-1}{N(Js + c)}\tau_d + \frac{K_mK_v}{R(Js + c)}\omega_d \tag{6.15}$$

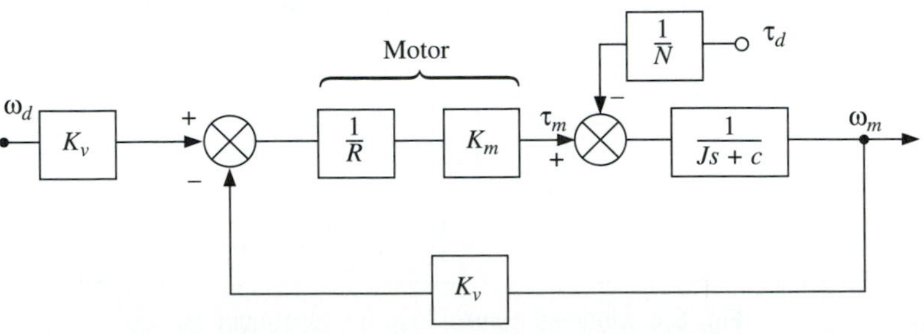

Fig. 6.5 Motor speed control system

$$\omega_m[R(Js + c) + K_mK_v] = -\frac{R}{N}\tau_d + K_mK_v\omega_d \tag{6.16}$$

resulting in

$$\omega_m = \frac{K_mK_v}{R(Js + c) + K_mK_v}\omega_d - \frac{R/N}{R(Js + c) + K_mK_v}\tau_d \tag{6.17}$$

In the above equation, the contribution due to τ_d may be regarded as a drop in the steady-state speed generated by ω_d, the magnitude of the drop being a measure of the influence of the disturbance. If τ_d is assumed to be a unit step function, the speed drop may be determined from the final-value theorem as

$$(\Delta\omega_m)_{\mathrm{CL}} = \lim_{s\to 0} \frac{R/N}{R(Js + c) + K_mK_v} = \frac{1}{Nc + NK_mK_v/R} \tag{6.18}$$

The equivalent open-loop system is now shown in Fig. 6.6. This time the output speed is obtained directly from the two inputs as:

$$\omega_m = \frac{K_mK_v}{R(Js + c)}\omega_d - \frac{1}{N(Js + c)}\tau_d \tag{6.19}$$

Again, the effect of the disturbance results in a decrease in speed from that established by the input voltage ω_d and may be found using the final-value theorem if the form of τ_d is again assumed to be a unit step:

$$(\Delta\omega_m)_{\mathrm{OL}} = \lim_{s\to 0} \frac{1}{N(Js + c)} = \frac{1}{Nc} \tag{6.20}$$

The disturbance rejection ratio (DRR) for open- and closed-loop systems may be found from equations 6.20 and 6.18 as

$$\mathrm{DRR} = \frac{(\Delta\omega_m)_{\mathrm{OL}}}{(\Delta\omega_m)_{\mathrm{CL}}} = 1 + \frac{K_mK_v}{cR} \tag{6.21}$$

Figure 6.7 shows a typical laboratory feedback control system for which the following data apply: $K_v = 0.00025$ V/rad s, $K_m/R = 13.5$ N-m/V, $c = 0.05$ N-m/rad s. Substituting these values results in the disturbance rejection ratio

$$\mathrm{DRR} = 1.67 \tag{6.22}$$

indicating that the reduction in speed due to the disturbance applied to the closed-loop system is about 67% of the speed reduction due to the same disturbance applied to the open-loop system. Figure 6.8 shows experimental results obtained from the system shown in Fig. 6.7, in which the disturbance torque is generated by means of an eddy current brake applied to the output shaft. This brake applies a torque opposing the motion of the shaft and a magnitude proportional to the

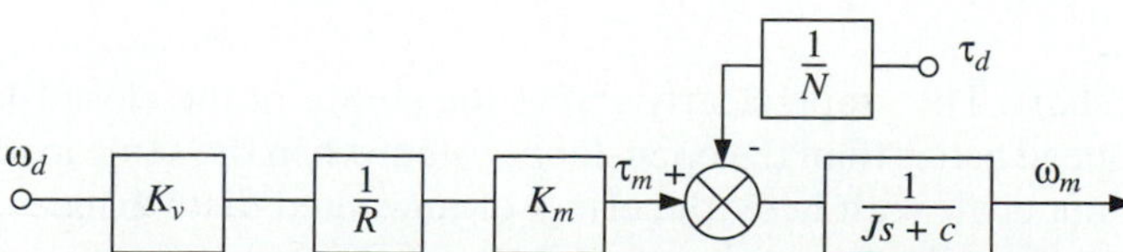

Fig. 6.6 Open-loop speed control system with disturbance

Fig. 6.7 Laboratory feedback control system

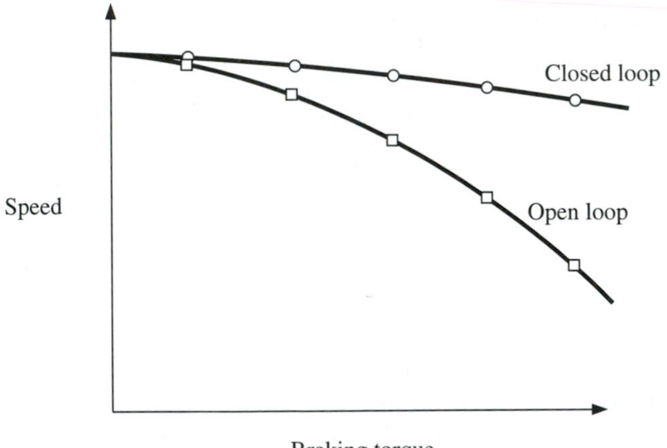

Fig. 6.8 Experimental results for open- and closed-loop disturbance rejection

speed of the shaft. The graph clearly shows the ability of the closed-loop system to maintain speed better than the open-loop system when the same load is applied.

Although not dealt with here, the effect of unwanted disturbances may be reduced by increasing the forward path gain (see SP6.1).

In order to understand the effect of velocity (rate) feedback, a particular example will be considered. The example is that of the position-controlled, motor-driven inertial load considered earlier in the module, except that for simplicity and without loss of generality, the effect of external disturbances will be ignored. The appropriate block diagram therefore is as shown in Fig. 6.3, without the disturbance. The closed-loop transfer function is

$$\frac{\theta_l}{K_p\,\theta_d} = \frac{K_aK_m/Rs(Js + c)N}{1 + K_aK_mK_p/Rs(Js + c)N} \tag{6.23}$$

which reduces to

$$\frac{\theta_l}{\theta_d} = \frac{K}{Js^2 + cs + K} \tag{6.24}$$

where

$$K = \frac{K_aK_mK_p}{RN} \tag{6.25}$$

The time-domain step response may be determined from the roots of the characteristic equation, which in turn is found from the location on the complex plane of the poles of the closed-loop transfer function, as shown in Module 5, Figs. 5.4 and 5.5. In this case there are two closed-loop poles located on the complex plane at

$$s = -\frac{c}{2J} \pm \frac{1}{2J}\sqrt{c^2 - 4JK} \tag{6.26}$$

Assuming the gain K to be a variable, the roots may be plotted on the complex plane for various gain values, as shown in Fig. 6.9. From the locus of the

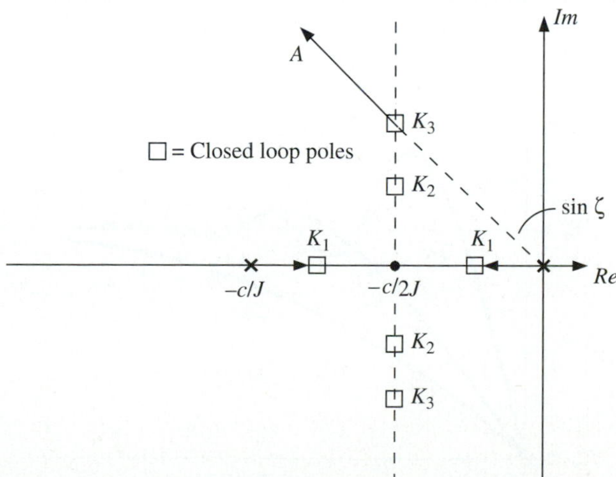

Fig. 6.9 Locus of closed-loop poles for servomechanism

closed-loop poles it is seen that as K increases, the roots move from the open-loop poles at

$$s = 0 \qquad \text{and} \qquad s = -c/J \tag{6.27}$$

toward each other, meeting at $s = -c/2J$. Increasing K further, the roots have a complex component indicating a damping ratio ζ less than unity (see Fig. 5.4.). Increasing K more still produces smaller damping ratios with increasing damped natural frequency ω_d.

Although it is possible to calculate a gain value that will give a required damping ratio, it is not possible to *completely* specify a desired response to a step input. For example, it may be necessary to have a damping ratio of 0.7 to limit the amount of overshoot the system has, but in addition it would be desirable to have the peak time as short as possible so that the system responds in the minimum time. Since the peak time is a function of ω_d, corresponding to the imaginary part of the closed-loop pole, it is not possible to *independently* select ζ and ω_d from the locus, since the poles are constrained to move along the vertical line. To increase ω_d while maintaining a constant damping ratio would require the closed-loop pole to move in the direction shown by the arrow A in Fig. 6.9; however, no locus exists in this part of the complex plane. Finally, the locus shows that the quantity $\zeta\omega_n$, which is the exponential envelope causing the decay of the oscillatory part of the response, is fixed and is given by

$$\zeta\omega_n = \frac{c}{2J} \tag{6.28}$$

and consequently cannot be changed by varying K alone. Figure 6.10 shows the type of step responses that may be obtained for the system by varying K and shows that a trade-off between speed of response and overshoot has to be made within the fixed-decay envelope. It will now be shown how the use of velocity feedback will allow independent specification of both damping ratio and speed of response. Recall Fig. 4.1, which shows a tachometer attached to the motor shaft of the servomechanism. This voltage is utilized in the feedback position control system, resulting in the block diagram shown in Fig. 6.11. In the figure, the transfer function K_v represents the tachometer constant (volts per radian per second) while the

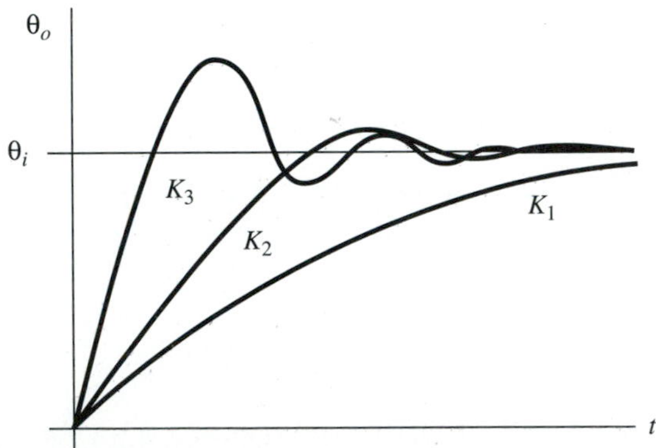

Fig. 6.10 Step responses for varying gain K

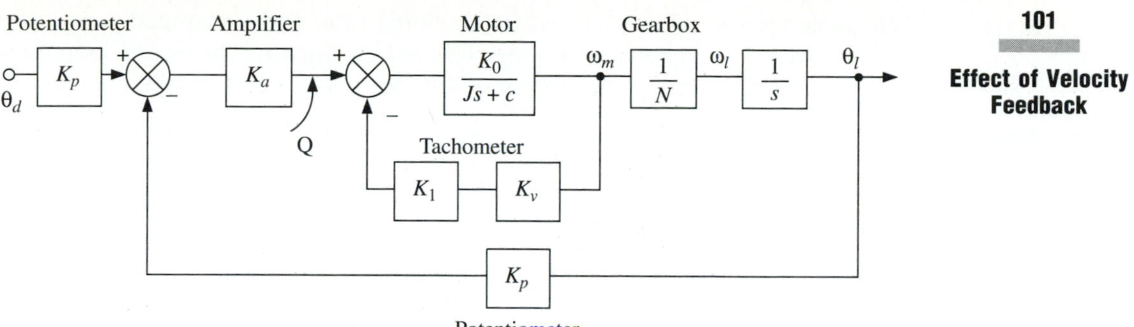

Fig. 6.11 Position control system with velocity feedback

parameter K_1 represents a gain allowing the tachometer signal to be amplified or attenuated before being fed back into the loop. In addition, the motor parameters have been grouped as

$$K_0 = \frac{K_m}{R} \tag{6.29}$$

Closing the inner loop first yields

$$\frac{\omega_m}{Q} = \frac{K_0/(Js + c)}{1 + K_0 K_v K_1/(Js + c)} \tag{6.30}$$

$$= \frac{K_0}{Js + c + K_0 K_v K_1} \tag{6.31}$$

$$= \frac{K_0}{Js + B} \tag{6.32}$$

where

$$B = c + K_0 K_v K_1 \tag{6.33}$$

Now closing the outer loop yields

$$\frac{\theta_l}{\theta_d} = \frac{K_p K_a K_0/Ns(Js + B)}{1 + K_p K_a K_0/Ns(Js + B)} \tag{6.34}$$

$$= \frac{K_p K_a K_0/N}{Js^2 + Bs + K_p K_a K_0/N} \tag{6.35}$$

$$= \frac{K'}{Js^2 + Bs + K'} \tag{6.36}$$

where

$$K' = K_p K_a K_0/N \tag{6.37}$$

The equation above is the same as the one developed for the case with position feedback only, except that the equivalent damping constant B is now a function of a selectable gain K_1 and the undamped natural frequency is also a function of a selectable gain K_a. The fact that the damping ratio and undamped natural frequency are now independent may be observed by equating the closed-loop transfer function with the standard form of second-order system to obtain

$$2\zeta\omega_n = \frac{B}{J} \qquad \omega_n^2 = \frac{K'}{J} \tag{6.38}$$

leading to

$$\omega_n = \sqrt{\frac{K_p K_a K_0}{NJ}} \tag{6.39}$$

and

$$\zeta = \frac{c + K_1 K_v K_0}{2\sqrt{JK_p K_a K_0/N}} \tag{6.40}$$

Using the above equations, ω_n may be selected using K_a while subsequent selection of ζ is made by adjusting K_1. Again, the closed-loop poles are given by the roots of the characteristic equation as

$$s = -\frac{B}{2J} \pm \frac{1}{2J}\sqrt{B^2 - 4JK'} \tag{6.41}$$

However, since B is given by equation 6.33, the locus may be changed by changing K_1 while the location of the roots on the locus is determined by K_a.

The particular relationship between the system parameters (K_1, K_a) and the required time-domain specifications must be determined for each particular problem. Since the undamped natural frequency and damping ratio have been determined above, they uniquely determine the location of the closed-loop poles on the complex plane, allowing other time-domain parameters to be obtained. The relationship between these parameters has already been discussed in earlier modules and is summarized in Fig. 5.4. In order to observe the determination of system parameters from the time-domain requirements, the student is referred to Sample Problem 6.3.

Sample Problem 6.1

As shown in Fig. SP6.1.1, a satellite tracking dish is positioned by a field-controlled DC motor using a unity-gain feedback control system. A disturbance torque acts on the dish due to a strong cross-wind. Determine the angular misalignment of the dish when the wind exerts a torque on the system of 175 N-m and the system gain is $K = 100$. What value of system gain is needed to reduce the positioning error to less than 1°?

Solution

For the purposes of determining the effect of the disturbance, the input θ_d will be assumed to be zero, allowing the transfer function from disturbance to dish output position to be written as

$$\frac{\theta_o}{\tau_d} = \frac{1 + 0.5s}{s(5s + 1)(1 + 0.5s) + 10K}$$

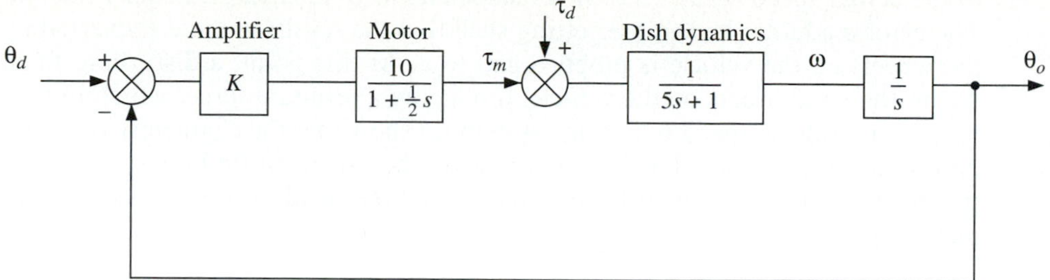

Fig. SP6.1.1

Taking the value $K = 100$ and writing the disturbance as

$$\tau_d = \frac{175}{s}$$

the steady-state output position θ_o may be obtained by using the final-value theorem

$$\lim_{t \to \infty} \theta_o = \lim_{s \to 0} \frac{175(1 + 0.5s)}{s(5s + 1)(1 + 0.5s) + 1000}$$

resulting in

$$\theta_{o(\mathrm{ss})} = 0.175 \text{ rad} \approx 10^\circ$$

In order to reduce the positioning error to less than 1° in the above equation, we write

$$\theta_{o(\mathrm{ss})} = \frac{175}{10K} = 0.0175$$

which gives

$$K = 1000$$

Sample Problem 6.2

Figure SP6.2.1 shows a speed control system for an automobile. The control speed v_c is converted to a voltage by means of the cruise control K_1 and compared

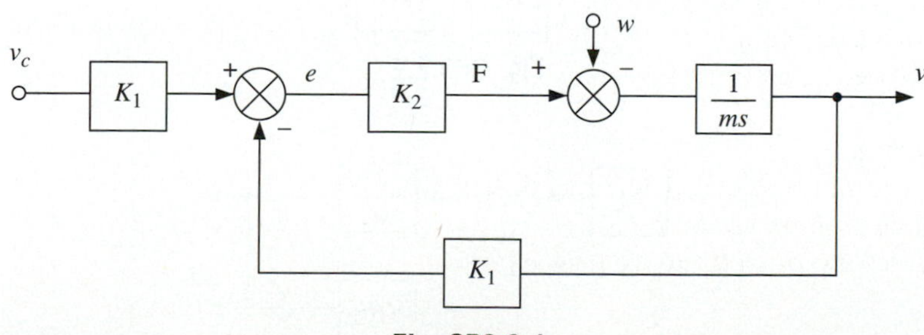

Fig. SP6.2.1

to the actual speed obtained from a tachometer with identical transfer function. The error e adjusts the throttle setting such that the resulting force F exerted by the wheels on the vehicle is proportional to e, At this point, a disturbing force due to the wind w acts on the vehicle, producing a resultant force, which in turn results in a vehicle speed v. For this system (a) show that the characteristic equation in the transfer functions v/v_c and v/w are the same, (b) find the steady-state speed when there is no disturbance, and (c) find the steady-state speed of the vehicle when w is a constant.

Solution

The transfer function between the output speed and the two inputs will be derived by using the superposition property of the system. Setting w to zero initially, the output speed due to v_c is

$$\frac{v}{K_1 v_c} = \frac{K_2/ms}{1 + K_1K_2/ms}$$

Hence

$$\frac{v}{v_c} = \frac{K_1K_2}{ms + K_1K_2}$$

In order to obtain the relationship between the disturbance and the output, we set $v_c = 0$ and rearrange the block diagram as shown in Fig. SP6.2.2 where the block containing the -1 corresponds to the summing junction with v_c. This block diagram may be simplified as in Fig. SP6.2.3 to get it in standard negative-feedback form. The output speed due to w becomes

$$\frac{v}{-w} = \frac{1/ms}{1 + K_1K_2/ms}$$

Rearranging yields

$$\frac{v}{w} = -\frac{1}{ms + K_1K_2}$$

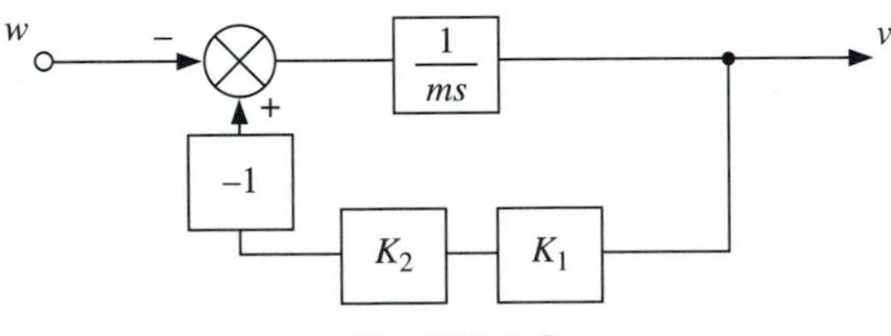

Fig. SP6.2.2

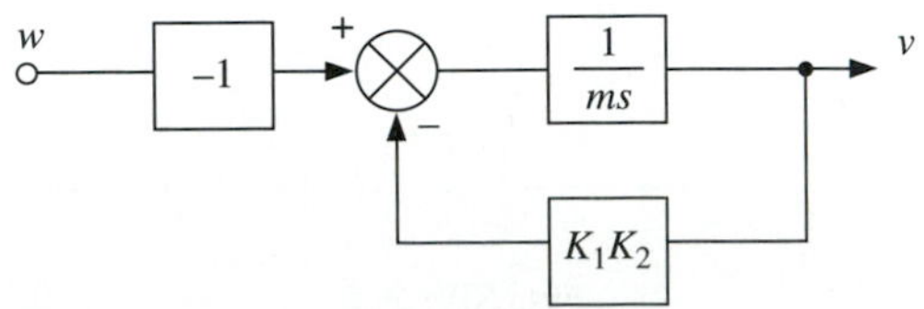

Fig. SP6.2.3

Adding these two results together,

$$v = \frac{K_1K_2}{ms + K_1K_2}v_c - \frac{1}{ms + K_1K_2}w$$

The above equations indicate that the characteristic equation for each transfer function is the same and is given by

$$ms + K_1K_2 = 0$$

It may be concluded, therefore, that the time constant of the response due to either input will be the same. When the disturbance is zero, the steady-state speed of the vehicle may be obtained by using the final-value theorem as

$$v_{ss} = \lim_{s\to 0} \frac{sK_1K_2}{ms + K_1K_2}v_c$$

When the commanded speed v_c is a constant V,

$$v_c = \frac{V}{s}$$

the steady-state speed becomes

$$v_{ss} = V$$

indicating that the vehicle travels at the requested speed. Now when the disturbance takes the form of a step input, the final-value theorem may again be used to obtain

$$v_{ss} = \lim_{s\to 0}\left(\frac{sK_1K_2}{ms + K_1K_2}\frac{V}{s} - \frac{s}{ms + K_1K_2}\frac{W}{s}\right)$$

The steady-state vehicle speed becomes

$$v_{ss} = V - \frac{W}{K_1K_2}$$

indicating a finite error for a constant speed. Means to eliminate such errors will be dealt with in later modules.

Sample Problem 6.3

A motor-driven position control system utilizes velocity feedback as shown in Fig. 6.11. The following data apply to the various components of the servomechanism:

Parameter	Value	Units
c	0.05	N-m s/rad
J	0.0125	N-m s^2/rad
K_0	13.5	N-m/V
K_v	0.02	V s/rad
K_p	6	V/rad
N	30	—

Find the values of the selectable gains K_1 and K_a that give the system a damped natural frequency $\omega_d = 5$ rad/s and a damping ratio of (a) $\zeta = 0.22$ and (b) $\zeta = 0.5$. Plot the resulting closed-loop poles on the complex plane and determine the 2% settling time for a step input.

Solution

CASE (A): $\zeta = 0.22$. The undamped natural frequency is found to be

$$\omega_n = \frac{\omega_d}{\sqrt{1 - \zeta^2}} = 5.13 \text{ rad/s}$$

Substituting the given data into equations 6.39 and 6.40 yields

$$\omega_n = \sqrt{K_p K_a K_0 / NJ} = 14.7\sqrt{K_a}$$

$$\zeta = \frac{c + K_1 K_v K_0}{2\sqrt{JK_p K_a K_0 / N}} = \frac{0.05 + 0.27K_1}{0.36\sqrt{K_a}}$$

Given ω_n, the value of K_a is found to be

$$K_a = 0.12$$

Knowing ζ and K_a, the value of K_1 is obtained from the above equation:

$$K_1 = -0.083$$

The negative sign implies changing the sign of the inner summing junction feedback path in Fig. 6.11, suggesting that damping has to be reduced from the natural value of c. Calculating the quantities B and K' from equations 6.33 and 6.37 gives

$$B = 0.0276 \qquad K' = 0.324$$

The value of B confirms that the system damping has been reduced. The closed-loop poles may be found from equation 6.41 as

$$s = -1.1 \pm 5.0j$$

The location of the real part of the roots is known to be

$$s = -\zeta\omega_n$$

which may be found directly from equation 6.28:

$$\zeta\omega_n = \frac{B}{2J} = 1.1$$

as before. The 2% settling time is obtained from equations 5.25 and 5.27:

$$T_s = \frac{4}{\zeta\omega_n} = 3.62 \text{ s}$$

CASE (B): $\zeta = 0.5$. The analysis proceeds as before. If

$$\omega_n = 5.78 \text{ rad/s}$$

then

$$K_a = 0.154 \qquad K_1 = 0.076 \qquad B = 0.0705 \qquad K' = 0.416$$

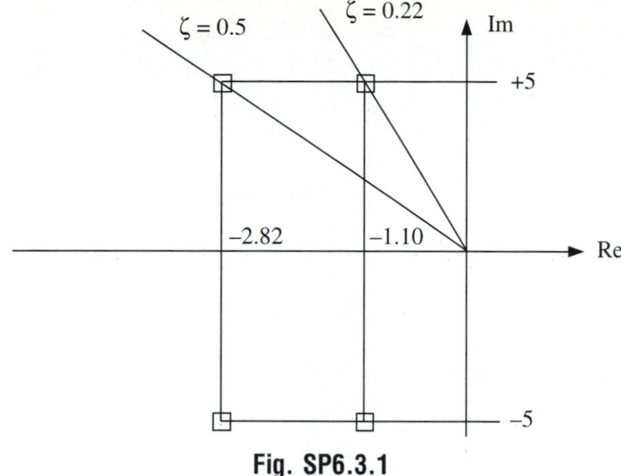

Fig. SP6.3.1

For this case the closed-loop poles are located at

$$s = -2.82 \pm 5.0j$$

and the corresponding settling time is calculated to be

$$T_s = 1.42 \text{ s}$$

Figure SP6.3.1 shows the location of the closed-loop poles for the two cases considered.

◆ PROBLEMS

6.1 Figure P6.1 shows a unity-feedback control system with one input R and two disturbances T_1 and T_2. Find the transfer function relating the output to each of the three inputs and determine the characteristic equation. What is the output if all inputs are present?

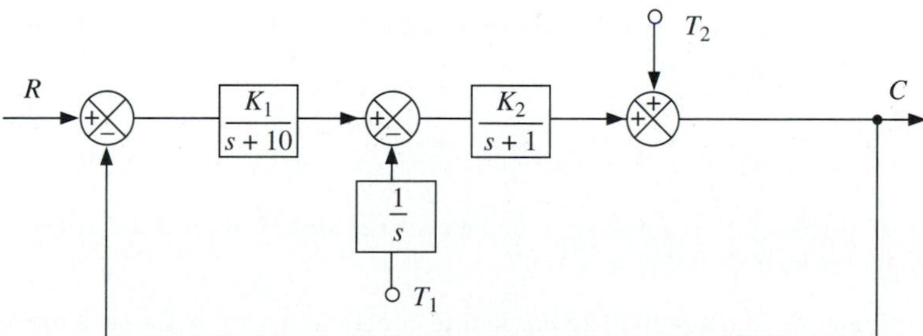

Fig. P6.1

6.2 Shown in Fig. P6.2 is a feedback control system with two disturbances: D_1 appears in the forward path while D_2 is introduced in the feedback path. Determine the output C as a function of all inputs.

6.3 A unity-feedback control system with a single disturbance input is shown in Fig. P6.3. Calculate the transfer function relating the error e to the two inputs R and D.

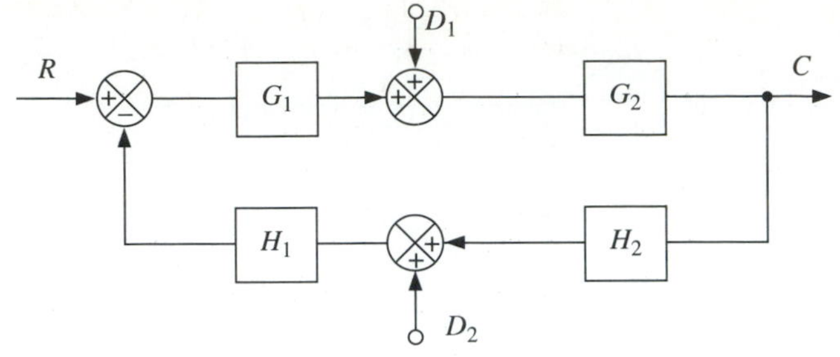

Fig. P6.2

Calculate the steady-state error e_{ss} when $r(t)$ is a unit ramp, $d(t)$ is a unit step, and $K = 10$.

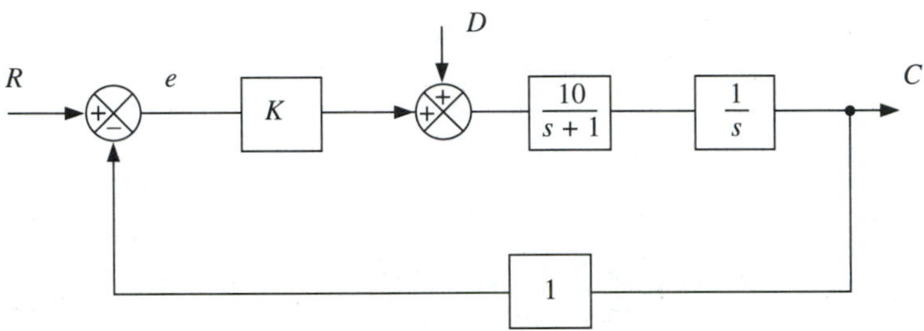

Fig. P6.3

6.4 Determine the disturbance rejection ratio (DRR) for the system shown in Fig. P6.4.

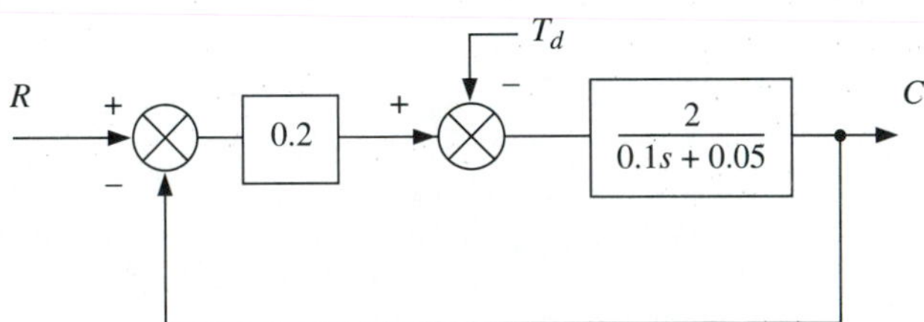

Fig. P6.4

6.5 A radar tracking antenna is shown in a simplified schematic form in Fig. P6.5. If the commanded input is a step,

$$\theta_d(s) = \frac{2}{s}$$

while the disturbance T_d is a unit step, determine the steady-state error e in the antenna position, assuming $J = 0.5$ N-m s^2/rad.

6.6 An automobile of mass 2000 kg has a drag coefficient of 10 N for each m/s of speed. The accelerator and engine may be considered to exert a force on the vehicle of 15 N for each degree of accelerator rotation u. If the vehicle is on a rising incline of grade G, show that the relationship between the vehicle speed v, the grade G, and accelerator rotation u is

$$v = \frac{0.0075u}{s + 0.005} - \frac{gG}{s + 0.005}$$

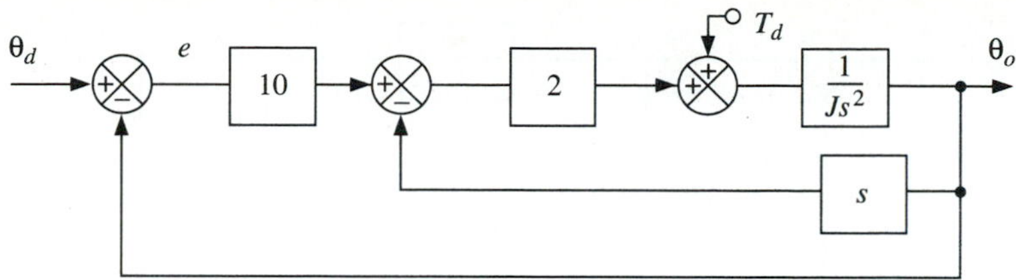

Fig. P6.5

where g is the acceleration due to gravity. Design a control system that will give a steady-state speed error of less than 2 m/s when the vehicle is on a 3% grade.

6.7 For the control system shown in Fig. P6.7, determine the percentage overshoot to a step input. Sketch the unit step response of $c(t)$, and estimate the percentage overshoot.

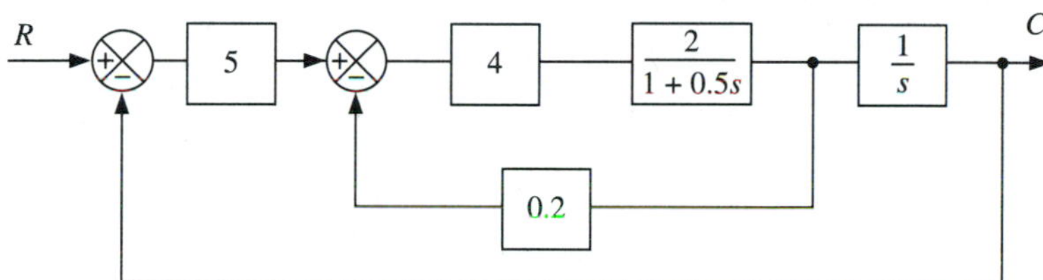

Fig. P6.7

6.8 Show that the system with rate feedback in Fig. P6.8a may be reduced to the system shown in Fig. 6.8b. Determine the values of K and K_f so that the overshoot to a step input is limited to 20% and the peak time is 1 s. What is the corresponding 5% settling time?

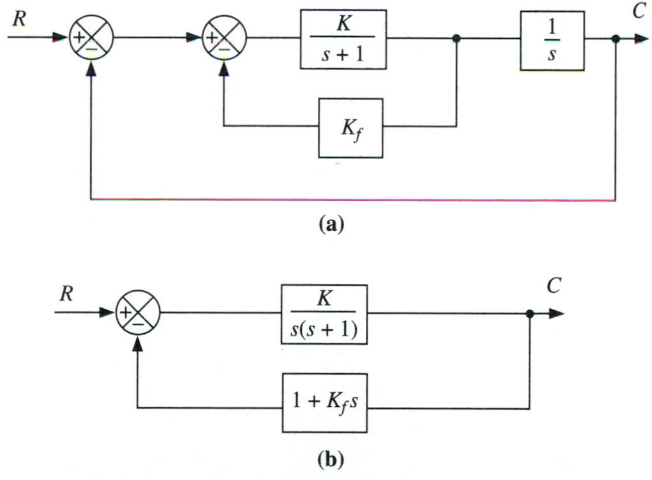

Fig. P6.8

6.9 A unity-feedback control system utilizes velocity feedback as shown in Fig. P6.9. Determine the characteristic equation for the system and find the relationship between the rate feedback parameter K, the damping ratio, and the undamped natural frequency. What value of K gives the system a 10% overshoot to a step input?

6.10 Consider the system discussed in Sample Problem 6.3. Represent graphically, in as compact a form as possible, the relationship between ζ, ω_n, and the system parameters K_a and K_1. What conclusions may be drawn from the graph regarding the possible

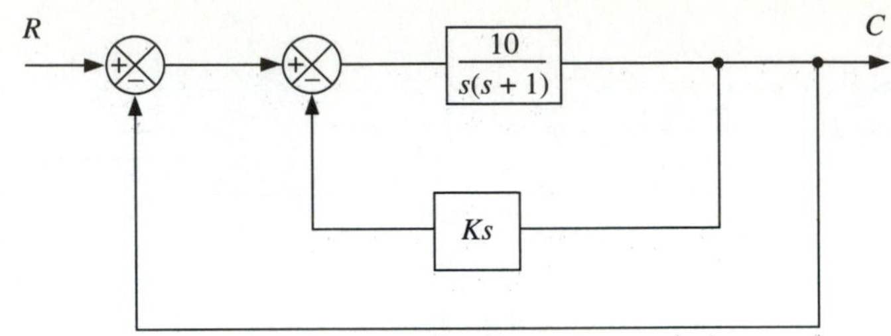

Fig. P6.9

range of ζ? If the sign of the rate feedback signal is changed from a minus to a plus (Fig. 6.11), what effect does this have on the graph?

Higher-Order Systems

◆ REDUCTION TO LOWER-ORDER SYSTEMS

It has been seen that higher-order systems may be reduced to lower-order systems through the identification of dominant poles. Some qualifications need to be specified, however, on the mechanism of reducing the appropriate transfer function, in order to ensure that it corresponds to the reduced pole-zero map. Suppose the higher-order system is given as

$$\frac{C}{R}(s) = \frac{100}{(s + 2)(s + 10)} \tag{7.1}$$

Expressing the transfer function in this way is sometimes called the Evans, or root locus, form. It is seen that the pole at $s = -2$ dominates the pole at $s = -10$, and the latter pole may be neglected in the resulting time-domain response. In order to obtain the correct transfer function corresponding to the reduced-order system, it is first necessary to express the transfer function in the so-called Bode form:

$$\frac{C}{R}(s) = \frac{5}{(1 + s/2)(1 + s/10)} \tag{7.2}$$

The reduced system now becomes

$$\frac{C}{R}(s) \approx \frac{5}{1 + s/2} \tag{7.3}$$

The transfer function must be in the Bode form before the reduction takes place; otherwise the DC gain of the transfer function changes as nondominant terms are deleted. For the system above

$$\frac{C}{R} = \frac{100}{(s + 2)(s + 10)} \tag{7.4}$$

if the input is

$$R(s) = \frac{1}{s} \tag{7.5}$$

then the steady-state output is given by

$$C_{ss} = \lim_{s \to 0} \frac{100}{(s + 2)(s + 10)} = 5 \tag{7.6}$$

If the nondominant term is ignored and it is assumed that

$$\frac{C}{R} \approx \frac{100}{s + 2} \tag{7.7}$$

then, for the same unit step input, the final value again yields

$$C_{ss} = \lim_{s \to 0} \frac{100}{s + 2} = 50 \tag{7.8}$$

If the deleted term is truly insignificant, a dramatic change in the steady-state output is clearly unexpected. Using the Bode form, however, the steady-state output for a unit step input is seen to be

$$C_{ss} = \lim_{s \to 0} \frac{5}{(1 + s/2)(1 + s/10)} \tag{7.9}$$

which yields the result

$$C_{ss} = 5 \tag{7.10}$$

whether the nondominant term is present or not. In conclusion, therefore, the Bode form must be used in order to correctly obtain the transfer function of the reduced-order system.

◆ THIRD-ORDER SYSTEMS

The advantage of reducing a third- (or higher-) order system to second order is that the time-domain response may be estimated quickly by determining the frequency of oscillation and overshoot from Figs. 4.16 and 4.17. This estimation becomes more difficult for higher-order systems. Before looking at the response of a third-order system, it is worthwhile addressing the obvious question of when may higher-order poles be neglected. Unfortunately, there is no simple answer to this question, since the effect of the neglected pole(s) may still have a catastrophic effect on system performance, even though the contribution to the response is small.

Previous discussion on the reduction of higher-order systems has focused on the criterion that the poles closest to the imaginary axis dominate those further

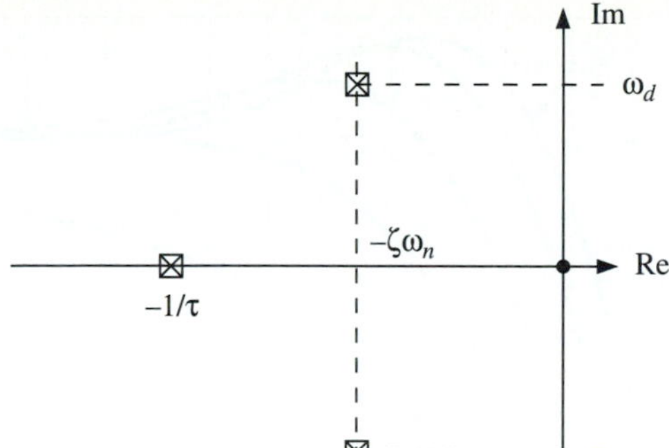

Fig. 7.1 Complex-plane representation of third-order system

away. A third-order system in the Bode form

$$\frac{C}{R} = \frac{1}{(1 + s\tau)(s^2/\omega_n^2 + 2\zeta s/\omega_n + 1)} \tag{7.11}$$

may be represented on the complex plane, as shown in Fig. 7.1. Whether the real pole may be neglected depends on the magnitude of $\zeta\omega_n$ and $1/\tau$. A common approach is to say that the real pole may be neglected if

$$\frac{|1/\tau|}{|\zeta\omega_n|} \geq 10 \tag{7.12}$$

although examples may be found where the ratio is quoted as 5 rather than 10. The effect of the real-axis pole contribution to the response will be to make it more *sluggish*. This effect is best observed by considering a specific example. Suppose that in equation 7.11

$$\omega_n = 2.0 \qquad \zeta = 0.5 \tag{7.13}$$

This places the complex poles at

$$s = -1 \pm 1.732j \tag{7.14}$$

The time-domain unit step response for various values of $1/\tau$ is shown in Fig. 7.2, including the case for the real-axis pole located at infinity. The influence of the real pole as it approaches the complex pair is clear. If the real-axis pole moves to the right of the complex pair, the real-axis pole dominates the response and no longer produces an oscillatory response. The results that have been generated for the particular case above may be generalized in order to determine transient response parameters for any third-order system and thereby act as an aid in system design. Writing the transfer function in equation 7.11 in the Evans form,

$$\frac{C}{R} = \frac{\omega_n^2/\tau}{(s + 1/\tau)(s^2 + 2\zeta\omega_n s + \omega_n^2)} \tag{7.15}$$

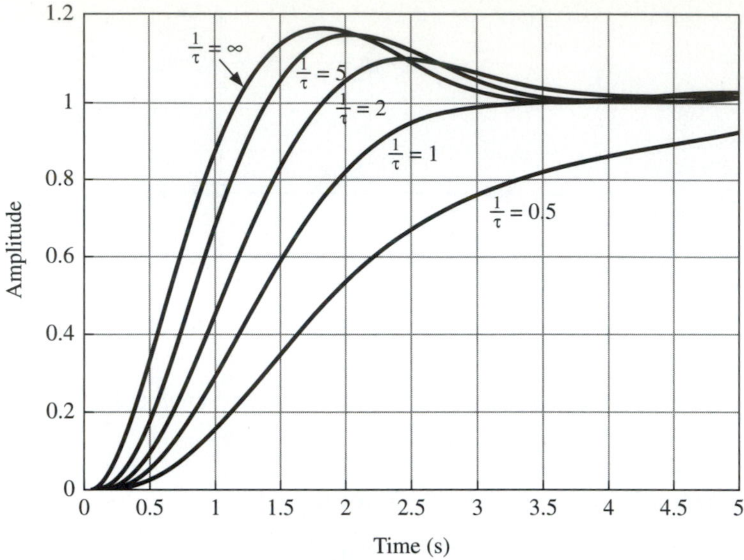

Fig. 7.2 Unit step response of third-order system

The location of the real-axis pole in relation to the real part of the complex poles is given by the parameter

$$\beta = \frac{1/\tau}{\zeta\omega_n} \tag{7.16}$$

When the real-axis pole moves to infinity along the negative axis, $\beta \to \infty$, but when the real pole is in close proximity to the complex poles, $\beta \to 1$. One of the principal transient response parameters of interest is the percentage overshoot (PO), which decreases when the effect of the real-axis pole becomes significant. The relationship between the percentage overshoot, damping ratio (of the complex closed-loop poles), and parameter β is shown in Fig. 7.3, which was obtained through computer modeling of the general third-order system given above. As $\beta \to \infty$, the PO approaches that of a second-order system with the specified damping ratio and corresponds to that obtained from Fig. 4.17.

◆ EFFECT OF A CLOSED-LOOP ZERO

Although the presence of a closed-loop zero with a pair of complex closed-loop poles does not increase the order of the system (it is still of degree 2), it will modify the time-domain response from that expected, by making it more *oscillatory*. Again consider the closed-loop transfer function

$$\frac{C}{R} = \frac{1 + s\tau}{s^2/\omega_n^2 + 2\zeta s/\omega_n + 1} \tag{7.17}$$

The closed-loop poles and zeros are shown on the complex plane in Fig. 7.4. Taking as an example the data from the previous section, where the closed-loop poles

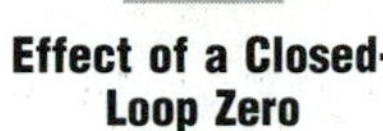

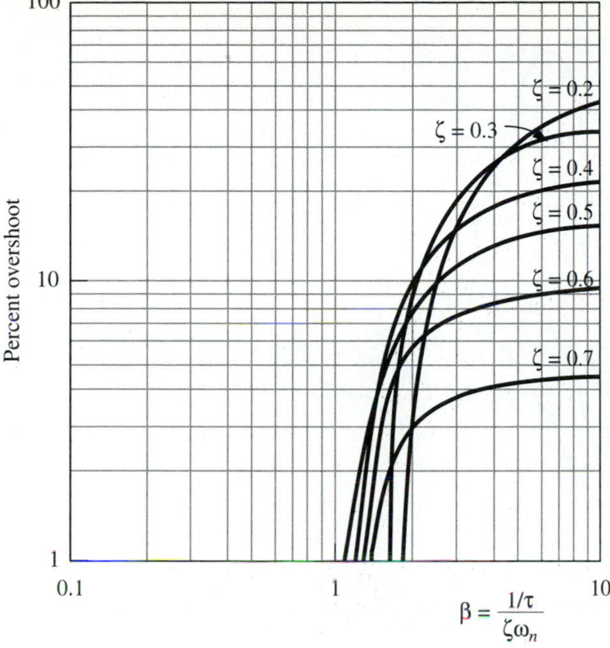

Fig. 7.3 Percentage overshoot for general third-order system

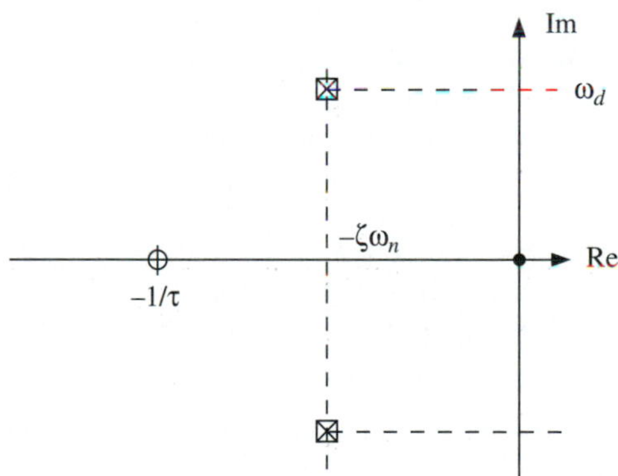

Fig. 7.4 Closed-loop zero

were located at

$$\omega_n = 2.0 \qquad \zeta = 0.5 \Rightarrow s = -1 \pm 1.732j \tag{7.18}$$

the unit step response of the system for various locations of the closed-loop zero is shown in Fig. 7.5. Note that when the effect of the zero becomes significant, the response becomes more oscillatory. Again the question arises of when the closed-loop zero may be neglected, and it is customary to allow this when

$$\frac{|1/\tau|}{|\zeta\omega_n|} \geq 10 \tag{7.19}$$

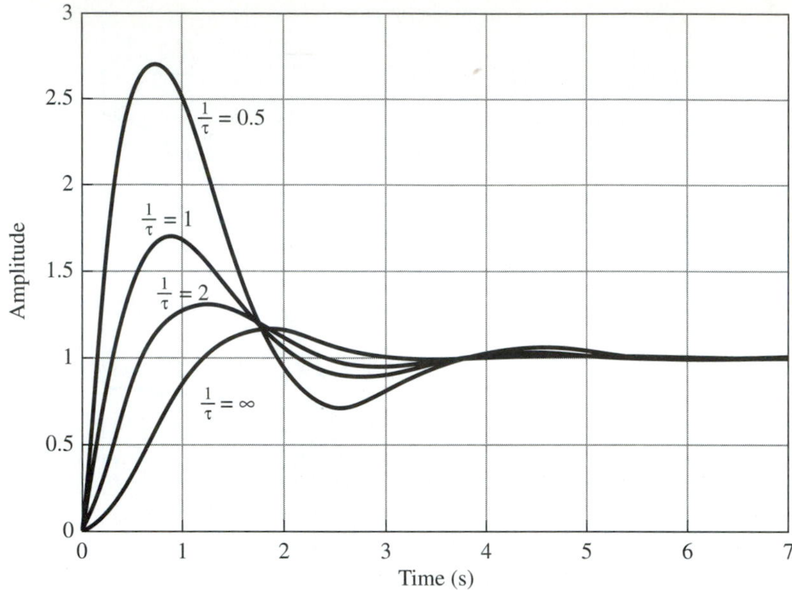

Fig. 7.5 Step response for various locations of closed-loop zero

As in the previous section, these results may be generalized so that the time-domain response parameters may be estimated if the location of the closed-loop zero is known. Since the response becomes more lightly damped, the effect of the zero on the percentage overshoot will be considered in detail, although note that other time-domain parameters such as the damped natural frequency and the settling time will also be affected. Again writing the closed-loop transfer function in the Evans form,

$$\frac{C}{R} = \frac{\omega_n^2 \tau (s + 1/\tau)}{s^2 + 2\zeta\omega_n s + \omega_n^2} \tag{7.20}$$

The location of the closed-loop zero in relation to the real part of the complex poles is given by the parameter

$$\gamma = \frac{1/\tau}{\zeta\omega_n} \tag{7.21}$$

The closed-loop zero materially affects the time-domain response when $\gamma \rightarrow 1$, while the condition $\gamma \rightarrow \infty$ leaves the response dominated by the complex poles. Using computer simulation, the relationship between the PO and the parameter γ may be plotted for several values of damping ratio, and the results are shown in Fig. 7.6. Although the closed-loop poles may lie on the negative real axis ($\zeta \geq 1$), the open-loop zero may still cause an oscillatory response. As the parameter γ approaches infinity, the effect of the closed-loop zero diminishes and the overshoot from Fig. 7.6 approaches that for a simple second-order system, as given in Fig. 4.17.

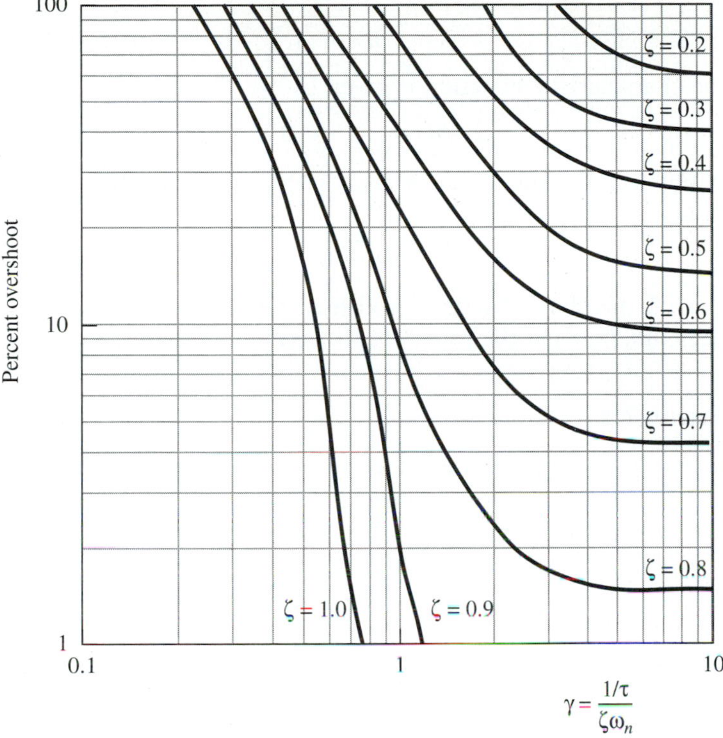

Fig. 7.6 Effect of zero on overshoot

◆ OCCURRENCE OF CLOSED-LOOP ZEROS

We have just seen that the presence of a closed-loop zero may affect the time-domain response of a closed-loop system considerably. It is worthwhile investigating where closed-loop zeros come from in terms of the forward-path and feedback transfer functions. Consider the general system shown in Fig. 7.7 where these transfer functions have been expressed in terms of the ratio of two polynomials in s. The closed-loop transfer function is given by

$$\frac{Y}{X} = \frac{A/B}{1 + AC/BD} \tag{7.22}$$

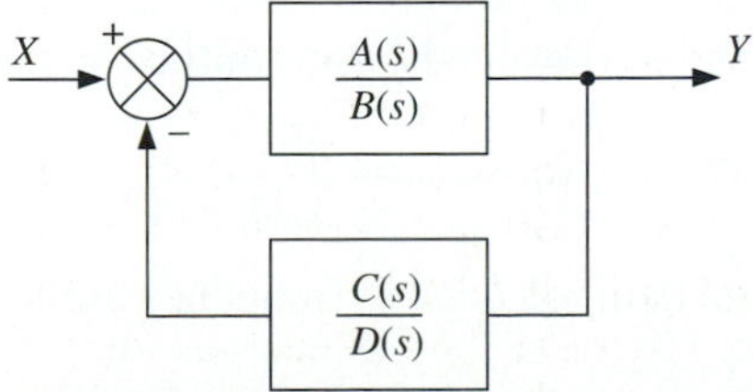

Fig. 7.7 General feedback system

which, when the fractions are cleared, becomes

$$\frac{Y}{X} = \frac{AD}{BD + AC} \tag{7.23}$$

Clearly, the characteristic equation is given by

$$BD + AC = 0 \tag{7.24}$$

but, more importantly, the zeros of the closed-loop transfer function are obtained from

$$AD = 0 \tag{7.25}$$

It may be stated, therefore, that the closed-loop zeros are the zeros of the forward-path transfer function and the poles of the feedback transfer function.

Sample Problem 7.1

Reduce the closed-loop transfer function of the system given below to second order, and sketch the unit step response:

$$\frac{C}{R} = \frac{600}{(s^2 + 22s + 120)(s^2 + 3s + 4)}$$

Solution

It is noticed that the solution of the two quadratic terms gives closed-loop poles at $s = -10, -12, -1.5 \pm 1.32j$. Factoring the first quadratic and writing the transfer function in the form

$$\frac{C}{R} = \frac{600}{(s + 10)(s + 12)(s^2 + 3s + 4)}$$

allows the transfer function to be written in the Bode form

$$\frac{C}{R} = \frac{600/480}{(1 + s/10)(1 + s/12)(s^2/4 + 3s/4 + 1)}$$

The closed-loop poles are shown in Fig. SP7.1.1. Although the two real-axis poles are closer than that required by the guideline given in equation 7.12, they will both be ignored in order to reduce the system to second order, which now may be done since the transfer function is in the Bode form. The reduced system becomes

$$\frac{C}{R} = \frac{1.25}{(s^2/4 + 3s/4 + 1)}$$

If the input is a unit step, the steady-state output becomes

$$C_{ss} = \lim_{s \to \infty} \frac{1.25}{(s^2/4 + 3s/4 + 1)} = 1.25$$

The damping ratio and damped natural frequency are found, from the complex poles, to be

$$\zeta = 0.75 \qquad \omega_d = 1.32 \text{ rad/s}$$

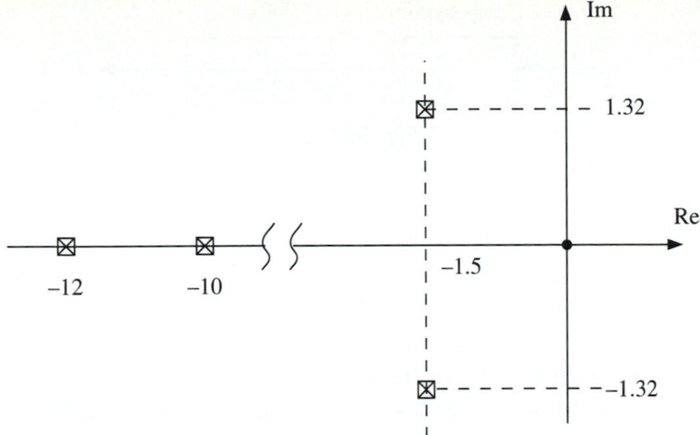

Fig. SP7.1.1

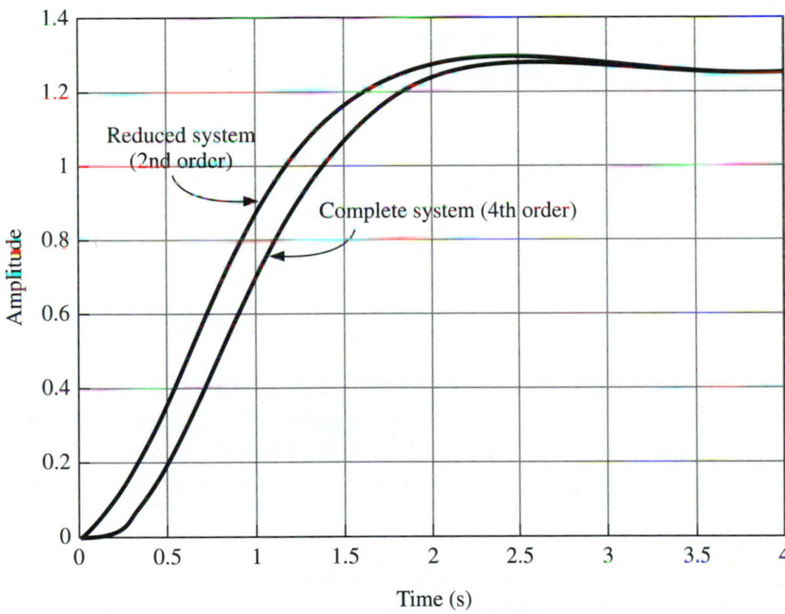

Fig. SP7.1.2

which give the percentage overshoot and peak time T_p:

$$\text{PO} = 2.84\% \qquad T_p = 2.38 \text{ rad/s}$$

The response is sketched in Fig. SP7.1.2, which also shows a computer simulation of the complete system including the two real-axis poles. As expected, the complete system is a little more sluggish in its response due to the presence of the additional poles.

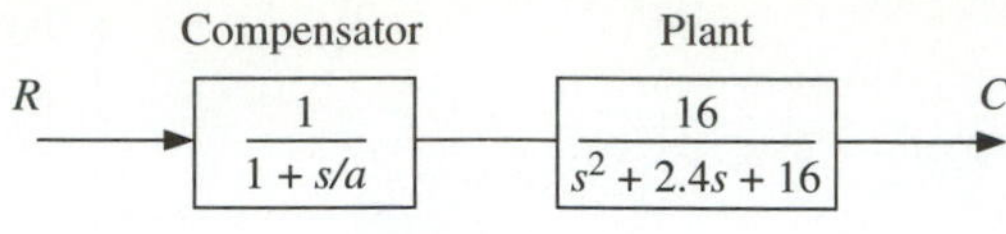

Fig. SP7.2.1

Sample Problem 7.2

Figure SP7.2.1 shows a plant represented by a second-order control system. It is required that the output of the closed-loop system be limited to no more than 10% overshoot, and this will be achieved through the use of a compensation element in the forward path. Determine the necessary value of the parameter a in the compensator, and locate the closed-loop poles on the complex plane.

Solution

Examination of the plant reveals the following time-domain parameters:

$$\omega_n = 4 \text{ rad/s} \qquad \zeta = 0.3$$

Clearly, the response to a step input without the compensator will have an overshoot of more than 10%, in fact from Fig. 4.17, PO = 37%. From Fig. 7.3, however, the third-order system with $\zeta = 0.3$ would have an overshoot of 10% if the additional pole were located such that the parameter $\beta = 2.1$. This is related to the parameter a by

$$\beta = \frac{1/\tau}{\zeta\omega_n} = \frac{a}{\zeta\omega_n} = 2.1$$

Hence

$$a = 2.1\zeta\omega_n = 2.52$$

The system poles are as shown in Fig. SP7.2.2. Moving the compensation pole nearer the imaginary axis decreases β, thereby reducing the overshoot, while moving the pole away from the imaginary axis has the reverse effect. In order to ensure no more than the specified overshoot, the constraint that applies is, therefore,

$$a < 2.52$$

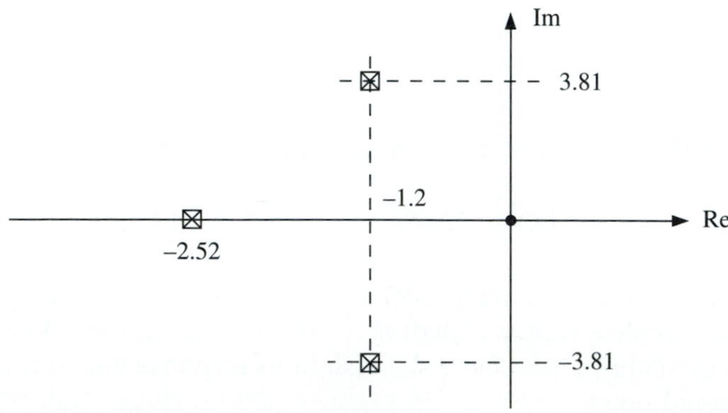

Fig. SP7.2.2

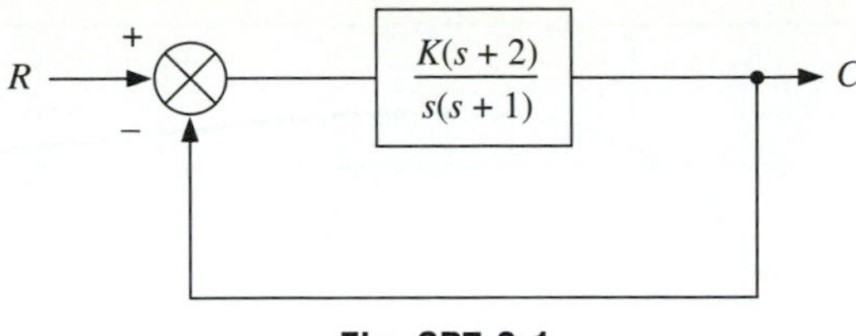

Fig. SP7.3.1

Sample Problem 7.3

For the unity-feedback system shown in Fig. SP7.3.1, determine the percentage overshoot for a system gain of $K = 4$. Determine the increase in overshoot attributable to the closed-loop zero.

Solution

The closed-loop transfer function is given by

$$\frac{C}{R} = \frac{K(s + 2)}{s^2 + s(1 + K) + 2K}$$

which for $K = 4$ yields

$$\frac{C}{R} = \frac{4(s + 2)}{s^2 + 5s + 8}$$

This system has two closed-loop poles located at

$$s = -2.5 \pm 1.32j$$

and a closed-loop zero at $s = -2$, as shown in Fig. SP7.3.2. Based on the characteristic equation, the time-domain response parameters may be calculated as

$$\omega_n = 2.82 \text{ rad/s} \qquad \zeta = 0.886$$

Examining Fig. 4.17 would suggest that for $\zeta \approx 0.9$, the percentage overshoot is negligible. However, the presence of a closed-loop zero modifies this figure.

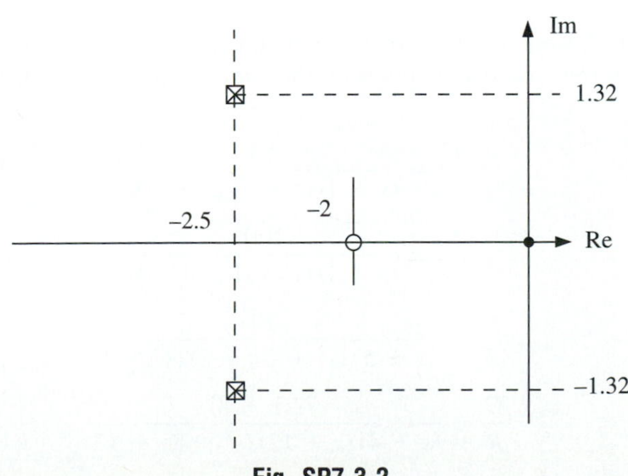

Fig. SP7.3.2

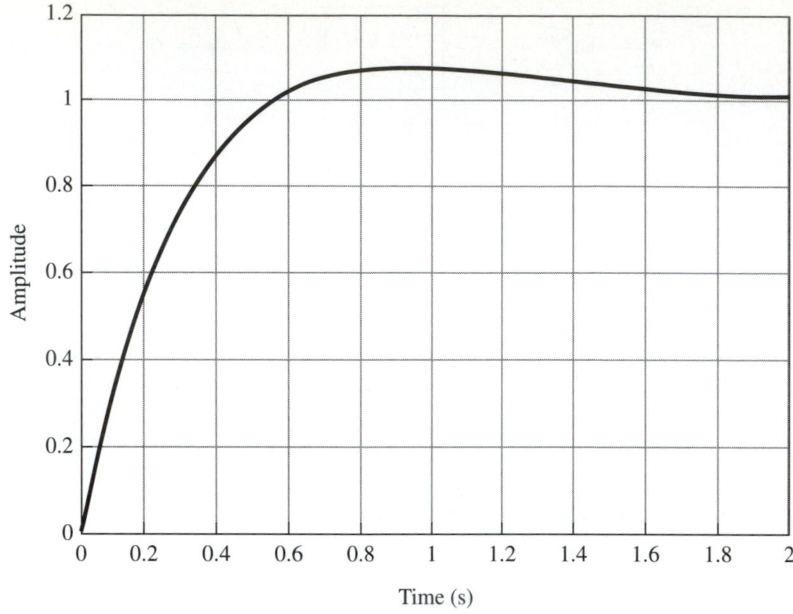

Fig. SP7.3.3

The value of the parameter γ is found to be

$$\gamma = \frac{1/\tau}{\zeta\omega_n} = \frac{2}{2.5} = 0.8$$

Examining Fig. 7.6 for $\gamma = 0.8$ and $\zeta = 0.9$ yields

$$\text{PO} = 7.0\%$$

A computer simulation of the unit step response of the system is shown in Fig. SP7.3.3, confirming the predicted value.

◆ PROBLEMS

For problems 7.1–7.4 reduce the transfer function to an approximate first- or second-order system, plot the closed-loop poles on the complex plane, and sketch the unit step response.

7.1

$$\frac{C}{R} = \frac{50}{(s+1)(s+5)(s+50)}$$

7.2

$$\frac{C}{R} = \frac{100}{(s+1)(s^2+12s+20)}$$

7.3

$$\frac{C}{R} = \frac{10}{(s+5)(s^2+2s+2)(s^2+4)}$$

7.4

$$\frac{C}{R} = \frac{72(s+8)}{(s+4)(s+12)(s^2+8s+12)}$$

7.5 Determine the closed-loop transfer function and the percentage overshoot for a system described by the pole-zero map shown in Fig. P7.5, assuming the steady-state gain of the closed-loop transfer function is unity.

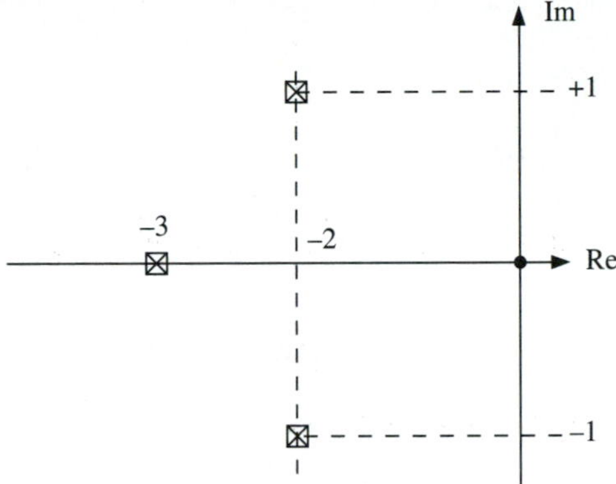

Fig. P7.5

7.6 For the system shown in Fig. P7.6, determine the maximum value of the output when the input takes the form

$$R(s) = \frac{4}{s}$$

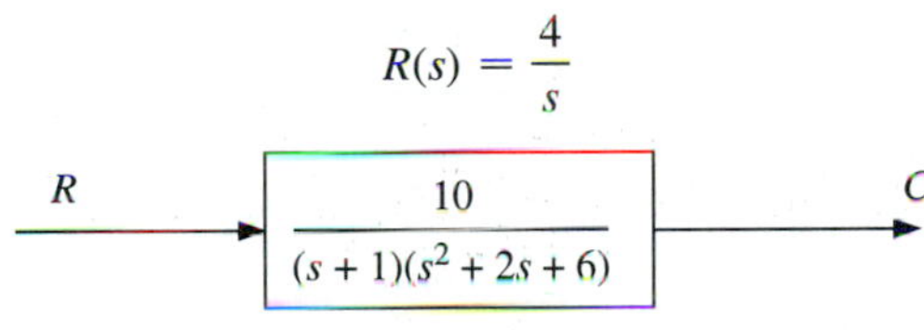

Fig. P7.6

7.7 A system has a transfer function that may be written in the form

$$\frac{C}{R} = \frac{1}{(s + 1)(s^2 + as + b)}$$

It is known that for the second-order term, $\zeta = 0.2$. Investigation of the unit step response reveals an overshoot of 5% of the input. Calculate the constants a and b, plot the closed-loop system poles on the complex plane, and comment on the reducability of the system.

7.8 Calculate the maximum value of output when a unit input is applied to the system described by the closed-loop transfer function

$$\frac{C}{R} = \frac{4(s + 3)}{2s^2 + 3s + 10}$$

7.9 Find a value of b that gives the following system a 5% overshoot to a unit step input. What is the maximum value of the system output:

$$\frac{C}{R} = \frac{s + b}{s^2 + 4s + 4}$$

7.10 Estimate, without reducing the order of the system, the percentage overshoot of the following system when the input is a unit step:

$$\frac{C}{R} = \frac{s + 2}{(s + 4)(s^2 + s + 1)}$$

7.11 For the systems described by the following forward-path and feedback transfer functions, determine the location of any closed-loop zeros:

a. $G(s) = \dfrac{K(s + 1)}{s}$, $H(s) = 1$

b. $G(s) = K$, $H(s) = s + 1$

c. $G(s) = \dfrac{(s + 1)(s + 2)}{s(s + 3)}$, $H(s) = \dfrac{s + 5}{s + 4}$

d. $G(s) = \dfrac{s}{s + 1}$, $H(s) = \dfrac{1}{s^2 + 2s + 10}$

System Type: Steady-State Errors

The concept of steady-state error has been introduced in Module 3 when the performance of first-order open- and closed-loop systems was discussed. It was shown that a closed-loop system, although less sensitive to parameter changes than the open-loop system, may have an inherent steady-state error for certain inputs. In this section, a more generalized view of the steady-state error of the closed-loop system is outlined and its relationship to the *system type* defined. As a result, by recognizing the system type and the kind of input the system is subjected to, the value of the steady-state error may be quickly identified.

The analysis begins by considering the generalized feedback control system shown in Fig. 8.1. The quantity C' is defined to be the measured output, and it is interesting to note that the true error $C - R$ is not observable, since the output has to be measured by H before its value is known. The quantity E_a is known as the actuating signal. Hence the true error is given by

$$E = R - C \tag{8.1}$$

while the actuating signal is

$$E_a = R - C' \tag{8.2}$$

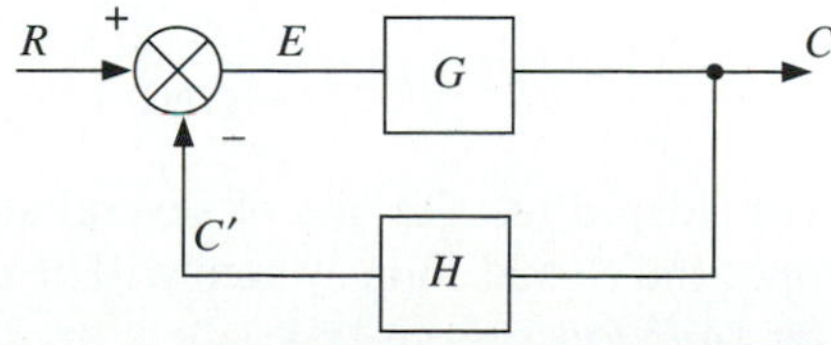

Fig. 8.1 Generalized feedback control system

The true error and the measured error are the same only for unity feedback. For now we will make the important assumption that the feedback transfer function is unity, i.e.,

$$H(s) = 1 \tag{8.3}$$

The basis for this assumption is that many control systems are indeed of this form, and therefore this constraint is not overly restrictive. Methods for dealing with non-unity-feedback systems will be discussed later in the module. From Fig. 8.1 with $H = 1$ we get the error to be

$$E = R - C = R - GE \tag{8.4}$$

yielding

$$E(1 + G) = R \tag{8.5}$$

and finally

$$E = \frac{R}{1 + G} \tag{8.6}$$

This equation will be used to determine the steady-state error of a closed-loop system to a given input by applying the final-value theorem. Before doing this, however, a generalized open-loop transfer function will be defined as

$$G(s) = \frac{K(s + z_1)(s + z_2)\cdots(s + z_m)}{s^n(s + p_1)(s + p_2)\cdots(s + p_q)} \tag{8.7}$$

$$= \frac{K \prod_{i=1}^{m}(s + z_i)}{s^n \prod_{k=1}^{q}(s + p_k)} \tag{8.8}$$

The system type is defined by the value of n. For example, a system with open-loop transfer function

$$G(s) = \frac{10(s + 1)}{s^2(s + 3)} \tag{8.9}$$

is a type 2 system. Note the distinction between the *type* number and the *order* of a system, the latter being the highest power of s in the characteristic equation. In the above example the type number is 2, while the system is of order 3. Applying the final-value theorem to equation 8.6 gives the steady-state error as

$$\lim_{t\to\infty} e(t) = \lim_{s\to 0} sE(s) \tag{8.10}$$

$$= \lim_{s\to 0} \frac{sR(s)}{1 + G(s)} \tag{8.11}$$

The input $R(s)$ will be considered to take one of several standard forms in turn, and the steady-state error of the closed-loop system will be evaluated as a function of the open-loop transfer function system type.

In this case

$$R(s) = 1 \tag{8.12}$$

and the steady-state error of the closed-loop system becomes

$$e_{ss}(t) = \lim_{s \to 0} \frac{s}{1 + G(s)} \tag{8.13}$$

For a type 0 system

$$G(s) \Rightarrow \frac{K \prod_{i=1}^{m} z_i}{\prod_{k=1}^{q} p_k} = K' \tag{8.14}$$

hence

$$e_{ss}(t) = \lim_{s \to 0} \frac{s}{1 + K'} = 0 \tag{8.15}$$

Now, considering a type 1 system,

$$G(s) \Rightarrow \frac{K \prod_{i=1}^{m} z_i}{s \prod_{k=1}^{q} p_k} = \infty \tag{8.16}$$

due to the s in the denominator; hence the steady-state error becomes

$$e_{ss}(t) = \lim_{s \to 0} \frac{s}{\infty} = 0 \tag{8.17}$$

For a system type greater than unity, it can be seen that the steady-state error will be zero due to the s in the numerator of equation 8.6 and due to $G(s) \Rightarrow \infty$ for $n > 0$. The conclusion from the above analysis is that a system of any type number will have a zero steady-state error for an impulsive input. An example of the closed-loop response to an impulsive input is shown in Fig. 8.2.

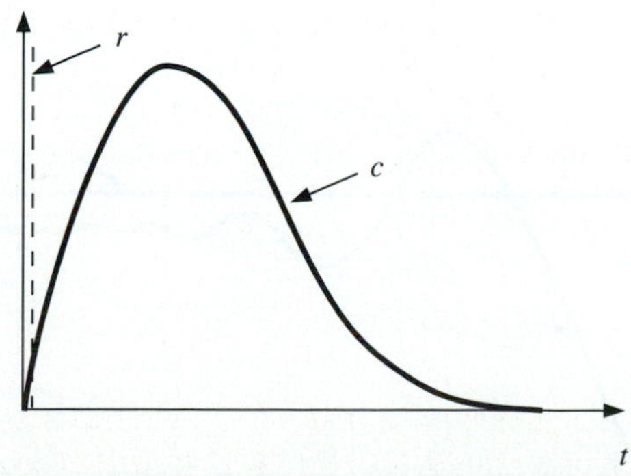

Fig. 8.2 System response to an impulsive input

◆ STEP INPUT

For a unit step input

$$R(s) = \frac{1}{s} \tag{8.18}$$

The steady-state error of the closed-loop system becomes

$$e_{ss}(t) = \lim_{s \to 0} \frac{1}{1 + G(s)} \tag{8.19}$$

Considering a type 0 system,

$$G(s) \Rightarrow \frac{K \prod_{i=1}^{m} z_i}{\prod_{k=1}^{q} p_k} = K_p \tag{8.20}$$

Because a step input to a system is usually associated with a demanded change in position, the constant K_p is called the position error constant. The steady-state error is now given by

$$e_{ss}(t) = \frac{1}{1 + K_p} \tag{8.21}$$

The response of a type 0 system to a unit step input is shown in Fig. 8.3 and is similar to Fig. 3.10. As previously discussed, the value of the steady-state error may be made smaller by increasing K_p, although, as we shall see later, this may have a serious effect on system stability. Now considering a type 1 system subject to a unit step input, it is seen that

$$G(s) \Rightarrow \frac{K \prod_{i=1}^{m} z_i}{s \prod_{k=1}^{q} p_k} = \infty \tag{8.22}$$

resulting in the steady-state error

$$e_{ss}(t) = \lim_{s \to 0} \frac{1}{1 + \infty} = 0 \tag{8.23}$$

For the same reasons discussed for the previous case of an impulse input

$$e_{ss}(t) = 0 \qquad n > 0 \tag{8.24}$$

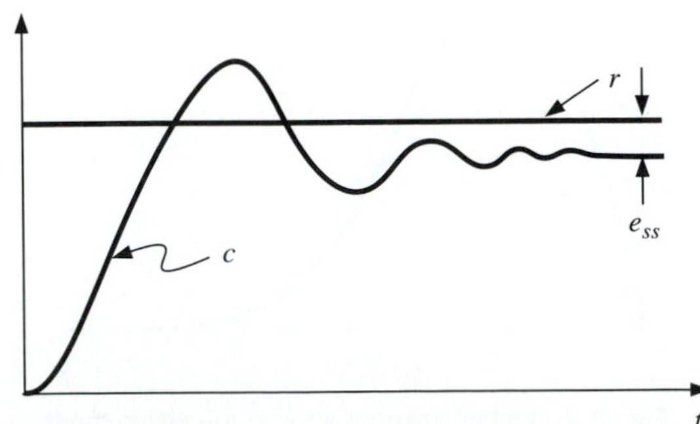

Fig. 8.3 Type 0 system response to a step input

For a unit ramp input

$$R(s) = \frac{1}{s^2} \tag{8.25}$$

Again the steady-state error is determined by the open-loop transfer function $G(s)$ from equation 8.6 as

$$e_{ss}(t) = \lim_{s\to 0} \frac{1/s}{1 + G(s)} \tag{8.26}$$

For a type 0 system

$$G(s) \Rightarrow \frac{K \prod_{i=1}^{m} z_i}{\prod_{k=1}^{q} p_k} = K_v \tag{8.27}$$

This time, the constant K_v is called the velocity error constant because a ramp input is usually associated with a position that changes uniformly with time, i.e., a velocity input. The steady-state error becomes

$$e_{ss}(t) = \lim_{s\to 0} \frac{1/s}{1 + K_v} = \infty \tag{8.28}$$

For a type 1 system,

$$G(s) \Rightarrow \frac{K_v}{s} \tag{8.29}$$

and the steady-state error becomes

$$e_{ss}(t) = \lim_{s\to 0} \frac{1/s}{1 + K_v/s} \tag{8.30}$$

$$= \lim_{s\to 0} \frac{1}{s + K_v} \tag{8.31}$$

$$= \frac{1}{K_v} \tag{8.32}$$

A typical response of a type 1 system to a unit ramp input is shown in Fig. 8.4. For a type 2 system

$$G(s) \Rightarrow \frac{K_v}{s^2} \tag{8.33}$$

resulting in the steady-state error

$$e_{ss}(t) = \lim_{s\to 0} \frac{1/s}{1 + K_v/s^2} \tag{8.34}$$

$$= \lim_{s\to 0} \frac{s}{s^2 + K_v} \tag{8.35}$$

$$= 0 \tag{8.36}$$

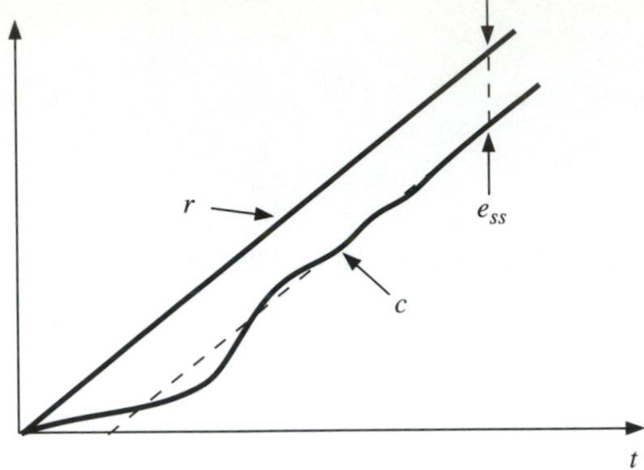

Fig. 8.4 Type 1 system response to a ramp input

Again, it is seen that

$$e_{ss}(t) = 0 \qquad n > 1 \tag{8.37}$$

◆ ACCELERATION INPUT

Finally, a unit acceleration input is considered where $r(t) = t^2$, or

$$R(s) = \frac{1}{s^3} \tag{8.38}$$

Equation 8.6 becomes

$$e_{ss}(t) = \lim_{s \to 0} \frac{1/s^2}{1 + G(s)} \tag{8.39}$$

For a type 0 system

$$G(s) \Rightarrow \frac{K \prod_{i=1}^{m} z_i}{\prod_{k=1}^{q} p_k} = K_a \tag{8.40}$$

This time, the constant K_a is called the acceleration error constant, since the input is a velocity that changes uniformly with time. The steady-state error now becomes

$$e_{ss}(t) = \lim_{s \to 0} \frac{1/s^2}{1 + K_a} = \infty \tag{8.41}$$

For a type 1 system

$$G(s) \Rightarrow \frac{K_a}{s} \tag{8.42}$$

producing the steady-state error

$$e_{ss}(t) = \lim_{s\to 0} \frac{1/s^2}{1 + K_a/s} \tag{8.43}$$

$$= \lim_{s\to 0} \frac{1/s}{s + K_a} \tag{8.44}$$

$$= \infty \tag{8.45}$$

For a type 2 system

$$G(s) \Rightarrow \frac{K_a}{s^2} \tag{8.46}$$

resulting in the steady-state error

$$e_{ss}(t) = \lim_{s\to 0} \frac{1/s^2}{1 + K_a/s^2} \tag{8.47}$$

$$= \lim_{s\to 0} \frac{1}{s^2 + K_a} \tag{8.48}$$

$$= \frac{1}{K_a} \tag{8.49}$$

The output of a type 2 system subject to a unit acceleration input is shown in Fig. 8.5. For a type 3 system

$$G(s) \Rightarrow \frac{K_a}{s^3} \tag{8.50}$$

and the steady-state error becomes

$$e_{ss}(t) = \lim_{s\to 0} \frac{1/s^2}{1 + K_a/s^3} \tag{8.51}$$

$$= \lim_{s\to 0} \frac{s}{s^3 + K_a} \tag{8.52}$$

$$= 0 \tag{8.53}$$

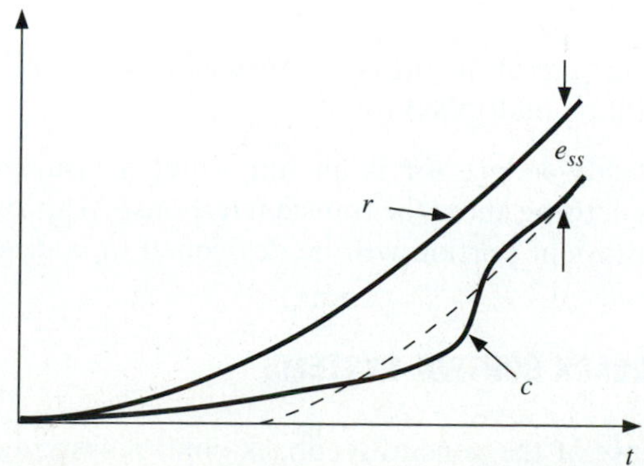

Fig. 8.5 Type 2 system response to an acceleration input

TABLE 8.1. Steady-state error as a function of system type and input

System type	Input: Impulse	Step	Ramp	Acceleration
0	0	$\dfrac{1}{1 + K_p}$	∞	∞
1	0	0	$\dfrac{1}{K_v}$	∞
2	0	0	0	$\dfrac{1}{K_a}$

As before, it is seen that systems with a higher type number also produce zero steady-state error:

$$e_{ss}(t) = 0 \qquad n > 2 \tag{8.54}$$

The results derived may be represented as in Table 8.1, showing the response of systems of various type numbers as a function of the nature of the input. All error constants are given by

$$\lim_{s \to 0} G(s) = \frac{K \prod_{i=1}^{m} z_i}{\prod_{k=1}^{q} p_k} = K' \tag{8.55}$$

but the constant K' is defined to be K_p, K_v, and K_a depending upon the input function.

In summary, therefore, if the system under consideration has unity feedback, the steady-state error present in the closed-loop system may be obtained from the open-loop transfer function $G(s)$ by determining the system type and inspecting Table 8.1 for the appropriate input. Note the following points:

1. Table 8.1 may be extended for higher-order inputs and systems of higher type numbers by observing the pattern in the table. The elements on the superdiagonal are finite, the upper triangular portion infinity, while the lower triangular portion is zero.

2. If any of the inputs are of magnitude A instead of unity, all the finite steady-state errors would be multiplied by A.

3. Although the steady-state error is an important measure of system performance, it tells us nothing about the transient response. Approximate methods to determine the transient portion will be developed in later sections.

◆ NON-UNITY-FEEDBACK CONTROL SYSTEMS

Returning to the case of the general feedback control system for $H(s) \neq 1$, it is seen that the transfer function from input to true error E is obtained from

$$E = R - C \tag{8.56}$$

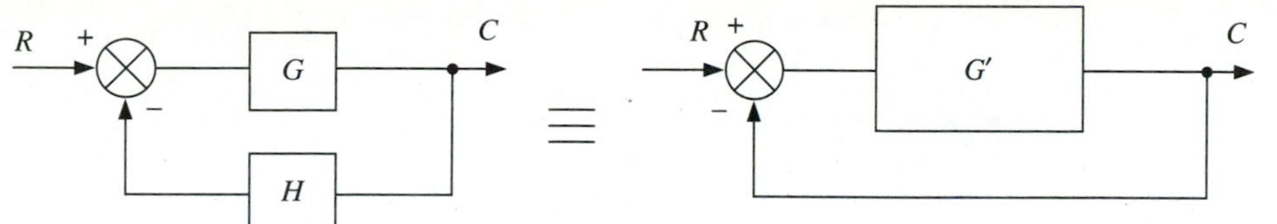

Fig. 8.6 Non-unity-feedback system and equivalent unity-feedback system

Dividing by the input R,

$$\frac{E}{R} = 1 - \frac{C}{R} \tag{8.57}$$

$$= 1 - \frac{G}{1 + GH} \tag{8.58}$$

$$= \frac{1 + G(H - 1)}{1 + GH} \tag{8.59}$$

This closed-loop system will be made equivalent to a unity-feedback system which has the forward-path transfer function G', as shown in Fig. 8.6. If the two systems are to be equivalent, their closed-loop transfer functions must be the same:

$$\frac{C}{R} = \frac{G}{1 + GH} = \frac{G'}{1 + G'} \tag{8.60}$$

Equating and multiplying out the two terms on the right yields

$$G(1 + G') = G'(1 + GH) \tag{8.61}$$

resulting in

$$G' = \frac{G}{1 + G(H - 1)} \tag{8.62}$$

This important result allows any non-unity-feedback system to be written in an equivalent unity-feedback form, thereby allowing the use of Table 8.1 in the determination of the steady-state error. It should be mentioned that in some cases, it is simpler to substitute the known input R into equation 8.59 and apply the final-value theorem directly in order to determine the steady-state error. Both methods will work, and the reader is encouraged to try both.

A final note to emphasize the difference between the error and the actuating signal for non-unity-feedback control systems. Suppose that

$$H(s) = s \tag{8.63}$$

It is seen that for *any* constant output C, the measured output C' is zero, resulting in the actuating signal taking the value

$$E_a = R \tag{8.64}$$

Clearly C may take any of an infinite number of constant values, having no effect at all on the value of E_a. In such a case, and in general when $H(s) \neq 1$, the true error and the actuating signal are quite different.

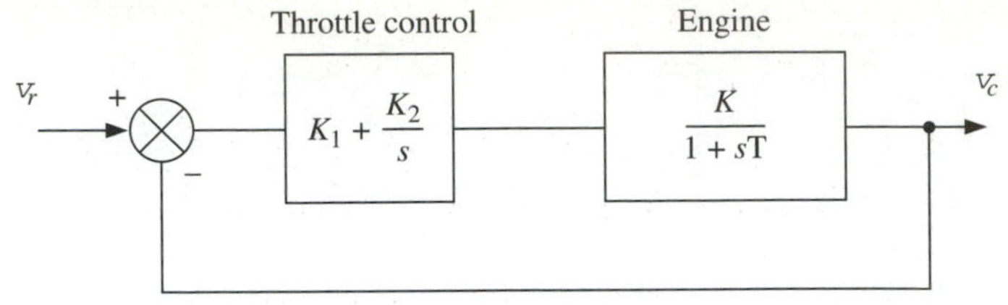

Fig. SP8.1.1

SAMPLE PROBLEM 8.1

An engine speed control system is shown in Fig. SP8.1.1. The engine itself is modeled as a first-order system with time constant T, while the electronic throttle controller may have the constants K_1 and K_2 set to arbitrary values.

1. What is the steady-state error for a step of magnitude A if $K_2 = 0$?
2. What is the steady-state error for a step of magnitude A when $K_2 \neq 0$?
3. Determine the steady-state error when the input is a ramp of slope A and (i) $K_2 = 0$, (ii) $K_2 \neq 0$.
4. Given $K_1 = 1.2$, $K_2 = 8.4$, and $T = 0.5$, what value of K gives a velocity error constant of 6 for a unit ramp input? Find the corresponding steady-state error, and sketch the input and output as functions of time for this case.

Solution

The system has unity feedback; therefore the various error constants and steady-state errors may be determined from Table 8.1. The open-loop transfer function is given by

$$G(s) = \frac{K(K_1 s + K_2)}{s(1 + sT)}$$

1. When $K_2 = 0$, this transfer function reduces to

$$G(s) = \frac{KK_1}{1 + sT}$$

which represents a system of type 0. The position error constant for a step input is obtained by writing the open-loop transfer function in the form of equation 8.8:

$$G(s) = \frac{KK_1/T}{s + 1/T}$$

hence

$$K_p = \lim_{s \to 0} G(s) = KK_1$$

This could also have been obtained directly from

$$G(s) = \frac{KK_1}{1 + sT}$$

without first writing the open-loop transfer function in the form of equation 8.8. The steady-state error for a unit step input, from table 8.1, becomes

$$e_{ss}(t) = \frac{1}{1 + K_p} = \frac{1}{1 + KK_1}$$

For a step of magnitude A, therefore

$$e_{ss}(t) = \frac{A}{1 + KK_1}$$

2. When $K_2 \neq 0$, the open-loop transfer function reverts to the form

$$G(s) = \frac{K(K_1 s + K_2)}{s(1 + sT)}$$

which represents a system of type 1. From Table 8.1, the steady-state error for any magnitude step is zero, i.e.,

$$e_{ss}(t) = 0$$

3. When $K_2 = 0$ and the input is a ramp, Table 8.1 indicates that because the system is of type 0, the steady-state error becomes infinite. Hence

$$e_{ss}(t) = \infty$$

This result does not necessarily imply that the system is unstable, but rather that the output is a ramp like the input but of different slope, as shown in Fig. SP8.1.2, where it is seen that the error becomes infinite as t becomes large.

When $K_2 \neq 0$, the system becomes type 1, and the open-loop transfer function may be written in the form

$$G(s) = \frac{K(K_1 s + K_2)}{s(1 + sT)} = \frac{KK_1(s + K_2/K_1)}{Ts(s + 1/T)}$$

The velocity error coefficient K_v is obtained as

$$K_v = KK_2$$

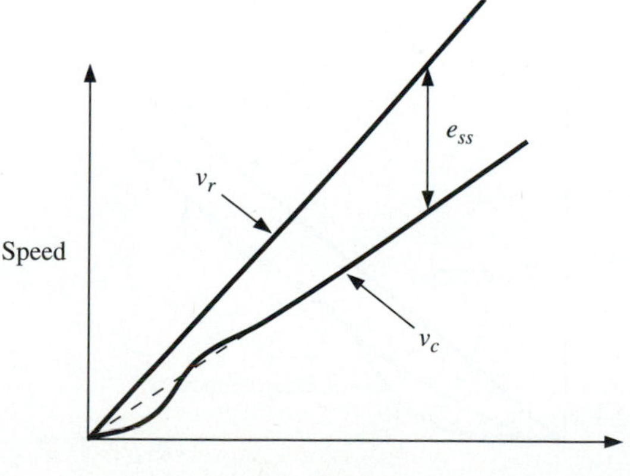

Fig. SP8.1.2

This result may also be obtained from either form of the open-loop transfer function $G(s)$. The steady-state error becomes

$$e_{ss}(t) = \frac{A}{K_v} = \frac{A}{KK_2}$$

4. Given $K_1 = 1.2$, $K_2 = 8.4$, and $T = 0.5$, it is required that

$$K_v = 6 = 8.4K$$

Hence the result

$$K = 0.714$$

The steady-state error of the closed-loop system becomes

$$e_{ss}(t) = \frac{1}{K_v} = 0.167$$

The output will be a unit ramp similar to the input but will lag behind it in the steady state. In order to sketch both the input and output, the transient portion of the output will also be determined. This may be necessary since the closed-loop transfer function will be of second order and the transient will be either underdamped or overdamped. This will lead to two possible types of responses as indicated in Fig. SP8.1.3. The response may be evaluated from the characteristic equation

$$1 + G(s) = 0$$

which gives

$$1 + \frac{K(K_1 s + K_2)}{s(1 + sT)} = 0$$

Substituting known values yields

$$s^2 + 3.716s + 12 = 0$$

giving the closed-loop poles

$$s = -1.858 \pm 2.94j$$

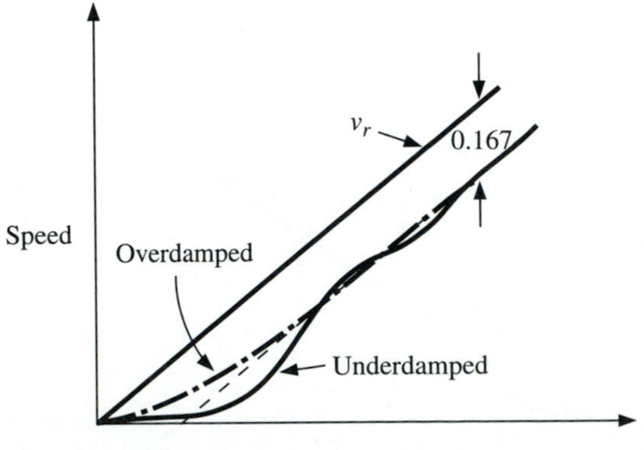

Fig. SP8.1.3

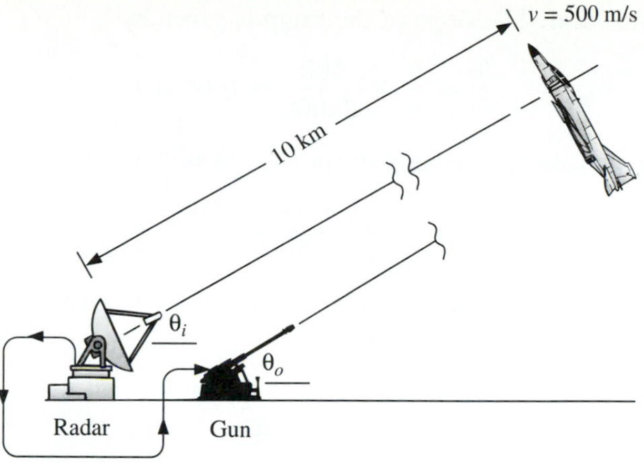

Fig. SP8.2.1

Since both closed-loop poles are complex, the system is underdamped, as indicated in Fig. 8.1.3.

Sample Problem 8.2

Figure SP8.2.1 shows an antiaircraft system consisting of a ground radar that measures the distance along the radar beam to an aircraft, together with the position relative to the tracking system based on the angle of the antenna with the ground. This information is passed to the gun-pointing system that attempts to track the aircraft, as shown in Fig. SP8.2.2, where $K = 240$. If the speed of the aircraft is $v = 500$ m/s at a range of 10 km and may be assumed perpendicular to the line of sight from the radar, by what distance do shells miss the target? If the required miss distance is to be no greater than 1 m, calculate the required value of system gain K.

Solution

Although the radar antenna provides an angular position input, the aircraft motion makes the angle θ change uniformly with time, at least for the instantaneous position shown. This subjects the gun-pointing system to a ramp input. Note that the system has unity feedback; therefore the results summarized in

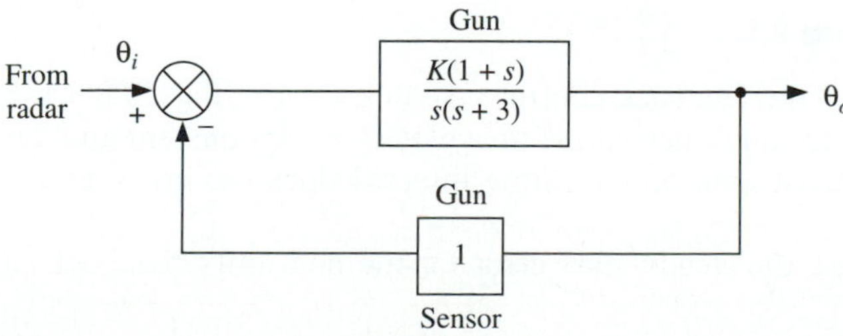

Fig. SP8.2.2

Table 8.1 are valid. The slope of the ramp is given by

$$\frac{d\theta}{dt} = \frac{v}{r} = \frac{500}{10{,}000} = 0.05 \text{ rad/s}$$

The open-loop transfer function of the system is

$$GH(s) = \frac{K(1 + s)}{s(s + 3)} = \frac{240(1 + s)}{s(s + 3)}$$

The system is of type 1 and therefore has a steady-state error for a unit ramp input (obtained from Table 8.1) of

$$e_{ss}(t) = \frac{1}{K_v}$$

where K_v, the velocity error constant, is obtained from the open-loop transfer function as

$$K_v = \frac{K \prod_{i=1}^{m} z_i}{\prod_{k=1}^{q} p_k} = \frac{240}{3}$$

The steady-state error for a ramp of slope 0.05 rad/s will be, therefore,

$$e_{ss}(t) = \frac{0.05}{240/3} = 0.000625 \text{ rad}$$

This represents the angular error between the radar antenna that is pointing directly at the aircraft and the gun. If the shells travel in a straight line, they will miss the target 10 km away by d, where

$$d = r\delta\theta = 10{,}000 \times 0.000625 = 6.25 \text{ m}$$

To reduce the miss distance to no more than 1 m, the value of K has to be increased. Since the miss distance is given by

$$d = r\delta\theta = \frac{0.05r}{K_v} = \frac{0.05r}{K/3}$$

which yields the required value of K to be

$$K = \frac{0.15r}{d} = 1500$$

Sample Problem 8.3

1. Consider the feedback control system shown in Fig. SP8.3.1. For $H(s) = 1$ and a unit ramp input, determine the velocity error constant and the steady-state error. After what approximate time interval does the error achieve its steady-state value?

2. Calculate the steady-state error for the non-unity-feedback case

$$H(s) = \frac{1}{s + 1}$$

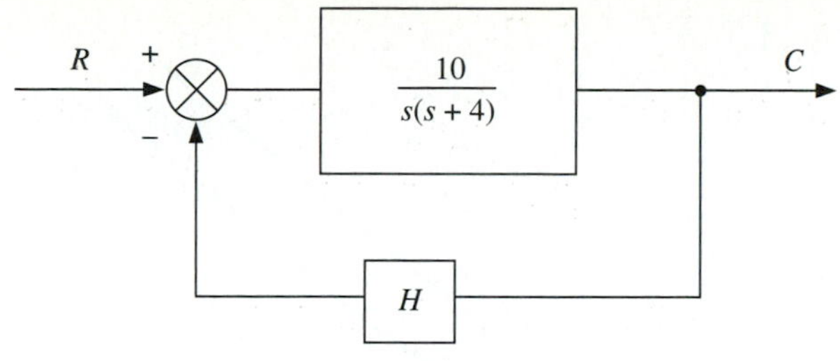

Fig. SP8.3.1

3. What is the steady-state error for $H(s) = 1$ and a time-domain input given by

$$r(t) = \begin{cases} 0 & t \leq 0 \\ 2t + 5 & t > 0 \end{cases}$$

Solution

1. For the unity-feedback case, the open-loop transfer function is given by

$$G(s) = \frac{10}{s(s + 4)}$$

and the system is of type 1. The velocity error coefficient is given by

$$K_v = \lim_{s \to 0} G(s) = \tfrac{10}{4} = 2.5$$

The steady-state error is therefore

$$e_{ss}(t) = \frac{1}{K_v} = 0.4$$

To determine the time taken for the transient to disappear, the 2% settling time introduced in Module 5 may be used. Although this was defined for a step-input response, we will assume it is a good measure of the disappearance of the transient for any input. The closed-loop transfer function is

$$\frac{C}{R} = \frac{10}{s(s + 4) + 10} = \frac{10}{s^2 + 4s + 10}$$

yielding

$$\zeta = 0.63 \qquad \omega_n = 3.16 \text{ rad/s}$$

The settling time is

$$T_s = \frac{4}{\zeta\omega_n} = 2 \text{ s}$$

It may be concluded, therefore, that the system achieves its steady-state error 2 s after the input is applied, and the total response will be as shown in Fig. SP8.3.2.

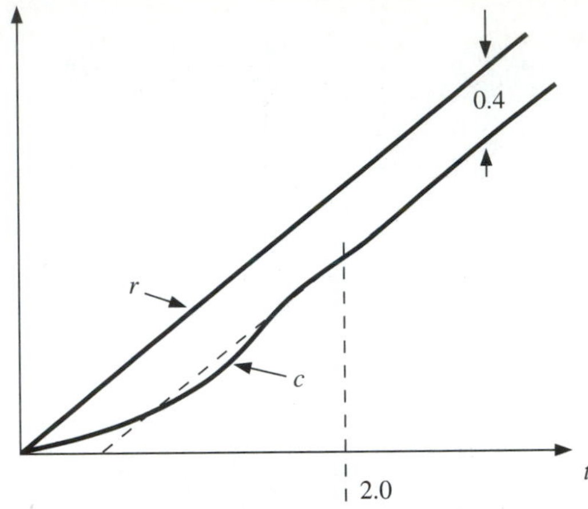

Fig. SP8.3.2

2. Considering the non-unity-feedback case now, the steady-state error will be calculated using equation 8.59. For this system we have

$$\frac{E}{R} = \frac{1 + G(H - 1)}{1 + GH}$$

$$= \frac{s(s + 1)(s + 4) - 10s}{s(s + 1)(s + 4) + 10}$$

Letting the input

$$R(s) = \frac{1}{s^2}$$

the steady-state error becomes

$$e_{ss}(t) = \lim_{s \to 0} sR(s)\frac{s(s + 1)(s + 4) - 10s}{s(s + 1)(s + 4) + 10} = -\frac{6}{10}$$

An alternative approach is to reduce the system to an equivalent one with unity feedback. The resulting forward-path transfer function becomes

$$G' = \frac{1 + G(H - 1)}{1 + GH} = \frac{10(s + 1)}{s(s^2 + 5s - 6)}$$

This system is of type 1 and third order, with a velocity error constant of

$$K_v = \lim_{s \to 0} G(s) = -\frac{10}{6}$$

The velocity error becomes the same as previously calculated:

$$e_{ss}(t) = \frac{1}{K_v} = -\frac{6}{10}$$

The significance of the negative error becomes clear by recalling that

$$e = r - c$$

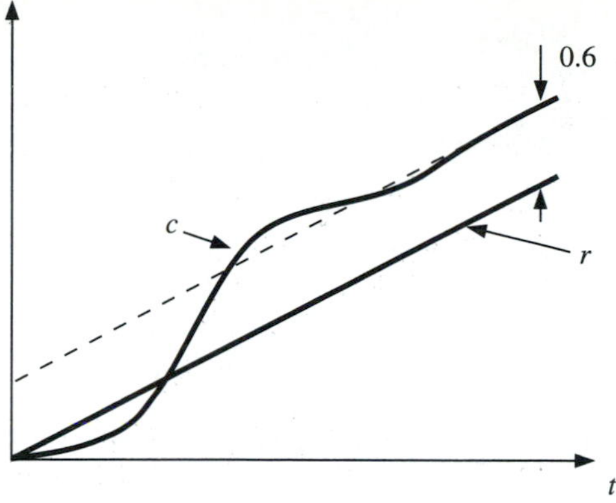

Fig. SP8.3.3

implying that, in this case, the steady-state output is larger than the input, as shown in Fig. SP8.3.3.

3. For the final part of the question, the input takes the form

$$r(t) = \begin{cases} 0 & t \leq 0 \\ 2t + 5 & t > 0 \end{cases}$$

Because the relationship between the input R and the output C is described by a *linear* differential equation, the principle of superposition may be applied. This means that the steady-state error will be given by the sum of the steady-state errors due to the individual inputs

$$r(t) = 2t$$

and

$$r(t) = 5$$

applied separately. Since the system is of order 1 and the feedback element is unity, Table 8.1 may be used to determine the steady-state error to each of these inputs, which may be identified as a ramp of slope 2 and a step of magnitude 5, respectively. As a system of order 1 has zero steady-state error for any step, only the ramp will contribute to the steady-state error. Because the ramp input has a slope twice that of a unit ramp, the steady-state error will be twice that calculated previously:

$$e_{ss}(t) = 2 \times 0.4 = 0.8$$

The total steady-state error is therefore 0.8.

◆ PROBLEMS

8.1 For unity-feedback systems described by the following open-loop transfer functions, determine the system type, any finite, nonzero error coefficients, and the corresponding steady-state errors:

a. $$G(s) = \frac{10}{s^2}$$

b. $$G(s) = \frac{s + 1}{s^4 + 2s^3 + 7s^2 + 5s}$$

c. $$G(s) = \frac{7}{s^5 + 2s^4 + 3s^3 + 3s^2 + 2s + 1}$$

d. $$G(s) = \frac{5}{(s + 1)^3}$$

e. $$G(s) = \frac{(s + 2)(s + 3)(s^2 + 4)}{s}$$

8.2 Repeat problem 8.1 but introduce a feedback element $H = 2$. For each case determine the forward-path transfer function G' of the equivalent unity-feedback system.

8.3 A feedback system is described by the forward-path and feedback transfer functions

$$G(s) = \frac{10}{s^2 + 2s + 5} \qquad H(s) = \frac{1}{1 + s}$$

Reduce the system to unity feedback by calculating the equivalent forward-path transfer function G'. Determine the steady-state error due to a unit step input by using Table 8.1, and confirm the result by direct calculation of the steady-state error from equation 8.59.

8.4 A system is described by the transfer functions

$$G(s) = \frac{5}{s(1 + s)} \qquad H(s) = 1$$

Determine the velocity error coefficient due to a ramp input of slope 4, and calculate the corresponding steady-state error. Determine the steady-state error for the same input but when $H(s)$ takes the following forms:

a. $$H(s) = 4$$

b. $$H(s) = s$$

c. $$H(s) = \frac{1}{1 + s}$$

d. $$H(s) = \frac{s + 1}{s + 2}$$

8.5 Confirm the results derived in Module 5 that, for the second-order system shown in Fig. P8.5, the steady-state error for a ramp input of slope P is given by $e_{ss}(t) = 2P\zeta/\omega_n$. Show also that the time delay between the input reaching a certain value and the steady-state output to achieve the same value is given by

$$\tau = \frac{2\zeta}{\omega_n}$$

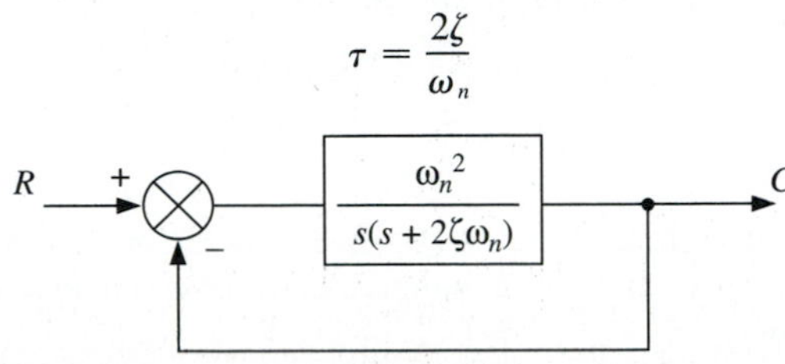

Fig. P8.5

8.6 For the system shown in Fig. P8.6, determine the system type and the steady-state error to a unit step, ramp, and acceleration inputs.

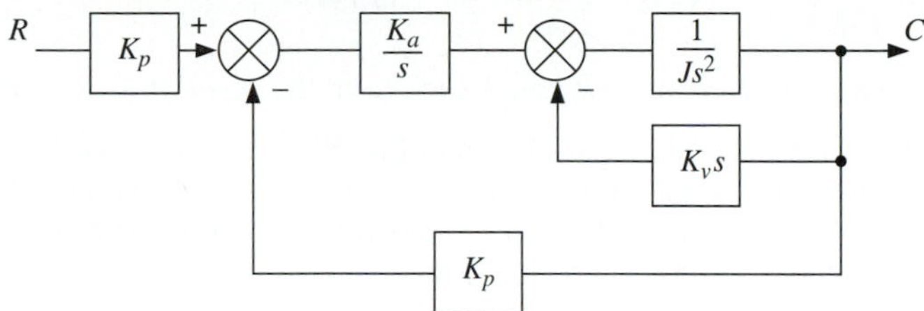

Fig. P8.6

8.7 Consider the unity-feedback control system shown in Fig. P8.7. Select the simplest values of the coefficients a, b, and c that will allow the closed-loop system to meet the following performance specifications:

a. Zero steady-state error for a step input.

b. A steady-state error of no more than 5% due to a ramp input.

c. No more than 10% overshoot due to a step input.

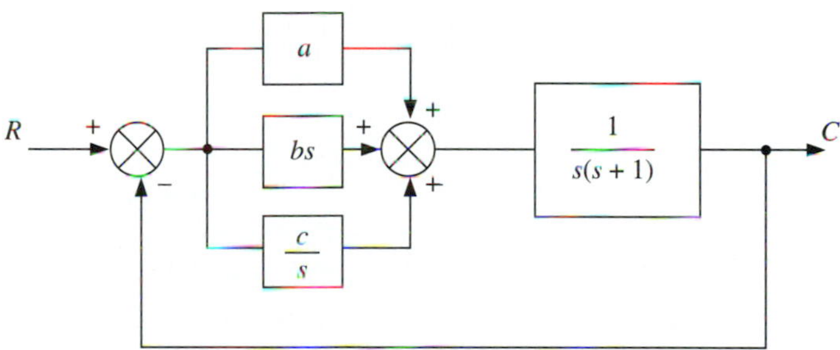

Fig. P8.7

8.8 Figure P8.8 shows a speed control system subject to a disturbance torque T. If the system transfer functions are given by

$$G_c(s) = \frac{K}{s+1} \qquad H(s) = 1$$

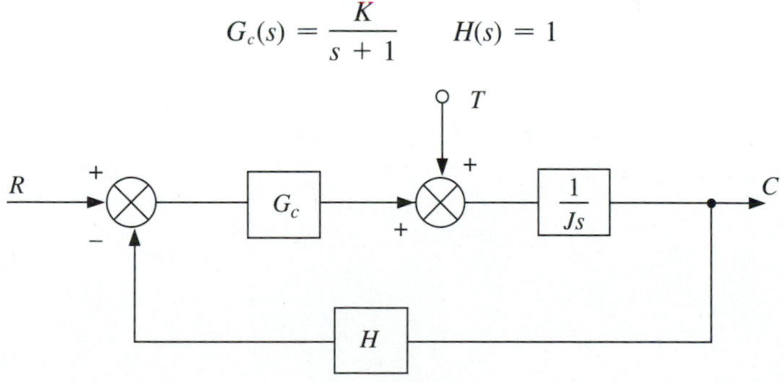

Fig. P8.8

determine the steady-state error due to a unit step change in input R (assuming T constant) and a unit step change in the disturbance torque T (assuming R constant), assuming $K = 1$. Determine K such that the resulting change in speed is less than 10% of the disturbance value.

8.9 Repeat problem 8.8 but with the following system transfer functions:

$$G_c(s) = \frac{K}{s(s+1)} \qquad H(s) = 2$$

What are the steady state errors if R and T each take the form of a unit ramp?

8.10 Consider the generalized system shown in Fig. P8.10. Derive an expression for the steady-state error $e(t)$ for when both R and T are unit step functions. Discuss the specifications of G_1, G_2, and H that make this steady-state error zero. Discuss your conclusions with respect to the results of the previous two problems.

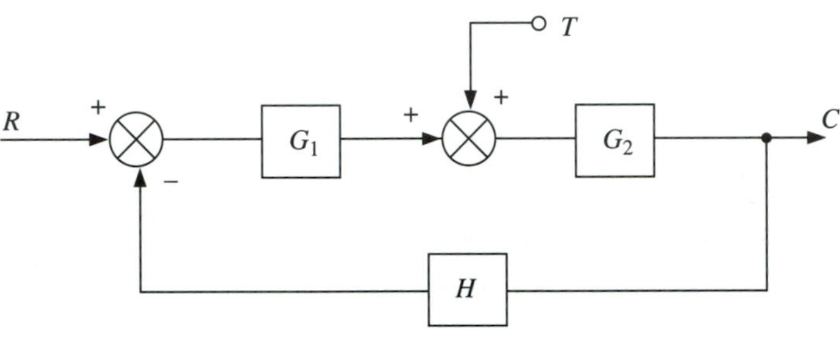

Fig. P8.10

MODULE 9

Routh's Method, Root Locus: Magnitude and Phase Equations

◆ ROUTH'S STABILITY CRITERION

Stability is one of the more important topics in which the control systems engineer is interested, and several modules will be devoted to the study of this subject. It may also be appreciated that system performance and stability are closely related topics, and a study of one invariably provides insight into the other. Our first priority, of course, is to determine whether a proposed design for a feedback control system will be stable, since catastrophic consequences may result if this is not done. Beyond this, however, we will determine how to calculate those system parameters that are under our control so as to achieve "good" response from the closed-loop system. It will be shown that this may be done using either the knowledge of the transfer functions of the forward and feedback path elements, or if these component transfer functions are unknown or too complex to calculate, it may be achieved through experimentally derived data on the open-loop system.

Detailed performance analyses can become quite involved, even using the techniques to be outlined in later modules, but if our only interest is in determining whether a system is stable or not, then the Routh method provides a rapid and simple test. What is not provided by this method is an indication of performance if the system is shown to be stable.

We have seen in earlier modules that the transient response of a closed-loop system may be qualitatively determined from the position of the closed-loop poles on the complex plane. In addition, it was apparent that the system became unstable as soon as one closed-loop pole was located in the right-hand half of the complex plane. Recall that the closed-loop poles may be found by solving the characteristic equation

$$1 + GH(s) = 0 \tag{9.1}$$

If the characteristic equation represents a fairly simple system and is of order 1 or 2, the closed-loop poles may be found by direct solution. For higher-order systems, however, solving the characteristic equation becomes tedious and usually requires numerical methods. Remember that it is not necessary to actually solve the characteristic equation to determine stability, only to find out if there are any roots with positive real parts. This is where Routh's method becomes useful, since it determines the number of roots located in the right-hand half of the complex plane using only the coefficients of the characteristic equation without actually solving the characteristic equation for the roots themselves. The method may be used by following these steps:

Step 1: Write the characteristic equation in the polynomial form

$$a_0 s^n + a_1 s^{n-1} + a_2 s^{n-2} + \cdots + a_{n-1} s + a_n = 0 \tag{9.2}$$

If any of the coefficients are negative or zero, and there is at least one positive root, then the system is not asymptotically stable.[1] If the characteristic equation has a negative or zero coefficient, it may be shown that there is either a closed-loop pole in the right-hand half plane (unstable) or one or more closed-loop poles on the imaginary axis (marginally stable).

Although it may be said that a negative coefficient will result in an unstable system, it is *not* possible to say that if all coefficients are positive, then the system is stable. The requirement for all coefficients to be positive is *necessary* but it is not *sufficient,* and further analysis is required.

Step 2: If there are no negative coefficients, the Routh array must be formed in the following manner:

$$\begin{array}{r|llllll}
s^n: & a_0 & a_2 & a_4 & a_6 & \cdot & \cdot\ \cdot \\
s^{n-1}: & a_1 & a_3 & a_5 & a_7 & \cdot & \cdot\ \cdot \\
s^{n-2}: & b_1 & b_2 & b_3 & b_4 & \cdot & \cdot\ \cdot \\
s^{n-3}: & c_1 & c_2 & c_3 & c_4 & \cdot & \cdot\ \cdot \\
s^{n-4}: & d_1 & d_2 & d_3 & d_4 & \cdot & \cdot\ \cdot \\
\cdot & \cdot & \cdot & \cdot & \cdot & \cdot & \cdot\ \cdot \\
\cdot & \cdot & \cdot & \cdot & \cdot & \cdot & \cdot\ \cdot \\
s^2: & e_1 & e_2 & e_3 & \cdot & \cdot & \cdot\ \cdot \\
s^1: & f_1 & f_2 & \cdot & \cdot & \cdot & \cdot\ \cdot \\
s^0: & g_1 & \cdot & \cdot & \cdot & \cdot & \cdot\ \cdot
\end{array}$$

The array is constructed one row at a time, with each row being labeled on the left-hand side with the highest power of s present in the characteristic equation for the first row, the next highest power for the second row, and so on. The first row consists of the coefficient of s^n and then alternate coefficients until all coefficients have been used. The next row begins with the coefficient of s^{n-1} and takes alternative coefficients until all have been used. Once all coefficients have been used up, zeros are inserted for the remainder of the row elements. While the first two rows of the array are obtained directly from the characteristic equation, each

[1]By asymptotically stable we mean that a system subjected to any disturbance will return to its equilibrium state in a finite time.

subsequent row is obtained from the two rows immediately preceding it. The remaining coefficients are derived as follows:

$$b_1 = \frac{a_1 a_2 - a_0 a_3}{a_1} \tag{9.3}$$

$$b_2 = \frac{a_1 a_4 - a_0 a_5}{a_1} \tag{9.4}$$

$$b_3 = \frac{a_1 a_6 - a_0 a_7}{a_1} \tag{9.5}$$

$$\vdots$$

All of the elements of the third row are computed until they become zero. The next row is computed from the previous two rows using the same pattern of cross-multiplying the elements:

$$c_1 = \frac{b_1 a_3 - a_1 b_2}{b_1} \tag{9.6}$$

$$c_2 = \frac{b_1 a_5 - a_1 b_3}{b_1} \tag{9.7}$$

$$c_3 = \frac{b_1 a_7 - a_1 b_4}{b_1} \tag{9.8}$$

$$\vdots$$

The next row is computed from the previous two rows in a similar manner:

$$d_1 = \frac{c_1 b_2 - b_1 c_2}{c_1} \tag{9.9}$$

$$d_2 = \frac{c_1 b_3 - b_1 c_3}{c_1} \tag{9.10}$$

$$d_3 = \frac{c_1 b_4 - b_1 c_4}{c_1} \tag{9.11}$$

$$\vdots$$

This process is continued until all rows have been completed down to the terms in row s^0. In practice, the array becomes triangular, reducing to a single element in the last row. Once completed, the stability of the system may be determined from Routh's criterion, which states that *the number of closed-loop poles in the right-hand half complex plane is equal to the number of sign changes of the elements of the first column of Routh's array.*[2]

[2]To be accurate, the complete test is known as the Routh-Hurwitz stability criterion, where the necessary examination of the coefficients of the characteristic equation is attributed to Hurwitz, while the sufficiency condition obtained from the array was developed by Routh.

EXAMPLE

Determine if the system described by the characteristic equation

$$s^4 + 6s^3 + 4s^2 + 3s + 1 = 0 \tag{9.12}$$

is stable. Inspection of the coefficients of the characteristic equation indicates that they are all positive, so we proceed to step 2 and construct the Routh array. It is seen to be

$$\begin{array}{lrrr} s^4: & 1 & 4 & 1 \\ s^3: & 6 & 3 & 0 \\ s^2: & 3.50 & 1 & \\ s^1: & 1.28 & 0 & \\ s^0: & 1 & & \end{array}$$

Observe how the array is triangular and trailing zeros are placed at the end of each row. Since there are no sign changes in the first column, there are no closed-loop poles to the right of the imaginary axis, and the system is stable.

◆ ROOT LOCUS METHOD: MAGNITUDE AND PHASE EQUATIONS

It was seen in the previous section that the characteristic equation need not be solved to determine the stability of a system. Placing the coefficients into the Routh array followed by simple arithmetic will be sufficient. In many cases, however, the determination of stability alone is insufficient, and some idea of how a stable system responds to a variety of inputs is required. Since this implies a knowledge of the time-domain response of the system, a solution of the characteristic equation is required. For all but very simple systems, solving the characteristic equation can be a formidable task. However, the root locus method provides a straightforward method of graphically "solving" the characteristic equation so that the closed-loop poles may be located on the complex plane. In some cases these locations may be found precisely, while in others they will be only approximate. In either case, the time-domain response of the system may be predicted with reasonable accuracy.

We begin by writing the characteristic equation for the generalized feedback control system as

$$1 + G(s)H(s) = 0 \tag{9.13}$$

The objective is to determine values of s that satisfy the above equation, remembering that s may be complex. Suppose a trial value of s is substituted into the characteristic equation. The two functions of s, G and H, will themselves become complex numbers. We rewrite equation 9.13 in the form

$$GH(s) = -1 + 0j \tag{9.14}$$

The left-hand side of the above equation may be regarded as a single complex number evaluated for the trial value of s. Recall that the complex number $s = \sigma + j\omega$ may be represented in the complex plane, as shown in Fig. 9.1, where the modulus, or magnitude of s, is the length of the vector from the origin to s, while the argument, or phase angle of s, is the angle between the vector from the

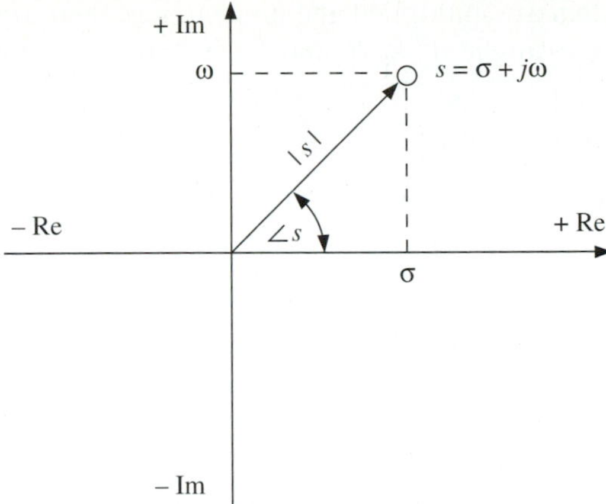

Fig. 9.1 Complex-number representation

origin to s, and the positive real axis. Since equation 9.14 simply equates two complex numbers, it may be inferred that both the magnitude and phase of each are equal and write

$$|GH(s)| = 1 \tag{9.15}$$

$$\angle GH(s) = -180^\circ \tag{9.16}$$

Note that the phase angle of $-1 + j0$ is strictly $-\pi, -3\pi, -5\pi, \cdots$; however, the principal value will be taken here. What has been accomplished so far is to say that a value of s satisfying the characteristic equation must also satisfy equations 9.15 and 9.16, known as the *magnitude* and *argument equations,* respectively. The root locus method is performed in two stages:

1. Finding all values of s satisfying the argument equation 9.16.
2. Finding particular values of s that satisfy the magnitude equation 9.15.

In order to see how this is done, a simple example will be considered. Suppose that

$$GH(s) = \frac{K}{s(s+1)} \tag{9.17}$$

Equation 9.16 implies that

$$\angle GH(s) = \angle \frac{K}{s(s+1)} = -180^\circ \tag{9.18}$$

The left-hand side of equation 9.18 may be thought of as a single complex number whose argument has to be -180°. However, in guessing a value of s that hopefully meets the argument requirement, the left-hand side of equation 9.18 is decomposed into three subcomponents K, s, and $s + 1$, each contributing to the overall argument. Actually, since K is independent of s, its associated argument is zero, and only s and $s + 1$ contribute to the argument.

For a given value of s, our guessed value, each of these subcomponents has its own argument. The total argument may be determined by using the usual rules of complex numbers that are multiplied and divided, i.e., their arguments are added and subtracted. For example, if A, B, and C are complex numbers, then

$$\left|\frac{A}{BC}\right| = \frac{|A|}{|B|\,|C|} \qquad \angle\frac{A}{BC} = \angle A - \angle B - \angle C \tag{9.19}$$

In this particular example, this results in

$$\angle\frac{K}{s(s+1)} = -\angle s - \angle(s+1) \tag{9.20}$$

Although $\angle s$ may be determined from the complex-plane representation of s, $\angle(s+1)$ has to be interpreted more carefully.

Suppose the trial guess for s is denoted by s_t. This complex number may also be represented on the complex plane, as shown in Fig. 9.2, where the open-loop poles ($s = 0, -1$) have been drawn. Consider for now just the vector drawn from the open-loop pole at $s = -1$ to the test point s_t. If this vector is denoted by $\bar{x}$, the vector equation may be written

$$\overline{-1} + \bar{x} = \bar{s}_t \tag{9.21}$$

Hence

$$\bar{x} = \overline{s_t + 1} \tag{9.22}$$

It may be seen, therefore, that the vector from the open-loop pole at $s = -1$ to the trial point s_t is given by $s_t + 1$ and will have a magnitude and argument associated with it, as defined in Fig. 9.2. Now that terms like $\angle(s+1)$ have been defined, it is possible to "solve" the argument equation for this example. In Fig. 9.3, the vectors from the open-loop poles have been drawn and their phase angles have been denoted by α_1 and α_2. From equations 9.16 and 9.20, s_t satisfies the angle equation if

$$-\alpha_1 - \alpha_2 = -180^\circ \tag{9.23}$$

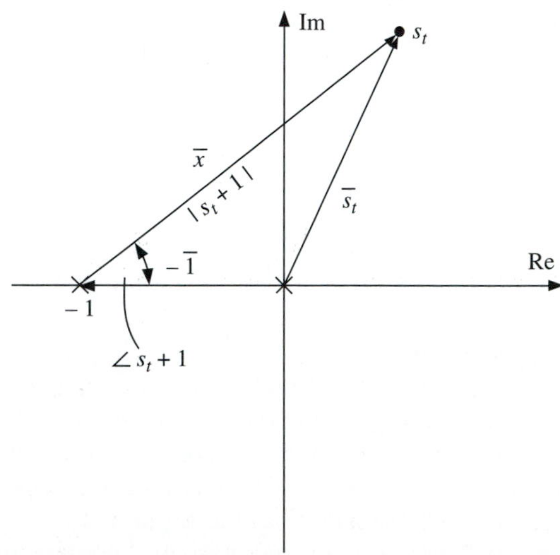

Fig. 9.2 Argument and magnitude of $s_t + 1$

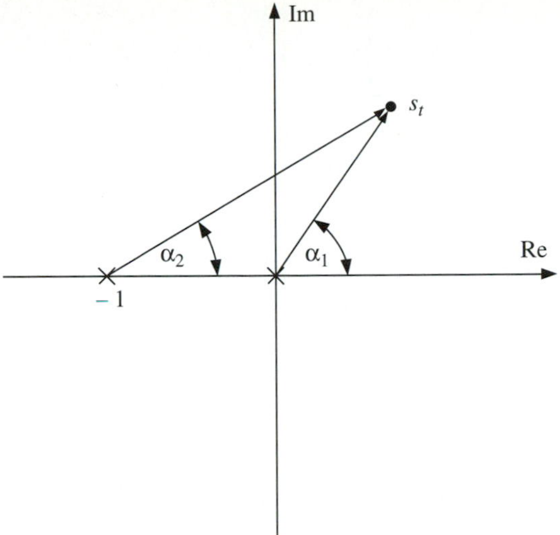

Fig. 9.3 Arguments of s and $s_t + 1$

The first part of the root locus method involves searching the whole of the complex plane for values of s_t satisfying the argument equation. With a little thought and some experience, this is not as daunting a task as it seems. For example, the search begins by inspecting only the real axis, where it may be seen that the total argument when $s_t > 0$ is zero. For $0 > s_t > -1$, $\alpha_1 = 180°$ and $\alpha_2 = 0$; hence equation 9.23 is satisfied. Further investigation of the real axis shows that for s_t to be the left of -1, $\alpha_1 = \alpha_2 = 180°$, which does not satisfy equation 9.23. Only the real axis for $0 > s_t > -1$ is, therefore, part of the root locus. The next part of the investigation examines locations of s_t off the real axis. By imagining the point s_t to be connected to the open-loop poles by "rubber band" lines, so that as s_t moves around the complex plane, these lines simply stretch, the student may spot that any value of s_t on a vertical line through $s = -0.5$ also satisfies the argument equation. This is shown in Fig. 9.4. The argument equation is satisfied by recalling the external angle properties of the triangle ABC. An alternative approach is to imagine a circle centered at the origin of infinite radius. An approximation to this situation is shown in Fig. 9.5, where the circle is of large, though not infinite, radius and has the test point s_t located on it. The problem to solve is: where on the circle should s_t be located in order to solve the argument equation? As the radius increases to infinity, the angles α_1 and α_2 become identical. From equation 9.23, therefore, the solution must be

$$\alpha_1 = \alpha_2 = 90° \tag{9.24}$$

Closed-loop poles must lie on a line inclined at 90° to the real axis, at infinity. This completes the solution of the argument equation, and the appropriate sections of the complex plane that satisfy it are marked with a heavy line in Fig. 9.6. In order to apply the second part of the root locus method, namely finding a particular value of s that satisfies the magnitude equation 9.15, the definitions of $|s|$ and $|s + 1|$ from Fig. 9.2 are recalled. Restricting s_t to those sections of the negative real axis known to be solutions of the argument equation, equations 9.15

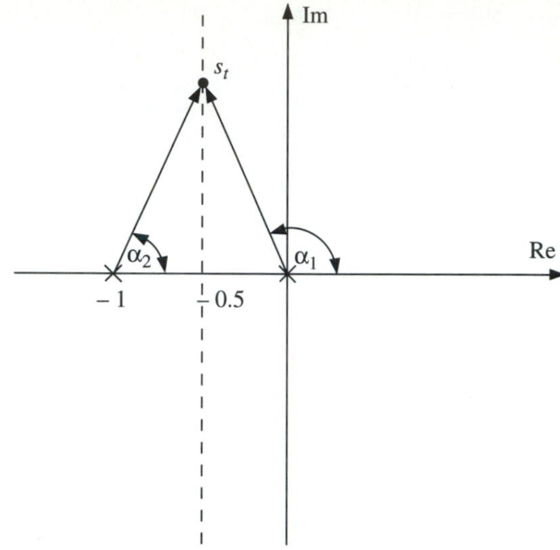

Fig. 9.4 Off-axis solutions of the argument equation

and 9.19 can be used for our example to write

$$\frac{|K|}{|s| \times |s_t + 1|} = 1 \tag{9.25}$$

This may be written as

$$K = |s| \times |s + 1| \tag{9.26}$$

Suppose $s_t = -1.5$. Then K may be evaluated from equation 9.26 above by measuring the lengths of the vectors on a root locus drawn to scale. Note, however,

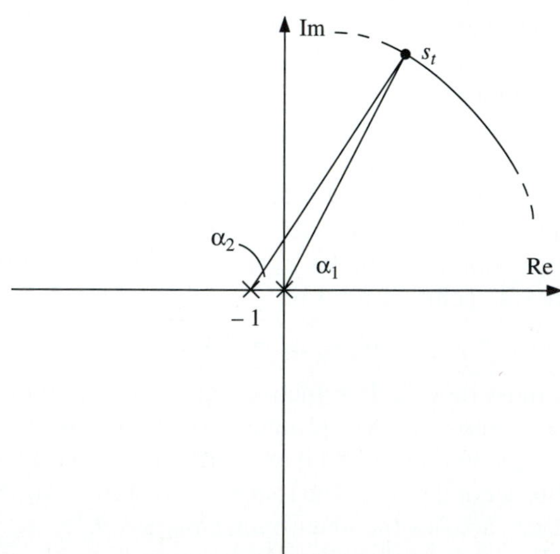

Fig. 9.5 Test point on an infinite circle

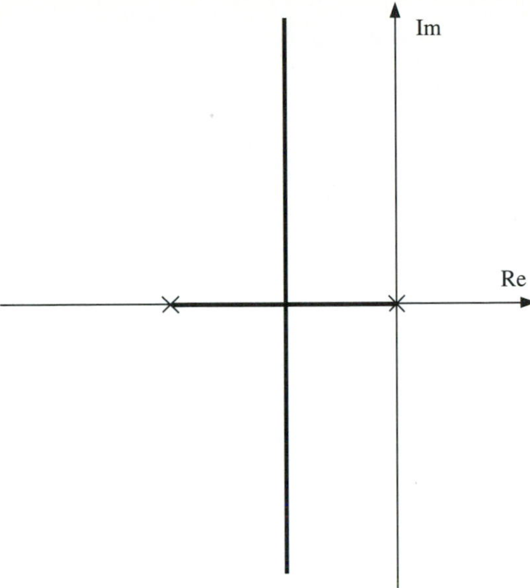

Fig. 9.6 Complex-plane solutions of the argument equation

that it is difficult to determine s_t given a value K. It can be seen from equation 9.26 that if a value of s_t is chosen very close to an open-loop pole, then one of the magnitude terms in the numerator will be close to zero, resulting in $K \approx 0$. Alternately, if the root is located a long way from the real axis, the corresponding gain is $K \approx \infty$. This relationship between the location of s_t and the corresponding value of K may be put in the form: As the value of K increases from zero to infinity, the corresponding location of s_t "moves" from an open-loop pole toward infinity. The direction the closed-loop poles move is indicated by the arrows in Fig. 9.7. It may be seen from Fig. 9.7 that, given a small value of K, there are in fact two corresponding locations of s_t, one close to the pole at $s = 0$ and another close to the pole at $s = -1$. As K increases, the location of s_t moves from the open-loop pole at $s = 0$ to the left, meeting the other closed-loop pole moving to the right from the open-loop pole at $s = -1$. They meet at $s = -0.5$ and then branch out along the vertical line, one going up while the other goes down. Increasing K further moves the closed-loop poles toward infinity.

We will calculate the value of gain at particular points on the locus. For closed-loop poles coincident with the open-loop poles, clearly $K = 0$. Using equation 9.26, when the two closed-loop poles are coincident with each other at $s = -0.5$, we obtain

$$K = 0.5 \times 0.5 = 0.25 \tag{9.27}$$

When the closed-loop poles are at $s = -0.5 \pm 0.5j$, as shown in Fig. 9.7, the gain may be found from

$$K = \frac{1}{\sqrt{2}} \times \frac{1}{\sqrt{2}} = 0.5 \tag{9.28}$$

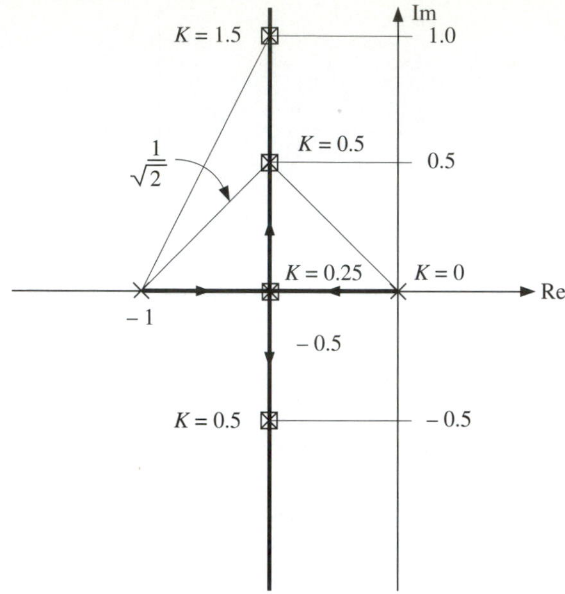

Fig. 9.7 Location of closed-loop poles with gain variation

For roots at $s = -0.5 \pm j$, the gain may be computed as 1.5, and so on. Recall that the purpose of the root locus method is to solve the characteristic equation. For this example, the characteristic equation may be solved directly for comparison with the methods just presented. The characteristic equation is

$$1 + GH(s) = s^2 + s + K = 0 \tag{9.29}$$

This has the solutions

$$s = \frac{-1 \pm \sqrt{1 - 4K}}{2} = -0.5 \pm \sqrt{0.25 - K} \tag{9.30}$$

Substituting $K = 0$, 0.25, 1.5 into this equation yields the same closed-loop poles shown in Fig. 9.7. The power of the root locus method is, of course, in solving higher-order characteristic equations for which analytical solutions are not available.

We will now look at another example, this time involving an open-loop zero:

$$GH(s) = \frac{K(s + 1)}{(s + 2)(s + 3)} \tag{9.31}$$

By substituting this particular open-loop transfer function into the characteristic equation, we get

$$1 + \frac{K(s + 1)}{(s + 2)(s + 3)} = 0 \tag{9.32}$$

The appropriate argument equation becomes

$$\angle \frac{K(s + 1)}{(s + 2)(s + 3)} = -180^\circ \tag{9.33}$$

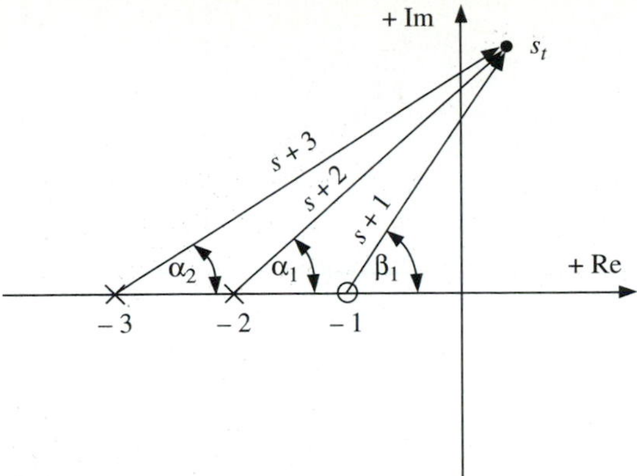

Fig. 9.8 Arguments of $s_t + 1$, $s_t + 2$, and $s_t + 3$

In this particular example, this reduces to

$$\begin{aligned}\angle\frac{K(s+1)}{(s+2)(s+3)} &= \angle(s+1) - \angle(s+2) - \angle(s+3) \\ &= \beta_1 - \alpha_1 - \alpha_2 \\ &= -180^\circ \qquad (9.34)\end{aligned}$$

Figure 9.8 shows the appropriate arguments of the components of the open-loop transfer function. Again, the search begins by inspecting only the real axis, where it may be seen that the total argument when $s_t > -1$ is zero. For $-2 < s_t < -1$, $\beta_1 = -180^\circ$, $\alpha_1 = \alpha_2 = 0$; hence equation 9.34 is satisfied. Further investigation of the real-axis segments reveals that the argument equation is also satisfied for $-\infty < s_t < -3$, but not for $-3 < s_t < -2$.

To determine if there are any values of s_t off the real axis, consider the infinite circle, centered on the origin. For s_t located anywhere on this circle, $\beta_1 = \alpha_1 = \alpha_2 = \theta$. The total argument from equation 9.34 will therefore be equal to θ. The only value that satisfies equation 9.34 is $\theta = -180^\circ$, which has already been determined as a solution. It is concluded, therefore, that there are no values of s_t satisfying the argument equation located far from the origin, except at $s_t = -\infty$. The solutions to the argument equation are shown in Fig. 9.9. Turning now to the magnitude equation, we may write

$$\frac{K|s_t + 1|}{|s_t + 2||s_t + 3|} = 1 \qquad (9.35)$$

which becomes

$$K = \frac{|s_t + 2||s_t + 3|}{|s_t + 1|} \qquad (9.36)$$

It may be seen from Fig. 9.9 that, given a small value of K, there are again two corresponding locations of s_t, one close to the pole at $s = -2$ and another close to the pole at $s = -3$. As K increases, the location of s_t moves from the pole

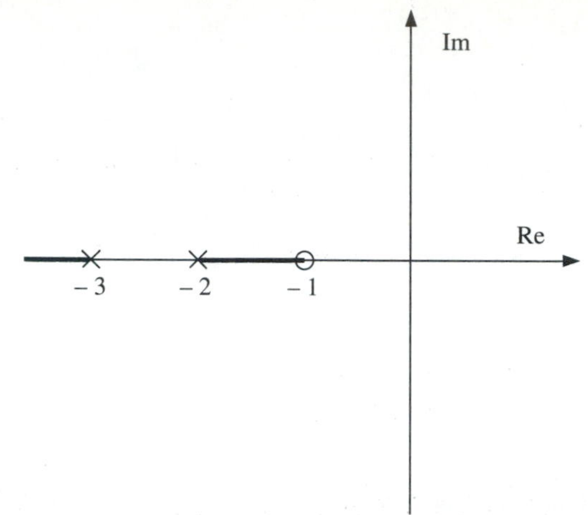

Fig. 9.9 Root locus for $GH(s) = K(s + 1)/(s + 2)(s + 3)$

at $s = -2$ toward the zero at $s = -1$. The other location of s_t moves from the pole at $s = -3$ along the negative real axis toward $-\infty$. An observation may be made that as K increases from zero to infinity, the closed-loop poles move from the open-loop poles to the open-loop zeros if there are a sufficient number or off to infinity if there are not a sufficient number. Calculation of the gain at specific closed-loop pole locations may be made as in the previous example, i.e., either by calculating the lengths geometrically or by measuring them, and then substituting into equation 9.36. The final root locus, with selected gain calculations, is shown in Fig. 9.10.

It will be seen in a later section that the root locus method provides much more insight into the time-domain response of the system as well as providing a powerful design capability, something not obtainable from a direct solution of the characteristic equation. After gaining some experience in sketching simple root loci,

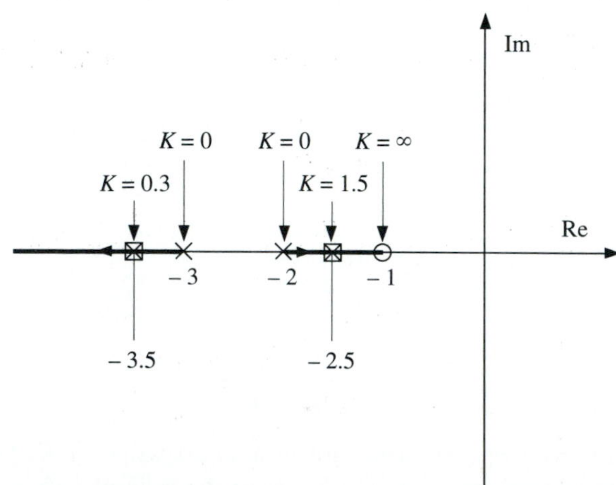

Fig. 9.10 Final root locus with selected gain calculations

some general guidelines will be developed to help in the rapid formulation of the loci for more complex systems.

Sample Problem 9.1

Draw the root locus for the system with open-loop transfer function

$$GH(s) = \frac{K}{s(s+2)(s+4)}$$

and determine the location of the closed-loop poles for $K = 10$.

Solution

We begin, as before, by drawing the complex plane and marking in the open-loop poles (by a cross) and the open loop zeros (by a circle). In this example there are three open-loop poles at $s = 0, -2, -4$ and no zeros. Inspecting the real axis, it may be seen that the argument equation is satisfied only for $-\infty < s_t < -4$ and $-2 < s_t < 0$. Figure SP9.1.1 shows the locus so far. It is also known that the locus begins at the open-loop poles and, as K increases, moves toward the open-loop zeros if there are any or to infinity if not. Since there are no open-loop zeros in this example, it is necessary to determine *how* the roots go to infinity. One of the three roots seems to go along the negative real axis to infinity, but what happens to the other two roots emanating from the poles at $s = -2$ and the origin?

Consider again a circle of infinite radius centered on the origin with the trial point s_t located somewhere on the circumference representing the closed-loop pole for $K \approx \infty$. Where on the circle does s_t have to be located to satisfy the argument equation? If each of the open-loop poles and s_t are connected with "rubber band" lines that expand and contract as s_t moves, it is seen that the arguments of s, $s + 2$ and $s + 4$, will be approximately equal, as shown in Fig. SP9.1.2. Since the phase angle equation requires

$$-\angle s - \angle(s+2) - \angle(s+4) \approx -3\theta = -180°$$

the argument from any of the poles θ has to be 60°. Note also that $\theta = -60°$ also satisfies the above equation. The locus from the two remaining poles

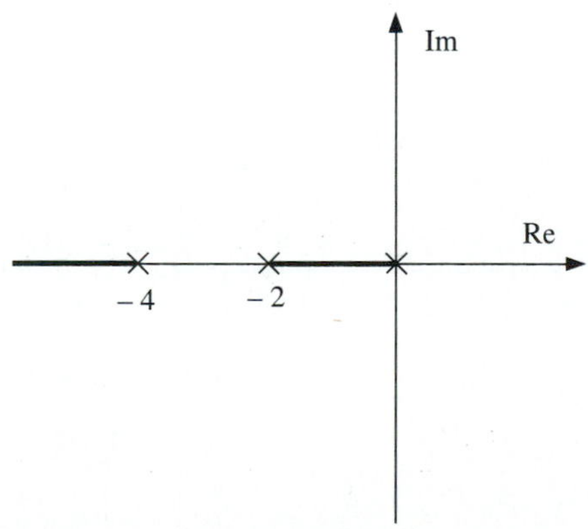

Fig. SP9.1.1

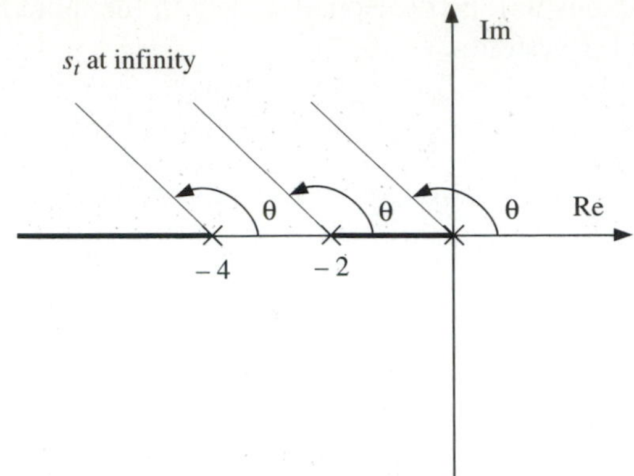

Fig. SP9.1.2

must therefore go to infinity along *asymptotes* inclined at ±60° to the positive real axis.

Two questions remain: (1) Where do the asymptotes intersect the real axis? (2) Where is the locus near the real axis? The intersection of the asymptotes with the real axis $\sigma_a + j0$ is given by the following formula (the derivation of which is beyond the scope of this text):

$$\sigma_a = \frac{\sum p_i - \sum z_i}{n - m}$$

where p_i is the real part of the open-loop poles, z_i is the real part of the open-loop zeros, n is the number of poles, and m is the number of zeros. Substituting values into this equation gives

$$\sigma_a = \frac{0 - 2 - 4}{3} = -2$$

Figure SP9.1.3 shows the asymptotes drawn on the complex plane. Two of the closed-loop poles start at the open-loop poles at $s = 0$ and $s = -2$ for $K = 0$

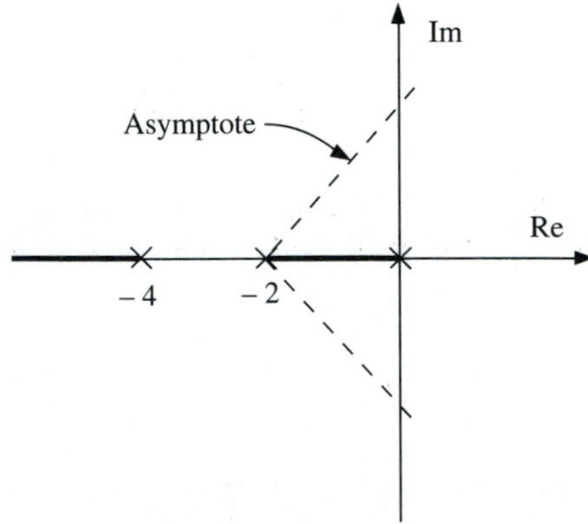

Fig. SP9.1.3

and move toward each other on the real axis as K increases. It may be argued that these closed-loop poles must then meet somewhere, break away from the real axis (one above and one below), and then move to infinity, becoming closer and closer to the asymptotes. The determination of the point where the locus leaves the real axis would be of considerable value in plotting the remaining locus, and we shall determine this point now. By considering the way the closed-loop poles move as the gain increases, it is readily seen that the point of departure from the real axis corresponds to the maximum gain for any closed-loop pole between the open-loop poles at $s = 0$ and $s = -2$. The point of departure is determined, therefore, by calculating the gain (using the magnitude equation) for several points on this real-axis segment and identifying the maximum value either graphically or by trial and error. In this example, points at 0.2 intervals will be selected between 0 and 2, the gain calculated graphically, and the results tabulated, as shown in the table below. This indicates the peak gain occurs at around $s = 0.8$. Further trial and error produces the more precise result of $s = 0.85$ as the breakaway point:

s	K
−0.0	0.000
−0.2	1.728
−0.4	2.304
−0.6	2.856
−0.8	3.072
−1.0	3.000
−1.2	2.688
−1.4	2.184
−1.6	1.536
−1.8	0.792
−2.0	0.000

Another way to determine the gain at the breakaway point is to write the characteristic equation

$$1 + \frac{K}{s(s + 2)(s + 4)} = 0$$

in the form where K is expressed directly as a function of s:

$$K = -s(s + 2)(s + 4)$$

or, upon expansion,

$$K = -s^3 - 6s^2 - 8s$$

Since we are looking for the maximum value of K, this equation may be differentiated and equated to zero:

$$\frac{dK}{ds} = -3s^2 - 12s - 8 = 0$$

Solving the quadratic yields the values

$$s = -2 \pm 1.15$$

which gives $s = -0.845$ as the only solution compatible with the real-axis segments. Note that this method is suitable only for characteristic equations of order 3 or less; otherwise the differentiated equation will be cubic or higher, requiring a numerical method to find the solution. In such a case, graphical determination of the maximum gain K as outlined before is to be preferred.

One more precise point may be found. It is known that the locus approaches the asymptote as the gain increases, but it is not known how quickly. The location of the closed-loop pole on the locus as it crosses the imaginary axis may be calculated by observing that at this point the closed-loop pole takes the particular form $s = j\omega$. If this value is substituted into the characteristic equation, ω can be determined. In this example the characteristic equation may be written as

$$s^3 + 6s^2 + 8s + K = 0$$

Substituting $s = j\omega$ gives

$$(K - 6\omega^2) + j(8\omega - \omega^3) = 0$$

Equating the imaginary part to zero yields $\omega = \sqrt{8} = 2.83$. The locus crosses the imaginary axis at $s = \pm 2.83$, enabling the remainder of the locus to be sketched, as shown in Fig. SP9.1.4. In order to determine the location of the closed-loop poles for $K = 10$, it is seen from the above equation that setting the real part of the left-hand side to zero yields $K = 6\omega^2 = 48$. The required root, therefore, must be before the interception of the locus with the imaginary axis. Note that for $K > 48$ the system becomes unstable.

A trial point s_t is selected and joined to all of the open-loop poles and zeros. Recalling the definition of the modulus equation from the previous section,

$$K = a \times b \times c$$

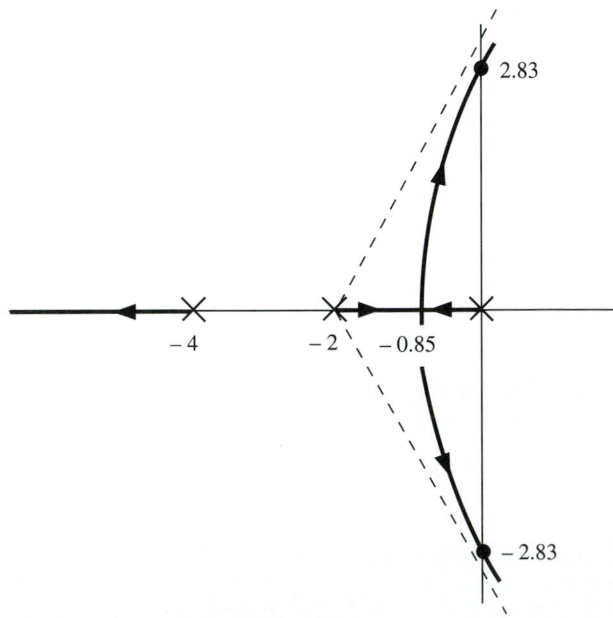

Fig. SP9.1.4

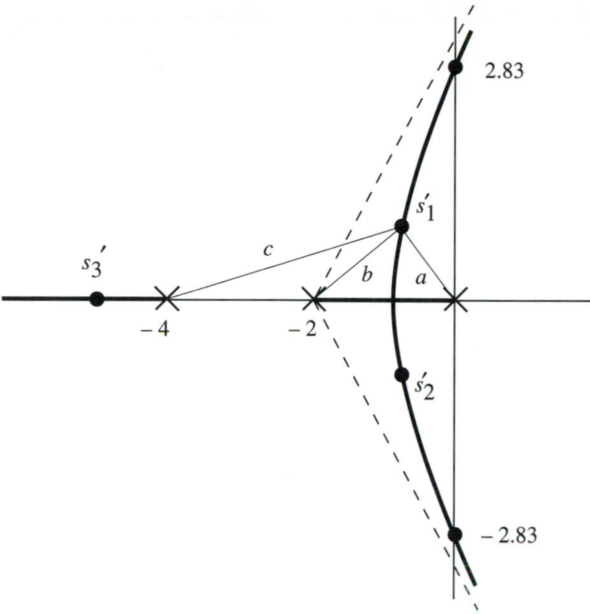

Fig. SP9.1.5

where a, b, and c are the distances between the open-loop poles and the trial point s_t. Estimating the gain at the real-axis breakaway point, $K \approx 1 \times 1 \times 3 = 3$. The roots for $K = 10$ must lie on the locus between the real-axis breakaway point and the imaginary-axis crossing. By trial and error, the first root is determined to be as shown by s_1' in Fig. SP9.1.5. Note that in determining the location of the first root, moving s_t toward the poles will reduce a, b, and c, thereby decreasing the gain value K, while moving s_t away from the poles will increase K. There are two other closed-loop poles, one being the complex conjugate s_2', the other being s_3', located on the negative real axis to the left of the open-loop pole at $s = -4$. Note that the third root may be located from the rule[3] that for $n - m \geq 2$ the sum of the real parts of the closed-loop poles is constant, independent of K, and is equal to the sum of the real parts of the open-loop poles. Figure SP9.1.6 shows a computer-generated plot of the root locus for this problem and may be compared with the sketch in Fig. SP9.1.5.

Sample Problem 9.2

Draw the root locus for the system with open-loop transfer function

$$GH(s) = \frac{K(s + 1)}{s^2(s + 2)}$$

Solution

Figure SP9.2.1 shows the complex plane with the open-loop poles and zeros marked. Since there are two more poles than zeros, there must be two branches

[3]To be proven later.

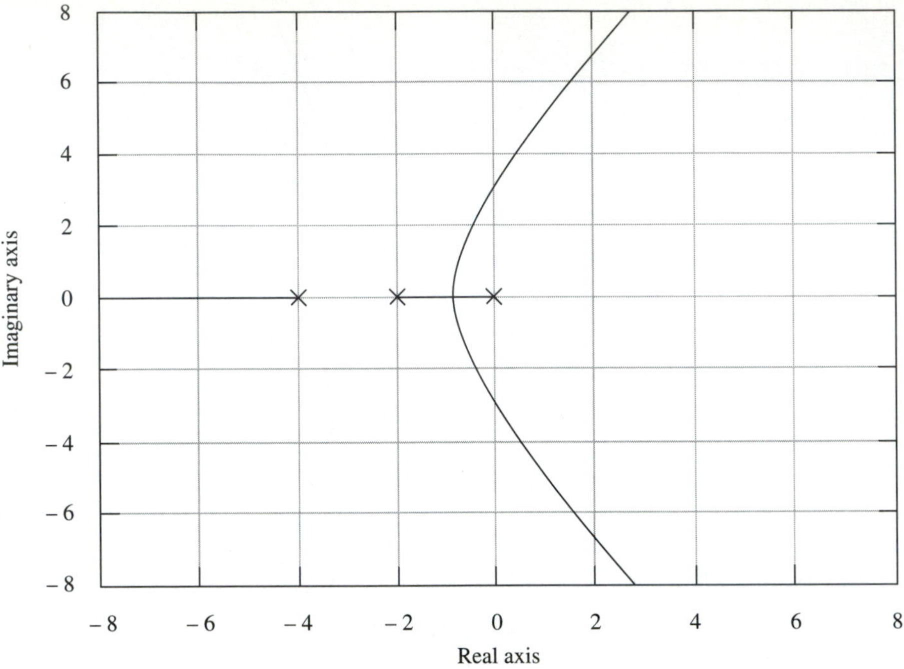

Fig. SP9.1.6

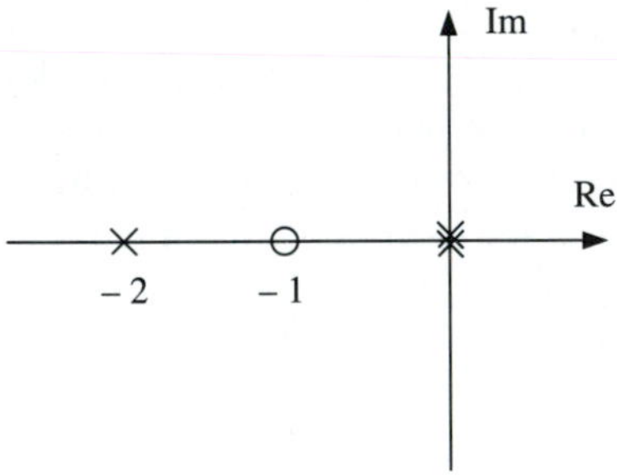

Fig. SP9.2.1

of the locus going to infinity. The construction of the locus begins by determining which sections of the real axis satisfy the argument equation and then finding out in which direction the branches go to infinity. A trial point s_t is indicated in Fig. SP9.2.2 and the argument of each component of the open-loop transfer function is shown. In this example, the argument equation becomes

$$\beta_1 - \alpha_1 - 2\alpha_2 = -180^\circ$$

Note the effect of the double pole at the origin. Imagining s_t to be on different portions of the real axis, α_1, α_2, and β_1 take the values of either 0 or 180°. It is straightforward to verify that the only section of the real axis satisfying the argument equation is when $-2 < s_t < -1$, and this is marked in Fig. SP9.2.3. In order to determine the direction that the branches go to infinity, a circle of

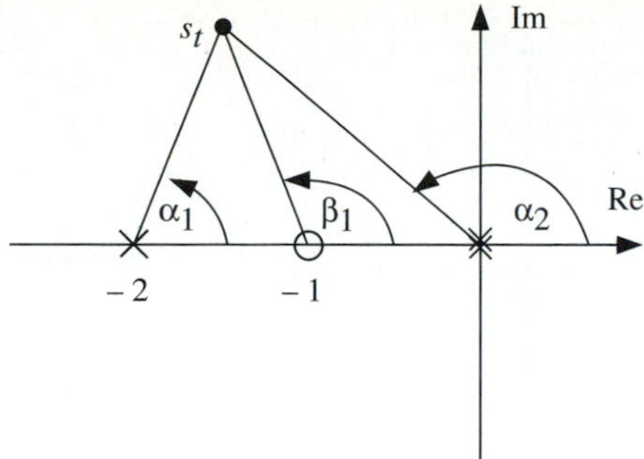

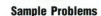

Fig. SP9.2.2

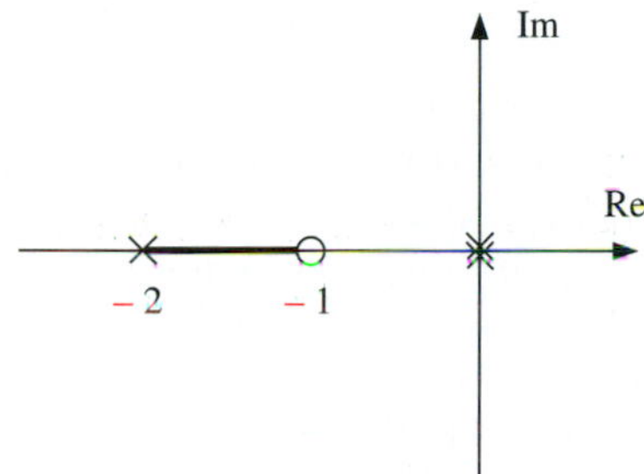

Fig. SP9.2.3

infinite radius centered at the origin is considered. For a closed-loop pole located on the circle and with each of the phase angles equal, it may be seen that in order to satisfy the phase angle equation above, the roots must go to infinity in a direction at 90° to the real axis. The interception of the asymptotes with the real axis is obtained from the equation

$$\sigma_a = \frac{\sum p_i - \sum z_i}{n - m}$$

as before, which yields $\sigma_a = -0.5$. Figure SP9.2.4 shows the locus construction so far. Two possibilities for the shape of the locus are considered. The locus

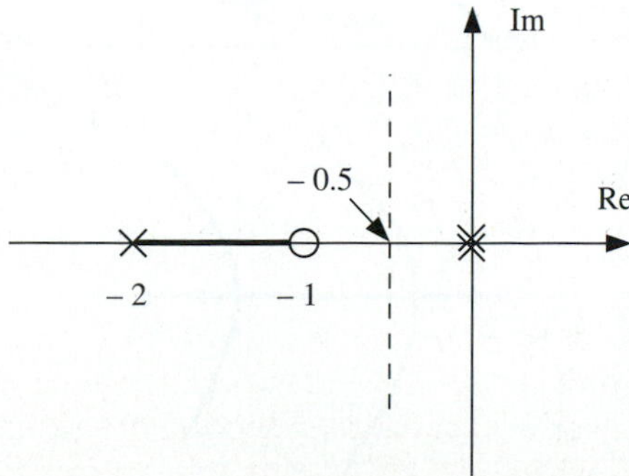

Fig. SP9.2.4

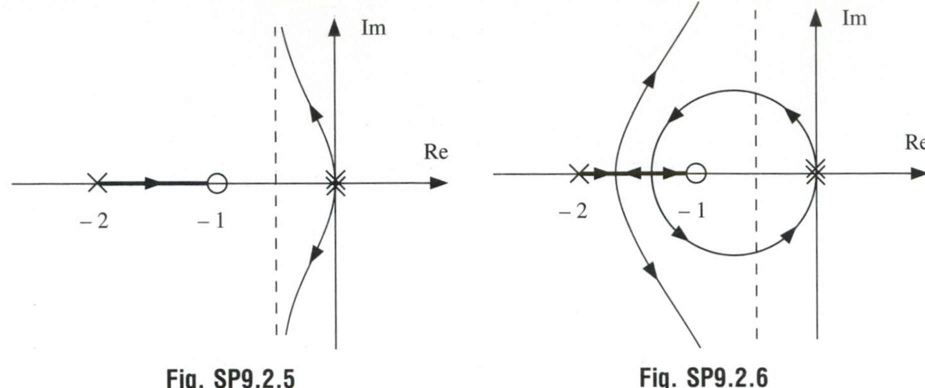

Fig. SP9.2.5

Fig. SP9.2.6

from the double pole at the origin may go directly to infinity along the asymptotes. Alternatively, the locus from the double pole at the origin may circle around the zero and join the real axis between the pole at $s = -2$ and the zero at $s = -1$. One of these poles then moves toward the other closed-loop pole coming from the open-loop pole at $s = -2$, branching away from the real axis when they meet and going to infinity along the asymptotes. The remaining closed-loop pole would then go to the zero at $s = -1$ as K approaches infinity. These two solutions are shown in Figs. SP9.2.5 and SP9.2.6, respectively. One way to determine which of these two options prevails is to use the method outlined in the previous example to calculate the points at which the locus enters and leaves the real axis. Although the previous example dealt with only a point where the locus leaves a real-axis segment, it is seen that the point where the locus enters the real axis will be characterized by a local minimum value of gain. If the locus does look like Fig. SP9.2.6, then the gain plotted as a function of s will look like Fig. SP9.2.7, whereas the locus shown in Fig. SP9.2.5 produces a gain plot more like that shown in Fig. SP9.2.8. Calculating sample

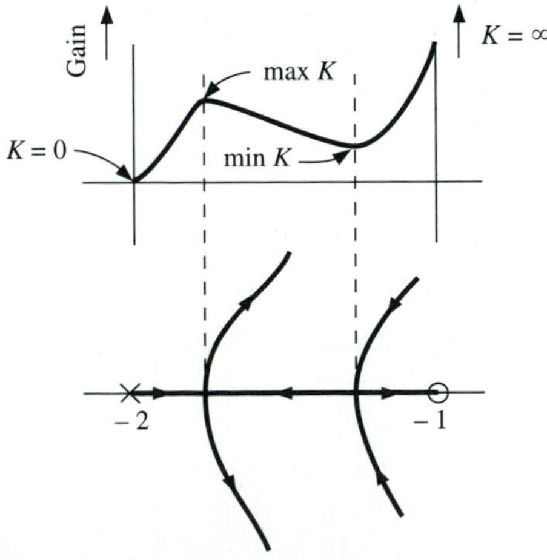

Fig. SP9.2.7

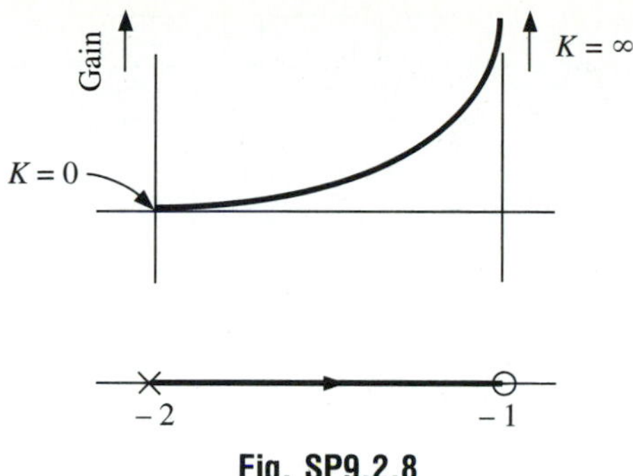

Fig. SP9.2.8

gains at locations between $s = -1$ and $s = -2$ indicates a variation like that in Fig. SP9.2.8, confirming Fig. SP9.2.5 as the correct plot. It is concluded, therefore, that the double pole at the origin moves directly to infinity along the asymptotes. Note, however, that the locus near the real axis for this branch can only be guessed, and the accurate locus determined by selecting trial points in an attempt to satisfy the argument equation. Figure SP9.2.9 shows a computer plot of the accurate solution.

Sample Problem 9.3

Figure SP9.3.1 shows a feedback control system. Draw the root locus for the system and determine the smallest value of gain K for which a closed-loop pole has a positive real part.

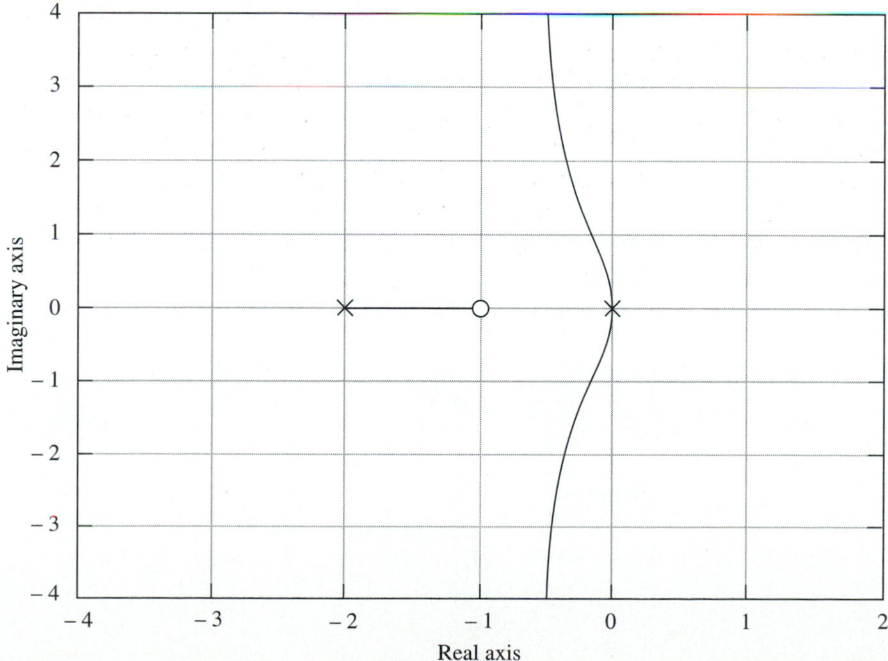

Fig. SP9.2.9

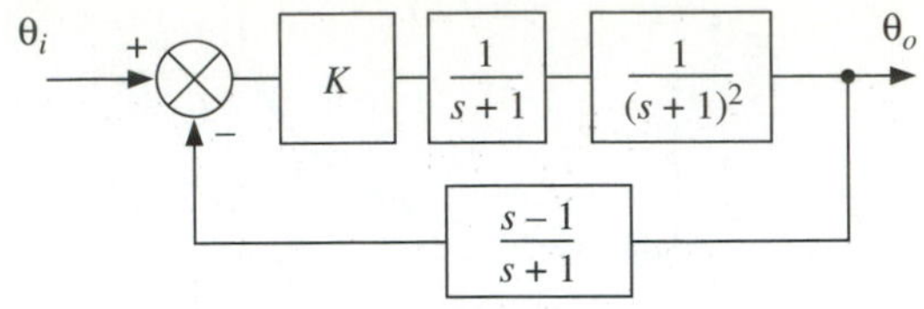

Fig. SP9.3.1

Solution

The open-loop transfer function is determined as

$$GH(s) = \frac{K(s - 1)}{(s + 1)^4}$$

Figure SP9.3.2 shows the complex plane with the open-loop zero at $s = +1$ and the four coincident poles at $s = -1$. Proceeding as before by denoting the argument from a zero by β and the argument from a pole by α, the argument equation becomes

$$\beta_1 - \alpha_1 - \alpha_2 - \alpha_3 - \alpha_4 = -180°$$

It is seen that the real axis for $s_t < +1$ is a solution of the above equation. There are four poles and one zero. Therefore, there will be three branches of the locus going to infinity. By considering the location of s_t at infinity, it may be seen that the argument equation is satisfied when the locus goes to infinity in a direction inclined at 60° to the real axis. It may also be seen that other solutions to the argument equation are −60° and −180°. The three asymptotes are inclined at ±60° and −180° to the positive real axis.

The interception of the asymptotes with the real axis may be calculated as

$$\sigma_a = \frac{(4 \times -1) - (+1)}{4 - 1} = -1.667$$

Figure SP9.3.3 shows the complex plane with the asymptotes marked. One closed-loop pole appears to move directly to the open-loop zero at $s = +1$, while another pole moves to infinity along the negative real axis. The two remaining closed-loop poles branch from the real axis and move to infinity along the two asymptotes inclined at ±60° to the real axis. The point at which the locus crosses the imaginary axis is determined by substituting $s = j\omega$ into the

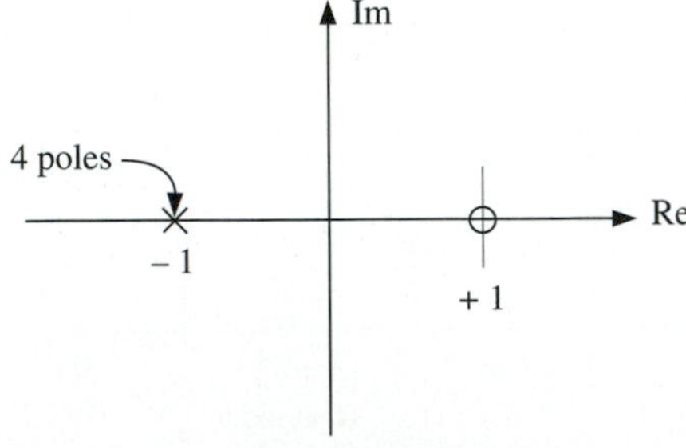

Fig. SP9.3.2

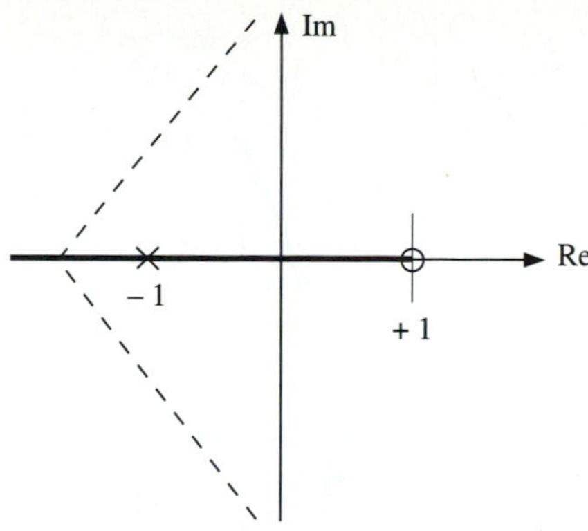

Fig. SP9.3.3

characteristic equation. The characteristic equation is

$$s^4 + 4s^3 + 6s^2 + s(K + 4) + (1 - K) = 0$$

After substituting, collecting terms, and equating to zero, two equations result:

$$\omega^4 - 6\omega^2 + 1 - K = 0 \qquad \omega(4 + K - 2\omega^2) = 0$$

Solving these equations yields $\omega = 0, \pm 3.08$. The first solution, $\omega = 0$, corresponds to the point at which the root from the pole at $s = -1$ crosses the imaginary axis on its way to the zero at $s = +1$. The other two solutions, $\omega = \pm 3.08$, correspond to the near-asymptote crossings of the imaginary axis. These points have been marked in Fig. SP9.3.4 and the remaining portions of the locus sketched. In order to determine the smallest value of gain necessary to give one of the closed-loop poles a positive real part, it is necessary to determine which segment of the locus crosses the imaginary axis first. For the segment crossing at $s = \pm 3.08j$, a trial point s_t may be placed at the crossing point and K found from

$$K = \frac{a^4}{b}$$

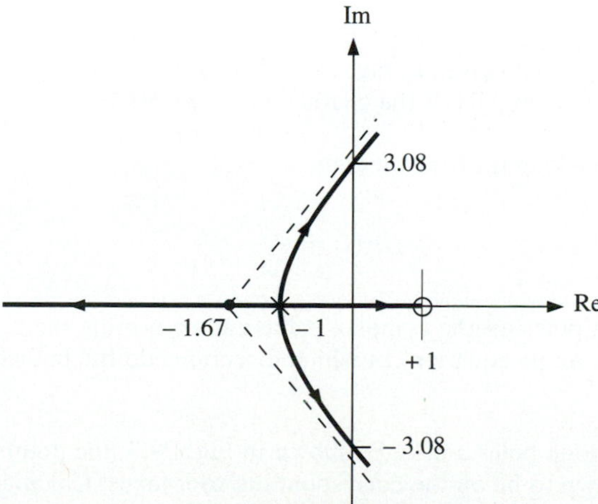

Fig. SP9.3.4

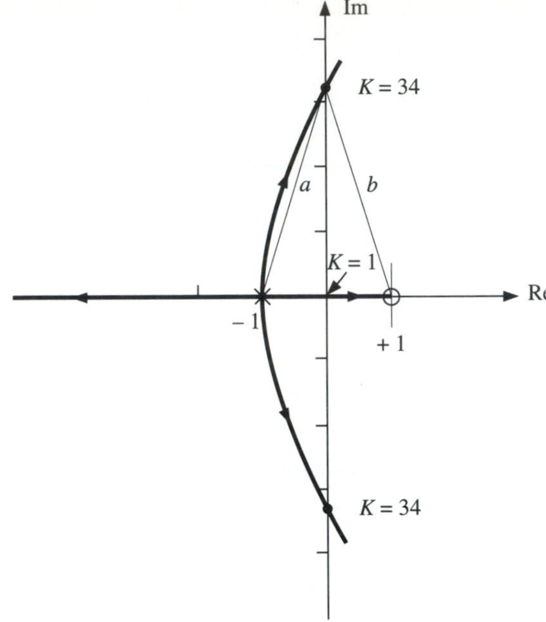

Fig. SP9.3.5

where a and b are defined in Fig. SP9.3.5. Alternatively, K may be calculated from the second of the two equations obtained by substituting $s = j\omega$ into the characteristic equation above. Either of these methods will give $K = 34$. For a trial point located at the origin, where that segment crosses the imaginary axis, the gain may be calculated directly as $K = 1^4/1 = 1$. The smallest value of K that makes the system unstable is $K = 1$. An accurate root locus plot is shown in Fig. SP9.3.6.

◆ PROBLEMS

9.1 For the pole-zero map given in Fig. P9.1, determine those segments of the real axis that satisfy the argument part of the characteristic equation.

9.2 Given the open-loop transfer function

$$GH(s) = \frac{K}{s(s + 4)}$$

plot the open-loop poles on the complex plane and determine the segments of the real axis satisfying the angle equation. In which directions do the branches of the locus go to infinity?

9.3 For the open-loop poles and zeros shown in Fig. P9.3, the points $s = -2 \pm 2j$ and $s = -2.5$ are known to be on the corresponding root locus. Calculate the gain at each of these points.

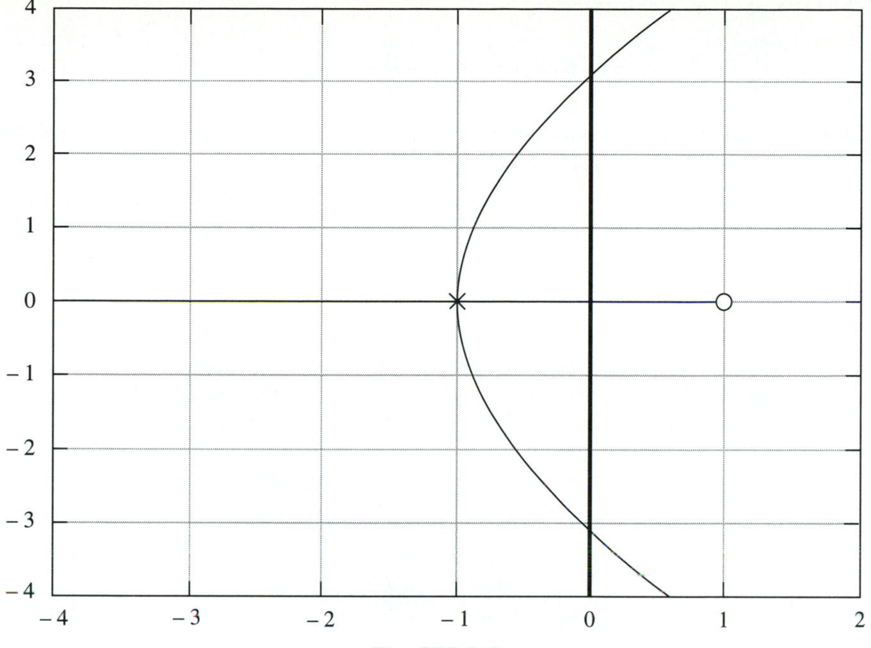

Fig. SP9.3.6

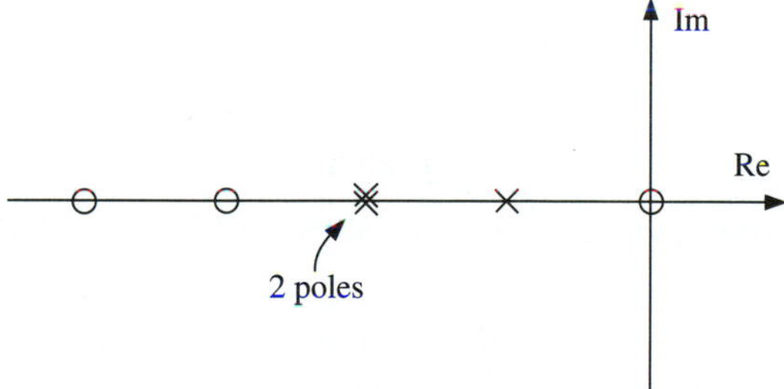

Fig. P9.1

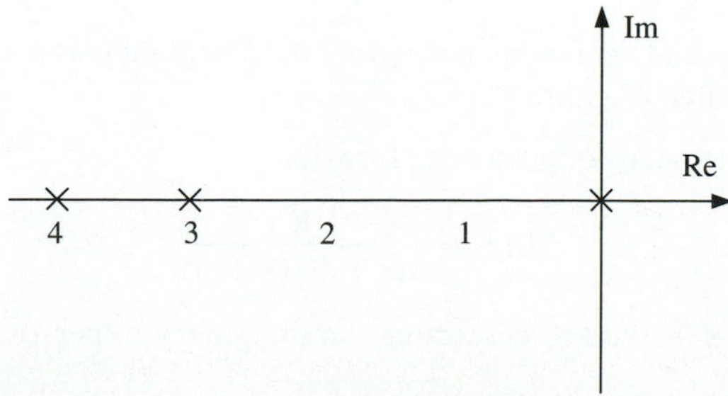

Fig. P9.3

9.4 For the open-loop transfer function

$$GH(s) = \frac{K}{s(s+5)(s+6)}$$

find the points on the root locus that have **(a)** a real part $\sigma = -2$ and **(b)** an imaginary part $\omega = \pm 1$. Calculate the gain at these locations.

9.5 A unity-feedback position control system shown in Fig. P9.5 has an actuator and load represented by the transfer function

$$G(s) = \frac{K}{s(s+2)}$$

Write down the characteristic equation of the system, solve it, and plot the locus of its roots as the gain K varies from zero to infinity. Compare the result with that obtained from the root locus approach utilizing the open-loop transfer function of the system.

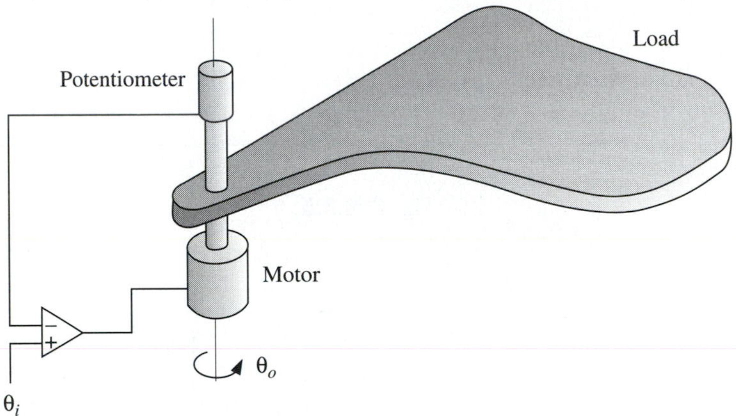

Fig. P9.5

9.6 For each of the open-loop poles and zeros shown in Fig. P9.6, *sketch* the corresponding root locus.

9.7 The open-loop pole-zero map of a satellite attitude control system is shown in Fig. P9.7. Draw the root locus and determine, as accurately as possible, the point at which the locus leaves the real axis.

9.8 A system has open-loop transfer function

$$GH(s) = \frac{K(s+0.5)^2}{s^3}$$

Sketch the root locus and determine accurately the imaginary-axis crossing points and the corresponding value of gain.

9.9 A system has the open-loop transfer function

$$GH(s) = \frac{K}{s(s+3)(s+4)}$$

a. Use the root locus method to determine the imaginary-axis crossing points.

b. Check the previous result graphically.

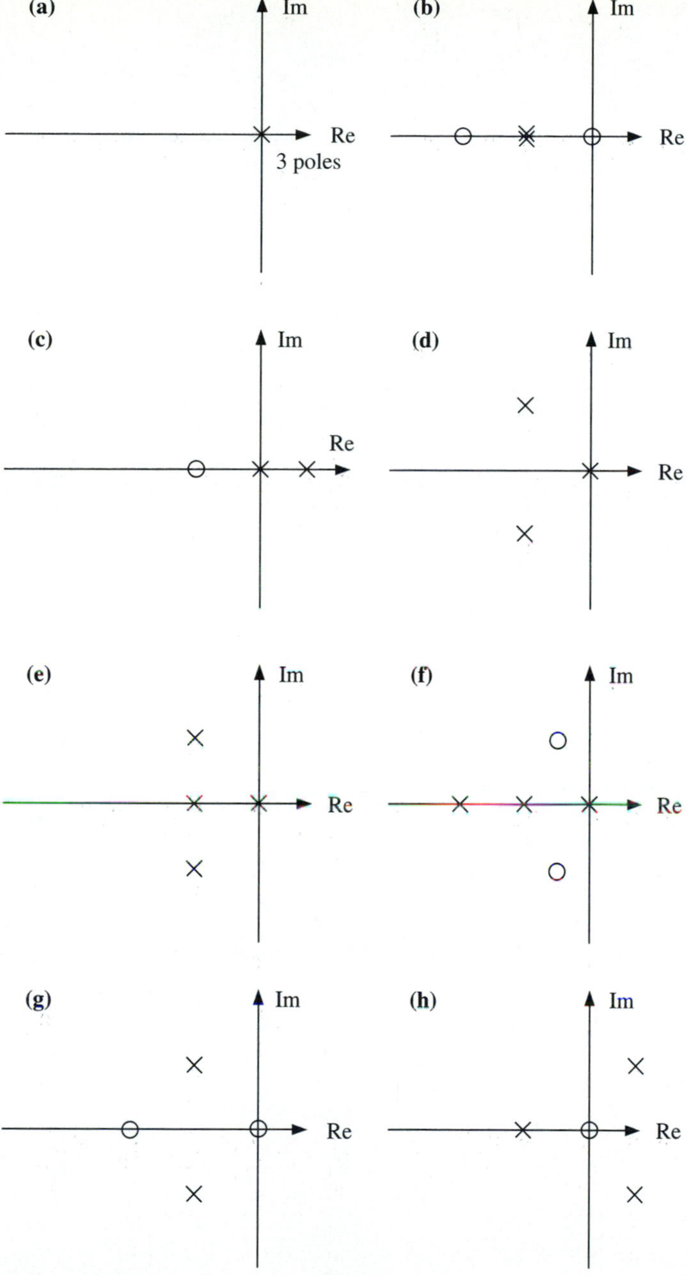

Fig. P9.6

c. Use Routh's method to determine the maximum value of K consistent with stability, thereby confirming the previous results.

9.10 For the open-loop poles and zeros shown in Fig. P9.10, determine the value of gain as *each* closed-loop pole crosses the imaginary axis. Determine also the points at which the crossings take place.

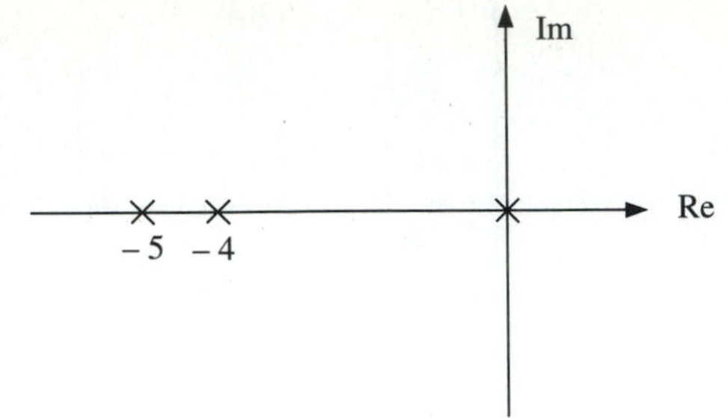

Fig. P9.7

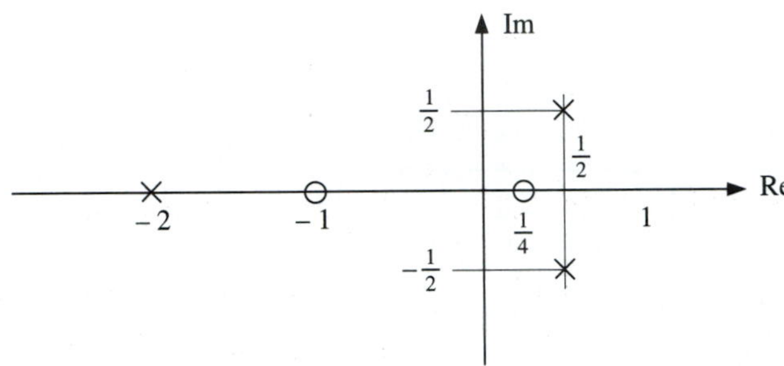

Fig. P9.10

MODULE 10

Rules for Plotting the Root Locus

In this section the concepts outlined previously will be developed further into some straightforward guidelines for plotting more complex root loci, which will be illustrated by focusing on a specific example. Suppose we are required to draw the root locus for the system with open-loop transfer function

$$GH(s) = \frac{K(s + 5)}{s(s^2 + 4s + 8)} \tag{10.1}$$

and to (1) calculate K for a closed-loop pole at $s = -1$ and (2) determine the locations of all other roots at this value of gain. The root locus will be drawn using the following rules.

> **Rule #1.** *Draw the complex plane and mark the n open-loop poles and m zeros. The locus starts at a pole for $K = 0$ and finishes at a zero or infinity when $K = \infty$. The number of segments going to infinity is therefore $n - m$.*

In this example we have one zero at $s = -5$, one real pole at $s = 0$, and a complex pair at $s = -2 \pm 2j$, where the complex open-loop poles come from the solution of the quadratic term $s^2 + 4s + 8$. Since $n = 3$ and $m = 1$, we have two branches of the locus going to infinity and one branch going from a pole to the zero.

> **Rule #2.** *Segments of the real axis to the left of an odd number of poles or zeros are segments of the root locus, remembering that complex poles or zeros have no effect.*

Part of the locus, therefore, is the negative real axis between the origin and the zero at $s = -5$. Figure 10.1 shows the locus so far, together with a trial point s_t on this part of the locus. The student may verify that the argument contribution from the pole at $s = -2 \pm 2j$ is α, while that from $s = -2 - 2j$ is $360° - \alpha$, resulting in a total contribution to the argument of zero.

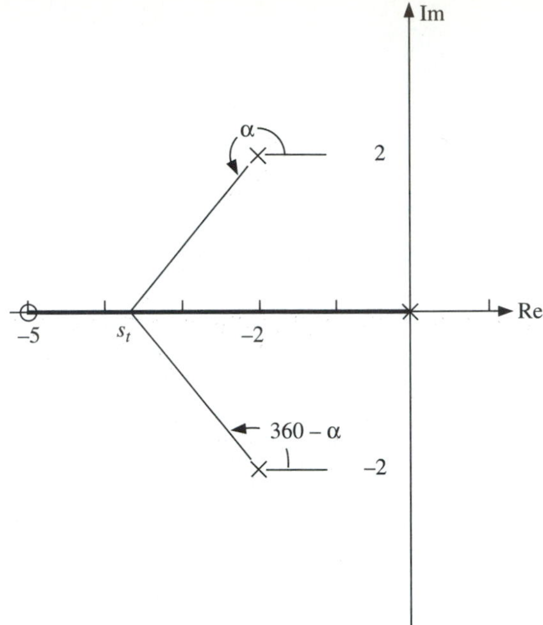

Fig. 10.1 Real-axis segments

Rule #3. *The loci are symmetrical about the real axis since complex roots are always in conjugate pairs. The angle between adjacent asymptotes is* 360°/$(n - m)$, *and to obey the symmetry rule, the negative real axis is one asymptote when* $n - m$ *is odd.*

Here, $n - m = 2$ and the angle between asymptotes is 180°, and the negative real axis is not an asymptote. This suggests the asymptotes must be parallel to the imaginary axis.

Rule #4. *The asymptotes intersect the real axis at* σ_a, *where*

$$\sigma_a = \frac{\sum p_i - \sum z_i}{n - m} \tag{10.2}$$

where $\Sigma\, p_i$ *is the sum of the real parts of the open-loop poles (including complex roots) and* $\Sigma\, z_i$ *is the sum of the real parts of the open-loop zeros (also including complex roots).*

Substituting into this equation yields

$$\sigma_a = \frac{(0 - 2 - 2) - (-5)}{3 - 1} = +0.5 \tag{10.3}$$

Since the orientation of the asymptotes and the point of intersection with the real axis are known, they can be drawn as shown in Fig. 10.2. It seems likely at this stage that the closed-loop pole moves from the open-loop pole at $s = 0$ to the zero at $s = -5$, and the two complex poles move toward the asymptotes and off to infinity. This may be checked by determining the angle of emergence from the complex open-loop poles.

Consider a trial point s_t located on a small circle drawn around the complex pole at $s = -2 + 2j$ in Fig. 10.3. As s_t moves around the circle, the argument angle

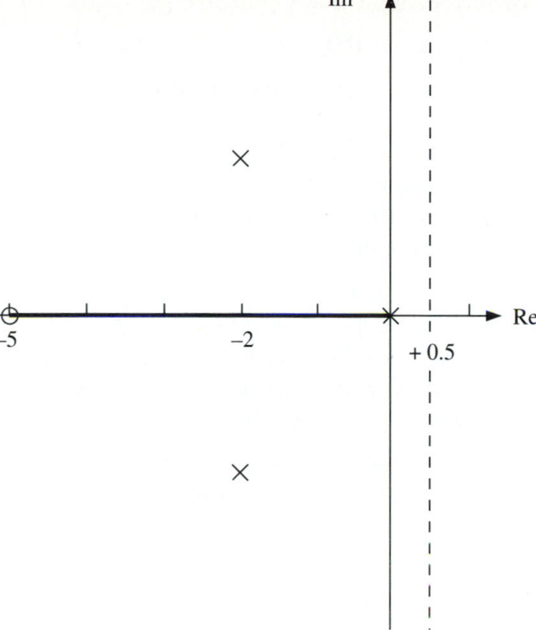

Fig. 10.2 Locus asymptotes

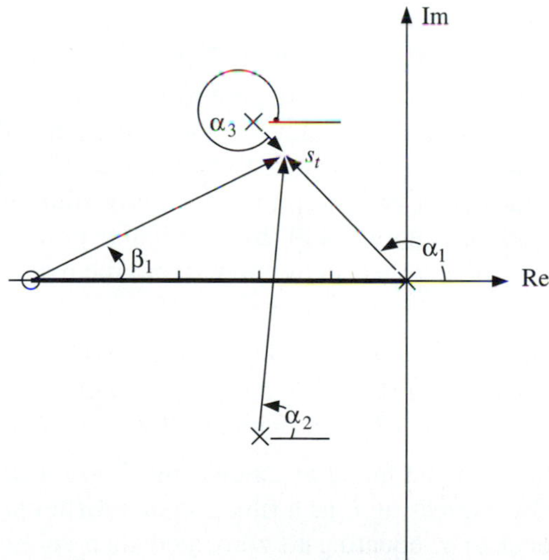

Fig. 10.3 Angle of emergence from complex poles

contributions from the other open-loop poles and zeros vary little, while the contribution from the encircled pole varies from zero to 2π. The question is, where on the circle does s_t satisfy the argument equation? Measuring α_1, α_2, and β_1 and substituting into the argument equation yields

$$\beta_1 - \alpha_1 - \alpha_2 - \alpha_3 = -180^\circ \tag{10.4}$$

As the radius of the circle approaches zero, we get

$$\begin{aligned}\alpha_3 &= 180 - \alpha_1 - \alpha_2 + \beta_1 \\ &= 180 - (135 + 90) + 33 \\ &= -12^\circ \end{aligned} \tag{10.5}$$

From considerations of symmetry, the locus leaves the complex poles to the right at an angle of $\pm 12^\circ$ to the positive real axis. Observing the grouping of the individual arguments in the above equation, we may write:

Rule #5. *The angle of emergence from complex poles is given by* $180^\circ - \Sigma$ *(angles of the vectors from all other open-loop poles to the pole in question)* $+ \Sigma$ *(angles of the vectors from the open-loop zeros to the complex pole in question). The angle of entry into a complex zero may be found from the same rule and then the sign changed to produce the final result.*

It has been seen that when the locus intersects the imaginary axis, we have $s = 0 + j\omega$ or simply $s = j\omega$, leading to the following rule for determining the point of intersection:

Rule #6. *The point where the locus crosses the imaginary axis may be obtained by substituting* $s = j\omega$ *into the characteristic equation and solving for* ω.

The characteristic equation for this problem is

$$s^3 + 4s^2 + s(K + 8) + 5K = 0 \tag{10.6}$$

Substituting $s = j\omega$ yields

$$(5K - 4\omega^2) + j[(K + 8)\omega - \omega^3] = 0 \tag{10.7}$$

Equating the real and imaginary parts to zero results in two equations in ω and K that solve to give $\omega = 6.32$ and $K = 32$. The locus intercepts the imaginary axis, therefore, at ± 6.32, and the final locus is shown in Fig. 10.4. Note that whatever the order of the characteristic equation, the above procedure will yield two equations in ω and K. Although not used in this particular problem, the rule relating to finding the point(s) of emergence (or entry) from the real axis may be defined as follows:

Rule #7. *The point at which the locus leaves a real-axis segment is found by determining a local maximum value of K, while the point at which the locus enters a real-axis segment is found by determining a local minimum value of K.*

The stationary values may be found by calculating K graphically using the magnitude equation or by expressing K as a function of s differentiating the resulting equation with respect to s, equating to zero, and then solving for s. The latter method is simpler to use only when the characteristic equation is of order less than 4.

Again, although not needed for this particular plot, the directions in which coincident closed-loop poles depart from (or enter) the real axis may be found from:

Rule #8. *The angle between the directions of emergence (or entry) of q coincident poles (or zeros) on the real axis is given by*

$$\psi = \frac{360}{q}$$

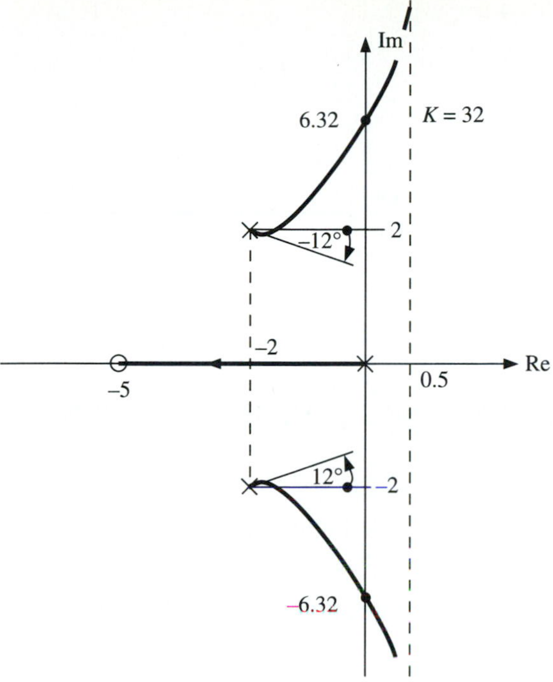

Fig. 10.4 Imaginary-axis crossing point

The directions are symmetrical about the real axis, and if q is odd, the negative real axis is one of the directions.

In the majority of cases this rule determines that two closed-loop poles ($q = 2$) that meet on the real axis and then break away do so in directions inclined at an angle of

$$\psi = \frac{360}{2} = 180° \tag{10.8}$$

to each other, as shown in Fig. 10.5. If the root locus of the system given by the open-loop transfer function

$$GH(s) = \frac{K}{(s + 1)^3} \tag{10.9}$$

is plotted, the locus starts at the three coincident ($q = 3$) open-loop poles at $s = -1$ and immediately breaks away from the real axis in directions inclined at

$$\psi = \frac{360}{3} = 120° \tag{10.10}$$

to each other. Since q is odd, the negative real axis is one of the directions, and the closed-loop poles begin to move in the directions shown in Fig. 10.6.

This completes the rules needed to draw those parts of the locus satisfying the argument equation, and attention may now be focused on the calculation of K for

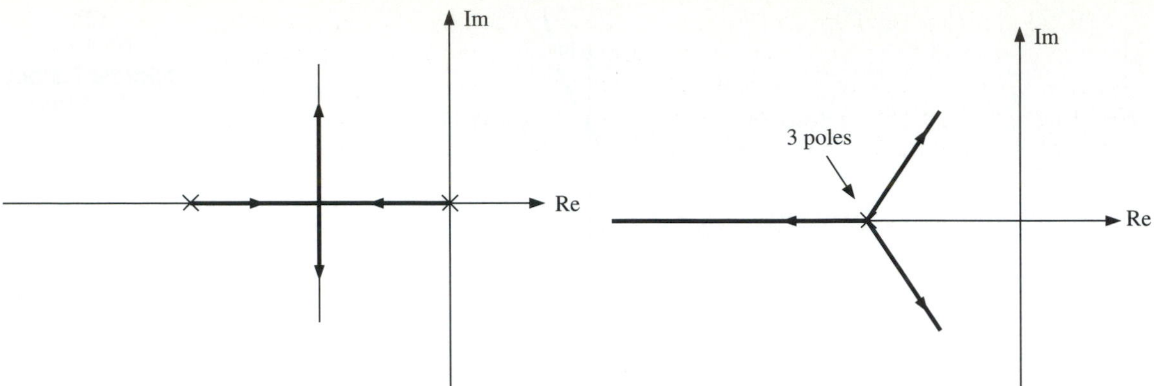

Fig. 10.5 Two poles breaking away from the real axis

Fig. 10.6 Three poles breaking away from the real axis

selected locations on the locus. Figure 10.7 shows the location of the required closed-loop pole at $s = -1$. The gain may be found from:

Rule #9. *The gain at a selected point s_t on the locus is obtained by joining the point to all open-loop poles and zeros and measuring the length of each line $|s_t + p_i|$, $|s_t + z_i|$. The gain is given by*

$$K = \frac{\prod |s_t + p_i|}{\prod |s_t + z_i|} \tag{10.11}$$

To locate a point with specified gain, use trial and error, remembering that moving s_t toward the poles reduces the gain; moving away from the poles increases the gain.

Using this rule, the individual moduli are marked in Fig. 10.7 as a, b, c, and d. The gain for a root at $s = -1$ is

$$K = \frac{a \times b \times c}{d} = 1.25 \tag{10.12}$$

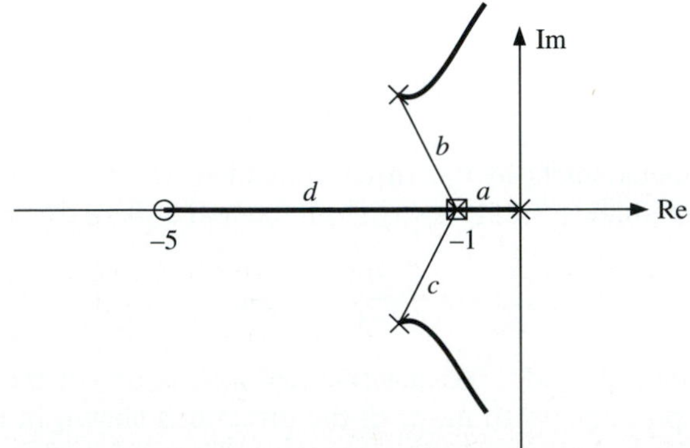

Fig. 10.7 Gain for specified root

The location of the remaining two roots will have to be found by trial and error by placing a test point on the branch from the complex pole and calculating the gain. The trial point is then moved until the calculated gain is equal to 1.25. If, however, there are at least two more open-loop poles than open-loop zeros, the location of the remaining closed-loop poles may be determined by using the following rule:

Rule #10. *If there are at least two more open-loop poles than open-loop zeros, the sum of the real parts of the closed-loop poles is constant, independent of K, and equal to the sum of the real parts of the open-loop poles.*[1]

If the two remaining roots are located at $-\sigma' \pm \omega'$, we can write from Rule #10 that

$$-1 - \sigma' - \sigma' = 0 - 2 - 2 \tag{10.13}$$

giving the result

$$\sigma' = 1.5 \tag{10.14}$$

These two roots are shown in Fig. 10.8. A computer-generated plot is given in Fig. 10.9 to compare with the estimated plot.

Sample Problem 10.1

Draw the root locus for the system with open-loop transfer function

$$GH(s) = \frac{K(s^2 + 2s + 5)}{s^2(s + 2)(s + 3)}$$

Find all of the closed-loop poles for $K = 1$.

Solution

We will solve this problem using the rules defined in the previous section.

Rule #1. *Draw the complex plane and mark the m open-loop poles and n zeros. The locus starts at a pole for $K = 0$ and finishes at a zero or infinity when $K = \infty$. The number of segments going to infinity is therefore $n - m$.*

[1]*Proof:* Writing the open-loop transfer function as

$$GH(s) = \frac{K(s^{n-2} + b_{n-3}s^{n-3} + \cdots + b_0)}{s^n + a_{n-1}s^{n-1} + \cdots + a_0}$$

the characteristic equation is

$$s^n + a_{n-1}s^{n-1} + (a_{n-2} + K)s^{n-2} + \cdots + (a_0 + b_0K) = 0$$

It is known from linear algebra that the sum of the roots of a polynomial is equal to the term a_{n-1}, which in this case is the sum of the closed-loop poles. Further, the denominator of the open-loop transfer function yields the open-loop poles as the solution of

$$s^n + a_{n-1}s^{n-1} + \cdots + a_0 = 0$$

Again, the sum of the roots of this equation are equal to a_{n-1}. It is concluded, therefore, that the sum of the closed-loop poles is equal to the sum of the open-loop poles, which is a constant and independent of the gain K.

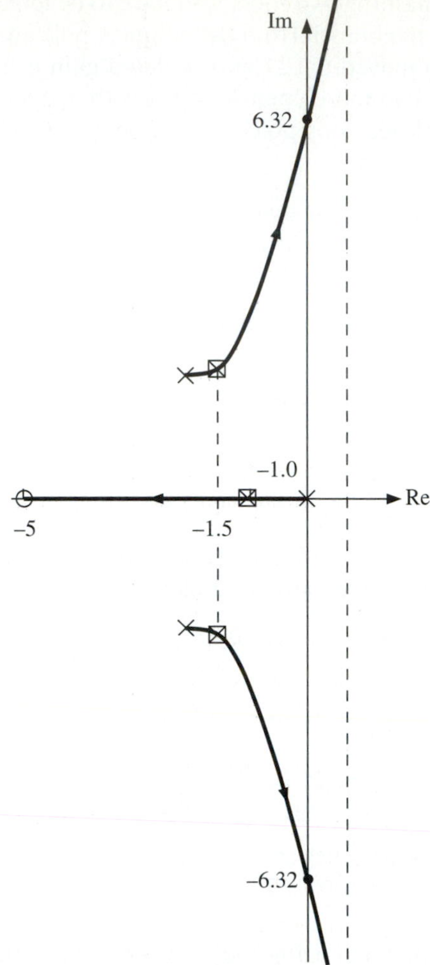

Fig. 10.8 Complete root locus with roots for $K = 5$

Here we have two complex zeros located at the roots of $s^2 + 2s + 5$, which are $s = -1 \pm 2j$. There are four open-loop poles, two at the origin and two at $s = -2, -3$.

Rule #2. *Segments of the real axis to the left of an odd number of poles or zeros are segments of the root locus, remembering that complex poles or zeros have no effect.*

The real-axis segment forming part of the locus is only between the poles at $s = -2$ and $s = -3$. The locus generated up to this point is shown in Fig. SP10.1.1.

Rule #3. *The loci are symmetrical about the real axis since complex roots are always in conjugate pairs. The angle between adjacent asymptotes is 360°/$(n - m)$ and to obey the symmetry rule, the negative real axis is one asymptote when $n - m$ is odd.*

There will be two asymptotes inclined at 180° to each other.

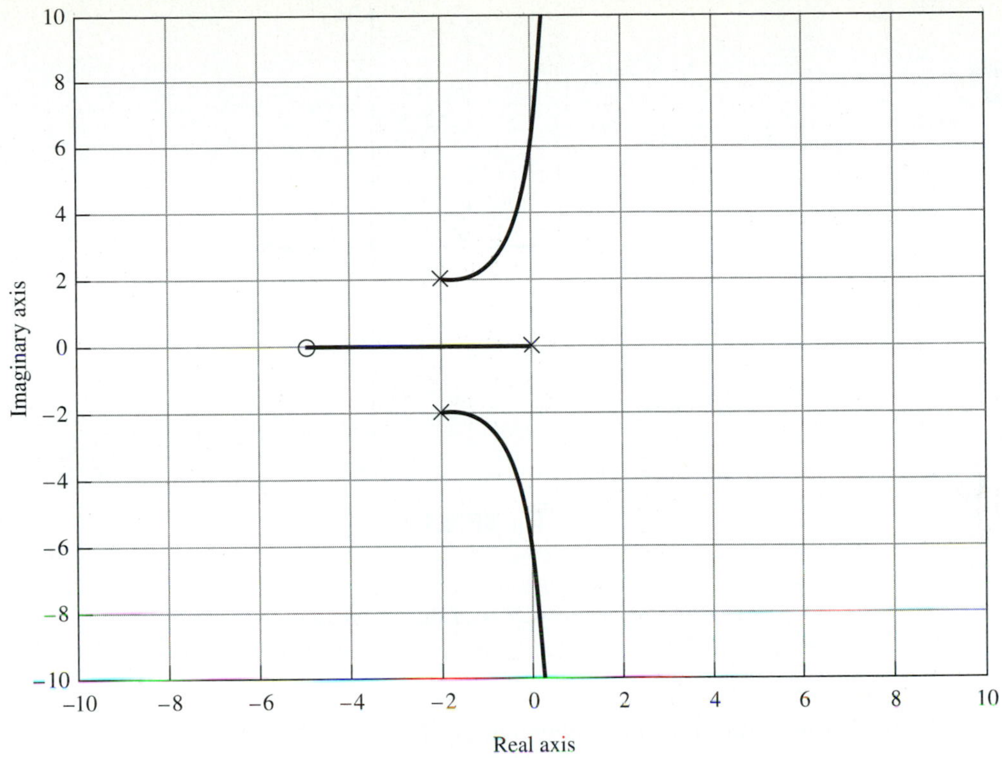

Fig. 10.9 Computer-generated root locus

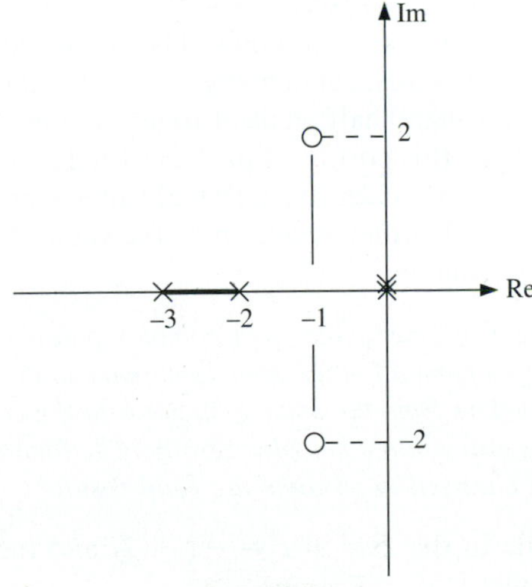

Fig. SP10.1.1

Rule #4. *The asymptotes intersect the real axis at σ_a, where*

$$\sigma_a = \frac{\sum p_i - \sum z_i}{n - m}$$

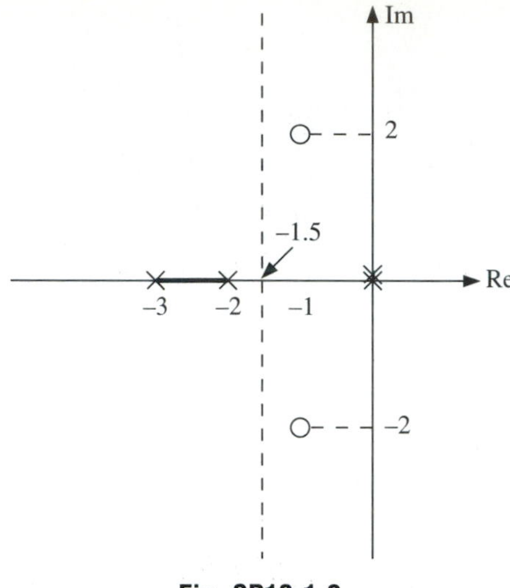

Fig. SP10.1.2

where Σp_i is the sum of the real part of the open-loop poles (including complex roots) and Σz_i is the sum of the real part of the open-loop zeros (also including complex roots).

Substituting into the above equation yields

$$\sigma_a = -1.5$$

Figure SP10.1.2 shows the location and direction of the asymptotes. At this stage it seems as though the two poles at $s = -2, -3$ move toward each other, meet, and branch out along the asymptotes. The double pole at the origin will move toward the complex zeros, but in doing so, it is not known if the locus will be partially on the right-hand half plane or totally in the left-hand half plane. Two factors will help plot this portion of the locus further: (1) the angle of entry into the complex zeros and (2) identifying a real value of ω corresponding to an imaginary axis crossing. In order to determine the angle of entry into the complex zeros, Rule #5 is used.

Rule #5. *The angle of emergence from complex poles is given by* 180° − Σ *(angles of the vectors from all other open-loop poles to the pole in question)* + Σ *(angles of the vectors from the open-loop zeros to the complex pole in question). The angle of entry into a complex zero may be found from the same rule and then the sign changed to produce the final result.*

Applying this rule to the zero at $s = -1 + 2j$ and measuring the angles from Fig. SP10.1.3 give

$$\beta_1 = -180 + (116 + 116 + 64 + 45) - (90) = 71°$$

Since, for this example, the angle of entry points almost in the opposite direction to the origin, the locus apparently makes a large sweep around the complex zero and probably crosses the imaginary axis.

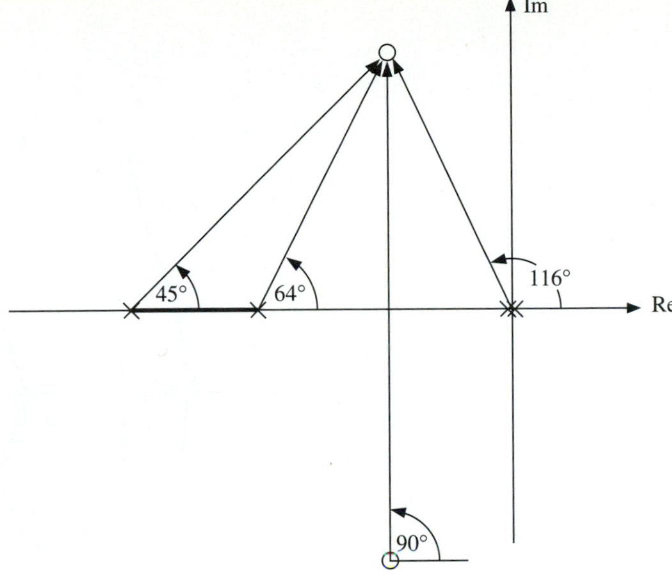

Fig. SP10.1.3

Rule #6. *The point where the locus crosses the imaginary axis may be obtained by substituting $s = j\omega$ into the characteristic equation and solving for ω.*

The characteristic equation for this problem becomes

$$s^4 + 5s^3 + s^2(K + 6) + 2Ks + 5K = 0$$

Substituting $s = j\omega$ yields the following equations:

$$2K - 5\omega^2 = 0$$

and

$$\omega^4 - (6 + K)\omega^2 + 5K = 0$$

which may be solved simultaneously to yield $\omega^2 = 4.33$ rad/s and $K = 10.83$. The locus crosses the imaginary axis at $\omega = 2.08$, at which point $K = 10.83$. Incidentally, another solution to the above equations is $\omega = 0$, corresponding to the two poles at the origin when $K = 0$.

Rule #7. *The point at which the locus leaves a real-axis segment is found by determining a local maximum value of K, while the point at which the locus enters a real-axis segment is found by determining a local minimum value of K.*

Clearly, there is a real-axis breakaway point between the open-loop poles at $s = -2$ and $s = -3$. Since the system characteristic equation is fourth order and there are two open-loop zeros, any stationary values of K will be determined graphically. Plotting K over the real-axis segment produces Fig. SP10.1.4, giving the approximate location of the break point as $s = -2.6$. This completes the process to determine the root locus, and the final plot is shown in Fig. SP10.1.5.

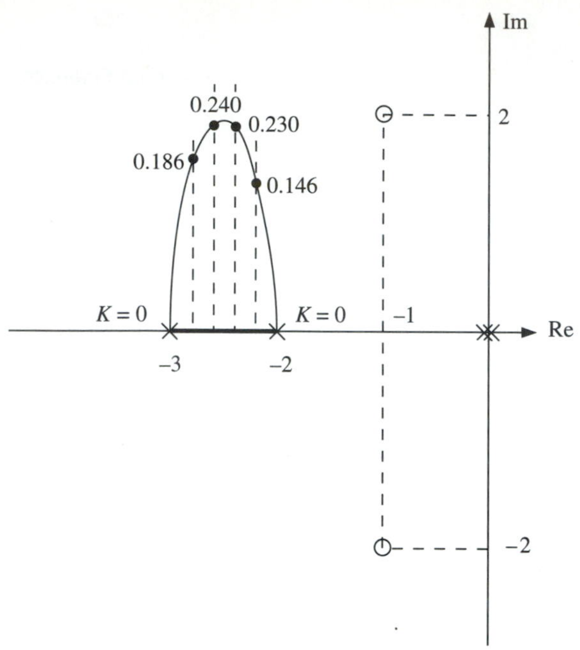

Fig. SP10.1.4

Fig. SP10.1.5

Rule #9. *The gain at a selected point s_t on the locus is obtained by joining the point to all open-loop poles and zeros and measuring the length of each line $|s_t + p_i|$, $|s_t + z_i|$. The gain is given by*

$$K = \frac{\prod |s_t + p_i|}{\prod |s_t + z_i|}$$

To locate a point with specified gain, use trial and error, remembering that moving s_t toward the poles reduces the gain; moving away from the poles increases the gain.

Knowing that the gain at the imaginary-axis crossing point is greater than the gain we are looking for limits the segment of the locus where s_t is placed. A few trials locates the two roots, as shown in Fig. SP10.1.6.

Rule #10. *The sum of the real parts of the closed-loop poles is constant, independent of K, and equal to the sum of the real parts of the open-loop poles.*

This rule is of little use in this problem because the real part of the roots does not vary much with K. In order to determine the location of the other two roots on the left-hand segment, Rule #9 is used, and after a few trials the location is identified, as shown in Fig. SP10.1.6. For comparative purposes, a computer-generated root locus plot of this problem is shown in Fig. SP10.1.7.

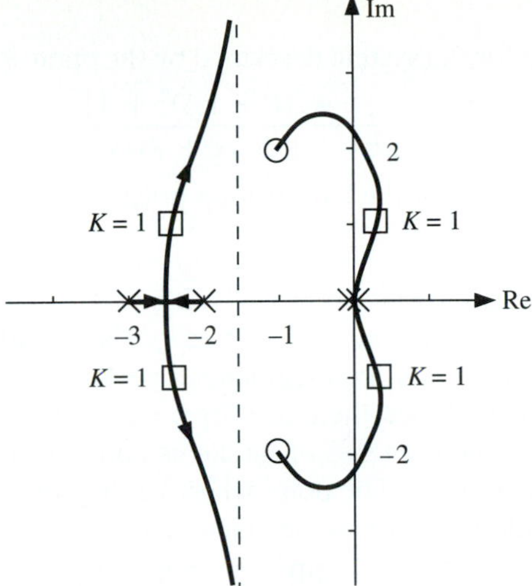

Fig. SP10.1.6

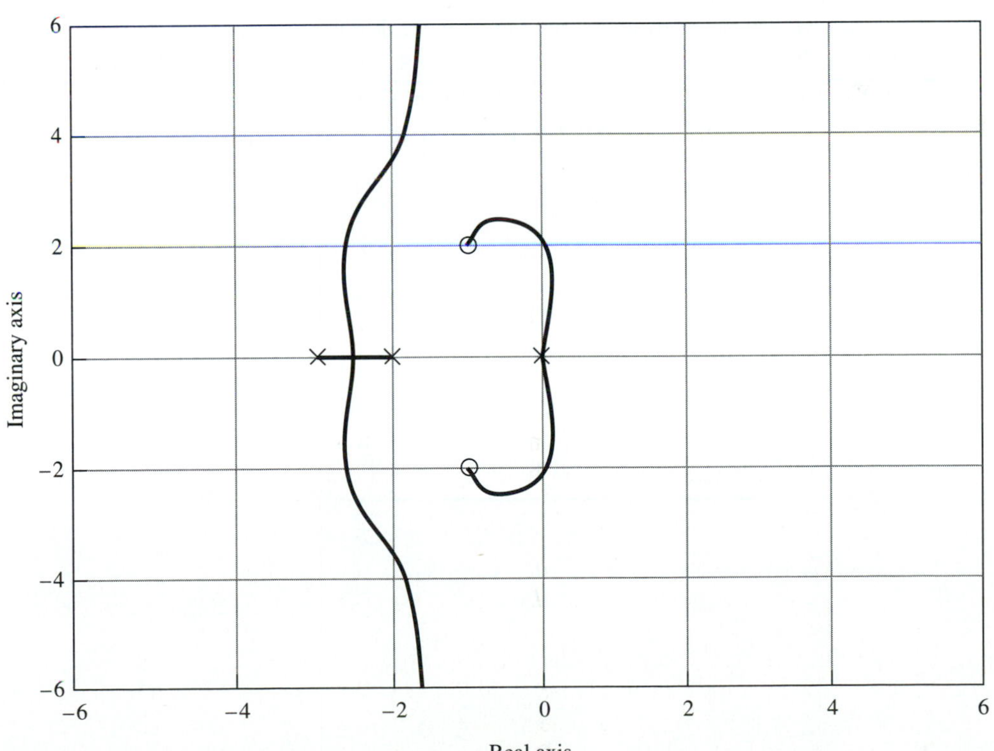

Fig. SP10.1.7

SAMPLE PROBLEM 10.2

Draw the root locus for the system described by the open-loop transfer function

$$GH(s) = \frac{K[(s + 1.5)^2 + 1]}{s^2(s + 0.5)(s + 8)(s + 10)}$$

and determine the location of the closed-loop poles when the gain $K = 50$.

Solution

For this system there are five poles at $s = -0.5, -8, -10$ and two at the origin. There are two complex zeros at $s = -1.5 \pm 1j$. Using Rules #1 and #2, it may be seen that the negative real-axis segments for $-8 < s < -0.5$ and $s < -10$ from part of the locus. Since there are three more poles than zeros, Rule #3 indicates there are three branches of the locus going to infinity and the negative real axis is one of them. The point where the asymptotes intersect the real axis is given by Rule #4 and may be calculated from

$$\sigma_a = \frac{(-0.5 - 8 - 10) - (-1.5 - 1.5)}{5 - 2} = -5.167$$

The locus up to this point is shown in Fig. SP10.2.1. It is reasonable to suppose that the two poles at the origin move out toward the asymptotes, the pole at $s = -10$ moves along the negative real axis to infinity, and the two roots from the poles at $s = -0.5$ and $s = -8$ come together and then branch into the complex zeros.

In order to determine the behavior of the locus from the double pole at the origin, it is necessary to find points where the locus crosses the imaginary axis, if there are any. Rule #6 applied to the characteristic equation

$$s^5 + 18.5s^4 + 89s^3 + (40 + K)s^2 + 3Ks + 3.25K = 0$$

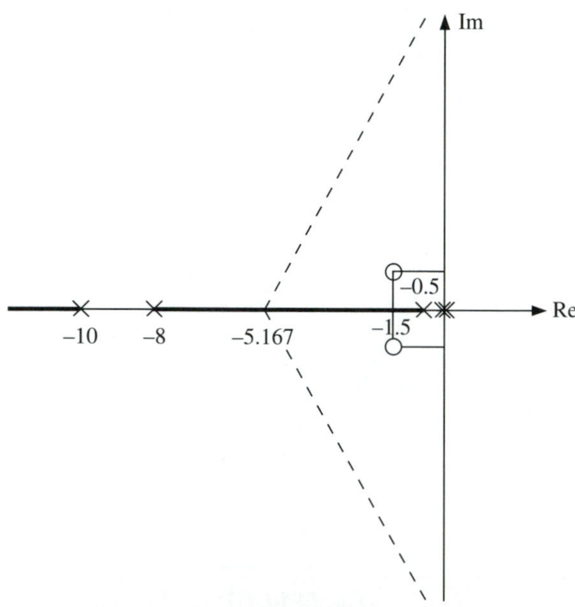

Fig. SP10.2.1

yields

$$[18.5\omega^4 - (40 + K)\omega^2 + 3.25K] + j[\omega^5 - 89\omega^3 + 3K\omega] = 0$$

Equating the real and imaginary parts of this equation and substituting $x = \omega^2$ give

$$x^2 - 89x + 3K = 0$$

$$18.5x^2 - (40 + K)x + 3.25K = 0$$

Solving these equations for x and consequently ω gives $\omega_1 = \pm 2.19$ at $K = 135$ and $\omega_2 = 5.65$ where $K = 607$. Since there are two imaginary-axis crossing points, the locus must move into the right-hand half plane upon leaving the double pole at the origin, cross into the left-hand half plane at $\omega = 2.19$, and recross the imaginary axis at $\omega = 5.65$ before moving toward the asymptotes. This section of the locus is shown in Fig. SP10.2.2. The angle at which the locus enters the complex zeros may be obtained from Rule #5,

$$\beta_k = -180 + (146 + 146 + 134 + 9 + 7) - 90 = 172°$$

where the angles from all other poles and zeros have been measured from Fig. SP10.2.3. The remaining sections of the root locus, particularly in the region of the real axis, are best determined by using Rule #7. Due to the complexity of the open-loop transfer function, the gain for the real-axis segment $-0.5 > s > -8.0$ is calculated graphically using the magnitude equation. The plot of K over this real-axis section indicates a maximum value at about $s = -2.5$, corresponding to the real-axis break point. The final locus is sketched in Fig. SP10.2.4 and may be compared with the exact plot shown in Fig. SP10.2.5. To determine the locations of all closed-loop poles for $K = 50$, it should be noted that the first imaginary-axis crossing point occurs when

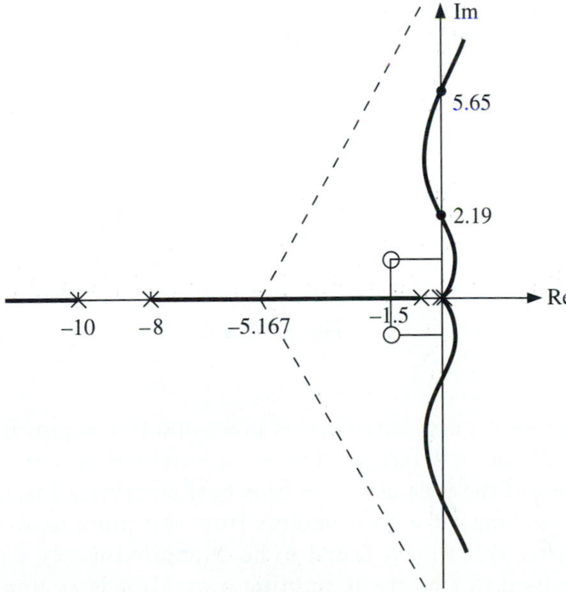

Fig. SP10.2.2

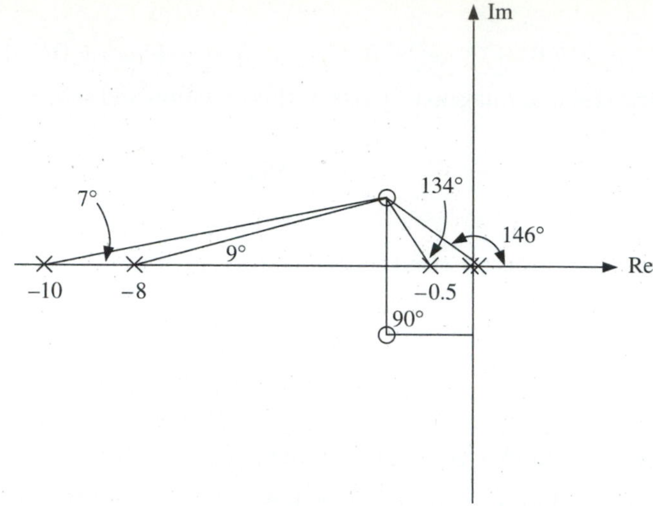

Fig. SP10.2.3

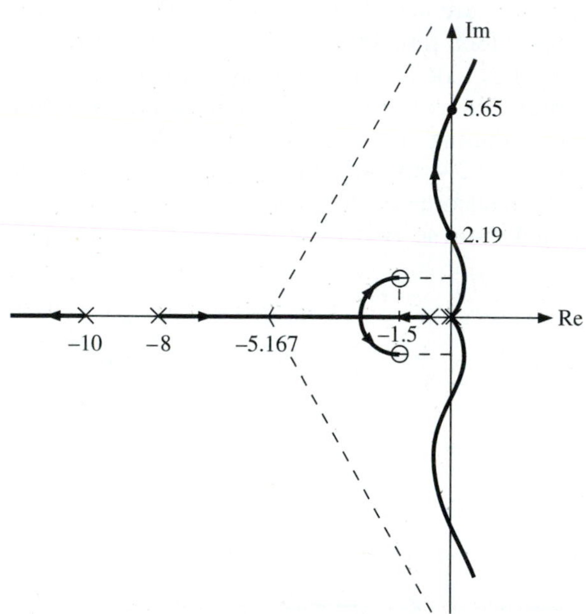

Fig. SP10.2.4

$K = 135$. A root occurs between this point and the origin. By using trial and error, this location is determined to be approximately $s = -0.2 \pm 1.5j$. The root to the left of the pole at $s = -10$ is best determined next. Again, using a trial point s_t and measuring the vectors from this point to all of the open-loop poles and zeros, this root is found to be at approximately $s = -11$. Rule #10 may now be used to find the remaining roots. If it is assumed that these two roots lie off the real axis at $s = -\sigma' + j\omega'$, then from Rule #10 and the roots

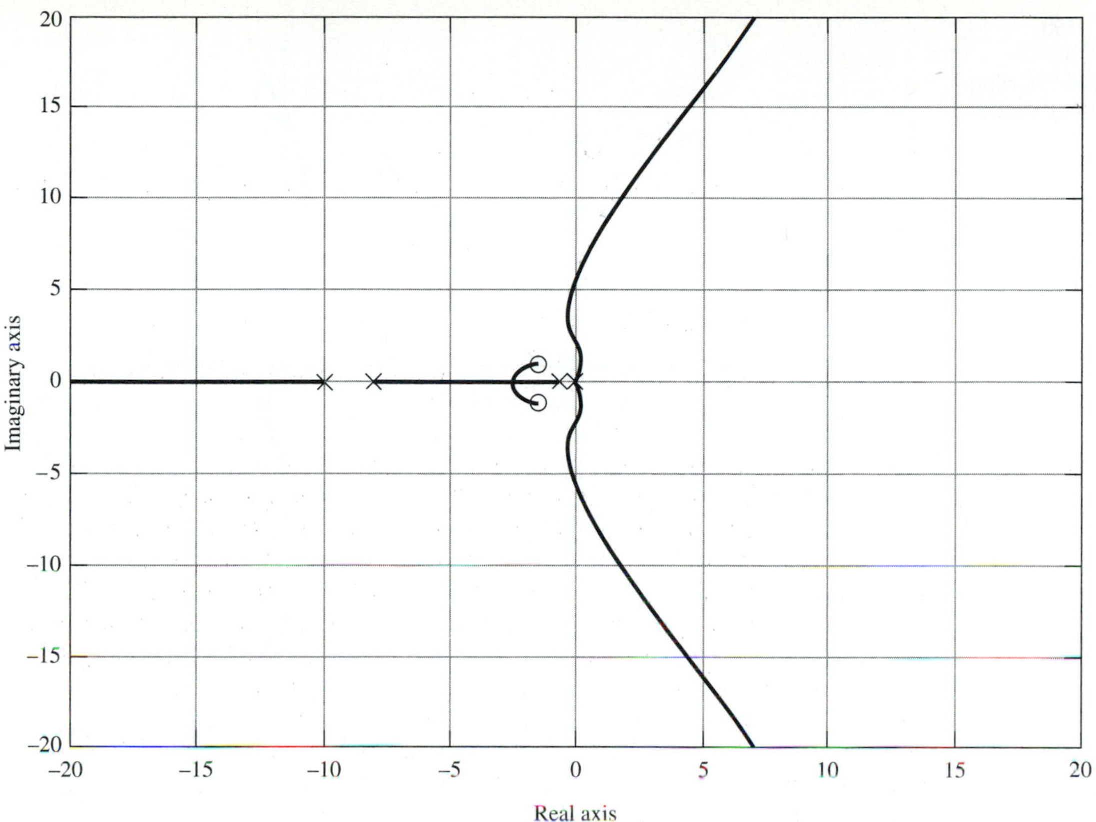

Fig. SP10.2.5

found so far

$$+0.2 + 0.2 - 11 - 2\sigma' = -0.5 - 10 - 8$$

which gives

$$\sigma' = +3.7$$

Since this cannot be the case, it is concluded that the two remaining roots do not lie off the real axis for this value of K. Further trial-and-error tests yield the roots at about $s = -1.1, -6.5$, as shown in Fig. SP10.2.6. It may be appreciated from this example that determining the roots from the locus for a given value of K may be somewhat tedious, particularly if there are several roots to find.

Sample Problem 10.3

Draw the root locus for the system shown in Fig. SP10.3.1 when (a) $\gamma = 0.5$, (b) $\gamma = 4$, and (c) $\gamma = 2$.

Solution

Case (a): $\gamma = 0.5$. The solution is started by determining the open-loop transfer function and plotting the open-loop poles and zeros on the complex

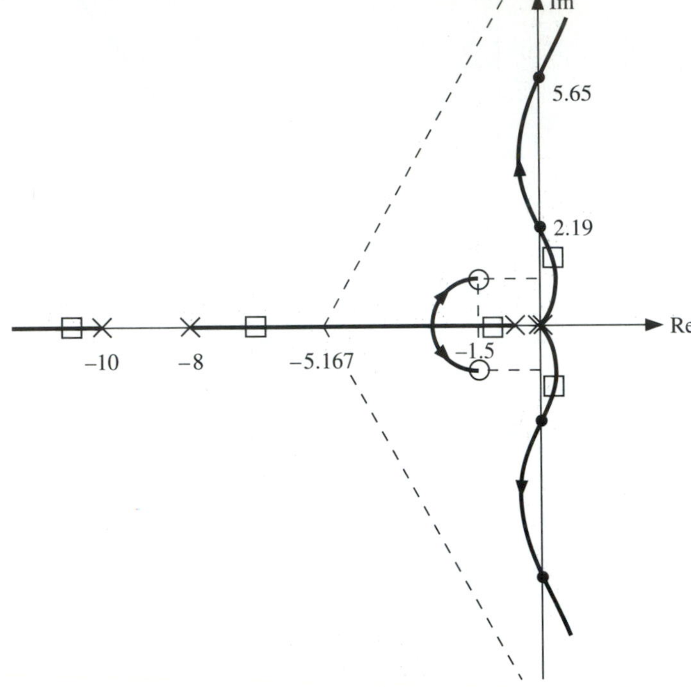

Fig. SP10.2.6

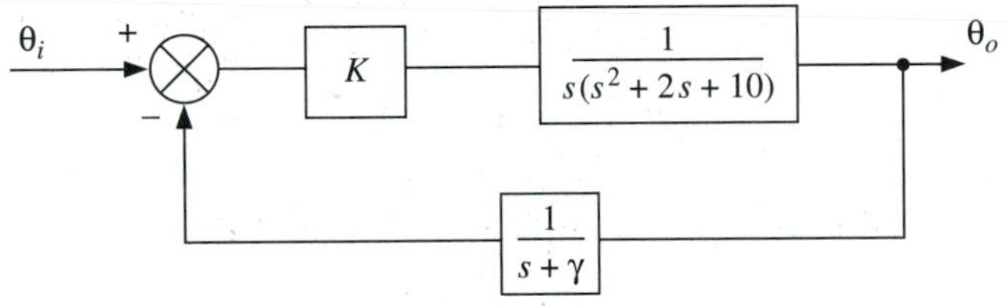

Fig. SP10.3.1

plane. For the system shown in Fig. SP10.3.1, the open-loop transfer function is given by

$$GH(s) = \frac{1}{s(s+0.5)(s^2+2s+10)}$$

There are no zeros, and the poles are located at $s = 0, 1, -1 \pm 3j$. From Rule #2 the negative real axis between the pole at $s = 0$ and the pole at $s = -0.5$ is part of the locus, resulting in Fig. SP10.3.2. Rule #3 indicates that there are four branches of the locus going to infinity and the asymptotes are inclined at 90° to each other. The point at which the asymptotes intersect the real axis is given by

$$\sigma_a = \tfrac{1}{4}(-0 - 1 - 1 - 0.5) = -0.625$$

These asymptotes are shown in Fig. SP10.3.3. Since there are asymptotes in the right-hand half plane, it appears that the locus crosses the imaginary axis at

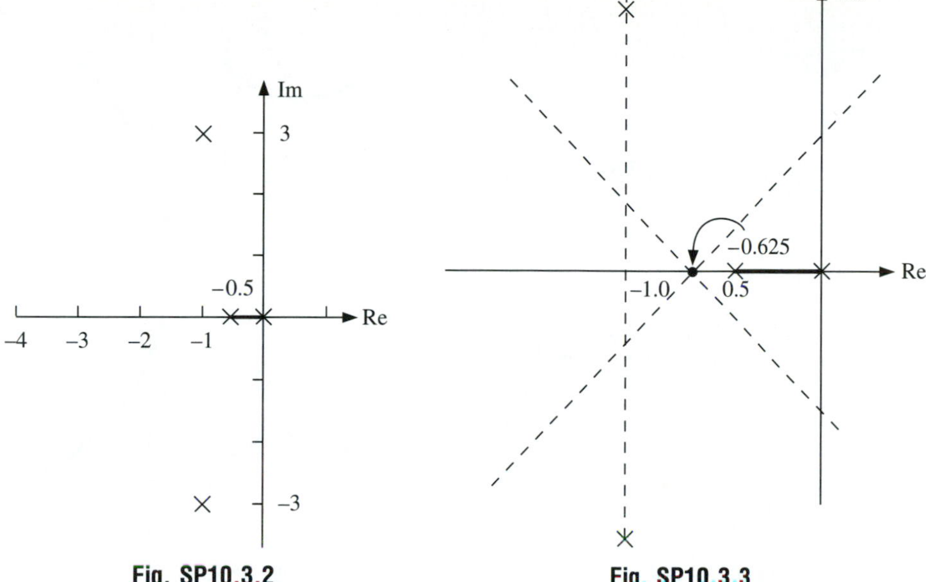

Fig. SP10.3.2

Fig. SP10.3.3

some point, although it is not clear whether the roots from the complex poles cross the axis or whether the two real-axis poles meet, branch, and then cross the axis. As before, the imaginary-axis crossing point is obtained by substituting $s = j\omega$ into the characteristic equation. Here the characteristic equation is seen to be

$$s^4 + 2.25s^3 + 11s^2 + 5s + K = 0$$

After substituting, the following equation results:

$$(\omega^4 - 11\omega^2 + K) + j\omega(5 - 2.5\omega^2) = 0$$

Equating the imaginary part to zero yields $\omega = 0, 1.414$, with corresponding values of $K = 0$ and $K = 18$, respectively. The first solution corresponds to the pole at $s = 0$ when $K = 0$, while the other solution is the imaginary-axis crossing for $K > 0$. Since the crossing point is closer to the pole at the origin than the complex pole, it is reasonable to suppose that the poles at $s = 0$ and $s = -0.5$ meet and then branch out into the right-hand half plane, leaving the complex poles to move toward the asymptotes in the left-hand half plane.

The angle of emergence from the complex pole located at $s = -1 + 3j$ may be evaluated by using Rule #5 and Fig. SP10.3.4 as follows:

$$\alpha_k = 180 - (108 + 99 + 90) = -117°$$

This confirms that the roots from the complex poles move toward the left-hand asymptotes. The point at which the locus leaves the real axis is given by Rule #7, which yields the result $s = -0.25$ for the break point. The final plot may be sketched as in Fig. SP10.3.5 and compared with the accurate plot shown in Fig. SP10.3.6.

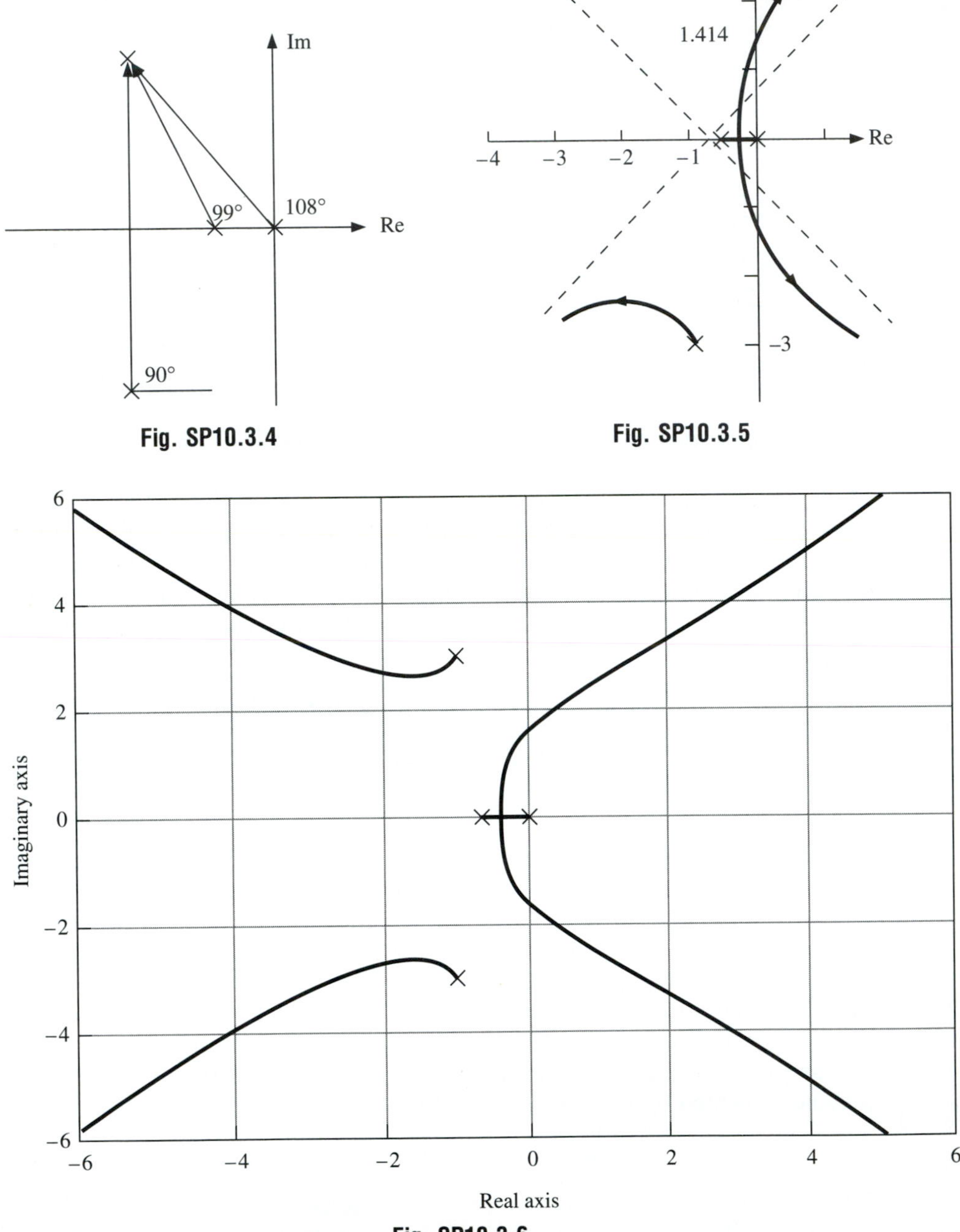

Fig. SP10.3.4

Fig. SP10.3.5

Fig. SP10.3.6

CASE (B): $\gamma = 4$. Moving the real-axis pole from $s = -0.5$ to $s = -4$ does not change the real-axis segments, the number of asymptotes, or their direction but does change the point of interception of the asymptotes with the real axis.

The new value of σ_a is given by

$$\sigma_a = \tfrac{1}{4}(-1 - 1 - 4) = -1.5$$

The point at which the locus crosses the imaginary axis will also change. The new characteristic equation becomes

$$s^4 + 6s^3 + 18s^2 + 40s + K = 0$$

Proceeding as before yields

$$(\omega^4 - 18\omega^2 + K) + j(40\omega - 6\omega^3) = 0$$

which gives a crossing value of $\omega = 2.58$, where $K = 75.6$. Since this puts the crossing point closer to the complex open-loop poles than the real poles, it is reasonable to suppose that the roots from the complex poles move to the right, while those from the two real open-loop poles come together and branch in the direction of the left-hand plane asymptotes. This may be confirmed by calculating the complex-pole departure angle α_k using Fig. SP10.3.7. This gives

$$\alpha_k = 180 - (90 + 108 + 45) = -63°$$

Calculating the point at which the locus leaves the real axis is achieved using Rule #7, applied for the real-axis segment $0 > s > -4$, yielding the break point at $s = -2.5$. This completes the calculations, and the plot may be sketched as in Fig. SP10.3.8 and compared with the accurate plot in Fig. SP10.3.9.

CASE (C): $\gamma = 2$. Again the asymptote directions and real-axis segments remain the same, but the asymptotes now intercept the real axis at

$$\sigma_a = \tfrac{1}{4}(-1 - 1 - 2) = -1$$

The new characteristic equation

$$s^4 + 4s^3 + 14s^2 + 20s + K = 0$$

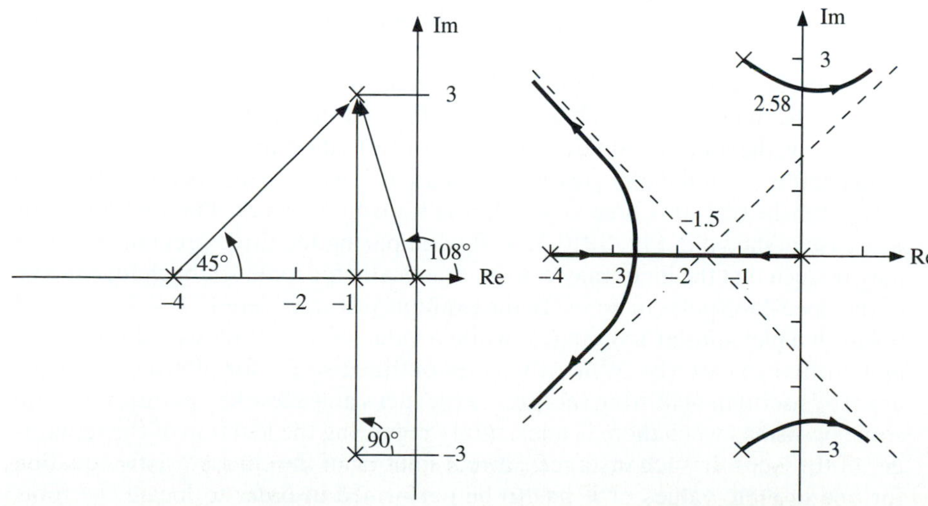

Fig. SP10.3.7 **Fig. SP10.3.8**

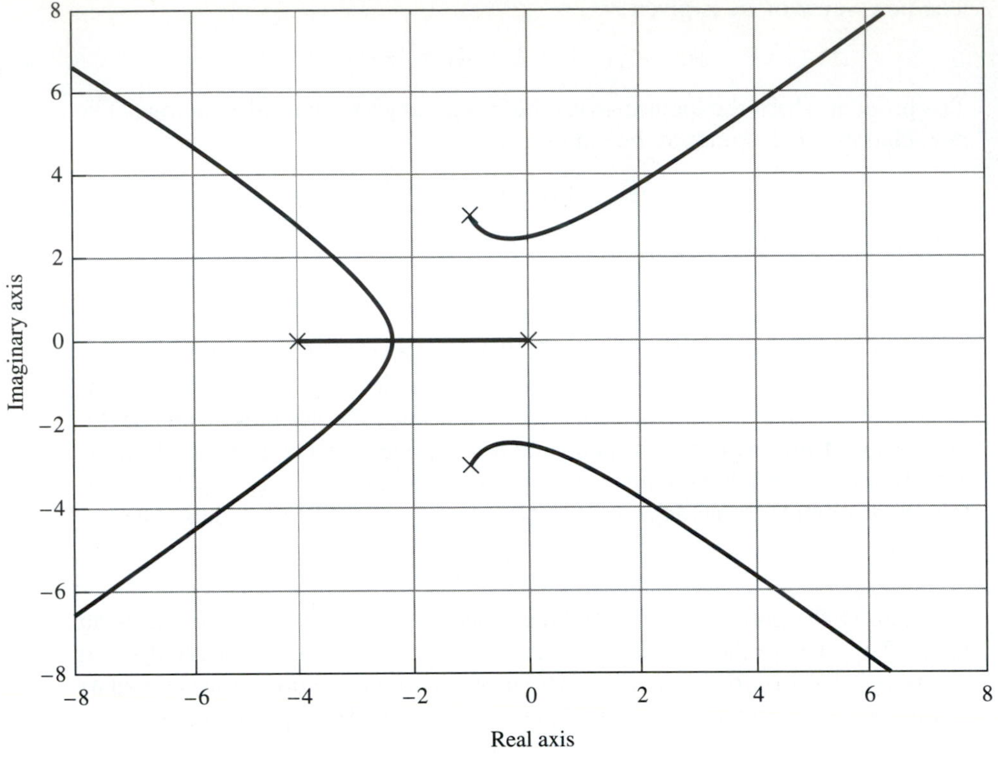

Fig. SP10.3.9

yields the imaginary-axis crossing point of $\omega = 2.23$. The angle of departure from the complex pole at $s = -1 + 3j$ is given by

$$\alpha_k = 180 - (108 + 72 + 90) = -90^\circ$$

The plot so far is shown in Fig. SP10.3.10, and it is not immediately obvious what the rest of the locus looks like. However, notice how, in case (a), the complex poles move toward the left-hand pair of asymptotes, while in case (b) they move to the right-hand pair. It follows that for some location of the variable pole at $s = -\gamma$, the locus must flip from one configuration to the other. The angle of departure from the complex poles in case (c) suggests this is the critical location for the real-axis pole at which this transition occurs. The root locus for this case is shown in Fig. SP10.3.11. By comparing the three previous cases, it may be seen that the locus may change drastically for some slight change in one of the open-loop poles or zeros. In the example just considered, $\gamma = 1.95$ would result in a plot similar to case (a), while a value of $\gamma = 2.05$ would produce a plot similar to case (b). While the rules outlined so far for plotting the locus are very useful in obtaining sections of the locus and a few key points, there are some occasions when there is uncertainty regarding the location of the remainder of the locus. In such instances, direct solution of the characteristic equation for one or more values of K has to be performed in order to locate the roots

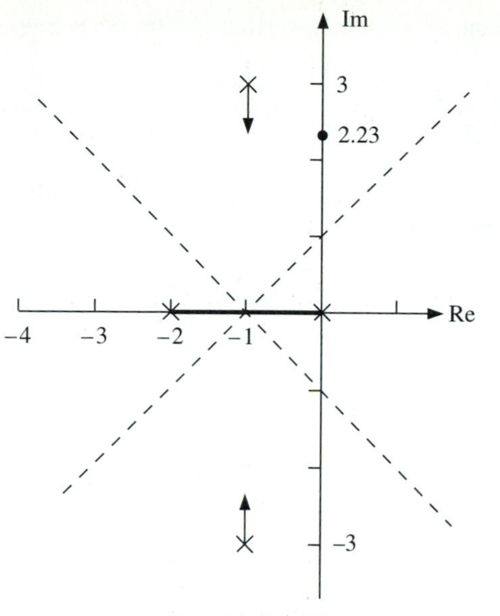

Fig. SP10.3.10

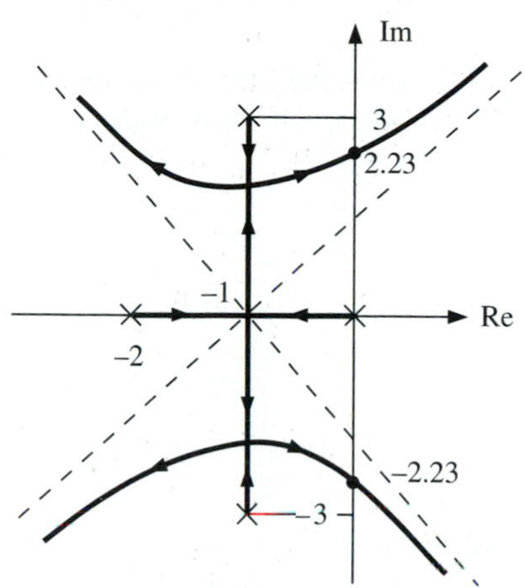

Fig. SP10.3.11

accurate enough to draw the locus. Fortunately, not many points are required, and some simple commercial computer tools, such as MATLAB, may be used to advantage.

◆ PROBLEMS

10.1 Figure P10.1 shows a pole-zero map of a feedback control system, where the location of one pole is denoted by the variable γ. Determine the angle of emergence of the locus from the complex poles for $\gamma = -1, -2, -3$, respectively.

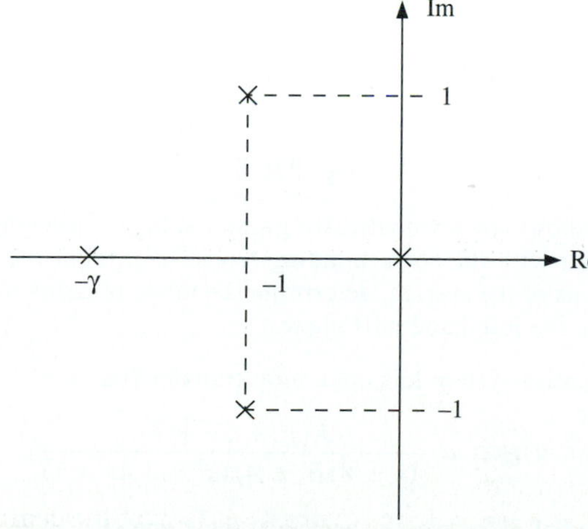

Fig. P10.1

10.2 Plot the root locus for the feedback control system specified by the open-loop transfer function

$$GH(s) = \frac{K(s^2 + 4s + 13)}{(s + 0.5)(s^2 + 2s + 2)}$$

10.3 Given the control system shown in Fig. P10.3, sketch the root locus and determine the points at which the locus crosses the imaginary axis.

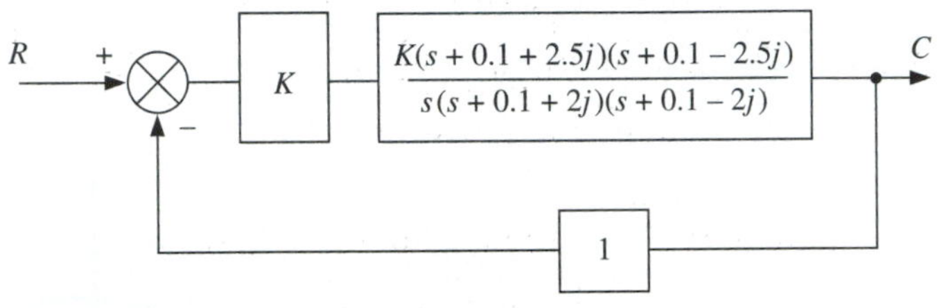

Fig. P10.3

10.4 For the system previously described in Fig. P10.1, what is the range of departure angle from the complex pole at $s = -1 + j$ as γ varies between $\pm\infty$? If there was a zero at $s = \gamma$ instead of a pole, what would the range of departure angles become?

10.5 Shown in Fig. P10.5 is an open-loop pole-zero map of a feedback control system in which it is necessary to ensure that all closed-loop poles have the same real part. Determine the required value of gain and the real part of all the closed-loop poles.

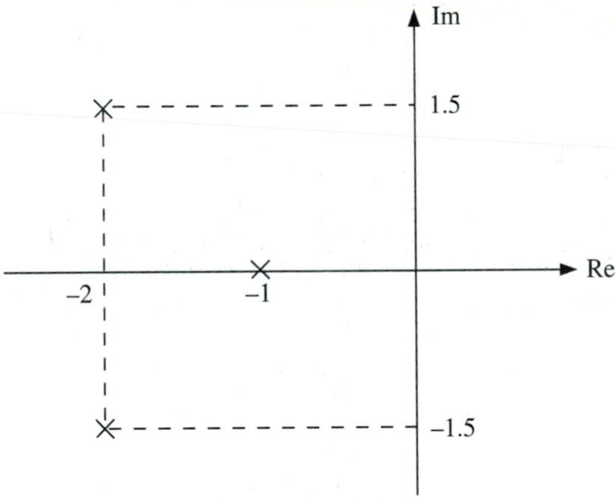

Fig. P10.5

10.6 Figure P10.6 shows an automatically guided vehicle following a buried track in the roadway, together with the corresponding block diagram of the control system. By plotting the root locus of the system, determine the range of gains for which the closed-loop poles are all in the left-hand half plane.

10.7 A feedback control system has open loop transfer function

$$GH(s) = \frac{K(s^2 + 2s + 4)}{s(s + 8)(s + 4)(s^2 + 1.4s + 1)}$$

Determine the range of K for which there is at least one closed-loop pole in the right-hand half plane.

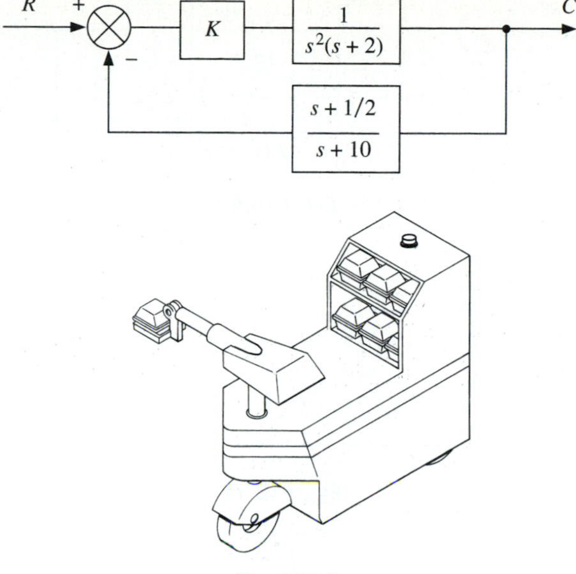

Fig. P10.6

10.8 Shown in Fig. P10.8 is the pole-zero map of a feedback control system in which the location of the real-axis pole is given by the variable ϕ. Determine the angles of emergence and entry into the complex poles and zeros when **(a)** $\phi = 0$ and **(b)** $\phi = -3$. Sketch the resulting root locus for each case.

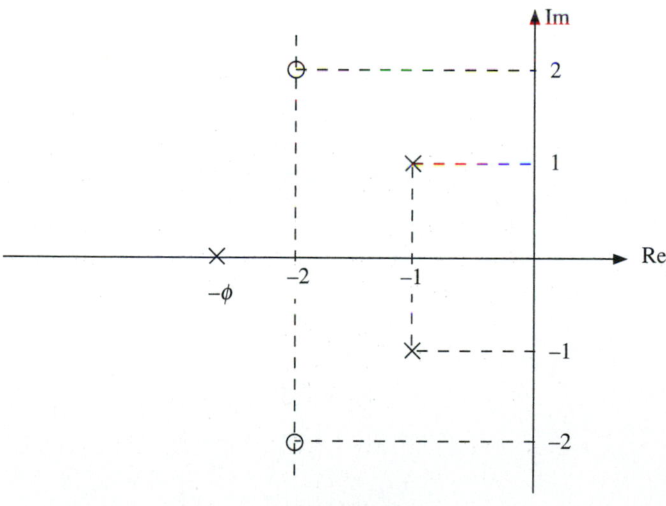

Fig. P10.8

10.9 Plot the root locus for the feedback control system shown in Fig. P10.9, and determine the location of all roots for $K = 20$.

10.10 The feedback control system shown in Fig. P10.10 has two feedback loops. By redrawing the block diagram into a more convenient form, plot the root locus and determine the location of all closed-loop poles for $K = 20$.

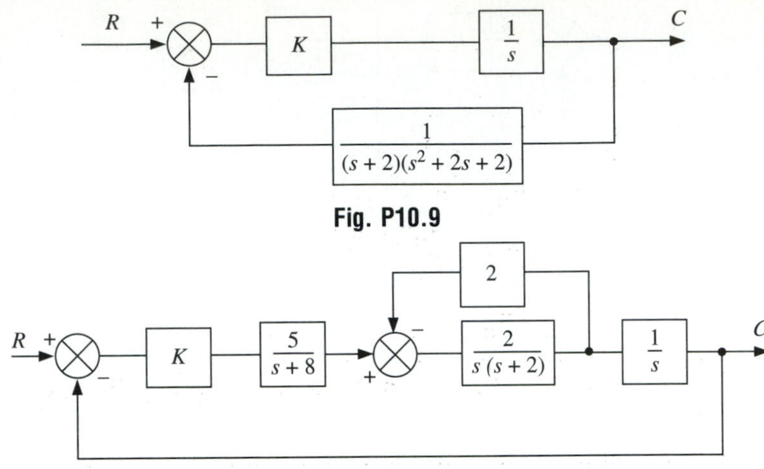

Fig. P10.9

Fig. P10.10

System Design Using the Root Locus

So far, we have used the root locus as a means to determine whether a system is stable or unstable or to determine the appropriate ranges of forward-path gain K for conditionally stable systems. However, if we review the material in Module 5 describing the relationship between the location of the system closed-loop poles in the complex plane and the resulting transient response, it should be clear that the root locus may be used as a powerful design tool in which the gain is set in order to meet given design constraints. Before discussing this in detail, we will look at another application of root locus methods to multiloop systems, particularly those where the block diagram reduction methods outlined in Module 2 fail to provide insight into the system performance.

◆ MULTILOOP SYSTEMS

Up to this point, only single-loop control systems have been addressed; yet many control problems involve several nested loops, such as the one shown in Fig. 11.1.

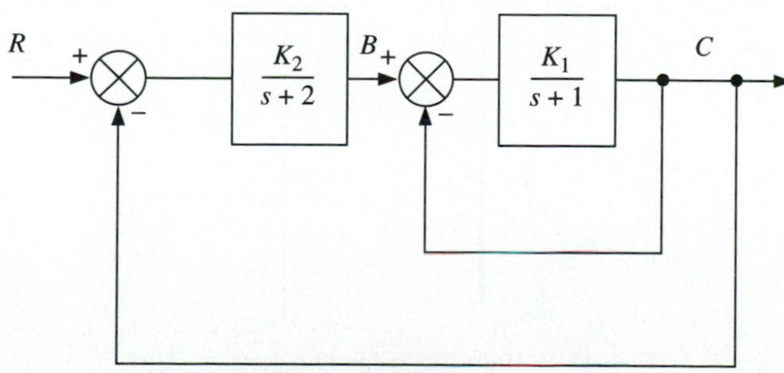

Fig. 11.1 Multiloop system

Block diagram algebra can always be used to reduce the system to one having a single loop and the root locus drawn from that point; however, later examples will show that this method becomes cumbersome for all but the simplest transfer functions. First the example shown in Fig. 11.1 be used to demonstrate an alternative technique for drawing the root locus of a multiloop system, and then the limitations of block diagram reduction will be demonstrated.

Since the transfer functions in the example system are very simple, the first approach will be to determine the open-loop transfer function of the complete system. To begin, the inner loop may be represented by the closed-loop transfer function

$$\frac{C}{B}(s) = \frac{K_1/(s+1)}{1 + K_1/(s+1)} \tag{11.1}$$

$$= \frac{K_1}{(s + 1 + K_1)} \tag{11.2}$$

Assume that for now $K_1 = 2$. The open-loop transfer function of the complete system may be written as

$$GH(s) = \frac{2K_2}{(s+2)(s+3)} \tag{11.3}$$

for which the root locus is as shown in Fig. 11.2. If K_2 is now chosen to be 5, the open-loop transfer function becomes

$$GH(s) = \frac{10}{(s+2)(s+3)} \tag{11.4}$$

which locates the closed-loop poles at $s = -2.5 \pm 3.123j$. Now consider an alternative method for locating the system closed-loop poles for $K_1 = 2$ and $K_2 = 5$.

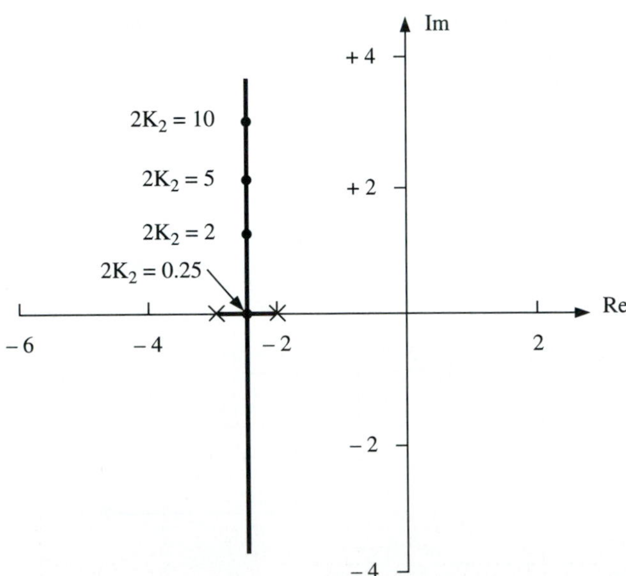

Fig. 11.2 Root locus for complete system

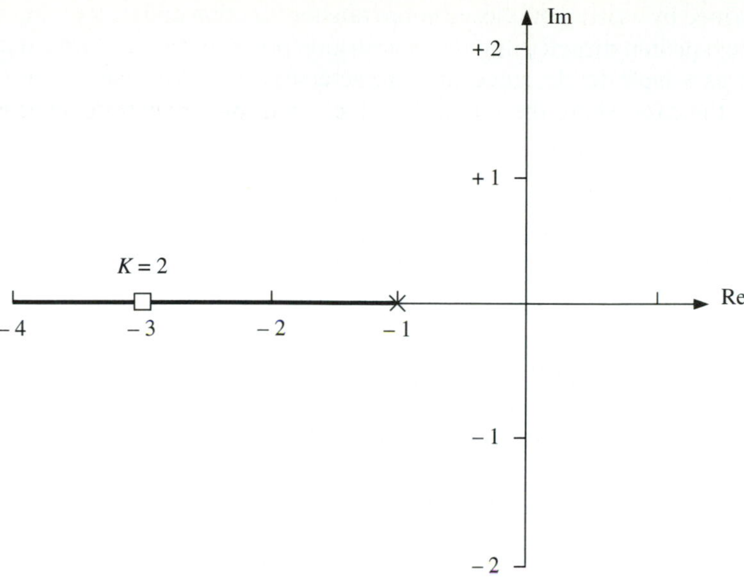

Fig. 11.3 Root locus for inner loop

Suppose the root locus is plotted just for the inner loop. The appropriate open-loop transfer function will be

$$GH(s) = \frac{K_1}{s + 1} \tag{11.5}$$

and the locus will be as in Fig. 11.3 with the closed-loop pole located at $s = -3$ for the required value of $K_1 = 2$. By plotting the root locus of the inner loop, the characteristic equation has been solved for the specified value of gain to determine the closed-loop pole. As a result, the closed-loop transfer function may be written as

$$\frac{C}{B} = \frac{K_1}{s + 3} = \frac{2}{s + 3} \tag{11.6}$$

The complete inner loop may then be replaced by this closed-loop transfer function as shown in Fig. 11.4, the root locus plotted, and closed-loop poles determined for $K_2 = 5$ as previously shown in Fig. 11.2. Obviously the two methods yield the same result; indeed the closed-loop transfer function of the inner loop is exactly

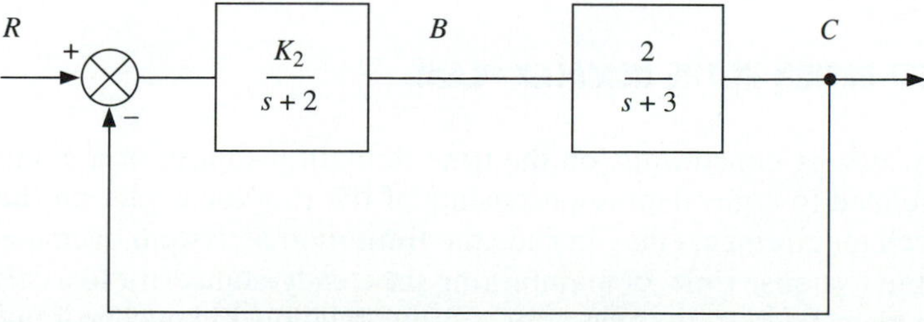

Fig. 11.4 Inner loop closed

that obtained by writing the closed-loop transfer function and factorizing the characteristic equation directly into the closed-loop poles. In this example the factorization was simple to do since the charactersitic equation was of order 1, but consider the case where the forward-path element in the inner loop becomes

$$G(s) = \frac{K_1(s + 4)}{s(s + 1)(s + 3)} \tag{11.7}$$

and the inner loop has unity feedback. The closed-loop transfer function of the inner loop becomes

$$\frac{C}{B}(s) = \frac{K_1(s + 4)}{s^3 + 4s^2 + (3 + K_1)s + 4K_1} \tag{11.8}$$

which is not easy to factorize into the closed-loop poles for a given value of K_1. It is easy, however, to plot the root locus and determine the closed-loop poles from the new open-loop transfer function

$$GH(s) = \frac{K_1(s + 4)}{s(s + 1)(s + 3)} \tag{11.9}$$

allowing the inner loop to be replaced by the transfer function

$$\frac{C}{B} = \frac{2(s + 4)}{(s + p_1)(s + p_2)(s + p_3)} \tag{11.10}$$

where p_1, p_2, and p_3 are the closed-loop poles of the inner loop obtained from the root locus for $K_1 = 2$. In the above analysis, the closed-loop poles of the inner loop become the open-loop poles of the outer loop, while the open-loop zeros of the inner loop become the open-loop zeros of the outer loop. Care needs to be taken in using this method since closing the inner loop using the root locus will only provide the closed-loop poles of the characteristic equation, and the student must determine the appropriate open-loop zeros using the analysis in Module 7, which specifies when closed-loop zeros occur and identifies their origin. For a given value of K, the characteristic equation may be factorized by simply replacing it by the locations of the closed-loop poles from the root locus. The numerator of the new open-loop transfer function to be used in the analysis of the next outermost loop contains the gain K, the open-loop zeros that come from the zeros of $G(s)$ and the poles of $H(s)$. Only when all these elements are considered will the correct transfer function for the inner loop be determined.

◆ SYSTEM DESIGN IN THE COMPLEX PLANE

Requirements, or constraints, on the time-domain response of a control system may be related to equivalent requirements of the root locus plot on the complex plane. Such requirements may include the limitation of system overshoot, having a minimum response time, or maintaining the steady-state error to a certain input to within given bounds. In some cases, a direct relationship may be found between each constraint and one or more parameters used in specifying the location of the closed-loop poles from the root locus.

It is important to realize that the root locus method is a means to graphically solve the characteristic equation so as to yield the closed-loop poles. *If there are no closed-loop zeros present or if they may be neglected,* the system response may then be estimated for a particular input from the location of the dominant closed-loop pole(s). Students should appreciate the subtle point that the closed-loop poles do indeed characterize the closed-loop response, but this same closed-loop response may appear very different depending upon the presence or absence of closed-loop zeros, as detailed in Module 7.

◆ PERFORMANCE REQUIREMENTS AS COMPLEX-PLANE CONSTRAINTS

Usually, any time-domain performance demand may be interpreted as a constraint on the damping ratio ζ and undamped natural frequency ω_n of an equivalent second-order system, as outlined in Module 5. For example, requiring specific values for the percentage overshoot, rise time, peak time, and settling time, all may be translated into required values of damping ratio and undamped natural frequency.[1] This may be illustrated by Figs. 11.5 and 11.6, which indicate those areas of the complex plane from which the closed-loop poles would be prohibited if the system were to have a damping ratio $\zeta > 0.4$ and a time constant (for a first- or second-order system) where $\tau < 0.5$ s.

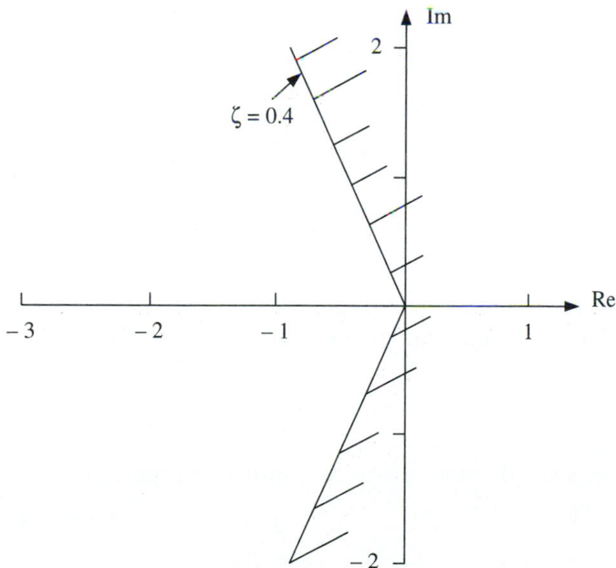

Fig. 11.5 Available complex plane for $\zeta > 0.4$

[1]Strictly speaking, some of these measures also require particular values of the damped natural frequency ω_d, but this may be derived from ζ and ω_n.

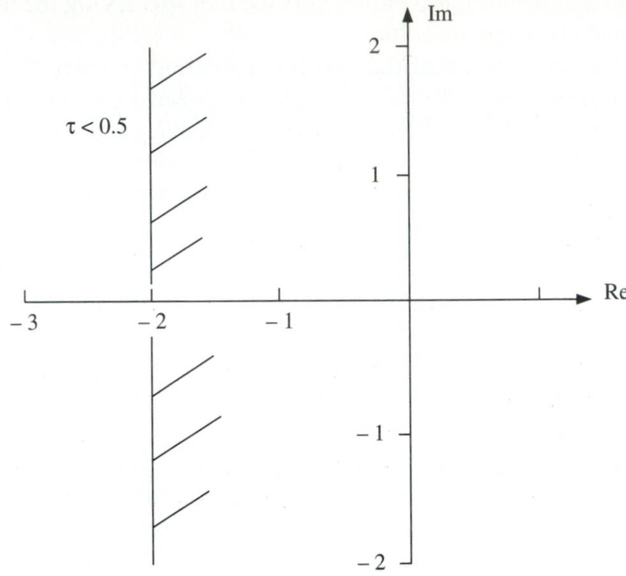

Fig. 11.6 Available complex plane for $\tau < 0.5$

◆ STEADY-STATE ERROR

The steady-state error of a particular system depends first on the system *type*, as discussed in Module 8. For a given system type, there will be a finite steady-state error due only to one particular input; inputs of order less than this will generate zero steady-state error, while inputs greater than this will produce infinite steady-state error. For example, a type 1 system will have zero steady-state error for a step input, a steady state error of $1/K_v$ for a ramp input, and infinite steady-state error for an acceleration input. Recall that K_v is known as the velocity error coefficient. Specification of a limiting steady-state error may be given either in absolute terms, i.e., radians per second, or in terms of the velocity error coefficient K_v. In either case, the error coefficient is seen to be simply the gain used in the root locus. Note, however, the difference in the specification of the transfer function for the analysis of system type and that used for plotting the root locus. For the former, a typical open-loop transfer function was given in the Bode form as

$$GH(s) = \frac{K}{1 + \frac{1}{2}s} \tag{11.11}$$

which results in a steady-state error, to a unit step input, of

$$e_{ss} = \frac{1}{1 + K_p} \tag{11.12}$$

where $K_p = K$, the position constant. For the root locus method, however, the transfer function is written in the Evans form

$$GH(s) = \frac{K'}{s + 2} \tag{11.13}$$

Note that $K \neq K'$, and care must be used in using the Evans gain to calculate the error constant. First, the transfer function used to calculate the root locus has to be written in Bode form, and then the error constant calculated by letting $s \rightarrow 0$.

It may be impossible to satisfy a requirement for zero steady-state error if the system is not of the correct type, since this feature is not controlled by the location of the closed-loop poles on the root locus plot. Other techniques, to be described in a later module, have to be used. Similarly, it is easy to overconstrain a design by specifying too many, or contradictory, requirements that cannot be met with a single gain value. As an example, suppose the system described by

$$GH(s) = \frac{K}{s(s + 2)} \tag{11.14}$$

has to satisfy the performance requirements

1. overshoot less than 10%,
2. steady-state error to a unit ramp less than 10%, and
3. dominant time constant less than 0.1 s.

It may be shown from the root locus (Fig. 11.7) that these translate into the following gain constraints:

1. $K < 2.89$ and
2. $K > 20$.

The third requirement is that the closed-loop poles lie to the left of $s = -10$, and clearly no part of the locus satisfies this constraint.

◆ DESIRABLE AREAS OF COMPLEX PLANE FOR "GOOD" RESPONSE

It would be convenient if there were particular areas of the complex plane where system designers could strive to place the closed-loop poles. Unfortunately, such

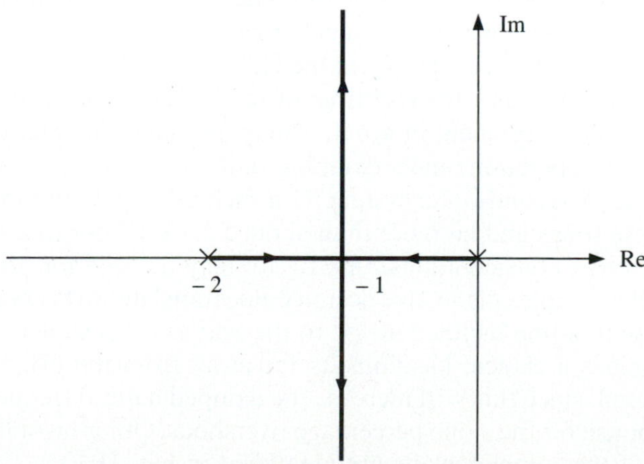

Fig. 11.7 Design example root locus

Fig. 11.8 Component placement robot

areas do not exist because each control system has its own special performance requirements. For example, Fig. 11.8 shows a manipulator placing delicate components onto a circuit card.

Considering the downward motion of the arm, it must be at least critically damped, or preferably overdamped, since any overshoot will damage the components. The designer might specify a damping ratio of $\zeta = 2$ to allow some factor of safety. On the other hand, Fig. 11.9 shows a ship's decoy flare launcher, designed as a last attempt to divert an incoming missile away from the thermal image of the ship, as seen by the missile guidance system, and toward the flare.

In rotating the launcher prior to firing the flare, the exact direction is not as important as the speed with which the launcher can rotate and fire. In such a case, perhaps $\zeta = 0.2$ would be acceptable, if the flare can be fired while the launcher is still rotating. In this case, the rise time of the system is minimized.

Keeping the above argument in mind, it may be stated that many systems are not critical in their operation, and a damping ratio of $\zeta \approx 0.7$ is commonly used as a design guide. A second-order system with such a damping ratio will have reasonable response times and no more than about 5% overshoot to a step input. In the absence of other considerations, this figure may be used for many systems.

In terms of the complex plane, this design guide translates to keeping the closed-loop poles close to a line inclined at 45° to the real axis, as shown in Fig. 11.10. If the designer has a choice, locations as far away from the origin as possible should be selected, since this will increase the damped natural frequency and lead to a faster response for the same percentage overshoot. Other favorable and unfavorable areas of the complex plane are identified in Fig. 11.10.

It was shown in the introductory sections on feedback how the natural response of the system could be changed through gain compensation. However, as perfor-

Fig. 11.9 Ship's decoy flare launcher (USNA)

mance requirements become more specific and numerous, it has now been illustrated that it may not be possible to meet all of the constraints put upon the system without resorting to some new methods not yet discussed. Such methods are available to assist in this task of meeting a greater range of control system performance

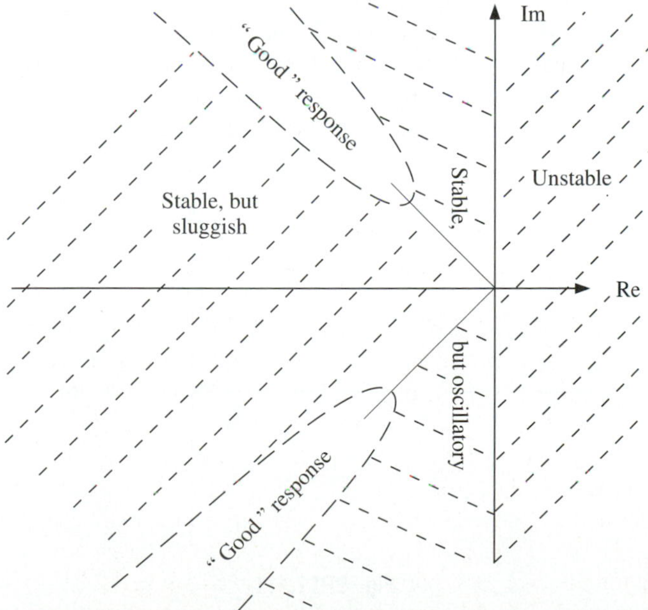

Fig. 11.10 Desirable areas of the complex plane

demands and involve the addition of new elements to the forward path of the loop. These techniques, known as *compensation* methods, will be introduced in later modules.

Sample Problem 11.1

A system has the following open-loop transfer function:

$$GH(s) = \frac{K}{s(s^2 + 6s + 10)}$$

Find the value of K so that the dominant system poles have a damping ratio of $\zeta = 0.5$, and sketch the resulting unit step response.

Solution

Factoring the quadratic in the denominator reveals a pair of complex poles at $s = -3 \pm j$, producing the pole-zero map shown in Fig. SP11.1.1. Applying the rules for plotting the root locus gives the following results:

1. The negative real axis is part of the locus.
2. There are three asymptotes inclined at 120° to each other, the negative real axis being one of them.
3. The asymptotes intersect the real axis at

$$\sigma_a = \tfrac{1}{3}(-3 - 3 - 0) = -2$$

These results are shown in Fig. SP11.1.2. At this stage, two possible forms of the root locus seem possible, as shown in Figs. SP11.1.3 and SP11.1.4, respectively. To determine which locus occurs, it is simplest to plot the gain of a test point s_t as it moves along the negative real axis from the origin to infinity, although in practice only the region $0 > s_t > -3$ needs to be investigated. If the locus is as shown in Fig. SP11.1.3, the gain of such a test point will increase monotonically from zero to infinity. On the other hand, the gain for Fig. SP11.1.4 would increase from zero to a local maximum at the breakaway point, through a local minimum at the break-in point, and then would increase

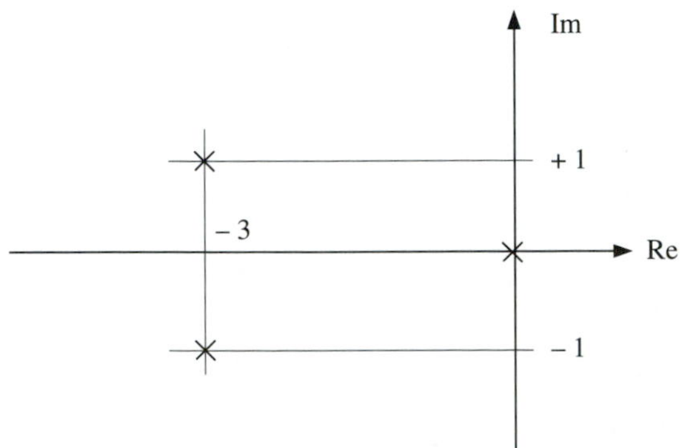

Fig. SP11.1.1

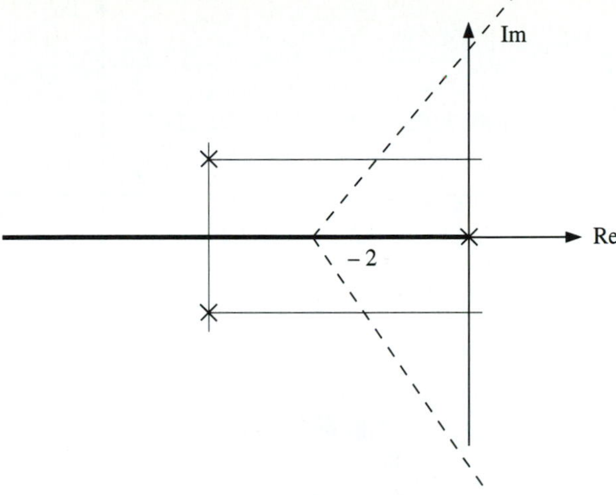

Fig. SP11.1.2

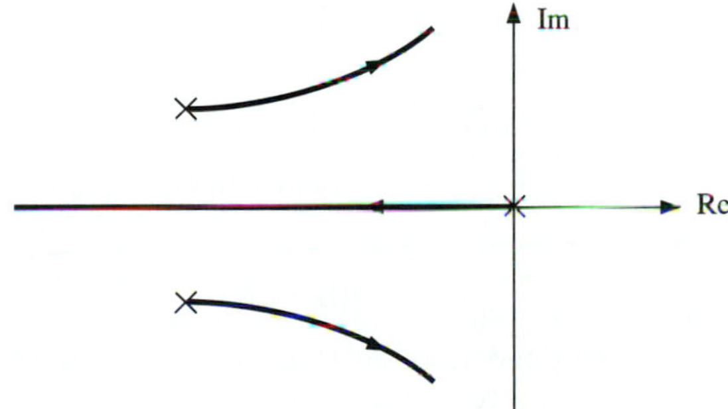

Fig. SP11.1.3

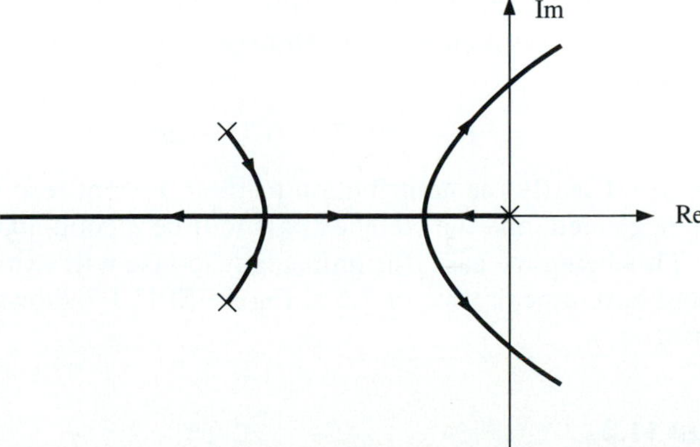

Fig. SP11.1.4

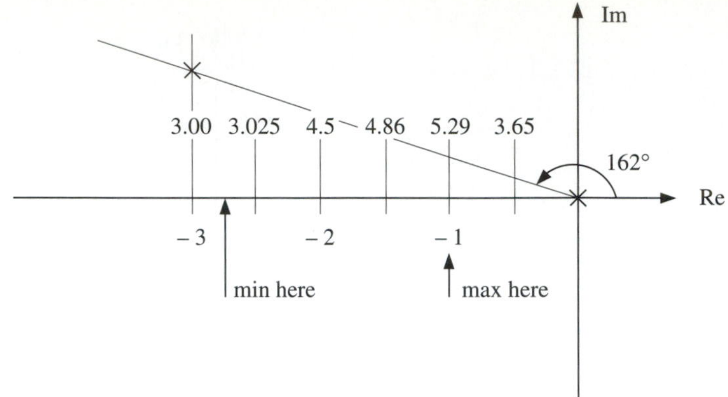

Fig. SP11.1.5

uniformly to infinity. Figure SP11.1.5 shows the gain calculated for various locations on the negative real axis and indicates the locus will be as shown in Fig. SP11.1.4, with a breakaway point located approximately at $s = -1$ and a break-in point around $s = -2.75$. Completing the data for plotting the root locus and using the angles shown in Fig. SP11.1.5, the departure angle from the complex pole at $s = -3 + j$ becomes

$$\alpha_k = 180 - (162 + 90) = -72°$$

The imaginary-axis crossing points are obtained from the characteristic equation, which is

$$s^3 + 6s^2 + 10s + K = 0$$

Substituting $s = j\omega$ and collecting real and imaginary parts give

$$(K - 6\omega^2) + j\omega(10 - \omega^2) = 0$$

The locus crosses the imaginary axis at $\omega = 3.16$, at which point $K = 60$. The root locus is shown in Fig. SP11.1.6. Marking the radial line corresponding to $\zeta = 0.5$ in Fig. SP11.1.6 shows that it intersects the locus at $s = -0.75 \pm 1.3j$, which is the location of the dominant complex pair of roots. Noting that the characteristic equation is a cubic, Rule #8 may be used to find the third root on the real axis:

$$-6 = -0.75 - 0.75 - \sigma$$

giving $\sigma = 4.5$. Clearly the contribution to the transient response from this root may be neglected, and the complex pair will be a good approximation to the system. This being the case, the unit step response will exhibit about 15% overshoot and have a peak time of 2.5 s. Figure SP11.1.7 shows the expected step response.

Sample Problem 11.2

Figure SP11.2.1 shows the block diagram representation of a satellite attitude control system that has unity feedback. Show that the system is unstable. The sys-

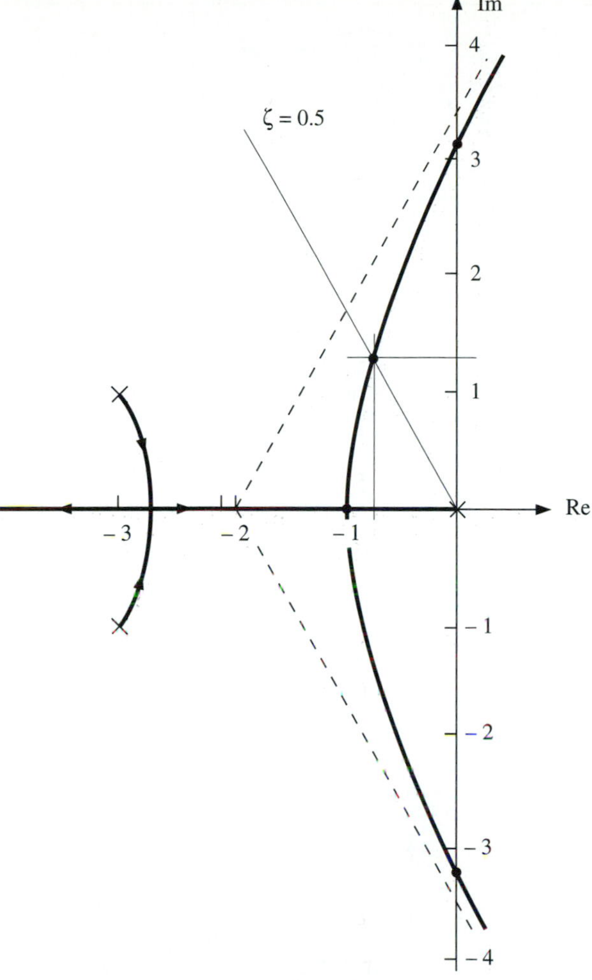

Fig. SP11.1.6

tem is then stabilized by means of an additional feedback path and the inclusion of a controller that has a transfer function of the form

$$Q(s) = \frac{K(s + a)}{s + b}$$

as shown in Fig. SP11.2.2. Suggest suitable values of a and b that stabilize the system, and for those values, find the range of K for which the system is stable.

Solution

The root locus for the system shown in Fig. SP11.2.1 is shown in Fig. SP11.2.3, from which it can be seen that the system is unstable for any value of gain. For the particular value $K = 4$, the closed-loop poles are located at $s = +0.5 \pm 1.94j$. Using the method outlined in the previous section for multiloop systems, the inner loop will be closed first using the root locus method, and these closed-

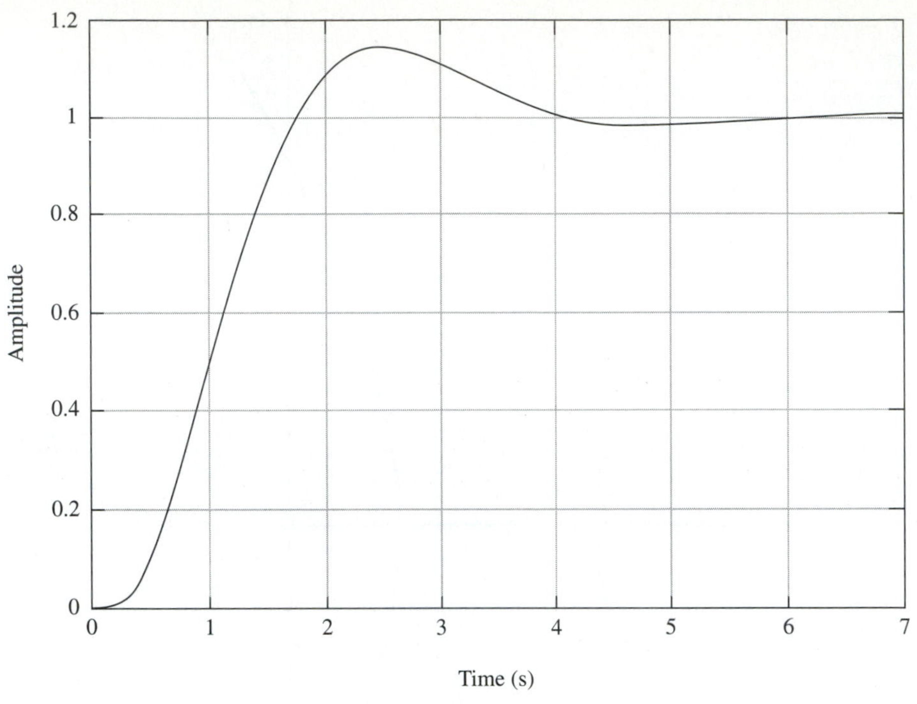

Fig. SP11.1.7

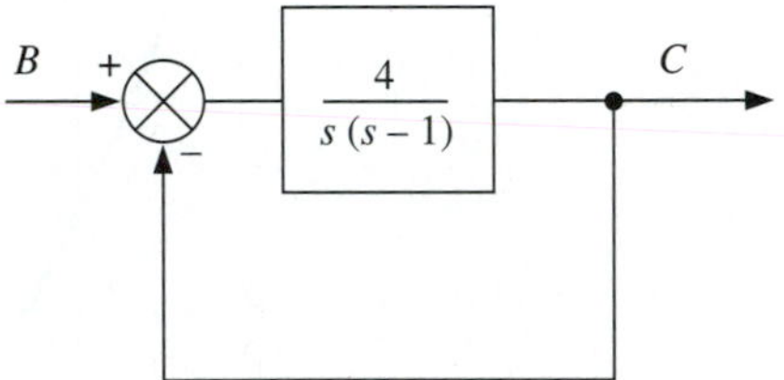

Fig. SP11.2.1

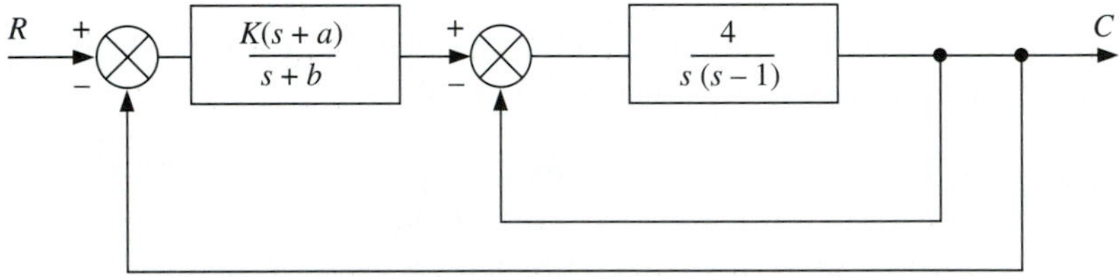

Fig. SP11.2.2

loop poles will become the open-loop poles of the outer loop. From the root locus shown in Fig. SP11.2.3, the closed-loop transfer function of the inner loop may be expressed as

$$\frac{C}{B}(s) = \frac{4}{(s - 0.5 + 1.94j)(s - 0.5 - 1.94j)}$$

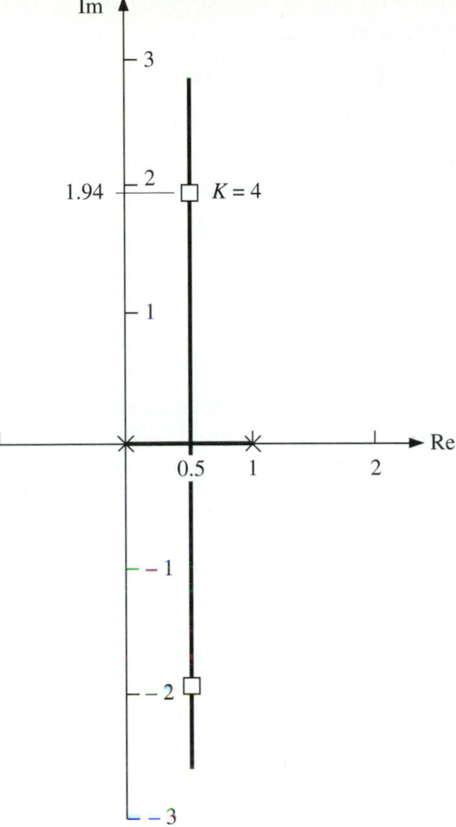

Fig. SP11.2.3

The system with the additional feedback loop and controller may be represented as in Fig. SP11.2.4, from which the open-loop transfer function becomes

$$GH(s) = \frac{4K(s + a)}{(s + b)(s - 0.5 + 1.94j)(s - 0.5 - 1.94j)}$$

In designing the controller, it is possible to introduce one pole and one zero into the root locus, and the only question to resolve is whether the pole should be closer to the imaginary axis than the zero, or vice versa. Suppose the root locus is sketched for the first case where $a = 5$ and $b = 1$. There is nothing special about these numbers; they are only initial working values. The inclusion of the new controller has no effect on the number of asymptotes; there

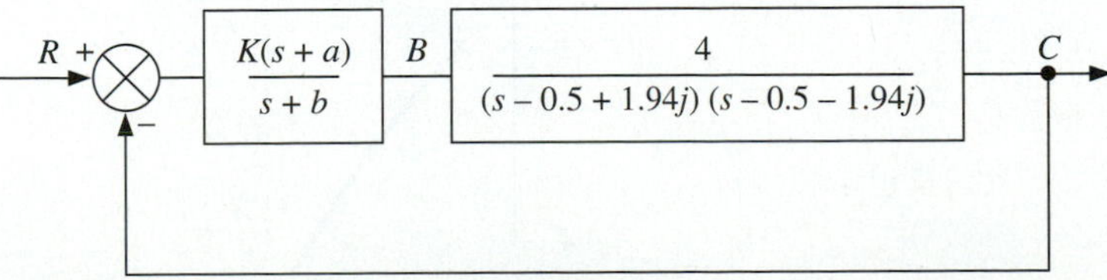

Fig. SP11.2.4

are still two of them inclined at 180° to each other. The new location of the asymptotes, however, is at

$$\sigma_a = \tfrac{1}{2}[(-1 + 0.5 + 0.5) - (-5)] = +2.5$$

The locus appears to take the form shown in Fig. SP11.2.5 and will be unstable for any value of K. If now the locations of the pole and zero are reversed, that is, $a = -1$ and $b = -5$, the open-loop transfer function becomes

$$GH(s) = \frac{K(s + 1)}{(s + 5)(s - 0.5 + 1.94j)(s - 0.5 - 1.94j)}$$

This time the asymptotes intersect the real axis at

$$\sigma_a = \tfrac{1}{2}[(-5 + 0.5 + 0.5) - (-1)] = -1.5$$

For this case, the root locus may be sketched as in Fig. SP11.2.6, observing that for gain above a critical value, the system will be stable, though lightly damped. The characteristic equation for the system becomes

$$s^3 + 4s^2 + s(K - 1) + (20 + K) = 0$$

Substituting $s = j\omega$ into this equation and solving yield the imaginary-axis crossing points at $\omega = 2.64$ rad/s for $K = 8$. The system is stable, therefore, when $K > 8$, although the damping ratio of the closed-loop poles never gets larger than about 0.2.

The situation may be improved by moving the zero closer to the origin, which has the effect of moving the point at which the asymptotes intersect the real axis to the left. For example, if the zero is placed at the origin, it results in the asymptotes being located at $\sigma_a = -2.0$. In practice there are difficulties involved in physically constructing a controller that has a zero located at the origin. Alter-

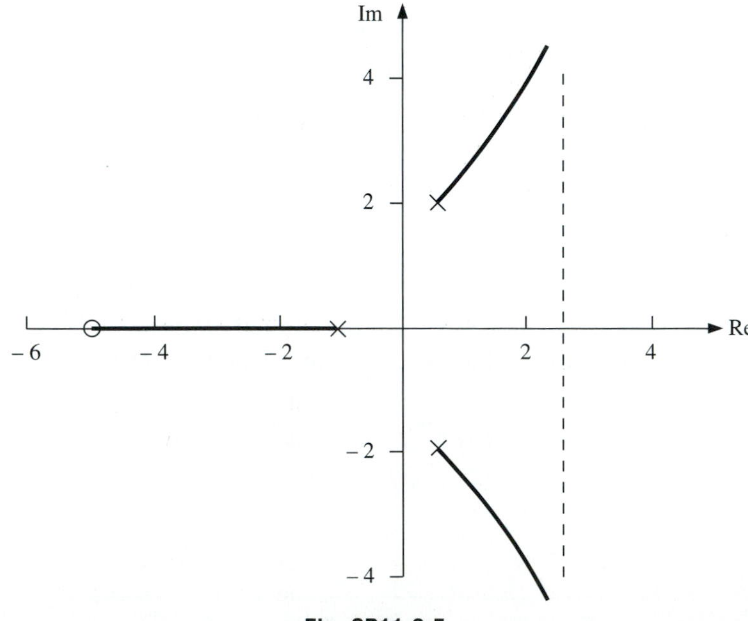

Fig. SP11.2.5

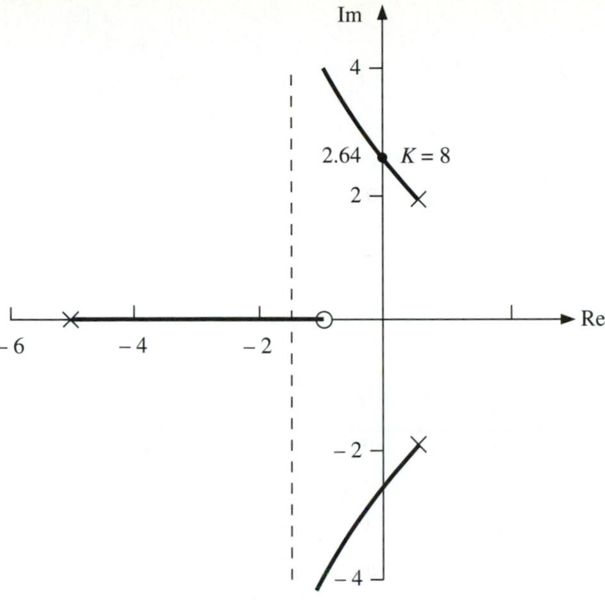

Fig. SP11.2.6

natively, the pole at $s = -5$ may be moved further to the left, which in turn moves the intersection of the asymptotes to the left. This will place more of the stable portion of the locus from the complex poles further to the left, allowing more choice in the selection of damping ratio. In addition, because the system has a closed-loop zero (same location as the open-loop zero), the closed-loop system response is not directly predictable from the location of the closed-loop poles alone.

Sample Problem 11.3

A ship's gun-pointing system is described by the transfer function

$$G(s) = \frac{K(s + 2)}{s(s + 1)(s + 15)} \qquad H(s) = 1$$

Find K for the fastest response with reasonable overshoot, and demonstrate the system response to a unit impulse.

Solution

To begin the design process, the root locus of the system will be plotted. The system has three poles and one zero, resulting in two asymptotes going to infinity inclined at 180° to each other. The asymptotes intersect the real axis at

$$\sigma_a = \tfrac{1}{2}[(-15 - 1) - (-2)] = -7$$

It seems as though the two poles near the origin loop around the zero, but it is not known whether they then move toward the asymptotes and off to infinity or whether they rejoin the real axis and then move along it. The issue is resolved

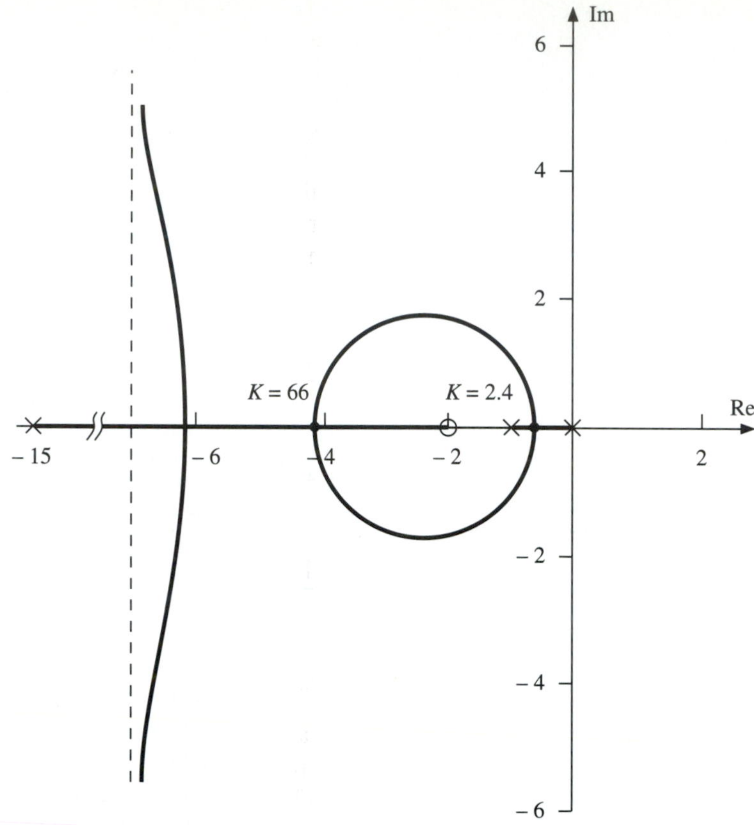

Fig. SP11.3.1

using Rule #7 in this case, indicating there is a local minimum and a local maximum on the real axis in the range $-2 > s > -15$. The locus may be sketched as in Fig. SP11.3.1 and indicates that the roots from the open-loop poles at $s = 0$ and $s = -1$ meet on the axis at about $s = -0.6$, loop around the zero at $s = -2$, and meet back on the real axis again at $s = -4.2$. At this point, one root moves toward the zero while the other moves to the left, meeting the remaining root coming in from the pole at $s = -15$. After meeting at about $s = -6.5$, these roots then move off to infinity along the asymptotes. For $K < 2.4$ the response is dominated by a single real pole located between the origin and $s = -0.6$. For $66 > K > 2.4$ the system may be represented by a pair of complex closed-loop poles. For $K > 66$, the system response is again dominated by a single pole on the real axis, the value of which approaches the zero at $s = -2$ as the gain approaches infinity. The minimum time constant therefore corresponds to the double pole at $s = -4.2$ when $K = 66$ and leads to a value of $\tau = 1/4.2 = 0.24$ s and a critically damped system. The significance of the double pole at $s = -4.2$ may be observed by writing the closed-loop transfer function as

$$\frac{C}{R}(s) = \frac{K(s + 2)}{s(s + 1)(s + 5) + K(s + 2)}$$

where the denominator, for $K = 66$, may be factorized into the closed-loop poles. Note that the third closed-loop pole has to be identified ($s = -8$) in order to write the closed-loop transfer function in the following form:

$$\frac{C}{R}(s) = \frac{66(s + 2)}{(s + 4.2)^2(s + 8)}$$

If the input is a unit step, $c(t)$ may be calculated by splitting into partial fractions and then taking the inverse Laplace transform. Although the system described by the double pole will exhibit monotonic (no overshoot) behavior for a given input, it represents a different physical system to that of a single pole. Consider a closed-loop system described by

$$\frac{C}{R}(s) = \frac{1}{(s + 4.2)^2}$$

If the input R is a unit impulse, the output is given by

$$c(t) = te^{-4.2t}$$

whereas the system described by

$$\frac{C}{R}(s) = \frac{1}{s + 4.2}$$

will have the unit impulse response of

$$c(t) = e^{-4.2t}$$

The system with the double pole has inertia and cannot immediately change its position, while that described by the single pole does not and is able to change position instantaneously. The concept of a time constant has to be interpreted somewhat carefully for the former. Representative forms of the impulse response are illustrated in Figs. SP11.3.2 and 11.3.3 for the second- and first-order systems, respectively. Because this system possesses an open-loop zero, it will also have a closed-loop zero, the effect of which has been neglected in the above analysis. This zero will have to be considered in order for the full closed-loop time response to be obtained. For an impulsive input, the output becomes

$$C(s) = \frac{66(s + 2)}{(s + 4.2)^2(s + 8)} = \frac{as + b}{(s + 4.2)^2} + \frac{c}{s + 8}$$

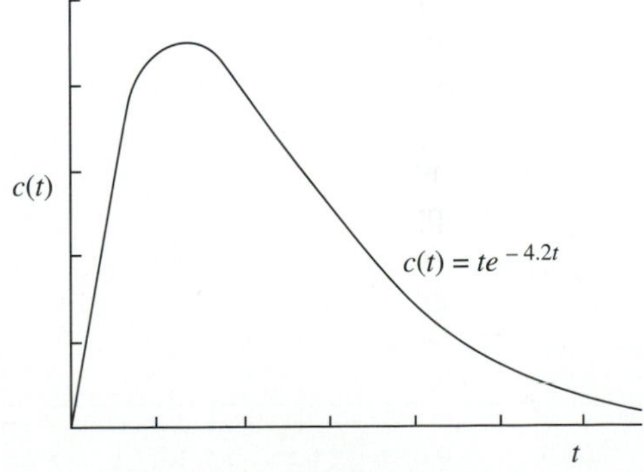

Fig. SP11.3.2

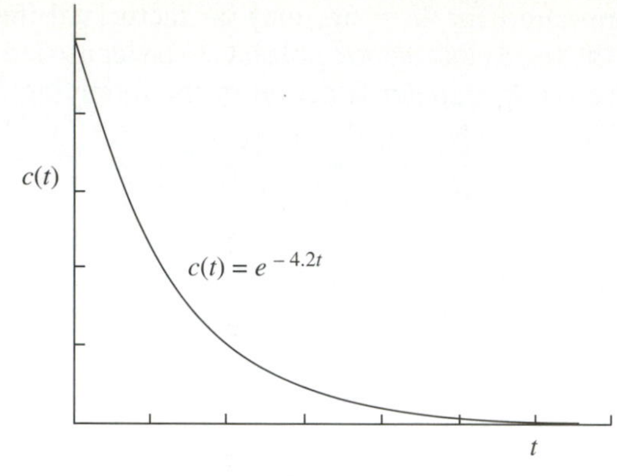

Fig. SP11.3.3

The student may verify that the partial-fraction coefficients give

$$C(s) = \frac{27.47s + 77.1}{(s + 4.2)^2} - \frac{27.47}{s + 8}$$

which may be written as

$$C(s) = 27.47\left[\frac{1}{s + 4.2} - \frac{1.4}{(s + 4.2)^2} - \frac{1}{s + 8}\right]$$

which yields the inverse transform

$$c(t) = 27.47[e^{-4.2t} - 1.4te^{-4.2t} - e^{-8t}]$$

Figure SP11.3.4 shows the impulse response. It may be seen that the effect of the closed-loop zero is to make the system response faster (peaks at 0.17 s instead of 0.22 s) and has a small overshoot. This is consistent with our previous discussion (Module 7) of the effect of closed-loop zeros on system response. The

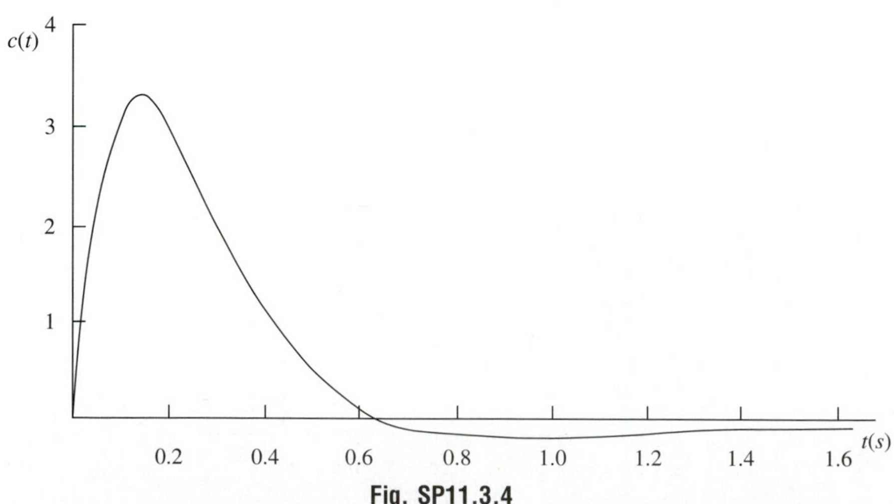

Fig. SP11.3.4

expected overshoot for a unit step may be obtained by considering the open-loop transfer function. If we assume the extra pole at $s = -8$ may be ignored but the zero at $s = -2$ cannot, Fig. 7.6 may be used. The dominant poles at $s = -4.2$ are critically damped, and the parameter γ used in Fig. 7.6 is computed as

$$\gamma = \frac{2}{4.2} = 0.48$$

Figure 7.6 suggests a 15% overshoot. The step response may be computed from MATLAB using the following code:

```
D1=[1 4.2];
D2=CONV(D1,D1);
D3=[1 8];
DEN=CONV(D2,D3);
NUM=66*[1 2];
STEP(NUM,DEN)
```

The resulting plot is shown in Fig. SP11.3.5. Note that becuase the system is of type 0, it has a steady-state error for a position input.

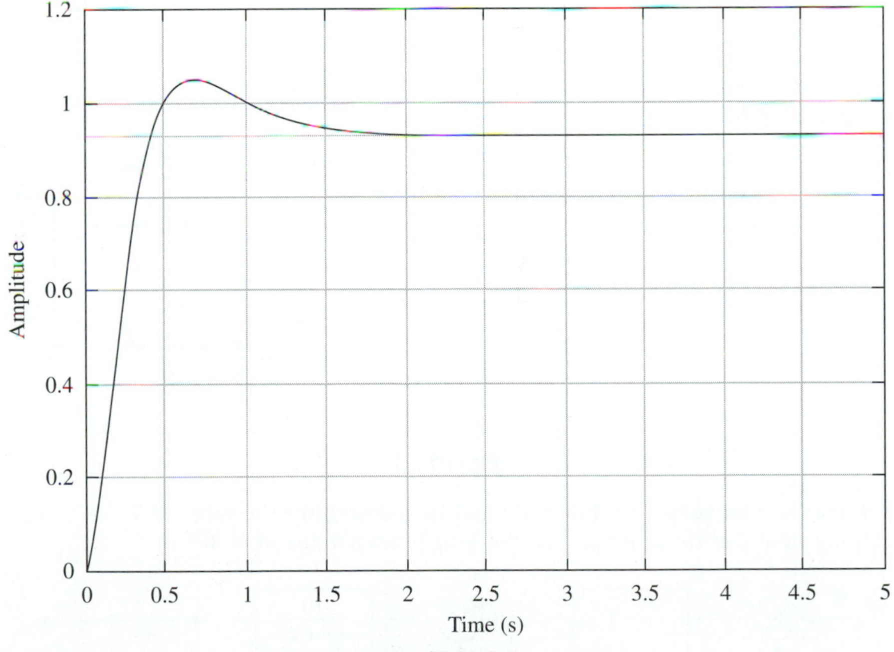

Fig. SP11.3.5

◆ PROBLEMS

11.1 Shown in Fig. P11.1 is a multiloop position control system. Determine analytically and by using the root locus method the system closed-loop poles and zeros when $K_1 = K_2 = 10$.

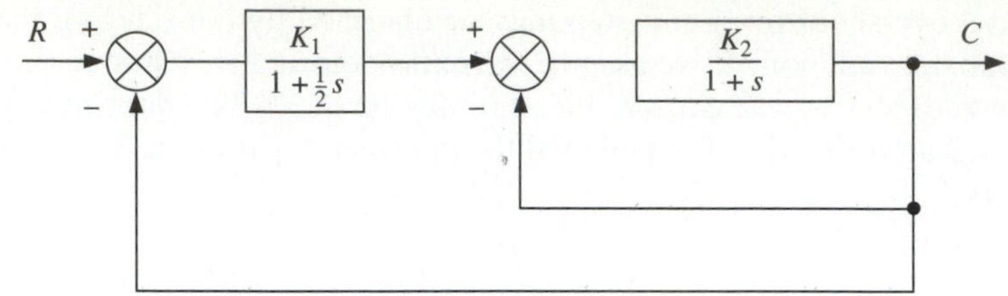

Fig. P11.1

11.2 Figure P11.2 illustrates an electromechanical position control system with tachometer as well as potentiometer feedback. Determine the range of damping ratio of the closed-loop poles available for $0 < K_v < 1$.

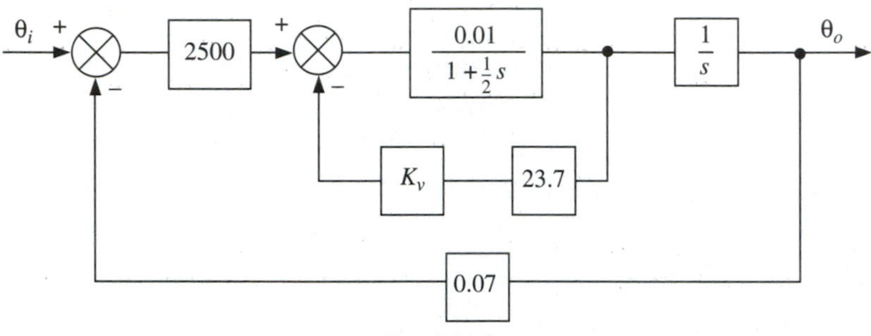

Fig. P11.2

11.3 For the multiloop system shown in Fig. P11.3, determine the location of all closed-loop poles for $K = 1$.

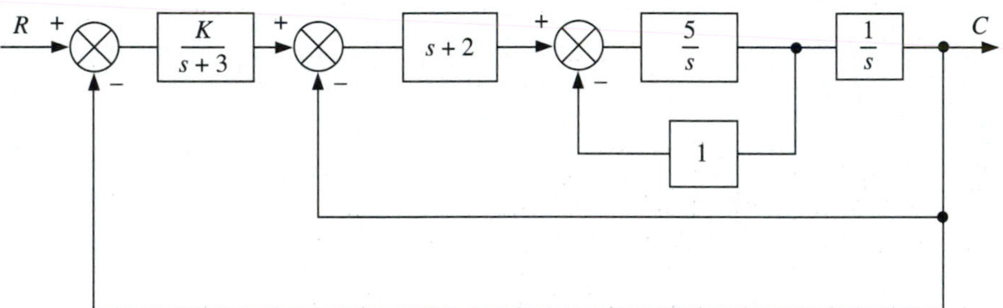

Fig. P11.3

11.4 For the control system shown in Fig. P11.4, determine the value of K that gives the damping ratio ζ of the dominant closed-loop poles a value of 0.707.

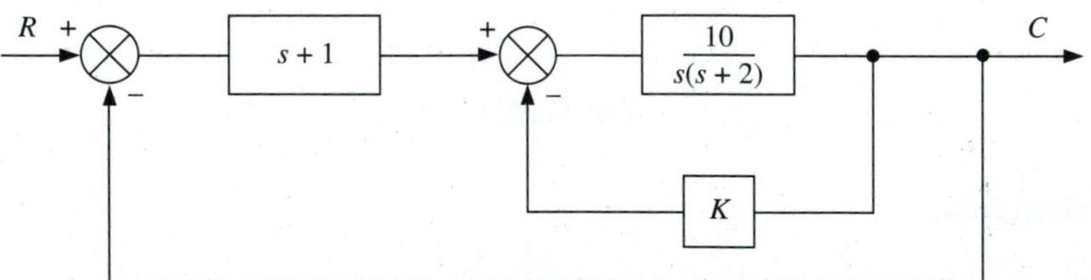

Fig. P11.4

11.5 A robot takes television tubes from a conveyer belt and places them in a test fixture. The vertical axis of the robot is controlled by a system with open-loop transfer function

$$GH(s) = \frac{K(s^2 + 16)}{s(s^2 + 10s + 120)}$$

Determine the "best" value of K to accomplish this task, stating reasons for the choice.

11.6 A control system has open-loop transfer function given by

$$GH(s) = \frac{K}{(s + 2)^2(s + 3)}$$

Determine the value of K such that the time response to a step input has a damping ratio $\zeta = 0.5$. What is the damped natural frequency for this value of K? An element of the form

$$D(s) = \frac{s + 4}{s + 12}$$

is now included in the forward path. Find the value of K that gives the closed-loop poles the same damping ratio. What is the value of ω_d now?

11.7 Figure P11.7 shows the open-loop poles and zeros of a feedback control system. Determine the value of gain that makes the dominant closed-loop complex poles have a damping ratio of $\zeta = 0.35$, and calculate the steady-state error for a unit step input.

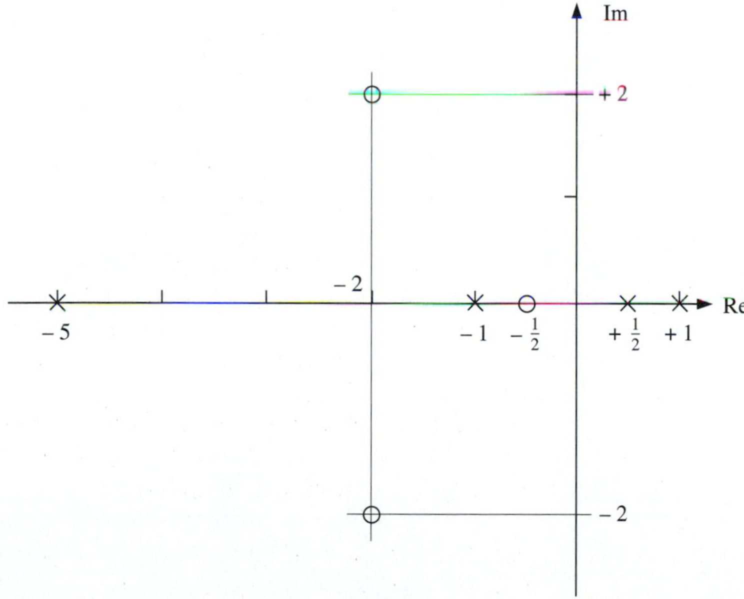

Fig. P11.7

11.8 Consider the control system shown in Fig. P11.8. Sketch the root locus and estimate the range of K for which the closed-loop poles satisfies the following design constraints:

a. Not more than 15% overshoot.

b. Dominant time constant less than 2 s.

c. Damped natural frequency of oscillation less than 2 rad/s.

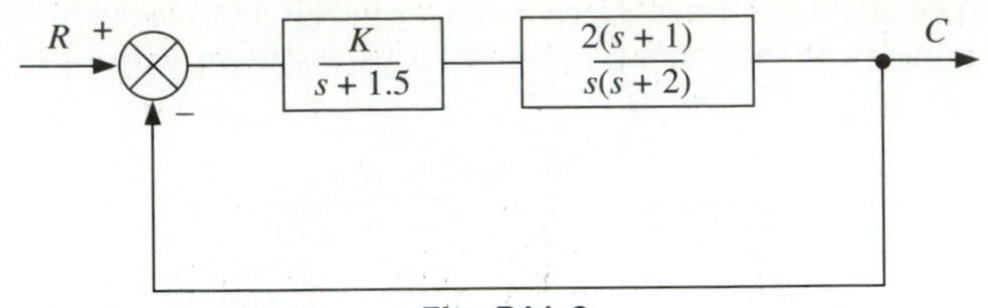

Fig. P11.8

11.9 For a control system described by the open-loop transfer function

$$GH(s) = \frac{K(s + 2)(s^2 + s + 1.69)}{s(s + 1)(s^2 + 2s + 5)}$$

calculate if a range of K is possible such that *all* closed-loop poles satisfy the conditions $\zeta > 0.7$ and $\omega_n > 0.6$ rad/s. If not, what is the best compromise available?

11.10 Consider the control system given by the open-loop transfer function

$$GH(s) = \frac{K(s + 2)(s + 3)}{s^2(s + 1)(s + 10)}$$

Estimate a suitable value of K such that the dominant closed-loop poles satisfy the following time-domain constraints:

a. $\zeta > 0.25$.

b. $\omega_d < 8$ rad/s.

c. Steady-state error for an acceleration input less than 15%.

Frequency Response and Nyquist Diagrams

The student may rightly question why more methods need to be introduced in order to study stability and system performance, since the root locus approach discussed previously seems to be more than adequate in predicting the transient behavior of a closed-loop system, given the open-loop transfer function. In addition, it provides a ready way in which the stability of a system may be determined.

The weakness of the root locus method is that it relies on the existence of the open-loop transfer function, from which the locus itself is drawn. Suppose we are designing a control system to regulate the power output from a nuclear power plant by positioning control rods in the reactor core. The proposed closed-loop system is shown in Fig. 12.1, where the reactor is represented as a single transfer function $G(s)$, taking control rod position x as input and generated power as output. If the transfer function $G(s)$ is too complex to determine or is simply unknown, the root locus method cannot be used to determine if the closed-loop system will be stable or not. Clearly, for this case, it will be advisable to determine the system stability before the loop is closed, since if the controller causes instability, the result would be catastrophic. The question is, can stability be determined in the absence

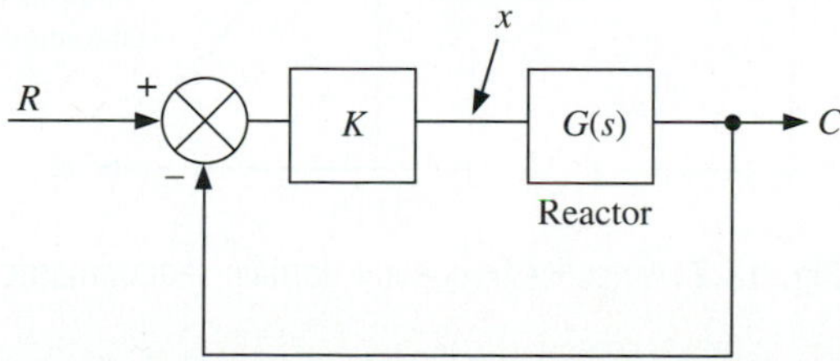

Fig. 12.1 Proposed closed-loop system

of a transfer function? The answer is yes, provided the open-loop frequency response may be found. In the case of the reactor, this could be done experimentally by applying a small harmonic signal to x and observing the power fluctuation. This fluctuation would also be harmonic, where the magnitude of the power and the phase relative to the input signal would be functions of the frequency ω of the input signal.

Performance requirements may sometimes be specified in the frequency domain, rather than in terms of transient response parameters. For example, if we wish to define performance requirements of an audio (hi-fi) amplifier, we may say that it should be capable of transmitting, without attenuation, all frequencies within the human audible spectrum.[1] Given that the human audible frequency spectrum extends up to about 20 kHz, Fig. 12.2 shows the required amplifier frequency requirements. There is, therefore, a need to investigate the performance and stability of systems whose transfer functions are not known, and this may be done by using data relating to the open-loop frequency response of the system. In many cases these data are obtained experimentally.

◆ FREQUENCY RESPONSE

Recall from Modules 3 and 5 that to obtain the frequency response of a function $F(s)$, the value

$$s = j\omega \tag{12.1}$$

is substituted in the function, which may then be evaluated in terms of the magnitude and phase angle:

$$M = |F(j\omega)| \qquad \phi = \angle F(j\omega) \tag{12.2}$$

There are three parameters involved in the above equation: (1) the independent variable frequency ω, (2) the magnitude M of the function, and (3) the phase angle ϕ between the input and the output.

There are several analytical techniques for investigating the stability and performance of systems, which may be classed as frequency response methods. They all involve the study of the relationships between the three parameters described

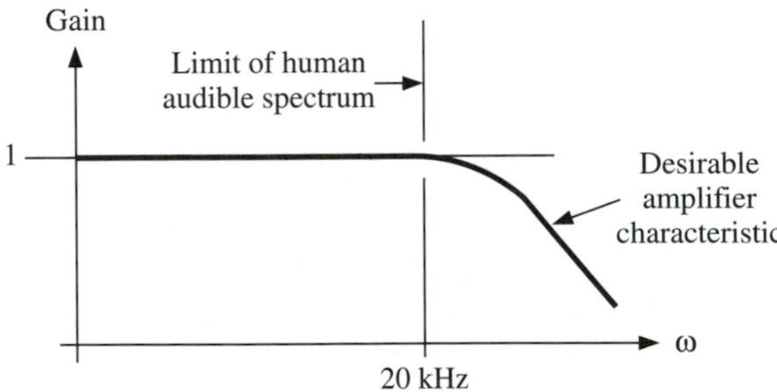

Fig. 12.2 Amplifier frequency-domain requirements

[1]This requirement may be obtained by realizing that all finite signals (such as music) may be reduced to a distribution of frequencies using Fourier analysis.

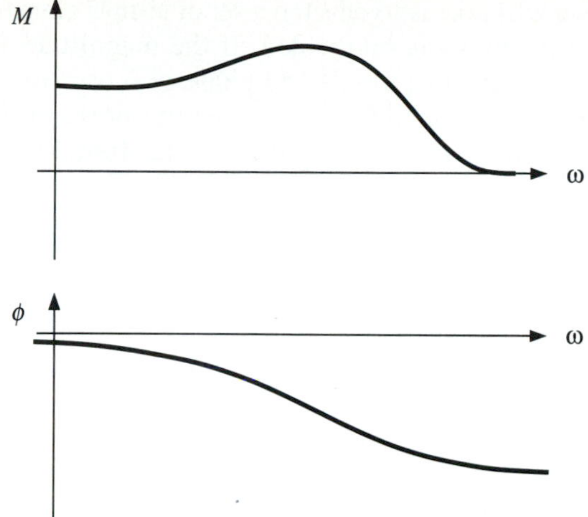

Fig. 12.3 Bode representation of frequency response

above. One way of representing these variables is shown in Fig. 12.3, where the $M-\omega$ and $\phi-\omega$ plots are drawn separately. This representation is known as a Bode diagram and will be studied in later modules.

Another representation of the same data is shown in Fig. 12.4, in which M and ϕ define a vector for a particular frequency ω. This representation is known as a Nyquist diagram and will be investigated in more detail.

◆ NYQUIST DIAGRAMS FROM TRANSFER FUNCTIONS

Although it was just stated that frequency response methods have the advantage over the root locus that experimental data may be used to predict the closed-loop response, we will begin the study of one particular frequency response method, the Nyquist diagram, by analyzing open-loop transfer functions. Obviously, any open-loop transfer function $GH(s)$ may have $j\omega$ substituted for s and the magnitude and phase subsequently calculated for $0 < \omega < \infty$. It may also be obvious that this process can become tedious and sometimes difficult.

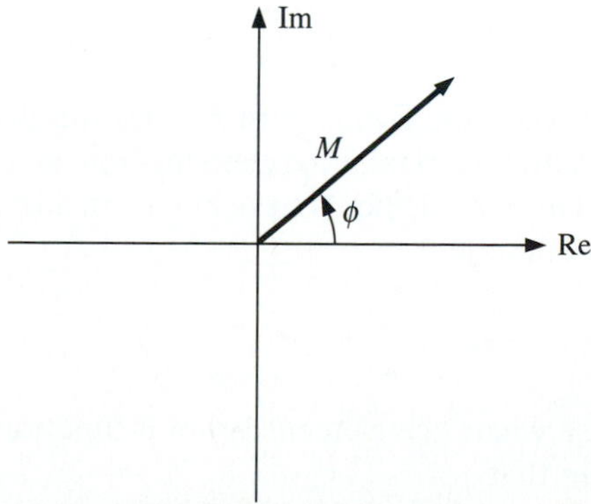

Fig. 12.4 Nyquist representation of frequency response

The approach we will take is to develop a set of simple elements into which all common transfer functions can be divided. If the magnitude and phase of each element is known, then the magnitude and phase of a transfer function made up of the elements may be computed from combinations of the elemental values. For example, if the magnitude and phase of the transfer function

$$GH(s) = \frac{1}{s(s+a)} \tag{12.3}$$

is required and the magnitude and phase of the elements are known in the forms

$$\left|\frac{1}{s}\right| = M_1 \qquad \angle\left[\frac{1}{s}\right] = \phi_1 \tag{12.4}$$

and

$$\left|\frac{1}{s+a}\right| = M_2 \qquad \angle\left[\frac{1}{s+a}\right] = \phi_2 \tag{12.5}$$

then using the usual rules for complex-number multiplication of multiplying the moduli and adding the phase angles, the modulus and phase of $GH(s)$ are

$$|GH(s)| = M_1 \times M_2 \tag{12.6}$$

and

$$\angle GH(s) = \phi_1 + \phi_2 \tag{12.7}$$

If the general open-loop transfer function is written in the form

$$GH(s) = \frac{K_E \prod(s+z_i)\prod(s^2+b_js+c_j)}{s^n \prod(s+p_k)\prod(s^2+d_ls+e_l)} \tag{12.8}$$

it becomes clear what the elements are that need to be dealt with. Before proceeding, however, the open-loop transfer function will be written in a different form. The form shown above is particularly useful for extracting the open-loop poles and zeros for the root locus and is referred to as the Evans form. Instead, we write the open-loop transfer function in the Bode form

$$GH(s) = \frac{K_B \prod(1+sT_i)\prod[1+(2\zeta_j/\omega_{n_j})s+s^2/\omega_{n_j}^2]}{s^n \prod(1+sT_k)\prod[1+(2\zeta_l/\omega_{n_l})s+s^2/\omega_{n_l}^2]} \tag{12.9}$$

Note the difference between the Evans gain K_E and the Bode gain K_B.

The procedure, therefore, is to take the elements one at a time and evaluate the magnitude and phase. Once all elements have been considered, a table summarizing the results will be compiled.

$GH(s) = K$

This is a simple case where the element is not a function of s, but a real number. It follows therefore that

$$|K| = K \qquad \angle K = 0 \tag{12.10}$$

$GH(s) = 1/1 + sT$

Substituting $s = j\omega$ into the transfer function yields

$$GH(j\omega) = \frac{1}{1 + j\omega T} \tag{12.11}$$

Rationalizing this complex number produces

$$GH(j\omega) = \frac{1 - j\omega T}{1 + \omega^2 T^2} \tag{12.12}$$

The magnitude and phase become

$$M = |GH(j\omega)| = \frac{1}{\sqrt{(1 + \omega^2 T^2)}} \tag{12.13}$$

and

$$\phi = \angle GH(j\omega) = -\tan^{-1}(\omega T) \tag{12.14}$$

It will be useful for later work if the asymptotic values of M and ϕ are calculated. It is seen that as

$$\omega \rightarrow 0: \qquad M \rightarrow 1 \qquad \phi \rightarrow 0 \tag{12.15}$$

and as

$$\omega \rightarrow \infty: \qquad M \rightarrow 0 \qquad \phi \rightarrow -90 \tag{12.16}$$

$GH(s) = 1 + sT$

This element is simply the inverse of the previous one. Substituting $s = j\omega$ into the transfer function gives

$$GH(j\omega) = 1 + j\omega T \tag{12.17}$$

the magnitude and phase of which are

$$M = \sqrt{1 + \omega^2 T^2} \qquad \phi = +\tan^{-1}(\omega T) \tag{12.18}$$

The asymptotic values are

$$\omega \rightarrow 0: \qquad M \rightarrow 1 \qquad \phi \rightarrow 0 \tag{12.19}$$

$$\omega \rightarrow \infty: \qquad M \rightarrow \infty \qquad \phi \rightarrow +90 \tag{12.20}$$

$GH(s) = 1/s$

Again substituting $s = j\omega$ results in

$$GH(j\omega) = \frac{1}{j\omega} = \frac{j\omega}{-\omega^2} = -\frac{j}{\omega} \tag{12.21}$$

yielding

$$M = \frac{1}{\omega} \qquad \phi = -90 \tag{12.22}$$

Here, the asymptotic values are

$$\omega \to 0: \qquad M \to \infty \qquad \phi \to -90 \tag{12.23}$$

$$\omega \to \infty: \qquad M \to 0 \qquad \phi \to -90 \tag{12.24}$$

Although rarely occurring in practice, the student may easily verify that the magnitude and phase of $GH(s) = s$ are

$$M = \omega \qquad \phi = +90 \tag{12.25}$$

with the asymptotic values

$$\omega \to 0: \qquad M \to 0 \qquad \phi \to +90 \tag{12.26}$$

$$\omega \to \infty: \qquad M \to \infty \qquad \phi \to +90 \tag{12.27}$$

$GH(s) = 1/s^n$

Although it is possible to determine the magnitude and phase in the usual way, this particular transfer function lends itself to the modular complex-number approach. Consider

$$\frac{1}{s^n} = \frac{1}{s}\frac{1}{s}\frac{1}{s}\cdots \tag{12.28}$$

Since each quotient on the right-hand side of the above equation has magnitude and phase of

$$M = \frac{1}{\omega} \qquad \phi = -90 \tag{12.29}$$

then, from the usual rules for multiplying complex numbers, the magnitude and phase of $1/s^n$ will be given by

$$M = \frac{1}{\omega^n} \qquad \phi = -90n \tag{12.30}$$

Similar arguments may be applied to obtain the magnitude and phase of $GH(s) = s^n$ as

$$M = \omega^n \qquad \phi = +90n \tag{12.31}$$

$GH(s) = \dfrac{1}{[1 + (2\zeta/\omega_n)s + s^2/\omega_n^2]}$

Substituting $s = j\omega$ produces

$$GH(j\omega) = \frac{1}{(1 - \omega^2/\omega_n^2) + 2\zeta j\omega/\omega_n} \tag{12.32}$$

Rationalizing this complex number results in

$$GH(j\omega) = \frac{1}{(1 - \omega^2/\omega_n^2)^2 + 4\zeta^2\omega^2/\omega_n^2}\left[\left(1 - \frac{\omega^2}{\omega_n^2}\right) - j\frac{2\zeta\omega}{\omega_n}\right] \tag{12.33}$$

producing the magnitude and phase

$$M = \frac{1}{\sqrt{\left(1 - \frac{\omega^2}{\omega_n^2}\right)^2 + \frac{4\zeta^2\omega^2}{\omega_n^2}}} \tag{12.34}$$

and

$$\phi = \tan^{-1}\left[\frac{-2\zeta\omega/\omega_n}{1 - \omega^2/\omega_n^2}\right] \tag{12.35}$$

If M and ϕ are plotted against the independent forcing frequency ω, the results are shown in Figs. 12.5 and 12.6, respectively. Note that both magnitude and phase graphs are functions of the damping ratio ζ. Asymptotic values may be obtained from the above plots, or by observing the open-loop transfer function immediately after $s = j\omega$ has been substituted into it,

$$GH(j\omega) = \frac{1}{\left[1 + \frac{2\zeta j\omega}{\omega_n} + \left(\frac{j\omega}{\omega_n}\right)^2\right]} \tag{12.36}$$

It is seen that as $\omega \to 0$, the second and third terms in the denominator become small compared to 1; hence

$$GH(j\omega) \to 1 \tag{12.37}$$

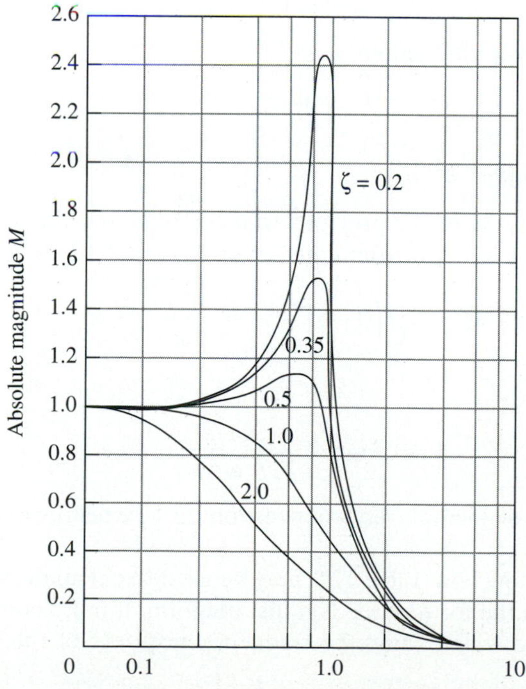

Fig. 12.5 Magnitude-frequency plot.

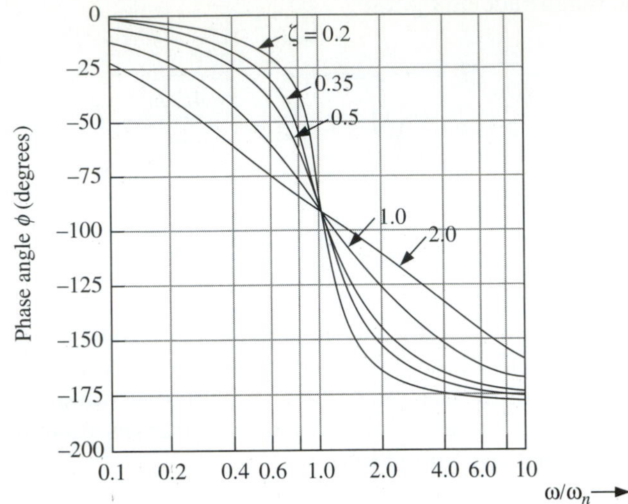

Fig. 12.6 Phase-frequency plot.

resulting in

$$M \to 1 \qquad \phi \to 0 \tag{12.38}$$

Alternatively, as $\omega \to \infty$, the last term in the denominator will dominate the others, the transfer function approximating to

$$GH(j\omega) \approx \frac{1}{(j\omega/\omega_n)^2} = -\left(\frac{\omega_n}{\omega}\right)^2 \tag{12.39}$$

This produces the asymptotic values

$$M \to 0 \qquad \phi \to -180 \tag{12.40}$$

$GH(s) = 1 + (2\zeta/\omega_n)s + s^2/\omega_n^2$

Since this is the inverse of the previous transfer function, the magnitude and phase may be obtained directly from equations 12.34 and 12.35 as

$$M = \sqrt{\left(1 - \frac{\omega^2}{\omega_n^2}\right)^2 + \frac{4\zeta^2\omega^2}{\omega_n^2}} \tag{12.41}$$

and

$$\phi = -\tan^{-1}\frac{-2\zeta\omega/\omega_n}{1 - \omega^2/\omega_n^2} \tag{12.42}$$

This completes all the elemental frequency responses. The results may be summarized in Table 12.1.

An example will show how Table 12.1 may be used to calculate and sketch the frequency response in the form of the Nyquist diagram of more complex transfer functions. Suppose we wish to find the frequency response of the system

$$GH(s) = \frac{K}{s(1 + sT)} \tag{12.43}$$

Transfer function	Magnitude M	Phase ϕ	Asymptotes M_0	M_∞	ϕ_0	ϕ_∞
K	K	0	K	K	0	0
$\dfrac{1}{1+sT}$	$\dfrac{1}{\sqrt{1+\omega^2T^2}}$	$-\tan^{-1}(\omega T)$	1	0	0	-90
$1+sT$	$\sqrt{1+\omega^2T^2}$	$+\tan^{-1}(\omega T)$	1	∞	0	$+90$
$\dfrac{1}{s}$	$\dfrac{1}{\omega}$	-90	∞	0	-90	-90
$\dfrac{1}{s^n}$	$\dfrac{1}{\omega^n}$	$-90n$	∞	0	$-90n$	$-90n$
$\dfrac{1}{1+(2\zeta/\omega_n)s+s^2/\omega_n^2}$	$\dfrac{1}{\sqrt{\left(1-\dfrac{\omega^2}{\omega_n^2}\right)^2+\dfrac{4\zeta^2\omega^2}{\omega_n^2}}}$	$\tan^{-1}\dfrac{-2\zeta\omega/\omega_n}{1-\omega^2/\omega_n^2}$	1	0	0	-180
$1+\dfrac{2\zeta}{\omega_n}s+\dfrac{s^2}{\omega_n^2}$	$\sqrt{\left(1-\dfrac{\omega^2}{\omega_n^2}\right)^2+\dfrac{4\zeta^2\omega^2}{\omega_n^2}}$	$-\tan^{-1}\dfrac{-2\zeta\omega/\omega_n}{1-\omega^2/\omega_n^2}$	1	∞	0	180

The open-loop transfer function is considered to consist of three elements:

$$K, \quad \frac{1}{s}, \quad \frac{1}{1+sT} \tag{12.44}$$

If the magnitude and phase of each element are obtained from the appropriate entry in Table 12.1, then the magnitude of $GH(s)$ is found by multiplying the elemental magnitudes and the phase by adding the elemental phases to obtain

$$\left|\frac{K}{s(1+sT)}\right| = \frac{K}{\omega\sqrt{1+\omega^2T^2}} \tag{12.45}$$

and

$$\angle\frac{K}{s(1+sT)} = 0 - 90 - \tan^{-1}(\omega T) \tag{12.46}$$

In a similar manner, the asymptotic values may be obtained from the elemental values in Table 12.1 as

$$\omega \to 0: \qquad M \to K \times \infty \times 1 \to \infty \tag{12.47}$$

$$\omega \to 0: \qquad \phi \to 0 - 90 - 0 \to -90 \tag{12.48}$$

whereas for high frequencies

$$\omega \to \infty: \qquad M \to K \times 0 \times 0 \to 0 \tag{12.49}$$

$$\omega \to \infty: \qquad \phi \to 0 - 90 - 90 \to -180 \tag{12.50}$$

A sketch of the Nyquist diagram is shown in Fig. 12.7. Note that the precise frequency response for $0 < \omega < \infty$ can only be determined by knowing the values

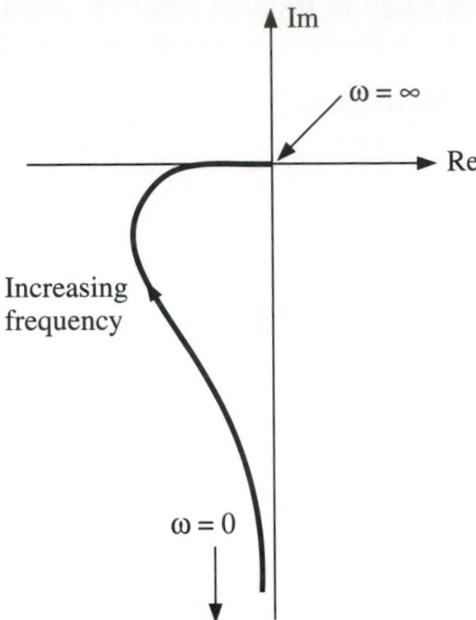

Fig. 12.7 Nyquist diagram

of K and T and evaluating equations 12.45 and 12.46 for particular values of ω. Fortunately, like the root locus, a sketch is all that is usually required, together with a few important points in the complex plane.

Sample Problem 12.1

An open-loop system has a transfer function of the form

$$GH(s) = \frac{0.1}{s(1 + 2s)(1 + 3s)}$$

Sketch the Nyquist diagram for the system and find the frequency that makes the phase angle $-180°$. What is the value of the magnitude at this frequency?

Solution

Using Table 12.1, the magnitude and phase of $GH(s)$ may be written directly as

$$M = \frac{0.1}{\omega\sqrt{1 + 4\omega^2}\sqrt{1 + 9\omega^2}}$$

and

$$\phi = -90 - \tan^{-1}(2\omega) - \tan^{-1}(3\omega)$$

Asymptotic values of the magnitude and phase may be obtained as

$$\omega \to 0: \quad M \to \infty \quad \phi \to -90$$

$$\omega \to \infty: \quad M \to 0 \quad \phi \to -270$$

A sketch of the Nyquist diagram is shown in Fig. SP12.1.1. The frequency response curve crosses the negative real axis ($\phi = -180$) for a particular value of frequency, at which point the phase equation gives

$$-180 = -90 - \tan^{-1}(2\omega) - \tan^{-1}(3\omega)$$

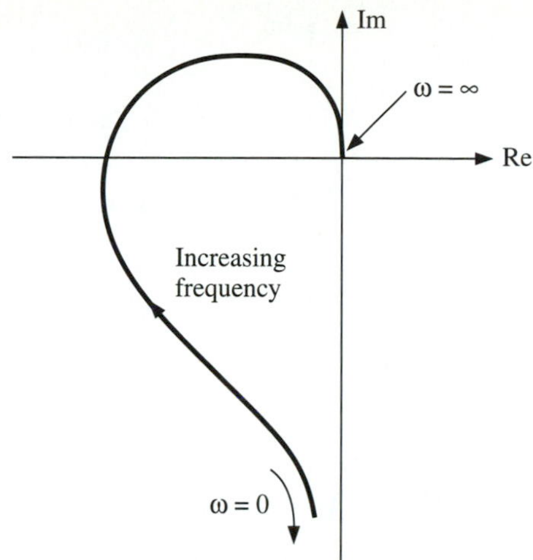

Fig. SP12.1.1

or

$$90 = \tan^{-1}(2\omega) + \tan^{-1}(3\omega)$$

Taking tangents of both sides of this equation yields

$$\infty = \tan\left[\tan^{-1}(2\omega) + \tan^{-1}(3\omega)\right]$$

Considering

$$\tan^{-1}(2\omega) = A \qquad \tan^{-1}(3\omega) = B$$

we can write the above phase equation as

$$\infty = \tan(A + B) = \frac{\tan A + \tan B}{1 - \tan A \tan B} = \frac{2\omega + 3\omega}{1 - 6\omega^2}$$

Hence

$$\omega = \frac{1}{\sqrt{6}}$$

Substituting this value of ω into the magnitude equation reveals

$$M = \frac{0.1}{(1/\sqrt{6})\sqrt{1 + 4/6}\sqrt{1 + 9/6}} = 0.12$$

Sample Problem 12.2

Sketch the Nyquist diagram for the system described by the open-loop transfer function

$$GH(s) = \frac{0.1s + 1}{s(0.2s + 1)}$$

and find the frequency and phase such that the magnitude is unity.

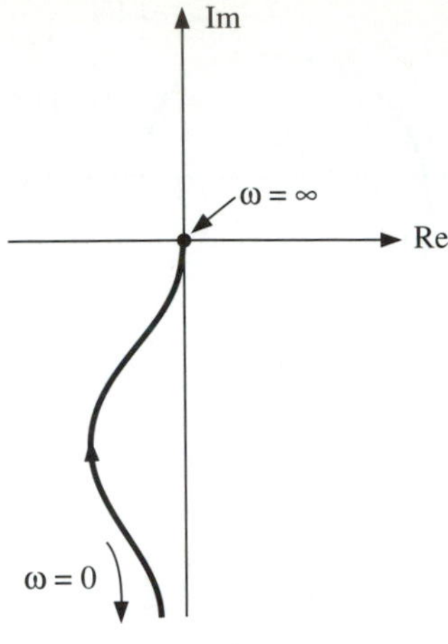

Fig. SP12.2.1

Solution

Using Table 12.1 and noting that the time constants of the numerator and denominator are 0.1 and 0.2 s, respectively, the magnitude and phase may be written directly as

$$M = \frac{\sqrt{1 + 0.01\omega^2}}{\omega\sqrt{1 + 0.04\omega^2}}$$

and

$$\phi = \tan^{-1}(0.1\omega) - 90 - \tan^{-1}(0.2\omega)$$

The asymptotic values are found to be

$\omega \to 0$: $\quad M \to 1 \times \infty \times 1 = \infty \qquad \phi \to 0 - 90 - 0 = -90$

$\omega \to \infty$: $\quad M \to 0 \times 0 \times \infty = 0$[2] $\qquad \phi \to +90 - 90 - 90 = -90$

The Nyquist diagram is as shown in Fig. SP12.2.1, noting that because the magnitude decreases from infinity to zero, it must be unity for some value of frequency. This frequency is obtained by setting the magnitude equal to unity:

$$1 = \frac{\sqrt{1 + 0.01\omega^2}}{\omega\sqrt{1 + 0.04\omega^2}}$$

or

$$\omega^2(1 + 0.04\omega^2) = 1 + 0.01\omega^2$$

leading to

$$0.04\omega^4 + 0.99\omega^2 - 1 = 0$$

[2]Observe how the terms in ω^2 go to infinity at the same rate, but the term in ω results in an overall zero magnitude.

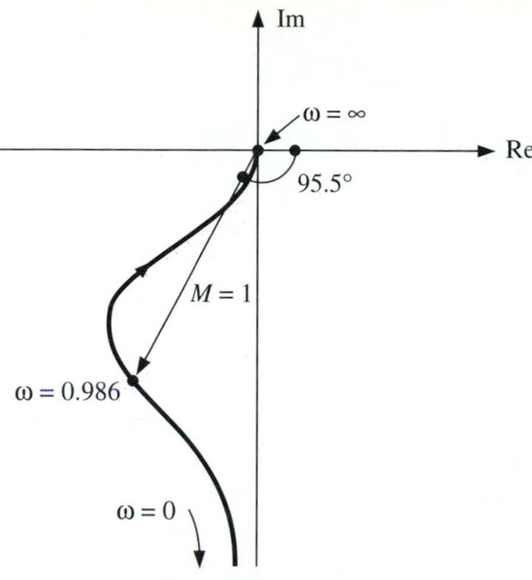

Fig. SP12.2.2

Putting $x = \omega^2$ yields the quadratic

$$0.04x^2 + 0.99x - 1 = 0$$

which has the solutions

$$x = -25.73, +0.972$$

Clearly the frequency has to be positive, leading to $\omega = 0.986$ rad/s. Substituting into the phase angle equation leads to

$$\phi = \tan^{-1}(0.0986) - 90 - \tan^{-1}(0.1972) = -95.5°$$

This vector is shown in Fig. SP12.2.2. Note that a vector with the phase angle of 95.5° intersects the open-loop frequency response twice, once for the point just found ($\omega = 0.986$ and $M = 1$) and again for a higher frequency and smaller magnitude. These two points could be found by solving the argument equation by setting $\phi = -95.5°$ and identifying the two frequencies that result. One would be 0.986 rad/s corresponding to $M = 1$, while the higher frequency would give the smaller magnitude.

Sample Problem 12.3

Plot the Nyquist diagram for the system described by the open-loop transfer function

$$GH(s) = \frac{1}{s(1 + 0.5s/10 + s^2/100)}$$

Solution

Note that the second-order element is already in the Bode form, and by comparing it with the standard form

$$GH'(s) = \frac{1}{1 + (2\zeta/\omega_n)s + s^2/\omega_n^2}$$

it is seen that

$$\omega_n = 10 \qquad \zeta = 0.25$$

Use of Table 12.1 will be made in determining the high- and low-frequency asymptotes for the Nyquist diagram:

$$\omega \to 0: \qquad M \to 1 \times \infty \times 1 = \infty \qquad \phi \to 0 - 90 - 0 = -90$$

$$\omega \to \infty: \qquad M \to 0 \times 1 \times 0 = 0 \qquad \phi \to 0 - 90 - 180 = -270$$

The general form of the Nyquist diagram looks as shown in the sketch in Fig. SP12.3.1. It is uncertain, however, how the magnitude varies with frequency. For example, it is not certain that the magnitude will decrease steadily from infinity to zero as the frequency increases, as suggested in Fig. SP12.3.1. In fact, since the damping ratio is quite low at 0.25, it is likely that a pronounced resonance peak will occur near the natural frequency ω_n, as suggested by Figs. 12.5 and 12.6. Under these circumstances, the Nyquist diagram may take the form shown in Fig. SP12.3.2, with the resonance peak occurring near $\omega = \omega_n$, at which frequency the phase ϕ is exactly $-180°$. For this example, we will plot the Nyquist diagram accurately using a computer program. First, Table 12.1 is used to determine the magnitude and phase to be

$$M = \frac{1}{\omega\sqrt{(1 - \omega^2/\omega_n^2)^2 + 4\zeta^2\omega^2/\omega_n^2}}$$

and

$$\phi = -90 + \tan^{-1}\left[\frac{-2\zeta\omega/\omega_n}{1 - \omega^2/\omega_n^2}\right]$$

where in this case $\zeta = 0.25$ and $\omega_n = 10$. The computer program listed below is written in BASIC and takes as input a value of ω and calculates the values of M and ϕ given above. Observing from Fig. SP12.3.2 that the phase is $-180°$ at

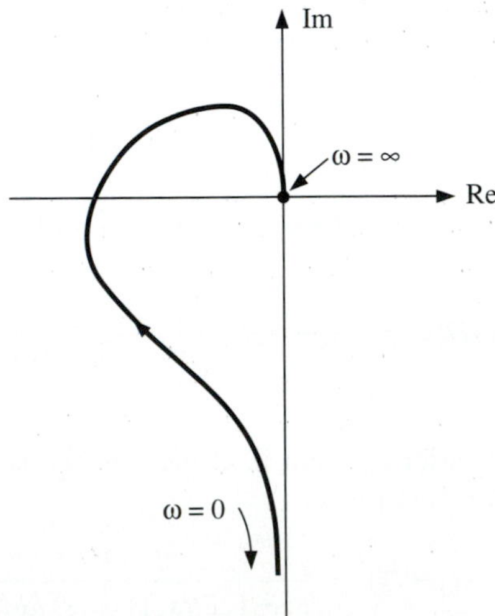

Fig. SP12.3.1

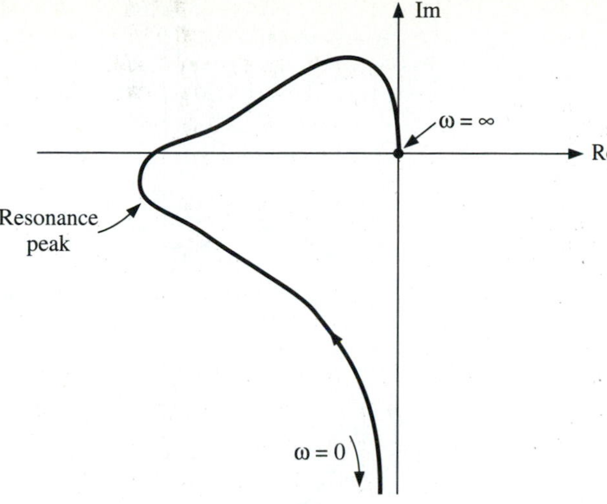

Fig. SP12.3.2

the natural frequency of 10 rad/s, values of ω in the range 1–100 rad/s seem appropriate in order to calculate the magnitude and phase:

```
1000 OMEGAN=10
1010 ZETA=0.25
2000 INPUT OMEGA
2010 WR=OMEGA/OMEGAN
2020 T1=WR*WR
2030 T2=1-T1
2040 M=1/(OMEGA*SQR(T2*T2 + 4*ZETA*ZETA*WR*WR))
2050 T4=ATN(2*ZETA*WR/T2)
2060 PHI=-90-180*T4/3.14159
2070 PRINT OMEGA,M,PHI
2080 GOTO 2000
```

Obtaining values of the magnitude and phase in the required range of ω produces the accurate Nyquist diagram shown in Fig. SP12.3.3.

◆ PROBLEMS

12.1 For the following open-loop transfer functions, sketch the Nyquist diagrams:

a. $GH(s) = \dfrac{K}{1 + 2s}$

b. $GH(s) = \dfrac{10}{s(1 + 4s)}$

c. $GH(s) = \dfrac{1}{s(s + 2)^2}$

d. $GH(s) = \dfrac{4}{s^2 + s + 1}$

e. $GH(s) = \dfrac{5}{s(s + 1)(s + 2)}$

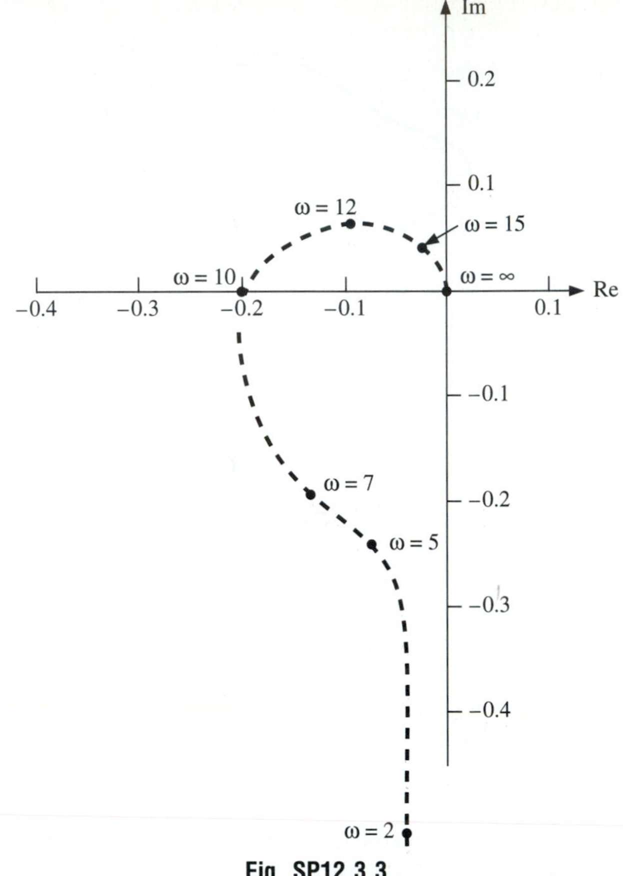

Fig. SP12.3.3

12.2 For problems 12.1 a–d, find the frequency that gives unity magnitude. For 12.1e, find the frequency that makes the phase −180°.

12.3 A system is described by the transfer function

$$GH(s) = \frac{2}{s + 5}$$

Calculate the frequencies necessary to make the phase angles −30°, −45°, and −60°, and calculate the corresponding magnitudes. Sketch the corresponding Nyquist diagram.

12.4 Sketch the Nyquist diagram for the standard second-order system described by

$$GH(s) = \frac{1}{1 + (2\zeta/\omega_n)s + s^2/\omega_n^2}$$

choosing values of damping ratio equal to 1.0, 0.5, 0.35, and 0.2.

12.5 Sketch the closed-loop frequency response of the unity-feedback system described by the open-loop transfer function

$$GH(s) = \frac{2(s + 10)}{s(s + 1)}$$

12.6 A system is described by the transfer function

$$GH(s) = \frac{5}{s(1 + 0.5s)(1 + 0.125s)}$$

Find the frequency for which the phase is $-180°$ and the corresponding magnitude. Use trial and error to determine the frequency at which the magnitude is unity and the corresponding phase angle.

12.7 Consider the system described by the *closed-loop* transfer function

$$GH(s) = \frac{1}{1 + 0.5s + 0.05s^2}$$

Draw as accurately as possible the closed-loop frequency response curve (Nyquist diagram) of this transfer function. Derive the open-loop transfer function using

$$\frac{C}{R} = \frac{G}{1 + GH}$$

and, assuming $H(s) = 1$, sketch the resulting Nyquist diagram.

12.8 Sketch the open-loop frequency response of the system described by

$$GH(s) = \frac{s + a}{s + b}$$

for the cases **(a)** $a > b$ and **(b)** $a < b$.

12.9 Figure P12.9 shows some experimentally determined open-loop frequency response plots. Match each plot with one of the transfer functions listed below:

a. $GH(s) = \dfrac{1}{(1 + sT_1)(1 + sT_2)(1 + sT_3)}$

b. $GH(s) = \dfrac{K}{s(s + a)(s + b)}$

c. $GH(s) = \dfrac{5}{s^3(s + 1)}$

d. $GH(s) = \dfrac{1}{s}$

e. $GH(s) = \dfrac{10}{(s + a)(s + b)(s + c)(s + d)}$

f. $GH(s) = 1 + sT$

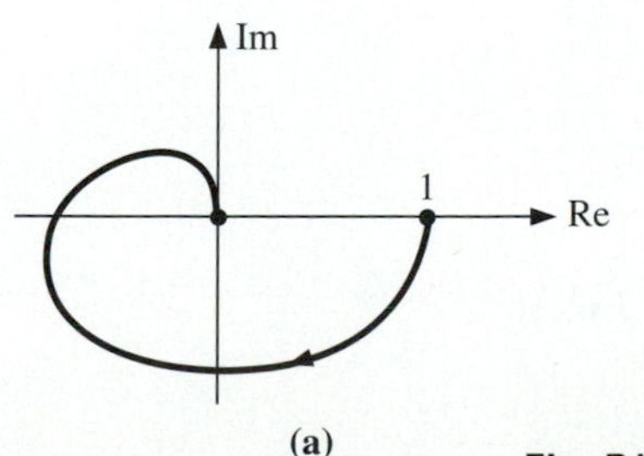

(a)

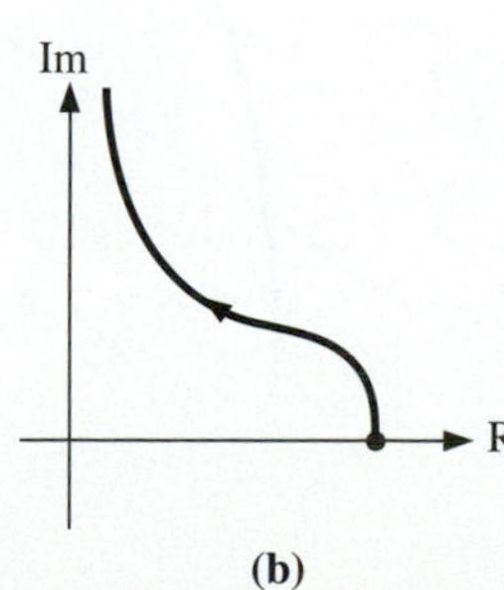

(b)

Fig. P12.9

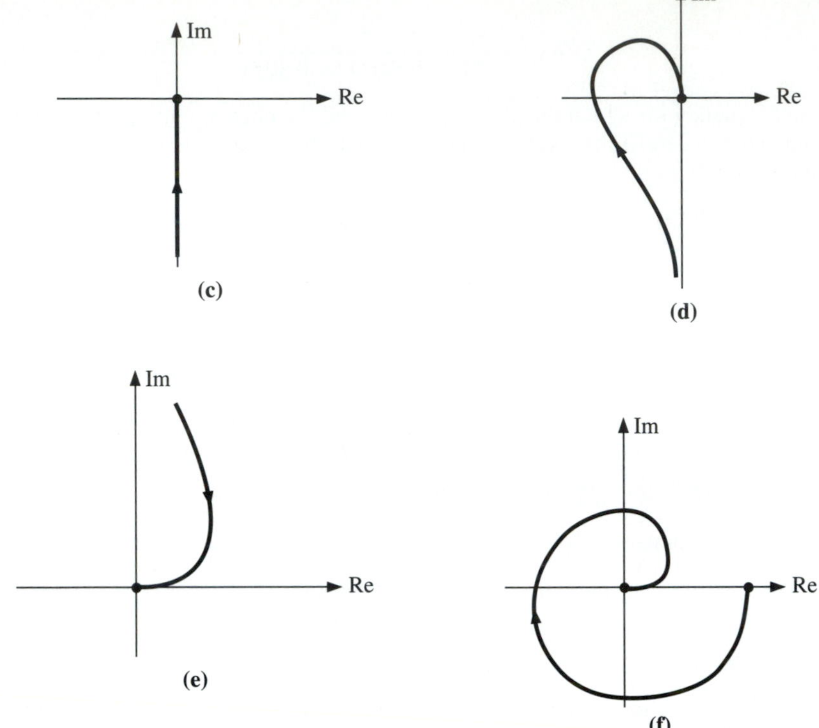

Fig. P12.9 (continued)

12.10 Figure P12.10 shows part of a Nyquist diagram for a system known to have a transfer function of the form

$$GH(s) = \frac{a}{s^3 + bs^2 + cs}$$

Evaluate a, b, and c, assuming the diagram is not drawn to scale.

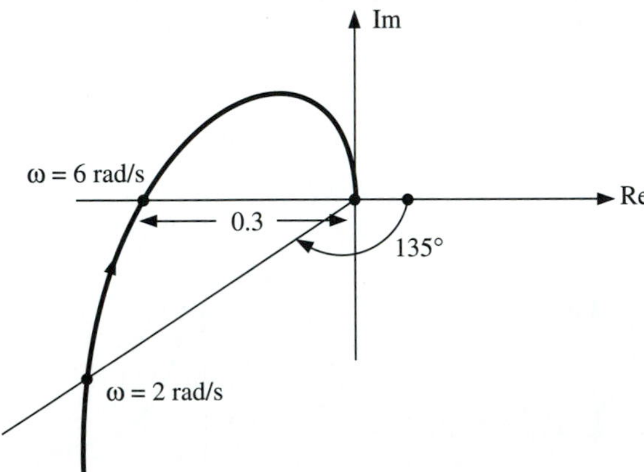

Fig. P12.10

MODULE 13

Nyquist Stability Criterion

Now that we know how to determine the frequency response of a system or if the frequency response is provided from experimental data, we are able to investigate system stability.

◆ CONFORMAL MAPPING: CAUCHY'S THEOREM

In order to see the relationship between the open-loop frequency response and the system stability, some rather abstract analysis needs to be studied. Consider a rational function $F(s)$ where s is a complex number. If this function is expressed in the manner

$$F(s) = \frac{\prod(s + z_i)}{\prod(s + p_j)} \tag{13.1}$$

then the z_i zeros and p_j poles of $F(s)$ may be drawn on the complex plane. To take a specific example, suppose that

$$F(s) = \frac{s + 1}{(s + 2)(s + 3)} \tag{13.2}$$

The poles and zeros of $F(s)$ are shown in Fig. 13.1. Suppose that an arbitrary contour Γ_s is defined in the complex plane, the s plane, as shown in Fig. 13.2. Now allow the variable s to take test values s_p as it traverses the contour. For any value of s_p on the contour, the magnitude and phase of $F(s)$ is calculated. From previous work with the root locus, it may be remembered that the magnitude and argument may be determined from Fig. 13.3 as

$$M = \frac{a \times b}{c} \tag{13.3}$$

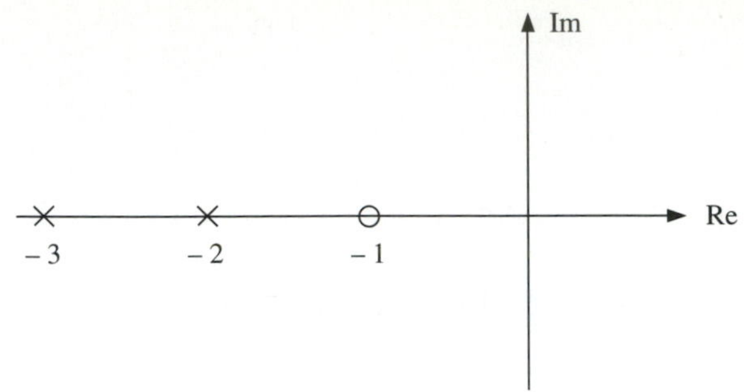

Fig. 13.1 Poles and zeros of $F(s)$

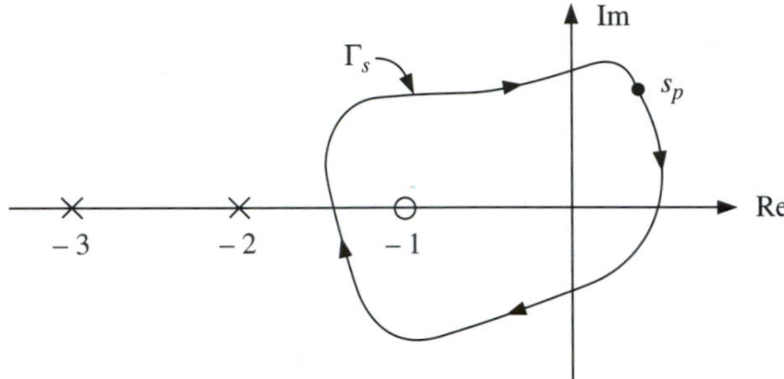

Fig. 13.2 Arbitrary contour Γ_S in the s plane

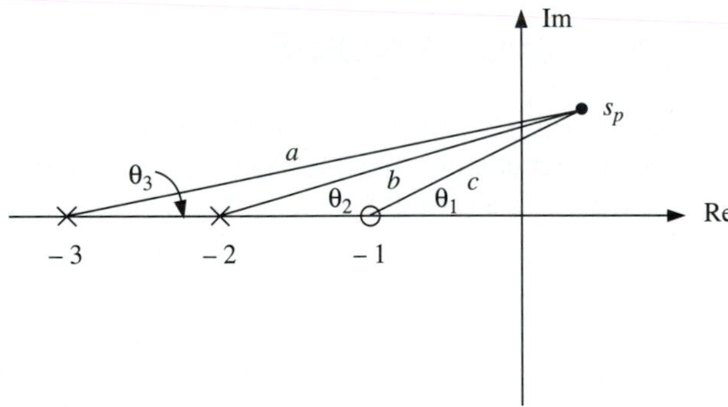

Fig. 13.3 Magnitude and phase calculation

and

$$\phi = \theta_1 - \theta_2 - \theta_3 \tag{13.4}$$

Now the magnitude and argument are plotted on another complex plane, the $F(s)$ plane, as a vector, the length of which is M and the angle with the positive real axis is ϕ. The original test point s_p is said to *map* to another point located at the tip of this vector. As s_p moves around the contour, there is a corresponding contour in the $F(s)$ plane. This contour is denoted by Γ_F and is shown in Fig. 13.4 for a rect-

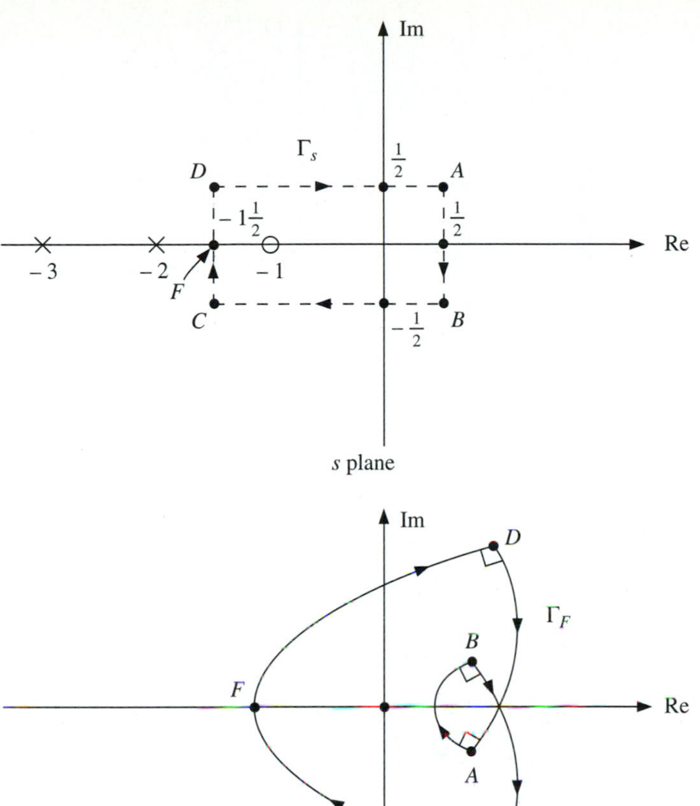

Fig. 13.4 Mapping the Γ_s contour to the Γ_F contour

angular Γ_s path.[1] Although the shape of the contour in the $F(s)$ plane may appear different from the contour in the s plane, the angles between adjacent lines are preserved.[2] Transformations that do not change these angles are called *conformal* transformations. Notice how the contour in the s plane makes one clockwise encirclement of a zero, producing a clockwise encirclement of the origin in the $F(s)$ plane. This may be understood from Fig. 13.3 by considering how the angles θ_1,

[1]The Γ_F contour may be evaluated for a particular transfer function using MATLAB. For example, the following function call is used here:

```
function [mag,phase]=conform(a,b)
s=a+b*j;
f=(s+1)/((s+2)*(s + 3))
```

[2]In the case of two adjacent curves, the angle between the tangents to the curves at the point of intersection is preserved.

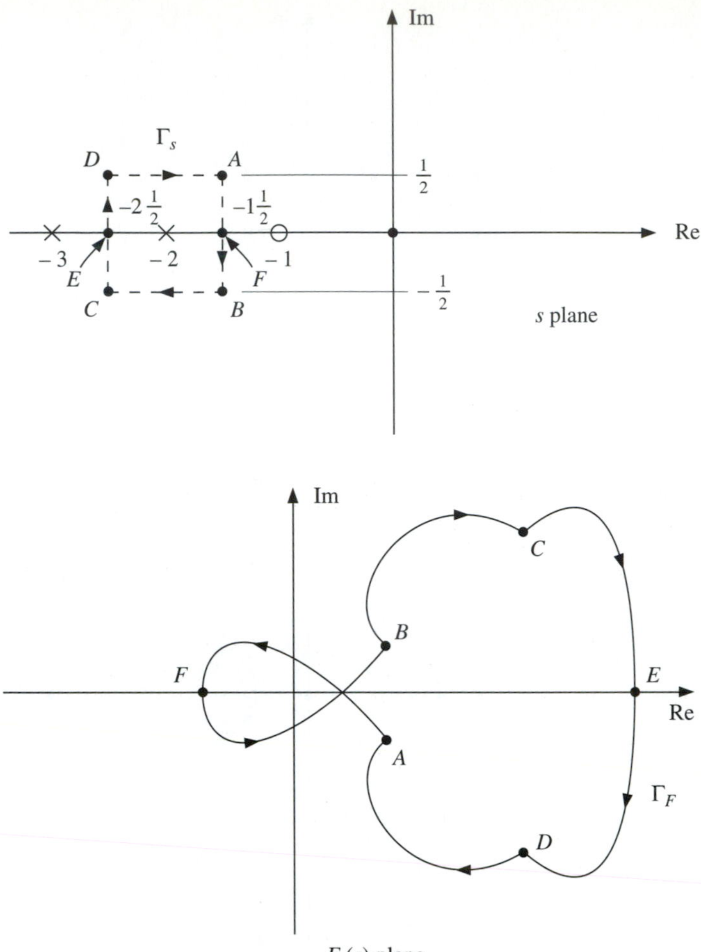

Fig. 13.5 Trajectory encircling a pole of $F(s)$

θ_2, and θ_3 vary as s_p traverses the contour. The contributions to the argument from the poles outside the contour will vary a little but will remain small relative to the contribution from the zero inside the contour. This causes the vector defining the contour in the $F(s)$ plane to perform one revolution in the clockwise direction. Figure 13.5 shows a different contour in the s plane, this time one that encircles a pole of $F(s)$ once in the clockwise direction. Shown also in the figure is the resulting contour in the $F(s)$ plane. In this case, the origin of the $F(s)$ plane is encircled once in the counterclockwise direction. Finally, shown in Fig. 13.6 is a trajectory that does not encircle either a pole or a zero, indicating that in this case, no encirclement of the origin of the $F(s)$ plane occurs.

These results may be summarized in Cauchy's theorem, which states:

> For a given contour in the s plane that encircles P poles and Z zeros of the function $F(s)$ in a clockwise direction, the resulting contour in the $F(s)$ plane encircles the origin a total of N times in a clockwise direction, where
>
> $$N = Z - P \tag{13.5}$$

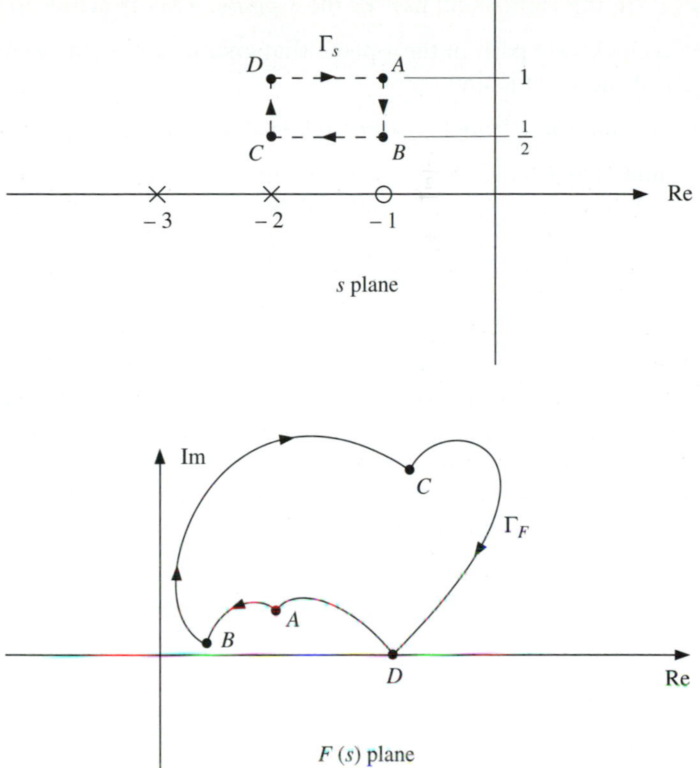

Fig. 13.6 Trajectory encircling no singularities of $F(s)$

One condition imposed on the contour in the s plane is that it must not pass through a pole or a zero, since this will cause indeterminate values for the argument. This analysis is certainly interesting, but how does it apply to control systems?

◆ APPLICATION TO STABILITY

Suppose we define the function $F(s)$ previously discussed to be equal to the characteristic equation

$$F(s) = 1 + GH \tag{13.6}$$

The roots of the characteristic equation are the closed-loop poles, which are the zeros of $F(s)$:

$$F(s) = 1 + GH = 0 \tag{13.7}$$

To determine the stability of the system described by the above characteristic equation, it will be necessary to determine whether there are any closed-loop poles [zeros of $F(s)$] in the right-hand half of the s plane. This is achieved by

1. defining a clockwise path in the s plane that encircles the whole of the right-hand half of the s plane and
2. observing the number of clockwise encirclements of the origin of the $F(s)$ plane.

Suppose for now that P in equation 13.5 is assumed zero. Then if the contour of $F(s)$ encircles the origin at least once in the clockwise direction, there exists a zero (a closed-loop pole) of $F(s)$ in the right-hand half of the s plane, and as a result the system will be unstable. Two questions remain: (1) What contour encompasses the right-hand half of the s plane? (2) How is the contour in the $F(s)$ plane obtained?

The first of these questions is answered by examining Fig. 13.7, which defines such a contour. This contour, known as the Nyquist contour, begins at the point $s = -j\infty$ and moves toward the origin out to the point $s = +j\infty$, then makes a circular path of infinite radius centered at the origin, back to the starting point. Clearly, such a contour encompasses all of the right-hand half plane. Although this contour appears somewhat artificial, it does have a physical significance, particularly if it is taken as three separate segments:

1. $0 < s < +j\infty$: In this segment, s takes the values of $j\omega$ where $0 < \omega < \infty$, which may be interpreted as a frequency response function for $F(s)$, namely $F(j\omega)$.
2. $-j\infty < s < 0$: Since the function

$$F(-j\omega) = -F(j\omega) \tag{13.8}$$

the frequency response function of $F(s)$ for negative frequencies will be the mirror image about the real axis of the positive frequency response function $F(j\omega)$.

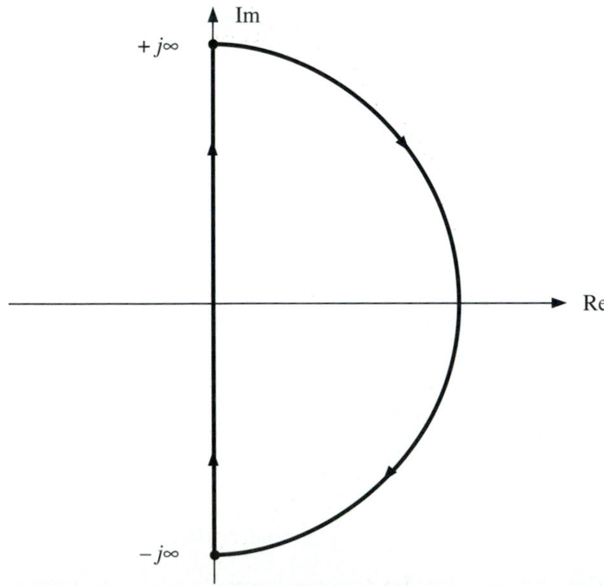

Fig. 13.7 Nyquist contour

3. Infinite semicircle: For all real systems, the magnitude of the frequency response function will approach zero as the frequency approaches infinity. Therefore, the two points in the s plane at $+j\infty$ and $-j\infty$ will map to the same point, the origin, in the $F(s)$ plane. The traversal of the infinite semicircle will not produce any movement in the $F(s)$ plane.

To address the second question regarding the interpretation of the contour in the $F(s)$ plane, recall that the Nyquist contour may be obtained solely from letting $s = j\omega$ and letting ω vary from zero to infinity. It is then necessary to calculate the magnitude and phase of $F(s)$. This is nothing more than calculating the frequency response of $F(s)$. Remembering that by definition

$$F(s) = 1 + GH(s) \tag{13.9}$$

it is seen that there is a connection between the frequency response of $F(j\omega)$ and the open-loop frequency response $GH(j\omega)$. In fact, if it was not for the 1 in the equation, the two frequency responses would be identical. This minor problem is circumvented by realizing that encirclement of the origin in the $1 + GH$ plane [the $F(s)$ plane] is the same as encirclement of the point $-1 + j0$ in the GH plane, as shown in Fig. 13.8. There are two loose ends to address:

1. It has been assumed that in Cauchy's theorem, $N = Z - P$, that the number of poles encircled in the s plane is zero. What if $P \neq 0$?
2. It was further assumed that the Nyquist path in the s plane did not pass through any singularities (poles or zeros). What happens if there are open-loop poles that lie on the Nyquist path?

To answer the first point, consider the characteristic equation in the form

$$F(s) = 1 + GH(s) = 1 + \frac{K\prod(s + z_i)}{\prod(s + p_j)} \tag{13.10}$$

As stated before, our prime interest is in the closed-loop poles, which are the zeros of the characteristic equation. Where do the poles of the characteristic equation come from? If a pole has a value of s that makes the function $F(s)$ infinite, then it is seen that the poles of the characteristic equation take the values $s = -p_j$, i.e., the open-loop poles. If the open-loop frequency response is obtained from the open-loop transfer function (using techniques described in the last module), then the open-loop poles are known. Note, however, that from the definition of the Nyquist

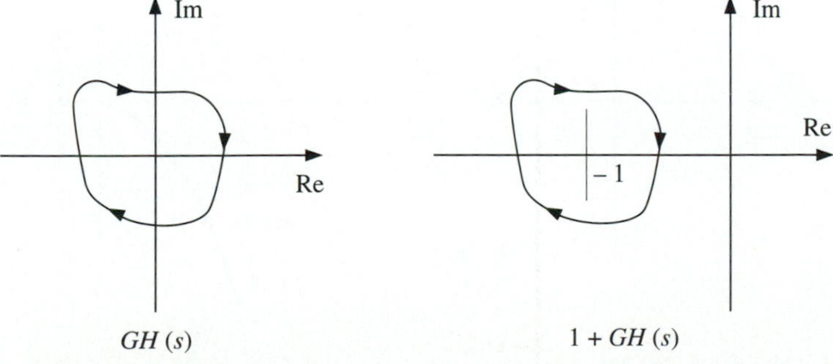

Fig. 13.8 Changing encirclement of the origin to the point $-1 + j0$

path, only the number of poles in the right-hand half of the s plane has to be known, and this number is equal to P in Cauchy's theorem. Systems with at least one open-loop pole or zero in the right-hand half of the s plane are known as *non-minimum-phase* systems, while those that do not have a pole or zero in the right hand half plane are called *minimum-phase* systems.

In this text, we will be concerned only with minimum-phase systems since these are the most common and non-minimum-phase systems require modifications to some of the stability methods outlined so far, such as the root locus. In addition, if a system has an open-loop pole in the right-hand half plane, an experimental determination of the open-loop frequency response will not be possible due to the system's unstable characteristics, thereby detracting from one of the chief advantages of using open-loop frequency response to predict closed-loop performance. We conclude, therefore, that for non-minimum-phase systems, $P = 0$.

To address the second issue, it is necessary to determine the effect of a singularity lying on the Nyquist path. Consider the particular example of

$$GH(s) = \frac{1}{s(s + 1)} \tag{13.11}$$

The s plane with the open-loop poles and zeros of $GH(s)$ plotted is shown in Fig. 13.9a. Note how the open-loop pole located at the origin lies on the Nyquist path. Using the techniques outlined in the previous module, the magnitude and argument of the open-loop system are

$$|GH(j\omega)| = \frac{1}{\omega\sqrt{1 + \omega^2}} \qquad \angle GH(j\omega) = -\tan^{-1}\omega - 90 \tag{13.12}$$

The frequency response in Nyquist form for $0 < \omega < \infty$ is shown in Fig. 13.9b. In order to avoid the Nyquist contour intersecting the pole at the origin, it is modified to include a small semicircle drawn around it. This semicircle is of radius ϵ, where $\epsilon \to 0$, and extends into the right-hand half s plane, as shown in Fig. 13.10. The open-loop frequency response for negative frequencies is obtained by reflecting the frequency response for positive frequencies about the real axis, as shown in Fig. 13.11. By substituting values of ω between $-\infty$ and $+\infty$ in equation 13.12,

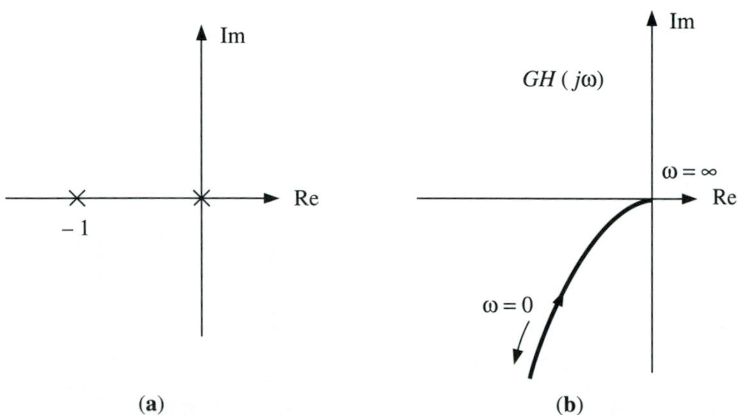

Fig. 13.9 (a) Open-loop poles and zeros and (b) frequency response of example system

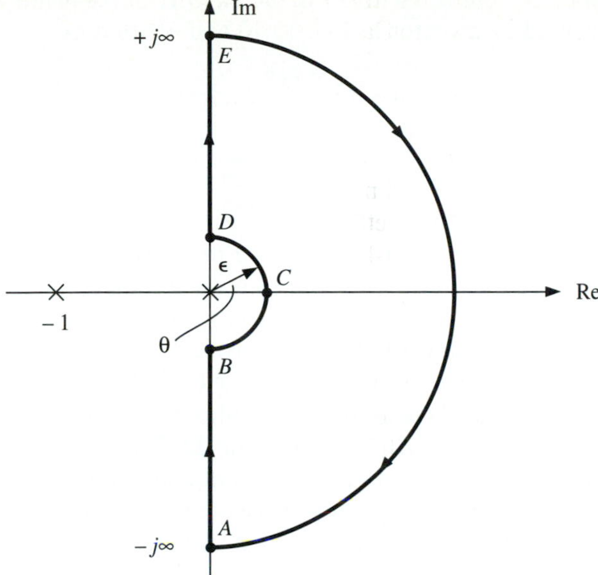

Fig. 13.10 Modified Nyquist contour

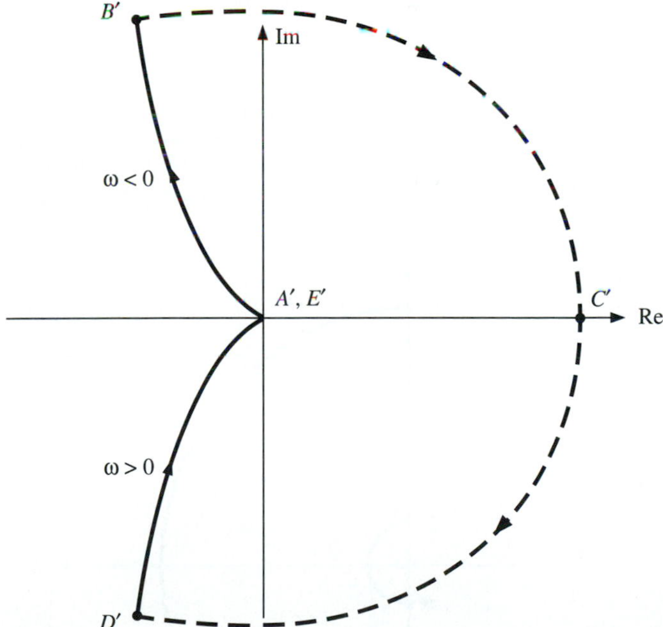

Fig. 13.11 Frequency response including a pole at the origin

the student will see that points A, B, C, D, and E map into A', B', C', D', and E' in the $GH(s)$ plane, shown in Fig. 13.11. Since the modulus is zero for infinite frequencies (positive and negative), the infinite semicircle in the s plane produces no change in the magnitude, and the contour in the GH plane remains at the origin. What happens to the small semicircle around the pole at the origin, or in other words, how is point B' connected to D'? In the vicinity of the origin of the s plane, s may be represented as a vector in Fig. 13.10 and written as

$$s = \epsilon e^{j\theta} \tag{13.13}$$

and

$$GH(s) = \frac{1}{s(s+1)} \approx \frac{1}{\epsilon e^{j\theta}} = \frac{1}{\epsilon} e^{-j\theta} \tag{13.14}$$

Clearly, whatever value the angle θ takes, the magnitude of the open-loop frequency response is infinity as $\epsilon \to 0$. Consider the argument of $GH(s)$ when s is located at point C. The total argument is zero, corresponding to point C' in Fig. 13.11. As s moves from C to D, θ varies from 0 to 90, but since it is a pole at the origin, the argument ϕ decreases from zero to -90 when s reaches D. In the $GH(s)$ plane, this corresponds to an infinite-radius quadrant from C' to D'. Similarly, when s moves from C to B, the contour in the $GH(s)$ plane moves along an infinite-radius quadrant from C' to B'. Note that a counterclockwise movement around the semicircle in the s plane produces a clockwise movement in the $GH(s)$ plane and how the direction would be reversed if there were a zero at the origin. Since the open-loop frequency response contour in the $GH(s)$ plane does not enclose the critical point $(-1, 0)$, the system is stable, as may be seen from a sketch of the root locus for this simple system.

Now consider the case of a double pole at the origin, as represented by the open-loop transfer function

$$GH(s) = \frac{1}{s^2(s+1)} \tag{13.15}$$

Proceeding as before, the Nyquist contour and open-loop frequency response are as shown in Figs. 13.12 and 13.13, respectively. In deciding how B' is connected to

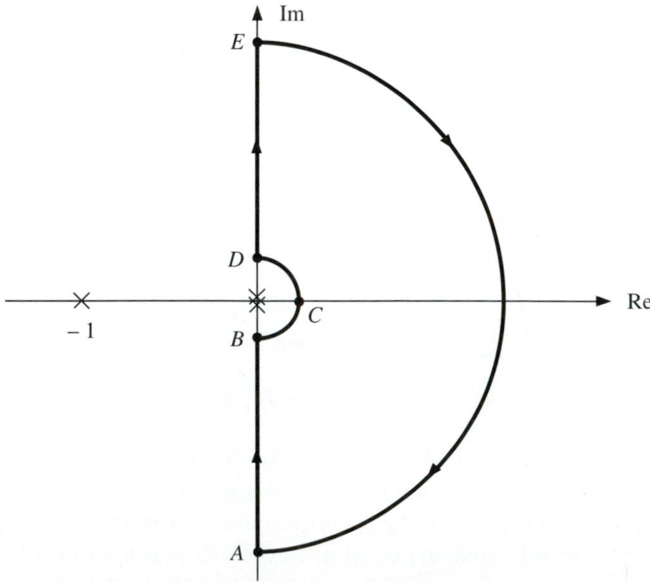

Fig. 13.12 Double-pole system

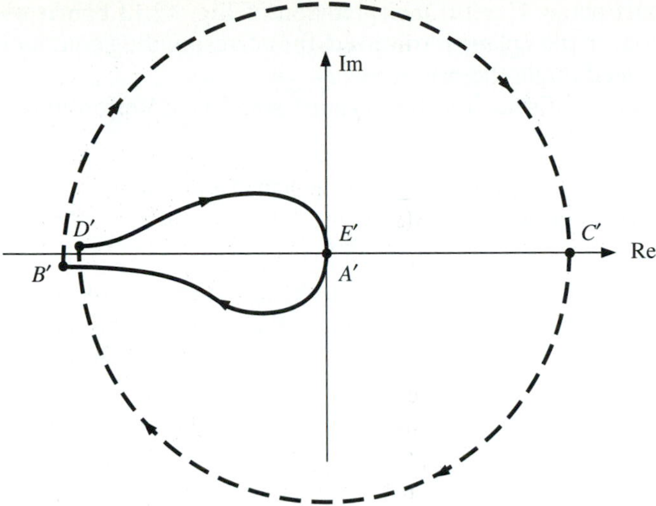

Fig. 13.13 Frequency response for double-pole system

D' in this case, note that two poles at the origin would produce an approximate open-loop frequency response

$$GH(s) = \frac{1}{\epsilon^2} e^{-2j\theta} \tag{13.16}$$

For each radian swept out by s_p as it moves from B to D, equation 13.16 indicates that the corresponding point in the $GH(s)$ plane sweeps out 2 rads; therefore as s moves from B to D (counterclockwise), the contour in the $GH(s)$ plane is an infinite-radius circle (two semicircles) in the clockwise direction, as shown in Fig. 13.13.

Since this contour in the $GH(s)$ plane encircles the point $(-1, 0)$ and there are no open-loop poles inside the Nyquist contour, there must be at least one closed-loop pole, or zero, of the characteristic equation inside the Nyquist contour, resulting in an unstable system. Sketching the root locus for this system confirms this observation; in fact the root locus shown in Fig. 13.14 shows two closed-loop poles in the

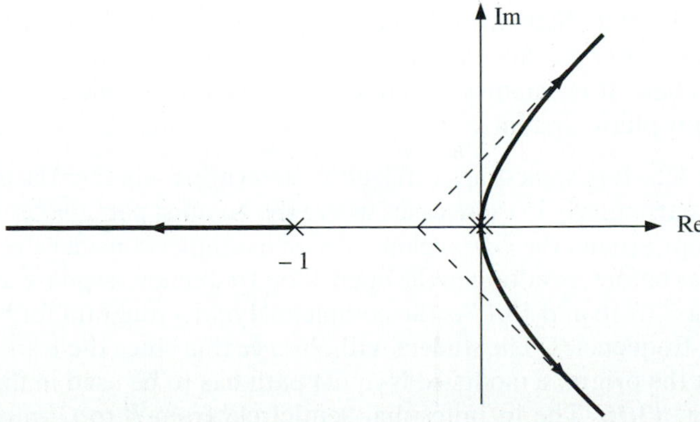

Fig. 13.14 Root locus for system with a double pole at the origin

right-hand half plane. Careful interpretation of Fig. 13.13 confirms that as the Nyquist contour in the s plane is traversed, the corresponding contour in the $GH(s)$ plane does indeed circle the origin *twice.*

Given the above discussion, the Nyquist stability criterion may be stated as follows:

> For a system having P open-loop poles in the right-hand half of the s plane to be stable, the open-loop frequency response must encircle the point $(-1, 0j)$ P times in a counterclockwise direction in the $GH(s)$ plane.

It is understood that if the open-loop transfer function places a singularity on the Nyquist contour, a modified Nyquist contour that bypasses the singularity must be used. Alternatively, the criterion may be written as: A system is stable if $Z \leq 0$, where

$$Z = N + P \tag{13.17}$$

and
Z = number of roots of $1 + GH(s)$ in right-hand half plane
N = number of clockwise encirclements of the critical point.
P = number of open-loop poles of $GH(s)$ in right-hand half plane

◆ SOME COMMENTS ON NYQUIST STABILITY

COMMENT #1. One simple physical interpretation of the Nyquist criterion is to imagine walking along the open-loop frequency response curve for $\omega > 0$ in the direction of increasing frequency. If the critical point $(-1, +0j)$ passes on the *right,* then the system is unstable; however, if the critical point passes to the *left,* the system is stable. The critical point must be directly viewable, i.e., no other part of the response curve should be between the observer and the critical point. Consider the open-loop transfer function

$$GH(s) = \frac{(1 + 0.34s)(1 + 0.2s)}{s^3(1 + 0.01s)(1 + 0.005s)} \tag{13.18}$$

This has an open-loop frequency response plot for positive frequency, as shown in Fig. 13.15. Imagine walking along this curve from $\omega = 0$ toward $\omega = \infty$ in the direction of the arrow. Since the critical point passes to the left of the observer, the system is stable. Note that when the observer is at point a, the critical point is to the right of the observer; however, it is not directly in view and therefore is not a valid observation. It is important to note that this simple rule does not apply to non-minimum-phase systems.

COMMENT #2. It is sometimes difficult to determine whether the critical point is encircled, particularly in those cases where the Nyquist path makes several total or partial loops around the $GH(s)$ plane. As an example, consider the same transfer function as before, resulting in the open-loop frequency response as Fig. 13.15. Shown in Figs. 13.16 and 13.17 is the complete Nyquist diagram for both positive and negative frequencies. The student will observe that since there are three open-loop poles at the origin, a modified Nyquist path has to be used in the s plane, as shown in Fig. 13.16. The infinitesimal semicircle from B to C traversed in the

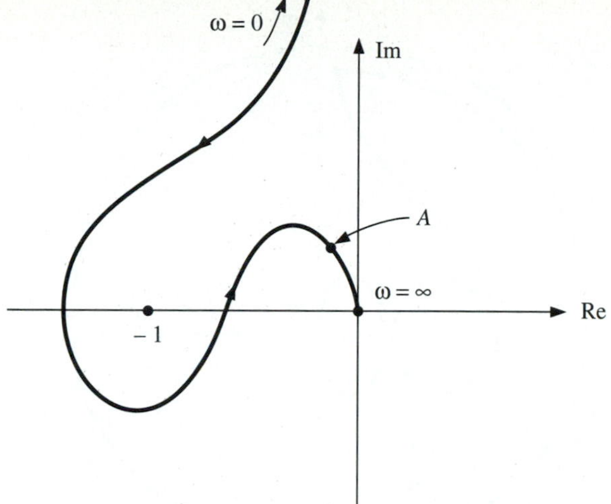

Fig. 13.15 Open-loop frequency response

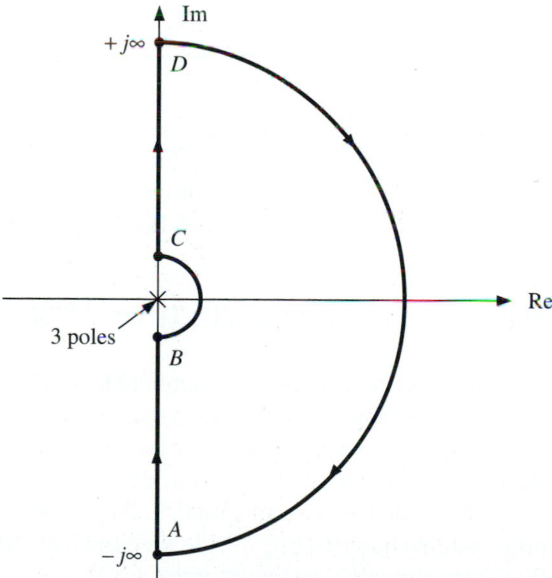

Fig. 13.16 Modified Nyquist path in the *s* plane

counterclockwise direction produces three semicircles in the *GH*(*s*) plane traversed in the clockwise direction. In order to calculate the number of encirclements of the critical point, draw a straight line radially outward from the critical point cutting all paths of the Nyquist diagram. Although this line is arbitrary, it should not be tangential to any part of the path. Such a line is *CX* in Fig. 13.17. The number of clockwise encirclements is obtained as the difference between N_{cw} and N_{ccw}, where N_{cw} is the number of times the path crosses the line in a clockwise sense and N_{ccw} is the number of times the path crosses the line in the counterclockwise sense. The path crosses the line twice in the counterclockwise sense at

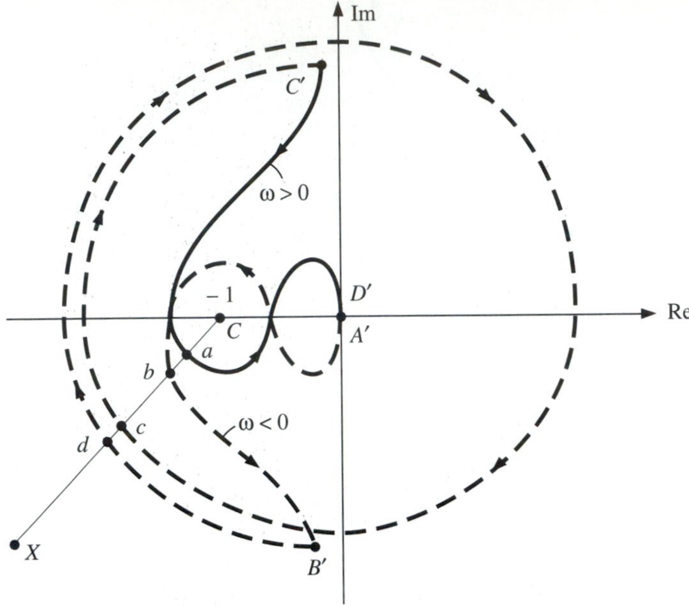

Fig. 13.17 Complete Nyquist diagram

a, b and twice in the clockwise sense at c, d. There are, therefore, no encirclements of the critical point, and since there are no open-loop poles or zeros in the right-hand half of the s plane, the system is stable, confirming the previous result obtained from the simple, physical interpretation of the Nyquist stability criterion.

◆ ALTERNATIVE APPROACH TO NYQUIST STABILITY CRITERION

The previous discussion on Nyquist stability is applicable to all systems, although we have seen that for minimum-phase systems[3] it may be reduced to examining the magnitude of the frequency response at the frequency that makes the phase $-180°$. This conclusion may be arrived at without dealing with the mappings of the previous sections, although it is done in a qualitative, rather than quantitative, manner. Consider the system shown in Fig. 13.18. Note the labeled signals E and B. We will assume that, initially, all signals are zero, i.e.,

$$r = e = c = b = 0 \tag{13.19}$$

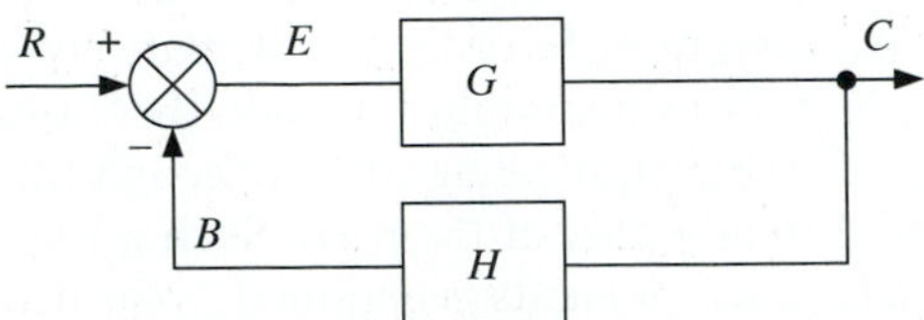

Fig. 13.18 General system frequency response

[3]Those without open-loop poles or zeros in the right-hand half of the complex plane.

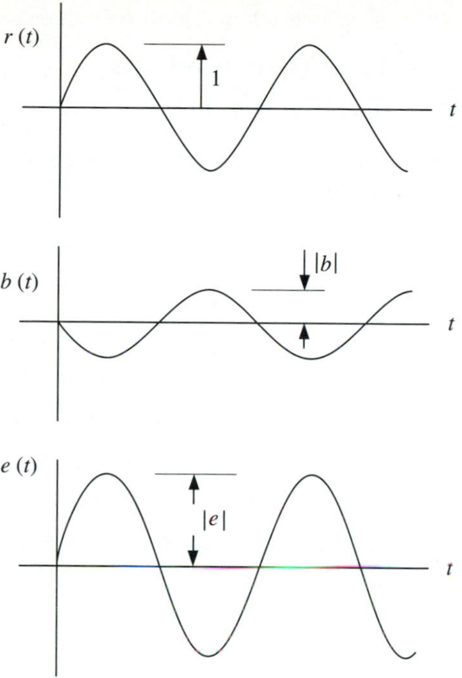

Fig. 13.19 Time-domain signals for harmonic input

Now suppose r is sinusoidal and of unity magnitude. Suppose further that the frequency ω' of r is such that the phase of b relative to r is exactly $-180°$. Figure 13.19 shows the signals, r, b, and e.

Since r and b are antiphase, the differencing junction will *add* them to produce e. Assuming the transient response disappears quickly, we will view the magnitudes of e and b starting at $t = 0$, as the signals propagate around the loop in a discrete-time fashion:

1. At $t = 0(+)$, $|r| = 1$ but $|b| = 0$; hence $|e| = 1$.
2. $GH(j\omega')$ will modify the magnitude of e. We know it changes the phase by $-180°$, but suppose $|GH(j\omega')| = 0.5$ also. Now $|b| = 0.5$.
3. Next time around the loop, r and b are added, making $|e| = 1.5$.
4. This now makes $|b| = 0.75$, which when added to r makes $|e| = 1.75$, and so on

It should be apparent that e successively takes the values

$$e = 1,\ 1\tfrac{1}{2},\ 1\tfrac{3}{4},\ 1\tfrac{7}{8},\ 1\tfrac{15}{16},\ \cdots \tag{13.20}$$

Clearly the system is stable since the error is approaching the value 2. Now consider the case where $|GH(j\omega')| = 2$:

1. At $t = 0(+)$, $|r| = 1$ but $|b| = 0$; hence $|e| = 1$.
2. Since $|GH(j\omega')| = 2$, $|b| = 2$.
3. Next time around the loop, r and b are added, making $|e| = 3$.

4. This now makes $|b| = 6$, which when added to r makes $|e| = 7$, and so on.

For this case, the values of e successively become

$$e = 1, 3, 7, 15, 31, \cdots \tag{13.21}$$

In this case the system is unstable, since the error increases indefinitely. Based on the above observations, the Nyquist stability criterion could be stated in the following form:

> For a system to be stable, the magnitude of $GH(j\omega)$ must be less than unity when evaluated at the frequency that makes the phase of $GH(j\omega) = -180°$.

This statement is consistent with the Nyquist criterion obtained earlier for nonminimum-phase systems. We shall, however, restate this observation as the physical interpretation of the Nyquist criterion discussed previously, namely:

> For a system to be stable, the critical point must pass to the left of the frequency response, as viewed in the direction of increasing frequency.

The correspondence between these two approaches is apparent, but as we shall see in later modules, only the latter one gives reliable information regarding system stability.

Sample Problem 13.1

Figure SP13.1.1 shows a contour Γ_s in the s plane. Calculate and plot the corresponding contour Γ_F in the GH plane when

(a) $$GH(s) = s + 2 - j$$

(b) $$GH(s) = \frac{1}{s + 1}$$

(c) $$GH(s) = \frac{1}{(s + 1)(s + 2)(s + 3)}$$

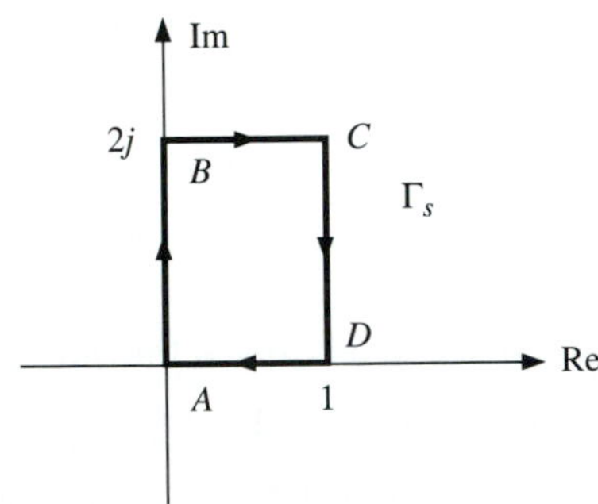

Fig. SP13.1.1

Solution

Case (a). Figure SP13.1.1 shows the contour with the corners labeled and the direction of the contour defined to be clockwise. In the first case considered, the transformation may be identified as *linear;* therefore straight lines in the s plane will be straight in the $GH(s)$ plane. The quickest way to draw the $GH(s)$-plane contour is to tabulate points in the s plane and calculate the corresponding points in the $GH(s)$ plane:

Point	s	$GH(s)$
A	$0 + 0j$	$2 - j$
B	$0 + 2j$	$2 + j$
C	$1 + 2j$	$3 + j$
D	$1 + 0j$	$3 - j$

The resulting contour is shown in Fig. SP13.1.2.

CASE (B). This mapping may be undertaken in a similar fashion to the previous one, except that it will involve the rationalization of some complex numbers. Each point on the s-plane contour will be taken in turn. The symbol $\Rightarrow$ will be used to denote "maps to":

$$s \Rightarrow \frac{1}{s + 1}$$

$$A \Rightarrow \frac{1}{0 + 0j + 1} = 1$$

$$B \Rightarrow \frac{1}{0 + 2j + 1} = \frac{1 - 2j}{1 + 4} = 0.2 - 0.4j$$

$$C \Rightarrow \frac{1}{1 + 2j + 1} = \frac{2 - 2j}{4 + 4} = 0.25 - 0.25j$$

$$D \Rightarrow \frac{1}{1 + 0j + 1} = 0.5$$

These points are plotted in Fig. SP13.1.3.

Note that these points should not necessarily be joined with straight lines, since the mapping is not linear (reciprocal function). This may be confirmed by mapping point $E = 0 + j$ located halfway between A and B:

$$E \Rightarrow \frac{1}{0 + j + 1} = 0.5 - 0.5j$$

which is not on the straight line joining A and B. The precise contour may only be obtained by calculating more points using a technique provided in the following case.

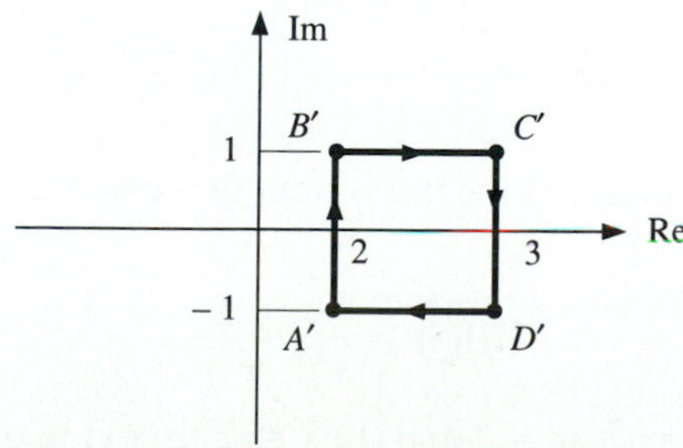

Fig. SP13.1.2

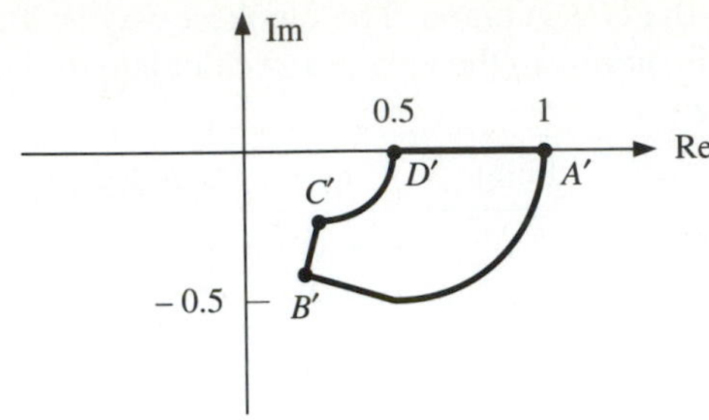

Fig. SP13.1.3

CASE (C). If we were to proceed as before, it is obvious that the rationalization of the resulting complex numbers for A', B', C', and D' in the $GH(s)$ plane will be cumbersome and is best achieved using computer methods. This may be accomplished using any language or software tool that supports complex-number data types. Given below is a FORTRAN program that takes a user-supplied complex number and calculates the corresponding point in the $GH(s)$ plane:

```
      COMPLEX S
1000  WRITE(6,*) 'INPUT S IN (REAL,IMAGINARY) FORMAT'
      READ(5,*) S
      WRITE(6,*) 1/((S+1)*(S+2)*(S+3))
      GOTO 1000
      END
```

Using this program to calculate the value of $GH(s)$ gives the contour shown in Fig. SP13.1.4.

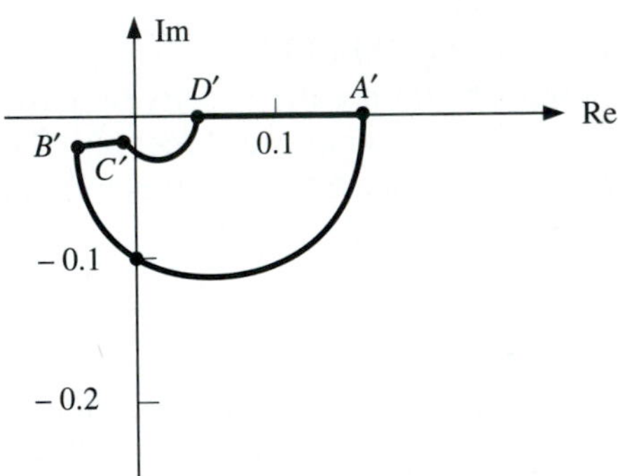

Fig. SP13.1.4

SAMPLE PROBLEM 13.2

Sketch the Nyquist diagram for the system described by the open-loop transfer function

$$GH(s) = \frac{s + z}{s^2(s + p)}$$

for the two cases (a) $z = 1, p = 2$ and (b) $z = 2, p = 1$. Comment on the system stability for each case.

Solution

CASE (a): z = 1, p = 2. The problem solution begins by drawing the s plane with the open-loop poles and zeros marked on it, together with the Nyquist path suitably modified so as to avoid the double open-loop pole at the origin. This is shown in Fig. SP13.2.1. It now remains to find the corresponding $GH(s)$ contour and determine if the critical point $(-1, 0j)$ is enclosed. The $GH(s)$ contour is begun by recognizing that the portion corresponding to s moving from D to E is the open-loop frequency response plot. It becomes necessary to determine the open-loop frequency response plot using the methods outlined in the previous module:

$$GH(s) = \frac{s+1}{s^2(s+2)} = \frac{1+s}{2s^2(1+\frac{1}{2}s)}$$

The magnitude and phase may be written directly as

$$M = \frac{0.5\sqrt{1+\omega^2}}{\omega^2\sqrt{1+\omega^2/4}}$$

and

$$\phi = -180 + \tan^{-1}(\omega) - \tan^{-1}(0.5\omega)$$

Checking asymptotic values yields

$$\omega \to 0: \qquad M \to \infty \qquad \phi \to -180$$

$$\omega \to \infty: \qquad M \to 0 \qquad \phi \to -180$$

It may be seen from the phase angle equation that for any positive value of ω between zero and infinity

$$\tan^{-1}(\omega) > \tan^{-1}(0.5\omega)$$

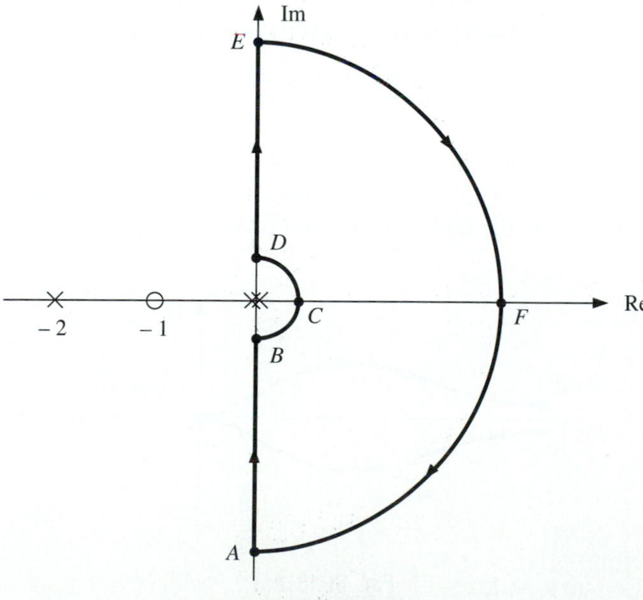

Fig. SP13.2.1

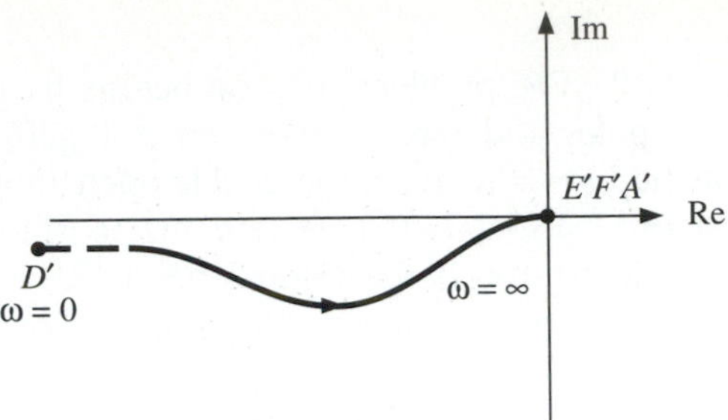

Fig. SP13.2.2

hence the resulting phase angle will be of the form

$$\phi = -180 + \Delta$$

ensuring that ϕ is always in the third quadrant. This portion of the contour is shown as $D'E'$ in the $GH(s)$ plane in Fig. SP13.2.2. The infinite semicircle EFA in the s plane maps to a single point, the origin, in the $GH(s)$ plane. Next, the contour in the s plane from B to A is the negative frequency response function $GH(-j\omega)$ and is plotted in the $GH(s)$ plane as the mirror image of $GH(j\omega)$, producing the contour segment $A'B'$, as shown in Fig. SP13.2.3. Finally, the traversal of the infinitesimal semicircle bypassing the double open-loop pole at the origin of the s plane has to be considered. Moving along the path BCD in the s plane produces a clockwise circular path of infinite radius in the $GH(s)$ plane connecting B' with D'. The two poles at the origin determine that this path is one complete revolution. The complete $GH(s)$ plane contour is shown in Fig. SP13.2.4. Since the critical point $(-1, 0j)$ is not encircled, it may be concluded that there are no zeros of the characteristic equation (closed-loop poles) within the Nyquist contour (right-hand half plane) and that the system is unconditionally stable. The student may check this result by considering the root locus of the system sketched in Fig. SP13.2.5. This confirms unconditional stability since the asymptotes will intersect the real axis at

$$\sigma_a = \frac{(-2) - (-1)}{3 - 1} = -\frac{1}{2}$$

CASE (b): z = 2, p = 1. This case is solved by proceeding in the same manner as before, marking the Nyquist contour on the s plane, plotting the corresponding contour on the $GH(s)$ plane, and investigating the encirclement of the

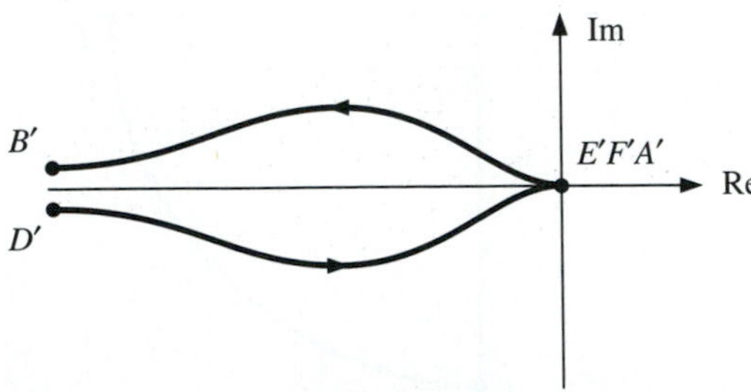

Fig. SP13.2.3

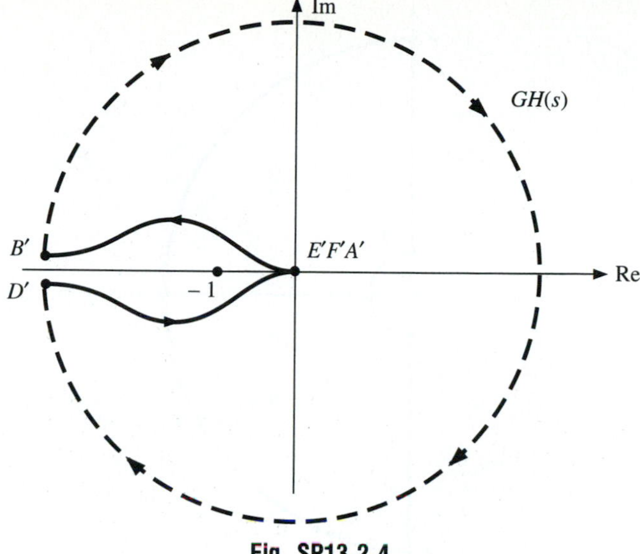

Fig. SP13.2.4

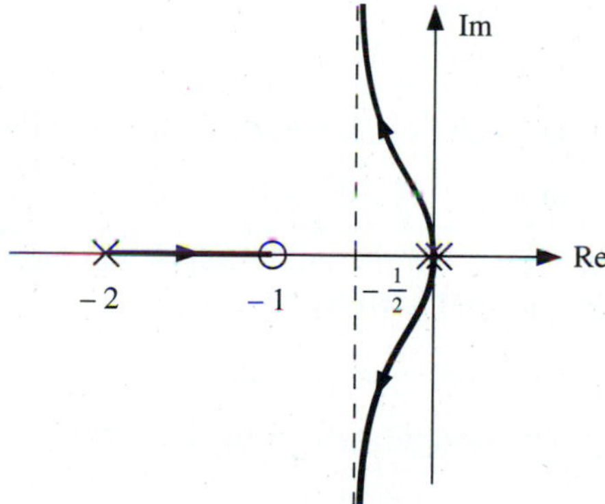

Fig. SP13.2.5

$(-1, +0j)$ point. The Nyquist contour, together with the open-loop poles and zeros, is shown in Fig. SP13.2.6. Again, the first part of the contour *DE* is obtained from the open-loop frequency response function $GH(j\omega)$. In this case

$$GH(j\omega) = \frac{2(1 + 0.5j\omega)}{(j\omega)^2(1 + j\omega)}$$

from which the modulus and phase are seen to be

$$M = \frac{2\sqrt{1 + \frac{1}{4}\omega^2}}{\omega^2\sqrt{1 + \omega^2}}$$

and

$$\phi = -180 + \tan^{-1}(0.5\omega) - \tan^{-1}(\omega)$$

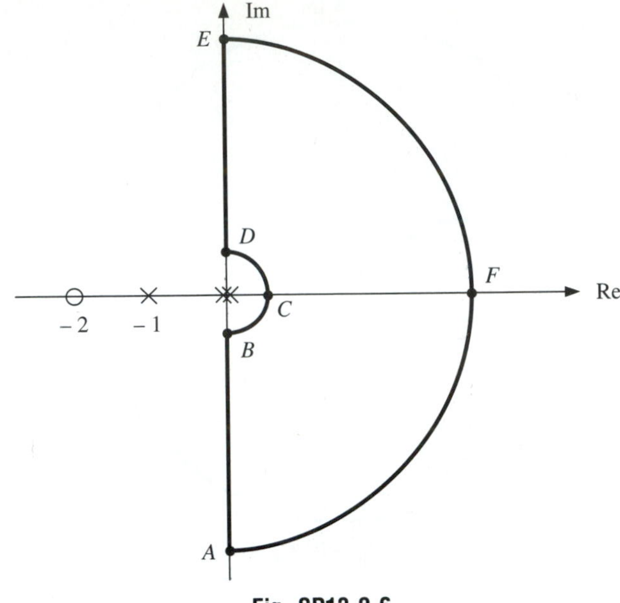

Fig. SP13.2.6

Asymptotic values of ω yield

$$\omega \to 0: \quad M \to \infty \quad \phi \to -180$$

$$\omega \to \infty: \quad M \to 0 \quad \phi \to -180$$

This time it may be seen from the phase angle equation that for any positive value of ω between zero and infinity

$$\tan^{-1}(\omega) > \tan^{-1}(0.5\omega)$$

Hence the resulting phase angle will be of the form

$$\phi = -180 - \Delta$$

ensuring that ϕ is always in the second quadrant, marked as $D'E'$ in the $GH(s)$ plane in Fig. SP13.2.7. Also shown in Fig. SP13.2.7 is the negative open-loop frequency response function as the image of $GH(j\omega)$ about the real axis, noting that the contour EFA still maps to the origin of the $GH(s)$ plane. The last part of the contour is the counterclockwise path around the two open-loop poles at

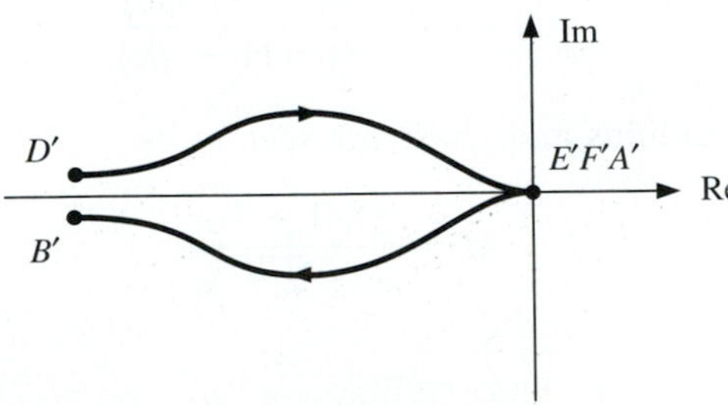

Fig. SP13.2.7

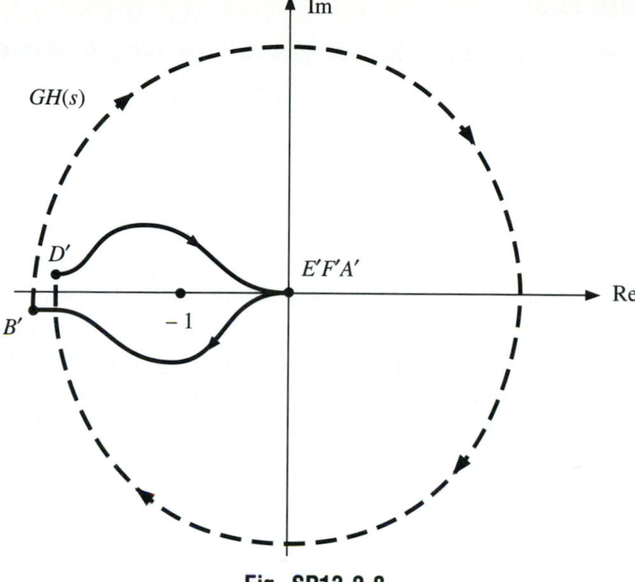

Fig. SP13.2.8

the origin, starting at B and finishing at D. It is known that this produces a complete infinite-radius circle in the clockwise direction, starting at B' and finishing at D'. This final contour is shown in Fig. SP13.2.8.

From this contour, it is seen that the critical point $(-1, 0j)$ is encircled twice, indicating that there are two closed-loop poles in the right-hand half of the s plane, resulting in an unstable system. Again, consideration of the root locus for the system confirms the result, since the intersection of the two asymptotes with the real axis is given by

$$\sigma_a = \frac{(-1) - (-2)}{3 - 1} = +\frac{1}{2}$$

The complete locus is shown in Fig. SP13.2.9.

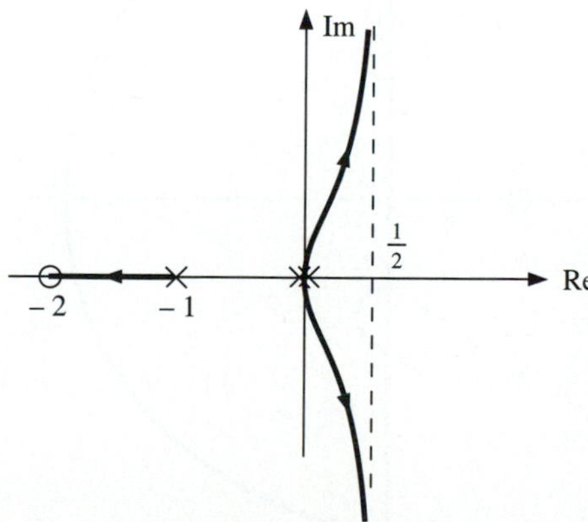

Fig. SP13.2.9

Sample Problem 13.3

Examine the stability of the system described by the open-loop transfer function

$$GH(s) = \frac{(s + 1)(s + 2)}{s^2 + 4}$$

Solution

The analysis is started by drawing the s plane with the open-loop poles and zeros marked. This is shown in Fig. SP13.3.1. In this case, there are two open-loop poles located on the imaginary axis, which is part of the proposed Nyquist path. It is necessary, therefore, to use a modified Nyquist path that avoids these poles with infinitesimal semicircles. This modified Nyquist path is shown in Fig. SP13.3.2 with some important points marked. In order to evaluate the open-loop frequency response, so as to evaluate the $GH(s)$ contour correspond-

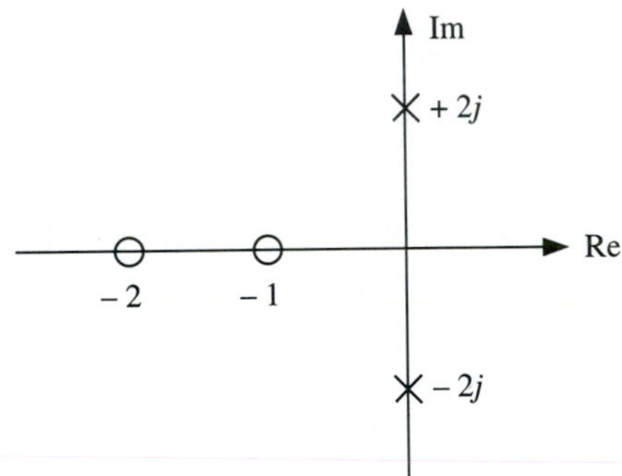

Fig. SP13.3.1

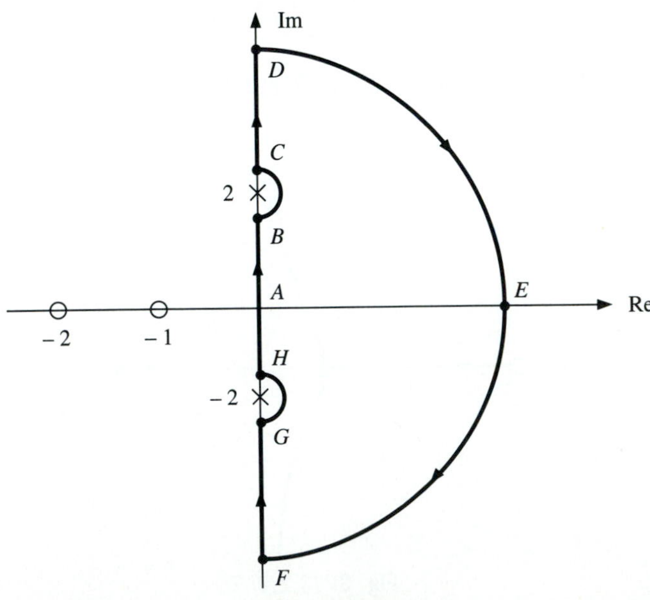

Fig. SP13.3.2

ing to path AD in Fig. SP13.3.2, the open-loop transfer function is first written in Bode form as

$$GH(s) = \frac{2(1 + s)(1 + 0.5s)}{4(1 + 0.25s^2)}$$

Then Table 12.1 is used to write the magnitude and phase as

$$M = \frac{0.5\sqrt{(1 + \omega^2)(1 + 0.25\omega^2)}}{1 - \omega^2/4}$$

and

$$\phi = \tan^{-1}(\omega) + \tan^{-1}(0.5\omega) - \Psi$$

where $\Psi = 0$ when $\omega < 2$, $\Psi = -180$ when $\omega > 2$, and $\Psi = -90$ when $\omega = 2$. Note the interpretation of the second-order system magnitude and phase for the special case of $\zeta = 0$.

Rather than investigate the asymptotic values of the open-loop frequency response, the diagram will be sketched for increasing frequency. Initially, $\omega = 0$ and the magnitude and phase become

$$M \to 0.5 \qquad \phi \to 0$$

As ω increases, the phase contribution from the second-order element remains zero, while the other two terms for ϕ increase, resulting in a positive phase angle. In addition, the magnitude increases. At the undamped natural frequency ($\omega_n = 2$ rad/s), the magnitude is infinite, at which point the phase increases from

$$\phi = \tan^{-1}(2) + \tan^{-1}(0.5) = 107.4$$

to the value

$$\phi = 107.4 - 180.0 = -73.6$$

These two points are shown in Fig. SP13.3.3. Since the increase in frequency from $2 - \delta$ to $2 + \delta$ corresponds to moving counterclockwise around BC in

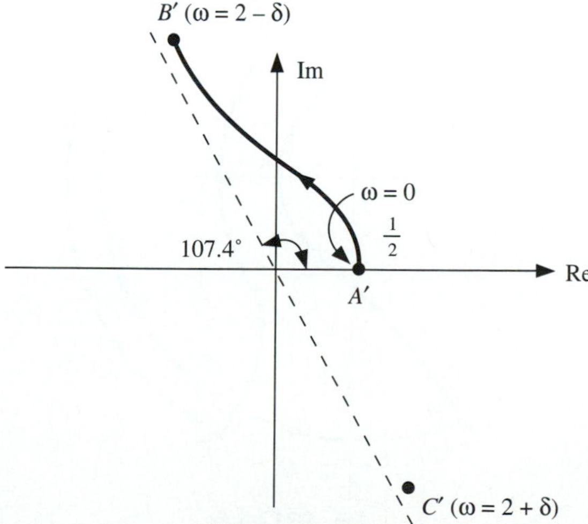

Fig. SP13.3.3

the s plane (Fig. SP13.3.2), it is seen that B' is joined to C' in the $GH(s)$ plane by an infinite-radius *clockwise* semicircle. Increasing ω further causes the magnitude to get smaller and the phase to increase in a positive sense. In the limit as $\omega \rightarrow \infty$,

$$M \rightarrow 1 \qquad \phi \rightarrow 0$$

The complete open-loop frequency response is shown in Fig. SP13.3.4.

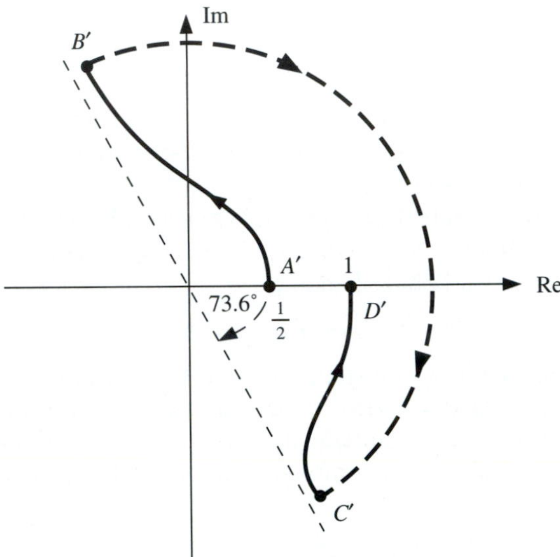

Fig. SP13.3.4

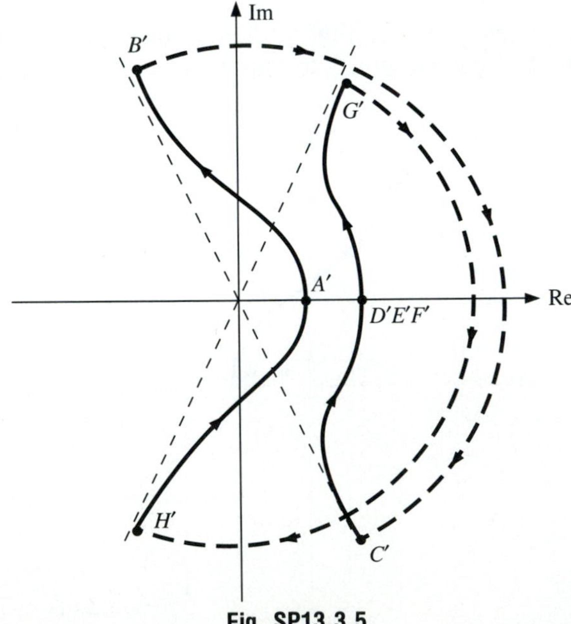

Fig. SP13.3.5

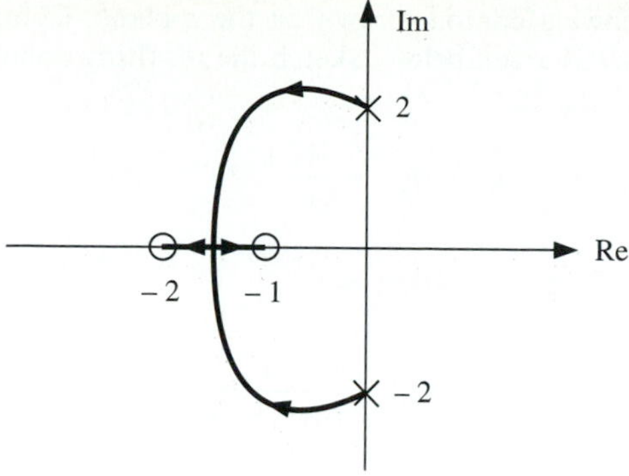

Fig. SP13.3.6

Since the negative frequency response will be required to determine closed-loop system stability, the open-loop frequency response is reflected about the real axis to provide the required plot. The complete Nyquist diagram is shown in Fig. SP13.3.5. The infinite-radius contour in the s plane DEF reduces to a single point in the $GH(s)$ plane. As before, the counterclockwise contour GH produces a clockwise contour $G'H'$. Since there are no open-loop poles in the right-hand half of the s plane, there should be no encirclements of the critical $(-1, +0j)$ point on the $GH(s)$ plane, and there are none. It may be concluded that the system is unconditionally stable and is confirmed from the root locus of the system, as shown in Fig. SP13.3.6.

◆ PROBLEMS

13.1 From first principles, calculate and plot the $GH(s)$ contour corresponding to the s-plane contour shown in Fig. P13.1 for the following transfer functions:

a. $GH(s) = s - 1$

b. $GH(s) = 2s + 1$

c. $GH(s) = s + 1 - 2j$

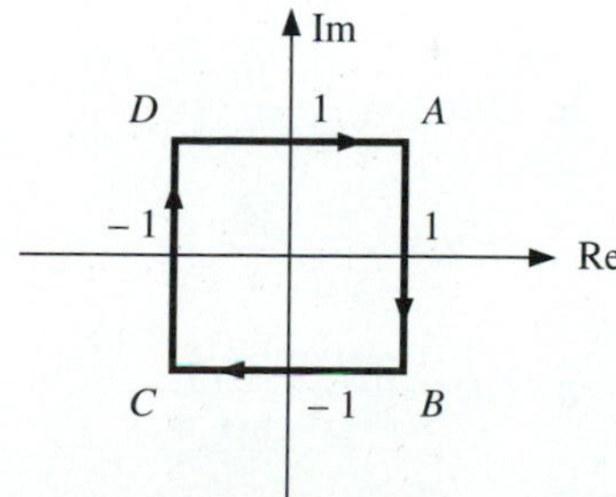

Fig. P13.1

13.2 Figure P13.2 shows a contour drawn on the s plane. Compute the mappings of points A–H for the $GH(s)$ given below. Sketch the resulting contour for the first transform only:

a. $GH(s) = \dfrac{1}{s}$

b. $GH(s) = \dfrac{1}{s^3}$

c. $GH(s) = \dfrac{1}{s(s+1)}$

d. $GH(s) = \dfrac{1}{(s+1)(s+2)}$

e. $GH(s) = \dfrac{1}{s^2 + 0.25}$

f. $GH(s) = \dfrac{(s+0.5)}{s(s+1)^2}$

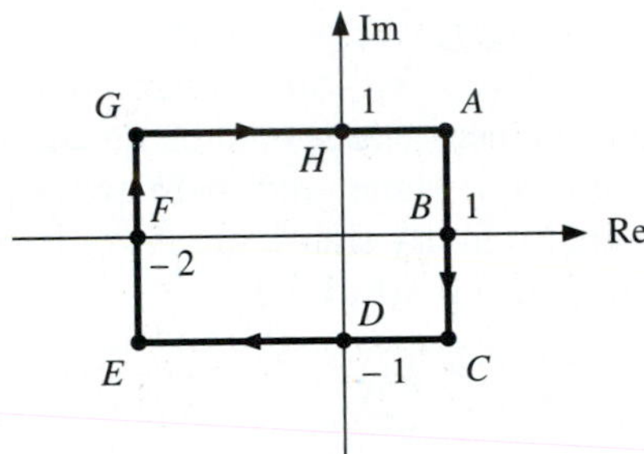

Fig. P13.2

13.3 a. In Fig. P13.3a, match each s-plane contour with the corresponding $GH(s)$ contour.
b. In Fig. P13.3b, how many zeros are enclosed by the s-plane contour?
c. In Fig. P13.3c, are any of the points A–E encircled by the contour? If so, how many, and in which direction do the encirclements occur?

13.4 For the following open-loop transfer functions, sketch the Nyquist diagrams for both positive and negative frequency and comment on the system stability:

a. $GH(s) = \dfrac{K}{1 + 2s}$

b. $GH(s) = \dfrac{10}{s(1 + 4s)}$

c. $GH(s) = \dfrac{10}{s(s+2)^2}$

d. $GH(s) = \dfrac{4}{s^2 + s + 1}$

e. $GH(s) = \dfrac{5}{s(s+1)(s+2)}$

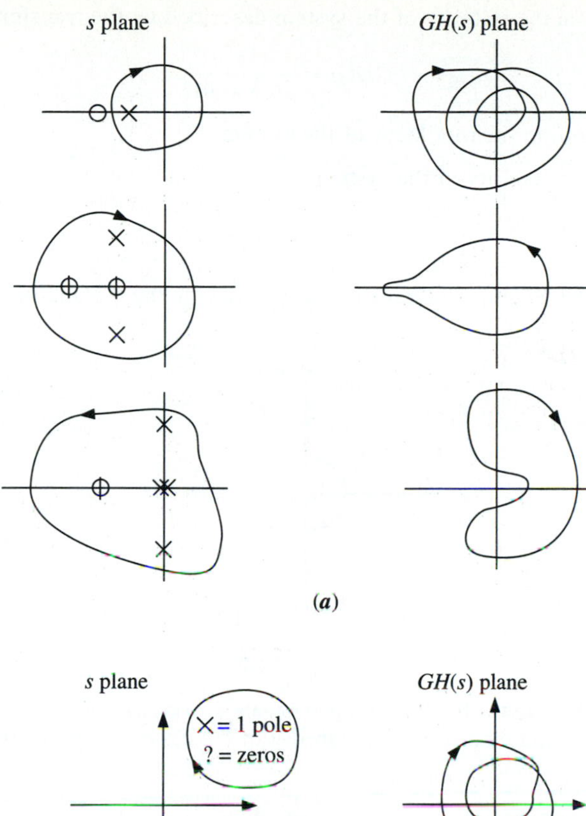

(*a*)

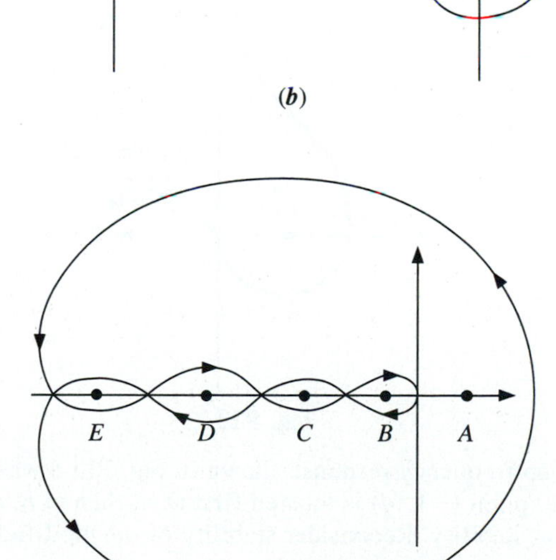

(*b*)

E D C B A

(*c*)

Fig. P13.3

13.5 Comment on the stability of the system described by the transfer function

$$GH(s) = \frac{1}{s(s^2 + 9)}$$

and check by plotting the root locus of the system.

13.6 Examine the stability of the system

$$GH(s) = \frac{s + 1}{s^2}$$

by modifying the Nyquist diagram in the vicinity of the s-plane origin as shown in Fig. P13.6.

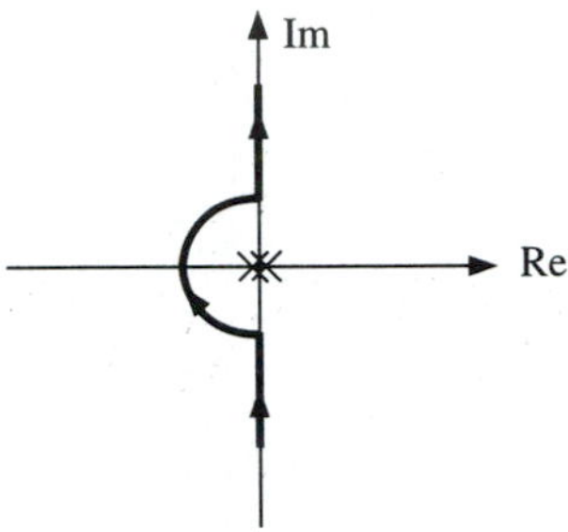

Fig. P13.6

13.7 Figure P13.7 shows the open-loop frequency response of a system. Estimate the transfer function, and determine the stability of the closed-loop system.

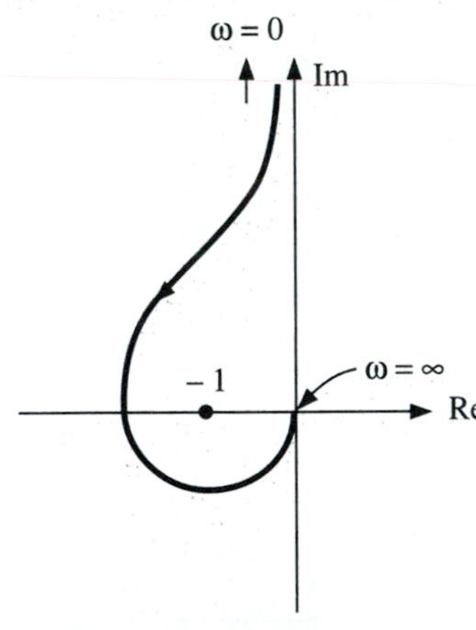

Fig. P13.7

13.8 The open-loop frequency response shown in Fig. P13.8 was obtained experimentally. If the critical point $(-1, 0j)$ is located first at A, then at B, and finally at C, comment on the system stability. Reconsider stability of the modified contour as shown in Fig. P13.6.

13.9 Consider a contour in the s plane such that it will encircle any closed-loop pole that has a time constant less than 1 s. Use this contour to determine if the closed-loop system obtained from

$$GH(s) = \frac{5}{s(s + 10)}$$

does indeed have a closed-loop pole with a time constant of less than 1 s.

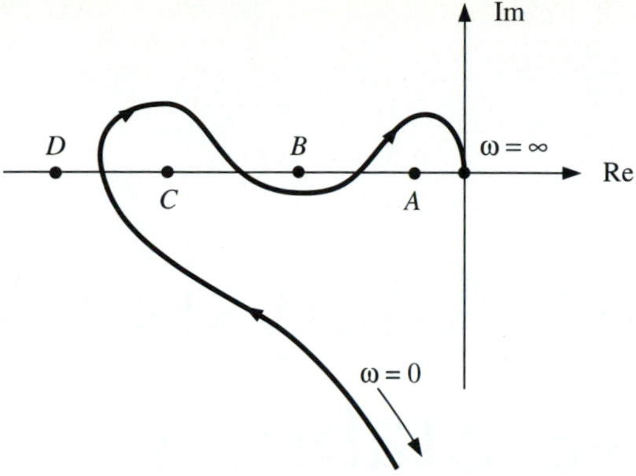

Fig. P13.8

13.10 Define a modified Nyquist path in the s plane, and find a corresponding path in the $GH(s)$ plane, to determine if the closed-loop response of the system

$$GH(s) = \frac{K}{s(s+2)}$$

has a closed-loop pole with $\zeta > 0.707$ when **(a)** $K = 1$ and **(b)** $K = 10$.

MODULE 14

Nyquist Analysis and Relative Stability

Now that the Nyquist stability criterion has been discussed, we will use it to study the conditions under which systems may be stable or unstable. In addition, measures will be developed that describe how close a system is to instability, thereby giving the designer a safety margin for stable operation.

◆ CONDITIONAL STABILITY

So far, we have drawn Nyquist diagrams and determined closed-loop system stability assuming a constant value of forward-path gain. It is known, however, from root locus analysis that some systems that have variable forward-path gain may be stable, marginally stable, or unstable depending upon the value of the gain. In this section we will investigate how this conditional stability appears in the Nyquist diagram.

Consider the system examined in Sample Problem 12.1, except that the constant forward-path gain of 0.1 is replaced by a variable gain K to give

$$GH(s) = \frac{K}{s(1 + 2s)(1 + 3s)} \tag{14.1}$$

Again using Table 12.1, the magnitude and phase may be written directly as

$$M = \frac{K}{\omega\sqrt{1 + 4\omega^2}\sqrt{1 + 9\omega^2}} \tag{14.2}$$

and

$$\phi = -90 - \tan^{-1}(2\omega) - \tan^{-1}(3\omega) \tag{14.3}$$

It follows that for a given value of frequency, the phase will be unchanged but the magnitude will be proportional to K. As the gain K increases, the resulting Nyquist diagram will be radially expanded about the origin. It is easily seen that the Nyquist diagram for $K = 0.3$ will have each magnitude three times that for $K = 0.1$. Further, the value $K = 0.05$ will produce a diagram where each magnitude is half that corresponding to $K = 0.1$. Example plots are shown in Fig. 14.1. Note, in particular, that as K increases, the diagram crosses the negative real axis at increasing values (in a negative sense). To check system stability, we will use the interpretation of the Nyquist stability criterion of observing the critical point $(-1, 0j)$ as we move along the $GH(s)$ contour in the direction of increasing ω. It is obvious, however, that if K is increased sufficiently, the intercept with the negative real axis will be to the left of the critical point, resulting in an unstable system. Since the phase is independent of K, the frequency at which the phase is $-180°$ is still $1/\sqrt{6}$, which allows the critical value of K to be obtained by equating the magnitude to unity:

$$1 = \frac{K}{(1/\sqrt{6})\sqrt{1 + 4/6}\sqrt{1 + 9/6}} \tag{14.4}$$

yielding the value

$$K_c = 0.832 \tag{14.5}$$

Values of K less than this value make the closed-loop system stable, values greater than K_c make the closed-loop system unstable, and $K = K_c$ gives the closed-loop

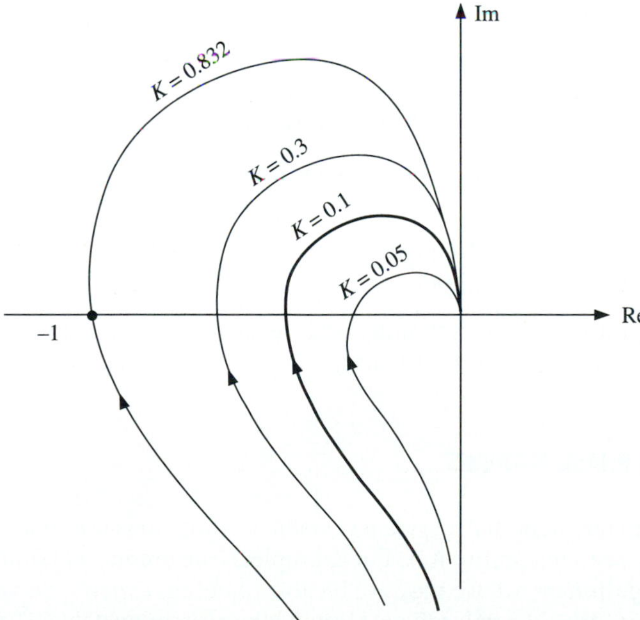

Fig. 14.1 Nyquist diagrams for different values of K

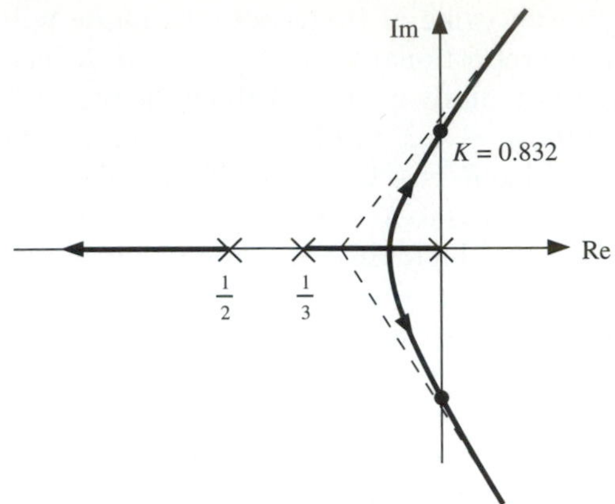

Fig. 14.2 Root locus for example system

system marginal stability, i.e., an oscillating response with no damping. In order for the system to be stable, therefore, we require

$$K < 0.832 \tag{14.6}$$

The corresponding root locus for this particular system is shown in Fig. 14.2. Comparing the root locus and the Nyquist diagram, it may be seen that the relationship between the imaginary axis of the root locus and closed-loop system stability is the same as the critical point in the Nyquist diagram and stability. In order to check the stability of a system, the following procedure may be followed:

1. Write the open-loop transfer function in Bode form, and write down expressions for the magnitude and phase.
2. Sketch the Nyquist diagram for an arbitrary value of gain to determine whether the critical point passes to the left or right of an observer moving along the frequency response curve in the direction of increasing frequency.
3. Using computer methods, if necessary, determine the frequency that makes the phase angle $-180°$.
4. Substitute this value of frequency into the magnitude equation and determine the corresponding magnitude.
5. If the magnitude is less than unity and the critical point passes to the left, the system is stable; otherwise it is not.

◆ GAIN AND PHASE MARGINS

Although the statement that a system is stable is of obvious use, it is also useful to know how close to instability it is. For example, if we predict a system is stable for the value of gain $K = 10$, we may not be able to obtain *exactly* the value 10 when we set the gain. Would a value $K = 11$ make the system unstable? To help answer such questions, the *gain margin* and the *phase margin* are sometimes quoted in order to specify how close to instability a system is. These margins both express

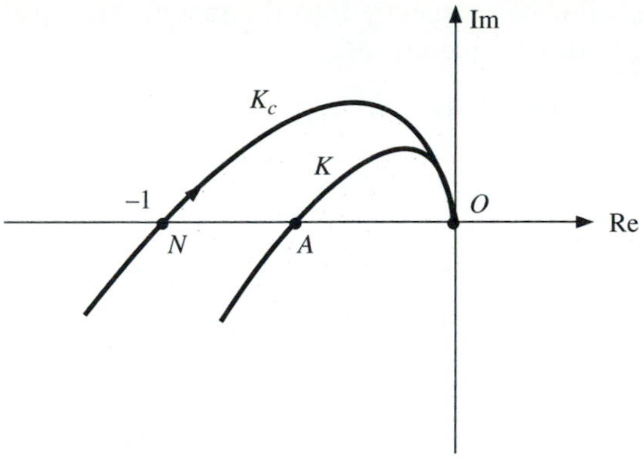

Fig. 14.3 Definition of gain margin

the "closeness" of the locus to the critical point, although they each express closeness differently. The gain margin is best explained in terms of Fig. 14.3. Suppose a system has the Nyquist diagram drawn for a particular value of K, as shown in Fig. 14.3. From the Nyquist stability criterion, it is clearly stable. Now consider the system Nyquist diagram for $K = K_c$, also drawn in Fig. 14.3:

> The *gain margin* is defined to be the gain needed to just make the system frequency response pass through the critical point divided by the actual system gain.

This may be interpreted as

$$\text{GM} = \frac{K_c}{K} \tag{14.7}$$

Since the magnitude is proportional to the value of K, the gain margin may be stated in terms of the following lengths in Fig. 14.3

$$\text{GM} = \frac{ON}{OA} = \frac{1}{OA} \tag{14.8}$$

Clearly for stable systems, the gain margin is greater than unity. Returning to the example of the system where it is proposed to make $K = 10$, suppose the gain margin were

$$\text{GM} = 1.5 \tag{14.9}$$

The gain could be set as high as

$$K_c = \text{GM} \times K = 15 \tag{14.10}$$

before system instability would occur; a value of $K = 11$ would not cause instability. The gain margin may be obtained from an accurately plotted Nyquist diagram. However, a procedure for calculating the gain margin is as follows:

1. From the system transfer function, derive expressions for the magnitude and phase as functions of frequency.
2. Using computer methods if necessary, determine the frequency at which the phase is $-180°$.

3. Substitute this value of frequency into the magnitude equation to determine the magnitude at this frequency, M_π.

4. Calculate the gain margin as

$$\text{GM} = \frac{1}{M_\pi} \tag{14.11}$$

5. If GM > 1 the system is stable; otherwise it is not.

Although gain margin is a measure of the closeness of the Nyquist diagram to the critical point, there are occasions when it is ambiguous and misleading. The system shown in Fig. 14.4a is a second-order system (root locus shown in Fig. 14.4b) that never crosses the negative real axis and is never unstable. As the forward-path gain is increased, the Nyquist diagram does get closer to the critical point. For all values of gain the gain margin is infinity. The root locus confirms that the system becomes more oscillatory even though it is not unstable. As another example, Fig. 14.5 shows two systems with the same gain margin. However, intuitively, system A is closer to instability than system B. The concept of gain margin becomes very unclear when the Nyquist diagram crosses the negative real axis more than once, as shown in Fig. 14.6. Clearly, another measure of closeness to the critical

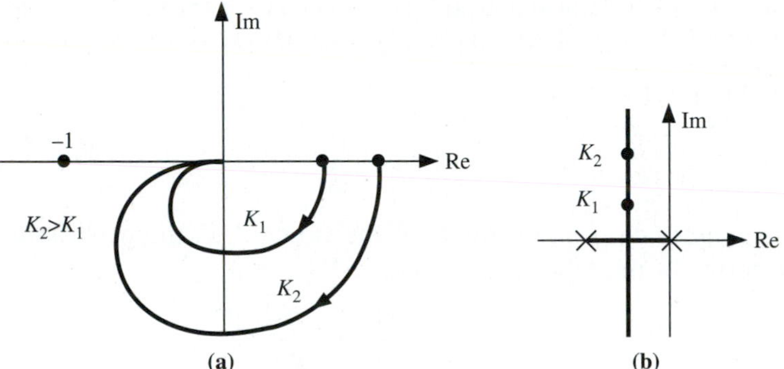

Fig. 14.4 Ambiguous gain margin

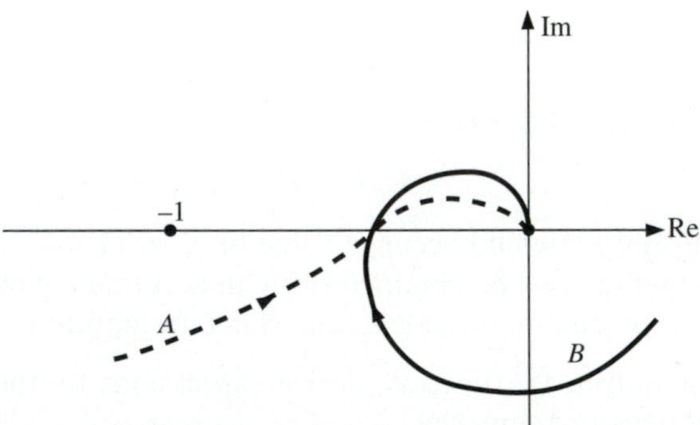

Fig. 14.5 Different systems having the same gain margin

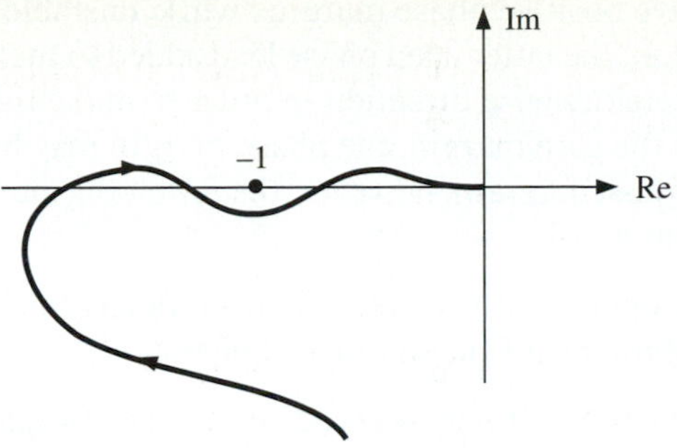

Fig. 14.6 Nyquist diagram crossing the real axis several times

point is required, and this is called the *phase margin*. This quantity is defined as follows:

> The phase margin is the amount of pure phase lag that has to be introduced to a system in order to make its frequency response pass through the critical point.

Again, this is best illustrated graphically. Consider Fig. 14.7, which shows a conditionally stable system. Consider a general vector *OB* for which the magnitude and phase have been evaluated for $\omega = \omega_1$. If a pure phase lag of, say, 10° is introduced to the system, the magnitude *OB* remains the same but the phase angle will increase in a counterclockwise sense by 10°. It may be seen that introducing a pure phase lag of θ degrees to the system will simply rotate the Nyquist diagram θ degrees in the clockwise direction. The Nyquist diagram shown in Fig. 14.7 will intersect the negative real axis at points closer to the critical point as this rotation occurs, and it may be appreciated that the system will become unstable when point *A*, which is the intersection of the Nyquist plot with the unit circle, rotates until it is coincident with the critical point. It follows that the angle ψ is the amount of pure phase that will cause instability and is, therefore, the phase margin.

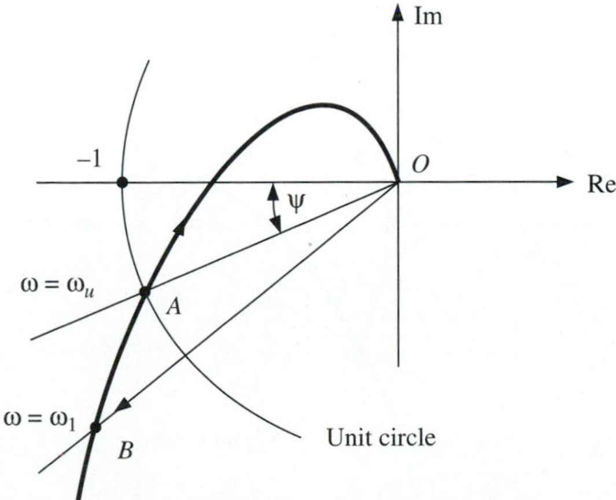

Fig. 14.7 Definition of phase margin

Stable systems have positive phase margins while unstable systems have negative phase margins, i.e., the latter need phase lead added to them, corresponding to rotation in the counterclockwise direction in order to make them stable, as shown in Fig. 14.8. As with the gain margin, the phase margin may be measured directly from an accurate Nyquist diagram; however, the following analytical method does not require a diagram at all:

1. From the system open-loop transfer function, determine expressions for the magnitude and phase as a function of frequency.
2. Using computer methods if necessary, calculate the frequency ω_u that makes the magnitude unity.
3. Substitute this frequency into the phase equation to determine the corresponding phase ϕ.
4. Evaluate the phase margin from

$$\text{PM} = 180 + \phi \tag{14.12}$$

5. If PM is positive, the system is stable; otherwise it is not.

In the expression for the phase margin, the correct sign for ϕ must be taken into account, i.e., a phase lag must be negative.

It will be shown later that although the gain margin test for stability (GM > 1) is not reliable, the test for the phase margin (PM > 0) does apply to all systems and should be considered the best test for system stability when using the Nyquist diagram. Although the relationships between gain and phase margins and system stability apply to some non-minimum-phase systems, they do not apply to all. It is best to regard the use of gain and phase margins to determine stability as applicable only to minimum-phase systems. It is not possible to say what the "best" values for gain and phase margins should be to obtain optimum response, but in

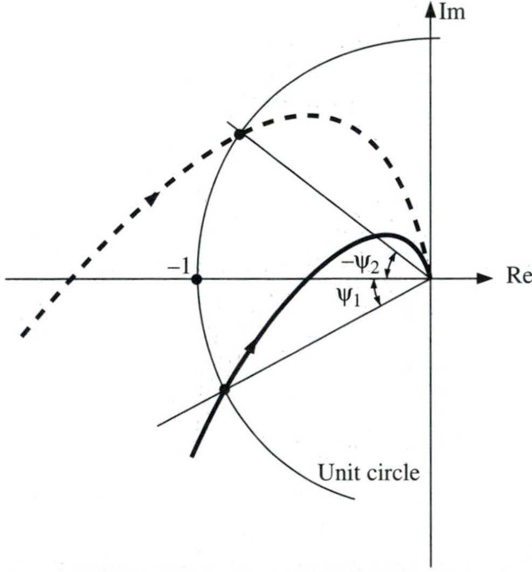

Fig. 14.8 Phase margin for stable and unstable systems

some published work a general rule of thumb is to attempt to obtain a gain margin of about 2 and a phase margin of between 40° and 60°. These are obviously rough guidelines, but they do produce a response with adequate margins of stability. These design guidelines correspond to specifying the "best" value of damping ratio when using the root locus method.

Sample Problem 14.1

Figure SP14.1.1 shows a unity-feedback control system. By sketching the Nyquist diagram of the system, determine the maximum value of K consistent with stability, and check the result using Routh's array. Sketch the root locus for the system.

Solution

The open-loop transfer function is first written in Bode form to facilitate the Nyquist analysis

$$GH(s) = \frac{K}{s(s^2 + 4s + 5)} = \frac{K/5}{s(1 + \frac{4}{5}s + \frac{1}{5}s^2)}$$

Comparing the second-order portion to the standard form

$$GH'(s) = \frac{1}{1 + (2\zeta/\omega_n)s + s^2/\omega_n^2}$$

it is seen that

$$\omega_n = \sqrt{5}$$

and the damping ratio is given by

$$\frac{2\zeta}{\omega_n} = \frac{4}{5} \qquad \zeta = 0.4\omega_n = 0.89$$

Using Table 12.1, the magnitude and phase may be written as

$$M = \frac{0.2K}{\omega\sqrt{(1 - \omega^2/\omega_n^2)^2 + 4\zeta^2\omega^2/\omega_n^2}}$$

and

$$\phi = -90 + \tan^{-1}\frac{-2\zeta\omega/\omega_n}{1 - \omega^2/\omega_n^2}$$

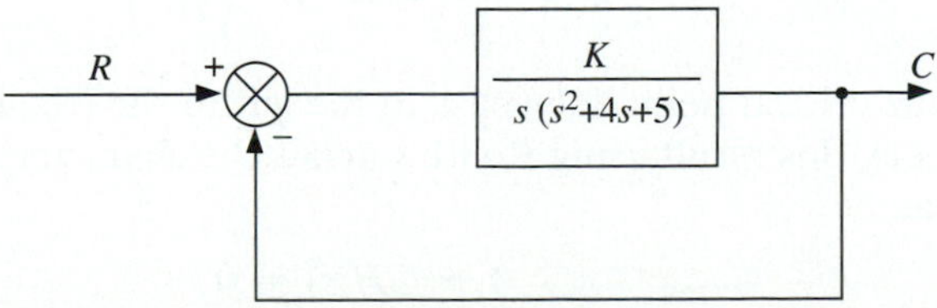

Fig. SP14.1.1

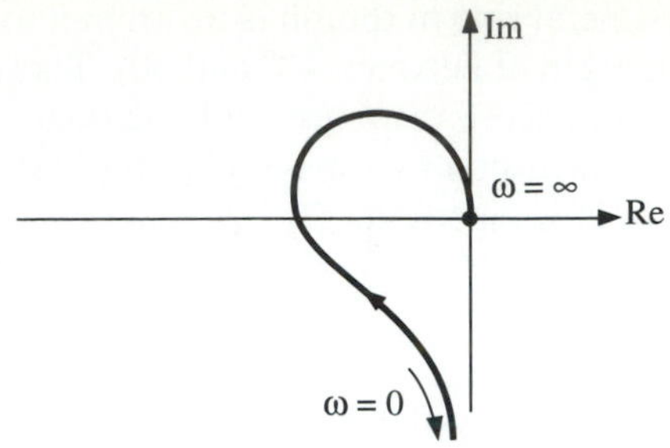

Fig. SP14.1.2

In order to sketch the Nyquist diagram, asymptotic values of the magnitude and phase are estimated:

$$\omega \to 0: \qquad M \to \infty \qquad \phi \to -90$$

$$\omega \to \infty: \qquad M \to 0 \qquad \phi \to -270$$

The Nyquist diagram is shown in Fig. SP14.1.2 and clearly crosses the negative real axis. To determine maximum gain K consistent with stability, we follow the method recommended earlier:

1. Find the frequency that makes the phase $-180°$.
2. Calculate the magnitude at this frequency.
3. Determine $\hat{K}$ from the equation $M = 1$, and check that the critical point passes on the left.

From the phase angle equation

$$-180 = -90 - \tan^{-1} \frac{2\zeta\omega/\omega_n}{1 - \omega^2/\omega_n^2}$$

Clearly, the second term on the right-hand side of the above equation has to be $-90°$, which will only be true when the argument of the arctangent function is infinity. This occurs when

$$\omega = \omega_n$$

Substituting this value back into the magnitude equation,

$$M = \frac{0.2K}{\sqrt{5} \times 2\zeta}$$

Setting $M = 1$ and substituting for ζ yields

$$\hat{K} = \frac{2\zeta\sqrt{5}}{0.2} = 20$$

For $K < \hat{K}$, the critical point does pass to the left of the frequency response. In order to check this result using Routh's array, the characteristic equation is first derived as

$$1 + GH(s) = 0$$

$$s^3 + 4s^2 + 5s + K = 0$$

Routh's array takes the form

$$\begin{array}{llll} s^3: & 1 & 5 & \cdots \\ s^2: & 4 & K & \cdots \\ s^1: & \frac{1}{4}(20 - K) & 0 & \cdots \\ s^0: & K & & \cdots \end{array}$$

Examination of the first column of the array generates the stability condition as $0 < K < 20$, resulting in $\hat{K} = 20$. The root locus may be plotted from the open-loop transfer function as given in Evans form in the problem statement. The student may verify that the locus is as shown in Fig. SP14.1.3. Interestingly, the locus from the complex poles does not move directly toward the asymptotes but enters the negative real axis first and then breaks away. The gain where the locus passes through the imaginary axis is indeed

$$\hat{K} = 20$$

Sample Problem 14.2

A feedback control system has the open-loop transfer function given by

$$GH(s) = \frac{K}{s(1 + s)(1 + 0.25s)}$$

Find the phase margin and gain margin corresponding to the value of forward-path gain $K = 1$.

Solution

Using Table 12.1 and noting that the transfer function is in Bode form, the magnitude and phase angle may be written as

$$M = \frac{1}{\omega\sqrt{(1 + \omega^2)(1 + \omega^2/16)}}$$

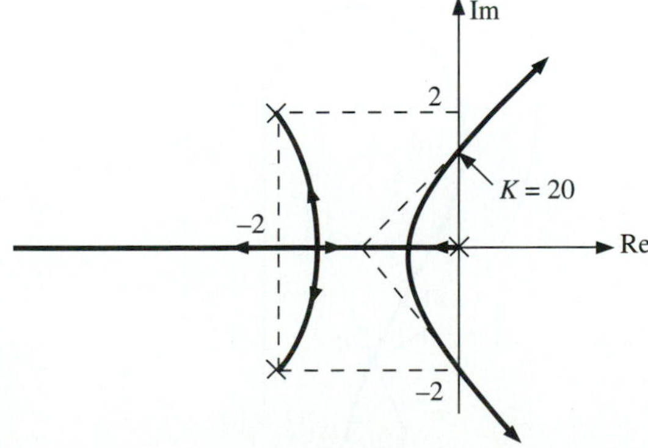

Fig. SP14.1.3

and

$$\phi = -90 - \tan^{-1}(\omega) - \tan^{-1}\left(\frac{\omega}{4}\right)$$

Taking asymptotic values, we get

$$\omega \to 0: \qquad M \to \infty \qquad \phi \to -90$$

$$\omega \to \infty: \qquad M \to 0 \qquad \phi \to -270$$

The Nyquist diagram takes the form shown in Fig. SP14.2.1, where it is assumed that the system is stable. This will be verified once the gain margin has been evaluated.

GAIN MARGIN. In order to determine the gain margin, it is necessary to determine the magnitude of the vector for which the phase is $-180°$. This vector is OA in Fig. SP14.2.2. Setting ϕ equal to $-180°$ yields

$$-180 = -90 - \tan^{-1}(\omega) - \tan^{-1}(\tfrac{1}{4}\omega)$$

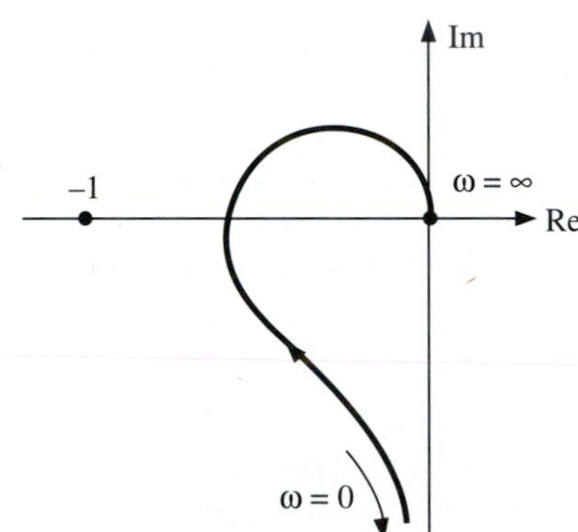

Fig. SP14.2.1

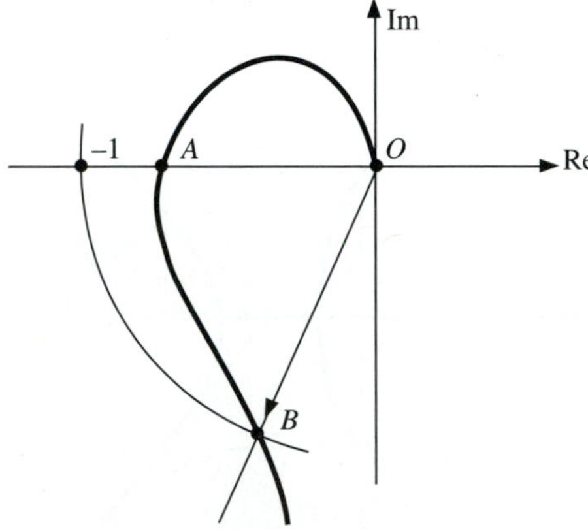

Fig. SP14.2.2

or, simplifying,

$$90 = \tan^{-1}(\omega) + \tan^{-1}(\tfrac{1}{4}\omega)$$

Using the technique outlined in Sample Problem 12.1, we take tangents of each side of this equation to yield

$$\infty = \frac{\omega + \frac{1}{4}\omega}{1 - \frac{1}{4}\omega^2}$$

This equation is only satisfied when

$$\omega = 2$$

Substituting into the magnitude equation gives

$$M = \frac{1}{2\sqrt{(1 + 4)(1 + 0.25)}} = 0.2$$

The gain margin is given by

$$\mathrm{GM} = \frac{1}{OA} = \frac{1}{M} = 5$$

From this value, it may be concluded that the system is stable, and the Nyquist diagram is as drawn in Figs. SP14.2.1 and SP14.2.2.

PHASE MARGIN. To determine the phase margin, we find the phase angle for the frequency that makes the magnitude unity. This vector is *OB* in Fig. SP14.2.2. Setting *M* equal to unity in the magnitude equation and squaring both sides yields

$$M^2 = 1 = \frac{1}{\omega^2(1 + \omega^2)(1 + \frac{1}{16}\omega^2)}$$

It may be seen that substituting $x = \omega^2$ and expanding the right-hand side would yield a cubic in x, which may be solved using numerical methods. In this case, however, we will solve the equation using trial-and-error methods based on a close approximation for x. It may be seen from Fig. SP14.2.2 that the magnitude is 0.2 when $\omega = 2$ rad/s. It follows that ω will be less than 2 when the magnitude is unity, probably around $\omega = 1$. We need to solve for x from

$$x(1 + x)(1 + \tfrac{1}{16}x) = 1$$

It is reasonable to assume that the third term $(1 + \frac{1}{16}x)$ will be close to unity for $x \approx 1$. The above equation approximates to

$$x(1 + x) = 1$$

which yields

$$x^2 + x - 1 = 0$$

We are obviously only interested in a positive solution for x, and this proves to be

$$x = 0.62$$

giving a value of frequency $\omega = \sqrt{0.62} = 0.79$. Using this value as a starting point, a few iterations give the exact value of ω that makes the magnitude unity as

$$\omega = 0.775 \text{ rad/s}$$

Substituting into the phase angle equation gives

$$\phi = -90 - \tan^{-1}(0.775) - \tan^{-1}(0.194) = -139°$$

The phase margin is obtained from

$$\text{PM} = 180 + \phi = 41°$$

For $K = 1$, this stable system has a gain margin of 5 and a phase margin of 41°.

Sample Problem 14.3

A system is described by feedback and forward-path transfer functions

$$G(s) = \frac{K(1 + 4s)}{s^2(1 + 0.25s)(1 + 0.1s)}$$

and

$$H(s) = 1$$

Find the values of gain that give (a) a gain margin of 2 and (b) a phase margin of 40°.

Solution

The open-loop transfer function is

$$GH(s) = \frac{K(1 + 4s)}{s^2(1 + 0.25s)(1 + 0.1s)}$$

Again, Table 12.1 is used to write expressions for the magnitude and phase angle as

$$M = \frac{K\sqrt{1 + 16\omega^2}}{\omega^2\sqrt{(1 + \frac{1}{16}\omega^2)(1 + \frac{1}{100}\omega^2)}}$$

and

$$\phi = -180 + \tan^{-1}(4\omega) - \tan^{-1}(\tfrac{1}{4}\omega) - \tan^{-1}(\tfrac{1}{10}\omega)$$

If we were to proceed as before to determine the gain margin by setting $\phi = -180$ and attempting to solve for ω_π, it can be seen from the phase angle equation that it would be difficult to obtain a solution. This method works well when there are only two factors in the denominator, but not when there are more or, as in this case, when there is an open-loop zero present.

To solve this problem, we will attempt to plot an accurate Nyquist diagram using a simple computer program to calculate the magnitude and phase. It will be useful first to check the asymptotic values of these quantities:

$$\omega \to 0: \quad M \to \infty \quad \phi \to -180$$

$$\omega \to \infty: \quad M \to 0 \quad \phi \to -270$$

In addition, it is important to notice that, for small values of ω, the term

$$\tan^{-1}(4\omega) > \tan^{-1}(\tfrac{1}{4}\omega) - \tan^{-1}(\tfrac{1}{10}\omega)$$

so that as ω increases from zero, the phase will initially increase in a positive sense. The Nyquist diagram will look similar to that shown in Fig. SP14.3.1, where the value of K has been selected so that the system is stable. The following listing is a BASIC program that requests a value of ω and calculates the magnitude and phase for this transfer function. The operation of the program should be self-evident. Note that the modulus expression assumes that $K = 1$:

```
1000 PI=3.14159265
1010 INPUT"TYPE IN THE VALUE OF OMEGA";W
1020 M1=SQR(1+16*W*W)
1030 M2=W*W*SQR((1+W*W/16)*(1+W*W/100))
1040 M=M1/M2
1050 PHI=-PI+ATN(4*W)-ATN(W/4)-ATN(W/100)
1060 PRINT W,M,PHI
1070 GOTO 1010
```

Running this program to obtain sufficient values of M and ϕ to plot a smooth curve produces the accurate Nyquist diagram shown in Fig. SP14.3.2, for $1 < \omega < 6$, while a more detailed plot in the region where the phase is $-180°$ is shown in Fig. SP14.3.3. For the value $K = 1$, the magnitude when $\phi = 180°$ is

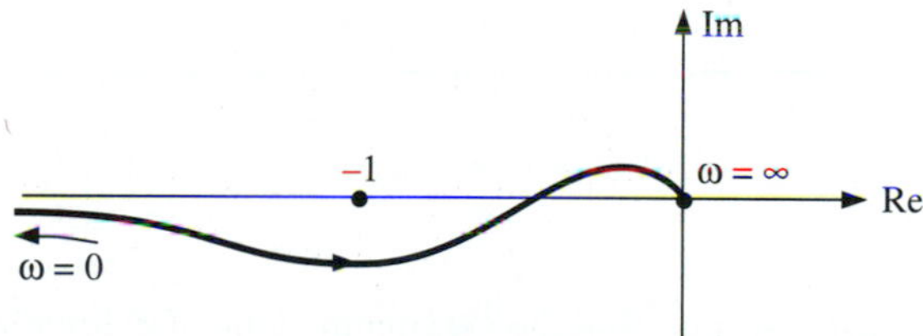

Fig. SP14.3.1

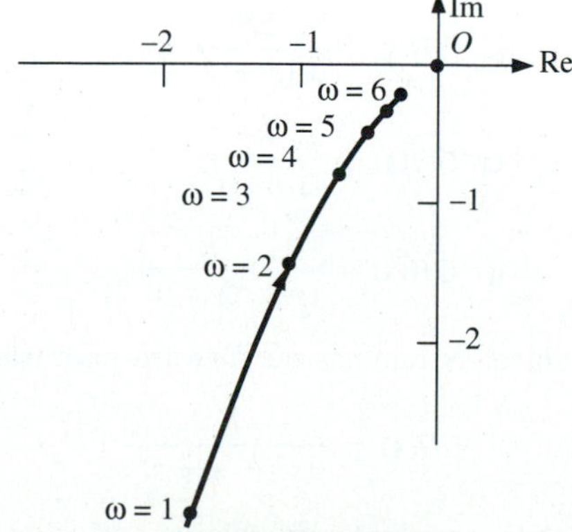

Fig. SP14.3.2

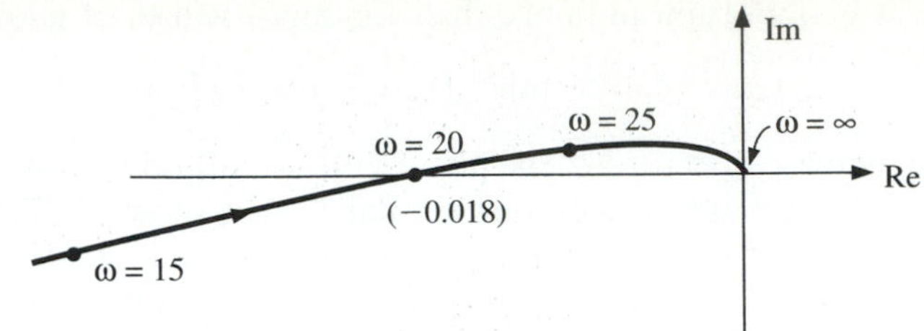

Fig. SP14.3.3

evaluated from Fig. SP14.3.3 as 0.018; hence in order for the gain margin to be 2, the magnitude needs to be 0.5. The required value of K is obtained from

$$\mathrm{GM} = \frac{1}{OA} = \frac{1}{0.018K}$$

This results in

$$K_{\mathrm{GM}=2} = \frac{1}{0.018\ \mathrm{GM}} = 27.8$$

From the program we find for $\omega = 0.22$ rad/s, $\phi = -140°$ (PM $= 40$), and the corresponding magnitude (for K $= 1$) is M $= 0.75$. In order to achieve the required phase margin, the value of K has to be

$$K_{\mathrm{PM}=40} = \frac{1}{0.75} = 1.34$$

◆ PROBLEMS

14.1 From the Nyquist diagram, find the maximum value of K for which the following systems are stable:

a. $GH(s) = \dfrac{K}{s(s + a)}$

b. $GH(s) = \dfrac{5K}{s^2(s + 2)}$

c. $GH(s) = \dfrac{K}{(s + 1)^3}$

d. $GH(s) = \dfrac{K}{(s + 1)(s + 2)^2}$

14.2 A unity-feedback control system has the forward-path transfer function

$$GH(s) = \frac{K}{s(s^2 + s + 9)}$$

Find the maximum value of K consistent with stability and check the result using Routh's method and by drawing the root locus.

14.3 Discuss the stability of the system for which the Nyquist diagram is shown in Fig. P14.3 as the forward-path gain **(a)** increases and **(b)** decreases. Sketch the imaginary-axis region of the corresponding root locus.

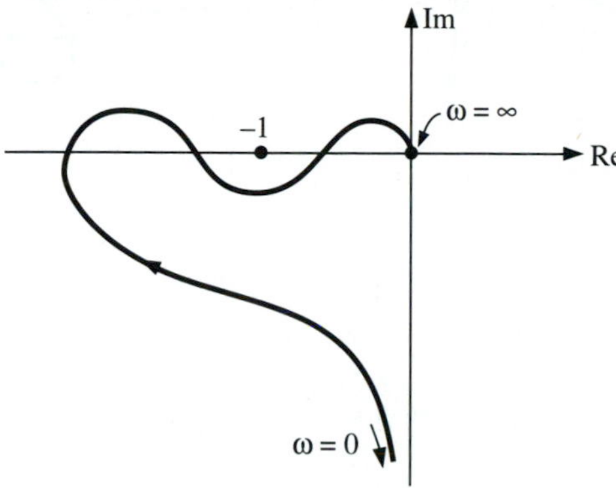

Fig. P14.3

14.4 Find the maximum value of K for which the following system is stable:

$$G(s) = \frac{K}{(1+s)(1+2s)(1+\frac{1}{2}s)} \qquad H(s) = 1$$

14.5 For the systems described by the Nyquist diagrams shown in Fig. P14.5 estimate the gain margin and phase margin.

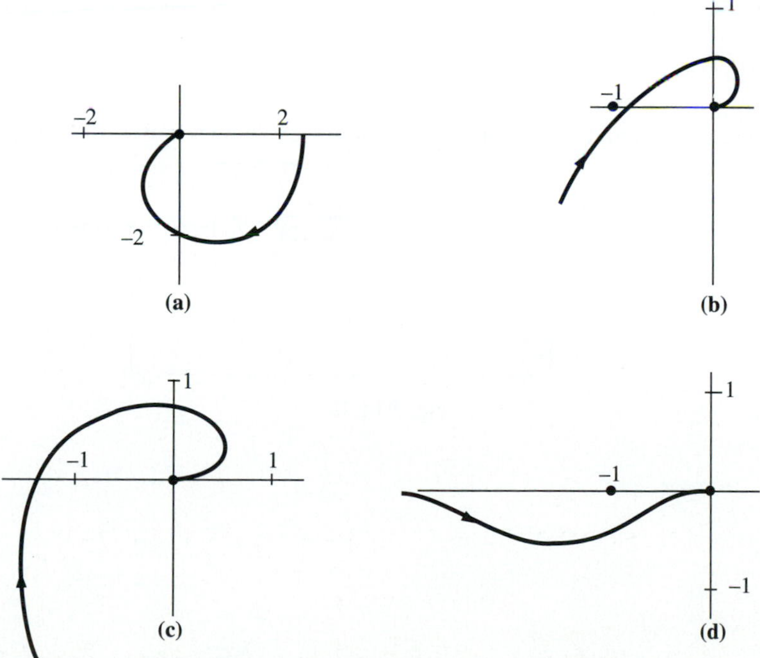

Fig. P14.5

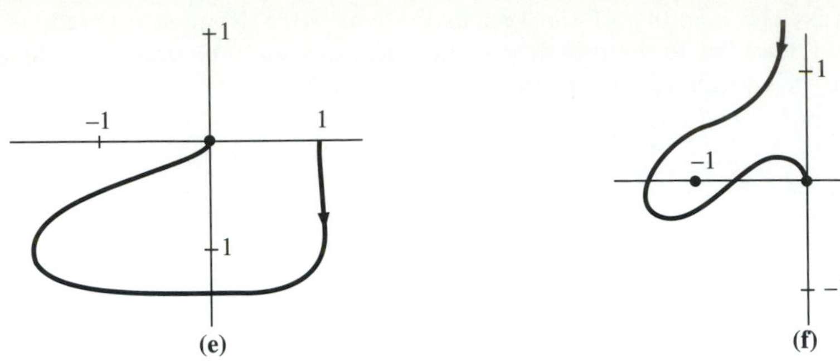

Fig. P14.5 (continued)

14.6 Determine the maximum K consistent with stability when

$$GH(s) = \frac{K(s+1)^2}{s^2(s+5)^2}$$

Find also the gain and phase margins when $K = 100$.

14.7 Find the gain margin and phase margin of the system

$$G(s) = \frac{5}{(s+1)(s^2+s+5)} \qquad H(s) = 1$$

14.8 What value of K will give the system described by

$$GH(s) = \frac{K(s+2)}{s^2(s^2+4s+20)}$$

a gain margin of 2?

14.9 Find the value of K that gives a phase margin of 60° to the system described by

$$GH(s) = \frac{2K}{s(s^2+2s+16)}$$

14.10 Find the gain and phase margins for the system shown in Fig. P14.10.

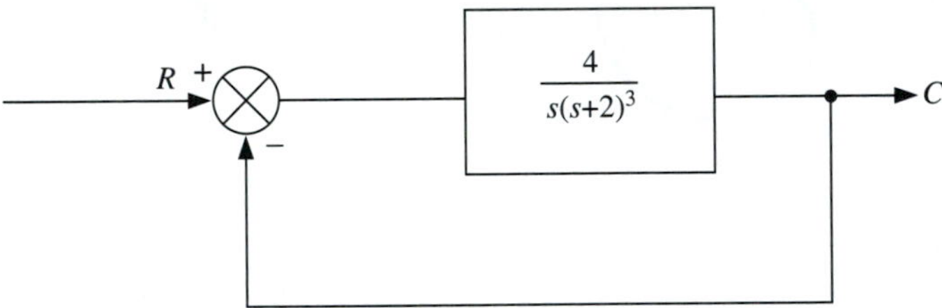

Fig. P14.10

Bode Diagrams

In the development of the Nyquist diagram, the open-loop frequency response was discussed in detail (Module 12). The frequency response was obtained by substituting $s = j\omega$ into the open-loop transfer function and then obtaining the magnitude M and phase ϕ. The Nyquist diagram consisted partly of the vectors defined by M and ϕ as ω varies from zero to infinity.

The Bode diagram contains the same information except that it consists of two separate graphs, one plot of M against ω and another plot of ϕ against ω. The student may ask why the Bode diagram needs to be studied if it contains no more information than the Nyquist diagram. For simple systems there is no clear advantage of one method over the other, but the student may recall that for more complex systems, such as those with three or more poles and/or open-loop zeros, it was necessary to utilize computer methods in order to plot the Nyquist diagram to sufficient accuracy so that the gain and phase margins may be obtained. The Bode method enables us to obtain the gain and phase margins without recourse to such computational techniques since the method allows fairly accurate frequency response data to be obtained, irrespective of the complexity of the system.

◆ BODE DIAGRAMS OF SIMPLE TRANSFER FUNCTIONS

We begin the study of Bode diagrams by means of a simple example. Suppose the Bode diagram of the open-loop transfer function

$$GH(s) = \frac{1}{1 + sT} \tag{15.1}$$

is required. Note first that the transfer function is written in the Bode form, rather than the Evans form used in the root locus method. It is known from previous work that the magnitude and phase are

$$M = \frac{1}{\sqrt{1 + \omega^2 T^2}} \tag{15.2}$$

and

$$\phi = -\tan^{-1} \omega T \tag{15.3}$$

In the Bode diagram, the modulus-frequency plot does not involve simply M, but the modulus expressed in decibels, M_{db}. The reasons for this will become clear in a moment. The magnitude in decibels may be calculated from the magnitude as a simple number by means of the following definition:

$$M_{db} = 20 \log_{10} M \tag{15.4}$$

so that if at a particular frequency the magnitude is 2, then

$$M_{db} = 20 \log_{10} 2 = 6.02 \tag{15.5}$$

In order to plot M_{db} against ω, we write

$$M = (1 + \omega^2 T^2)^{-1/2} \tag{15.6}$$

Hence

$$M_{db} = 20 \log_{10} (1 + \omega^2 T^2)^{-1/2} \tag{15.7}$$

$$= -10 \log_{10} (1 + \omega^2 T^2) \tag{15.8}$$

Now consider the low- and high-frequency asymptotes of the M_{db}–ω plot:

$$\omega \to 0: \qquad M_{db} \to -10 \log_{10} 1 = 0 \tag{15.9}$$

and

$$\omega \to \infty: \qquad M_{db} \to -10 \log_{10} (\omega^2 T^2) = -20 \log_{10} (\omega T) \tag{15.10}$$

This equation may be written as

$$M_{db} = -20 \log_{10} \omega - 20 \log_{10} T \tag{15.11}$$

The magnitude part of the Bode diagram consists of M_{db} plotted against $\log_{10} \omega$ for which the high-frequency asymptote (equation 15.11) may be compared with

$$y = mx + c \tag{15.12}$$

to indicate that this asymptote on the Bode diagram is a straight line of slope -20 and intercept $-20 \log_{10} T$. The high- and low-frequency asymptotes appear in Fig. 15.1. It should be noted that ω is plotted on a logarithmic scale, so that ω has the values of 1, 10, and 100 rad/s, corresponding to 0, 1, 2 on the horizontal scale, respectively. Further, M_{db} changes by -20 db when ω changes, say, from 1 to 10, or from 100 to 1000. Each of these changes is known as a decade (the larger frequency is 10 times the smaller frequency), so the slope of the high-frequency asymptote is -20 db/decade.

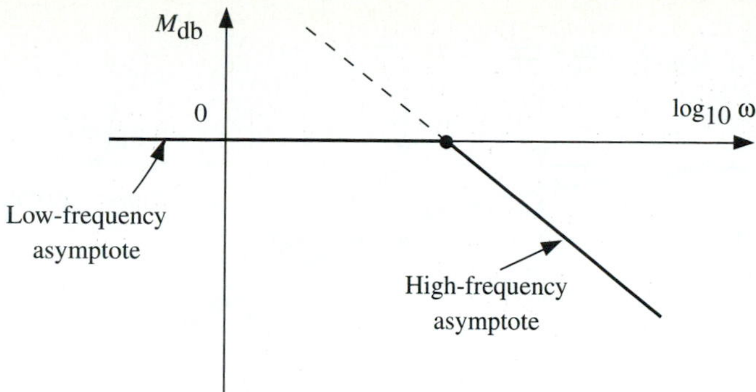

Fig. 15.1 Bode diagram asymptotes

The high- and low-frequency asymptotes meet at a frequency given by the solution of

$$0 = -20 \log_{10} \omega - 20 \log_{10} T \tag{15.13}$$

which yields

$$\omega = \frac{1}{T} \tag{15.14}$$

It is reasonable to assume that the high-frequency asymptote is valid for $\omega > 1/T$ and the low-frequency asymptote valid for $\omega < 1/T$. The frequency where the low- and high-frequency asymptotes meet is called the *break frequency*. The actual modulus will be a smooth curve approaching the asymptotes at very high and very low frequencies. It is useful to calculate the discrepancies between the true modulus and the asymptotic approximations at a few key frequencies. For example, at the break frequency, the true modulus has a value of

$$M_{db} = -10 \log_{10} (1 + 1) \approx -3 \text{ db} \tag{15.15}$$

Since the straight-line approximations suggest a modulus of 0 db at this frequency, the error is -3 db. At twice the break frequency, where

$$\omega = \frac{2}{T} \tag{15.16}$$

the exact modulus is given by

$$M_{db} = -10 \log_{10} (1 + 4) = -7 \text{ db} \tag{15.17}$$

while the approximate modulus is

$$M_{db} = -20 \log_{10} 2 = -6 \text{ db} \tag{15.18}$$

giving an error of 1 db. It is left to the student to verify that at one-half the break frequency ($\omega = 1/2T$) the error is also 1 db. The final form of this Bode diagram is shown in Fig. 15.2.

All that remains to do for this element is to plot the phase angle against frequency. The Bode diagram actually consists of plotting phase angle against $\log_{10} \omega$

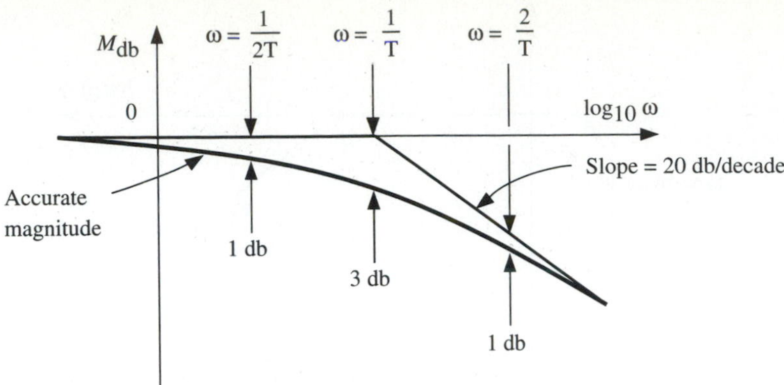

Fig. 15.2 Magnitude part of Bode diagram

so that the phase and magnitude may be shown to the same frequency scale. It is known that the phase angle for this element varies from zero at low frequencies to $-90°$ at high frequencies, as shown in Fig. 15.3. Also shown in the figure is a straight-line approximation for the phase angle starting one decade below the break frequency and changing uniformly by $-90°$ in two decades to one decade above the break frequency. Both accurate and straight-line approximations pass through $\phi = -45°$ at the break frequency, and the error between the true phase and approximate phase is never greater than about 5°. The Bode diagram for the element

$$GH(s) = \frac{1}{1 + sT} \tag{15.19}$$

may be summarized in Fig. 15.4 where the magnitude and phase are plotted together, and straight-line approximations to the true curves are used for both plots. Now consider another simple element for which we will plot the Bode diagram:

$$GH(s) = \frac{1}{s} \tag{15.20}$$

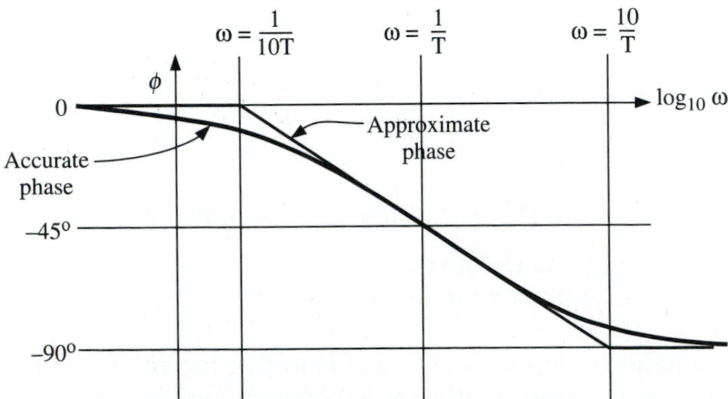

Fig. 15.3 Phase part of Bode diagram

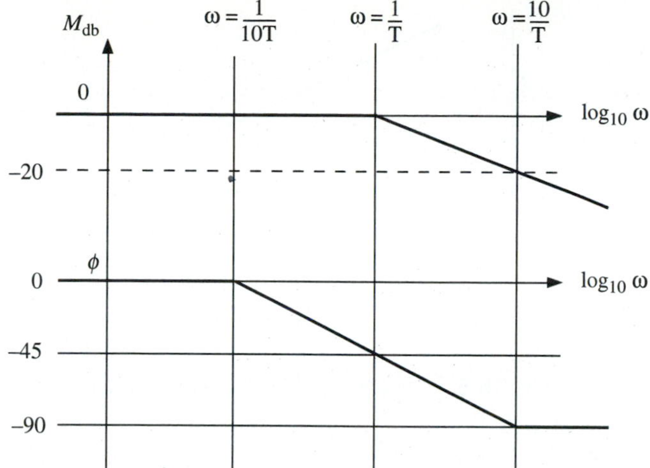

Fig. 15.4 Complete Bode diagram for first-order element

For this element, it is known from Table 12.1 that the magnitude and phase are given by

$$M = \frac{1}{\omega} \qquad \phi = -90 \tag{15.21}$$

The Bode magnitude is

$$M_{db} = 20 \log_{10} (\omega)^{-1} = -20 \log_{10} \omega \tag{15.22}$$

A plot of M_{db} against ω is shown in Fig. 15.5. Note that this plot is an exact straight line in that no approximations are necessary, as in the previous example, and consists of a straight line of slope -20 db/decade passing through the frequency $\omega = 1$ when $M_{db} = 0$. The phase plot for this element is simple, since it has the value of $-90°$ and is independent of frequency. The complete Bode diagram for this element is shown in Fig. 15.6.

◆ BODE DIAGRAMS OF COMPOUND TRANSFER FUNCTIONS

In this section it will become apparent why M_{db} is used in the Bode diagram, rather than simply M. Suppose we wish to draw the Bode diagram of the open-loop

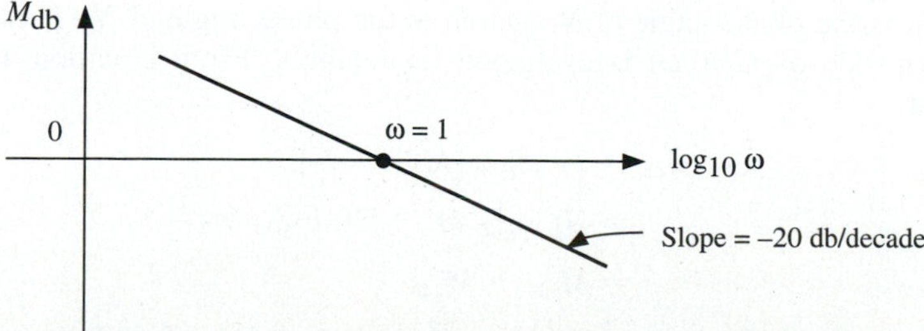

Fig. 15.5 Magnitude part of Bode diagram for integrator

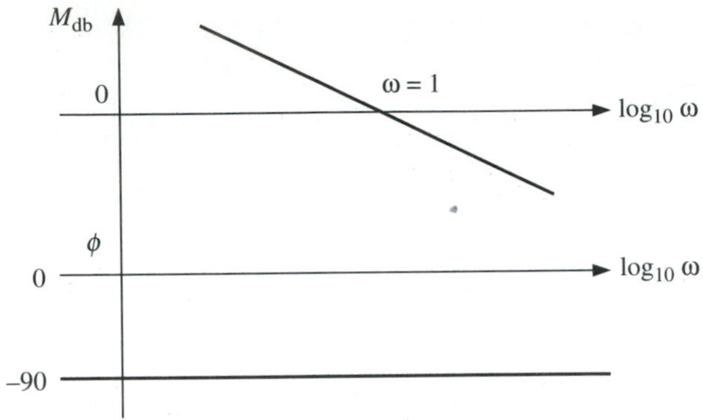

Fig. 15.6 Bode diagram for integrator element

transfer function

$$GH(s) = \frac{1}{s(1 + sT)} \tag{15.23}$$

We could proceed as in the Nyquist analysis by substituting $s = j\omega$ and evaluating the magnitude and phase and plotting them. However, consider the open-loop transfer function to consist of two separate transfer functions corresponding to

$$GH(s) = N_1 \times N_2 \tag{15.24}$$

where

$$N_1 = \frac{1}{s} \tag{15.25}$$

$$N_2 = \frac{1}{1 + sT} \tag{15.26}$$

Each transfer function could be regarded as having a magnitude and phase, each of which is a function of frequency ω. Using the usual rules of complex-number algebra, the magnitude of $GH(j\omega)$ is given by M, where, in this case,

$$M = M_1 \times M_2 \tag{15.27}$$

where M_1 is the magnitude of N_1 and M_2 is the magnitude of N_2. Similarly, for the phase angles

$$\phi = \phi_1 + \phi_2 \tag{15.28}$$

where ϕ_1 is the phase angle of N_1 and ϕ_2 is the phase angle of N_2. Now, suppose the magnitude of $GH(j\omega)$ is evaluated in decibels. From equation 15.27 it is seen that

$$M_{db} = 20 \log_{10} [M_1 \times M_2] \tag{15.29}$$

$$= 20 \log_{10} M_1 + 20 \log_{10} M_2 \tag{15.30}$$

$$= M_{1db} + M_{2db} \tag{15.31}$$

The composite magnitude portion of the Bode diagram may be obtained by adding *graphically* the individual magnitude curves for the component parts. Figure 15.7

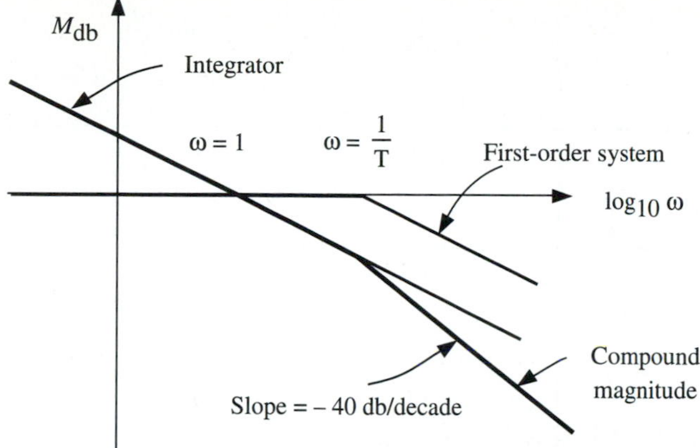

Fig. 15.7 Magnitude part of the Bode diagram for compound element

shows the straight-line representations of the two components of the transfer function, together with the resulting magnitude plot. One way to generate the magnitude plot is to begin at low frequency (left-hand part of the graph), and as we move to the higher frequency (right-hand part), we sum graphically the individual magnitude plots. Note how the slope of the straight line changes from -20 db/decade to -40 db/decade at the break frequency. Since it is known that the error between the true and straight-line approximations to the magnitude is 3 db at the crossover frequency, a better sketch of the magnitude plot would be as shown in Fig. 15.8. The resultant phase angle may also be obtained graphically from equation 15.28 directly. Figure 15.9 shows the approximate phase for each element of the transfer function and also the sum of these two, producing the phase of $GH(j\omega)$. The total phase angle varies between $-90°$ and $-180°$, passing through the value $-135°$ at the break frequency. Although it is common to obtain an accurate magnitude plot by marking the deviations from the straight-line approximations and plotting a smooth curve through them, it is usual not to do this for the phase angle plot. The resulting, complete Bode diagram for this example is shown in Fig. 15.10. The basic method for drawing Bode diagrams consists, therefore, of decomposing the transfer function of interest into component form and plotting the indi-

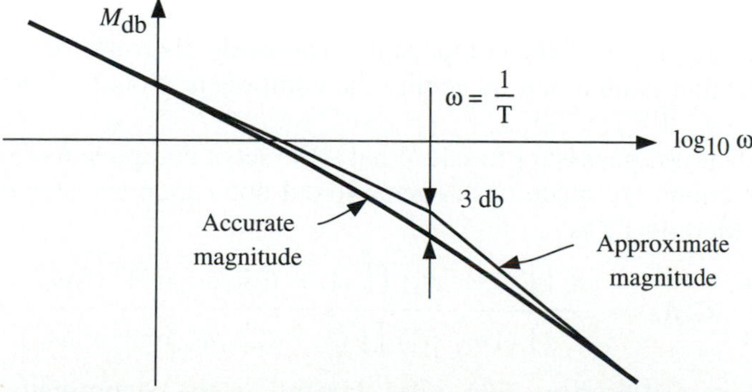

Fig. 15.8 Accurate magnitude plot for compound element

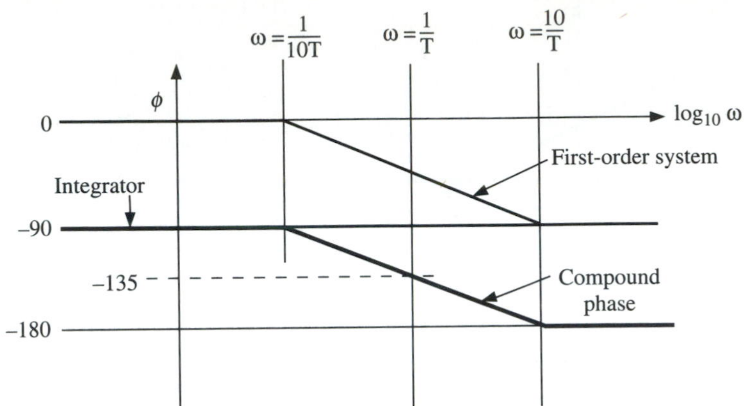

Fig. 15.9 Phase angle plot for compound element

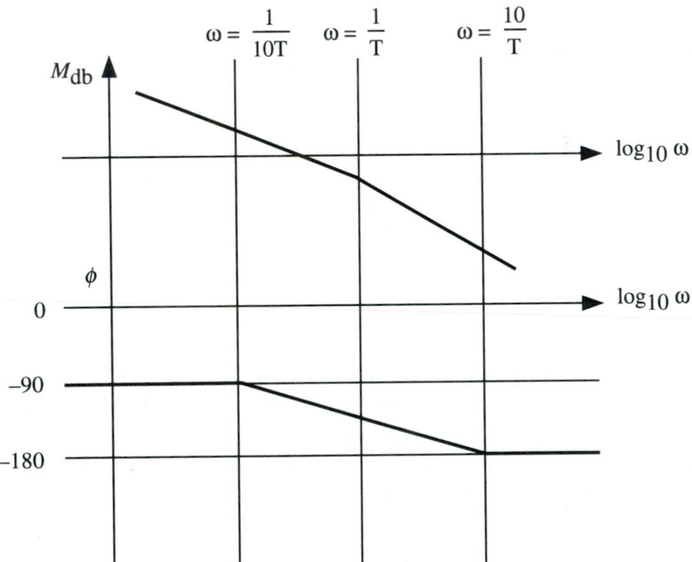

Fig. 15.10 Complete Bode diagram for compound element

vidual Bode diagrams of the components. The Bode diagram for the compound transfer function is obtained by adding the component plots for both magnitude and phase.

At this stage it is necessary to ask: What is the set of components from which all transfer functions are made up? A generalized open-loop transfer function was defined in Module 12 as

$$GH(s) = \frac{K \prod (1 + sT_i) \prod (1 + (2\zeta_j/\omega_{n_j})s + s^2/\omega_{n_j}^2)}{s^n \prod (1 + sT_k) \prod (1 + (2\zeta_l/\omega_{n_l})s + s^2/\omega_{n_l}^2)} \tag{15.32}$$

Usually, there will be more first-order elements in the denominator than in the numerator, as is the case for second-order elements. The rest of this module will develop the individual Bode diagrams for these elements.

$GH(s) = K$

In this case, the magnitude in decibels may be calculated directly as

$$M_{db} = 20 \log_{10} K = K_{db} \tag{15.33}$$

and is obviously independent of frequency. Similarly the phase angle of the real number K is zero for all ω. The Bode diagram is shown in Fig. 15.11.

$GH(s) = 1 + sT$

This element is the inverse of the first-order lag. Table 12.1 reveals

$$M = (1 + \omega^2 T^2)^{1/2} \tag{15.34}$$

Hence

$$M_{db} = 20 \log_{10} (1 + \omega^2 T^2)^{1/2} \tag{15.35}$$

$$= 10 \log_{10} (1 + \omega^2 T^2) \tag{15.36}$$

Considering the low- and high-frequency asymptotes

$$\omega \to 0: \qquad M_{db} \to 10 \log_{10} 1 = 0 \tag{15.37}$$

and

$$\omega \to \infty: \qquad M_{db} \to 10 \log_{10} (\omega^2 T^2) = 20 \log_{10} (\omega T) \tag{15.38}$$

This equation may be written as

$$M_{db} = 20 \log_{10} \omega + 20 \log_{10} T \tag{15.39}$$

Comparing this equation to equation 15.11 and its resulting magnitude plot, it is seen that the plot for this transfer function is 0 db for ω less than the break frequency and is represented by a straight line of slope +20 db/decade above the break frequency, as shown in Fig. 15.12. The true magnitude curve lies above the straight lines and has a maximum error of +3 db at the break frequency. Similarly, the phase angle is given from Table 12.1 by

$$\phi = +\tan^{-1} (\omega T) \tag{15.40}$$

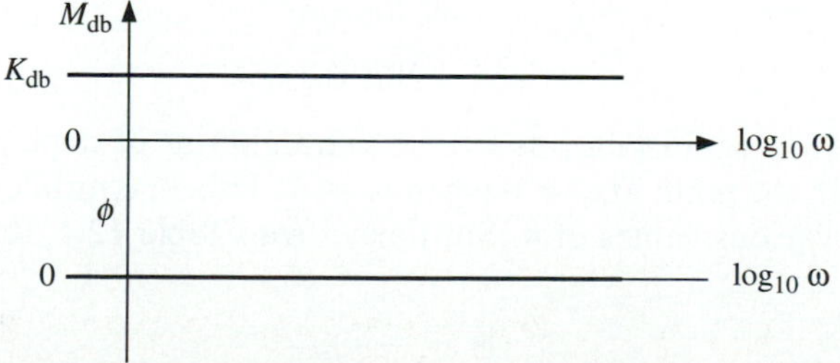

Fig. 15.11 Bode diagram for gain K

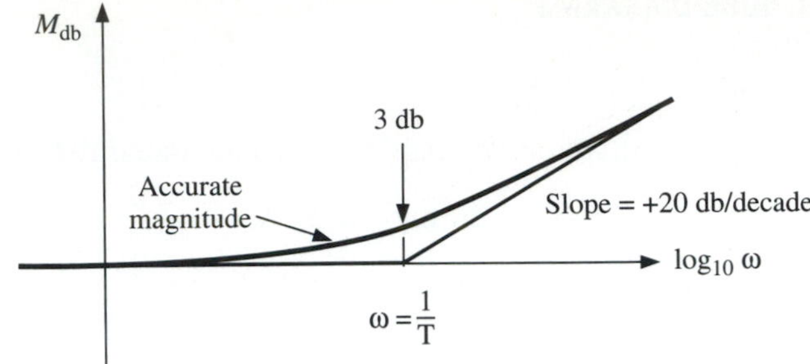

Fig. 15.12 Magnitude plot for $GH(s) = 1 + sT$

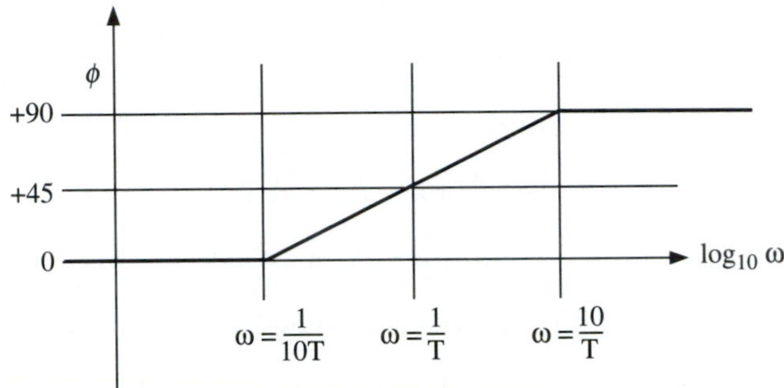

Fig. 15.13 Phase plot for $GH(s) = 1 + sT$

producing the phase plot shown in Fig. 15.13. Note how the Bode diagram for this element may be obtained from that of

$$GH(s) = \frac{1}{1 + sT} \tag{15.41}$$

simply by reflecting the magnitude and phase plots about the frequency axis. Use will be made of this feature later.

$GH(s) = 1/s^n$

Again from Table 12.1, the magnitude is given by

$$M_{db} = 20 \log_{10} \frac{1}{\omega^n} \tag{15.42}$$

$$= 20 \log_{10} \omega^{-n} \tag{15.43}$$

$$= -20n \log_{10} \omega \tag{15.44}$$

Hence a plot of M_{db} against $\log_{10} \omega$ will be a straight line of slope -20n db/decade passing through the point $M_{db} = 0$ when $\omega = 1$. This magnitude plot is shown in Fig. 15.14 for various values of n. Similarly, from Table 12.1, the phase angle is given by

$$\phi = -90n \tag{15.45}$$

resulting in Fig. 15.15.

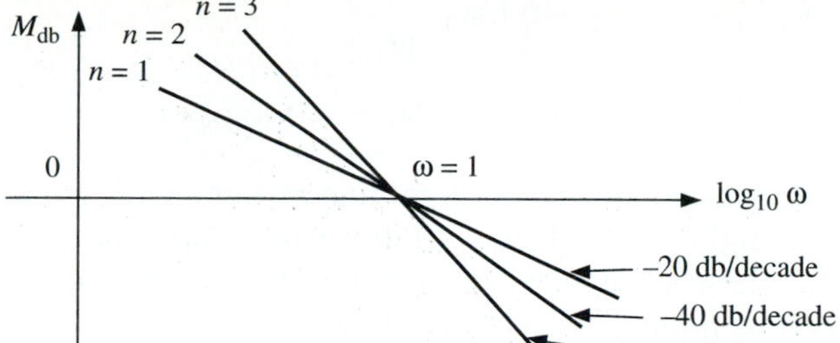

Fig. 15.14 Magnitude for $GH(s) = 1/s^n$

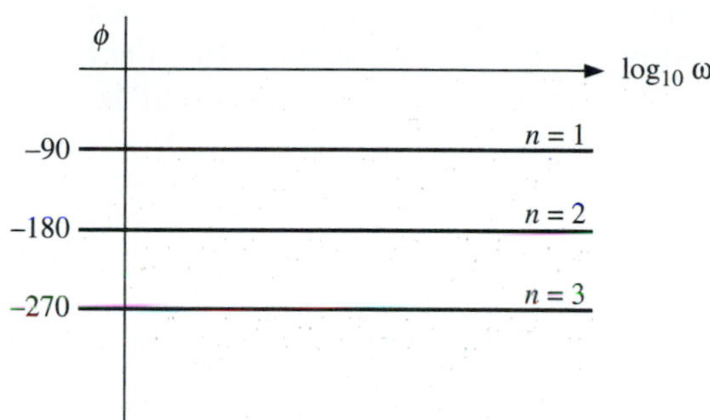

Fig. 15.15 Phase angle for $GH(s) = 1/s^n$

$GH(s) = \dfrac{1}{1 + (2\zeta/\omega_n)s + s^2/\omega_n^2}$

This element is a little more difficult to deal with; however, the results obtained in Module 12 may still be used, namely,

$$M = \frac{1}{\sqrt{(1 - \omega^2/\omega_n^2)^2 + 4\zeta^2\omega^2/\omega_n^2}} \tag{15.46}$$

and

$$\phi = \tan^{-1} \frac{-2\zeta\omega/\omega_n}{1 - \omega^2/\omega_n^2} \tag{15.47}$$

The asymptotic values of the magnitude may be obtained from Module 12 as

$$\omega \rightarrow 0: \qquad M \rightarrow 1 \qquad M_{db} \rightarrow 0 \tag{15.48}$$

$$\omega \rightarrow \infty: \qquad GH(j\omega) \rightarrow -\left(\frac{\omega_n}{\omega}\right)^2 \tag{15.49}$$

Hence the magnitude of $GH(j\omega)$ approaches

$$M = \left(\frac{\omega_n}{\omega}\right)^2 \tag{15.50}$$

The high-frequency magnitude in decibels becomes

$$M_{db} = 20 \log_{10} \left(\frac{\omega_n}{\omega}\right)^2 \tag{15.51}$$

$$= 40 \log_{10} \frac{\omega_n}{\omega} \tag{15.52}$$

$$= 40 \log_{10} \omega_n - 40 \log_{10} \omega \tag{15.53}$$

The low-frequency asymptote is a straight line at 0 db, while the high-frequency asymptote is a straight line of slope −40 db/decade. This line intersects the 0-db line at $\omega = \omega_n$. These asymptotes are shown in Fig. 15.16. As was seen in Module 12, the exact magnitude-frequency plots were multivalued, depending upon the damping ratio ζ. This will obviously be the case for M_{db}–$\log_{10} \omega$ plots, although they will look a little different. Such plots for different damping ratios are shown in Fig. 15.17. Unlike a first-order element which has a single-valued deviation between the approximate and accurate moduli, the discrepancy for the second-order element depends upon the damping ratio. Note that the true magnitude may be below or above the straight-line approximate magnitude.

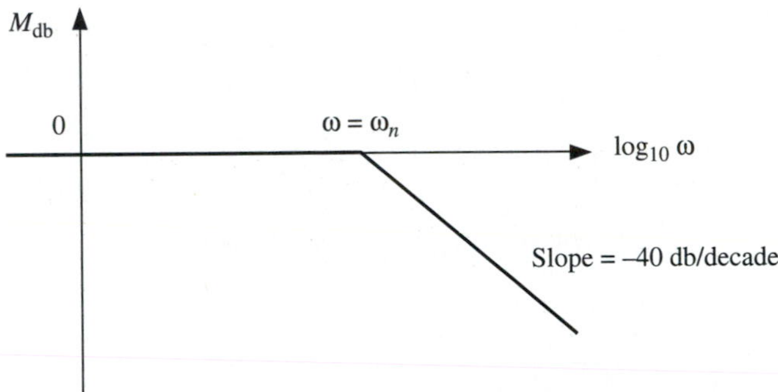

Fig. 15.16 Approximate magnitude for second-order element

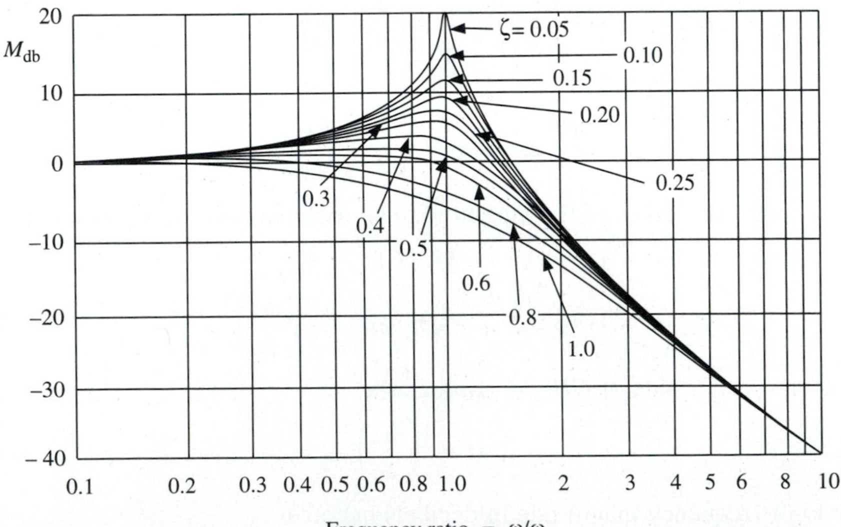

Fig. 15.17 Accurate magnitude for second-order element

A similar situation governs the phase-frequency plot. At low frequencies, the phase approaches

$$\phi \rightarrow \tan^{-1} 0 = 0 \tag{15.54}$$

while at high frequencies

$$\phi \rightarrow \tan^{-1} \frac{-2\zeta\omega\omega_n}{-\omega^2} \tag{15.55}$$

Taking the appropriate quadrant into consideration, the high-frequency phase approaches

$$\phi \rightarrow -180° \tag{15.56}$$

However, the rate at which the phase makes the transition from zero to $-180°$ also depends upon the damping ratio ζ, as shown in Fig. 15.18. Note also that all phase lines pass through the value $-90°$ at the frequency $\omega = \omega_n$. Because of the similarity of the first- and second-order elements, at least as far as the straight-line approximations are concerned, the value $\omega = \omega_n$ is sometimes referred to as the break frequency for the second-order element. Although not accurate in all cases, a straight-line approximation for the phase curve is sometimes used. This is defined as a linear change from 0° to $-180°$ beginning one decade below ω_n and finishing one decade above ω_n.

$GH(s) = 1 + (2\zeta/\omega_n)s + s^2/\omega_n^2$

Instead of reworking the analysis for this element, we will use the observation that this is the inverse of the element previously discussed. The Bode diagram of the element will be, therefore, the inverse of the previous element. The magnitude and phase will be as shown in Fig. 15.19.

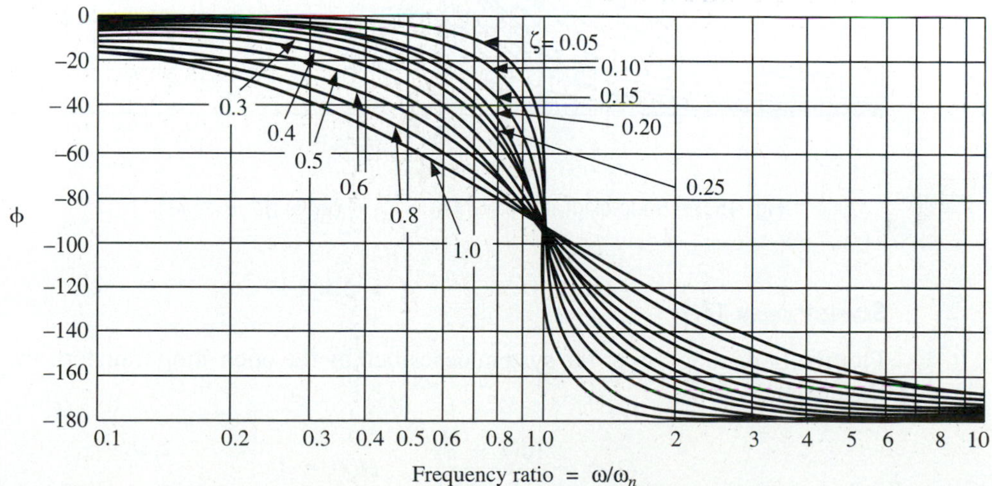

Fig. 15.18 Phase plot for second-order element

40
M_{db}
20
0
−20
1.0
0.8
0.6
0.5
0.4
0.3
0.25
0.20
0.15
0.10
$\zeta = 0.05$
$\omega = \omega_n$

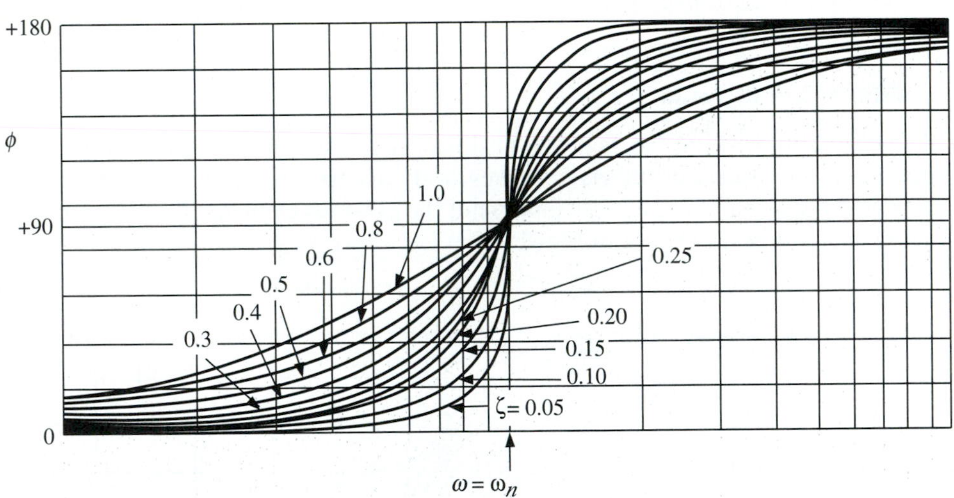

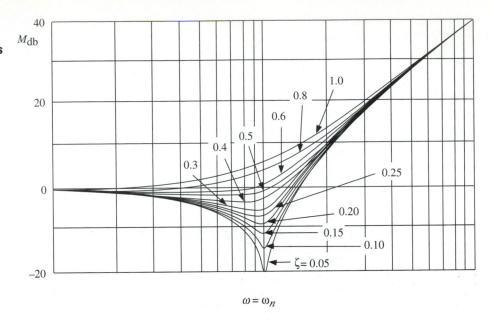

Fig. 15.19 Bode diagram for $GH(s) = 1 + (2\zeta/\omega_n)s + s^2/\omega_n^2$

Sample Problem 15.1

Plot the Bode diagram of the system described by the open-loop transfer function elements

$$G(s) = \frac{10(1 + s)}{s(1 + \frac{1}{2}s)} \qquad H(s) = 1$$

Solution

The transfer function consists of a constant gain of 10, a zero, a pole, and an integrator. The plotting of the Bode diagram will be performed as a series of simple steps.

STEP 1: CALCULATE THE BREAK FREQUENCIES. Each element, whether first or second order, will have a break point. For this system, the break points are as follows:

$$1 + s: \qquad \omega_b = 1 \text{ rad/s}$$

$$\frac{1}{s}: \qquad \omega_b = 1 \text{ rad/s}$$

$$\frac{1}{1 + \frac{1}{2}s}: \qquad \omega_b = 2 \text{ rad/s}$$

The gain K does not have a break point, although its value in decibels has to be calculated:

$$K_{\text{db}} = 20 \log_{10} 10 = 20 \text{ db}$$

STEP 2: DETERMINE THE FREQUENCY RANGE TO BE PLOTTED. All break frequencies are in the decade 1–10 rad/s. It is usual to plot at least one decade on either side of this range, so the Bode plot will be drawn for

$$0.1 < \omega < 100$$

It is customary to plot the Bode diagram on semilog paper so that ω appears directly on the horizontal axis and M_{db} is plotted on the linear, vertical scale. There are no general guidelines to determine the range of M_{db} that will be plotted; only experience will help here! A piece of semilog graph paper with at least three cycles will be required.

STEP 3: PLOT THE STRAIGHT-LINE MAGNITUDE APPROXIMATIONS. The straight-line approximations to each element making up the transfer function are plotted on the same set of axes. Figure SP15.1.1 shows the Bode diagram of the elements of the transfer function.

STEP 4: GRAPHICALLY ADD ALL ELEMENT MAGNITUDES. We begin at low frequencies and move to the right adding the magnitudes as we proceed. Be especially careful to allow for all elements when they overlap, as is the case of the two elements with poles at $s = 0$ and $s = -1$, for frequencies greater than the common break frequency of $\omega_b = 1$ rad/s. The compound magnitude line may be interpreted as a line whose gradient changes, usually by ± 20 db/decade, as each break frequency is passed. For example, in this plot, the low-frequency line is falling at -20 db/decade until the zero is encountered, at which point the line will become horizontal. When the frequency $\omega = 2$ rad/s is reached, the line slope changes to -20 db/decade again. The result of this summation is shown in Fig. SP15.1.2.

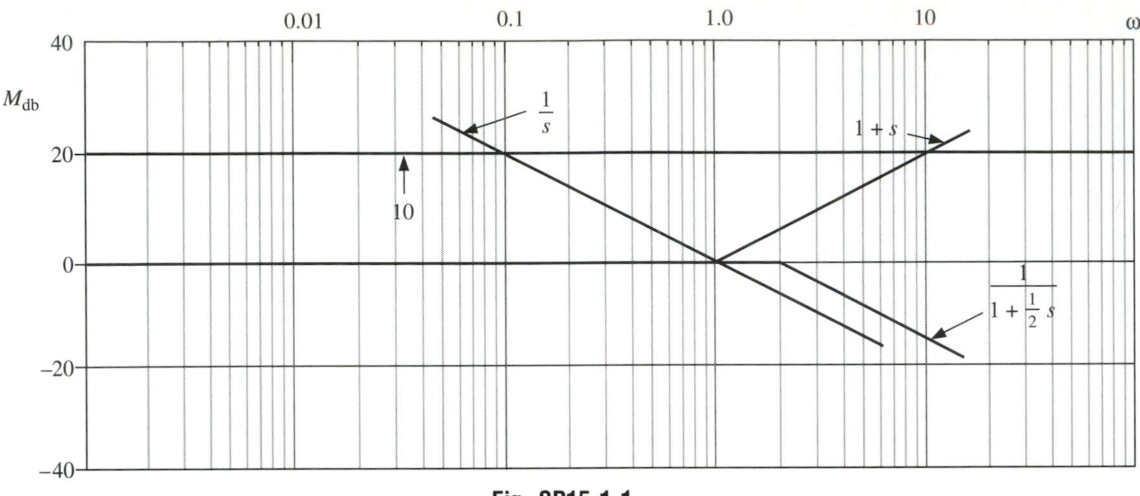

Fig. SP15.1.1

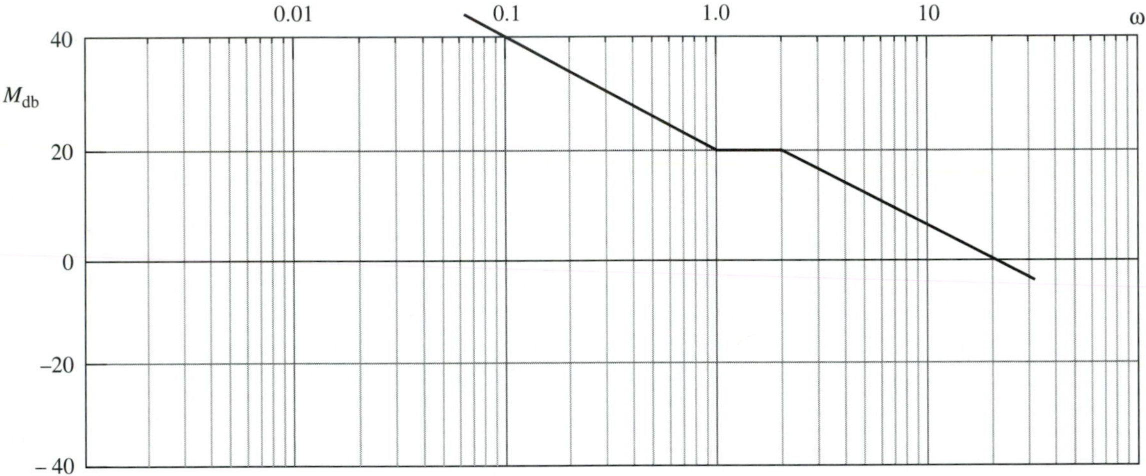

Fig. SP15.1.2

STEP 5: PLOT DEVIATIONS BETWEEN TRUE AND APPROXIMATE MAGNITUDES. The magnitudes of the gain and the pure integrator are accurate and do not involve any approximations. The error for the zero is +3 db at $\omega = 1$ rad/s, so we might be tempted to mark this error on the diagram. However, note that there is a deviation due to the pole at $s = -1$ of -3 db at $\omega = 2$ rad/s. Figure SP15.1.3 shows the two accurate magnitudes in this frequency range, and it can be appreciated that the net error at $\omega = 1$ rad/s will be somewhat less than +3 db. Similarly, the deviation at $\omega = 2$ rad/s will be a little less than -3 db. Because of the symmetry of the situation, it will be estimated that the deviation is closer to ±2 db at these frequencies. Note that if the two break frequencies were more spread out, they would interact less, and the accurate magnitude would be closer to ±3 db deviation from the approximate magnitude. If the two break frequencies were closer, they would interact more and the corresponding deviations would be less.

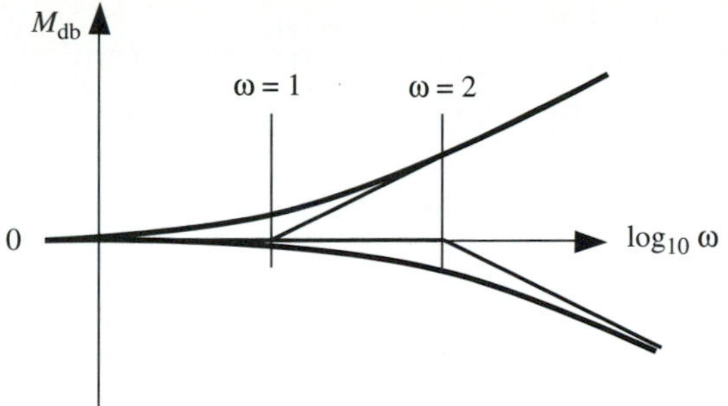

Fig. SP15.1.3

STEP 6: COMPLETE THE MAGNITUDE CURVE. A smooth curve is drawn through the estimated points in the vicinity of the break frequencies and approaching asymptotically the high- and low-frequency straight-line approximations to the magnitude. The completed figure is shown in Fig. SP15.1.4. This completes the magnitude part of the plot, and the phase part may now be considered. The phase-frequency part of the Bode diagram is drawn on semilog graph paper, with the frequency plotted on the same scale as the magnitude.

STEP 7: PLOT THE STRAIGHT-LINE PHASE APPROXIMATIONS. First-order elements may have their phase angles approximated by straight lines from one decade below to one decade above the break frequency. For the example given, the straight-line approximations for each component are shown in Fig. SP15.1.5. Note that the phase of the integrator is $-90°$ and the phase of the gain is zero.

STEP 8: GRAPHICALLY ADD ALL ELEMENT PHASES. This is best done the same way as the magnitudes, moving from low to high frequencies, noting the

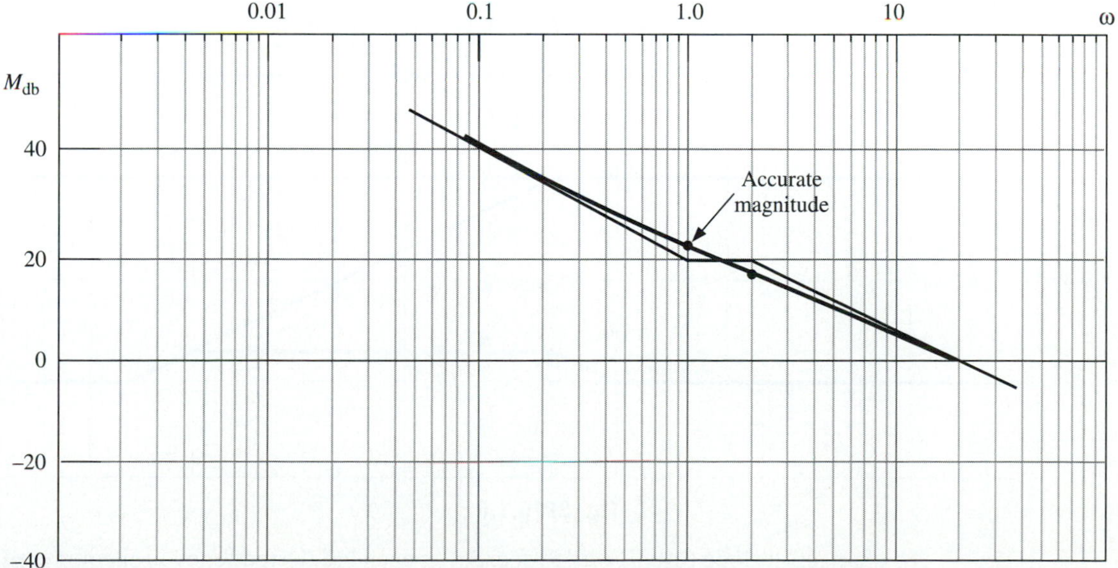

Fig. SP15.1.4

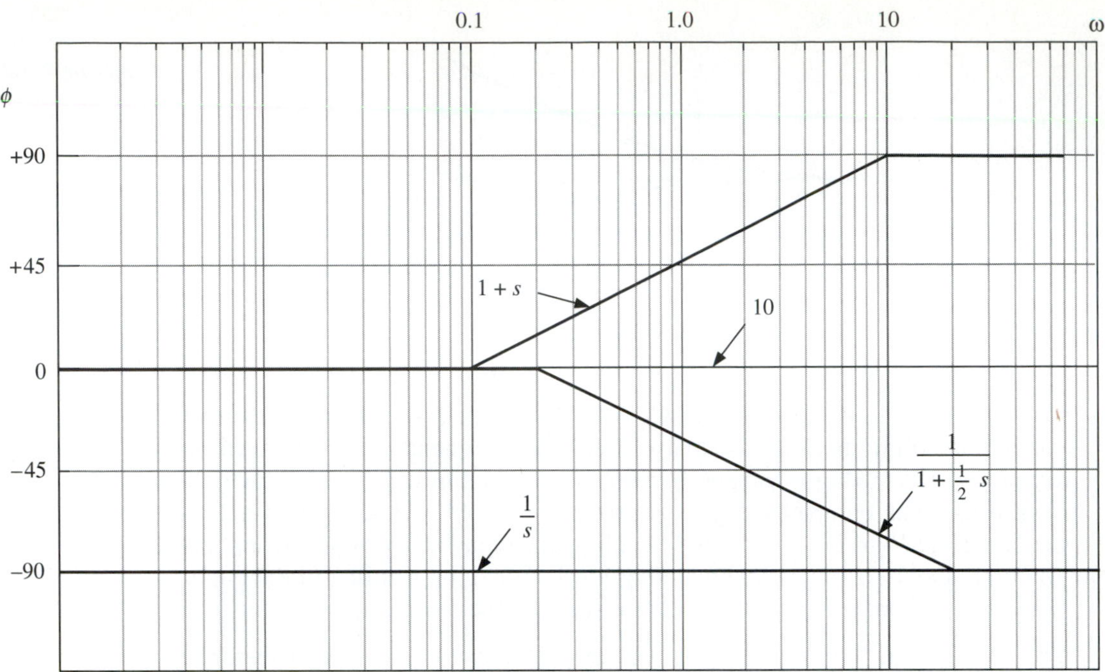

Fig. SP15.1.5

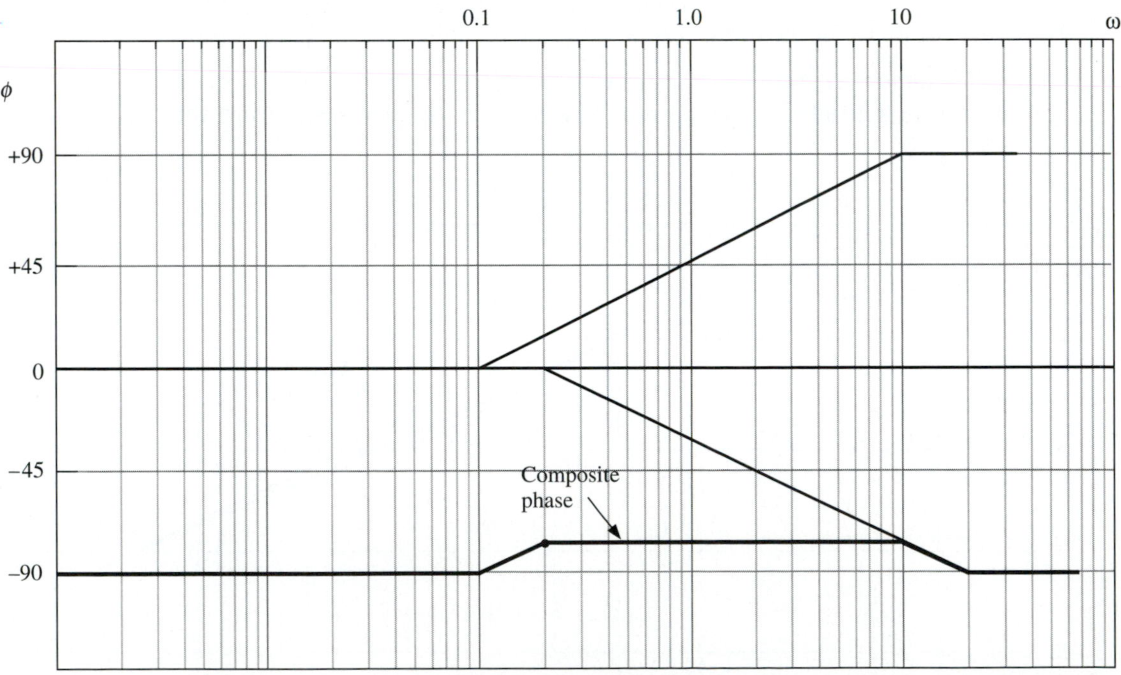

Fig. SP15.1.6

changes in slope (usually ±45°/decade) as each break frequency is encountered. This summation is shown in Fig. SP15.1.6. It is not usual to plot accurate phase data since the deviation between the actual and approximate values of phase is

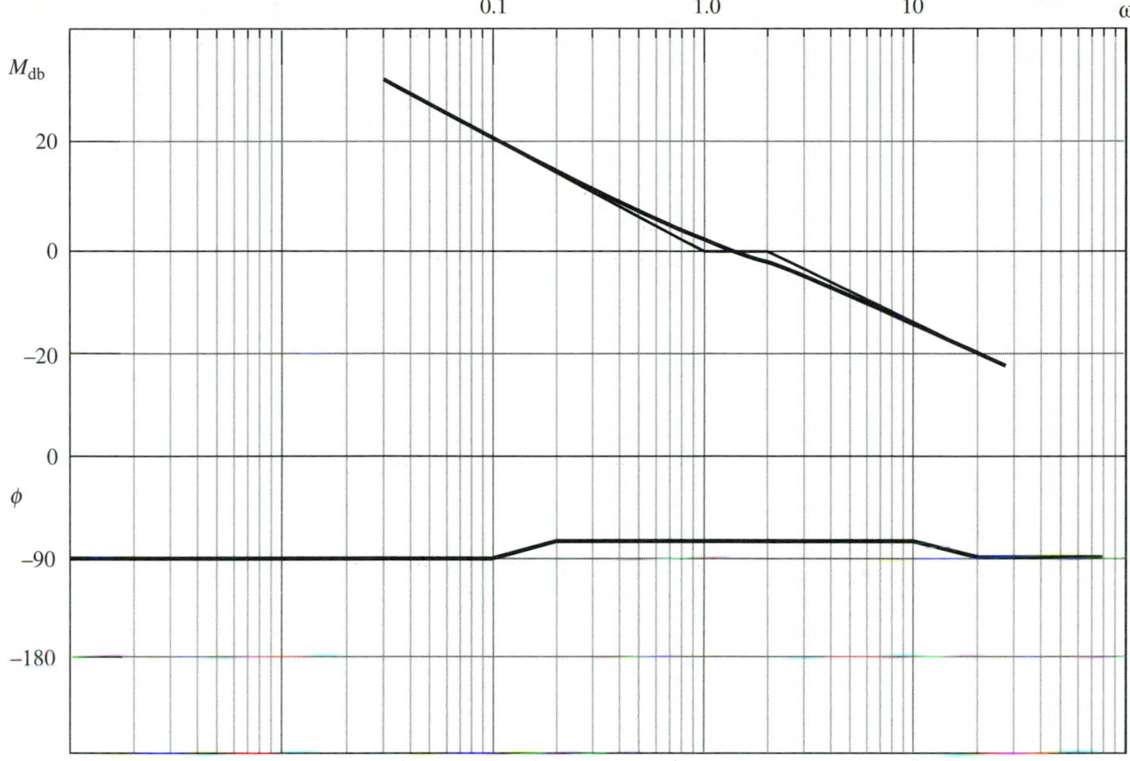

Fig. SP15.1.7

small. The complete Bode diagram for the example is shown in Fig. SP15.1.7, where the magnitude and phase are plotted together.

Sample Problem 15.2

Plot the Bode diagram for the unity-feedback system shown in Fig. SP15.2.1 when $K = 500$.

Solution

We begin by first writing the open-loop frequency response in Bode form:

$$GH(s) = \frac{500(s + 2)}{s(s + 20)(s^2 + 4s + 100)}$$

$$= \frac{500 \times 2(1 + s/2)}{2000s(1 + s/20)(1 + 4s/100 + s^2/100)}$$

$$= \frac{0.5(1 + s/2)}{s(1 + s/20)(1 + 4s/100 + s^2/100)}$$

Step 1: Calculate the Break Frequencies. Before this can be done, the second-order element has to be investigated to determine the damping ratio and the undamped natural frequency. Comparing the standard second-order system

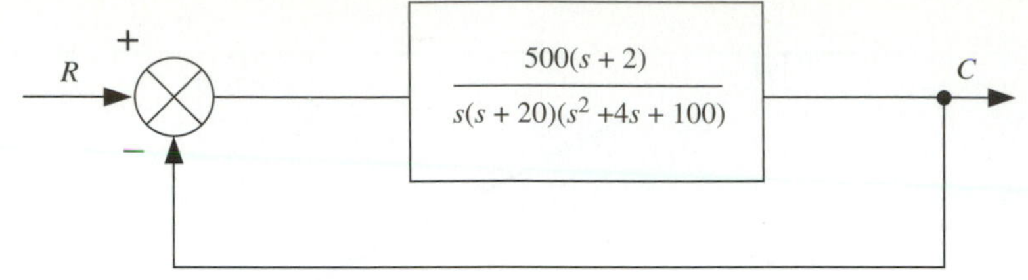

Fig. SP15.2.1

with the example

$$1 + \frac{2\zeta s}{\omega_n} + \frac{s^2}{\omega_n^2} = 1 + \frac{4s}{100} + \frac{s^2}{100}$$

reveals

$$\omega_n = 10$$

and

$$\frac{2\zeta}{\omega_n} = \frac{4}{100} \qquad \zeta = 0.2$$

For this problem, the break frequencies are 1 rad/s (integrator), 2 rad/s (zero), 10 rad/s (second-order element), and 20 rad/s (first-order pole). In addition, the Bode gain is

$$K_{\text{db}} = 20 \log_{10} 0.5 = -6 \text{ db}$$

STEP 2: DETERMINE THE FREQUENCY RANGE TO BE PLOTTED. An appropriate frequency range to plot will be

$$0.1 < \omega < 1000$$

since this will allow at least one decade on each side of the break frequencies.

STEP 3: PLOT THE STRAIGHT-LINE MAGNITUDE APPROXIMATIONS. Figure SP15.2.2 shows each of the straight-line representations of the elements comprising the open-loop transfer function. The slopes of these lines are ±20 db/decade, except for the second-order element, which has a slope of −40 db/decade.

STEP 4: GRAPHICALLY ADD ALL ELEMENT MAGNITUDES. Beginning at the frequency $\omega = 0.1$ rad/s, the magnitude is due to the $1/s$ term (+20 db) and the K_{db} term (−6 db) alone, resulting in a net magnitude of +14 db. Moving to the right in the direction of increasing frequency, the magnitude will fall at a rate of −20 db/decade due to the integrator.

When the next break frequency at $\omega = 2$ rad/s is reached, the effect of the zero will increase the magnitude at +20 db/decade, offsetting the decrease due to integrator causing the net magnitude to be horizontal. This continues until the next break frequency at $\omega = 10$ rad/s is reached. The effect of the second-order element at this frequency causes the magnitude to drop at −40 db/decade, until the final break point at $\omega = 20$ rad/s is encountered. The effect of this first-order pole is to cause the composite magnitude to decrease at a rate of −60 db/decade for all frequencies greater than 20 rad/s. The resulting magnitude plot is shown in Fig. SP15.2.3. Some students may prefer to visualize the

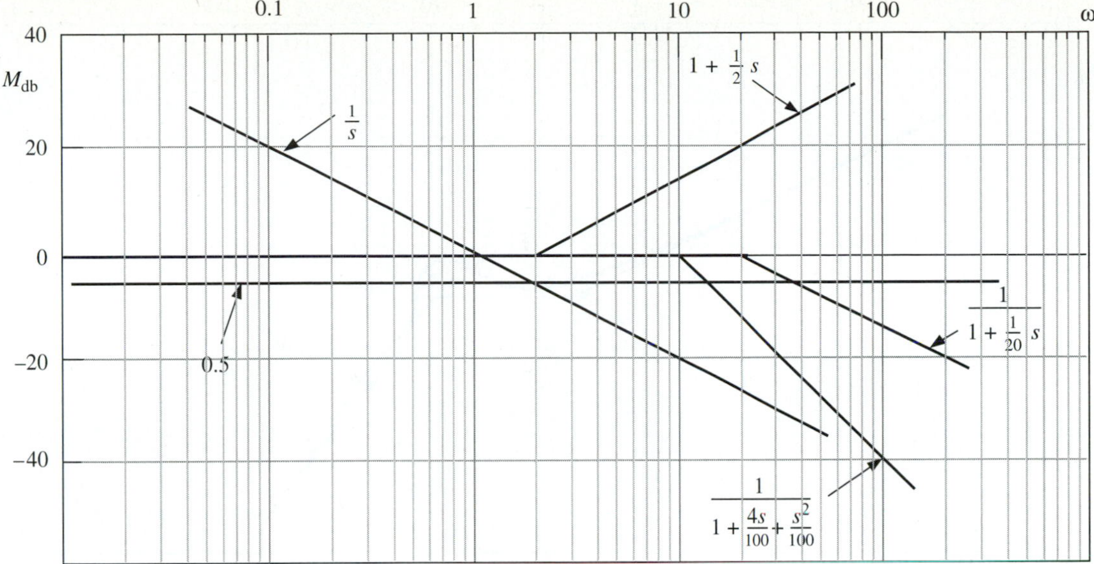

Fig. SP15.2.2

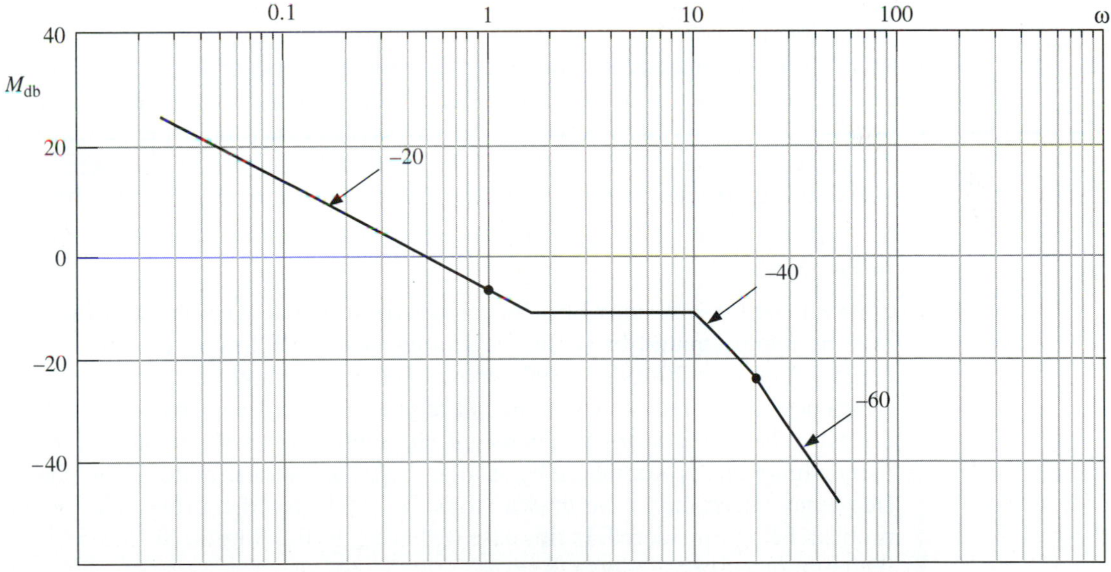

Fig. SP15.2.3

plot by first ignoring the effect of the gain K_{db} and then adding the magnitudes of all the other components of the transfer function. The gain is finally incorporated by seeing that its effect is to lower the entire plot by 6 db, as shown in Fig. SP15.2.4.

STEP 5: PLOT DEVIATIONS. The zero at $\omega = 2$ rad/s will have a deviation of +3 db. However, there is a possible contribution from the second-order element at $\omega = 10$ rad/s also. Inspection of Fig. 15.17 indicates that for $\zeta = 0.2$, the

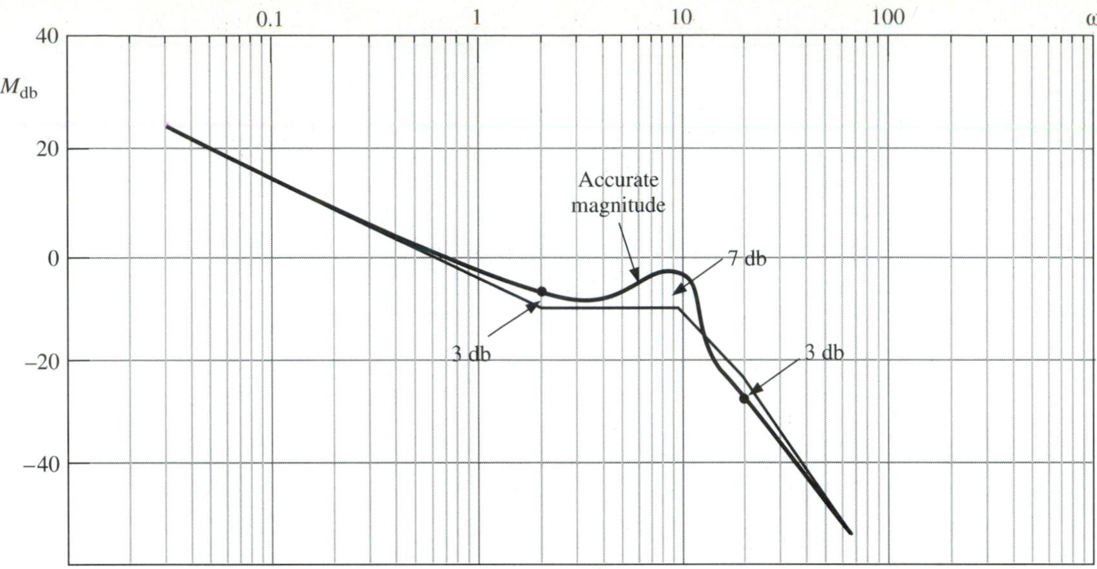

Fig. SP15.2.4

size of the resonance peak is about +8 db above the low-frequency asymptote and negligible deviation at half the break frequency. We might expect, therefore, a deviation of about +3 db at $\omega = 2$ rad/s due to the zero and about +7 db at $\omega = 10$ rad/s due to the combined effect of the second-order resonance peak and the pole. At $\omega = 20$ rad/s, the effect of the second-order system deviation is negligible, so a −3-db deviation due to the pole would be expected. At low frequencies, the true magnitude will be asymptotic to that of the integrator, while at high frequencies it will be asymptotic to the −60-db/decade line.

STEP 6: COMPLETE THE MAGNITUDE CURVE. The complete magnitude is shown in Fig. SP15.2.4.

STEP 7: PLOT THE STRAIGHT-LINE PHASE APPROXIMATIONS. The phase of the zero may be approximated by a line beginning at $\omega = 0.2$ rad/s (0° phase) and finishing at $\omega = 20$ rad/s (+90° phase), passing through +45° at the break frequency $\omega = 2$ rad/s. A similar phase plot may be drawn from the pole at 20 rad/s. The integrator has a constant phase of −90°. The second-order element cannot be approximated by straight lines, and an estimate of the phase-frequency curve has to be drawn from Fig. 15.18 for the value of ζ for the problem being considered. In this case we have $\zeta = 0.2$. Figure SP15.2.5 shows the component phase plots.

STEP 8: GRAPHICALLY ADD ALL ELEMENT PHASES. If a second-order element is involved in the transfer function, it is not possible to draw the composite phase-frequency plot as a series of straight-line segments, as was done in the previous problem. Instead, the phase is estimated at a few critical frequencies and a smooth curve is drawn through them. It will be seen in the next few modules that this will be sufficient for most Bode diagrams, and an accurate plot is rarely required. The estimation is performed in the following manner:

1. Observing Fig. SP15.2.5, it is seen that at low frequencies the phase will be −90°.

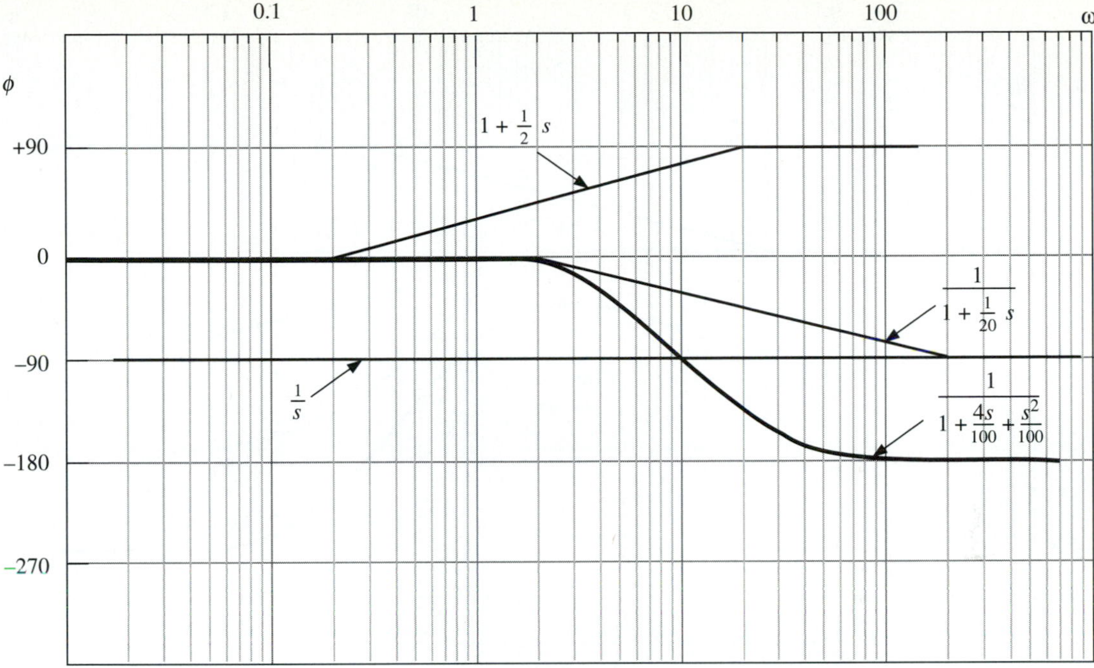

Fig. SP15.2.5

2. At $\omega = 2$ rad/s, the total phase will be $\phi = -90 + 45 - \delta$, where δ is the contribution from the second-order element. Estimating this to be 10°, $\phi = -55°$.

3. At $\omega = 10$ rad/s, $\phi = -90 - 90 + 80 - 40$, where the last two numbers are the estimated contributions from the zero and pole, respectively, at this frequency. This results in $\phi = -140°$.

4. At $\omega = 20$ rad/s, the phase is estimated to be $\phi = -90 + 90 - 45 - 160°$, where the last number is the estimated contribution from the second-order element. This results in $\phi = 205°$.

5. Above $\omega = 200$ rad/s, the phase asymptotically approaches the value $\phi = -270°$.

Figure SP15.2.6 shows these particular values of phase marked, and a composite phase plot is sketched through them. Finally, the complete Bode diagram showing magnitude and phase is given in Fig. SP15.2.7.

Sample Problem 15.3

Draw the Bode diagram of the system described by the open-loop transfer function

$$GH(s) = \frac{1.6(s^2 + 4s + 25)}{s(s^2 + 0.8s + 4)}$$

0.1
1
10
100
ω
φ
+90
0
−90
−180
−270
Composite
phase

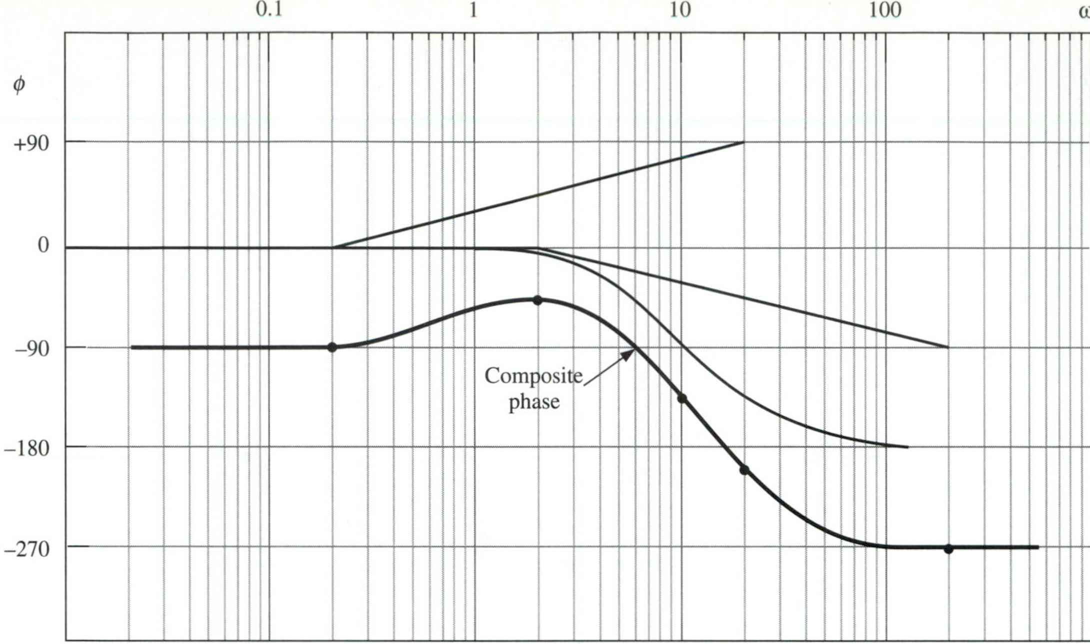

Fig. SP15.2.6

0.1
1
10
100
ω
20
M_{db}
0
−20
−40
0
φ
−90
−180
−270

Fig. SP15.2.7

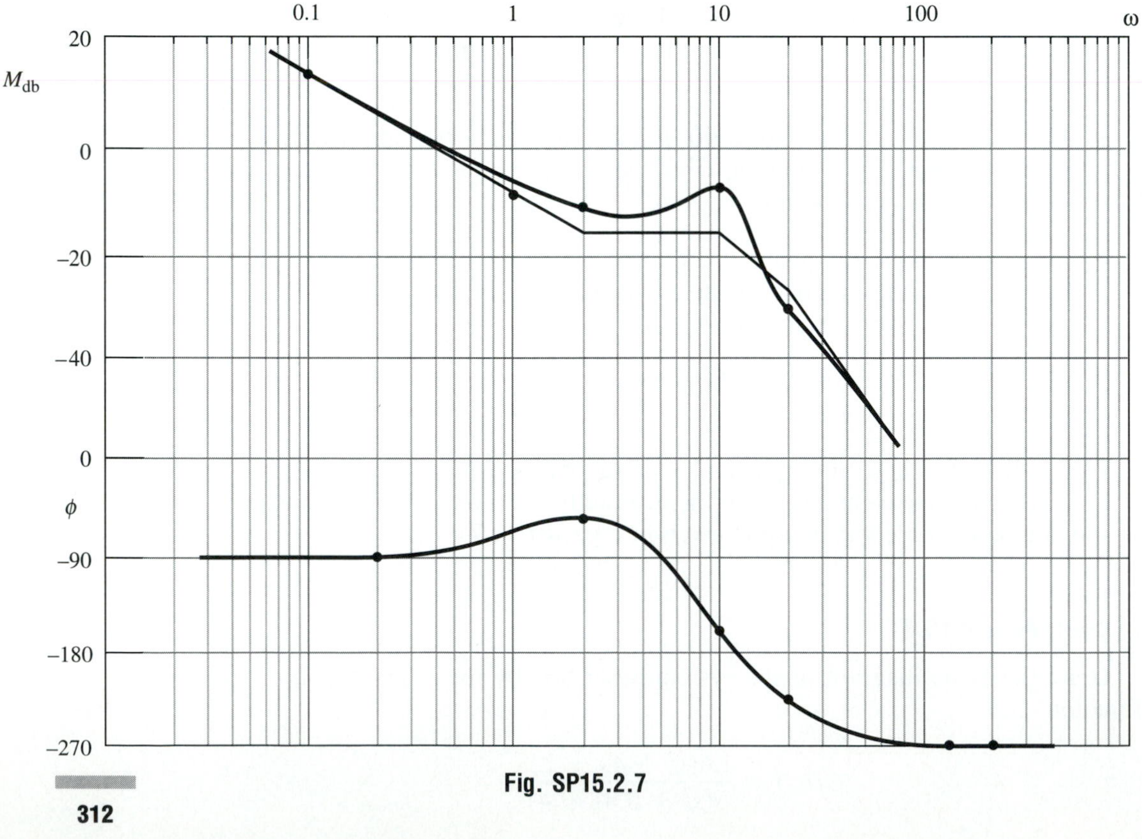

Solution

The solution is started by writing the open-loop transfer function in Bode form as

$$GH(s) = \frac{1.6 \times 25(s^2/25 + (4/25)s + 1)}{4s(s^2/4 + (0.8/4)s + 1)}$$

$$= \frac{10(s^2/25 + (4/25)s + 1)}{s(s^2/4 + (0.8/4)s + 1)}$$

In this system, we have two zeros and three poles. The second-order elements may yield complex or real roots depending on the value of the quadratic coefficients; therefore these quadratics are first solved to find the damping ratios and natural frequencies.

1. *Zeros.* $\omega_n = 5$ rad/s, and the damping ratio is given by

$$\frac{2\zeta}{\omega_n} = \frac{4}{25} \qquad \zeta = \frac{4}{2 \times 5} = 0.4$$

2. *Poles.* $\omega_n = 2$ rad/s, and the damping ratio is given by

$$\frac{2\zeta}{\omega_n} = \frac{0.8}{4} \qquad \zeta = \frac{0.8}{2 \times 2} = 0.2$$

Both poles and zeros are, therefore, complex. The various steps outlined in the two previous examples will not be listed in such rigorous form, since the plotting of the diagram proceeds in exactly the same manner. The break frequencies ω_b for the elements are seen to be 1, 2, and 5 rad/s. The straight-line segments are plotted in Fig. SP15.3.1 for unity Bode gain ($K_{db} = 0$). Figure SP15.3.2

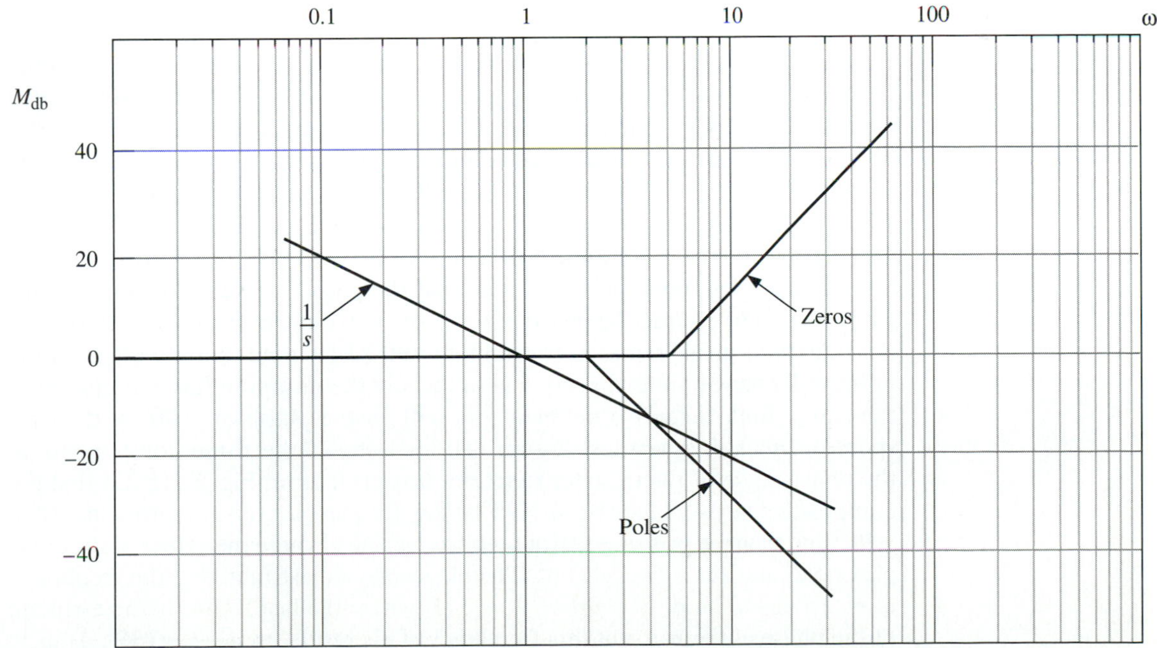

Fig. SP15.3.1

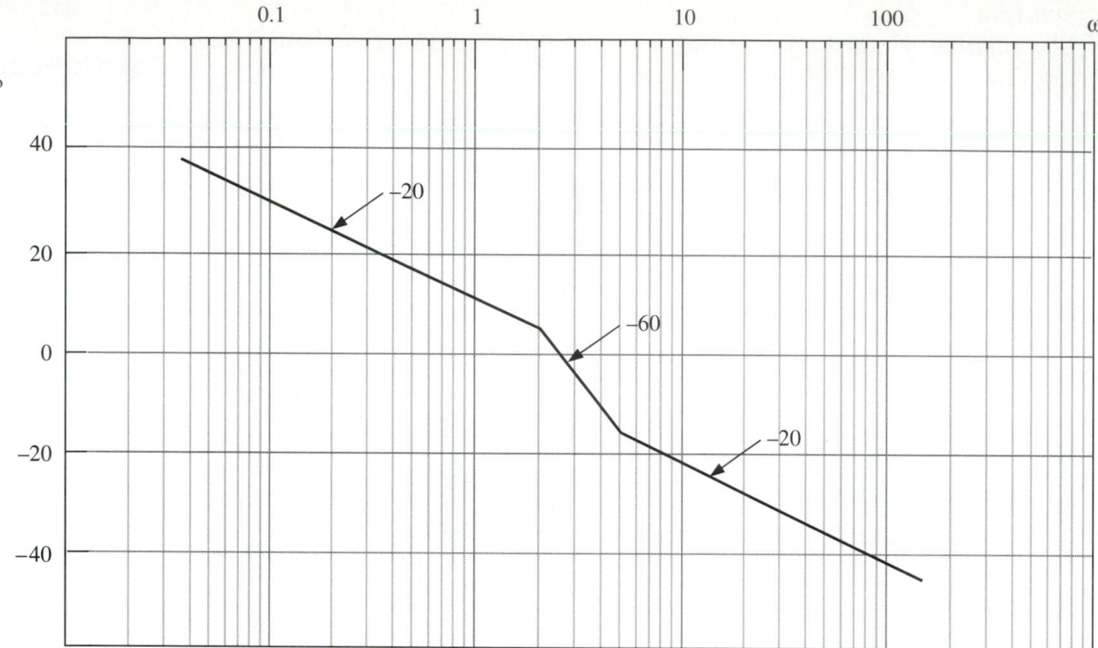

Fig. SP15.3.2

shows the summation of the straight-line segments to which is added the 20 db resulting from the Bode gain of

$$K_{db} = 20 \log_{10} 10 = 20 \text{ db}$$

In order to estimate the deviations from the straight-line segments, Fig. 15.17 has to be inspected for the values of $\zeta = 0.2$ and $\zeta = 0.4$. Taking the lowest frequency first, the poles with break frequency $\omega = 2$ rad/s have a damping ratio of 0.2 and will have a peak magnitude about 8 db above the low-frequency asymptote, although this will be reduced somewhat due to the zeros. The zeros, however, will have a deviation of -3 db below the low-frequency asymptote at $\omega = 5$ rad/s, and this will also be reduced due to the poles. It will be assumed that the deviation is $+7$ db at $\omega = 2$ rad/s and -2 db at $\omega = 5$ rad/s. The estimated true magnitude will therefore be as shown in Fig. SP15.3.3. For the phase angle–frequency plot, note that the phase of the zeros will change from 0 to $+180°$, while the phase of the poles will change from 0 to $-180°$, and the integrator phase remains constant at $-90°$. Also note that the phase of the poles will change more sharply due to the smaller damping ratio compared to the zeros. Both second-order elements will have a phase of $-90°$ at their respective break frequencies. Figure SP15.3.4 shows the three components to the phase angle of the transfer function. It is clear from Fig. SP15.3.4 that the composite phase changes from $-90°$ at low frequencies to something less than $-90°$ and then returns to $-90°$ at high frequencies, where the effect of the complex poles and zeros cancels out. The phase may be estimated at the frequency $\omega = 2$ rad/s to be $\phi = -90 - 90 + 45$, where the last figure is the estimate of the phase of the zeros at this frequency. This results in $\phi = -135°$. At $\omega = 5$ rad/s, the phase may be estimated as $\phi = -90 + 90 - 140$, where this time the last figure is the estimate of the phase contribution from the poles.

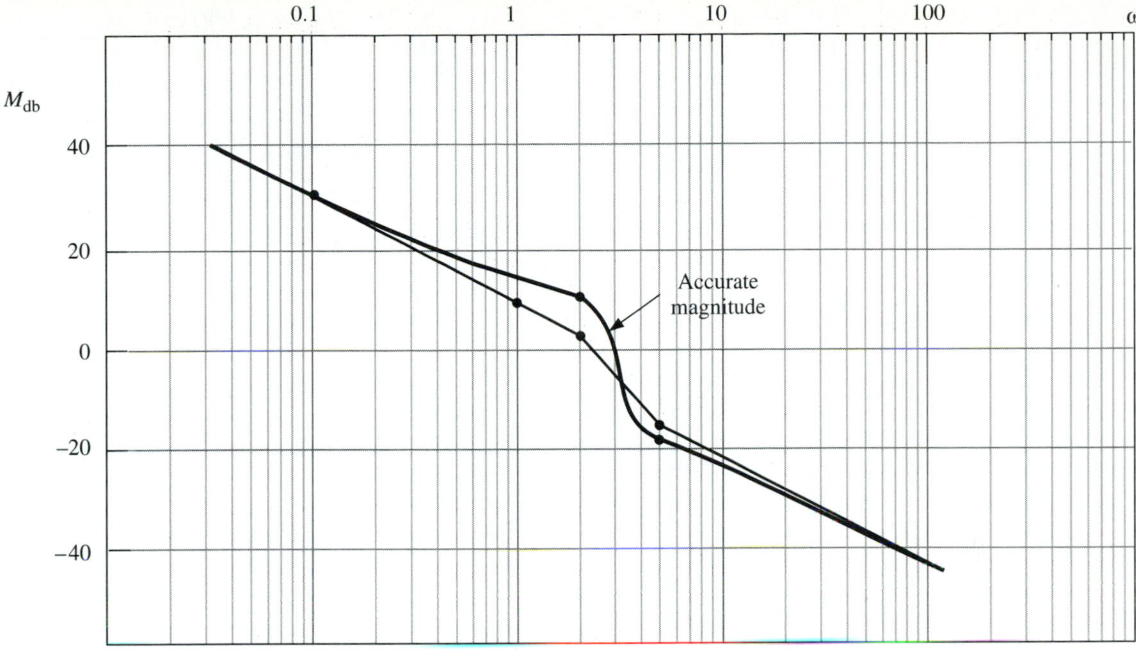

Fig. SP15.3.3

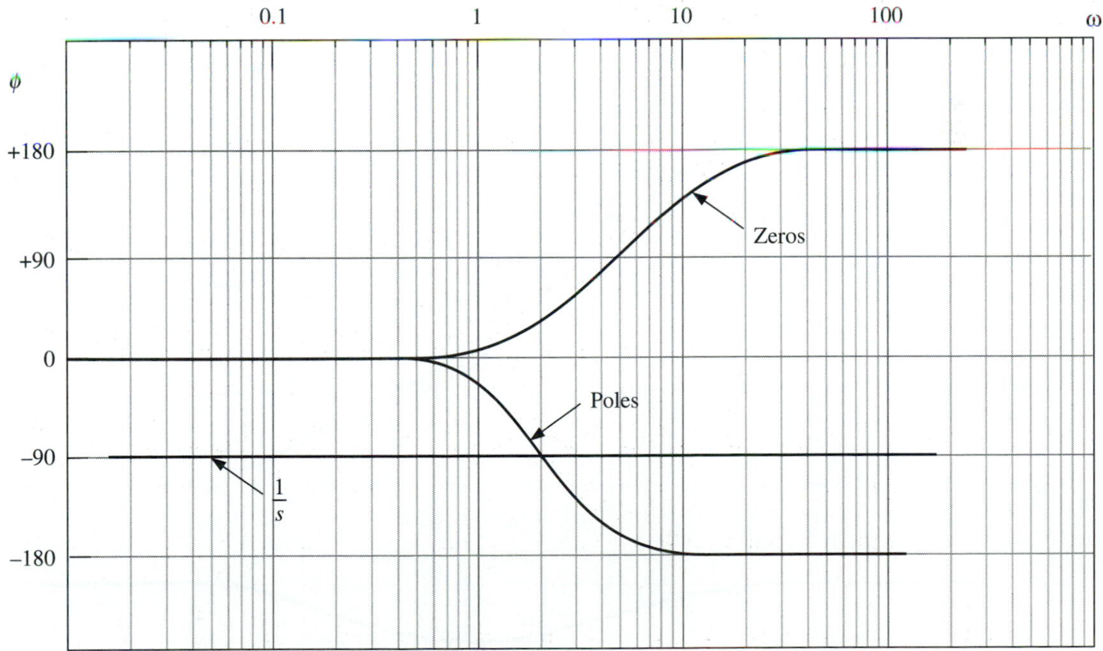

Fig. SP15.3.4

These two points are sufficient to sketch the phase-frequency plot, which is shown in Fig. SP15.3.5. The final Bode diagram for this system is shown in Fig. SP15.3.6.

0.1
1
10
100
ω
φ
+180
+90
0
–90
–180
Zeros
Poles
1/s
Accurate phase

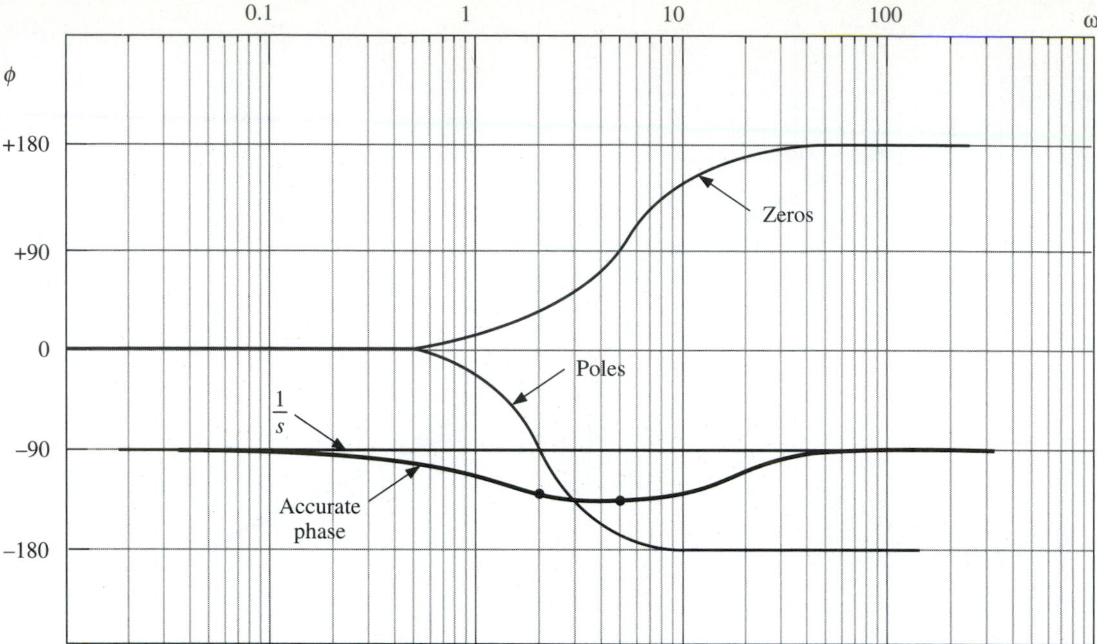

Fig. SP15.3.5

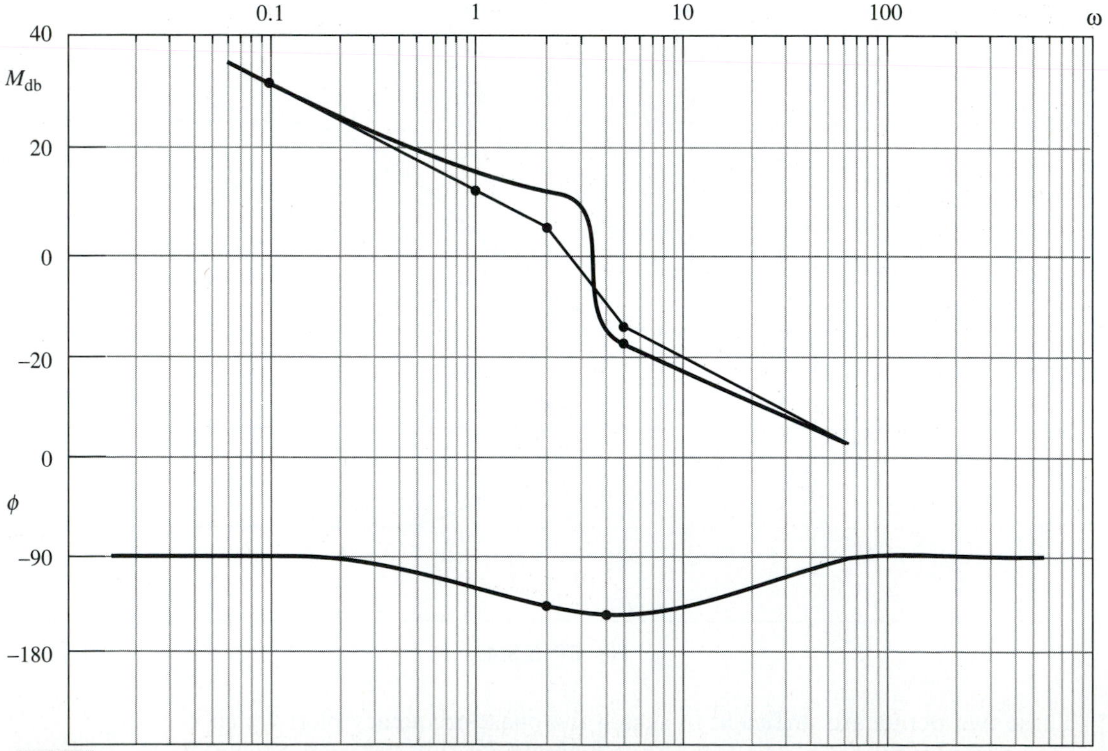

0.1
1
10
100
ω
40
M_{db}
20
0
–20
0
φ
–90
–180

Fig. SP15.3.6

15.1 Using straight-line approximations, plot the Bode diagrams of the systems described by the following open-loop transfer functions:

a. $GH(s) = \dfrac{10}{(1 + \frac{1}{3}s)(1 + \frac{1}{2}s)}$

b. $GH(s) = \dfrac{16s}{(s + 1)(s + 2)}$

c. $GH(s) = \dfrac{25(s + 1)}{s(s + 5)}$

d. $GH(s) = \dfrac{2(1 + \frac{1}{10}s)}{s(1 + \frac{1}{100}s)(1 + \frac{1}{1000}s)(1 + \frac{1}{2000}s)}$

e. $GH(s) = \dfrac{0.4s(s + 2)}{(s + 0.2)(s + 1)(1 + \frac{1}{10}s)}$

15.2 The following systems contain second-order components. By estimating the magnitude and phase contributions from these elements, plot the Bode diagram for each system:

a. $GH(s) = \dfrac{15}{s^2 + s + 64}$

b. $GH(s) = \dfrac{1 + 0.2s}{s(s^2 + 0.5s + 0.1)}$

c. $GH(s) = \dfrac{100(s + 3)}{s^2(1 + 0.2s + 0.01s^2)}$

d. $GH(s) = \dfrac{20}{(s^2 + 2s + 17)(s^2 + 3s + 50)}$

e. $GH(s) = \dfrac{0.625(s^2 + 0.8s + 4)}{s^2 + 4s + 25}$

15.3 Plot the Bode diagram of the system shown in Fig. P15.3.

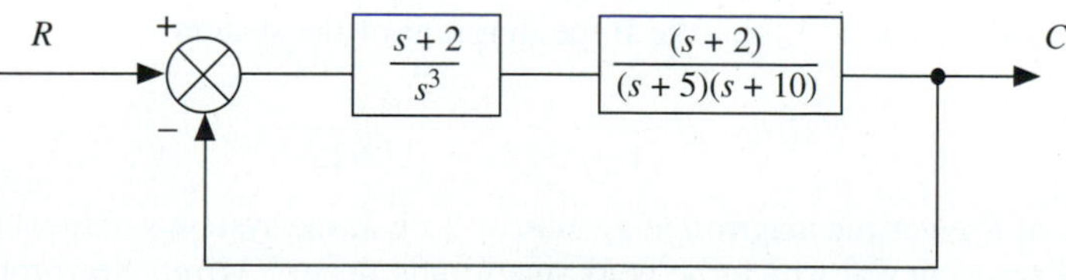

Fig. P15.3

15.4 Consider the open-loop transfer function elements

$$GH(s) = \frac{10}{(s + 1)(s + 3)} \qquad H(s) = 1$$

Sketch the Bode diagram of this system by **(a)** considering the denominator of G to be the product of two first-order elements and **(b)** multiplying out and then considering the denominator to be a second-order element.

15.5 What value of K gives the system described by

$$G(s) = \frac{K(s + 0.5)}{s^2 + 12s + 100}$$

a magnitude of -10 db at a frequency $\omega = 5$ rad/s?

15.6 For the following systems, calculate the phase angle when the magnitude is unity:

a. $GH(s) = \dfrac{10}{s^2(s + 3)}$

b. $GH(s) = \dfrac{25(1 + 0.1s)}{s(1 + 0.2s)(1 + 0.3s)}$

c. $GH(s) = \dfrac{12s(s + 1)}{(s + 2)^3}$

15.7 For the systems given in problem 15.6, determine the magnitude when the phase is $-225°$ for system (a), $-100°$ for system (b), and $0°$ for system (c).

15.8 For the standard second-order system described by

$$GH(s) = \frac{1}{1 + (2\zeta/\omega_n)s + s^2/\omega_n^2}$$

determine the frequency ratio ω/ω_n at which the peak of the magnitude curve (Fig. 15.17) occurs and the magnitude at this frequency. Plot a graph of $\hat{M}_{db}$ against the damping ratio ζ.

15.9 Plot the Bode diagram of the system shown in Fig. P15.9 for the value of forward-path gain $K = 1000$. What are the exact deviations between the asymptotic and true magnitude curves at the break frequencies?

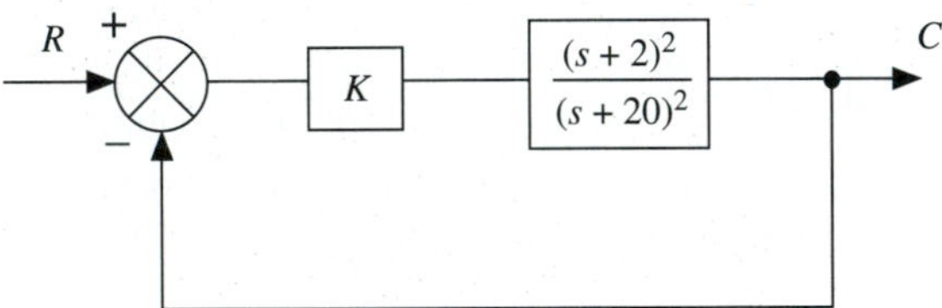

Fig. P15.9

15.10 For a value of $K = 1$, plot the Bode diagram of the system

$$GH(s) = \frac{K}{(1 + 0.15s + 0.0625s^2)^2}$$

What value of K gives the magnitude a value of 2 db at the system's natural frequency? What is the frequency at which the peak magnitude occurs? (*Hint:* See problem 15.8.)

Bode Analysis, Stability, and Gain and Phase Margins

◆ CONDITIONAL STABILITY

It is now well known from our studies of the root locus and Nyquist methods that a closed-loop system may be stable or unstable, depending upon the value of the forward-path gain K. This may be observed from the Bode diagram also. It may be seen from the previous module that the magnitude M of the forward-path transfer function with harmonic input is directly proportional to the Bode gain M_{db}, while the phase angle is independent of the gain. The effect of changing the gain is to shift the magnitude curve up (if K is increased) or down (if K is decreased), leaving the phase curve unchanged. For example, if a Bode diagram is plotted for a certain value of $K = K'$, then the Bode gain is

$$20 \log_{10} K' = K'_{db} \tag{16.1}$$

If the gain is doubled, the new Bode gain is

$$20 \log_{10} 2K' = 20 \log_{10} 2 + 20 \log_{10} K' = K'_{db} + 6 \tag{16.2}$$

while if the gain is halved,

$$20 \log_{10} 0.5K' = 20 \log_{10} 0.5 + 20 \log_{10} K' = K'_{db} - 6 \tag{16.3}$$

The effect of changing K may be thought of as adding or subtracting decibels from the initial magnitude plot.

In order to examine the conditional stability of a system using the Bode diagram, the same example will be used as in Module 14, where this issue was investigated using the Nyquist diagram. Recall that the system under study is

$$GH(s) = \frac{K}{s(1 + 2s)(1 + 3s)} \tag{16.4}$$

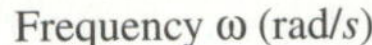

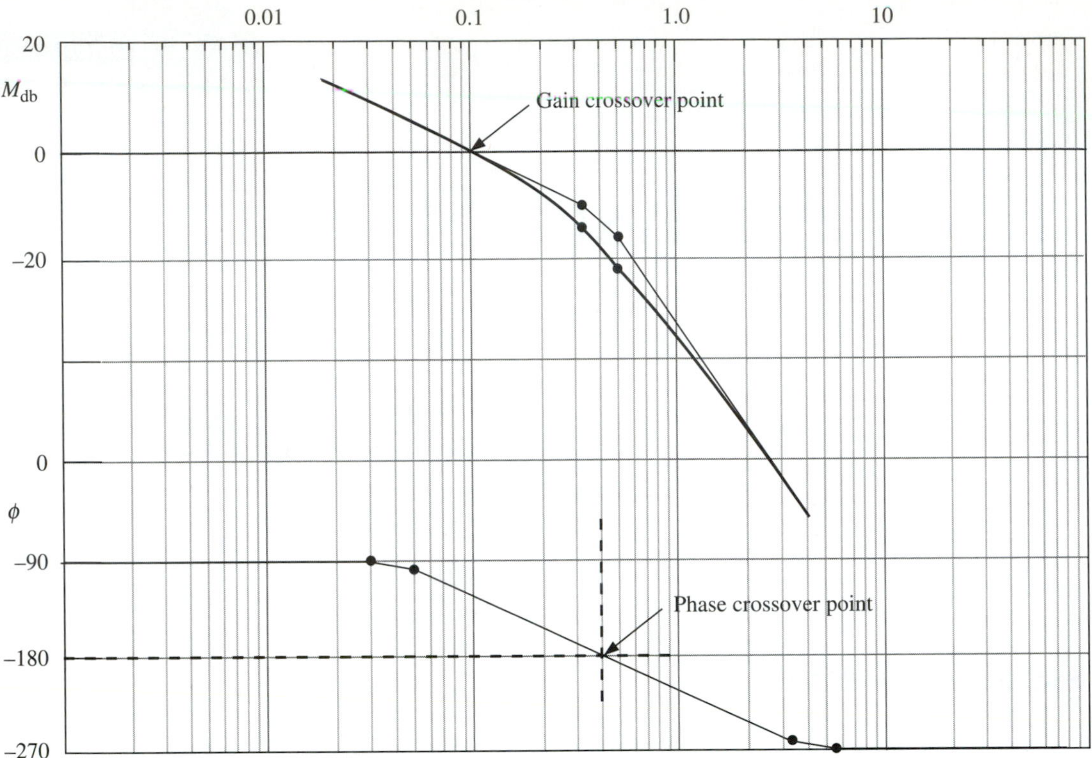

Fig. 16.1 Bode diagram for sample system.

where $K = 0.1$. The Bode diagram may be quickly plotted, recognizing that the break frequencies are 1, 0.50, and 0.34 rad/s. The complete diagram is shown in Fig. 16.1. To obtain a stability criterion for the Bode diagram, we shall begin with the simplified Nyquist stability criterion, as outlined in Module 14, namely:

> A system is stable if the magnitude is less than or equal to unity when the phase angle is $-180°$.

We shall see shortly that this criterion will not give reliable information about system stability; however, we will use it for now. In terms of the variables plotted in the Bode diagram, unity magnitude corresponds to $M_{db} = 0$. The Bode diagram for the sample system indicates that the phase is $-180°$ at a frequency of about 4 rad/s, and at this frequency, the magnitude is about -18 db. The magnitude is clearly much less than unity and the system may be identified as stable.

It is now apparent why the phase angle is plotted directly below the magnitude curve, since it is not necessary to determine the frequency at which the phase is $-180°$ (as was the case for the Nyquist method), since we can read the magnitude directly above the phase crossover point.

It was seen in the analysis of Nyquist diagrams that the system may be made unstable by increasing K, thereby expanding the diagram about the origin until the critical point is enclosed. Increasing K will move the magnitude curve upward in Fig. 16.1, moving the gain crossover point to the right. Note that the phase crossover point remains unchanged. Eventually the gain crossover point and the phase crossover point occur at the same frequency, implying that the magnitude is 0 db when the phase is $-180°$, and the system is on the point of instability. Increasing K further results in positive magnitude ($M > 1$), producing an unstable system. Figure 16.2 shows the Bode diagram plotted for several values of K, producing stable, marginally stable, and unstable systems. For comparative purposes, the corresponding Nyquist diagrams and positions of the closed-loop poles on the root locus are shown for the same system. The question arises, what is the gain for the sample system shown in Fig. 16.1 that just causes system instability? Recall that the system shown in the figure was plotted for $K = 0.1$. Therefore it is apparent that this critical value of gain K_c is greater than this. It is also clear that increasing K_{db} by about 18 db will result in K_c. From equation 16.2 it may be seen that

$$20 \log_{10} NK = 20 \log_{10} K + 20 \log_{10} N \tag{16.5}$$

where N is the value that K has to be *multiplied by* in order to make it equal to K_c. This equation may be written as

$$20 \log_{10} NK = K_{db} + N_{db} \tag{16.6}$$

where N_{db} is the number of decibels that the magnitude curve has to be shifted up in order to obtain the required plot. Since

$$N_{db} = 20 \log_{10} N \tag{16.7}$$

and $N_{db} = 18$, we get

$$N = 7.94 \tag{16.8}$$

This means that the initial value of K for which the magnitude curve was plotted has to be multiplied by 7.94 in order to make the magnitude curve pass through 0 db at the phase crossover frequency. This results in $K_c = 0.794$, which is close to the value obtained (0.832) using the Nyquist analysis.[1] If the system were already unstable, the gain would have to be reduced in order to stabilize it. This corresponds to subtracting decibels from the magnitude, which in turn implies multiplication of the original gain by a fraction, i.e., reducing it. This is what would be expected.

◆ GAIN AND PHASE MARGINS IN THE BODE DIAGRAM

Recalling the definition of gain margin from Module 14 (equation 14.7), as the ratio of the critical gain to the gain for which the frequency response is initially plotted

$$\mathrm{GM} = \frac{K_c}{K} \tag{16.9}$$

Taking $\log_{10}$ of each side of this equation results in

$$\mathrm{GM}_{db} = 20 \log_{10} K_c - 20 \log_{10} K \tag{16.10}$$

[1]The difference is due to the straight-line approximation to the phase angle plot in the Bode diagram, Fig. 16.1.

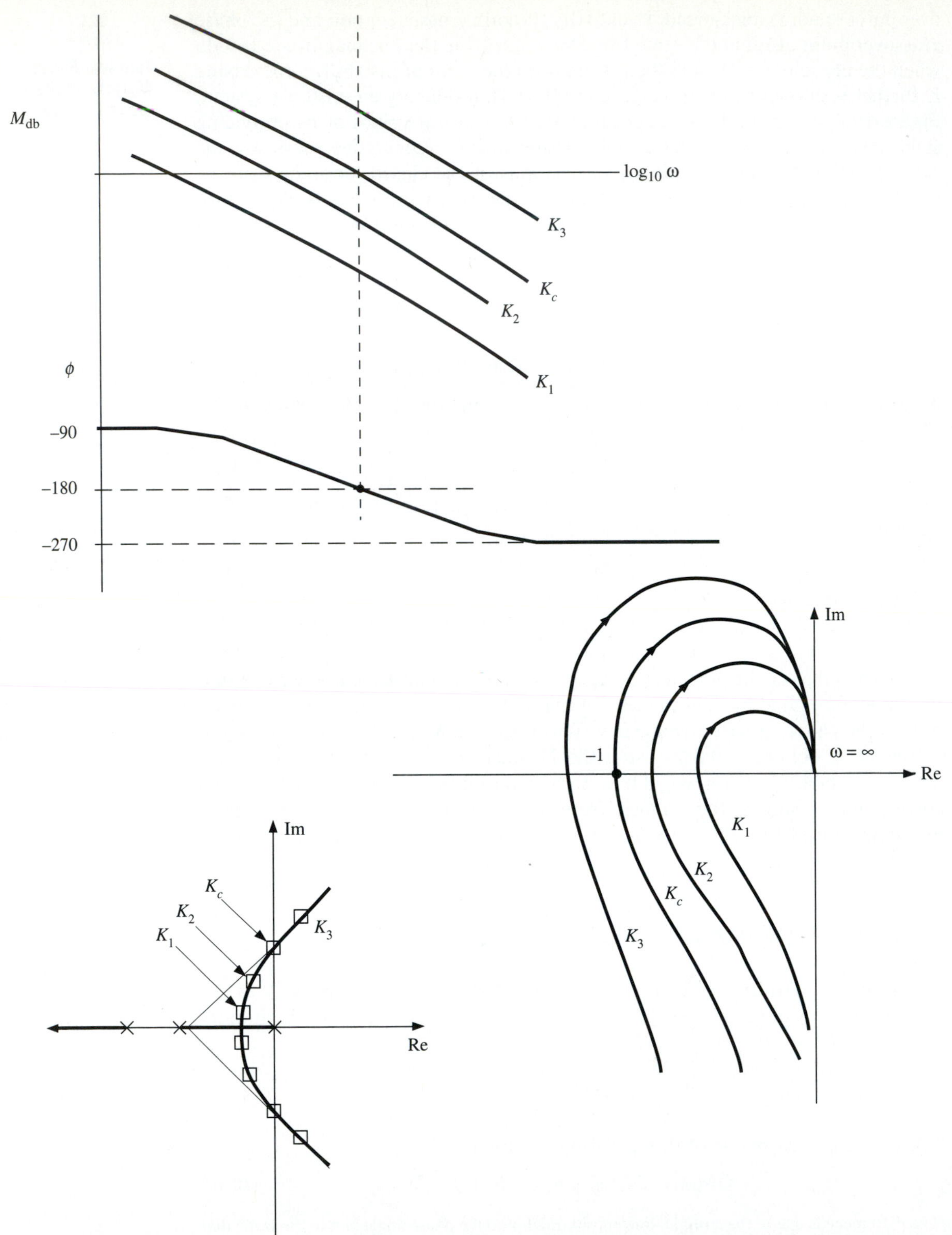

Fig. 16.2 Bode, Nyquist, and root locus of system with variable *K*.

The gain margin (in decibels) is the difference between K_c and the original gain K in decibels. It may be seen from Fig. 16.2 that, for the example chosen, the gain margin is 18 db.

Consider the definition of phase margin as being the amount of pure phase lag that may be introduced in order to make the frequency response pass through the critical point.[2] In terms of the Bode diagram, introducing pure phase lag corresponds to moving the phase curve down without changing the magnitude curve. As this is done, the phase crossover point moves to the left until the gain and phase crossover points occur at the same frequency, at which point the system becomes unstable. Referring to Fig. 16.1, this will happen when a total of 54° of additional phase lag is introduced. This value is, therefore, the phase margin; hence

$$\text{PM} = 54^\circ \tag{16.11}$$

Estimates of the gain and phase margins may be readily obtained from the Bode diagram, as indicated in Fig. 16.3. To summarize, the gain margin is obtained by investigating the phase crossover frequency. It is the difference, in decibels, between the magnitude at this frequency and the 0-db line. A stable system usually has positive gain margin while an unstable system has a negative gain margin. The phase margin is obtained by investigating the gain crossover frequency and is the difference between the phase at this frequency and the −180° line.

◆ SYSTEM TYPE AND STEADY-STATE ERROR FROM BODE DIAGRAMS

Recall the definition of a general open-loop transfer function from Module 8 as

$$G(s) = \frac{K(s + z_i)(s + z_2)\cdots(s + z_m)}{s^n(s + p_1)(s + p_2)\cdots(s + p_q)} \tag{16.12}$$

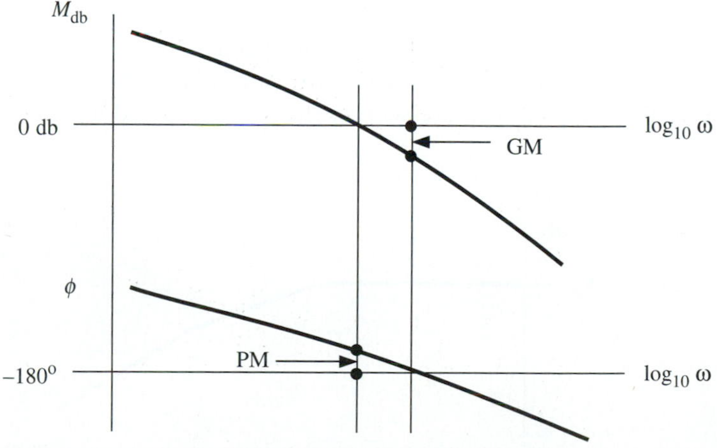

Fig. 16.3 Gain and phase margins.

[2] This will be interpreted as making the phase −180° when the magnitude is 0 db.

For plotting the Bode diagram, $H(s)$ is not limited to unity; hence we may write this general transfer function as

$$GH(s) = \frac{K \prod_{i=1}^{m} z_i(1 + s/z_1)(1 + s/z_2)\cdots(1 + s/z_m)}{\prod_{k=1}^{q} p_k s^n(1 + s/p_1)(1 + s/p_2)\cdots(1 + s/p_k)} \tag{16.13}$$

It may be seen that the gain K is the Evans gain, and when multiplied by the product of the zeros and divided by the product of the poles, it becomes simply the Bode gain, i.e.,

$$GH(s) = \frac{K_b(1 + s/z_i)(1 + s/z_2)\cdots(1 + s/z_m)}{s^n(1 + s/p_1)(1 + s/p_2)\cdots(1 + s/p_k)} \tag{16.14}$$

In the Bode diagram, allowing s to approach zero is the same as allowing $j\omega$ to approach zero, or simply letting ω approach zero. This corresponds to examining the low-frequency behavior of the system. This implies that at low frequencies, the above system approximates to

$$GH(s) \approx \frac{K_b}{s^n} \tag{16.15}$$

Also note that since the error coefficients are obtained by allowing $s \to 0$, the *Bode gain and the error constants are identical.* Since we will be mostly interested in obtaining the position, velocity, and acceleration constants, these are finite and nonzero for type 0, 1, and 2 systems, respectively. Each such system will now be examined to determine the value of the appropriate error constant from the Bode diagram.

Type 0 System

A type 0 system has the value $n = 0$; hence the system, at low frequencies, approximates to

$$GH(s) = K_b \tag{16.16}$$

In the M_{db}–$\log_{10} \omega$ plane this becomes

$$GH(j\omega) = K_p \tag{16.17}$$

since it is the position error constant that is finite and nonzero for a type 0 system. It is concluded that type 0 systems will have the Bode diagram consisting of a horizontal line at low frequencies, intersecting the M_{db} axis at $20 \log_{10} K_p$, as shown in Fig. 16.4.

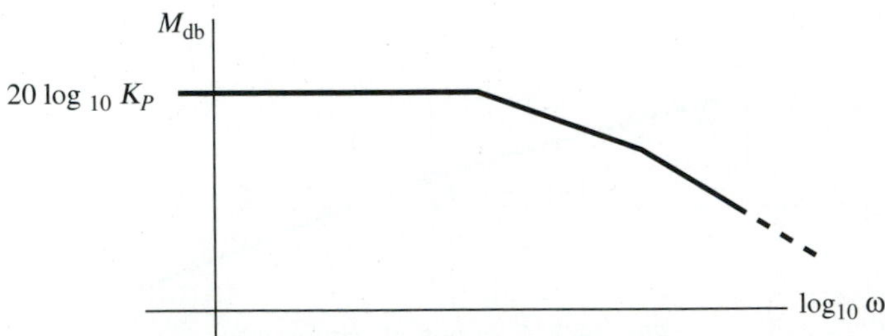

Fig. 16.4 Bode diagram of type 0 system.

Type 1 System

A type 1 system approximates at low frequencies to

$$GH(s) = \frac{K_b}{s} = \frac{K_b}{j\omega} \tag{16.18}$$

Also, for a type 1 system, it will be the velocity error constant that is equal to the Bode gain, i.e.,

$$GH(j\omega) = \frac{K_v}{\omega} \tag{16.19}$$

A type 1 system will behave (at low frequencies) as a combined integrator and gain. It will have a slope of -20 db/decade and would pass through $M_{db} = 0$ when $\omega = 1$ if $K_v = 1$. If the low-frequency part of the magnitude does not pass through 0 db, the value of K_v (in decibels) may be obtained from measuring the gain at $\omega = 1$, as shown in Fig. 16.5. If other factors of $GH(s)$ contribute to a composite magnitude plot before the frequency $\omega = 1$, the extension of the low-frequency asymptote should be used to find K_v, as indicated in Fig. 16.6.

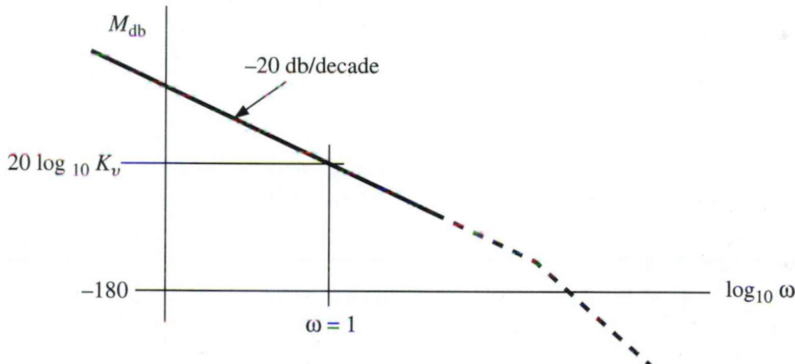

Fig. 16.5 Bode diagram of type 1 system.

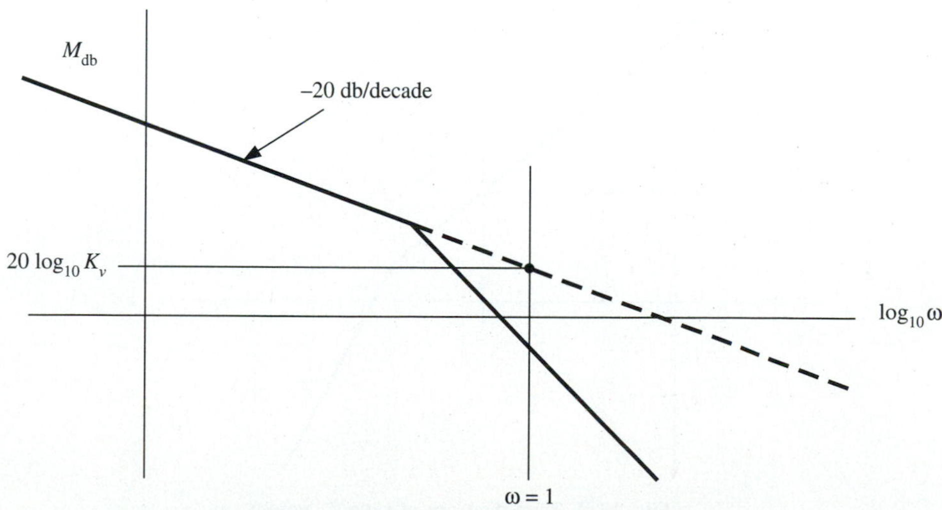

Fig. 16.6 Alternative Bode diagram of type 1 system.

Type 2 System

At low frequencies, a type 2 system ($n = 2$) may be approximated by

$$GH(s) = \frac{K_b}{s^2} = \frac{K_a}{(j\omega)^2} \tag{16.20}$$

since it is the acceleration constant K_a that is equal to the Bode gain K_b. The low-frequency part of the magnitude plot, or its extension, will have a slope of -40 db/decade and would pass through 0 db at $\omega = 1$ if $K_a = 1$. The value of K_a (in decibels) may be found by measuring the gain at $\omega = 1$, as shown in Fig. 16.7.

◆ FURTHER DISCUSSION OF GAIN AND PHASE MARGINS

Although gain and phase margins have been utilized to determine system stability, it is important to realize that the *phase margin* is the only reliable parameter on which to base such determinations. A few examples will indicate why this is so.

Simple Case

A simple case of system stability is provided by the example given earlier in this module, as shown in Fig. 16.3. The phase angle plot crosses the $-180°$ line from top to bottom, i.e., in a decreasing manner. In such a case, the gain margin is clearly positive, as is the phase margin. Recall, however, Fig. 14.5, which shows two systems with the same gain margin but differing phase margins, to realize why the phase margin is a more reliable measure of closeness to instability.

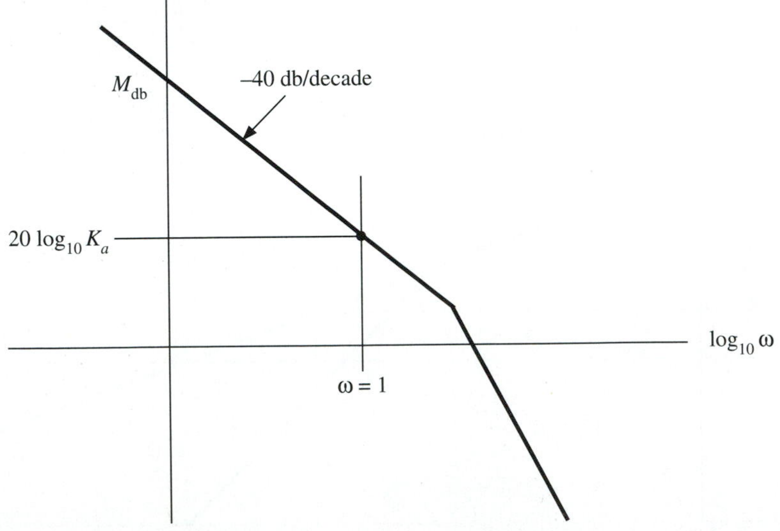

Fig. 16.7 Bode diagram of type 2 system.

No Frequency Crossover Points

Consider the Bode diagram of the system described by the open-loop transfer function

$$GH(s) = \frac{K(1 + s)}{s^2(1 + s/10)} \tag{16.21}$$

The student may verify that the Bode diagram is as shown in Fig. 16.8. The phase angle never crosses the $-180°$ line; therefore it is not possible to determine the gain margin at all. Do not attempt to extrapolate the phase curve to $\omega = 0$ or $\omega = \infty$ as conflicting gain margins will result for each case! Instead, consider the phase margin obtained from the gain crossover point, which indicates a stable system with a phase margin of about 45°. In fact, this system is stable for all gains, although it will be oscillatory for large K. This may be confirmed from the root locus, which is sketched in Fig. 16.9.

Multiple Frequency Crossover Points

For some transfer functions, the gain margin and phase margin may give conflicting indications of stability. Consider the system

$$GH(s) = \frac{K(1 + s)(1 + s/10)}{s^3(1 + s/100)(s + 1/1000)} \tag{16.22}$$

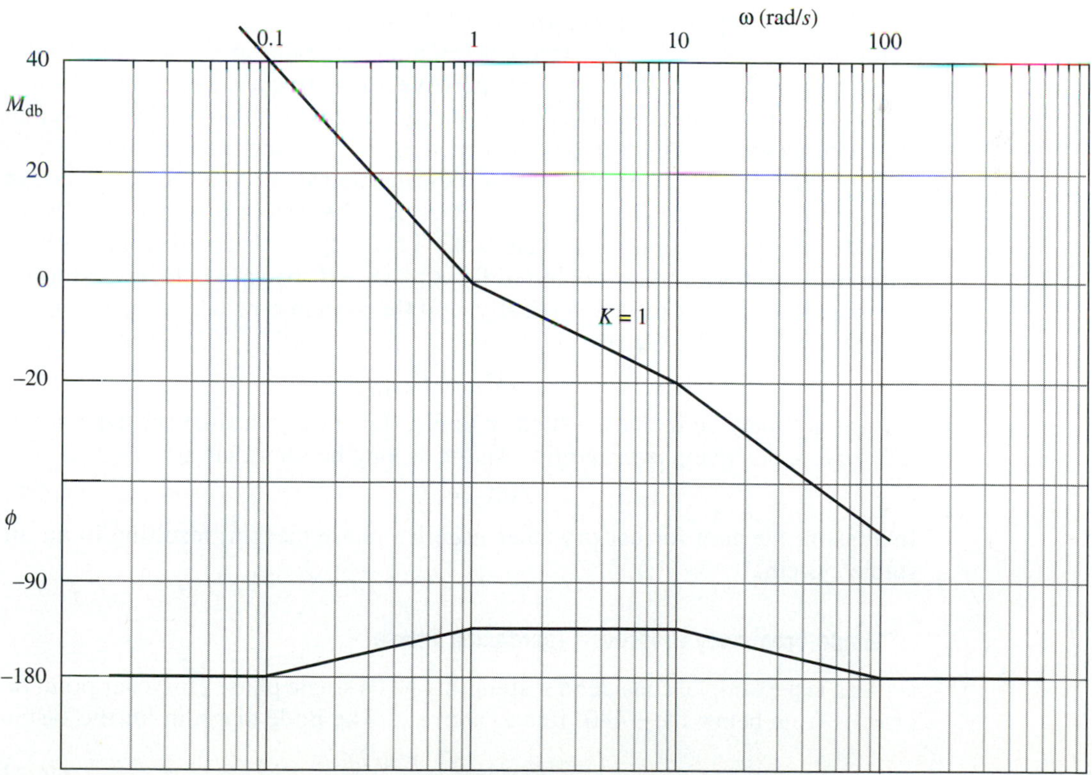

Fig. 16.8 Bode diagram with no phase crossover points.

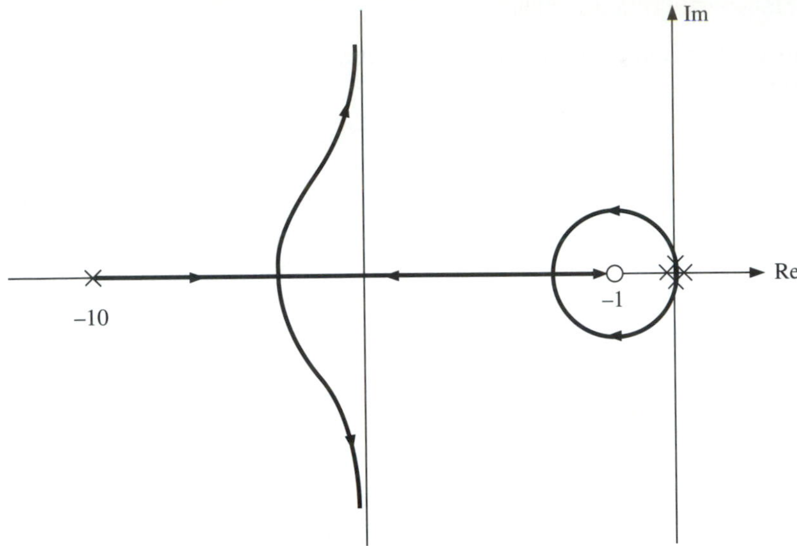

Fig. 16.9 Root locus of system.

The Bode diagram is shown in Fig. 16.10. Using straight-line approximations for both the magnitude and phase, it is seen that the gain crossover frequency is at $\omega = 1$, indicating a phase margin of $-90°$, suggesting the system is unstable. There are two phase crossover frequencies, at $\omega = 3$ rad/s and at $\omega = 300$ rad/s. At each frequency, the gain margin is positive, indicating a stable system! In such a case, it is important to rely only on the phase margin, concluding that the system is indeed unstable. This may be confirmed again by using the root locus, which is sketched in Fig. 16.11. The root locus indicates that at low gains $K < K_1$ the system is unstable. For larger gains, $K_1 < K < K_2$, the system is stable. Finally, for $K > K_2$ the system is once more unstable. This may be observed from the Bode diagram also. Figure 16.10 is plotted for $K < K_1$. If the gain is increased to K_1, the magnitude plot is moved vertically, and the system now has a positive phase margin and is stable. Clearly

$$\mathrm{GM}_1 = K_{1\mathrm{db}} \tag{16.23}$$

The gain K may be increased further to K_2, during which time the system will still have a positive phase margin. Again, it may be seen that

$$\mathrm{GM}_2 = K_{2\mathrm{db}} \tag{16.24}$$

Increasing the gain further produces negative phase margin, resulting in an unstable system.

Single-Frequency Crossover: Increasing Phase

As a last example, consider a system that has a single phase crossover point but crosses from below the $-180°$ line to above it. The Bode diagram for the system

$$GH(s) = \frac{(1 + s/2)^2}{s^3} \tag{16.25}$$

is shown in Fig. 16.12. Again we see conflicting stability predictions from a posi-

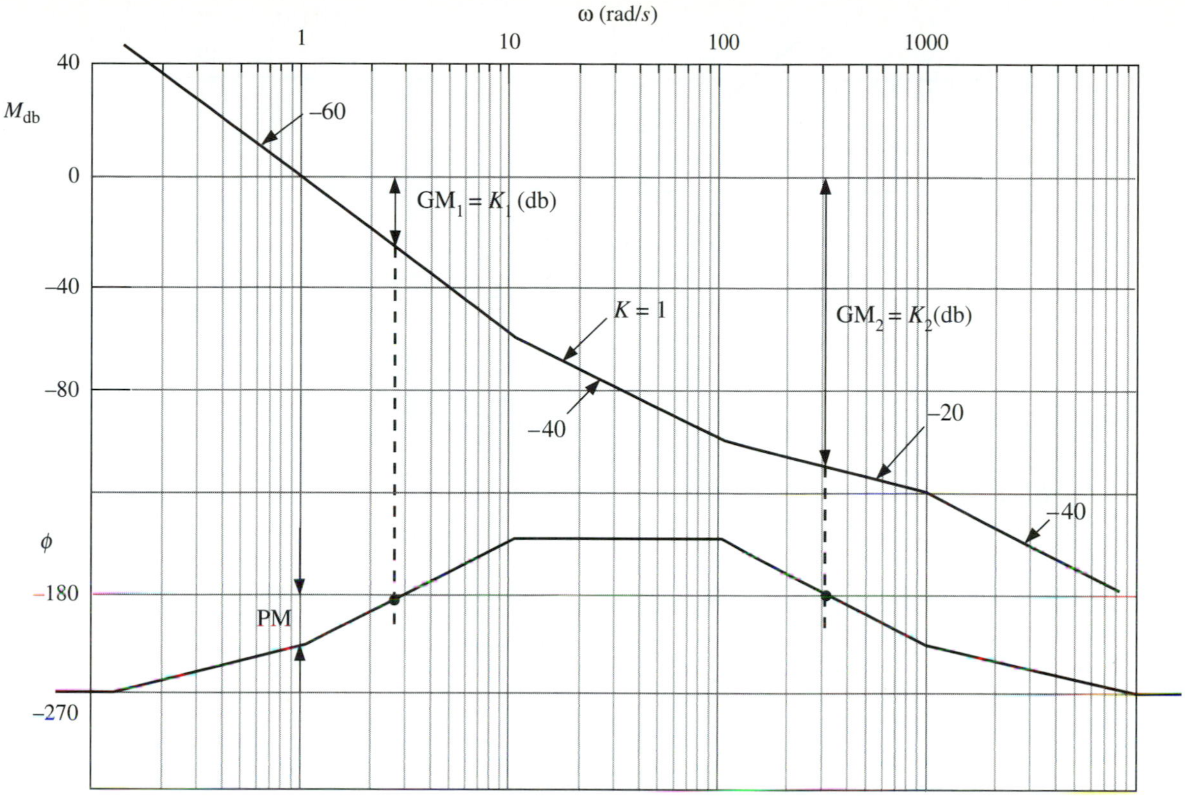

Fig. 16.10 System with two phase crossover points.

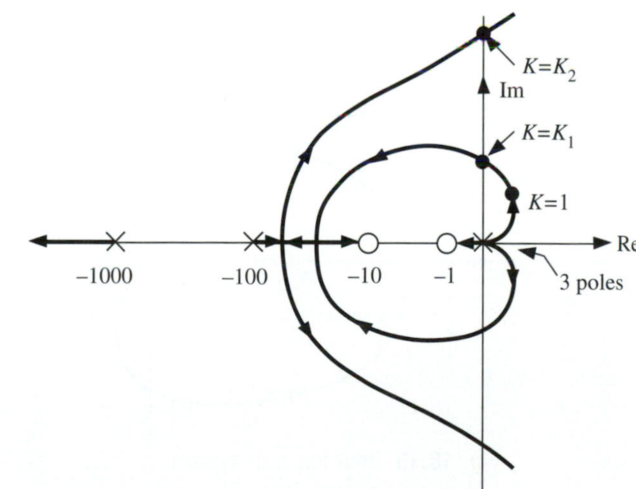

Fig. 16.11 Root locus of system.

tive gain margin but a negative phase margin. The system is unstable, as suggested by the phase margin, as the root locus shown in Fig. 16.13 confirms. Also, the system is seen to be conditionally stable, being unstable for $K < K_1$ and stable for $K > K_1$.

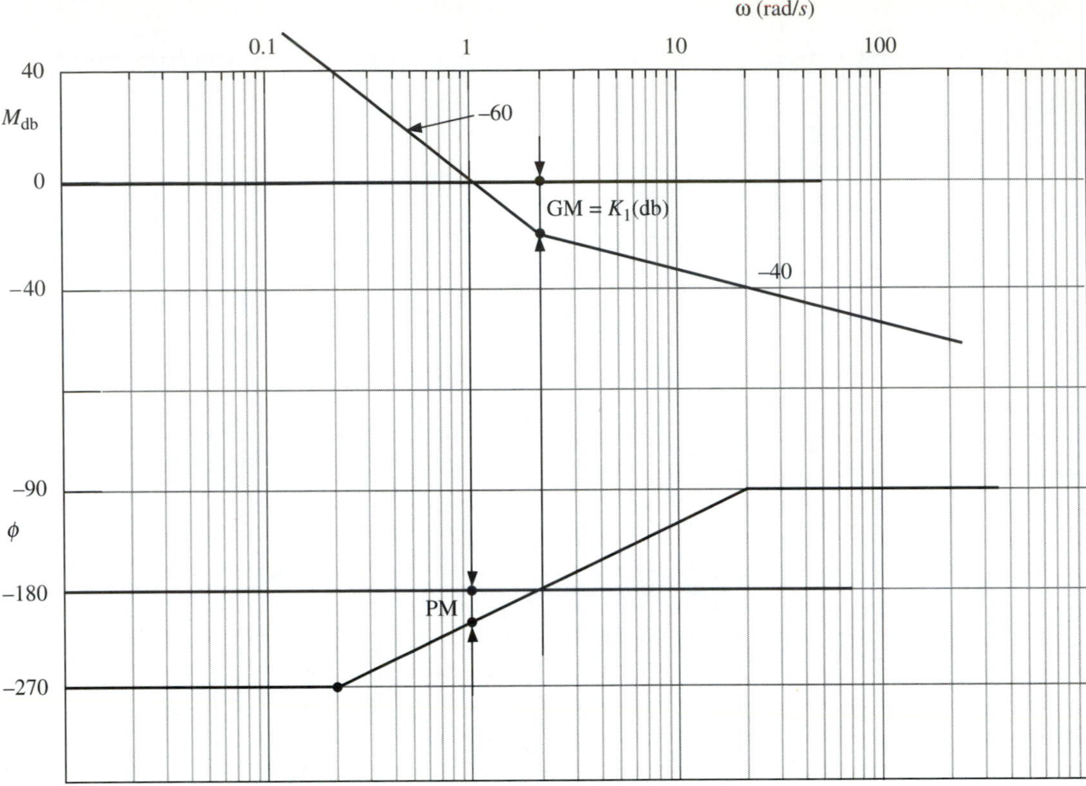

Fig. 16.12 Bode diagram of system with single increasing phase crossover point.

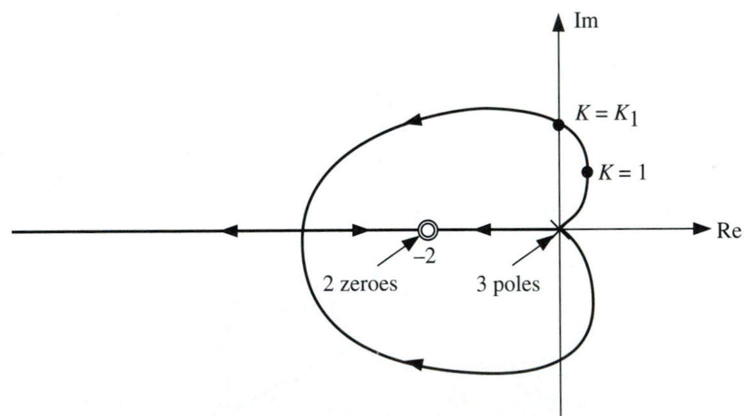

Fig. 16.13 Root locus of system.

It is apparent that if both gain and phase margins are used to predict system stability, contradictions may arise. The only reliable test for stability is through the use of the phase margin. In fact, the only reliable test of stability from the Bode diagram may be stated as:

A system is stable if it has a positive phase margin.

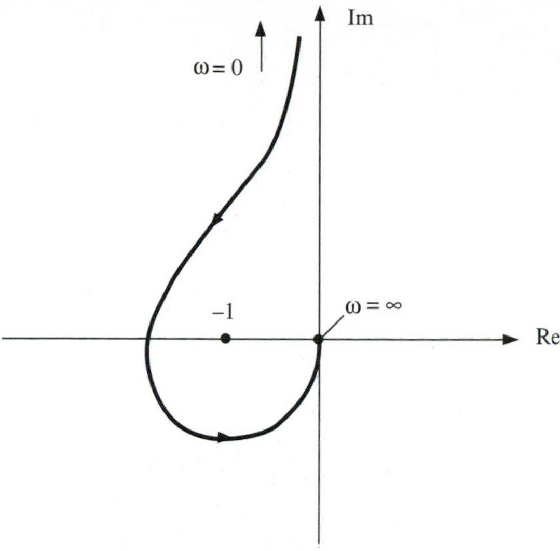

Fig. 16.14 Nyquist diagram of stable system.

This stability criterion is better than the modified Nyquist stability criterion in which the magnitude is inspected at the frequency that makes the phase angle $-180°$, which we can now see is simply determining stability based on the gain margin. Proof that this is incorrect may be observed from Fig. 16.14, which shows the Nyquist diagram of the previous example given by equation 16.25, plotted for $K > K_1$, i.e., for a stable system. Inspecting the magnitude when the phase is $-180°$ shows it to be greater than unity, suggesting an unstable system; however, the critical point passes the Nyquist plot on the left, suggesting a stable system. This demonstrates the fact that gain margin assessments of stability are not reliable. We may state our observations from the discussion as follows:

1. The phase margin is the only reliable measure of system stability from the Bode diagram.
2. The modified Nyquist criterion, based on traversing the frequency response in the direction of increasing frequency and observing whether the critical point passes on the left or right, is the only reliable means of checking stability from the polar plot, unless the complete Nyquist contour is evaluated.

Sample Problem 16.1

The system shown in Fig. SP16.1.1 has a forward-path gain K and unity-feedback gain. Plot the Bode diagram for $K = 45$, and determine the gain and phase margins. Calculate the maximum value of K consistent with stability, and check the answer using Routh's array.

Solution

Writing the open-loop transfer function in Bode form gives

$$GH(s) = \frac{45}{(s+2)(s+3)^2} = \frac{2.5}{(1+\frac{1}{2}s)(1+\frac{1}{3}s)^2}$$

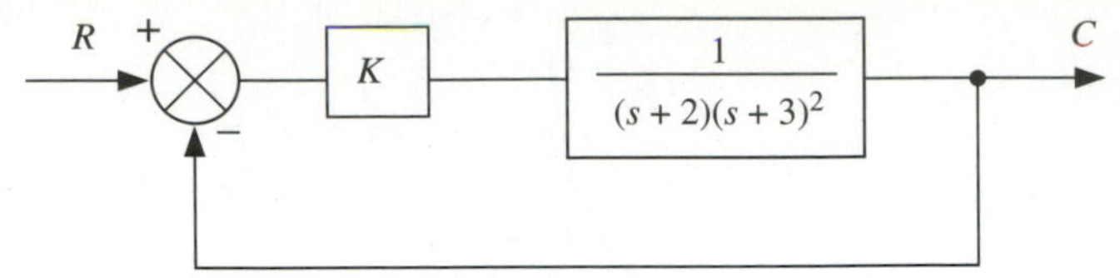

Fig. SP16.1.1

The break frequencies will occur at $\omega = 2$ and 3 rad/s. The Bode diagram will be plotted in the frequency range

$$0.1 < \omega < 100$$

The gain, expressed in decibels, is

$$K_{\text{db}} = 20 \log_{10} 2.5 = 8 \text{ db}$$

The second term in the denominator will be treated as two identical first-order elements, since it will be easier to draw the straight-line approximations to the phase curve than a second-order element. Figure SP16.1.2 shows the Bode diagram first using straight-line approximations and then allowing corrections of about -8 db at $\omega = 3$ rad/s and -7 db at $\omega = 2$ rad/s. It is seen that the system is stable, since at the phase crossover frequency of 5.4 rad/s the magnitude

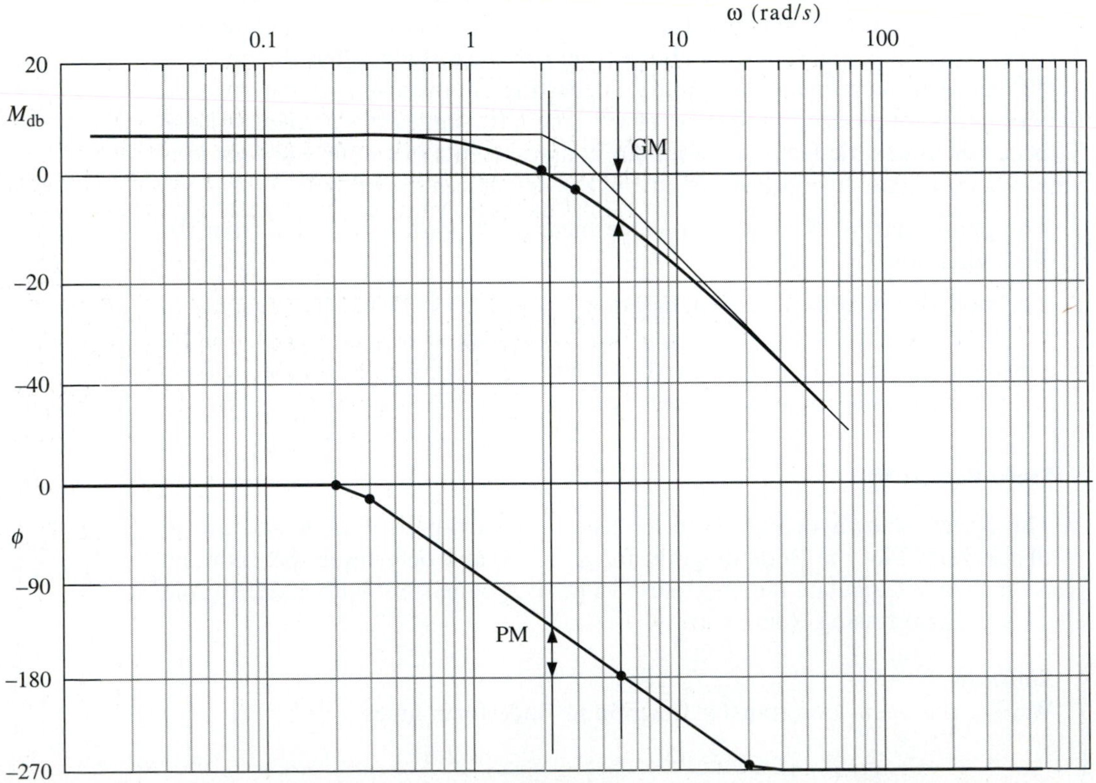

Fig. SP16.1.2

plot is less than 0 db. The gain and phase margins may be read directly from the Bode diagram as

$$\text{GM} = 11 \text{ db} \qquad \text{PM} = 50°$$

In order to place the system just on the verge of instability the magnitude plot would need to be moved up 11 db, which means multiplying K by some factor N that is greater than unity. Since the magnitude in Fig. SP16.1.2 is plotted for a gain $K = 45$, this means that the multiplier N may be obtained from

$$11 = 20 \log_{10} N$$

as shown in equation 16.7. This results in $N = 3.55$, resulting in a maximum value of gain for stability of

$$K = 45 \times 3.55 = 160$$

This value will now be checked using Routh's method. First, the characteristic equation may be written from the denominator of the closed-loop transfer function as

$$(s + 2)(s + 3)^2 + K = s^3 + 8s^2 + 21s + 18 + K = 0$$

This yields Routh's array:

s^3:	1	21
s^2:	8	$18 + K$
s^1:	$\frac{1}{8}(168 - 18 - K)$	0
s^0:	$18 + K$	0

For a stable system, the critical value of K is found from the left-hand column of the array:

$$168 - 18 - K > 0$$

yielding

$$K < 150$$

This is smaller than the value obtained from the Bode diagram due to the accuracy with which the gain margin can be obtained. For example, if the gain margin had been estimated as 12 db instead of 11 db, the resulting value of N would be 4, and the maximum value of K becomes 180. Clearly, the Bode diagram has to be used with some caution in determining critical-gain values if high accuracy is required.

Sample Problem 16.2

Consider the unity-feedback control system described by the open-loop transfer function

$$G(s) = \frac{K(s + 5)}{s(s + 1)(s^2 + s + 4)}$$

Plot the Bode diagram for the system when $K = 0.8$ and determine the gain and phase margins. What value of K gives the system a phase margin of about 60°?

Solution

Plotting the Bode diagram proceeds in the usual manner by first writing the open-loop transfer function in Bode form:

$$GH(s) = \frac{5K(1 + s/5)}{4s(1 + s)(1 + s/4 + s^2/4)} = \frac{1.25K(1 + s/5)}{s(1 + s)(1 + s/4 + s^2/4)}$$

The break frequencies for the first-order elements occur at $\omega = 1$ (twice) and 5 rad/s, while the second-order system break frequency is the undamped natural frequency of $\omega = 2$ rad/s. The damping ratio for the second-order system may be found from

$$\frac{2\zeta}{\omega_n} = \frac{1}{4}$$

leading to

$$\zeta = 0.25$$

The decade $1 < \omega < 10$ contains all the break points, so the Bode diagram will be plotted in the frequency range $0.1 < \omega < 100$ rad/s. In this example, $1.25K = 0$ db. The straight-line segments for both the magnitude and phase are shown in Fig. SP16.2.1. Note how the second-order phase curve is estimated

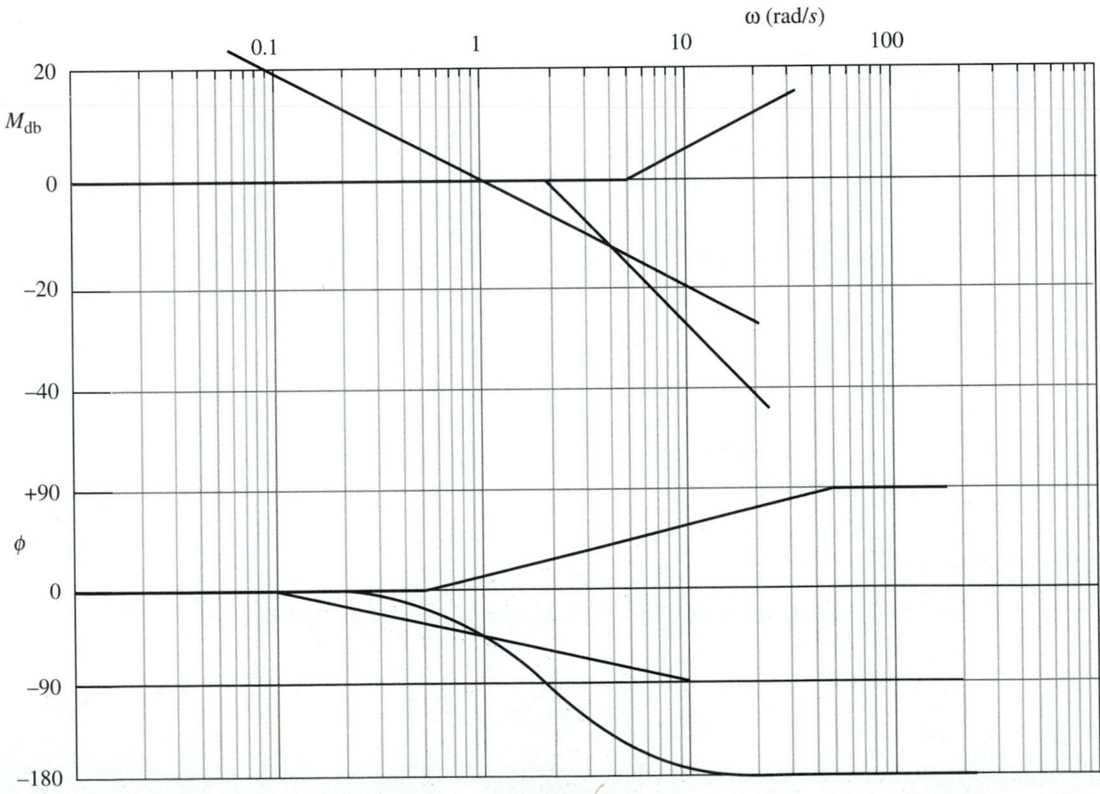

Fig. SP16.2.1

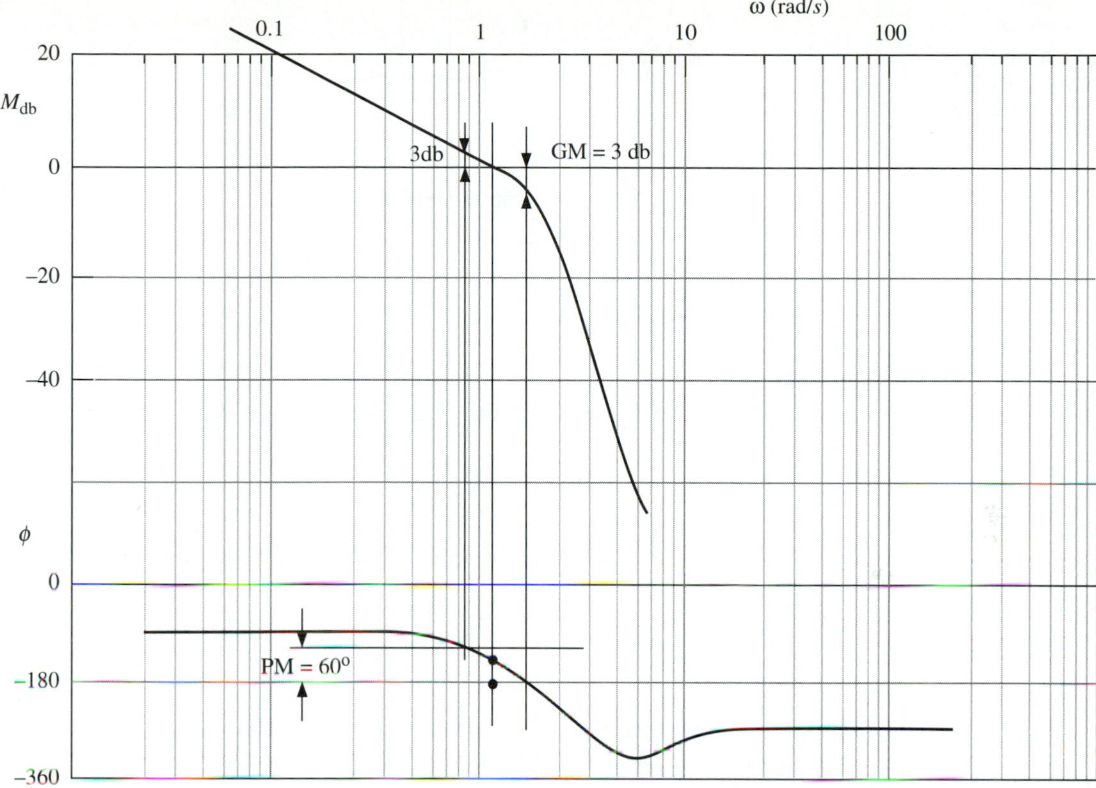

Fig. SP16.2.2

from Fig. 15.18. The size of the resonance peak for the second-order system with damping ratio of 0.25 is about 5 db, resulting in the more accurate plots shown in Fig. SP16.2.2. From Fig. SP16.2.2 it may be seen that the gain margin is 3 db and the phase margin is about 35°. In order to provide the system with a phase margin of 60°, the magnitude curve needs to be moved down about 3 db; hence using equation 16.6,

$$20 \log_{10} NK = K_{db} + N_{db}$$

Setting $N_{db} = -3$ results in

$$-3 = 20 \log_{10} N$$

or

$$N = 0.7$$

Hence, the new value of K to give the required phase margin is 0.7 multiplied by the value of K for which the original Bode diagram was plotted, i.e.,

$$K = 0.7 \times 0.8 = 0.56$$

Sample Problem 16.3

Consider the unity-feedback gain system shown in Fig. SP16.3.1. Determine the gain and phase margins when the Bode gain is unity and the maximum value of K

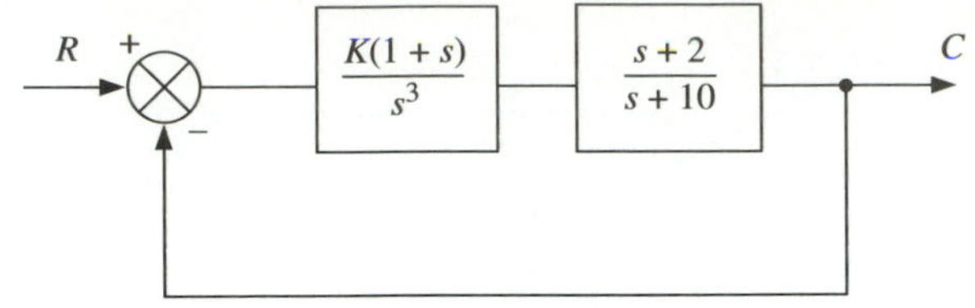

Fig. SP16.3.1

for which the system is stable. Confirm the result by using Routh's array and sketch the root locus for the system.

Solution

The open-loop transfer function may be written in the Bode form as

$$GH(s) = \frac{2K(1 + s)(1 + s/2)}{10s^3(1 + s/10)}$$

The break points will occur at ω values of 1, 2, and 10 rad/s, while setting the Bode gain to unity,

$$\frac{2K}{10} = 1$$

corresponds to putting $K = 5$. The Bode diagram for this value of gain is plotted in Fig. SP16.3.2. From the Bode diagram the gain and phase margins may be measured as

$$\text{GM} = 6 \text{ db} \qquad \text{PM} = -6°$$

This system is a case of a single-frequency crossover with increasing phase, where the gain and phase margin stability tests give conflicting information. Considering the phase margin only, it is confirmed that the system is in fact unstable for $K = 5$. To stabilize the system, K has to be increased. The phase crossover frequency will not change; so the magnitude plot has to be moved up by the gain margin value. To just make the closed-loop system stable, 6 db has to be added to the gain plot, which means multiplying the value of gain for which Fig. SP16.3.2 is plotted by N, where

$$20 \log_{10} N = 6$$

This produces $N = 2$, giving a minimum value of K consistent with stability of

$$\hat{K} = 5 \times 2 = 10$$

To check this with Routh's array, the characteristic equation may be written as

$$s^3(s + 10) + K(s + 1)(s + 2) = 0$$

leading to

$$s^4 + 10s^3 + Ks^2 + 3Ks + 2K = 0$$

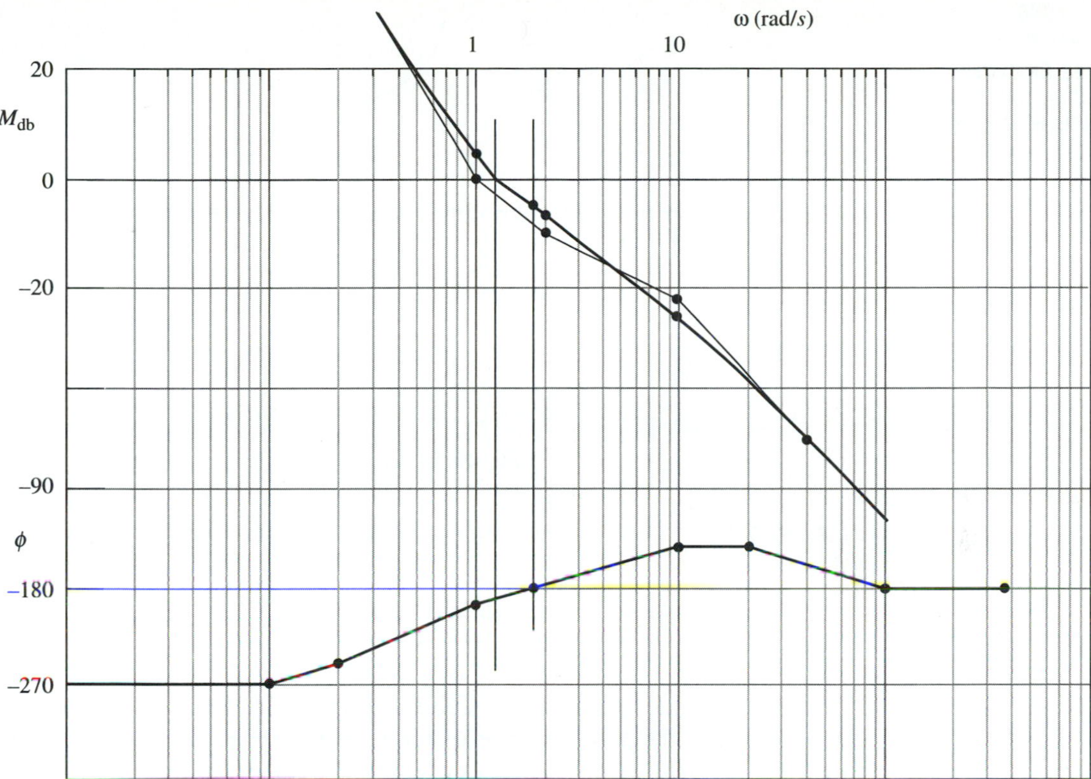

Fig. SP16.3.2

Routh's array takes the form

$$
\begin{array}{llll}
s^4: & 1 & K & 2K \\
s^3: & 10 & 3K & 0 \\
s^2: & 0.7K & 2K & 0 \\
s^1: & \dfrac{2.1K^2 - 20K}{0.7K} & 0 & \\
s^0: & 20K & 0 &
\end{array}
$$

The left-hand column element of the s^1 row yields

$$2.1K^2 - 20K > 0$$

or

$$K > 9.52$$

This is a little smaller than the value found from the Bode diagram, and again this discrepancy may be attributed to the accuracy of determining the phase crossover frequency and then estimating the gain margin. The student may verify that the root locus is as shown in Fig. SP16.3.3. Note that the three poles at the origin force the closed-loop poles to move out at 120° to each other, caus-

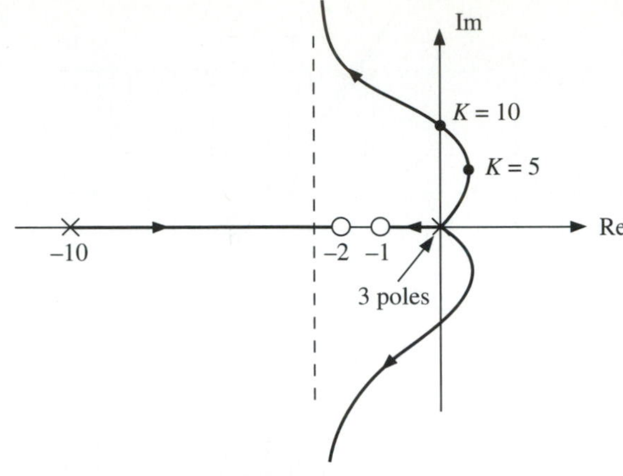

Fig. SP16.3.3

ing the system to be unstable at low gains. The asymptotes intersect the negative real axis at

$$\sigma_a = \frac{-10 + 2 + 1}{4 - 2} = -3.5$$

Therefore increasing the gain will stabilize the system. From the root locus it may be seen that the closed-loop poles for $K = 5$ are to the right of the imaginary axis, confirming the system is unstable for this value of gain.

◆ PROBLEMS

16.1 For the following systems, sketch the Bode diagram, and from the straight-line approximations to the gain and phase plots, estimate the maximum value of K for which the system is stable:

a. $GH(s) = \dfrac{K}{s(s + 1)(s + 4)}$

b. $GH(s) = \dfrac{K}{s^2(1 + s)}$

c. $GH(s) = \dfrac{Ks}{(s + 2)^4}$

d. $GH(s) = \dfrac{K}{s(s^2 + 2s + 16)}$

e. $GH(s) = \dfrac{5K(1 + s)}{s^2(1 + s/3)^2}$

16.2 For the system shown in Fig. P16.2, determine the maximum value of K for which the system is stable, and check the result using Routh's array.

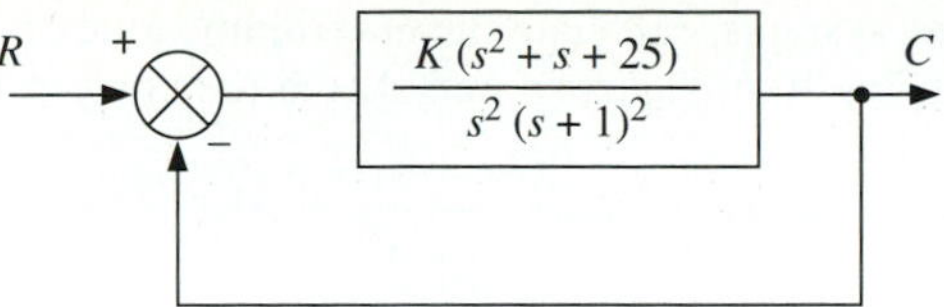

Fig. P16.2

16.3 Estimate for each system given below the gain and phase margins:

a. $GH(s) = \dfrac{15(s+2)}{s^2(s+9)}$

b. $GH(s) = \dfrac{20}{s^2(s+5)(s+2s+4)}$

c. $GH(s) = \dfrac{0.1(s^2+2s+9)}{s^2(1+s/10)^2}$

d. $GH(s) = \dfrac{2000(s+1)}{s(s+2)(s+5)(s+10)(s+20)}$

e. $GH(s) = \dfrac{0.5(s+10)(s+20)}{s^2(s+1)(s+100)}$

For each example, discuss system stability.

16.4 Figure P16.4 shows a unity-gain feedback control system. Calculate the value of K such that the system has a 20° phase margin.

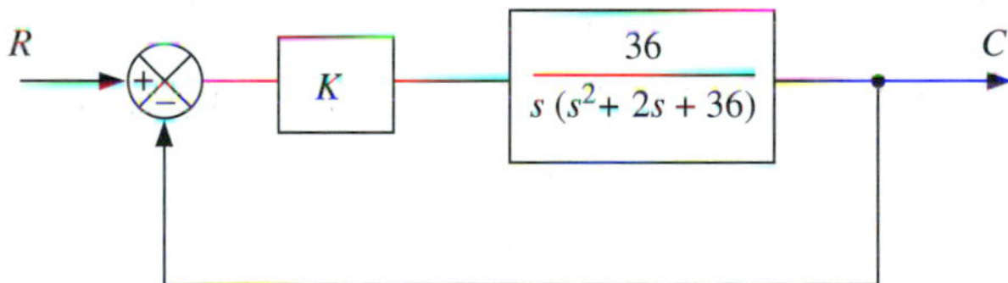

Fig. P16.4

16.5 The forward-path transfer function of a unity-feedback control system is

$$G(s) = \frac{K}{(s^2+2s+4)(s^2+s+12)}$$

Determine the stability of the system when **(a)** $K = 10$, **(b)** $K = 50$, and **(c)** $K = 200$. For each case estimate the phase margin.

16.6 For the system shown in Fig. P16.6 determine the value of K such that the phase margin is 60°. What is the corresponding value of gain margin?

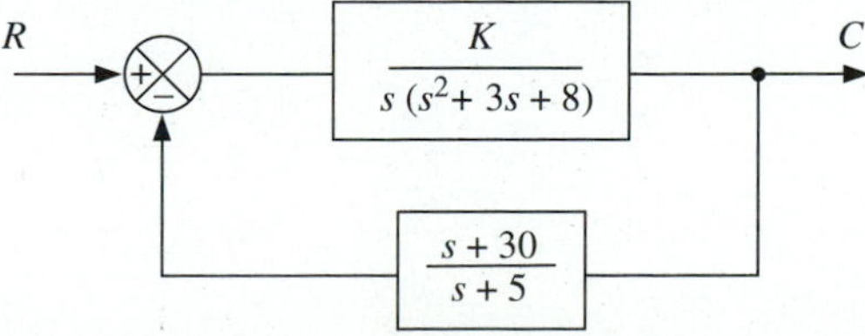

Fig. P16.6

16.7 For the following systems, calculate the maximum value of gain for which the system is stable, check using Routh's array, and sketch the corresponding root locus:

a. $GH(s) = \dfrac{K(s+1)(s+2)}{s^4}$

b. $GH(s) = \dfrac{K(1+0.1s)}{(s^2+3s+15)^2}$

c. $GH(s) = \dfrac{K}{s(s+0.01)(s+0.08)}$

16.8 In the following open-loop transfer function, determine the value of β that gives the system a phase margin of 50°:

$$GH(s) = \frac{(1+s/\beta)^2}{s^3}$$

16.9 For $K = 10$, draw the Bode diagram of the system shown in Fig. P16.9. From the Bode diagram construct the Nyquist diagram, and from this determine the gain and phase margins. Check the result using the Bode diagram. What value of K gives the system a velocity error constant of 2.0?

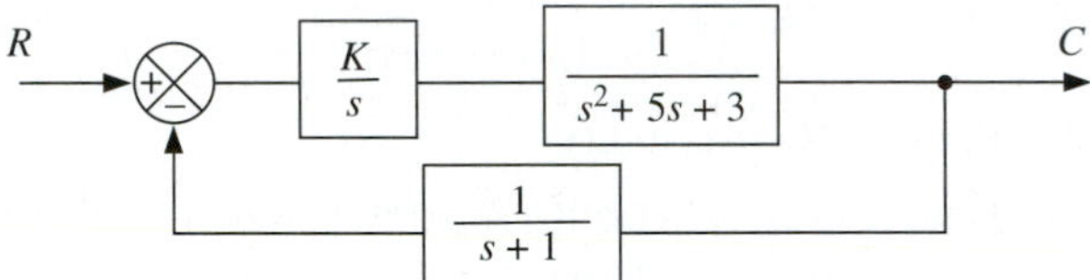

Fig. P16.9

16.10 Consider the system described by the open-loop transfer function

$$GH(s) = \frac{K}{s(1+s/3)^2}$$

Find the range of K such that the system is stable and the velocity error is less than 30%.

MODULE 17

Time Response from Frequency Response

So far, frequency response methods have been used only to study issues of stability. They may be considered somewhat inferior to the root locus method, for example, since it can give insight into the time-domain response based on the locations of the closed-loop poles on the complex plane. The question naturally arises whether frequency response methods can do the same. We will show that they can, by using parameters obtained from the Bode diagram to evaluate the damping ratio ζ and undamped natural frequency ω_n. From these, the characteristics of the time reponse for certain inputs may be determined.

Before this, however, we will look at a technique for obtaining the Bode diagram from the root locus plot, which will be useful in cases where both root locus and Bode plots are required.

◆ BODE DIAGRAM FROM THE ROOT LOCUS

The derivation of the Bode diagram from a root locus plot is performed graphically from the open-loop pole-zero map and is best explained with an example. Consider the system described by the open-loop transfer function

$$GH(s) = \frac{K(s+3)}{(s+1)(s+2)} \tag{17.1}$$

It is known that the frequency response is obtained by substituting $s = j\omega$ into GH and evaluating the magnitude and phase of the resulting complex number. This results in

$$GH(j\omega) = \frac{K(j\omega+3)}{(j\omega+1)(j\omega+2)} \tag{17.2}$$

To plot the Bode diagram, it is necessary, therefore, to calculate

$$M = |GH(j\omega)| \qquad \phi = \angle GH(j\omega) \tag{17.3}$$

How can these quantities be obtained from the open-loop poles and zeros? Consider Fig. 17.1, which shows the open-loop poles and zeros of the example on the complex plane. Consider the calculation of M and ϕ for a specific frequency $\omega = \omega_1$ rad/s. Setting $s = j\omega_1$ corresponds to the point $0 + j\omega_1$ on the complex plane, as shown in Fig. 17.1. The open-loop transfer function becomes

$$GH(j\omega_1) = \frac{K(j\omega_1 + 3)}{(j\omega_1 + 1)(j\omega_1 + 2)} \tag{17.4}$$

The magnitude and phase may be calculated by the usual rules for combining complex numbers using the angles and lengths defined in Fig. 17.2:

$$M = \frac{K|(j\omega_1 + 3)|}{|(j\omega_1 + 1)| \times |(j\omega_1 + 2)|} = \frac{Kl_1}{l_2 \times l_3} \tag{17.5}$$

and

$$\phi = \angle(j\omega_1 + 3) - \angle(j\omega_1 + 1) - \angle(j\omega_1 + 2) = \theta_3 - \theta_1 - \theta_2 \tag{17.6}$$

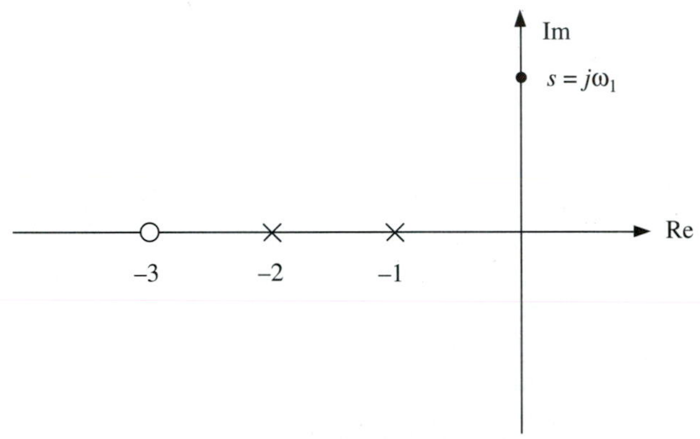

Fig. 17.1 Open-loop poles and zeros.

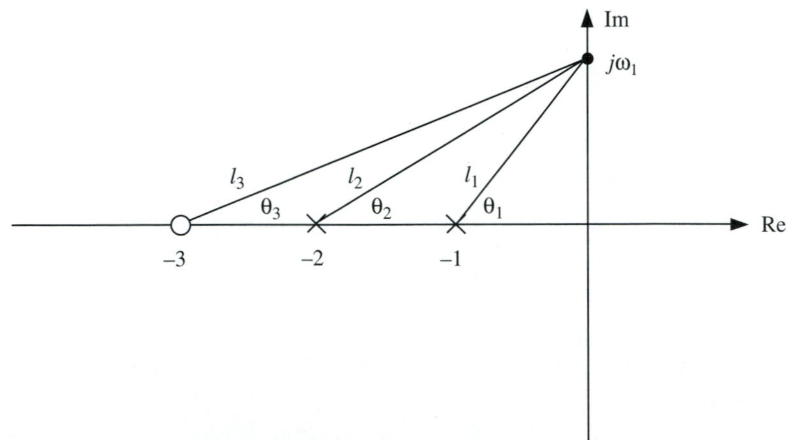

Fig. 17.2 Magnitude and phase calculation.

The three magnitudes and three angles would be measured and the above equations evaluated. The value of K may be set arbitrarily to 1 if desired and the magnitude converted to decibels. Then another point $s = j\omega_2$ would be selected and the procedure repeated. As s moves from the point $0 + 0j$ to the point $0 + j\infty$, i.e., as s moves out along the positive imaginary axis, the frequency at which the open-loop transfer function is evaluated increases from zero to infinity. This is exactly what happens when the Bode diagram of the system is determined. Plotting the results on appropriate axes will therefore produce the Bode diagram for this element, with the value of gain $K = 1$. This results in the plot shown in Fig. 17.3 and corresponds to the exact magnitude and phase plots. The presence of an open-loop pole on the imaginary axis will produce a magnitude of infinity and a change in the phase angle of $-180°$ as the pole is traversed. Since a pole on the real axis actually corresponds to a pair of undamped poles in the denominator of the open-loop transfer function, an infinite amplitude would be expected at this frequency, as shown for the $\zeta = 0$ case in Fig. 12.5. The common case of a pole at the origin will also produce an infinite magnitude for zero frequency. This is consistent with the Bode diagram for the element

$$GH(s) = \frac{K}{s} \tag{17.7}$$

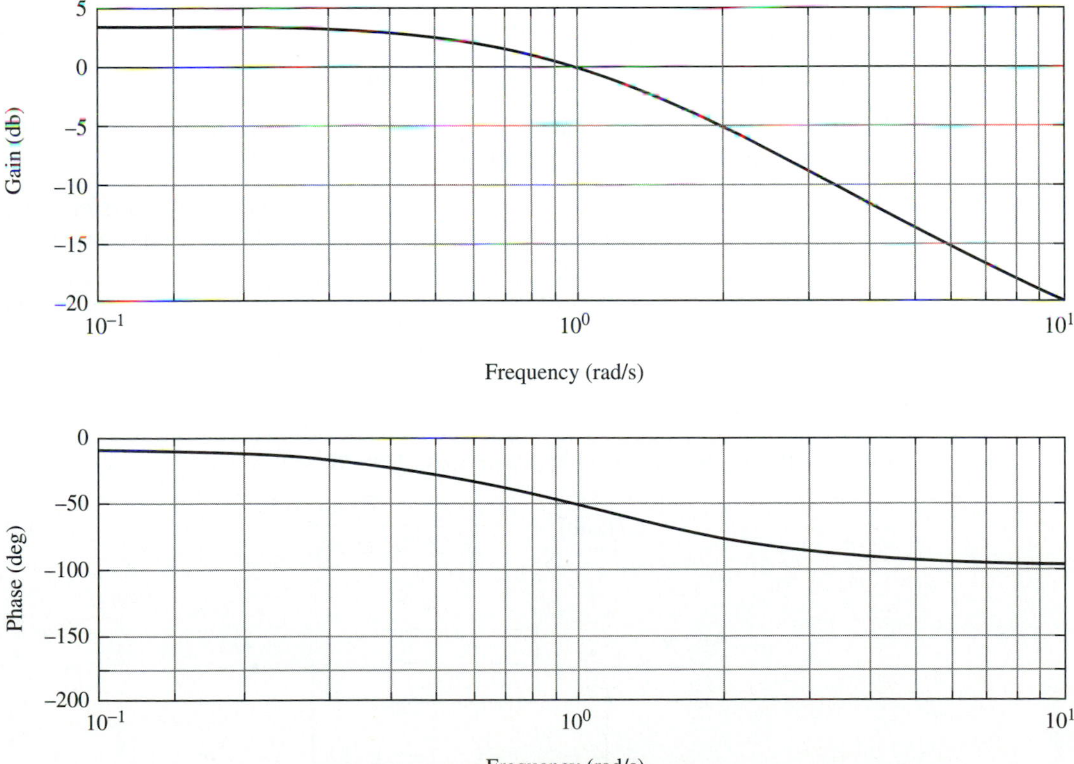

Fig. 17.3 Bode diagram from root locus.

As soon as the frequency increases to 0+, the phase changes from an indeterminant value to $-90°$, which is the phase expected from the Bode method.

◆ CLOSED-LOOP TIME RESPONSE FROM OPEN-LOOP PHASE MARGIN

We now turn our attention to obtaining some information about the closed-loop time-domain response of a system. We will consider the response of a standard, closed-loop second-order system. We start by determining what open-loop transfer function would, when a unity-feedback path is closed around it, produce a standard second-order closed-loop transfer function. Consider the system shown in Fig. 17.4. This system has unity feedback and a forward-path transfer function

$$G(s) = \frac{\omega_n^2}{s(s + 2\zeta\omega_n)} \tag{17.8}$$

The student may easily verify that the closed-loop system takes the form

$$\frac{C}{R} = \frac{\omega_n^2}{s^2 + 2\zeta\omega_n s + \omega_n^2} \tag{17.9}$$

and is of the required standard form with damping ratio ζ and undamped natural frequency ω_n. The strategy here will be as follows:

1. Plot the Bode diagram of the open-loop system, realizing that the closed-loop system is of standard second-order form.
2. Relate the phase margin to the damping ratio.
3. Relate the gain crossover frequency ω_{gc} (the frequency that makes $\phi = -180°$) to the undamped natural frequency ω_n.

We begin by investigating the frequency response of the open-loop transfer function that leads to a standard second-order system in the closed loop, namely

$$GH(j\omega) = \frac{\omega_n^2}{j\omega(j\omega + 2\zeta\omega_n)} \tag{17.10}$$

$$= \frac{\omega_n^2}{-\omega^2 + 2\zeta\omega_n\omega j} \tag{17.11}$$

After multiplying by the complex conjugate, this becomes

$$GH(j\omega) = \frac{\omega_n^2(-\omega^2 - 2\zeta\omega\omega_n j)}{\omega^4 + 4\zeta^2\omega^2\omega_n^2} \tag{17.12}$$

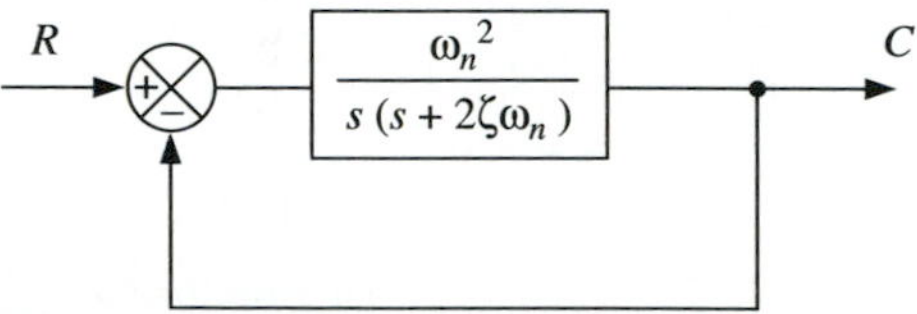

Fig. 17.4 Second-order closed-loop system.

$$= \frac{r^2 - 2\zeta rj}{r^4 + 4\zeta^2 r^2} \tag{17.13}$$

where r is the frequency ratio

$$r = \frac{\omega}{\omega_n} \tag{17.14}$$

This gives the magnitude and phase as

$$M = \sqrt{\frac{r^4}{(r^4 + 4\zeta^2 r^2)^2} + \frac{4\zeta^2 r^2}{(r^4 + 4\zeta^2 r^2)^2}} \tag{17.15}$$

$$= \frac{1}{\sqrt{r^4 + 4\zeta^2 r^2}} \tag{17.16}$$

and

$$\phi = \tan^{-1} \frac{-2\zeta}{r} = -\tan^{-1} \frac{2\zeta}{r} \tag{17.17}$$

It is necessary to find the frequency when the magnitude is unity, hence yielding

$$r^4 + 4\zeta^2 r^2 - 1 = 0 \tag{17.18}$$

which is a quadratic in r^2 having solutions

$$r^2 = \tfrac{1}{2}(-4\zeta^2 \pm \sqrt{16\zeta^4 + 4}) \tag{17.19}$$

Since r^2 must be positive, we get

$$r^2 = \sqrt{4\zeta^4 + 1} - 2\zeta^2 \tag{17.20}$$

Substituting into 17.17 for the corresponding phase (for $M = 1$) yields

$$\phi = -\tan^{-1} \frac{2\zeta}{(\sqrt{4\zeta^4 + 1} - 2\zeta^2)^{1/2}} \tag{17.21}$$

The phase margin is defined to be

$$\text{PM} = 180 + \phi \tag{17.22}$$

resulting in

$$\text{PM} = 180 - \tan^{-1} \frac{2\zeta}{(\sqrt{4\zeta^4 + 1} - 2\zeta^2)^{1/2}} \tag{17.23}$$

This equation shows that the phase margin is only a function of the damping ratio ζ and is plotted in Fig. 17.5. Also plotted in the figure is a straight-line approximation for the range $0 < \zeta < 0.6$. Neglecting the small error between the actual curve and the stright line over this region, it may be stated that the relationship between the phase margin and the damping ratio is given by

$$\text{PM} \approx \zeta \times 100 \tag{17.24}$$

This is the first relationship required and allows the damping ratio of the closed-loop system to be evaluated from the phase margin measured from the Bode dia-

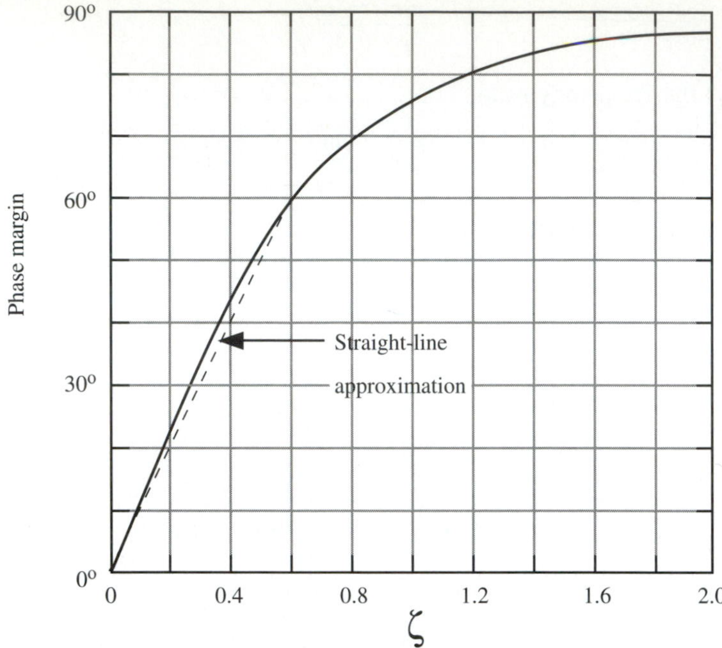

Fig. 17.5 Phase margin–damping ratio relationship.

gram of the open-loop system. The second relationship for ω_n is obtained from equation 17.20, which is valid only for a magnitude of unity, i.e.,

$$r = \frac{\omega_{gc}}{\omega_n} = [\sqrt{4\zeta^4 + 1} - 2\zeta^2]^{1/2} \tag{17.25}$$

Figure 17.6 shows this frequency ratio plotted as a function of damping ratio. This completes the identification of transient response parameters from the Bode diagram for an open-loop transfer function that leads to a standard second-order closed-loop form.

◆ TIME RESPONSE OF HIGHER-ORDER SYSTEMS

The previous analysis is valid only when the open-loop transfer function is of the form

$$GH(s) = \frac{\omega_n^2}{s(s + 2\zeta\omega_n)} \tag{17.26}$$

If the system is of higher order, it may be approximated to second order as long as the dominant closed-loop poles are significantly dominant.[1] Also, a closed-loop zero may cause considerable change in the transient response predicted by assuming the system to be represented by the closed-loop poles alone. There is no simple way to determine the transient response from the Bode diagram when a closed-

[1]A design guideline here is to say that the next most dominant poles are no less than five times closer to the imaginary axis of the complex plane compared to the dominant poles.

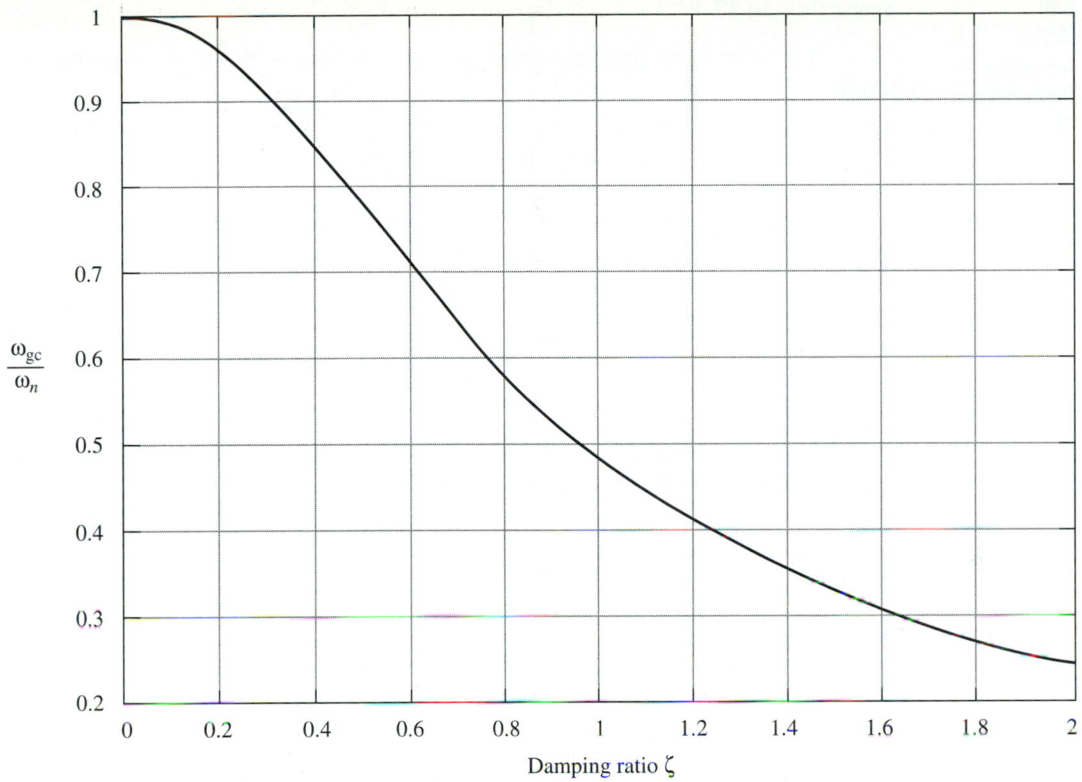

Fig. 17.6 Frequency ratio–damping ratio relationship.

loop zero is present or if the system is of order greater than 2. This is because the dominance, or otherwise, of a pair of poles over the remaining poles (or zero) is not apparent from the Bode or Nyquist plots, as it was for the root locus. However, there are a few guidelines that allow a reasonable approximation to the time-domain response to be made:

1. When a zero or pole is present, there is no unique relationship between the phase margin and the damping ratio, shown in the previous section; however, it is usually assumed that other system poles or zeros have little effect on the calculation of the damping ratio.
2. To determine the value of ω_n, the Bode diagram of the open-loop system should be plotted, and the gain crossover frequency ω_{gc} obtained. It is customary to equate the gain crossover frequency to the undamped natural frequency for systems that are of order higher than 2 or if there is a closed-loop zero present. The percentage overshoot or settling time may then be obtained in the usual way by allowing for the existence of a closed-loop zero, as indicated in Fig. 7.7. Other time-domain parameters such as peak times or settling times may also be estimated.

These guidelines will only allow an approximation to be made of the time-domain response of the closed-loop system, and if accurate response data are required, a computer solution should always be obtained.

SAMPLE PROBLEM 17.1

Draw the open-loop pole-zero map for the system with open-loop transfer function

$$GH(s) = \frac{K(s+3)}{s^2 + 2s + 5}$$

and from this map plot the Bode diagram for unity Evans gain. Compare the result with the conventionally sketched Bode diagram.

Solution

The solution of the denominator yields the open loop poles at

$$s = -1 \pm 2j$$

resulting in the pole-zero map shown in Fig. SP17.1.1. The magnitude and phase will be calculated graphically, as outlined in the previous section, by taking test points s_t along the positive imaginary axis. This technique will generate magnitude and argument data for unity Evans gain. We shall take points between $\omega = 0$ and $\omega = 5$ rad/s at intervals of 0.5 rad/s, for a total of 10 data points. Figure SP17.1.2 shows the calculation geometry for $\omega = 1$ rad/s in terms of the various lengths and angles.

ω	M	M_{db}	θ_1	θ_2	θ_3	ϕ
0	0.592	−4.6	0	−65	65	0
0.5	0.621	−4.1	10	−57	68	−1
1.0	0.693	−3.2	19	−46	72	−7
1.5	0.814	−1.8	26	−28	74	−20
2.0	0.865	−1.3	34	0	76	−42
2.5	0.750	−2.5	40	27	78	−65
3.0	0.589	−4.6	45	45	79	−79
3.5	0.456	−6.8	50	57	80	−87
4.0	0.365	−8.8	53	64	82	−93
4.5	0.304	−10.3	57	68	83	−94
5.0	0.259	−11.7	57	71	83	−93

TABLE 17.1. Magnitude and Phase of Transfer Function

The table shown calculates the magnitude and phase from the pole-zero map and the equations

$$M = \frac{l_1}{l_2 \times l_3} \qquad \phi = \theta_1 - \theta_2 - \theta_3$$

Note that at low frequencies, the magnitude is approximately equal to the value 0.6, corresponding to $K = 1$. The high frequency phase angle should approach −90°, but small errors in reading the three angles generate small errors in the total phase angle. Figure SP17.1.3, shows the Bode diagram obtained from the root locus plot.

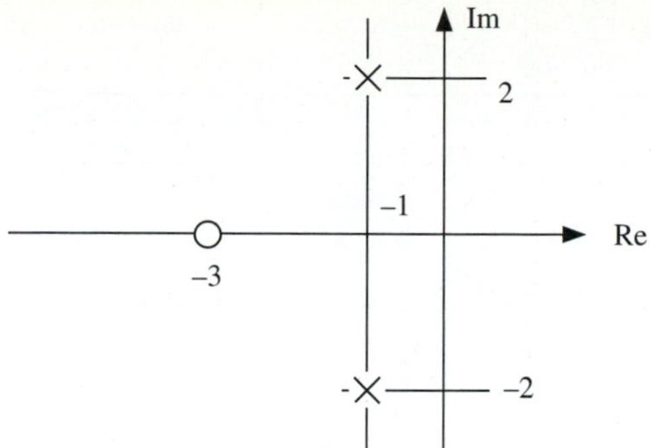

Fig. SP17.1.1

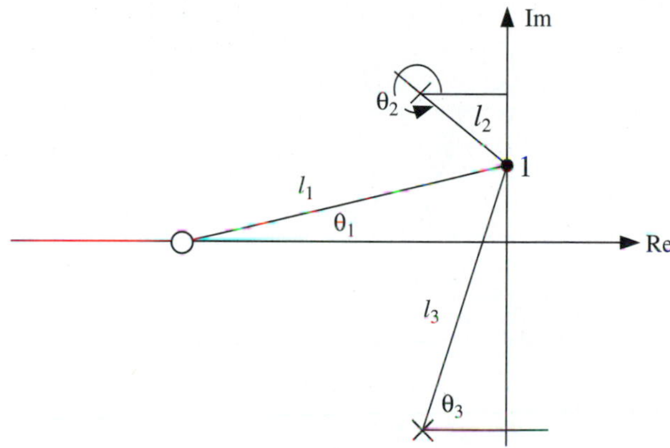

Fig. SP17.1.2

We now plot the Bode diagram directly from the transfer function. The transfer function is first written in Bode form as

$$GH(s) = \frac{0.6K(1 + s/3)}{1 + 2s/5 + s^2/5}$$

Since the Bode diagram obtained from the open loop pole-zero map is valid only for unity Evans gain, we must also here set $K = 1$ to obtain a comparative plot. Beginning with the second-order denominator, it may be seen that

$$\omega_n = \sqrt{5} = 2.23 \qquad \zeta = \frac{1}{\sqrt{5}} = 0.45$$

Similarly for the first order numerator, there will be a break point at

$$\omega_b = 3$$

The deviations from the asymptotic, straight line approximations to the magnitude will be approximately +3 db at $\omega = \omega_n$, and about +4 db at $\omega = 3$ rad/s. The asymptotes and accurate magnitude and phase plots are shown in Fig. SP17.1.4.

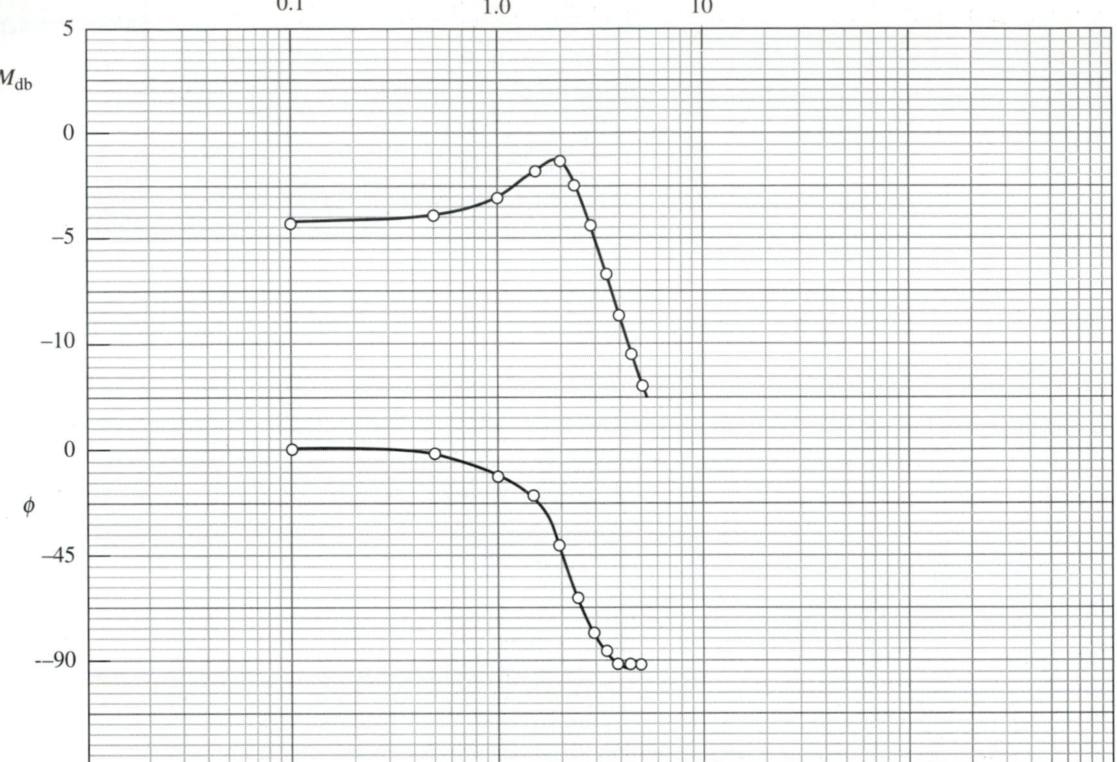

Fig. SP17.1.3

The two Bode diagrams appear to be very similar. Note that if the value of K had been given, the magnitude part of the Bode diagram could be moved up or down accordingly. However, if only the pole-zero map is provided, the value of K will be unknown, and the Bode diagram produced graphically from the open-loop poles and zeros will have the implicit assumption that $K = 1$.

Sample Problem 17.2

Consider the system described by the open-loop transfer function

$$GH(s) = \frac{K}{s(s + 10)}$$

Draw the Bode diagram of the system and estimate the unit step response of the closed-loop system when $K = 200$.

Solution

The transfer function is first put into Bode form:

$$GH(s) = \frac{K}{s(s + 10)} = \frac{K/10}{s(1 + s/10)} = \frac{20}{s(1 + s/10)}$$

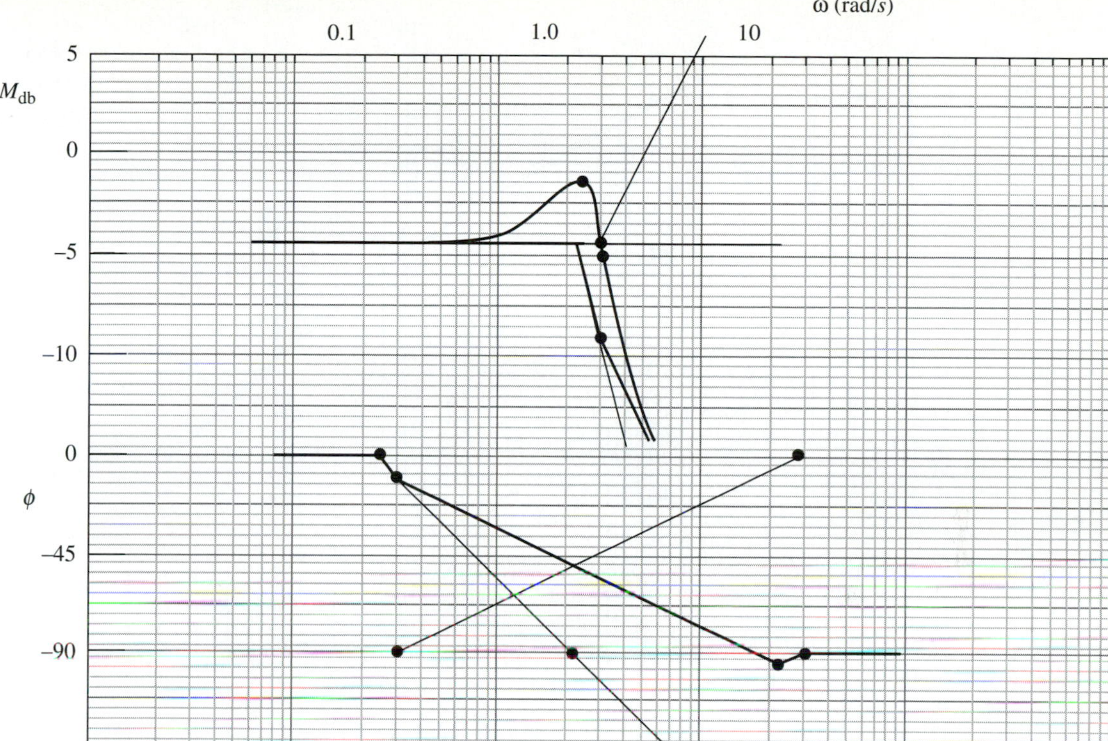

Fig. SP17.1.4

We will calculate the transfer function of the closed-loop system and determine the undamped natural frequency and damping ratio. In most cases this will not be easy to do, but for this simple system it is:

$$\frac{C}{R} = \frac{200}{s^2 + 10s + 200}$$

From this transfer function it may be shown that

$$\omega_n = 14.14 \text{ rad/s} \qquad \zeta = 0.354$$

We now plot the Bode diagram. The break frequencies are at 1 and 10 rad/s. The Bode gain is

$$K_{db} = 20 \log_{10} 20 = 26 \text{ db}$$

The Bode diagram is shown in Fig. SP17.2.1, where the phase is approximated by a straight-line segment spanning the break frequency, but a 3-db deviation from the magnitude curve has been allowed for. From the Bode diagram, the following data are obtained:

$$\text{PM} = 40° \qquad \omega_{gc} = 13 \text{ rad/s}$$

An approximate damping ratio is obtained as

$$\zeta = \frac{\text{PM}}{100} = 0.4$$

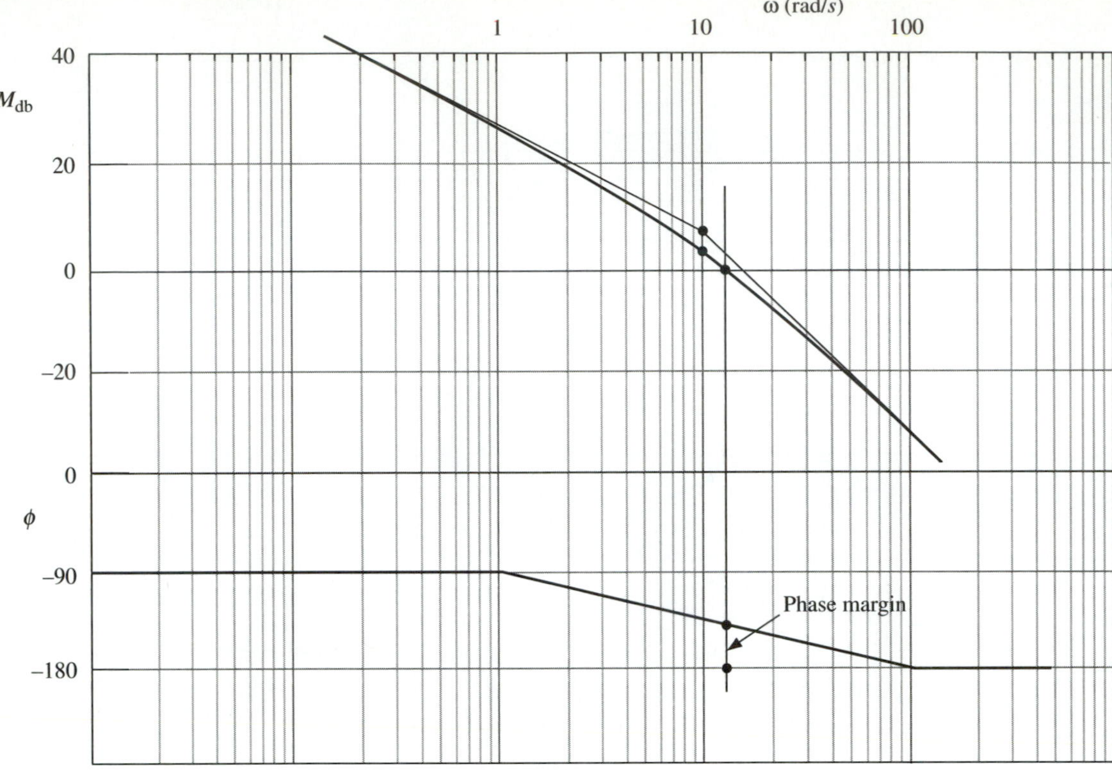

Fig. SP17.2.1

This allows the frequency ratio r to be obtained from equation 17.21:

$$r = \frac{\omega_{gc}}{\omega_n} = 0.85$$

yielding the undamped natural frequency as

$$\omega_n = 15.2 \text{ rad/s}$$

Comparing these results with those obtained by direct analysis of the closed-loop transfer function indicates good agreement. If better accuracy is required, the phase margin can be obtained from an accurate magnitude and phase plot or by using the `bode` and `margin` commands in MATLAB. If either of these is done, it results in

$$\text{PM} = 38.6^\circ \qquad \omega_{gc} = 12.55 \text{ rad/s}$$

Using Fig. 17.5 to get a better measure of the damping ratio, it is found to be $\zeta = 0.35$, exactly the value obtained by direct analysis. Substituting this value into equation 17.21 yields $r = 0.886$, giving an undamped natural frequency $\omega_n = 14.14$, again the same as given by direct analysis. It may be seen that the values obtained using approximations to the Bode diagram and the phase margin–damping ratio relationship are of acceptable accuracy. Now that the closed-loop undamped natural frequency and damping ratio have been determined, characteristics of the unit step response may be obtained. From Fig. 4.17, the percentage overshoot is

$$\text{PO} = 25\%$$

The damped natural frequency is calculated as

$$\omega_d = \omega_n \sqrt{1 - \zeta^2} = 14.0 \text{ rad/s}$$

The equivalent time constant is obtained from equation 5.25:

$$\tau_e = \frac{1}{\zeta \omega_n} = 0.165 \text{ s}$$

The peak time is obtained from equation 5.24:

$$T_p = \frac{\pi}{\omega_d} = 0.224 \text{ s}$$

Finally the 2% settling time is calculated from equation 5.27:

$$T_s = 4_e = 0.657 \text{ s}$$

These parameters will allow a reasonably accurate sketch of the time-domain step response to be made. However, another approach is to compare the above data with a simulation of the closed-loop system shown in Fig. SP17.2.2. If the accurate values of the undamped natural frequency and closed-loop damping

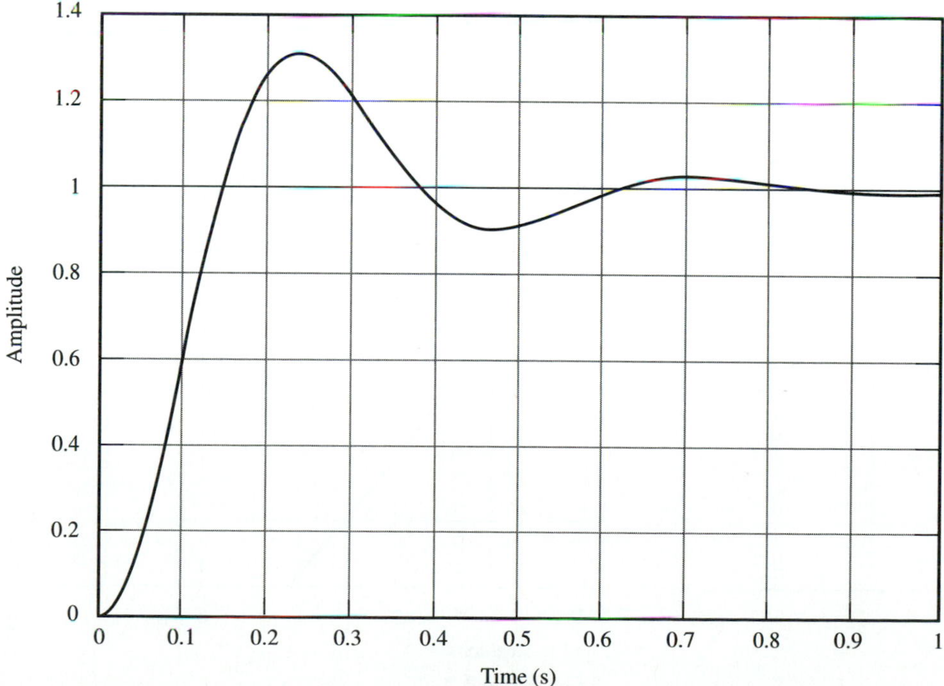

Fig. SP17.2.2

ratio had been used, an exact match to the time response shown in Fig. SP17.2.2 would have been obtained.

Sample Problem 17.3

From the Bode diagram, estimate the closed-loop time-domain response of the system given by

$$G(s) = \frac{K(s+3)}{s(s+1)(s+2)(s+5)} \qquad H(s) = 1$$

when the input is a unit step and K is selected from the root locus to give dominant closed-loop poles a damping ratio of $\zeta = 0.7$.

Solution

For this system, the root locus will be plotted to determine the required value of K and the time response estimated from the closed-loop pole locations in the complex plane. The same will then be done using the Bode diagram. The student may verify that the root locus takes the form shown in Fig. SP17.3.1 while a closeup of the region near the origin is shown in Fig. SP17.3.2. The value of gain required to give the dominant roots a damping ratio of 0.7 is calculated

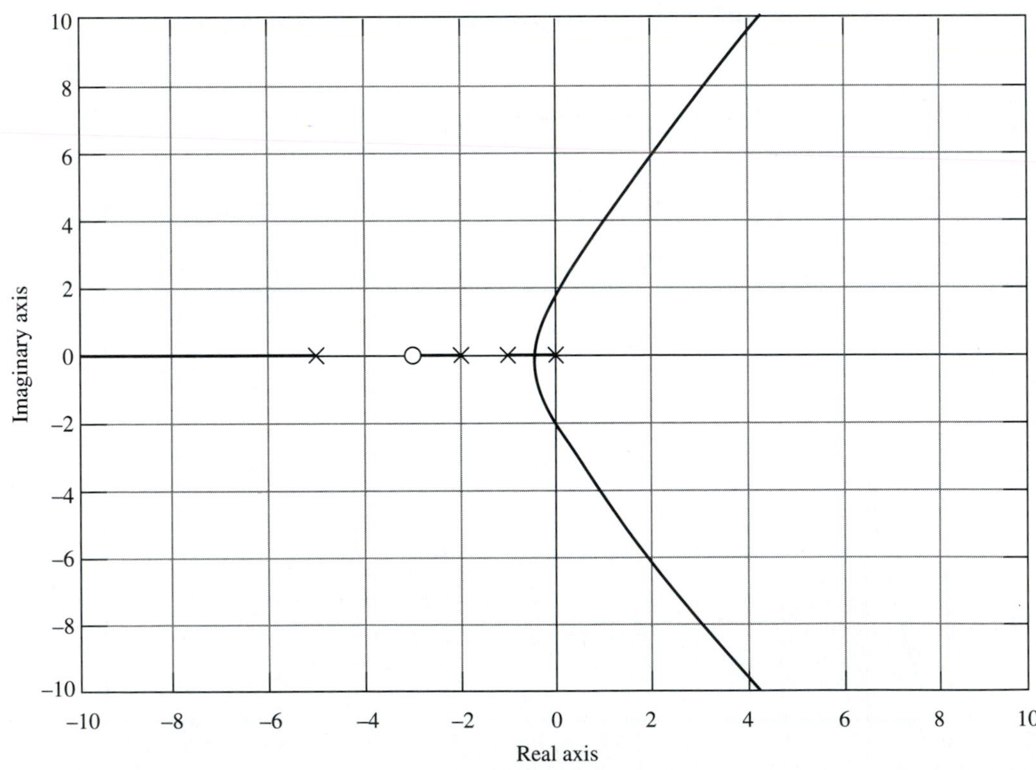

Fig. SP17.3.1

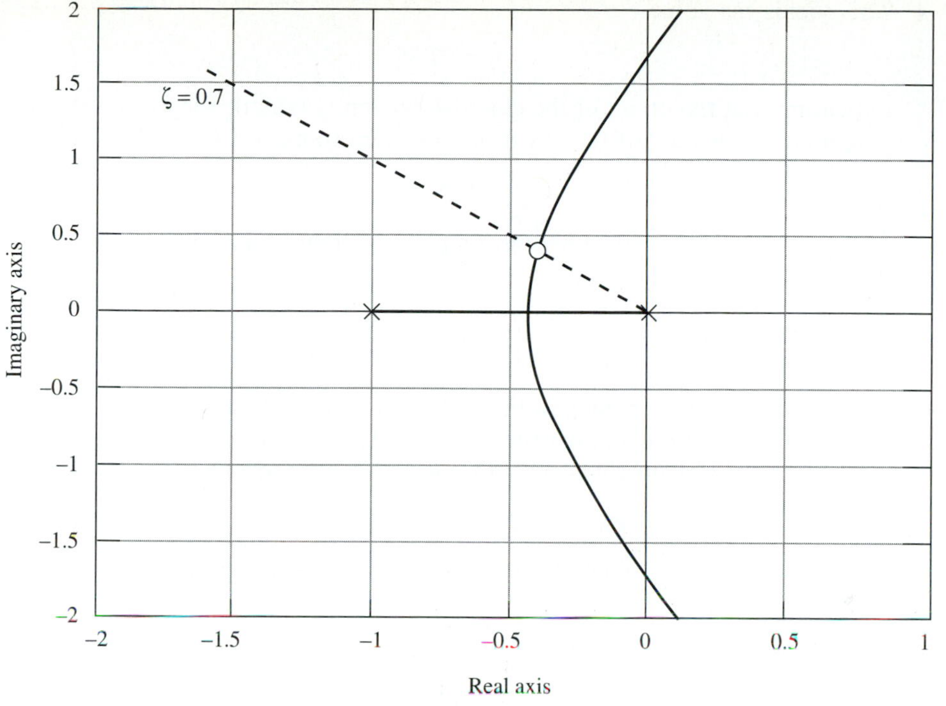

Fig. SP17.3.2

in the usual manner, the result being $K = 1.2$. The roots for this value of gain are at

$$s = -5.04, -2.15, -0.408 \pm 0.408j$$

Clearly, the system is dominated by the complex roots, although it is of order 4. The closed-loop transfer function yields

$$\frac{C}{R} = \frac{K(s+3)}{s(s+1)(s+2)(s+5) + K(s+3)}$$

The steady-state output for a unit step input will be obtained from the final-value theorem and is

$$c_{ss}(t) = 1$$

This system has a closed-loop zero. Therefore it must be taken into account when predicting the resulting time-domain response. Following the treatment of zeros given in Module 7, we note that the closed-loop zero will be located at the open-loop zero, which gives the following data for use in Fig. 7.7:

$$\omega_n = \sqrt{0.408^2 + 0.408^2} = 0.577 \text{ rad/s}$$

$$\zeta = 0.7 \qquad \frac{1}{\tau} = 3$$

This yields the value

$$\gamma = 7.35$$

indicating that the effect of the closed-loop zero is small, suggesting that the percentage overshoot of the closed loop system (obtained from Fig. 4.17) is

$$\text{PO} \approx 5\%$$

Turning now to the Bode diagram, we plot the open-loop transfer function for $K = 1.2$:

$$GH(s) = \frac{1.2(s + 3)}{s(s + 1)(s + 2)(s + 5)} = \frac{0.36(1 + s/3)}{s(1 + s)(1 + s/2)(1 + s/5)}$$

The Bode diagram is shown in Fig. SP17.3.3, where the accurate curves have been drawn in order to improve the accuracy of determining the system parameters. The variables obtained from the Bode diagram are

$$\text{PM} = 65^\circ \qquad \omega_{gc} = 0.34 \text{ rad/s}$$

Figure 17.5 gives the damping ratio from the phase margin to be

$$\zeta = 0.7$$

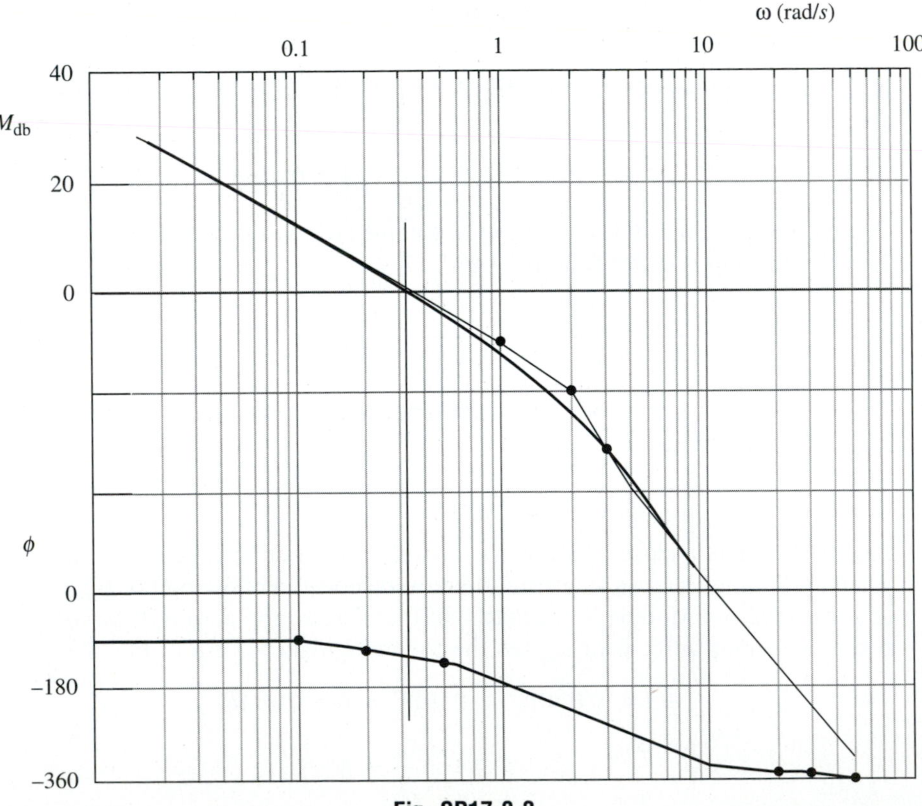

Fig. SP17.3.3

Surprisingly, this is an accurate value for the damping ratio, even though the system is fourth order! The student may be tempted to calculate the undamped natural frequency from equation 17.26, as was done for the previous example. This should not be done, since the equation is valid only for a second-order system. The undamped natural frequency is taken as the gain crossover frequency, i.e.,

$$\omega_n = \omega_{gc} = 0.34 \text{ rad/s}$$

To compare the actual system response to that predicted from the Bode diagram, we form the closed-loop system transfer function as

$$\frac{C}{R} = \frac{K(s+3)}{s(s+1)(s+2)(s+5) + K(s+3)}$$

For the Bode analysis, we have to represent the system as second order, even though we do not know for sure how valid an assumption this is. The Bode diagram indicates that the system may be approximated by a second-order system with damping ratio $\zeta = 0.7$ and undamped natural frequency $\omega_n = 0.34$ rad/s. The approximate closed-loop transfer function becomes

$$\frac{C}{R} = \frac{K(s+3)}{s^2 + 0.472s + 0.116}$$

Before we can compare the system responses, we note that the accurate system has a steady-state value for a unit step input of unity, while the approximated system has a steady-state value of

$$c_{ss}(t) = \frac{1.2 \times 3}{0.116} = 31$$

To compare the two responses, therefore, we divide the numerator by 31 to obtain a steady-state value of unity:

$$\frac{C}{R} = \frac{0.039s + 0.116}{s^2 + 0.472s + 0.116}$$

Figure SP17.3.4 shows the two system responses. Note how the percentage overshoot is about the same for both systems because the damping ratios are the same. The difference in undamped natural frequency causes the peak times to differ.

◆ PROBLEMS

17.1 For the following open-loop transfer functions, plot the Bode diagram from the pole-zero map:

a. $GH(s) = \dfrac{1}{s}$

b. $GH(s) = \dfrac{1}{1+s}$

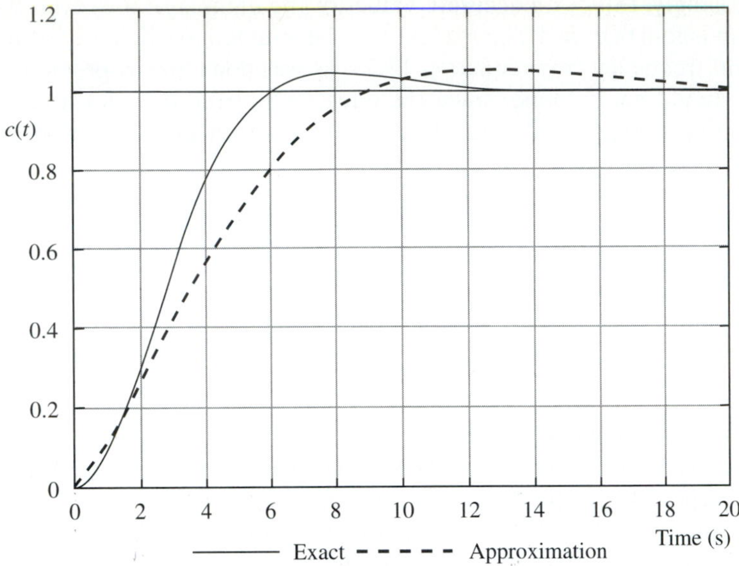

Fig. SP17.3.4

c. $GH(s) = \dfrac{s + 2}{s + 3}$

d. $GH(s) = \dfrac{s + 1}{s^2}$

e. $GH(s) = \dfrac{1}{s^2 + 9}$

17.2 Plot the Bode diagram for the following open-loop transfer functions directly from the open-loop poles and zeros. Then superimpose the asymptotic segments of the Bode diagram elements obtained in the usual manner and compare the results obtained:

a. $GH(s) = \dfrac{2(s + 4)}{s^2(s + 1)(s + 2)}$

b. $GH(s) = \dfrac{6(s + 1)}{s(s + 5)^2}$

c. $GH(s) = \dfrac{20(s^2 + 4s + 8)}{s^2 + 10s + 125}$

17.3 For the system shown in Figure P17.3, sketch the root locus, and from it determine the Bode diagram when $K = 1$.

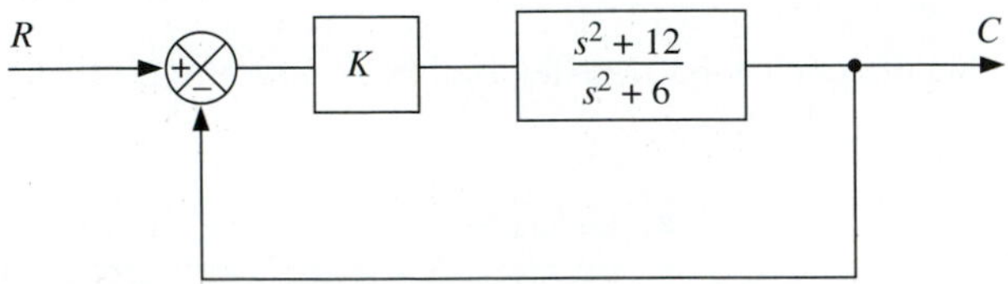

Fig. P17.3

17.4 From the Bode diagram, calculate the 2% settling time of the closed-loop system when

$$G(s) = \frac{K}{s(1 + s/4)} \qquad H(s) = 1$$

when **(a)** $K = 1$ and **(b)** $K = 10$. Check the result by direct analysis of the closed-loop transfer function.

17.5 Consider the unity-feedback system shown in Fig. P17.5. Plot the Bode diagram, and for the closed-loop system determine the damping ratio, peak time, 2% settling time, and damped and undamped natural frequencies.

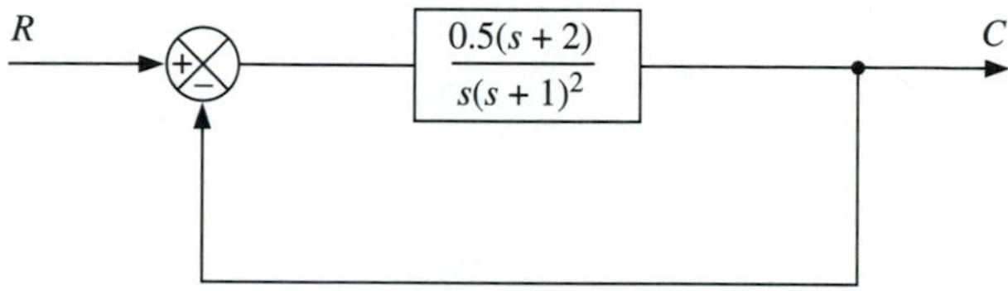

Fig. P17.5

17.6 For the system given in problem 17.4, use the Bode diagram to determine the value of K that will give the closed-loop system a maximum of 20% overshoot to a unit step input. At what time does this peak overshoot occur?

17.7 Consider the system described by the open-loop transfer function elements

$$GH(s) = \frac{5}{s} \qquad H(s) = \frac{1}{s + 1}$$

Draw the Bode diagram of the open-loop transfer function and write down an expression for the unit impulse response of the closed-loop system.

17.8 Sketch the root locus of the system shown in Fig. P17.8 and determine the value of K needed to give the dominant closed-loop poles a damping ratio of $\zeta = 0.707$. Estimate the closed-loop unit step response of the system by plotting the Bode diagram of the open-loop transfer function for this value of K, and discuss the two results. Compare the result with direct analysis of the closed-loop transfer function.

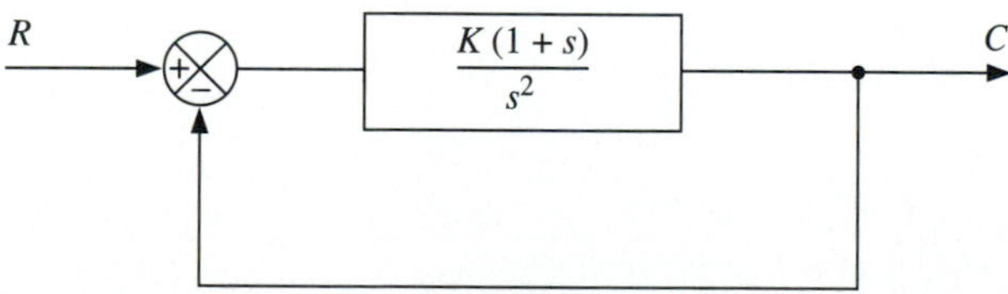

Fig. P17.8

17.9 Estimate the percentage overshoot of the closed-loop system

$$GH(s) = \frac{240(s + 1)}{(s + 2)(s + 3)(s + 4)}$$

by plotting the Bode diagram of the open-loop transfer function.

17.10 Consider the system shown in Fig. P17.10. Estimate the damping ratio of the closed-loop system when **(a)** $K = \frac{2}{3}$ and **(b)** $K = \frac{1}{3}$. For each case write down the transfer function that approximates the closed-loop system to second order.

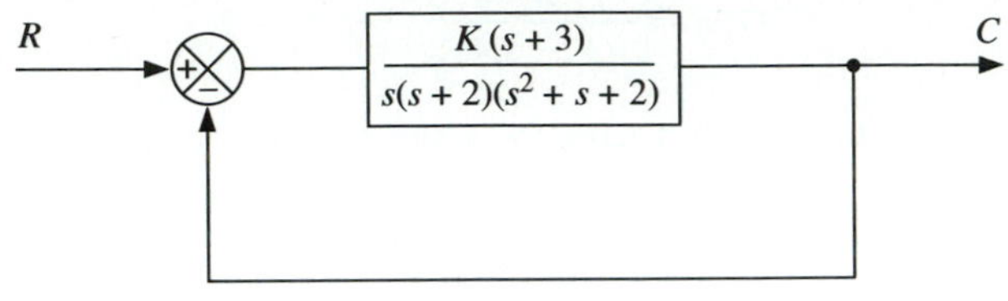

Fig. P17.10

MODULE 18

Frequency-Domain Specifications and Closed-Loop Frequency Response

Just as time-domain performance specifications allow different designs of feedback control systems to be compared, certain frequency-domain measures may also be defined and used to assess total system performance. These measures, which give information about the system not available from the transient response, will be defined in this module.

Although most of our stability analysis has been performed using the open-loop transfer function, it is important to keep in mind that ultimately the feedback loop will be closed, and the performance of the closed-loop system needs to be assessed. With this in mind, the closed-loop frequency response of a system will have to be determined. This may be done directly from the closed-loop transfer function, but if the open-loop frequency response has already been plotted for the purposes of checking stability, it seems desirable that the closed-loop frequency response be obtained from it, rather than duplicating the effort. Methods will be presented in this module for doing just this.

◆ FREQUENCY-DOMAIN SPECIFICATIONS

Before defining commonly used frequency-domain performance specifications, it is worthwhile looking at an example that illustrates why these measures are important and what they tell us about system performance. Suppose we are designing a missile that is to fly at a fixed height above the ground and is equipped with radar that measures this height, compares it to the desired height, and actuates the control surfaces to reduce the height error to zero. Figure 18.1 shows the physical system while Figure 18.2 shows it in block diagram form. The system plant will be assumed to be represented by the transfer function G shown. Suppose the missile

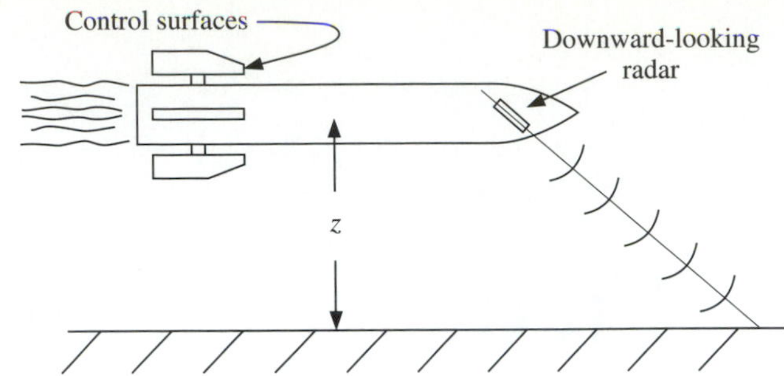

Fig. 18.1 Ground-skimming missile.

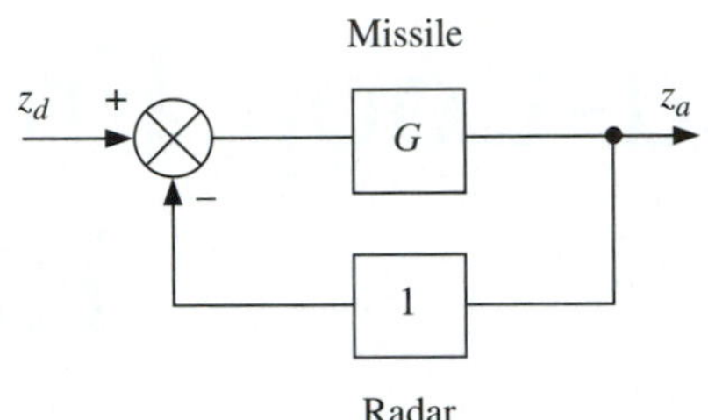

Fig. 18.2 Missile altitude control system.

flies over ground with buildings on it. We would like to design a control system that does not force the missile to follow the contour of the buildings but follows the contour of the ground on which the buildings are located, as indicated in Fig. 18.3.[1] How can we distinguish between the signal generated by the buildings and that generated by the ground? Assume the buildings are all the same height and spaced in equal distances along the path of the missile and that the speed of the missile is known and constant. The signal generated by the radar when flying over flat ground with buildings present is shown in Fig. 18.4a, while the signal obtained by

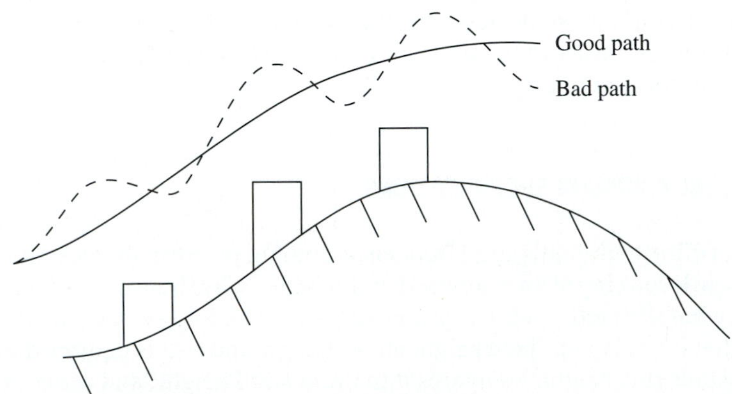

Fig. 18.3 Contour-following paths.

[1]Clearly, z_d must be greater than the expected building height.

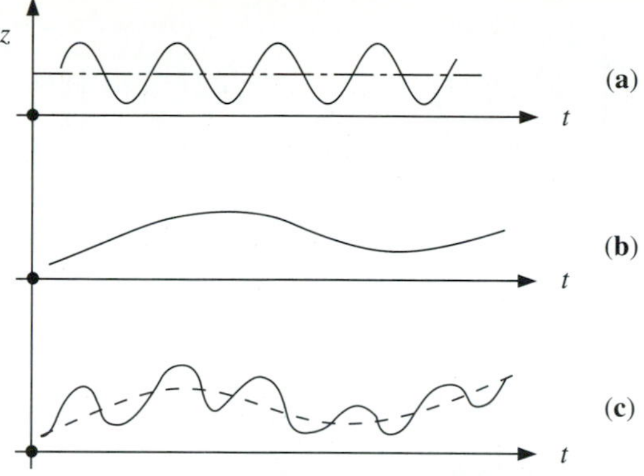

Fig. 18.4 Radar signal for different paths.

the radar when the missile flies over undulating ground with no buildings is shown in Fig. 18.4b. Figure 18.4c shows the signal expected when the missile flies over ground that is both undulating and with buildings and, as expected, is the sum of the two components.

In designing for a good path, as defined in Fig. 18.3, we are saying that the component of the signal due to the buildings should be ignored, or attenuated, while that due to the component of the ground should be emphasized. In this sense, the buildings generate an unwanted part of the total signal, usually referred to as noise, while the ground generates the true signal in which we are interested. How can we use frequency response to make the system respond to the signal but not the noise?

Notice that the signal is of much lower frequency than the noise. Also, if we consider the response of the missile height control system to signals of all frequencies, we may represent the situation as shown in Fig. 18.5, which is the closed-loop frequency response of the system. Suppose the missile control system happens to be represented by a second-order system. We know what its Bode diagram looks like. Notice that we are able to select the natural frequency and damping ratio, and we do so such that the natural frequency falls between the signal and the noise frequency, and perhaps select the damping ratio to be about critical, as shown in Fig. 18.6. Such a design has the desired effect, since it indicates that signals with the desired signal frequency are passed with slight magnification, while the noise is attenuated considerably, perhaps by 60 db or more. Many systems, particularly closed-loop systems, have a similar appearance to the one shown in Fig. 18.6, with a low-frequency magnitude of unity (0 db) and high-frequency attenuation. The range of frequencies passed by the system is usually referred to as the *bandwidth* ω_b, and this is commonly defined as the frequency range over which the magnitude does not drop more than 3 db. In some cases the low-frequency magnitude

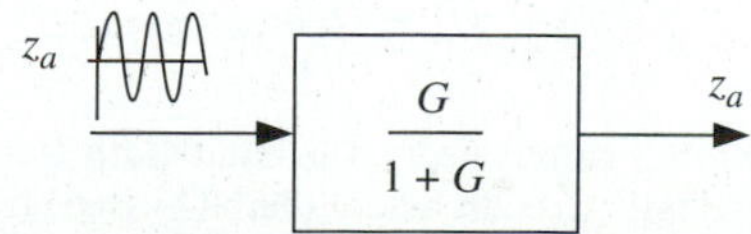

Fig. 18.5 Closed-loop frequency response.

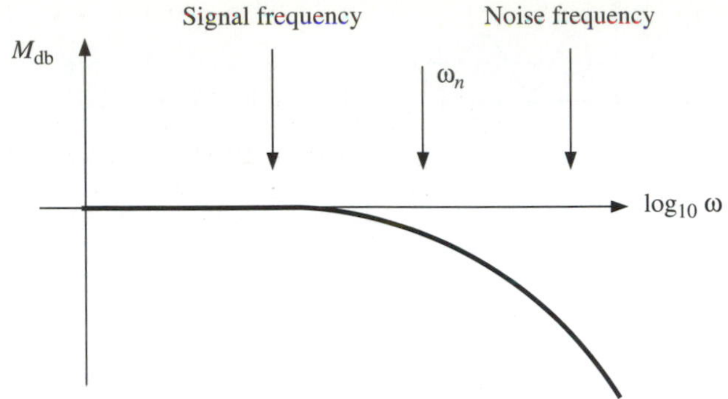

Fig. 18.6 Frequency response placement.

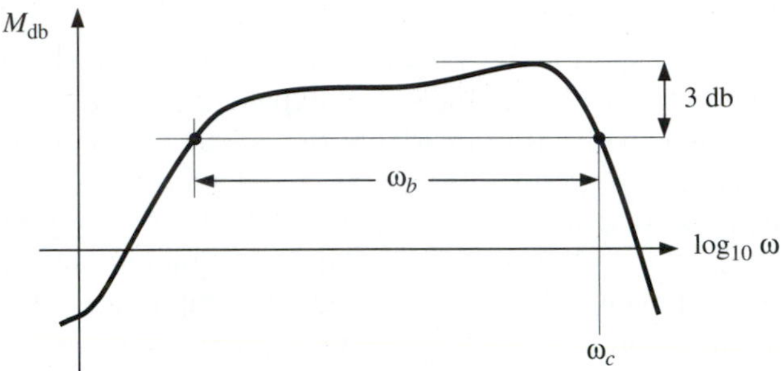

Fig. 18.7 Definition of bandwidth and cutoff frequency.

is not unity, and the *cutoff frequency* ω_c is used in conjunction with the bandwidth. Both definitions are shown in Fig. 18.7.[2] Another useful design parameter is the size of the resonance peak of the frequency response. In our design, we deliberately selected the damping ratio to be about the critical value, resulting in no resonance peak. It is important to examine this peak since if a large peak occurs, large excursions in missile height would be expected if the system were excited at a frequency close to the natural frequency ω_n. Such a response, which may be caused by flying over regularly spaced objects such as water waves, may cause excessive changes in missile altitude, ultimately resulting in the missile hitting the water.

Just as many systems may be approximated to second order for the purposes of determining time-domain response parameters (percentage overshoot, peak time, etc.), it is customary to approximate systems to second order in order to determine frequency-domain parameters. Recall such a system is described by

$$T(s) = \frac{\omega_n^2}{s^2 + 2\zeta\omega_n s + \omega_n^2} \tag{18.1}$$

[2]If the system low-frequency (DC) gain is unity, the bandwidth is usually taken as the frequency range over which the response falls to 3 db below the DC value. If the DC value is not defined, the bandwidth is defined as the frequency range over which the response falls to 3 db below the peak value.

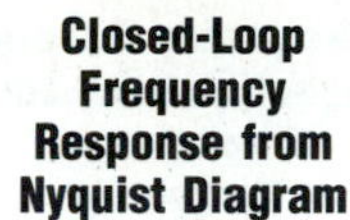

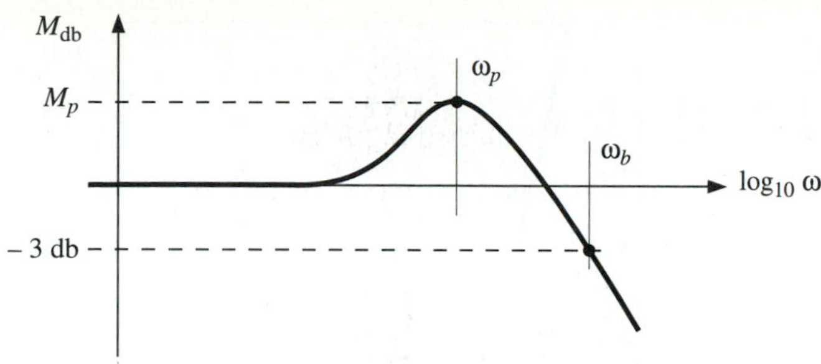

Fig. 18.8 Second-order system frequency response.

Note that here we are not distinguishing between open- or closed-loop transfer functions, since the definitions that follow apply equally to both. Since the frequency-domain parameters are all determined from the magnitude part of the Bode diagram for a second-order system, Fig. 15.17 is reproduced here in Fig. 18.8 for a particular value of damping ratio. For this system, three parameters are of interest:

1. the bandwidth ω_b, which in this case is the same as the cutoff frequency;
2. the magnitude of the resonance peak, M_p and
3. the frequency at which the resonance peak occurs, ω_p.

As can be seen from Fig. 18.8, all of these quantities will be functions of the damping ratio ζ. Although not derived here, these functional relationships are easily derived from the frequency response of a standard second-order system and are found to be

$$\omega_b = \omega_n[1 + \sqrt{4\zeta^4 - 4\zeta^2 + 2} - 2\zeta^2]^{1/2} \tag{18.2}$$

$$M_p = \frac{1}{2\zeta\sqrt{1 - \zeta^2}} \tag{18.3}$$

$$\omega_p = \omega_n\sqrt{1 - 2\zeta^2} \tag{18.4}$$

Instead of using the above equations, the parameters may be obtained from the graphs shown in Figs. 18.9 and 18.10. Knowing the frequency of signals to which we wish the system to respond and knowing the frequency of the noise, systems may be designed in the frequency domain to respond to the signal and reject the noise. Although the preceding analysis may be used on open-loop frequency response, it is just as applicable and is, in fact, more widely applied to closed-loop frequency response design. It remains, therefore, to develop means to derive the closed-loop response from the open-loop response, so that this design process may be performed.

◆ CLOSED-LOOP FREQUENCY RESPONSE FROM NYQUIST DIAGRAM

The closed-loop frequency response may be obtained in a graphical manner directly from the open-loop frequency response. We will first consider only unity-

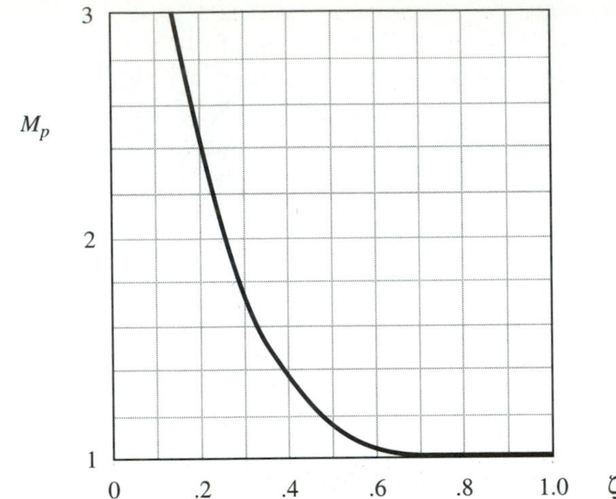

Fig. 18.9 Resonance peak magnitude and frequency as functions of damping ratio.

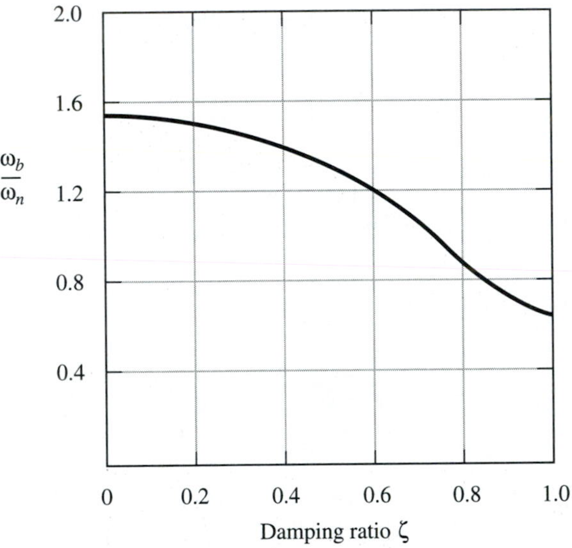

Fig. 18.10 System bandwidth as a function of damping ratio.

feedback systems and express the closed-loop transfer function as

$$\frac{C}{R}(s) = \frac{G}{1 + GH} \qquad H(s) = 1 \tag{18.5}$$

As before, the frequency response will be evaluated by substituting $s = j\omega$ into the above equation:

$$\frac{C}{R}(j\omega) = \frac{G(j\omega)}{1 + G(j\omega)} \tag{18.6}$$

The quantity on the left of this equation is what we are interested in, and it may be considered as a magnitude and phase, denoted as M_{cl} and α, respectively, corresponding to M and ϕ for the open-loop transfer function. These quantities are obtained by considering the right-hand side of the above equation to be the ratio of two complex numbers, G and $1 + G$, each of which has its own magnitude and phase. The magnitude and phase of G may be obtained directly from the open-loop Nyquist chart, since this is, by definition, what the Nyquist diagram portrays. It will be represented by a vector drawn from the origin to a point on the Nyquist contour corresponding to a frequency ω_1 in which we are interested. The magnitude and phase of G are as shown in Fig. 18.11. The quantity $1 + G$ may be obtained by recalling earlier work with the root locus where we defined the vector $s_t + a$ to be a vector drawn from $-a$ to the test point s_t. Similarly, the quantity $1 + G$ may be represented by a vector drawn from the point -1 to a point on the Nyquist contour at the frequency of interest. The magnitude and phase of the closed-loop frequency response are obtained from Fig. 18.11 as

$$M = \frac{|G|}{|1 + G|} = \frac{OA}{NA} \tag{18.7}$$

and

$$\alpha = \angle G - \angle(1 + G) = -\theta_2 + \theta_1 \tag{18.8}$$

It may be shown using simple geometry that the phase reduces to

$$\alpha = \angle OAN \tag{18.9}$$

The procedure would be, therefore, to pick points on the Nyquist contour at known frequencies and use the above equations to calculate the closed-loop magnitude and phase. If this is repeated for sufficient frequencies, a Nyquist or Bode diagram may be drawn for the closed-loop transfer function.

Suppose we wish to evaluate the size of the resonance peak for the closed-loop frequency response. We would have to evaluate sufficient points along the open-loop Nyquist contour to ensure that the maximum M_{cl} had been found. Alternatively, this task may be done by inspection by considering which points on the complex plane all have the same magnitude and which points all have the same phase. This process generates lines of constant magnitude and phase on the complex plane and is achieved in the following manner. Suppose we are interested in

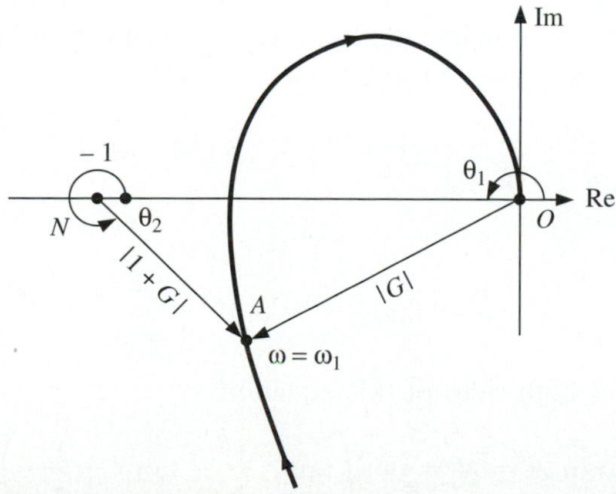

Fig. 18.11 Calculation of closed-loop frequency response.

a point on the Nyquist contour, the coordinates of which on the complex plane are (x, y). Then

$$G(j\omega) = x + yj \tag{18.10}$$

The magnitude of the closed-loop transfer function becomes

$$M_{\text{cl}} = \frac{|x + yj|}{|1 + x + yj|} = \frac{(x^2 + y^2)^{1/2}}{[(1 + x^2)^2 + y^2]^{1/2}} \tag{18.11}$$

Expanding yields

$$x^2(1 - M_{\text{cl}}^2) - 2M_{\text{cl}}^2 x - M_{\text{cl}}^2 + (1 - M_{\text{cl}}^2)y^2 = 0 \tag{18.12}$$

Dividing through by $M_{\text{cl}}^2 - 1$ and adding $M_{\text{cl}}^2/(M_{\text{cl}}^2 - 1)^2$ to both sides of the equation, the following is obtained:

$$\left(x + \frac{M_{\text{cl}}^2}{M_{\text{cl}}^2 - 1}\right)^2 + y^2 = \frac{M_{\text{cl}}^2}{(M_{\text{cl}}^2 - 1)^2} \tag{18.13}$$

If M_{cl} is regarded as a constant, the above equation represents a circle centered on the point (x_c, y_c), where

$$x_c = \frac{-M_{\text{cl}}^2}{M_{\text{cl}}^2 - 1} \tag{18.14}$$

and

$$y_c = 0 \tag{18.15}$$

and the circle will have radius

$$r = \left|\frac{M_{\text{cl}}}{M_{\text{cl}}^2 - 1}\right| \tag{18.16}$$

For $M > 1$, the circle will be centered to the left of the imaginary axis, while for $M < 1$ it will be centered to the right. The above equations become indeterminant when $M = 1$, so we go back to equation 18.11 to obtain

$$-2x - 1 = 0 \tag{18.17}$$

which has the solution $x = -\frac{1}{2}$ for any y. This is a straight vertical line through the point $(-0.5, 0)$. A complete family of circles is shown in Fig. 18.12. Turning now to lines of constant closed-loop phase, we write

$$\angle\frac{C}{R} = \angle(x + yj) - \angle(1 + x + yj) \tag{18.18}$$

or

$$\alpha = \tan^{-1}\frac{y}{x} - \tan^{-1}\frac{y}{1 + x} \tag{18.19}$$

Taking tangents of both sides of this equation

$$\tan\alpha = N = \tan\left(\tan^{-1}\frac{y}{x} - \tan^{-1}\frac{y}{1 + x}\right) \tag{18.20}$$

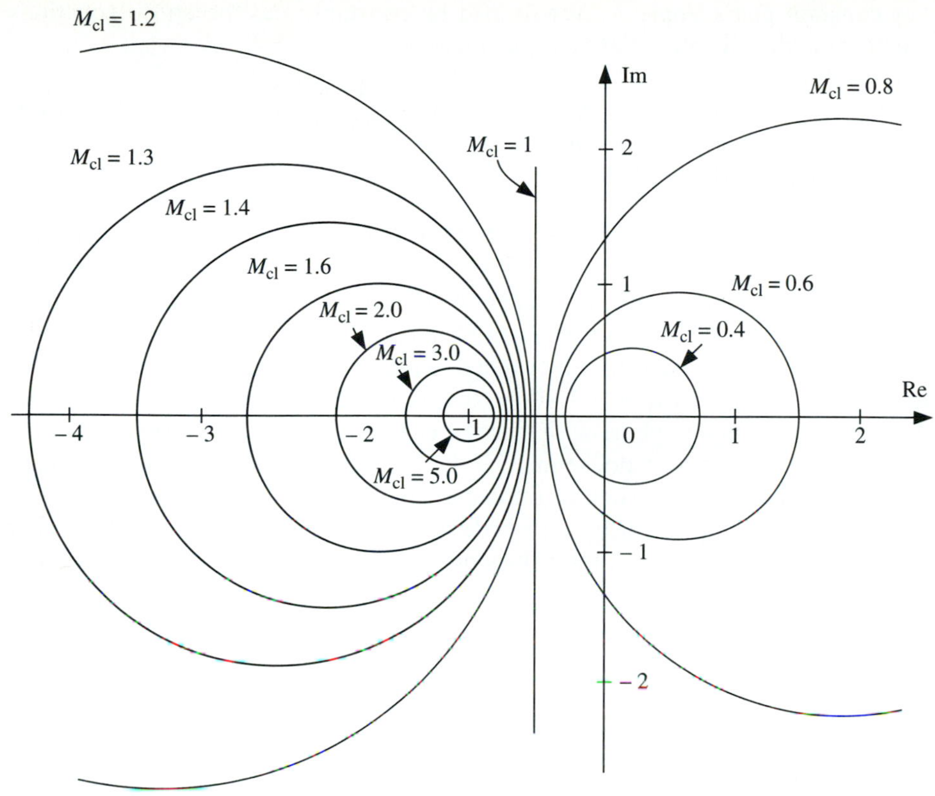

Fig. 18.12 Constant M_{cl} contours.

and using the identity

$$\tan (A - B) = \frac{\tan A - \tan B}{1 + \tan A \tan B} \tag{18.21}$$

we obtain, after some manipulation,

$$N = \frac{y}{x^2 + x + y^2} \tag{18.22}$$

or

$$x^2 + x + y^2 - \frac{y}{N^2} = 0 \tag{18.23}$$

Adding the term

$$\frac{1}{4} + \frac{1}{4N^2} \tag{18.24}$$

to each side of the equation gives

$$\left(x + \frac{1}{2}\right)^2 + \left(y - \frac{1}{2N}\right)^2 = \frac{1}{4} + \left(\frac{1}{2N}\right)^2 \tag{18.25}$$

For constant phase angle α, N will also be constant, and the above equations represent a circle centered at (x_c, y_c), where

$$x_c = -\tfrac{1}{2} \qquad y_c = \frac{1}{2N} \tag{18.26}$$

and the radius of the circle is

$$r = \sqrt{0.25 + \frac{1}{4N^2}} \tag{18.27}$$

Figure 18.13 shows the constant N circles on the complex plane for different values of α. Notice from the figure that a complete circle is made up of two constant N contours, the sum of each absolute value of which is 180°. For example, the N contours for $\alpha = 20°$ and $\alpha = -160°$ make up one circle. By plotting the Nyquist diagram on a complex plane with lines of constant M_{cl} and α marked on it, the closed-loop frequency response may be determined simply by reading the magnitude and phase for a particular frequency, without having to use the graphical procedure outlined previously. This is demonstrated in Figs. 18.14 and 18.15. For frequencies $\omega_1, \omega_2, \ldots, \omega_6$, the values of M_{cl} and α are estimated from the con-

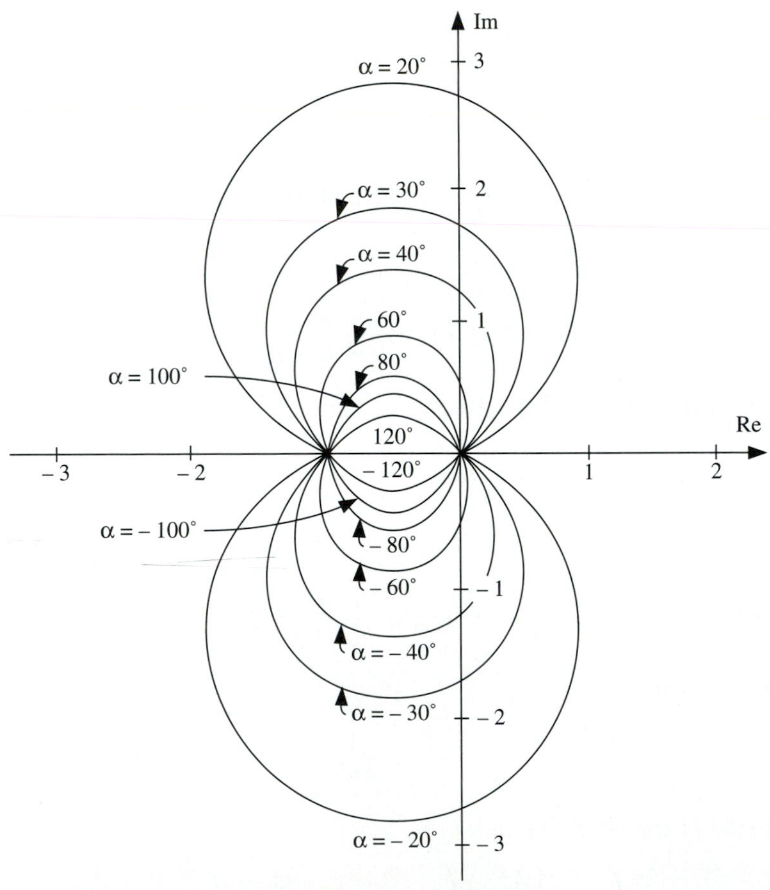

Fig. 18.13 Constant N contours.

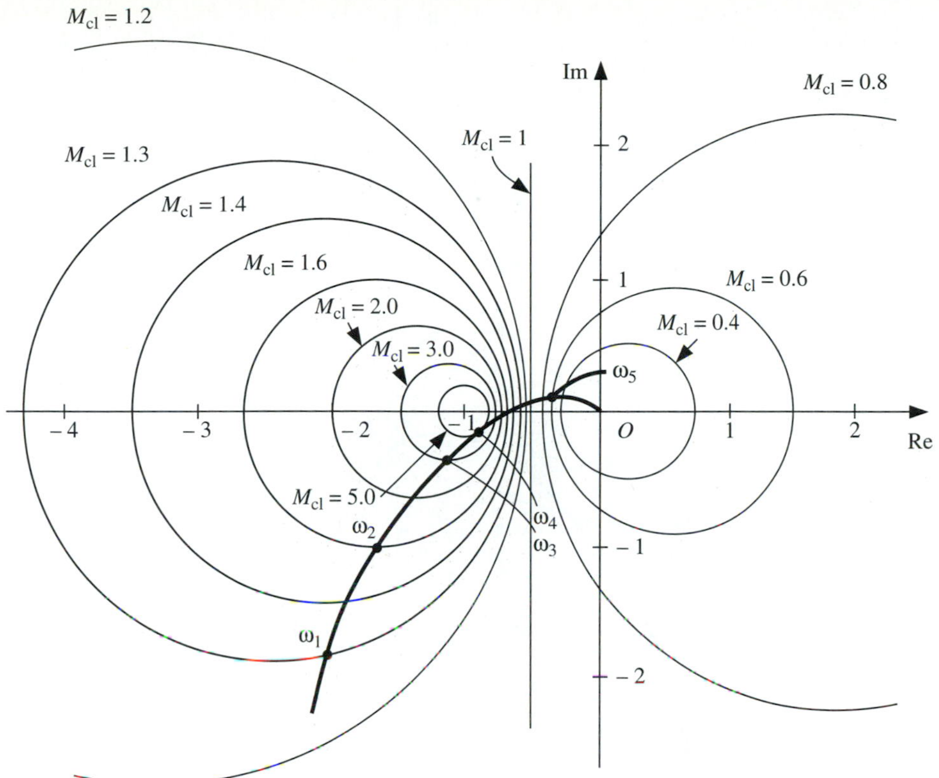

Fig. 18.14 Nyquist diagram and constant M_{cl} contours.

tours and replotted on a Bode-type plot, as shown in Fig. 18.16. Note that the size of the resonance peak is given by the largest value M_{cl} contour to which the open-loop Nyquist diagram is tangent.

◆ CLOSED-LOOP FREQUENCY RESPONSE FROM BODE DIAGRAM

Although it is possible to determine the closed-loop frequency response in the manner described, it is more common to derive it from the Bode diagram. The reasons are that experimental data are usually available in Bode diagram form, and for those cases where the open-loop frequency response is derived from a transfer function, over the frequency range studied, the Bode form is more accurate than the Nyquist form. Consider a point on the open-loop frequency response such as that shown in Fig. 18.11. Five parameters are represented by this one point on the $GH(j\omega)$ curve:

1. the frequency of interest, ω_1;
2. the open-loop magnitude M;
3. the open-loop phase ϕ;
4. the closed-loop magnitude M_{cl}; and
5. the closed-loop phase α.

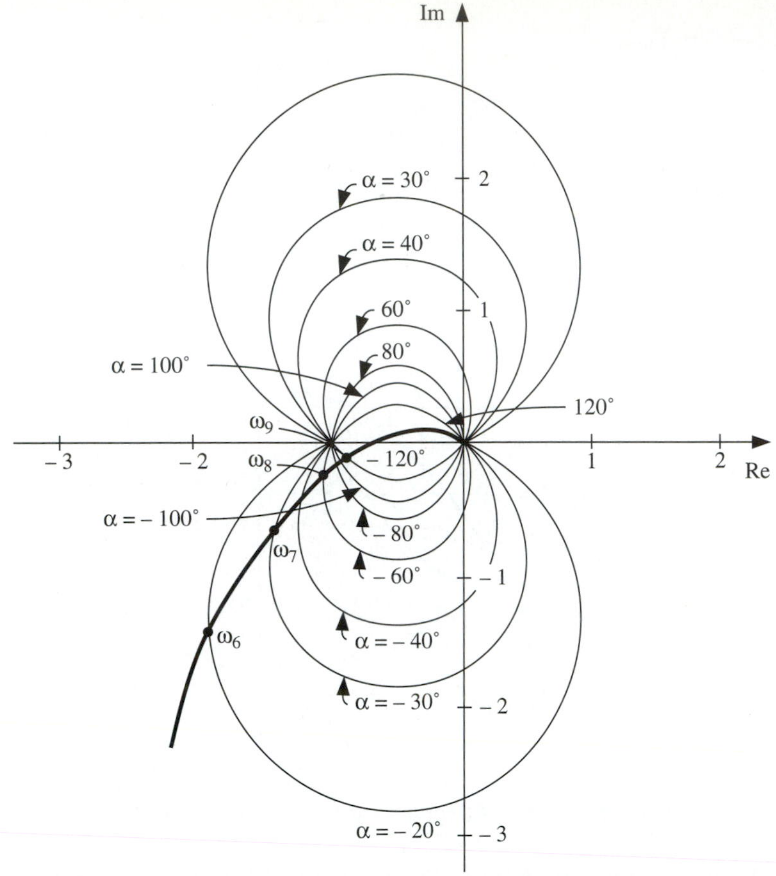

Fig. 18.15 Nyquist diagram and constant α contours.

Consider a simple mapping of these parameters such that we plot M_{db} vertically and ϕ horizontally. The point of interest will map to a corresponding point in this new diagram, as will the values of M_{cl} and α. The contours of constant M_{cl} and α in the complex plane, however, will not be circles in the new diagram and will look somewhat different, as indicated by Fig. 18.17, which shows the mapping. This mapping, with the new M_{cl} and α contours, is known as a Nichols chart and is shown in Fig. 18.18. In this diagram, note how both the open- and closed-loop magnitudes are in decibels. The Nichols chart is symmetrical about the $-180°$ line and repeats itself every 360°. Considerable information about both the open- and closed-loop frequency responses is available from the Nichols chart. For example, suppose Fig. 18.19 shows an open-loop frequency response plotted on the Nichols chart. The following data may be obtained:

1. The gain margin is 11 db.
2. The phase margin is about 32°.
3. The gain crossover frequency is 10 rad/s.

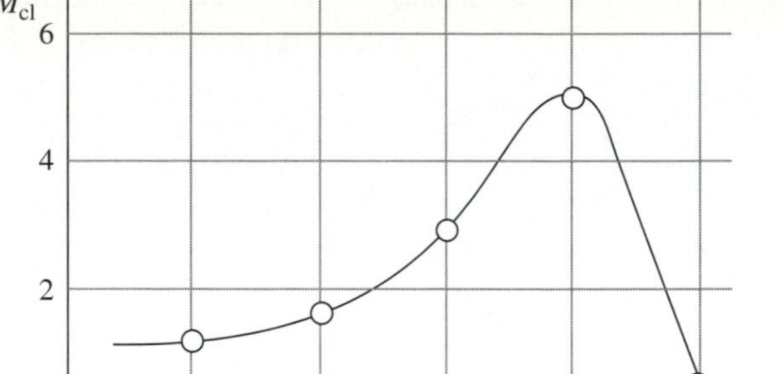

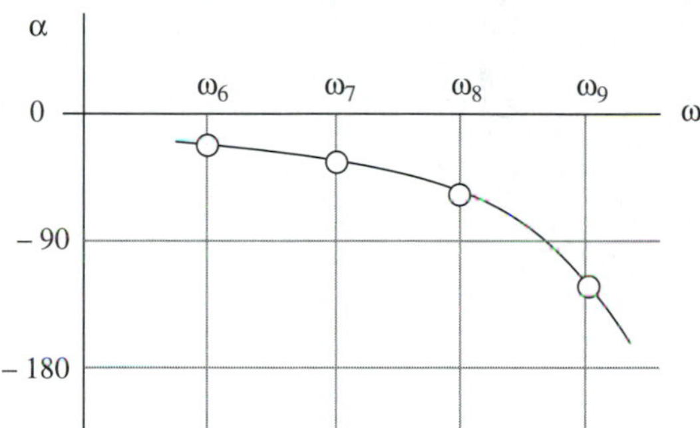

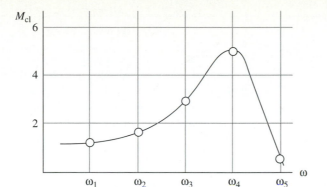

Fig. 18.16 Closed-loop frequency response.

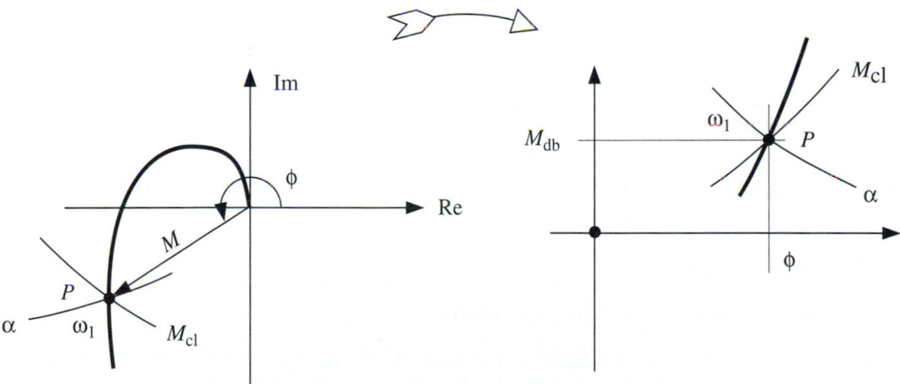

Fig. 18.17 Mapping of a point from Nyquist diagram to new axes.

4. The closed-loop bandwidth is 25 rad/s.
5. The size of the closed-loop resonance peak is 5 db.

Since the open-loop magnitude is plotted vertically, changing the forward-path gain K in the open-loop transfer function moves the Nichols plot up and down by K_{db}, just as in the Bode diagram. This fact will be used later in designing for par-

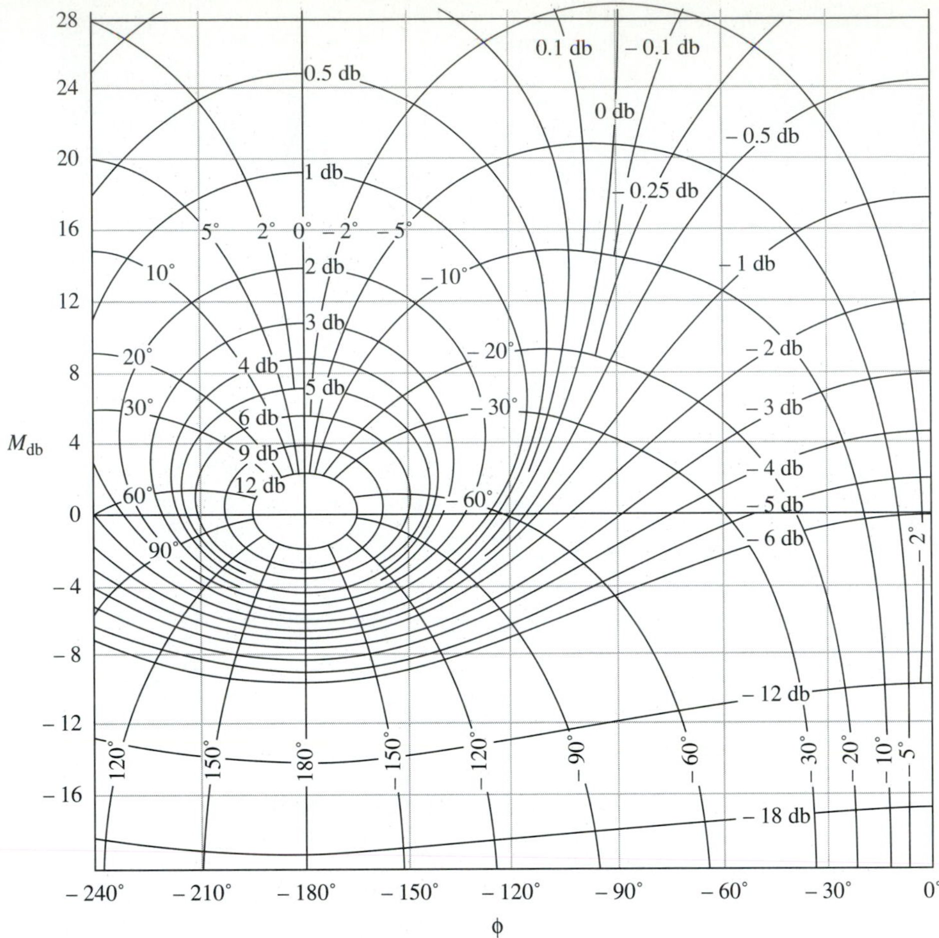

Fig. 18.18 Nichols chart.

ticular values of M_p. Using the Nichols chart to determine the closed-loop frequency response involves three steps:

1. Plot the Bode diagram.
2. Transfer the data to the Nichols chart.
3. From the M_{cl} and α contours on the Nichols chart, plot the closed-loop frequency response on another Bode diagram.

◆ GAIN FOR A DESIRED M_p FROM THE NYQUIST DIAGRAM

As we saw from the example of the missile altitude control system, it is desirable to design the feedback control system so that the resonance peak is within some specified value. This is known as designing for a maximum value of M_p, where

$$M_p = \hat{M}_{cl} \tag{18.28}$$

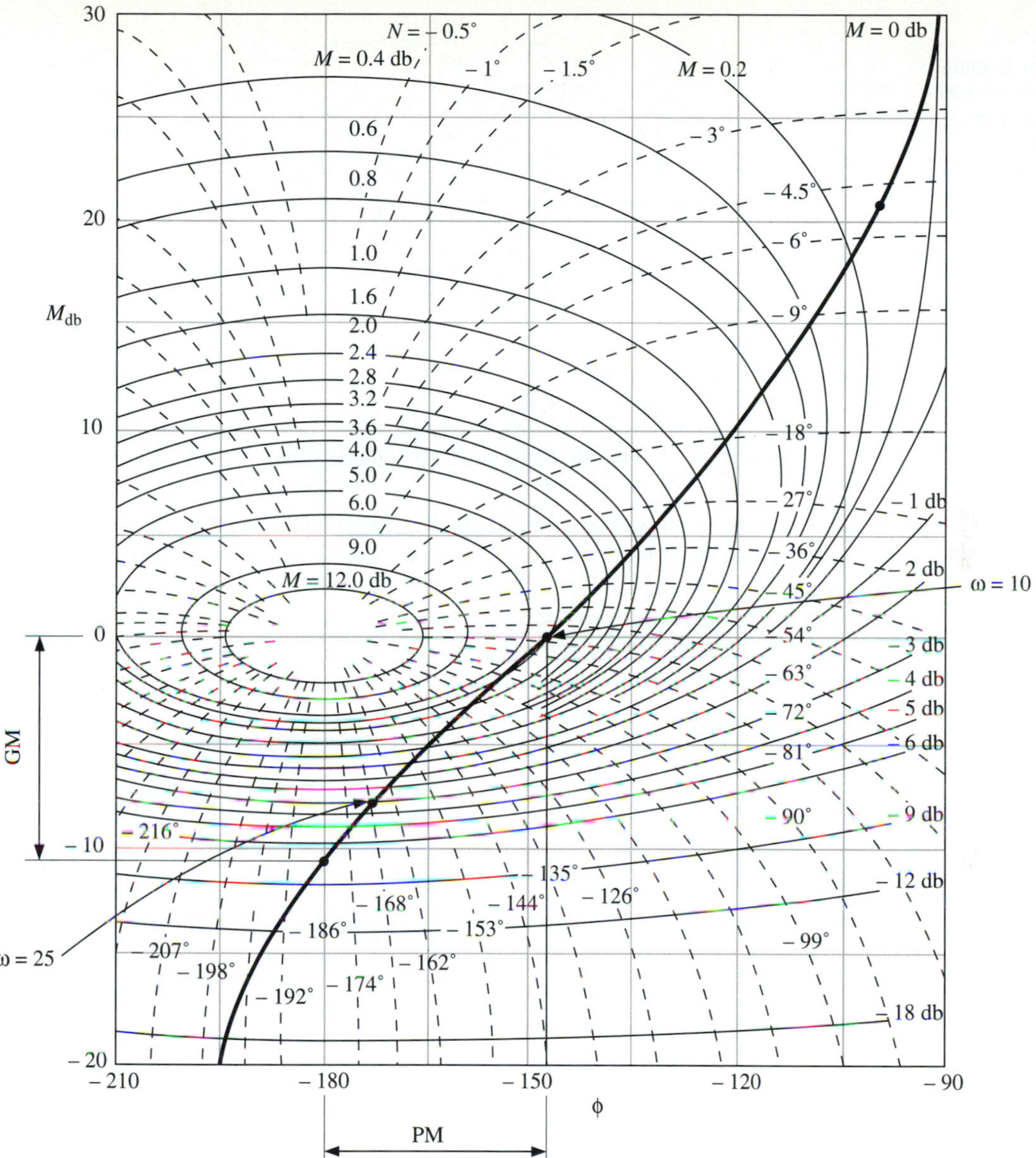

Fig. 18.19 Example system plotted on Nichols chart.

for the $GH(j\omega)$ contour. Consider a constant M_{cl} contour on the complex plane, as shown in Fig. 18.20. Previous analysis has determined the location of the center of the circle and the radius to be as marked in Fig. 18.20. It may be seen that the angle made by the line OT, tangent to the M_{cl} circle, makes an angle ψ with the real axis, where

$$\sin \psi = \frac{1}{M_{cl}} \tag{18.29}$$

It may also be shown that a line drawn parallel to the imaginary axis through T

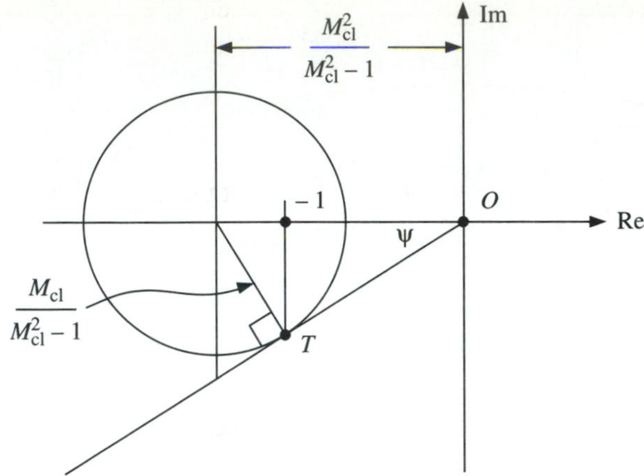

Fig. 18.20 *M* circle on complex plane.

passes through the critical point $(-1, 0j)$. It is known that increasing the forward-path gain K expands the open-loop frequency response radially, causing it to be tangent to different M_{cl} circles. The problem is to determine the value of K that just causes it to be tangent to the maximum desired value of closed-loop magnitude M_p without drawing many Nyquist diagrams. Without proof, the following procedure will determine this value of K:

1. From the open-loop transfer function $GH(j\omega)$, set the forward-path gain $K = 1$ and plot the Nyquist diagram. For a sample system, this results in Fig. 18.21.
2. Calculate the angle ψ using

$$\sin \psi = \frac{1}{M_p} \tag{18.30}$$

and draw a line ON at this angle from the origin of the complex plane.

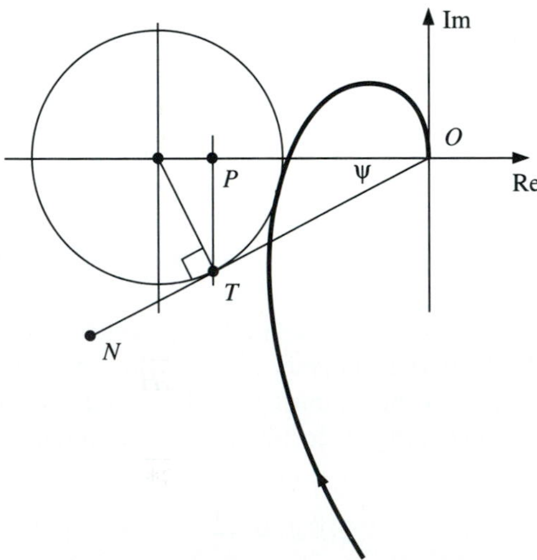

Fig. 18.21 Nyquist diagram of system.

3. Using trial-and-error methods, draw a circle with its center on the negative real axis, tangent to both the Nyquist plot and the line *ON*.
4. From the tangent point *T*, draw a vertical line to intersect the real axis at *P*, as shown in Fig. 18.21.
5. The gain required to make the Nyquist plot tangent to the desired M_p circle is given by

$$K = \frac{1}{OP} \tag{18.31}$$

◆ GAIN FOR A DESIRED M_p FROM THE NICHOLS CHART

The process of determining the required K to give a particular M_p is considerably easier using the Nichols chart. Consider the open-loop frequency response of the same system shown in Fig. 18.21, but this time plotted on the Nichols chart, as shown in Fig. 18.22. From this figure, it may be seen that in order to achieve tangency to a particular M_{cl} circle corresponding to M_p, the open-loop frequency response has to be moved up by increasing the gain by x_{db}. This is achieved by increasing the forward-path gain from unity (from which the open-loop frequency response was plotted) to the value given by solving

$$20 \log_{10} K = x \tag{18.32}$$

In some cases, the frequency response has to be moved down to achieve the required tangency condition, corresponding to a reduction in K. To determine exact tangency to the desired M_p contour, it may be necessary to trace the $GH(j\omega)$ contour onto a transparency and slide it up or down until it just touches the contour desired.

◆ NON-UNITY-FEEDBACK GAIN SYSTEMS

All the analysis described so far has been restricted to unity-gain feedback systems. If a system does not have unity feedback, some modifications of the system transfer function are required. Consider the closed-loop transfer function

$$\frac{C}{R} = \frac{G}{1 + GH} \tag{18.33}$$

$$= \frac{1}{H}\frac{GH}{1 + GH} \tag{18.34}$$

$$= \frac{1}{H}\frac{G^*}{1 + G^*} \tag{18.35}$$

where

$$G^* = GH \tag{18.36}$$

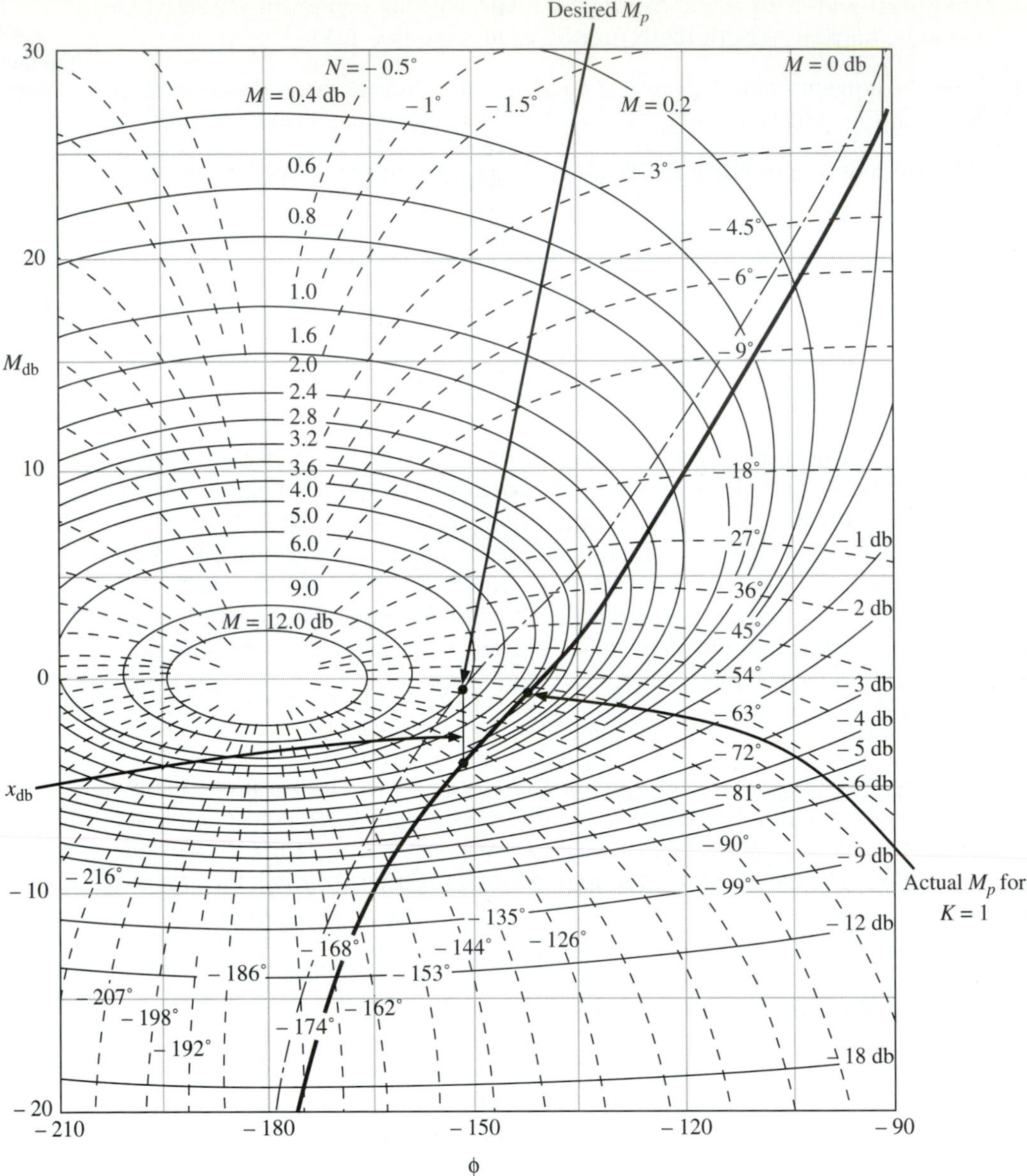

Fig. 18.22 Example frequency response plotted on Nichols chart.

The closed-loop frequency response is obtained in the following manner:

1. Plot the Bode diagram of the $G^*(j\omega)$ system.
2. Transfer the plot to the Nichols chart.
3. Calculate the closed-loop frequency response in the usual manner using the M_{cl} and α contours.
4. Plot the Bode diagram of $H(j\omega)$.

5. The closed-loop frequency response is obtained in Bode form by subtracting the $H(j\omega)$ plot from the closed-loop frequency response of the $G^*(j\omega)$ system.

Calculating the system forward-path gain K so that the closed-loop system has a maximum M_{cl} value may be performed in the same manner as for unity-feedback systems, since only the $G^*(j\omega)$ system will move on the Nichols chart when K is varied. It will be seen, however (in Sample Problem 18.3), that it is not possible to estimate what effect this has on the closed-loop frequency response from the Nichols chart, and a trial-and-error approach has to be used.

Sample Problem 18.1

A unity-feedback control system has a forward-path transfer function given by

$$G(s) = \frac{1}{s(1 + s)(1 + s/2)}$$

Using first the Nyquist diagram and then the Nichols chart, determine the value of the resonance peak M_p and the bandwidth of the closed-loop system.

Solution

Using the Nyquist Diagram. We begin by plotting the Nyquist diagram on the complex plane and then superimposing M_{cl} and α circles, enabling the closed-loop frequency response to be determined. The frequency response may be written as

$$GH(j\omega) = \frac{1}{j\omega(1 + j\omega)(1 + j\omega/2)}$$

from which the magnitude and phase of the open-loop system may be seen to be

$$M = |GH(j\omega)| = \frac{1}{\omega\sqrt{(1 + \omega^2)(1 + \omega^2/4)}}$$

and

$$\phi = \angle GH(j\omega) = -90 - \tan^{-1}\omega - \tan^{-1}\omega/2$$

A simple BASIC program may be written to calculate the magnitude and phase for any frequency. The following program performs this task:

```
1000 INPUT "W=";W
1010 M1=1+W*W
1020 M2=1+0.25*W*W
1030 M=1/(W*SQR(M1*M2))
1040 P1=ATN(W)
1050 P2=ATN(W/2)
1060 PHI=-90-180*(P1+P2)/3.14159
1070 PRINT W, M, PHI
1080 GOTO 1000
```

Considering the asymptotic values of the frequency response ($-90°$ and $-270°$), the complete Nyquist diagram will look as shown in Fig. SP18.1.1.

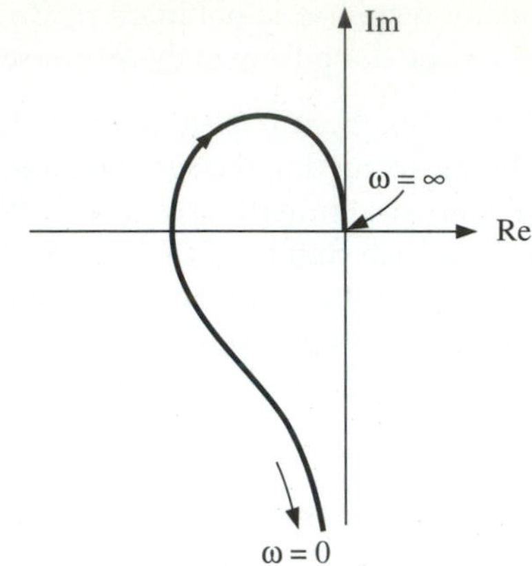

Fig. SP18.1.1

TABLE 18.1 Calculated magnitude and phase

ω	M	ϕ
0.5	1.75	−130
0.7	1.10	−144
0.9	0.75	−156
1.0	0.63	−161
1.4	0.33	−180
1.5	0.30	−183
2.0	0.16	−198
3.0	0.06	−217

Running the above program results in the data shown in Table 18.1. Plotting these data produces the Nyquist diagram shown in Fig. SP18.1.2. This diagram is now drawn on top of a template of constant M_{cl} and α circles, as shown in Figs. SP18.1.3 and SP18.1.4. In these figures, we estimate the closed-loop magnitude and phase for the known frequencies and replot M_{cl} against ω and α against ω. This produces the result shown in Fig. SP18.1.5, which is like a Bode diagram except that logarithm scales are not used. The resonance peak has a value of about 2.0. Since the bandwidth is defined as the frequency range over which the magnitude does not fall by more than 3 db, this value must be converted to a fraction for use in Fig. SP18.1.5. We have

$$20 \log_{10} x = -3$$

which yields the value $x = 0.707$. The bandwidth given from Fig. SP18.1.5 is about $\omega_b = 1.2$ rad/s.

USING THE NICHOLS CHART. The first stage in obtaining the closed-loop frequency response from the Nichols chart is to plot the Bode diagram of the open-

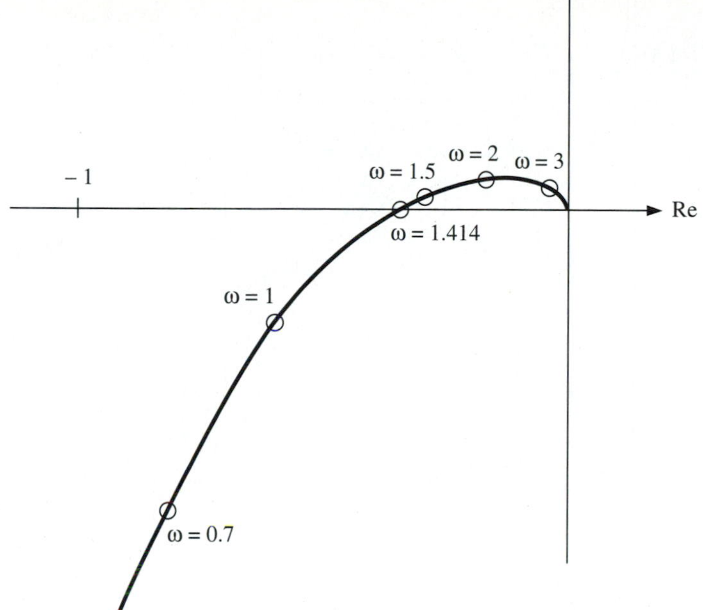

Fig. SP18.1.2

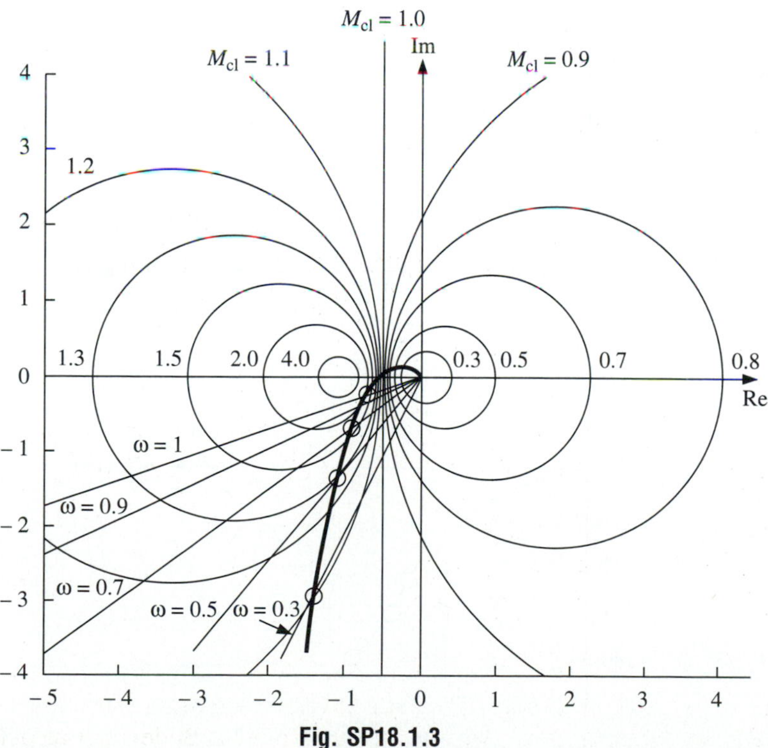

Fig. SP18.1.3

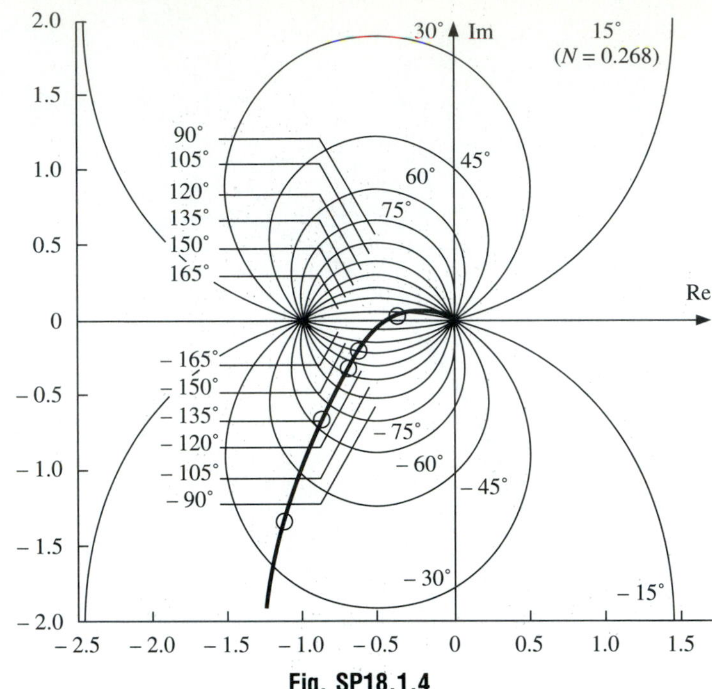

Fig. SP18.1.4

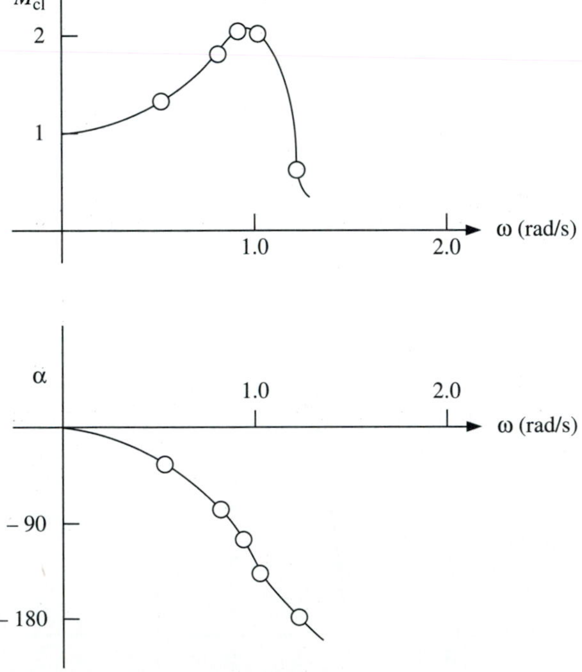

Fig. SP18.1.5

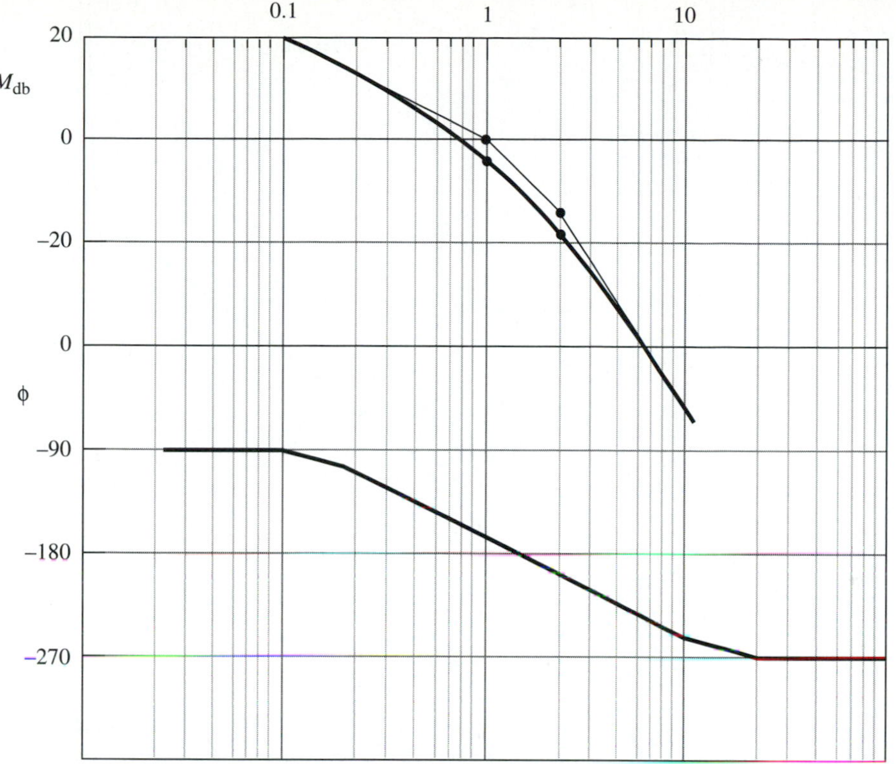

Fig. SP18.1.6

loop system. This is shown in asymptotic and accurate form in Fig. SP18.1.6. Note the estimated 5-db deviations at the two break points and that straight-line approximations have been used for the phase angle. This diagram is now replotted on the Nichols chart, as shown in Fig. SP18.1.7. Reading the constant M_{cl} and α contours directly from the Nichols chart, the closed-loop frequency response is plotted as in Fig. SP18.1.8. Note that the bandwidth is about the same, and the peak resonance is 9 db, or about 2.8. This difference between the two methods may be attributed to the errors introduced by approximating the phase to straight-line segments and the accuracy with which the M_{cl} and α contours may be interpolated. Notice, also, how the closed-loop magnitude is particularly sensitive to the accuracy with which the open-loop frequency response is drawn on the Nichols chart, as evidenced by the results obtained using asymptotic approximations to the magnitude in Fig. SP18.1.7.

Sample Problem 18.2

For the same system as that described in Sample Problem 18.1, assume that the forward-path gain is now K; hence

$$G(s) = \frac{K}{s(1 + s)(1 + s/2)} \qquad H(s) = 1$$

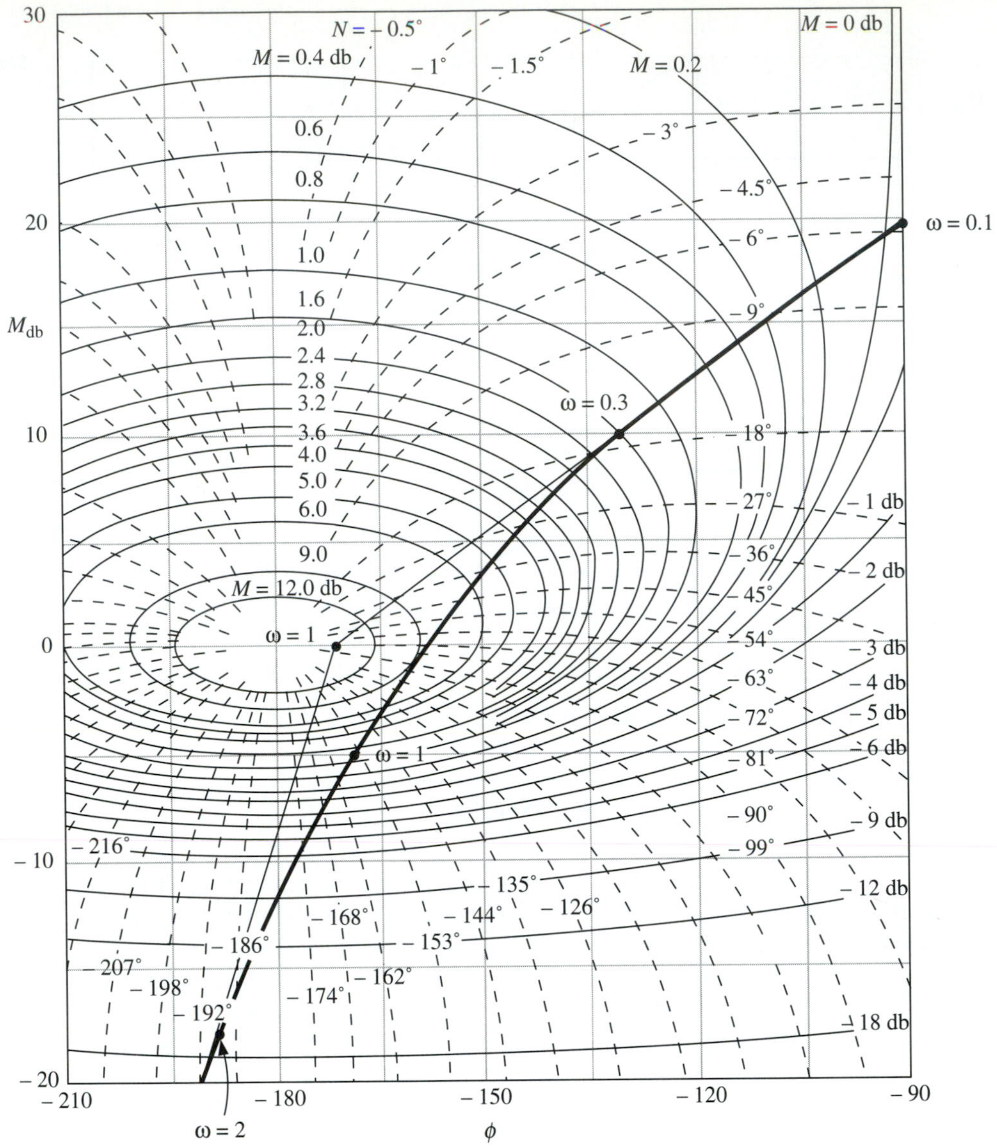

Fig. SP18.1.7

Using first the Nyquist diagram and then the Nichols chart, determine the value of K such that the closed-loop system has a peak resonance $M_p = 4.0$.

Solution

USING THE NYQUIST DIAGRAM. First, the Nyquist diagram of the system is drawn for $K = 1$. This has been done in the previous problem and is reproduced in Fig. SP18.2.1. A line inclined at ψ to the negative real axis is drawn, where

$$\sin \psi = \frac{1}{M_p} = 0.25$$

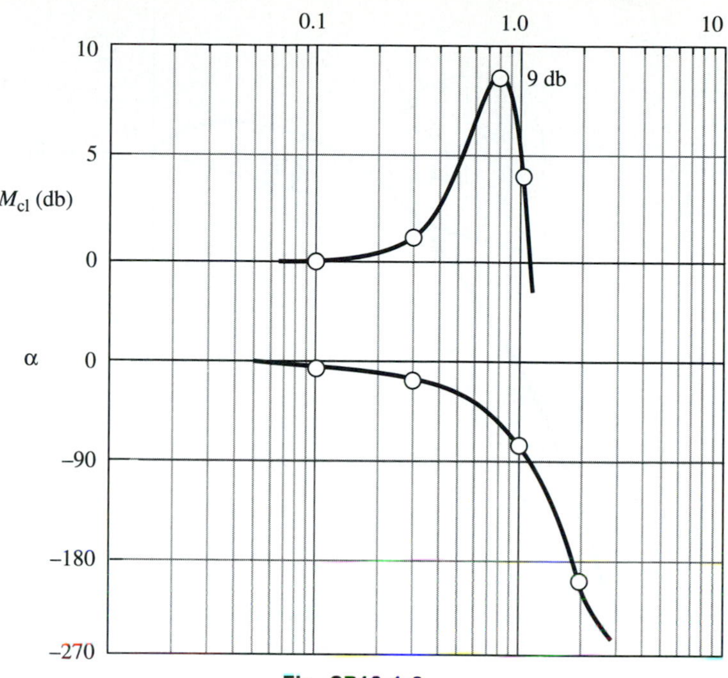

Fig. SP18.1.8

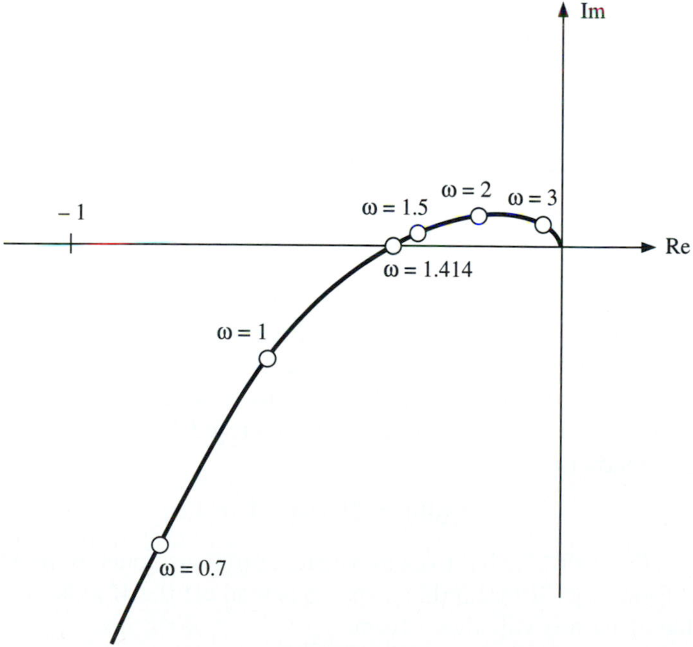

Fig. SP18.2.1

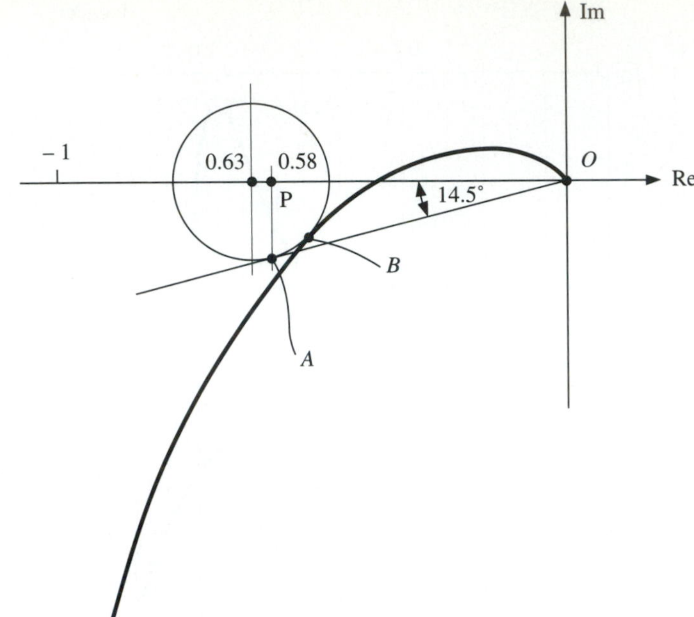

Fig. SP18.2.2

resulting in

$$\psi = 14.5^\circ$$

This line is shown in Fig. SP18.2.2. The next step is to use trial-and-error methods to find a circle that is tangent to both this line and the Nyquist plot $GH(j\omega)$. Such a circle is shown in Fig. SP18.2.2, has center (0.63, 0), and touches the line and plot at A and B, respectively. Next, a line is drawn from A to P on the negative, real axis such that angle $OPA = 90^\circ$. The distance OP is measured as 0.58. The required value of gain is obtained from

$$K = \frac{1}{OP} = \frac{1}{0.58} = 1.73$$

USING THE NICHOLS CHART. Again, the Bode diagram and Nichols chart have been drawn in the previous problem for this particular transfer function, and these have been reproduced in Fig. SP18.2.3. Increasing K simply moves the Nichols plot vertically upward, and it remains to be determined how far up it should be moved. If the desired value of the closed-loop resonance peak is 4, this corresponds to

$$M_p(\text{db}) = 20 \log_{10} 4 = 12$$

The plot in Fig. SP18.2.3 has to be moved up until it just touches the $M_p = 12$ db contour. From Fig. SP18.2.3 this distance is read off the M axis as 4.0 db. The new value of gain is calculated from

$$20 \log_{10} K = 4.0$$

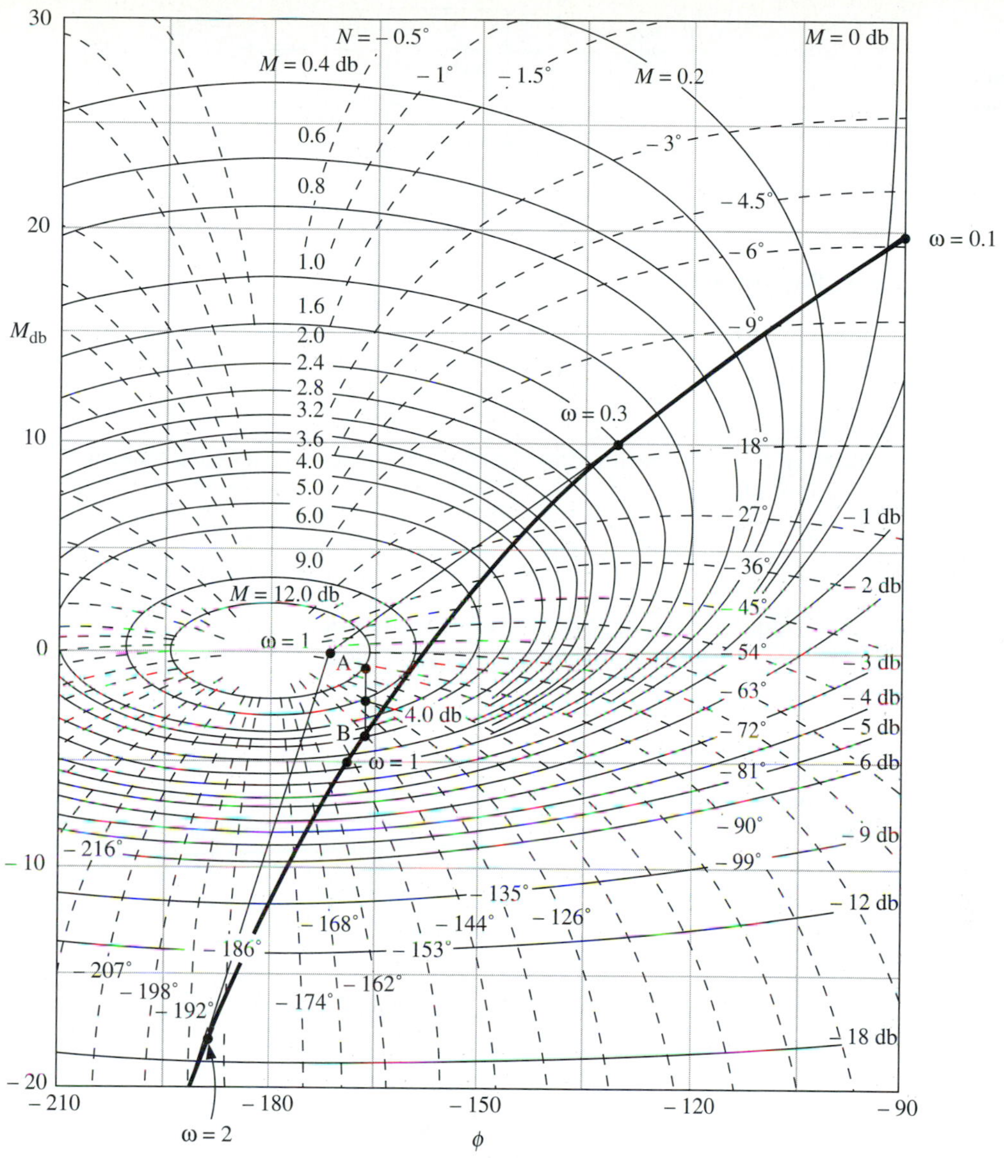

Fig. SP18.2.3

yielding

$$K = 1.60$$

The discrepancy between the two methods depends upon the accuracy with which the Nyquist and Nichols diagrams are plotted and read, and accurately drawn plots would be expected to give the same result. The student may observe that here the Nichols method is somewhat easier to apply, and this is true for most cases, particularly when the effort needed to estimate the open-loop frequency response is considered.

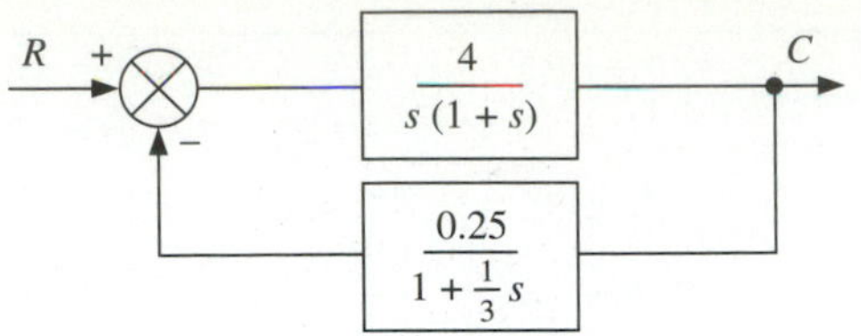

Fig. SP18.3.1

Sample Problem 18.3

Consider the feedback control system shown in Fig. SP18.3.1. Plot the closed-loop frequency response of the system and determine its bandwidth.

Solution

Following the steps outlined in the module for non–unity feedback, we form the system $G^*(s)$, where

$$G^*(s) = GH(s) = \frac{1}{s(1 + s)(1 + s/3)}$$

and the frequency response of $G^*(j\omega)$ is plotted in Bode form in Fig. SP18.3.2. In this plot, an accurate magnitude curve is plotted, but asymptotic approxi-

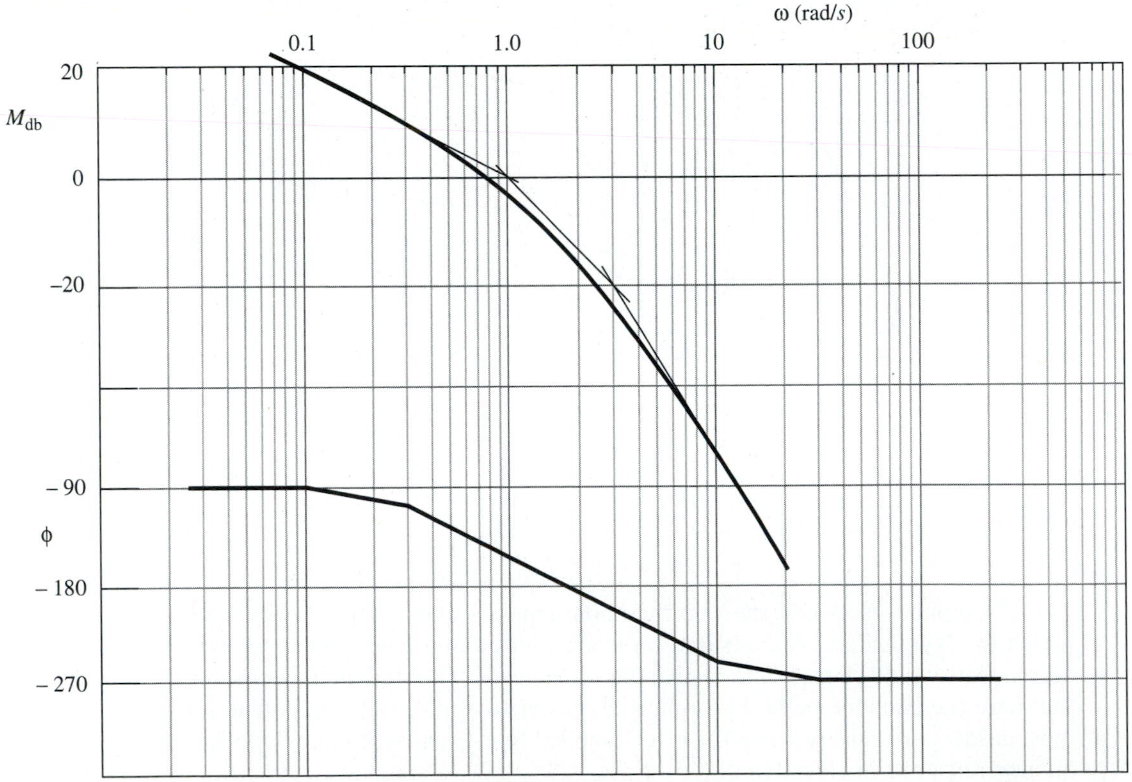

Fig. SP18.3.2

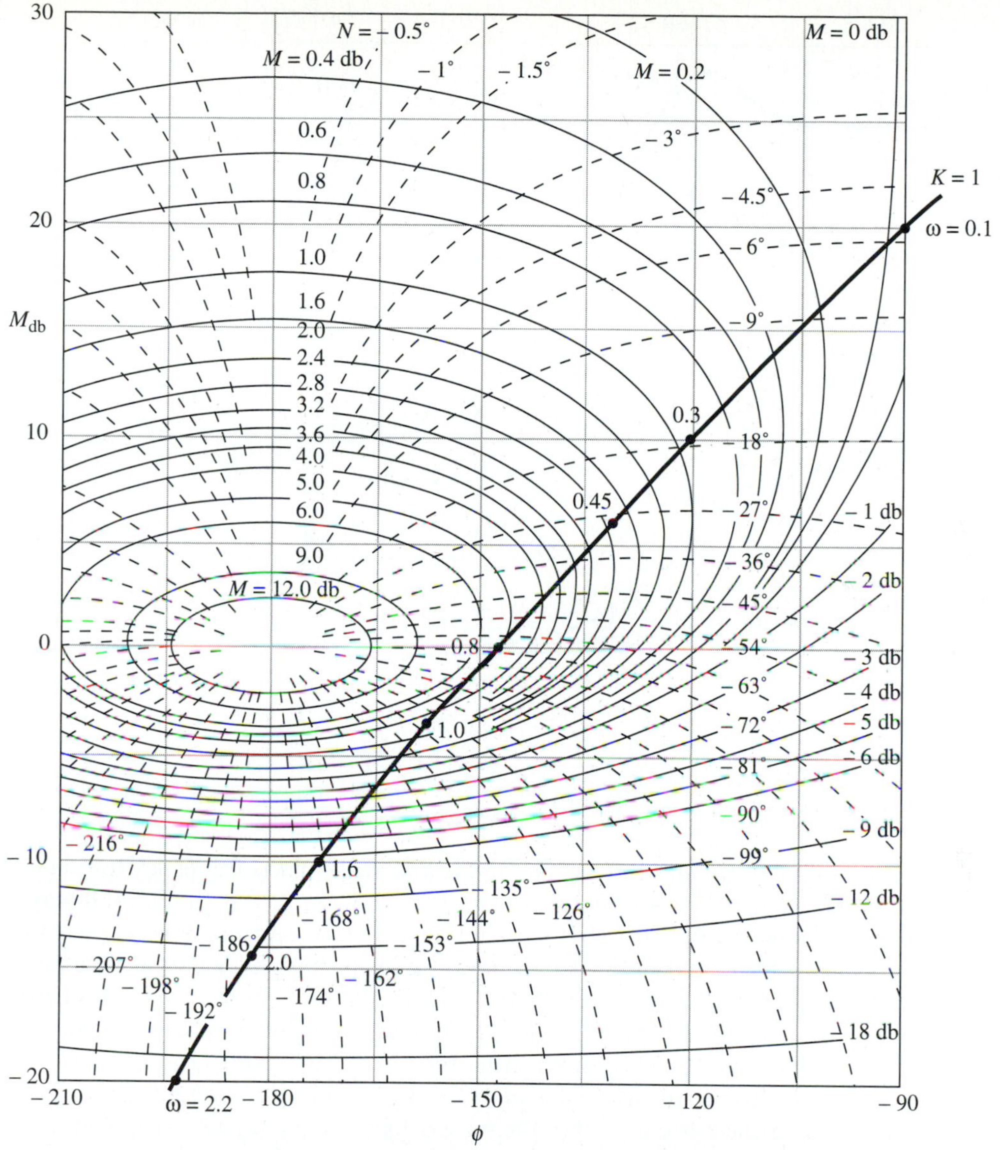

Fig. SP18.3.3

mations to the phase angle are used. This open-loop frequency response is now transferred to the Nichols chart, as shown in Fig. SP18.3.3. Using the constant M_{cl} and α contours, the closed-loop frequency response of the G^* system is plotted in Bode form, as shown in Fig. SP18.3.4. Also plotted in Fig. SP18.3.4 is the Bode diagram of $H(j\omega)$, which is a simple first-order system moved down by $20 \log_{10} 0.25$, or -12 db. Since

$$\frac{C}{R} = \frac{1}{H} \frac{G^*}{1 + G^*}$$

the closed-loop frequency response is obtained by subtracting the magnitude

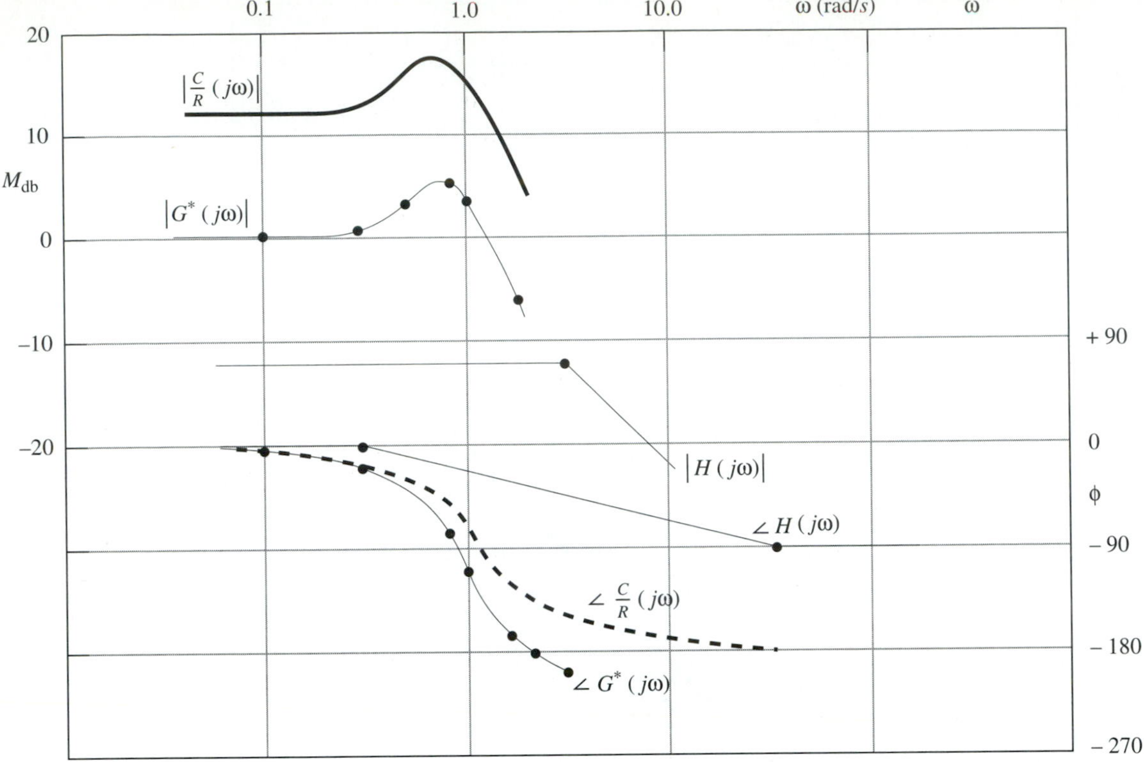

Fig. SP18.3.4

and phase of $H(j\omega)$ from the magnitude and phase of the frequency response of the G^* system. The result of doing this is also shown in Fig. SP18.3.4. It may be seen that below the break frequency of the feedback element ($\omega = 3$), the effect is to *add* 12 db to the G^* system, and since the bandwidth occurs at about $\omega_b =$ 2.1 rad/s, the effect of the high-frequency asymptote of the feedback element may be ignored. The phase angle is plotted only over the frequency range of interest and is obtained by subtracting the phase of $H(j\omega)$ from that of G^*.

It may be seen that the value of M_p of the closed-loop system is about 17 db. To design the system so that the closed-loop frequency response had a prescribed M_p would involve changing the forward-path gain from its current value of 4 to some other number. Suppose K is doubled to 8. The open-loop frequency response would shift up by 6 db, as would the Nichols plot, but the effect on the closed-loop frequency response of G^* cannot be predicted; it can only be plotted from the Nichols chart. Subtracting the Bode plot of $H(j\omega)$ would produce a new closed-loop frequency response with a new M_p. If this is not the value required, K would have to be changed once more and the process repeated. Through trial and error, the required value of M_p would be achieved.

In this case, and in many others, the closed-loop frequency response may be calculated directly from the closed-loop transfer function:

$$\frac{C}{R} = \frac{K/s(1 + s)}{1 + 0.25K/s(1 + s)(1 + s/3)}$$

which reduces to

$$\frac{C}{R} = \frac{K(1 + s/3)}{s^3/3 + 4s^2/3 + s + K/4}$$

Substituting $s = j\omega$ and collecting terms yield

$$\frac{C}{R} = \frac{K + jK\omega/3}{(0.25K - 4\omega^2/3) + j(\omega - \omega^3/3)}$$

which may be written in the form

$$\frac{C}{R} = \frac{a + jb}{c + jd}$$

It may be shown that in this form the magnitude and phase may be written as

$$M_{\text{cl}} = \frac{\sqrt{(ac + bd)^2 + (bc - ad)^2}}{c^2 + d^2}$$

and

$$\alpha = \tan^{-1} \frac{bc - ad}{ac + bd}$$

In our case, we have

$$a = K$$
$$b = K\omega/3$$
$$c = 0.25K - 4\omega^2/3$$
$$d = \omega - \omega^3/3$$

The magnitude and phase equations may be coded in a simple program and used to confirm the result shown in Fig. SP18.3.4 for $K = 4$. Alternatively, MATLAB may be used to plot the Bode diagram of the closed-loop transfer function directly, resulting in Fig. SP18.3.5. The use of such a program would speed up the trial-and-error process of changing K to design the closed-loop system for a specified M_p also.

◆ PROBLEMS

18.1 Figure P18.1 shows the magnitude part of the frequency response of different systems. For each case determine the bandwidth, the size of the resonance peak, and the frequency at which the peak occurs.

18.2 Figure P18.2 shows a model of an automobile suspension system, which may be considered as an open-loop system with input y_w and output y_a. If the suspension system is not to cause the automobile to move vertically when driving at speeds in excess of 40 m/s over bumps 1 m apart, what would a suitable bandwidth of the suspension system be? Select K and c if $m = 1000$ kg, to provide the selected bandwidth.

18.3 Figure P18.3 shows a rope bridge across a canyon. Through experiment the bridge has exhibited a natural frequency of 1 Hz. A man has a natural step length of 1 m. What

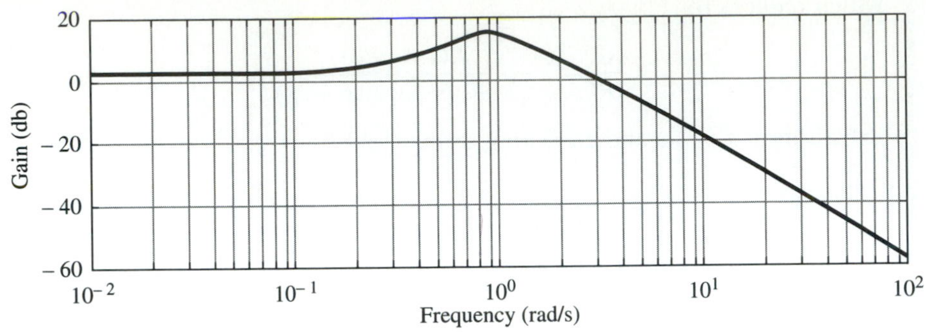

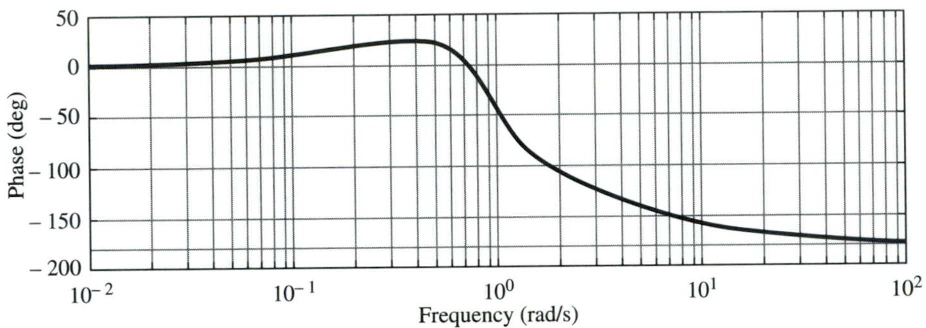

Fig. SP18.3.5

(a)

(b)

Fig. P18.1

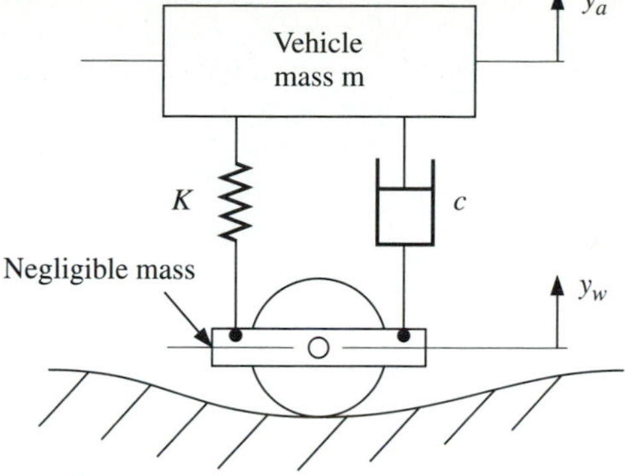

Fig. P18.2

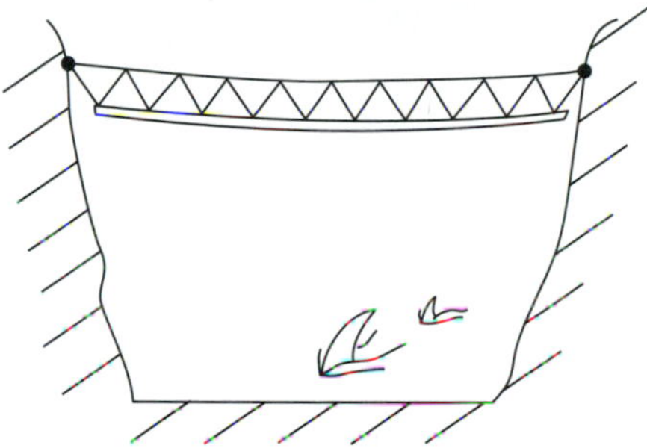

Fig. P18.3

is the most dangerous speed with which to cross the bridge, and why? Would it be better to walk more slowly or to run across the bridge? Give reasons for your answer. (Assume the step length remains invariant with speed.)

18.4 For the following system, use the Nyquist diagram to determine the closed-loop frequency response and measure the bandwidth:

$$G(s) = \frac{2}{s(1 + s)} \qquad H(s) = 1$$

18.5 From the Bode diagram and Nichols chart, plot the closed-loop frequency response of the following system and determine the bandwidth, the resonant frequency, and the size of the resonance peak:

$$G(s) = \frac{12(1 + s)}{s(s^2 + 2s + 10)} \qquad H(s) = 1$$

18.6 Use the Bode diagram to plot the closed-loop frequency response of the unity-feedback gain system

$$G(s) = \frac{8}{(s + 1)^2}$$

18.7 Consider the system shown in Fig. P18.7. Estimate the peak time and percentage overshoot for the closed-loop system if the input is a unit step.

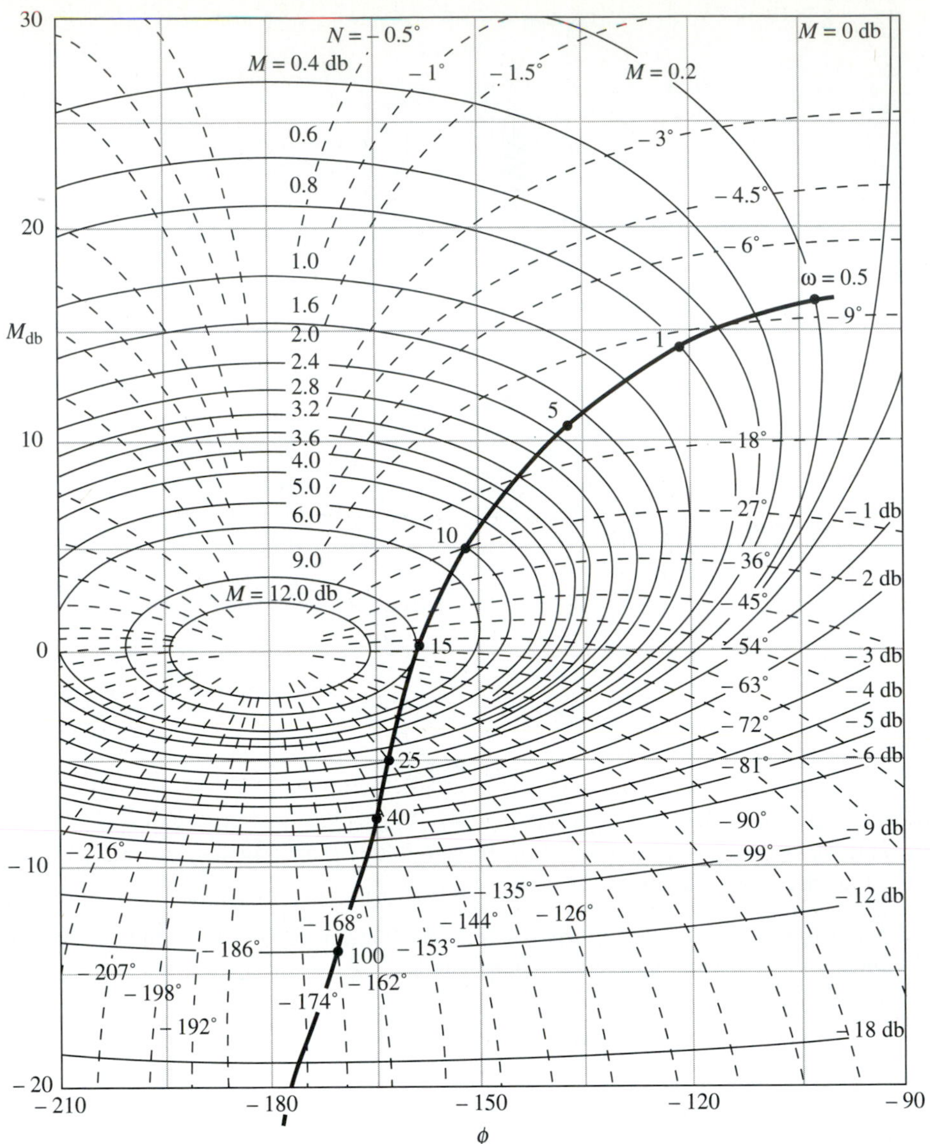

Fig. P18.7

18.8 Use the Nyquist diagram to determine the value of K that gives a closed-loop resonance peak of $M_p = 3$ for the unity-feedback system described by

$$G(s) = \frac{K(1 + s)}{s(1 + s/2)^3}$$

18.9 Consider the system given by the open-loop transfer function elements

$$G(s) = \frac{K(s^2 + 3s + 16)}{s^2(s + 16)^2} \qquad H(s) = 1$$

Use the Bode diagram and Nichols chart to determine K such that the closed-loop frequency response resonance peak satisfies $M_p = 1.6$.

18.10 For the system shown in Fig. P18.10, plot the closed-loop frequency response when $K = 1$.

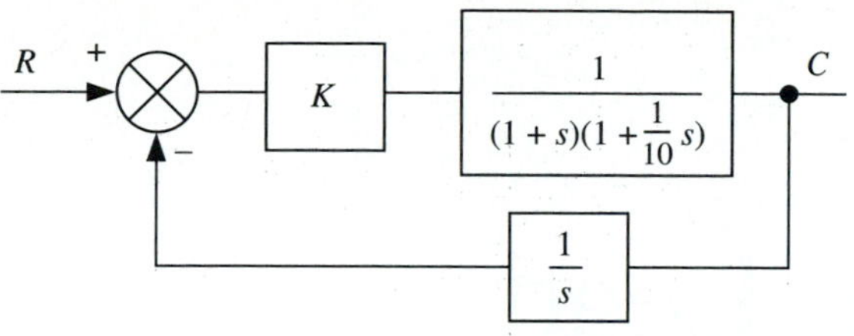

Fig. P18.10

MODULE 19

Phase Lead Compensation

◆ MULTIPLE-DESIGN CONSTRAINTS

In the design of control system performance up to this point, the forward-path gain K has been adjusted in order to meet some required objectives, such as

1. maximum allowable percentage overshoot or
2. maximum permitted steady-state error for a particular form of input.

When the achievement of a single performance requirement may be met by selecting a particular value of K, the process is called *gain compensation*. A question that arises is whether it is possible to meet more than one performance requirement with a single value of gain? To answer this question, consider the system described by the open-loop transfer function

$$GH(s) = \frac{K}{s(1 + 0.2s)} \tag{19.1}$$

Suppose we wish the closed-loop system to meet the following performance requirements:

1. The velocity error for a unit ramp is to be no more than 0.32% of the slope of the input.
2. The maximum percentage overshoot is to be no more than 20%.

For the first requirement, the velocity error constant may be calculated from equation 8.32 as

$$e_{ss}(t) = 0.0032 = \lim_{s\to 0} \frac{1/s}{1 + G(0)} = \frac{1}{K_v} \tag{19.2}$$

leading to

$$K = K_v = 316 \tag{19.3}$$

For the second requirement, we note that there are no closed-loop zeros, therefore it is necessary that the dominant closed-loop poles have a damping ratio obtained from Fig. 4.17 of $\zeta = 0.45$. In terms of frequency response, which we will use to study compensation techniques, the damping ratio requirement translates, approximately, into the phase margin requirement PM $= 45°$. We now plot the Bode diagram to determine whether both of these requirements may be met with a single value of K. Figure 19.1 shows the asymptotic Bode diagram of the system for $K = 316$, which satisfies the steady-state error requirement. It is seen that the gain crossover frequency is $\omega_{gc} = 40$ rad/s, and the phase margin is negligible. It may also be seen that the phase margin will be 45° at $\omega = 5$ rad/s; therefore, to meet the phase margin requirement, the magnitude must be 0 db at this frequency. From this example, this must occur when $K = 6.3$,[1] and the magnitude part of the Bode diagram is also shown in Fig. 19.1 for this value of gain.

Obviously, it is not possible to satisfy both system performance requirements with a single value of gain, and something new has to be done to the system in order to make it perform as required. The system needs to be modified in some way; i.e.,

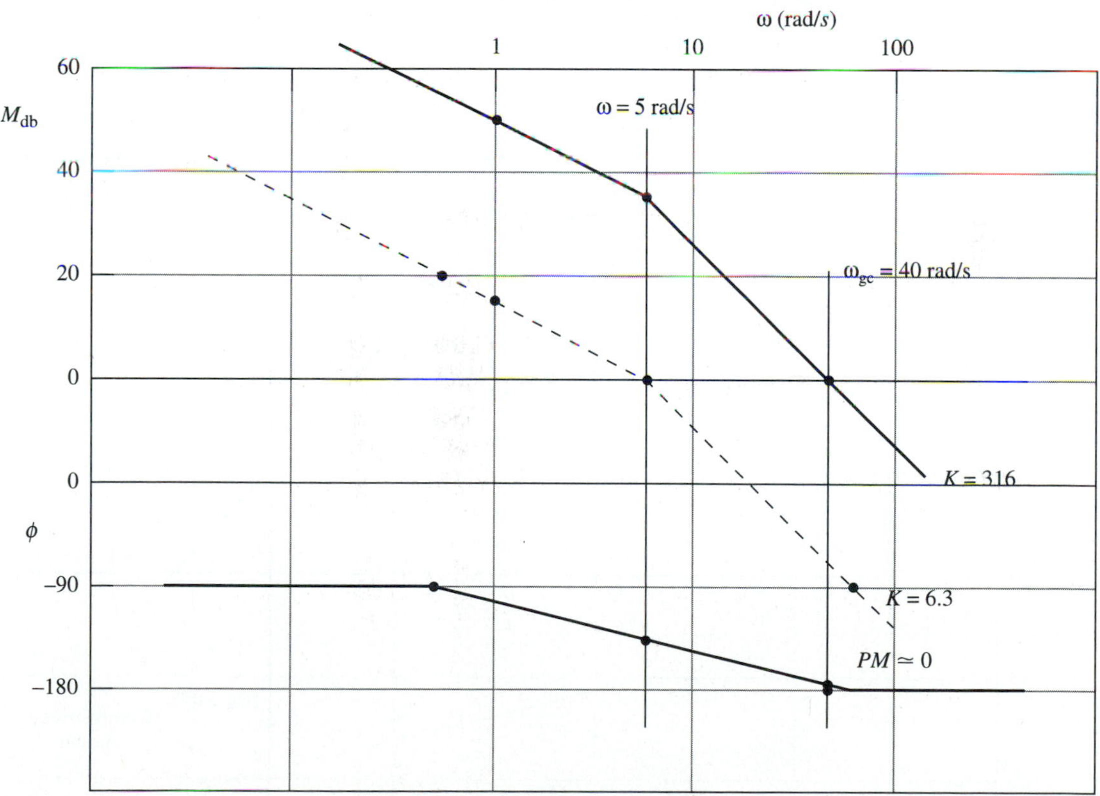

Fig. 19.1 Bode diagram for system.

[1]Since the magnitude at $\omega = 1$ is 16 db.

the shape of the Bode diagram has to be altered to allow it to achieve both performance requirements. This alteration of the Bode diagram is known as *system compensation*.

The system performance requirements stated in the example are typical of those found in many design cases; a steady-state error determines one value of gain while a desired transient response determines another. Note how each requirement relates to a different region of the frequency axis in the Bode diagram:

1. The steady-state error relates to the magnitude at low frequency, i.e., as $s(= j\omega) \to 0$ in equation 19.2.

2. The transient response requirement (damping ratio) relates to the gain cross-over frequency, which usually occurs at higher frequencies.

In order to meet both requirements, we could do one of two things:

1. Keep $K = 316$ so as to satisfy the steady-state error requirement but modify the phase angle in the region of $\omega = 40$ rad/s in order to meet the phase margin requirement. The suggested change is shown in Fig. 19.2.

2. Keep the phase angle plot the same, keep $K = 316$, but introduce another element in the forward path that attenuates higher frequencies in such a way that the modified magnitude plot passes through 0 db at $\omega = 5$ rad/s, as shown in Fig. 19.3.

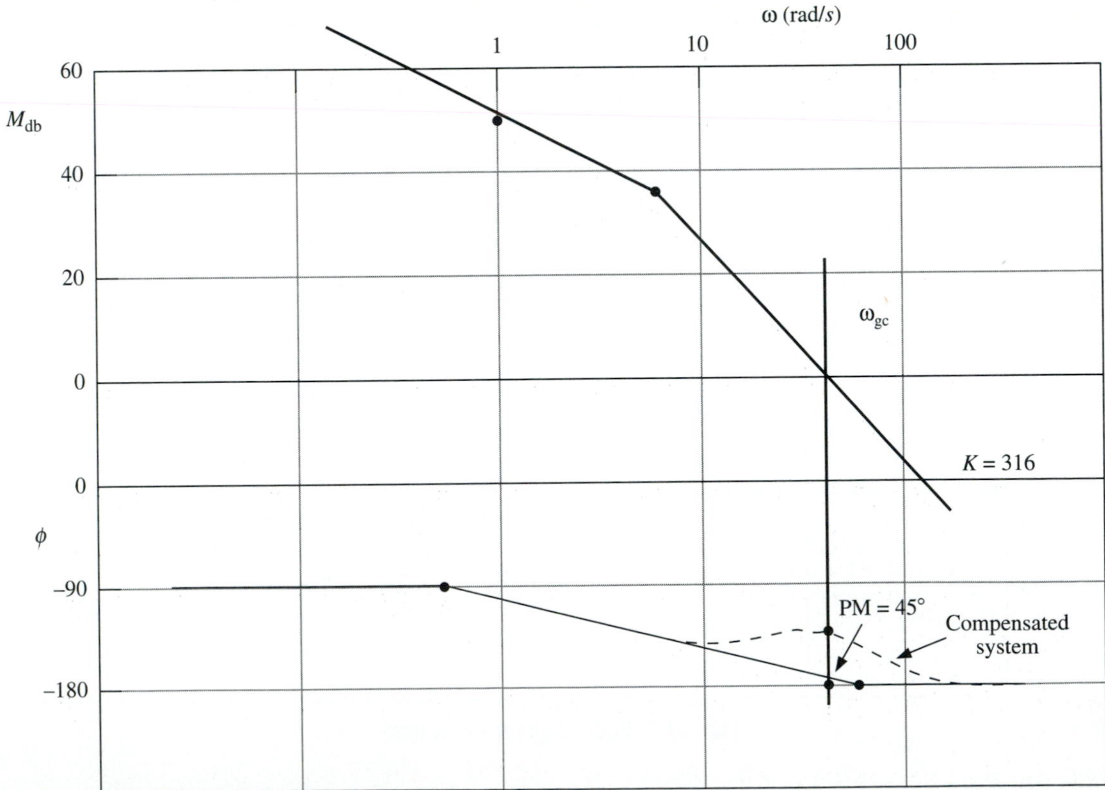

Fig. 19.2 Bode diagram with modified phase.

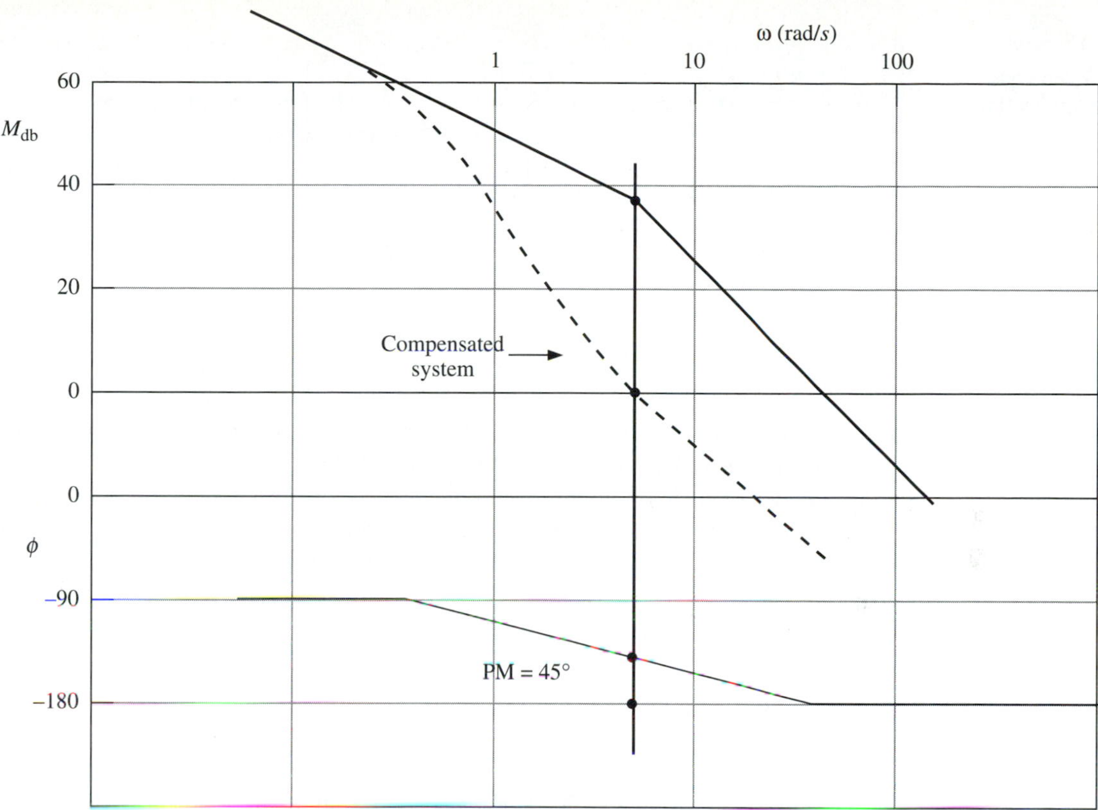

Fig. 19.3 Bode diagram with modified magnitude.

In the first case, the technique is known as phase advance, or phase lead compensation, since a positive phase angle is added to the uncompensated system. The second technique is commonly called phase lag compensation, mainly to contrast it to phase lead methods. In fact, the second technique uses magnitude attenuation to achieve the desired objective, but it is rarely described in these terms. Phase lead compensation will be dealt with in this module, while phase lag will be discussed in the next module, together with lead-lag compensation, which is a combination of both. In each case, the technique involves adding an additional compensation element $G_c(s)$ to the forward path of the system being compensated, as shown in Fig. 19.4. This approach may also be referred to as *cascade compensation,* since another element is included in series with the forward-path transfer function. Because the frequency response of the open-loop transfer function is altered, the compensation element that achieves the required performance is sometimes called a *filter.*

◆ TRANSFER FUNCTION OF PHASE LEAD ELEMENT

Although it is convenient to talk about altering the Bode diagram in some desirable manner, it is important to base compensation on the availability of physical systems that can accomplish the modification; otherwise the practical implemen-

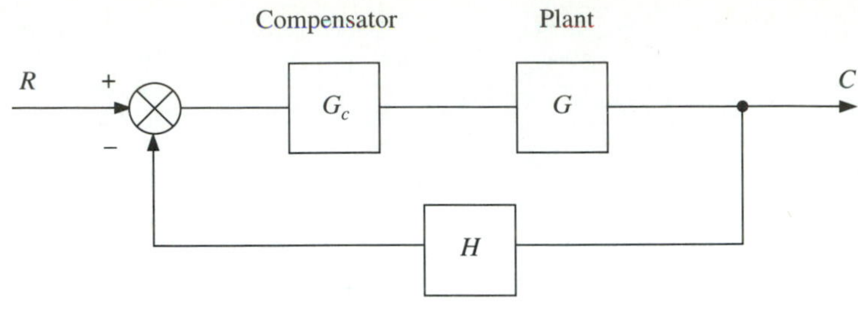

Fig. 19.4 Basic compensation technique.

tation of the technique will be very limited. One example of a compensation element that provides phase lead is shown in Fig. 19.5. In this case, the element is a simple passive electrical network, but phase lead elements may be constructed in other domains too. It may be shown that, for such a system, the transfer function takes the form

$$G_c(s) = \frac{1 + \alpha Ts}{\alpha(1 + Ts)} \tag{19.4}$$

where the constants T and α are given by

$$T = \frac{R_1 R_2 C}{R_1 + R_2} \qquad \alpha = \frac{R_1 + R_2}{R_2} \tag{19.5}$$

For a phase lead element we define $\alpha > 1$, which obviously places constraints on the values of resistance and capacitance used in the network.

The task of *designing* the compensation network involves the specification of T and α by means of a process to be described in a moment. First, the Bode diagram of the compensator will be plotted. We see that the transfer function comprises a zero at

$$\omega_z = \frac{1}{\alpha T} \tag{19.6}$$

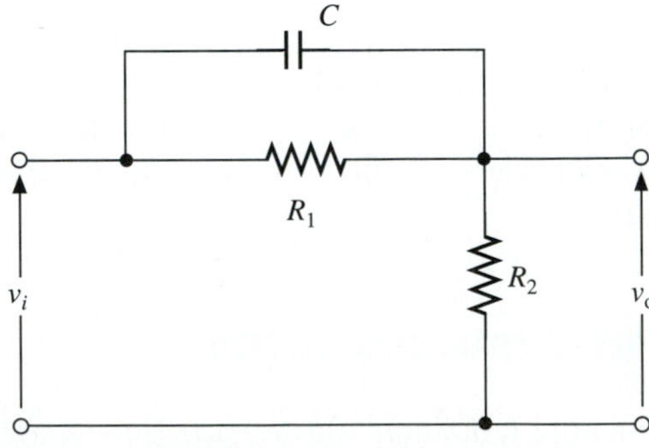

Fig. 19.5 Phase lead network.

and a pole at

$$\omega_p = \frac{1}{T} \tag{19.7}$$

It follows that

$$\omega_p = \alpha\omega_z \tag{19.8}$$

Since $\alpha > 1$, the zero is at a lower frequency than the pole, and the Bode diagram will be as shown in Fig. 19.6. From Fig. 19.6, the following points may be noted:

1. The element clearly introduces phase lead (positive phase angle), and it may be seen that the design approach will be to place the maximum phase lead in the region of the uncompensated gain crossover frequency in Fig. 19.1.
2. The amount of phase lead obtained from the element depends upon the separation of the numerator and denominator time constants, i.e., on α. If the time

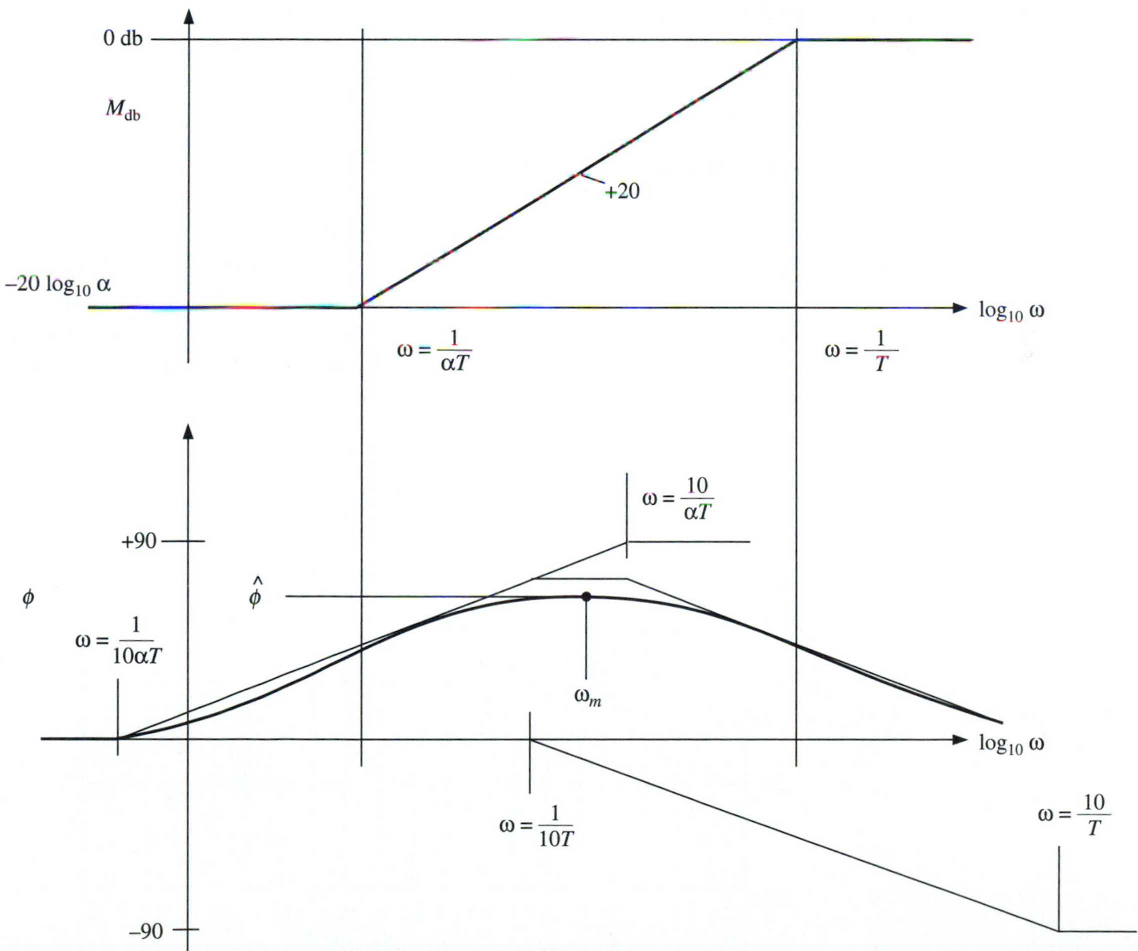

Fig. 19.6 Bode diagram of phase lead element.

constants are wide apart, the numerator is allowed to contribute virtually all of the available 90° before the denominator phase lag decreases the net phase down to the high-frequency value of zero. If, on the other hand, the time constants are close together, their effects cancel each other out, and very little phase lead is obtained.

The amount of phase lead provided may, therefore, be determined solely by α, and it may be shown that the maximum phase lead is related to α by

$$\sin \hat{\phi} = \frac{\alpha - 1}{\alpha + 1} \tag{19.9}$$

and that the frequency at which this maximum phase lead occurs is

$$\omega_m = \sqrt{\omega_z \omega_p} = \frac{1}{T\sqrt{\alpha}} \tag{19.10}$$

which is the geometric mean of the two break frequencies and may be identified as the halfway point between the break frequencies on a $\log_{10} \omega$ scale. Equations 19.9 and 19.10 are the basic design equations of the phase lead element used for the compensation process. For rapid estimation of α for a required phase lead, Fig. 19.7 may be used.

◆ PHASE LEAD COMPENSATION PROCESS

Before using the phase lead compensator, it is usual to redraw its Bode diagram so that the low-frequency asymptote has zero gain, rather than the $-20 \log_{10} \alpha$ shown in Fig. 19.6. This produces the modified Bode diagram shown in Fig. 19.8 and represents multiplying the transfer function by α. Since this corresponds to multiplying the compensator transfer function by α, the modified transfer function will now be of the form

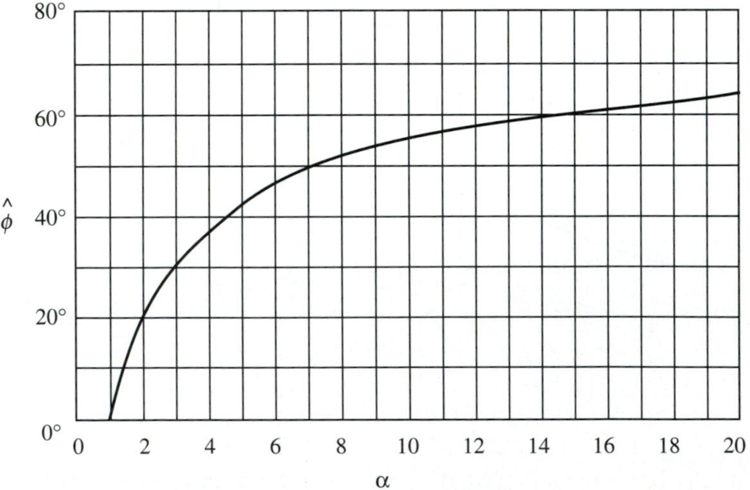

Fig. 19.7 Phase lead as a function of α.

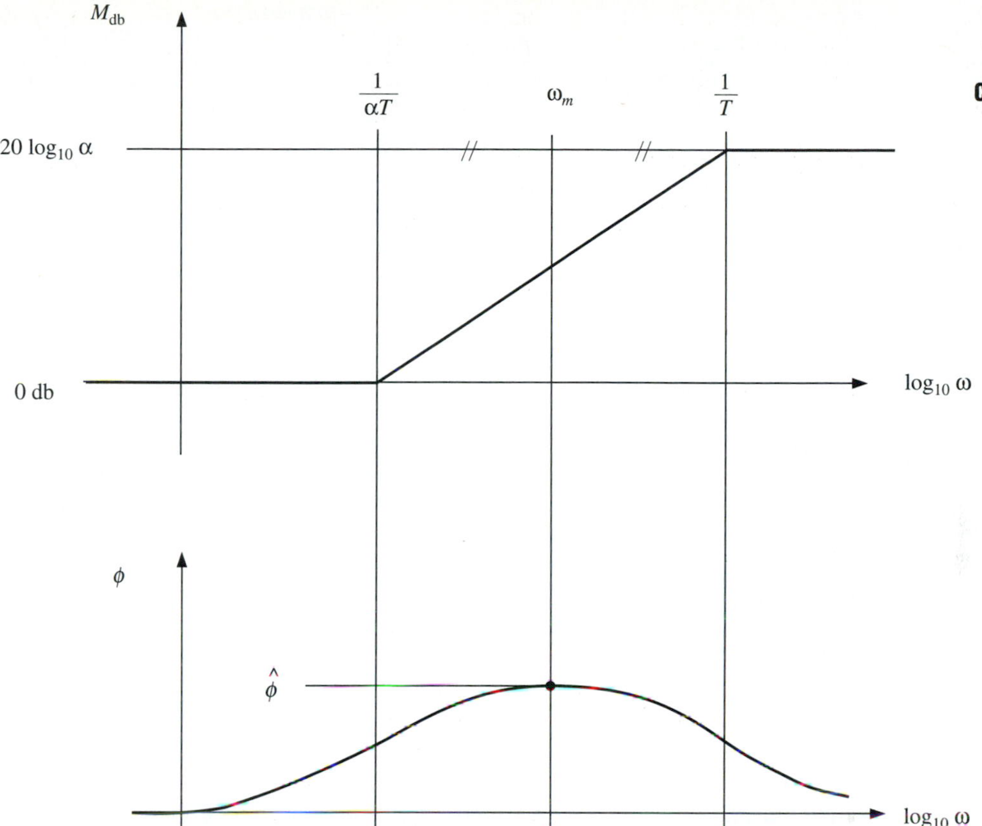

Fig. 19.8 Modified Bode diagram of phase lead compensator.

$$G_c(s) = \frac{1 + \alpha TS}{1 + Ts} \tag{19.11}$$

We will remember this operation after the compensator has been designed. To illustrate the use of the phase lead compensator, we will attempt to satisfy the two design requirements for the system described in the first part of this module. Recall the system was

$$GH(s) = \frac{K}{s(1 + 0.2s)} \tag{19.12}$$

and the two performance criteria could be met separately by values of $K = 6.3$ and $K = 316$. The first step is to plot the uncompensated magnitude such that the steady-state error requirement is met. This corresponds to $K = 316$, and the resulting plot is shown in Fig. 19.9. The next step is to estimate the additional phase lead required from the compensation element. The first approach might be to recognize that at the gain crossover frequency the actual phase margin is approximately zero. The required phase margin is 45°. The compensation element, therefore, has to supply an additional 45° of phase margin, which, from Fig. 19.7, indicates $\alpha = 6$. Our compensation network will be designed to produce this

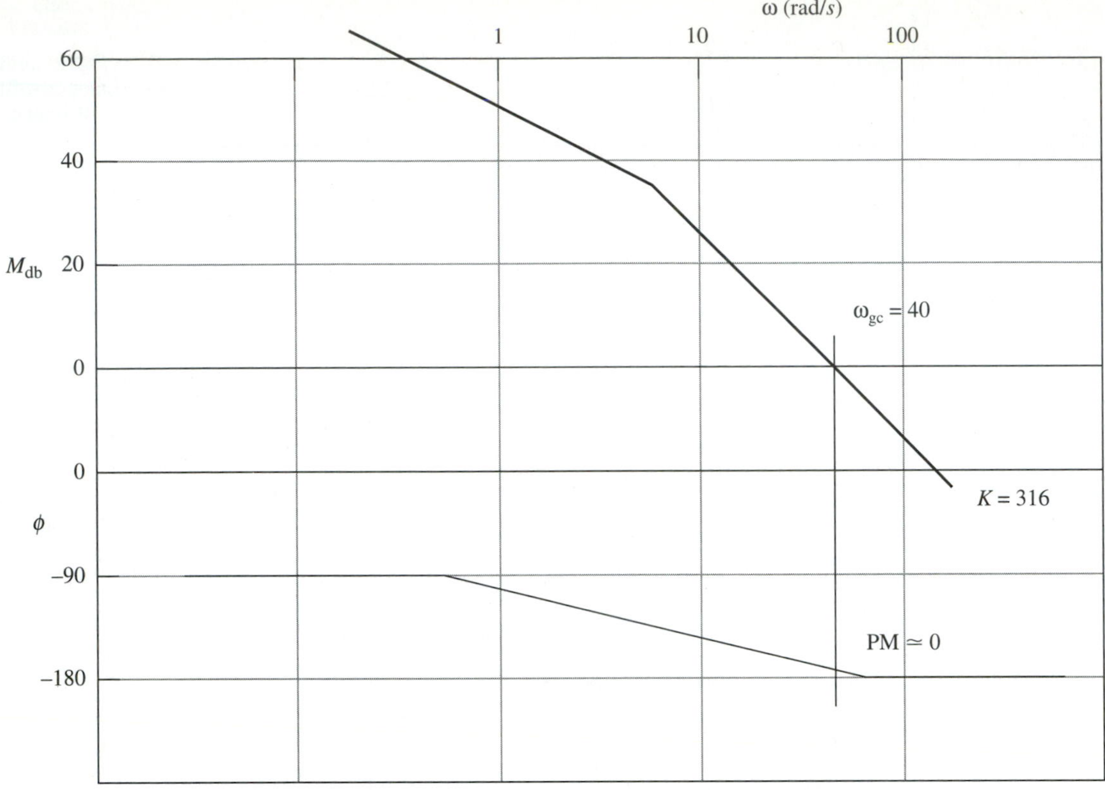

Fig. 19.9 Bode diagram satisfying steady-state error requirement.

amount of phase lead, and it will be placed at the uncompensated gain crossover frequency, as shown in Fig. 19.2. This appears to complete the first part of the compensator design, and it only remains to determine T. We can represent the phase part of our compensator as shown in Fig. 19.10.

Adding the phase of the uncompensated system to that of the compensation network, we obtain the total phase, which provides a phase margin of 45° at the uncompensated gain crossover frequency, as expected. However, we also have to add the magnitude part of the compensator to the uncompensated magnitude, and when this is done, as shown in Fig. 19.11 (assuming we known ω_z and ω_p), the gain crossover frequency has moved from the uncompensated value $\omega_{gc} = 40$ to a higher value $\omega'_{gc} = 150$ rad/s, with a corresponding decrease in phase margin to 10°. This means we need to place the maximum phase lead from the compensator further to the right, which in turn moves the new gain crossover frequency to the right, and so on. Can we avoid repeating the process indefinitely? First, note that, from Fig. 19.8, the magnitude has a value of half the total magnitude at the frequency ω_m, i.e.,

$$\Delta M = 10 \log_{10} \alpha \tag{19.13}$$

which, for our value of α, corresponds to $\Delta M = 7.8$ db. It is important to realize that since the phase margin has been fixed at the required value of 45° at ω_m, *this must also be the new (compensated) gain crossover frequency.*

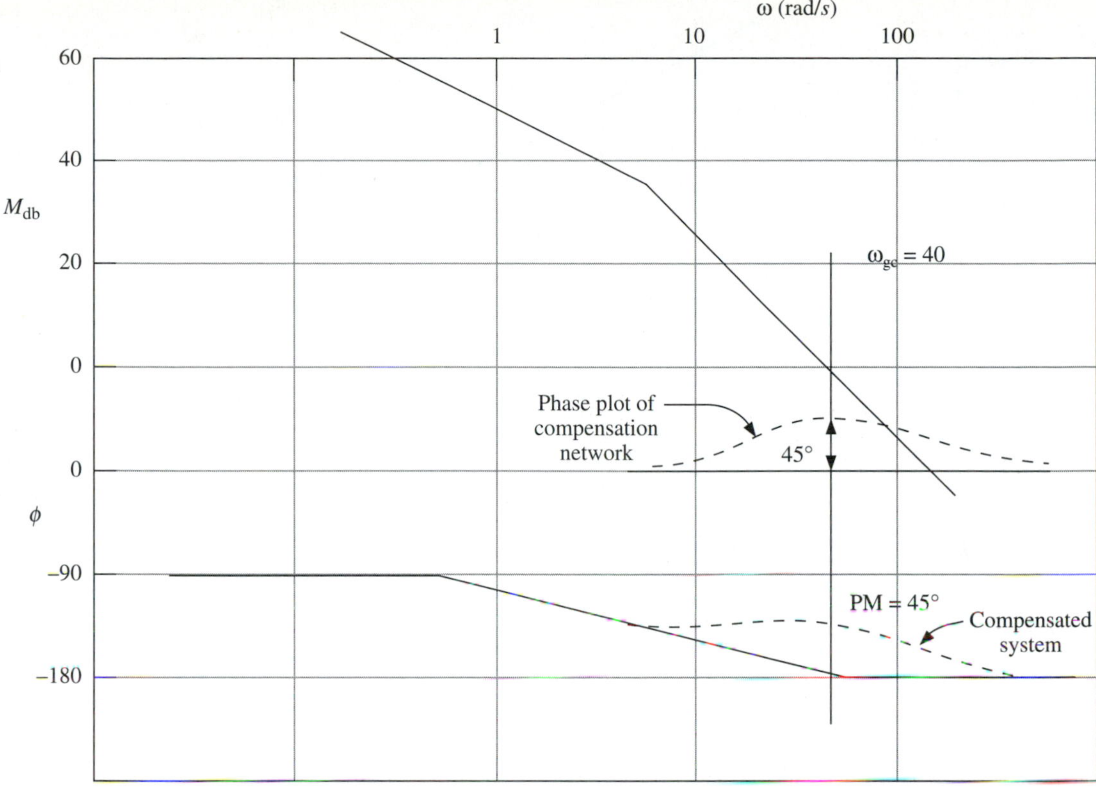

Fig. 19.10 Bode diagram with compensator phase angle added.

We do *not*, therefore, place the maximum phase at the old gain crossover frequency, as first attempted above, but at the frequency where the uncompensated magnitude has a value of $-10 \log_{10} \alpha$, thereby anticipating the subsequent increase in magnitude due to the compensation network. The determination of the compensated gain crossover frequency, which has the value 60 rad/s, is shown in Fig. 19.12. Two points remain:

1. The estimate of the required additional phase lead was obtained from the uncompensated gain crossover frequency. If this is no longer the new gain crossover frequency, does not this change the estimate for the additional phase lead? The answer is that it does, and although it is impossible to determine the exact amount required, a good estimate may be obtained from the following argument. Since the new gain crossover frequency will be higher than the old one, and since most systems have a falling phase-frequency relationship, the required phase advance will be greater than that estimated from the uncompensated system. Therefore add 10% as an estimate of the increase. In our case, the value of additional phase lead expected of the compensation network should be increased from 45° to 50°, leading to a new value of $\alpha = 7.5$. The new gain crossover frequency and ω_m will now be when the uncompensated magnitude plot has a value

$$\Delta M = -10 \log_{10} 7.5 = -8.75 \text{ db} \tag{19.14}$$

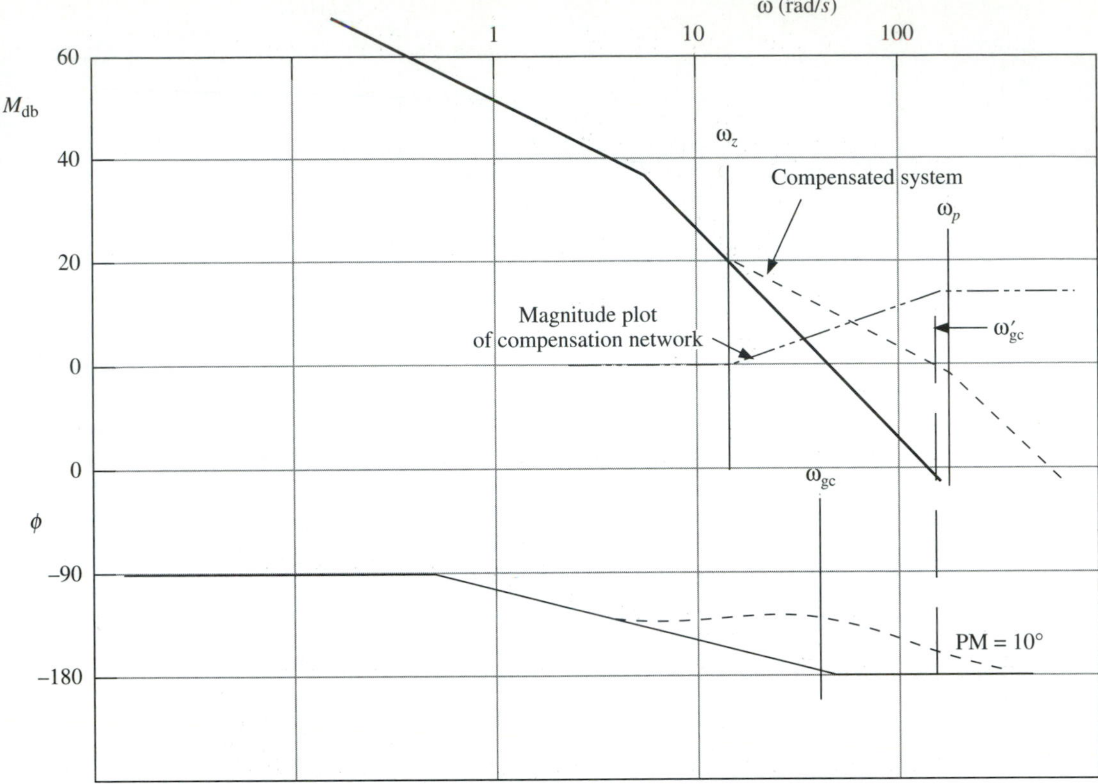

Fig. 19.11 Composite magnitude and phase of compensated system.

From Fig. 19.12, the accurate gain crossover frequency for the compensated system becomes 66 rad/s.

2. Having determined α and ω_m, how is T determined? This is obtained directly from one of the compensator design equations (19.10), which may be written in the form

$$T = \frac{1}{\omega_m \sqrt{\alpha}} \tag{19.15}$$

allowing T to be calculated directly as

$$T = 0.0055 \tag{19.16}$$

This completes the design of the compensation element, which now may be written in the form

$$G_c(s) = \frac{1 + \alpha Ts}{1 + Ts} = \frac{1 + 0.041s}{1 + 0.0055s} \tag{19.17}$$

From the above equation, the two break frequencies for the compensator network may be calculated as

$$\omega_z = \frac{1}{\alpha T} = 24 \text{ rad/s} \tag{19.18}$$

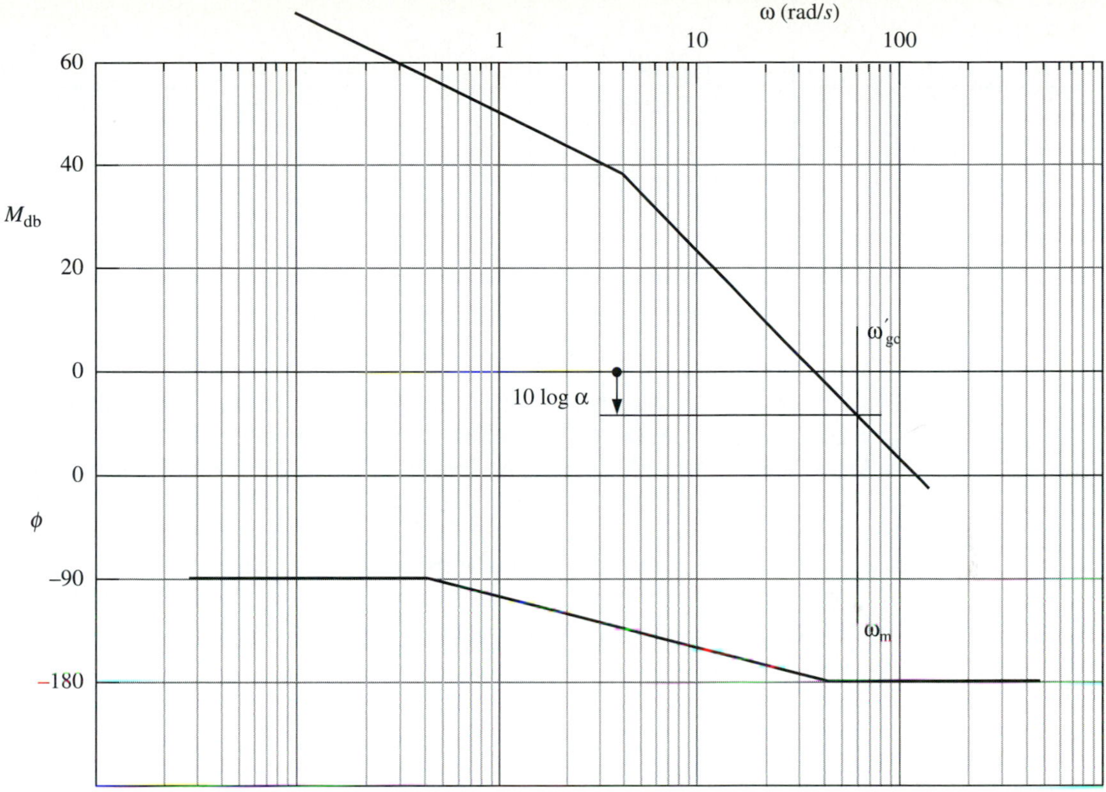

Fig. 19.12 Determination of compensated-system gain crossover frequency.

and

$$\omega_p = \frac{1}{T} = 182 \text{ rad/s} \tag{19.19}$$

Confirm from Fig. 19.12 that $\omega_m = 66$ rad/s is midway between these two frequencies on a logarithm scale.

One last point needs to be taken care of. All of the analysis so far has been based on the assumed transfer function for the compensator of

$$G_c(s) = \frac{1 + \alpha Ts}{1 + Ts} \tag{19.20}$$

A real compensator must have a transfer function

$$G_c(s) = \frac{1 + \alpha Ts}{\alpha(1 + Ts)} \tag{19.21}$$

To obtain the performance required, the original forward-path gain K must be multiplied by α in order to make the actual compensator transfer function equal to the assumed transfer function. This may be illustrated in Fig. 19.13, which shows the final form of the compensation network and the uncompensated forward-path transfer function. Note how K has been increased to $\alpha \times K$. Since a certain

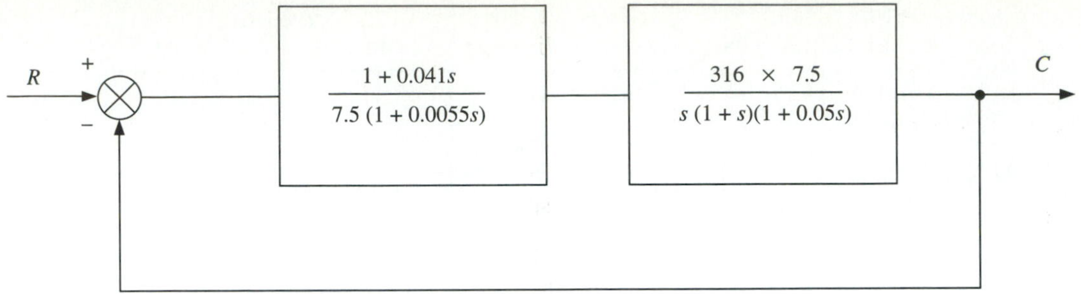

Fig. 19.13 Compensation network in forward path.

amount of estimation was used in determining the required amount of additional phase lead needed from the compensator, the final, compensated Bode diagram should be drawn and the phase margin checked. If the value is not sufficiently close to the desired value, the process may be repeated to obtain better accuracy. Given other assumptions made during the process, further iterations are occasionally required. The final Bode diagram of the compensated system is shown in Fig. 19.14.

The process just described may seem somewhat tedious, but it can be summarized in a few easy-to-follow rules that may be used to design the compensation

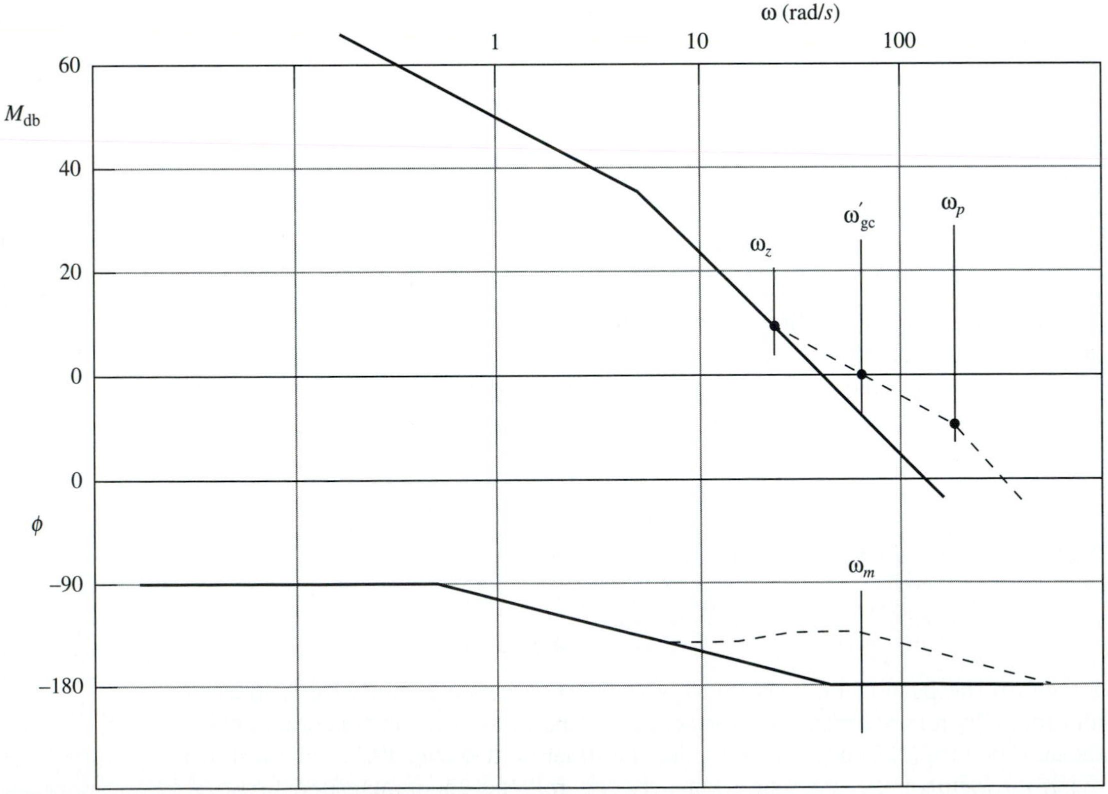

Fig. 19.14 Bode diagram of compensated system.

network. It is assumed that it is necessary to satisfy a steady-state error condition (low frequency) and a transient response condition (mid frequency):

1. Plot the uncompensated system for the gain value that satisfies the steady-state error constraint.
2. Determine the uncompensated phase margin, and estimate the phase margin required in order to satisfy the transient response performance constraint.
3. Calculate the extra phase advance required of the compensation network, add 10%, and determine the corresponding value of α.
4. Find the compensated gain crossover frequency, which equals the frequency of the peak phase lead contribution ω_m, by determining the frequency where the uncompensated magnitude satisfies

$$\Delta M = -10 \log_{10} \alpha \tag{19.22}$$

5. Calculate T from the equation

$$T = \frac{1}{\omega_m \sqrt{\alpha}} \tag{19.23}$$

6. Calculate ω_z and ω_p from

$$\omega_z = \frac{1}{\alpha T} \qquad \omega_p = \frac{1}{T} \tag{19.24}$$

7. Check that the compensated Bode diagram has the desired phase margin, and repeat the previous steps if necessary.
8. Determine the transfer function of the phase lead compensation network from

$$G_c(s) = \frac{1 + \alpha Ts}{\alpha(1 + Ts)} \tag{19.25}$$

9. Increase the forward-path gain K by a factor of α to ensure that the steady-state error constraint it still met.

For the example system just considered, the unit step response of the uncompensated and compensated systems are shown in Fig. 19.15. Phase lead compensation has some distinct advantages over other forms of compensation. It may also be difficult to use. These points are discussed briefly in the next section.

◆ COMMENTS ON THE APPLICABILITY AND RESULTS OF PHASE LEAD COMPENSATION

Some observations from the example just analyzed allow a few generalizations to be made regarding phase lead compensation:

1. The open-loop (and usually the closed-loop) bandwidths are increased. This is usually beneficial since the inclusion of higher frequencies in the response results in a faster response. It may cause problems, however, if noise exists at the higher, unattenuated frequencies.

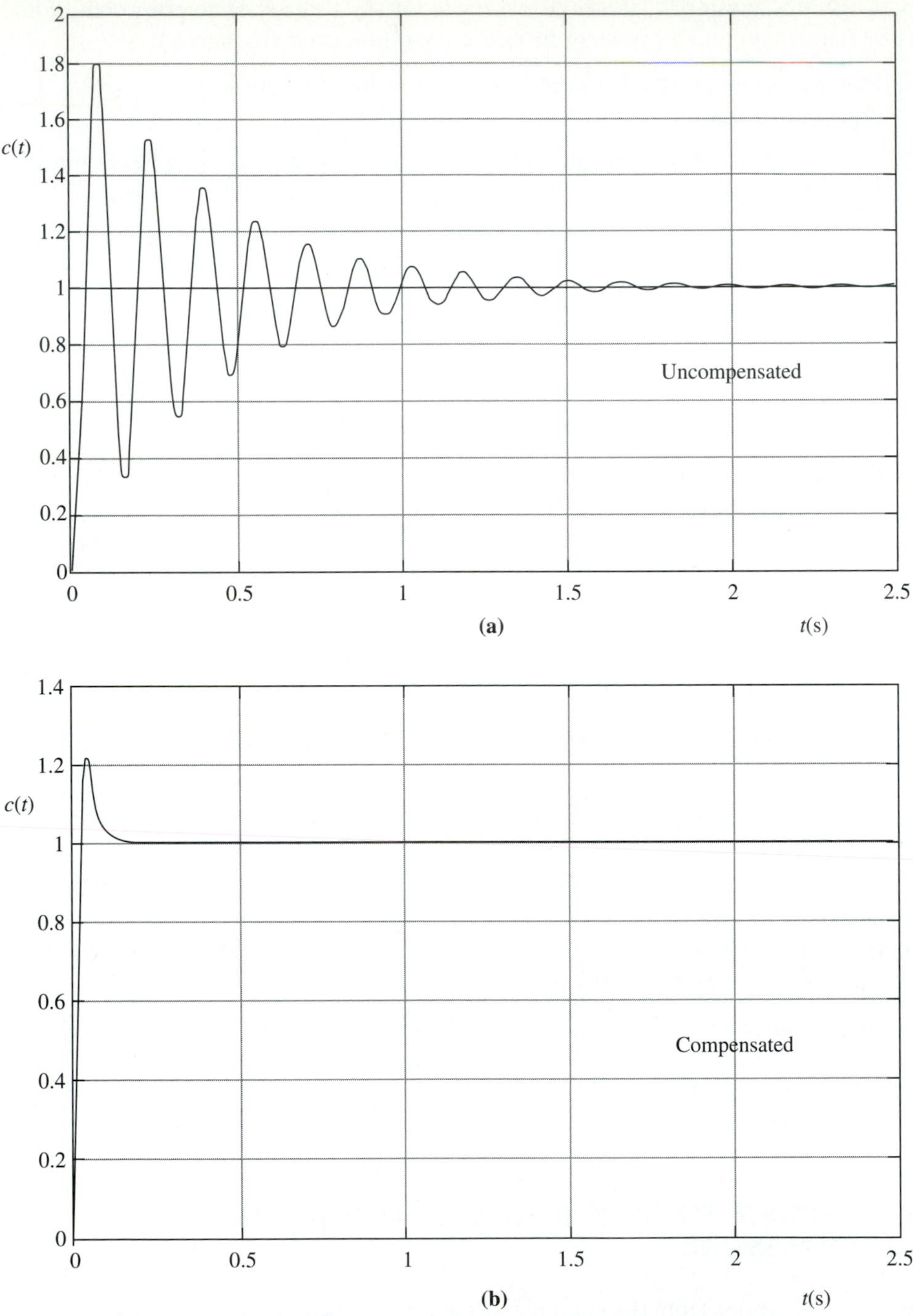

Fig. 19.15 Step response of uncompensated and compensated systems.

2. Usually the closed-loop resonance peak is reduced, as is the case in this example, confirmed by Fig. 19.16. This is also beneficial since the closed-loop system will be less excited by signals with significant components at, or near, the resonant frequency.

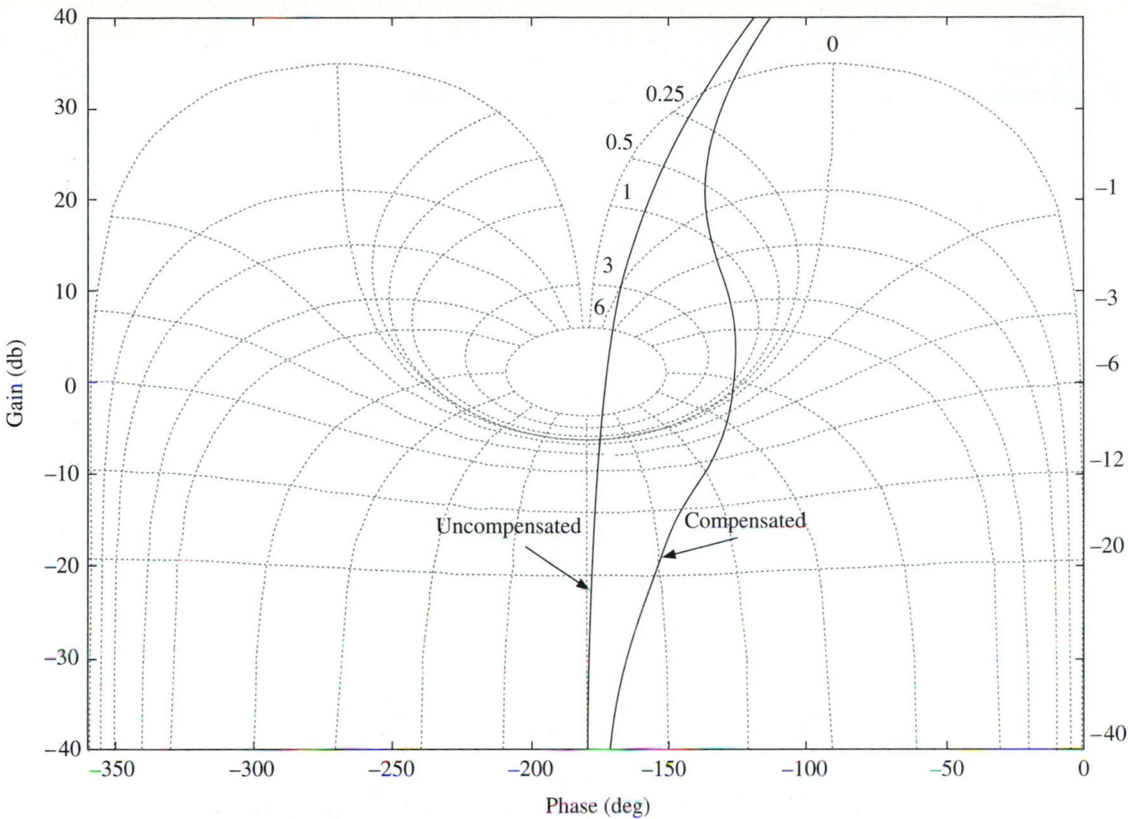

Fig. 19.16 Nichols chart of compensated and uncompensated systems.

3. Problems may occur when the phase plot has a steep slope in the vicinity of ω_m. This occurs because, as the new gain crossover point moves to the right, larger and larger phase lead is required from the compensator, demanding very large values of α. This is difficult to achieve when the compensator is realized with physical components. For this reason, values of $\alpha > 15$ should be avoided, and methods to compensate the system using other techniques, such as phase lag, should be investigated. One alternative approach is to add an additional 20% phase lag if the slope of the magnitude curve is -40 db/decade or steeper in the vicinity of ω_m.
4. Although the design of the compensation filter took place in the frequency domain, the change in system characteristics may also be observed from the complex plane using the root locus. Figures 19.17 and 19.18 show the uncompensated and compensated root loci, respectively. The closed-loop poles of the uncompensated system are located at

$$s = -2.5 \pm 40j \tag{19.26}$$

as shown in the figure, and the low damping ratio is apparent. If the charactersitic equation of the compensated system is solved, the roots become

$$s = -62 \pm 27j \qquad s = -63 \tag{19.27}$$

These roots have a much better damping ratio and a strong real-axis pole contribution, as shown in Fig. 19.18.

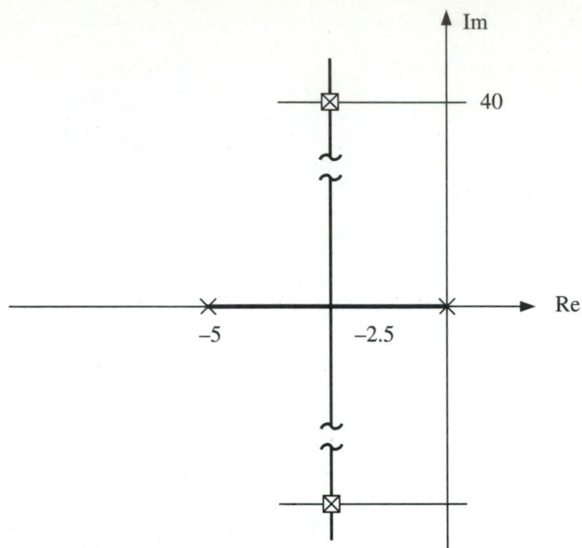

Fig. 19.17 Root locus of uncompensated system.

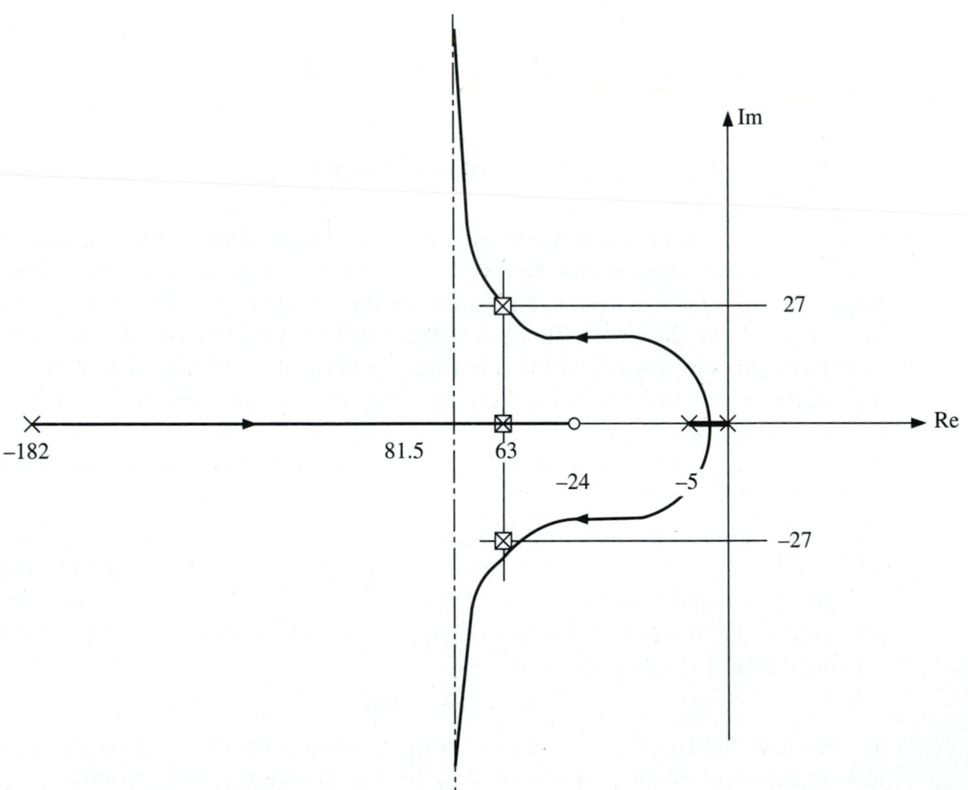

Fig. 19.18 Root locus of compensated system.

Sample Problem 19.1

Consider the system with the following open-loop transfer function:

$$GH(s) = \frac{K}{(1 + 10s)(1 + 25s)}$$

Design a phase lead compensator that allows the system to meet the following performance criteria:

1. steady-state error to a step input to be less than 5% of the step size and
2. less than a 15% overshoot to the same input.

Compare the unit step responses for the uncompensated and compensated systems.

Solution

This problem will be approached using the step-by-step procedure outlined in the module. First we calculate the gain K necessary to meet the steady-state performance requirement. This is seen to be

$$e_{ss}(t) = 0.05 = \frac{1}{K_p}$$

since the system is of type 0. This leads to

$$K = K_p = 20 = 26 \text{ db}$$

The remaining steps are as follows:

1. *Plot the uncompensated Bode diagram:* This plot is a straightfoward one to obtain, recognizing that the break points are at $\omega = 0.04$ rad/s and $\omega = 0.1$ rad/s. Straight-line approximations are used for both the magnitude and phase parts of the diagram, which is shown in Fig. SP19.1.1.

2. *Phase margins:* The uncompensated system has a gain crossover frequency of 2.8 rad/s and a corresponding phase margin of 31°. If the overshoot is to be limited to 15%, a damping ratio of 0.5 is required (Fig. 4.17).

3. *Calculate the phase advance required:* The required phase to be obtained from the compensation filter is

$$(50 - 31) + 10\% = 21°$$

From equation 19.9 we obtain

$$\sin 21 = \frac{\alpha - 1}{\alpha + 1}$$

yielding $\alpha = 2.5$.

4. *New gain crossover frequency:* This is obtained from the equation

$$\Delta M = -10 \log_{10} \alpha = -3.9 \text{ db}$$

Figure SP19.1.2 shows that the frequency where the uncompensated magnitude is -3.9 db is

$$\omega_m = \omega'_{gc} = 0.35 \text{ rad/s}$$

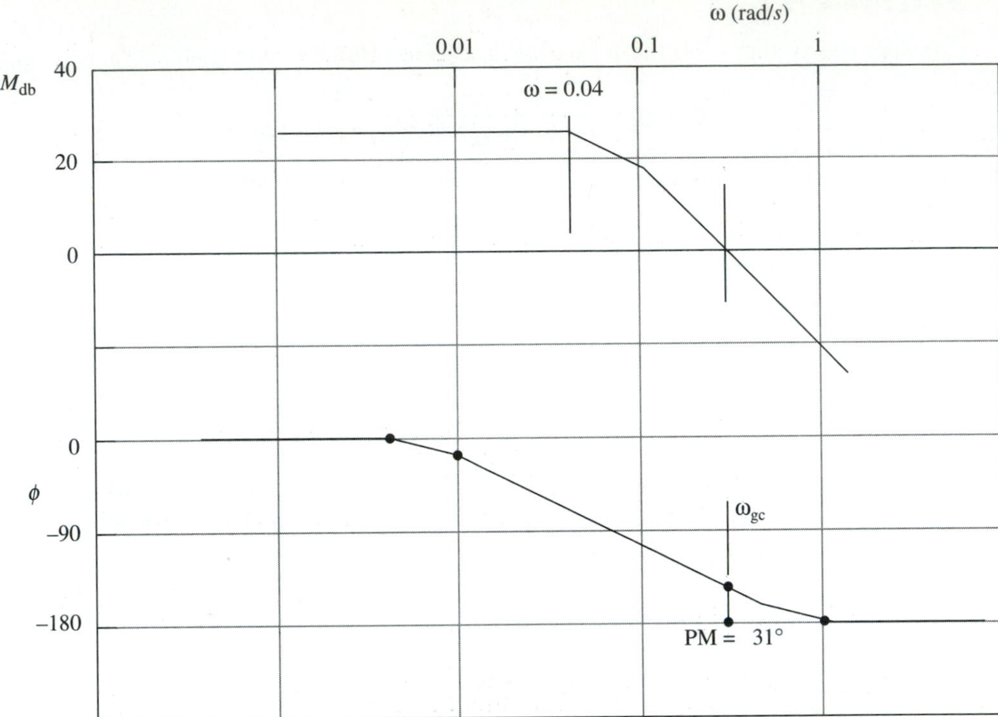

Fig. SP19.1.1

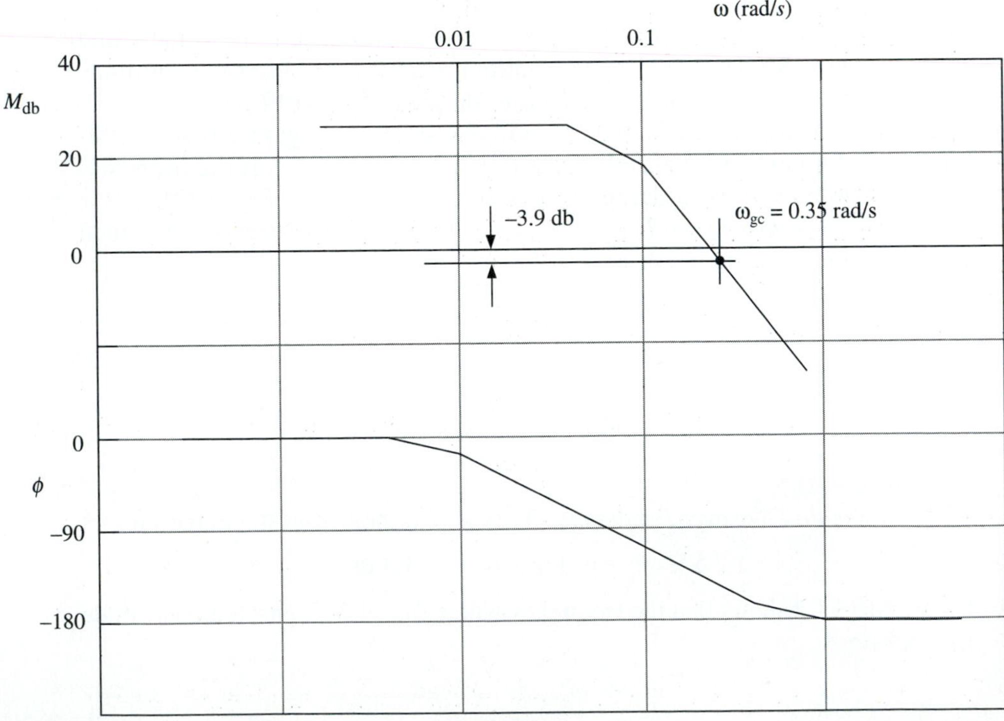

Fig. SP19.1.2

5. *Calculate T:* This is obtained from

$$T = \frac{1}{\omega_m \sqrt{\alpha}} = 1.8$$

giving $\alpha T = 4.5$. The compensation filter will have a zero break frequency of

$$\omega_z = \frac{1}{\alpha T} = 0.22 \text{ rad/s}$$

while the break frequency of the pole is

$$\omega_p = \frac{1}{T} = 0.56 \text{ rad/s}$$

6. *Draw compensated Bode diagram:* This is shown in Fig. SP19.1.3, where the straight-line approximations to the filter have been added to those of the uncompensated system. The compensated-magnitude part of the Bode diagram is drawn as follows. Mark the new gain crossover frequency ω_m as well as the zero and pole break frequencies ω_z, ω_p. The compensated magnitude departs from the original magnitude at ω_z, or point *A* in Fig. SP19.1.3. The compensated magnitude should have a slope of -20 db/decade less than that of the uncompensated system, passing through 0 db at ω_m, point *B*. When frequency ω_p is reached, the slope reverts to that of the uncompensated system. Note that if any uncompensated-system break points were encountered, the appropriate change in slope would have to be made. The phase is drawn in a similar

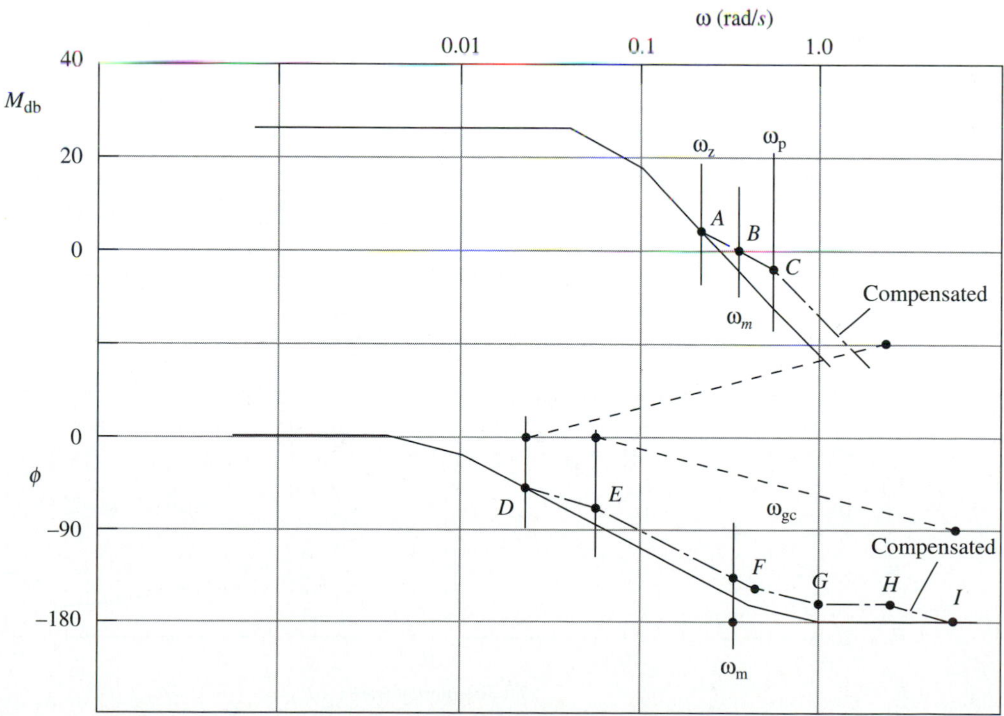

Fig. SP19.1.3

manner, although it is best to draw the two components of the filter as shown in Fig. SP19.1.3. Now work from low frequency to high frequency, making changes in the uncompensated slope of 45°/decade as break points are encountered. From Fig. SP19.1.3, the phase margin for the compensated system may be estimated as 43°, a little smaller than expected.

7. *Compensator transfer function:* Using the above data, the transfer function of the compensation filter may be written as

$$G_c(s) = \frac{1 + \alpha Ts}{1 + Ts} = \frac{1 + 4.5s}{1 + 1.8s}$$

where the attenuation due to the α in the denominator has been canceled by an assumed increase in K.

8. *System transfer function:* The final form of the compensated-system transfer function may be written as

$$G_cGH(s) = \frac{20(1 + 4.5s)}{(1 + 10s)(1 + 25s)(1 + 1.8s)}$$

Figure SP19.1.4 shows an accurate Bode diagram of the compensated system and suggests a phase margin in the region of 50° as required. The discrepancy between this value and the previous estimate is due to the use of straight-line approximations for the magnitude and phase parts of the Bode diagram. Finally, MATLAB has been used to generate the step response of the uncompensated and compensated systems, and these are shown in Fig. SP19.1.5. The

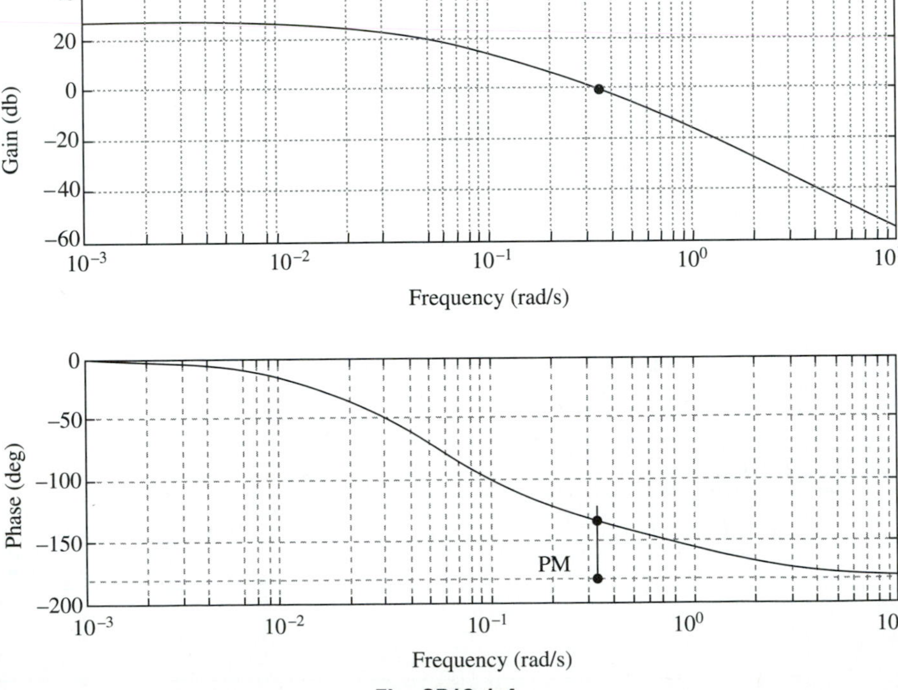

Fig. SP19.1.4

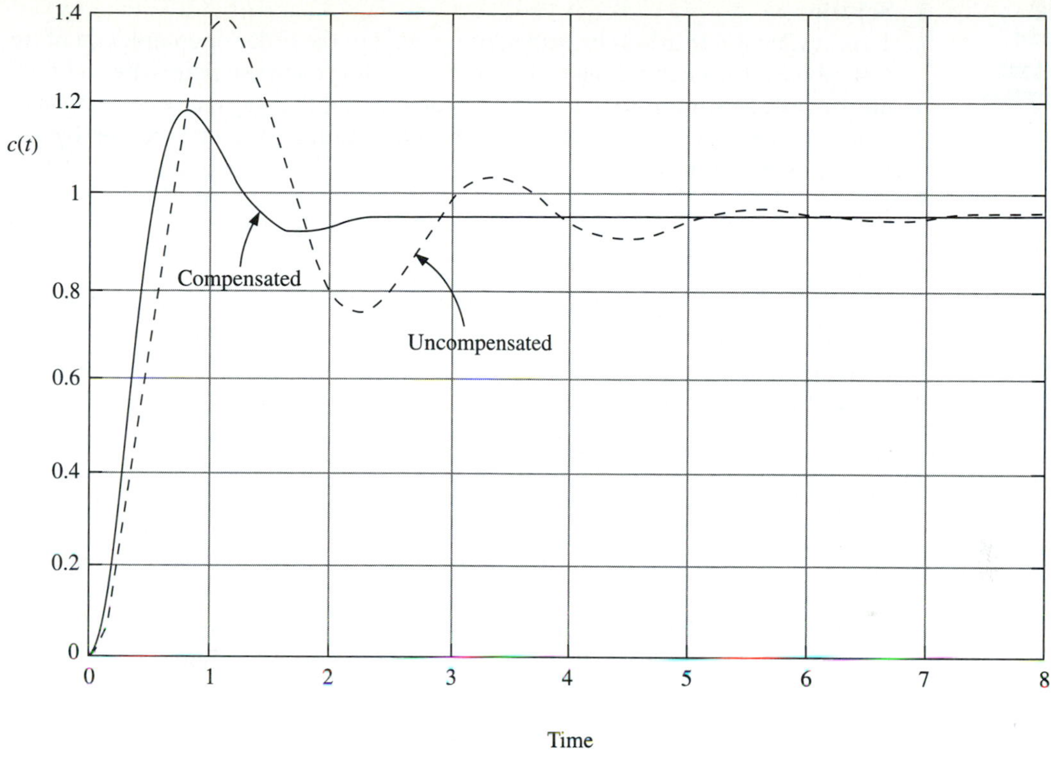

Fig. SP19.1.5

improvement in transient response is apparent, and the slight decrease in response time is due to the increased open-loop bandwidth of the system.

Sample Problem 19.2

Consider the system shown in Fig. SP19.2.1. Design a phase lead cascade compensation filter that will provide the corresponding closed-loop system with the following characteristics.

1. steady-state error due to a unit ramp input of less than 25% and
2. a closed-loop damping ratio of about 0.7.

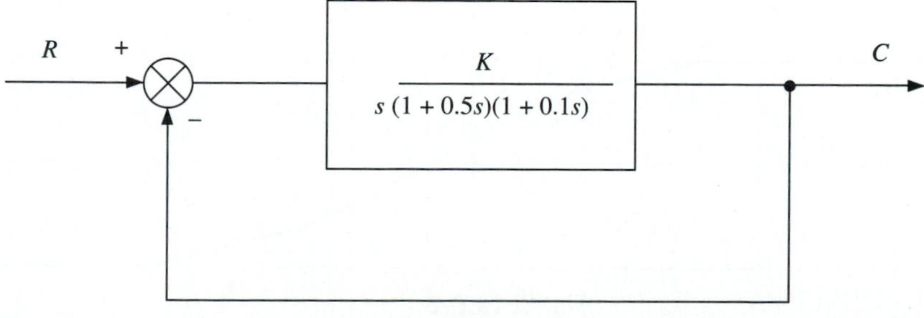

Fig. SP19.2.1

Solution

This system is a third-order system of type 1, but the order or complexity of the system has little bearing on the design of a suitable compensation filter. The approach will be similar to the previous problem, but the steps may be performed a little faster now. The steady-state performance measure may be met by setting the gain to the value

$$K = K_v = \frac{1}{e_{ss}} = \frac{1}{0.25} = 4$$

For this value of K, the uncompensated-system Bode diagram is as shown in Fig. SP19.2.2. From Fig. SP19.2.2 the existing phase margin is found to be 18°. Since we require a compensated phase margin around 70°, the compensator has to supply

$$\phi = (70 - 18) + 10\% = 57°$$

Substituting into

$$\sin \phi = \frac{\alpha - 1}{\alpha + 1}$$

yields $\alpha = 11.5$. The new gain crossover frequency is found from

$$\Delta M = -10 \log_{10} \alpha = -10.6 \text{ db}$$

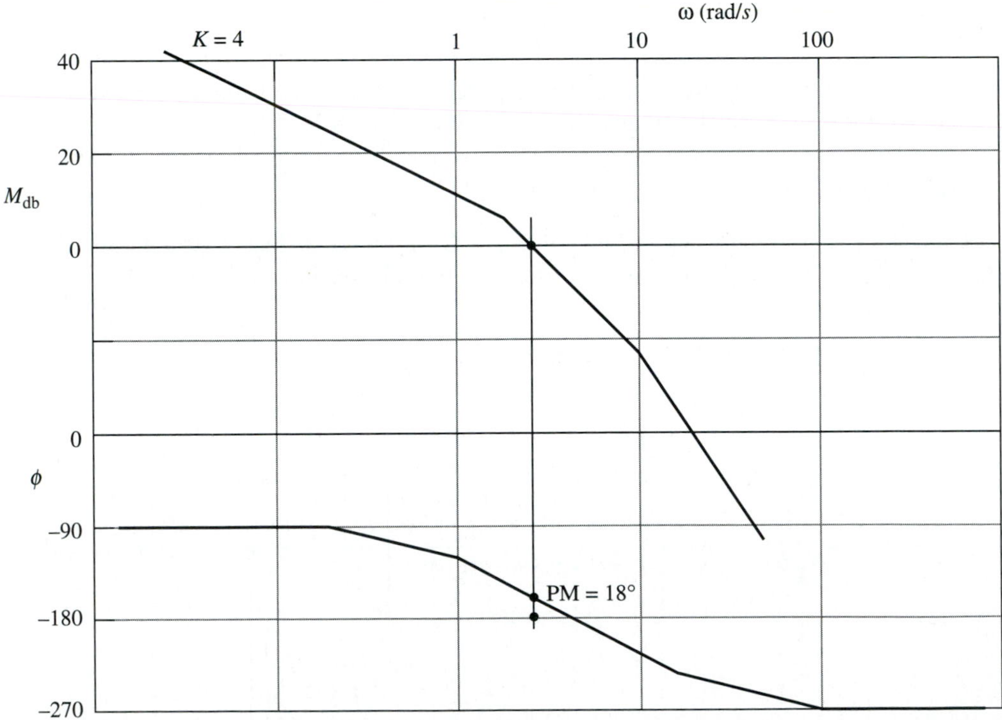

Fig. SP19.2.2

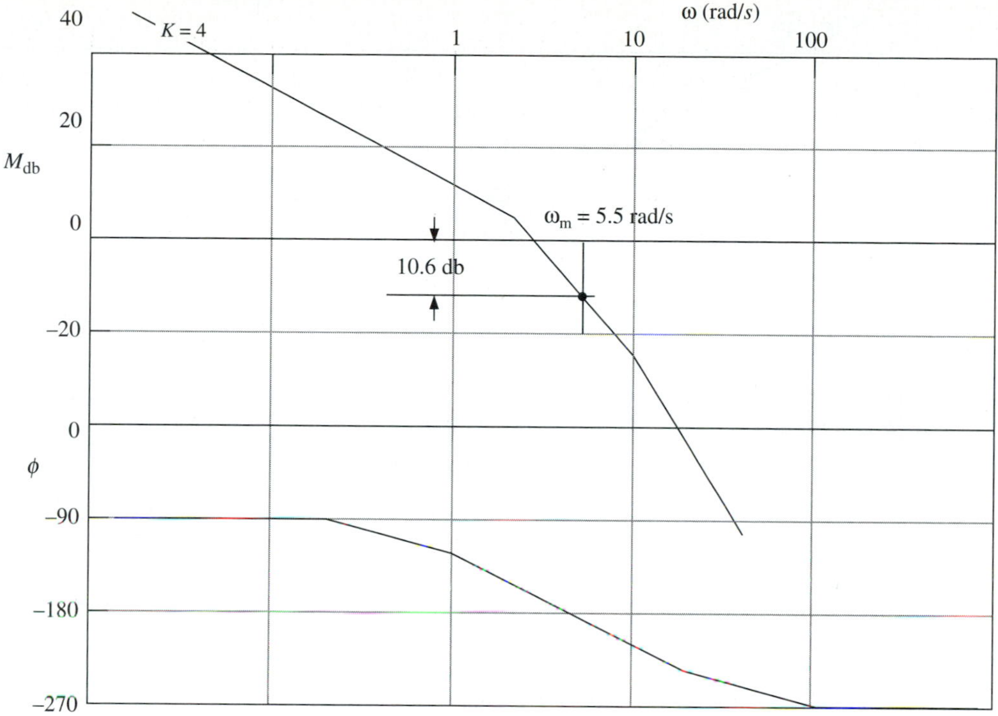

Fig. SP19.2.3

which, from Fig. SP19.2.3, is $\omega_m = 5.5$ rad/s. Next, the value of T is calculated as

$$T = \frac{1}{\omega\sqrt{\alpha}} = 0.054$$

producing $\alpha T = 0.617$. The compensation network (neglecting the attenuation) may be written as

$$G_c(s) = \frac{1 + 0.617s}{1 + 0.054s}$$

which has zero and pole break frequencies of $\omega_z = 1.62$ rad/s and $\omega_p = 18.5$ rad/s, respectively. Figure SP19.2.4 shows the magnitude part of the compensated Bode diagram. The diagram is plotted somewhat differently than before, due to the system break points located between ω_z and ω_p. The high-frequency gain of the filter is known to be

$$\Delta M = 20 \log_{10} \alpha = 21 \text{ db}$$

The high-frequency part of the magnitude is, therefore, parallel to the uncompensated magnitude, but 21 db above it. This time, moving from the high-frequency asymptote toward the low-frequency end, the break frequency at the pole of the compensator is reached first (point A) when the slope changes

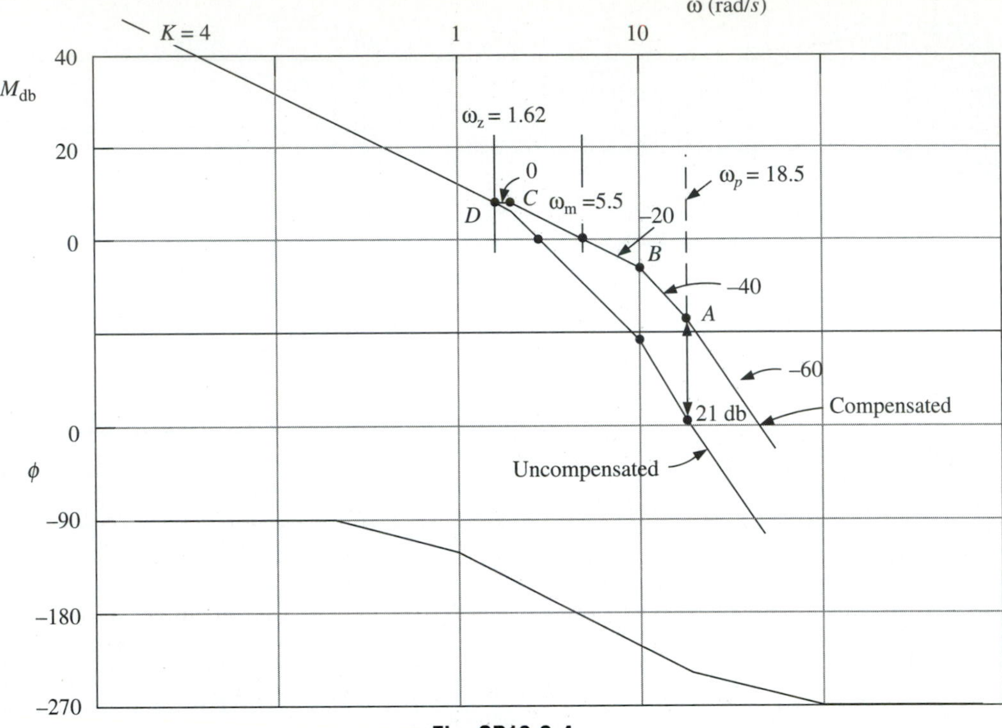

Fig. SP19.2.4

from -60 to -40 db/decade. Moving along this line, the system break point at $\omega = 10$ rad/s (point B) is reached when the slope changes to -20 db/decade. Moving along this line the next system break point is reaced at $\omega = 2$ rad/s (point C), where the slope of the magnitude plot becomes horizontal. This line should intersect the uncompensated magnitude at the break frequency of the zero of the compensation network, $\omega = 1.62$ rad/s (point D). The phase part of the Bode diagram is more cumbersome to plot, and so this will not be done here, except to note an expected increase in phase of 57° in the vicinity of $\omega_m = 5.5$ rad/s. Figure SP19.2.5 is a computer-generated Bode diagram and indicates that a phase margin of about 65° may be expected. For completeness, Fig. SP19.2.6 shows the unit step response of the system indicating that the closed-loop system has about 10% overshoot. This is a little higher than would be expected. However, it is sufficiently close to be acceptable.

Sample Problem 19.3

A unity-feedback control system has a forward-path transfer function given by

$$G(s) = \frac{K}{s^2(1 + 0.1s)}$$

Design a phase lead compensation filter that provides the closed-loop system step response with no more than a 25% overshoot.

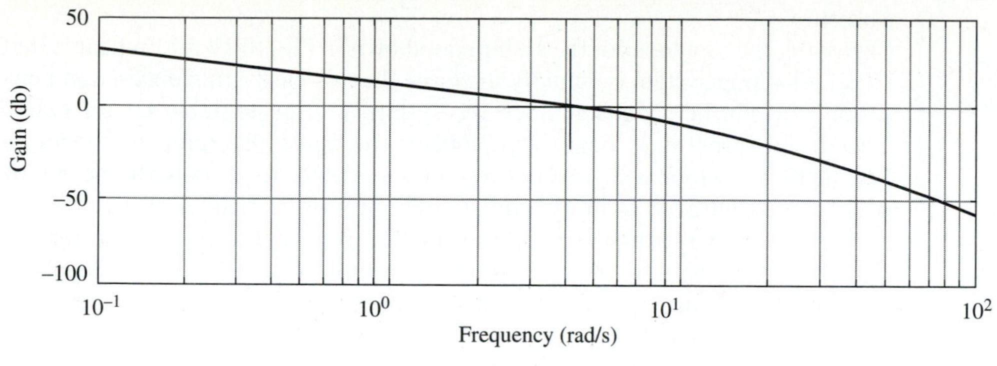

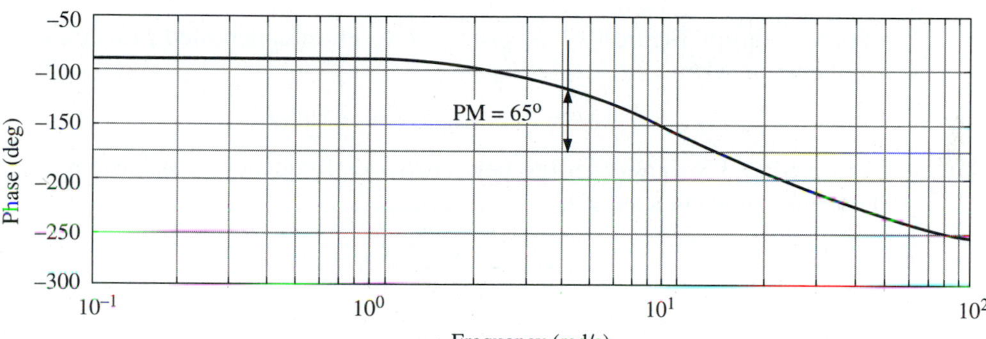

Fig. SP19.2.5

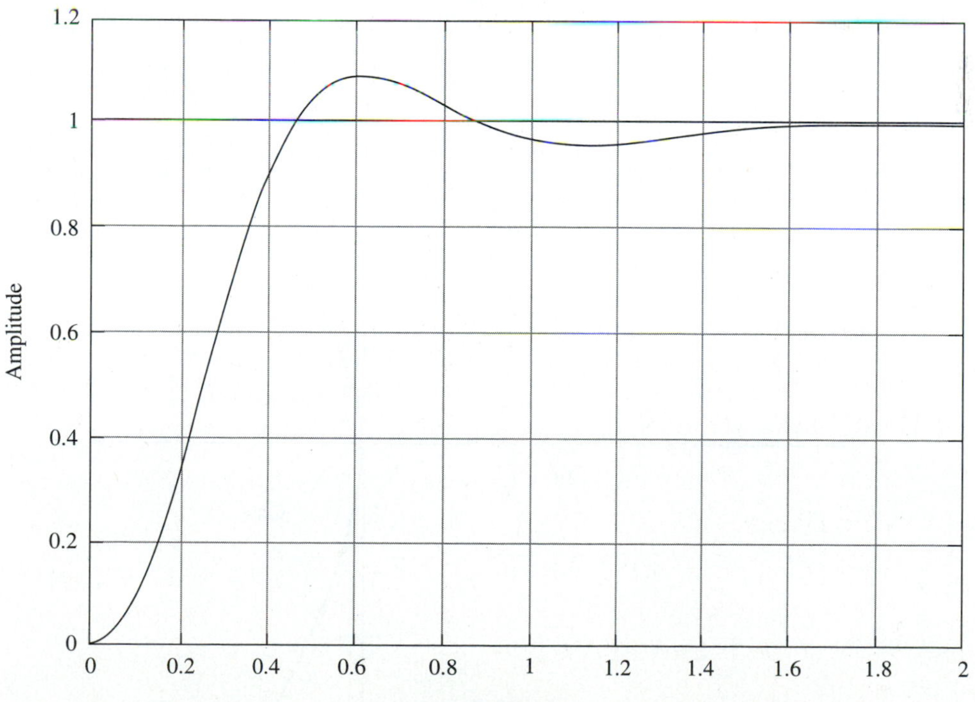

Fig. SP19.2.6

Solution

Sketching the root locus of the system, as shown in Fig. SP19.3.1, indicates that the closed-loop system is actually *unstable.* In this case, simple gain compensation will not be adequate, since the system will be unstable for all gains. The design process is begun by plotting the Bode diagram, as shown in Fig. SP19.3.2, which confirms the system is unstable. To begin with, the gain K will be set arbitrarily to unity, and an attempt will be made to design a filter that meets the performance requirements. The transient response requirement implies a damping ratio of $\zeta = 0.4$, or a phase margin of 40°. Since the phase margin at the uncompensated gain crossover frequency is about zero (for $K = 1$), our filter will have to provide

$$\phi = 40 + 10\% \approx 45^\circ$$

of phase margin. Figure 19.7 suggests $\alpha = 6$. The compensated gain crossover frequency is when

$$\Delta M = -10 \log_{10} \alpha = -7.8 \text{ db}$$

Figure SP19.3.3 shows that the new crossover frequency is $\omega_{gc} = 1.6$ rad/s. The filter time constant T is

$$T = \frac{1}{\omega_m \sqrt{\alpha}} = 0.26$$

making $\alpha T = 1.53$. The filter transfer function becomes

$$G_c(s) = \frac{1 + 1.53s}{1 + 0.26s}$$

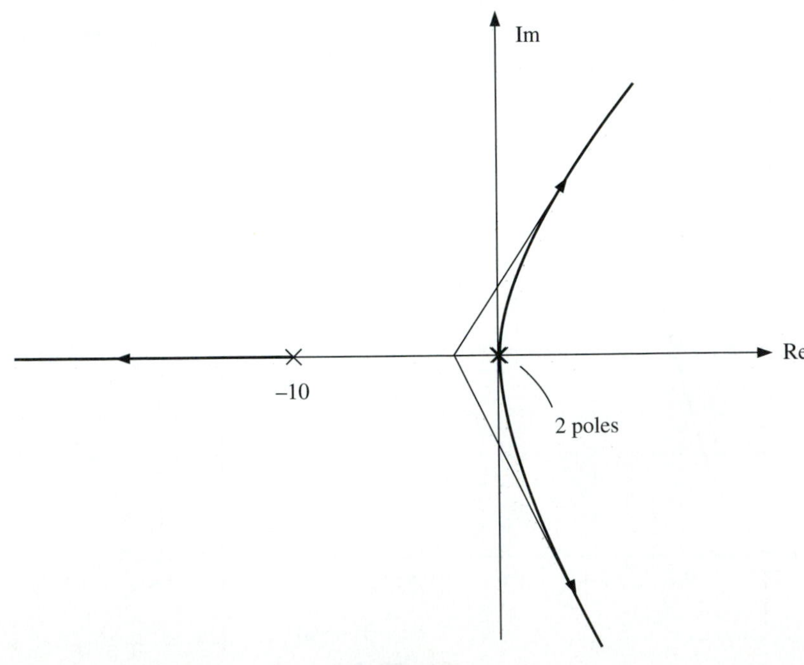

Fig. SP19.3.1

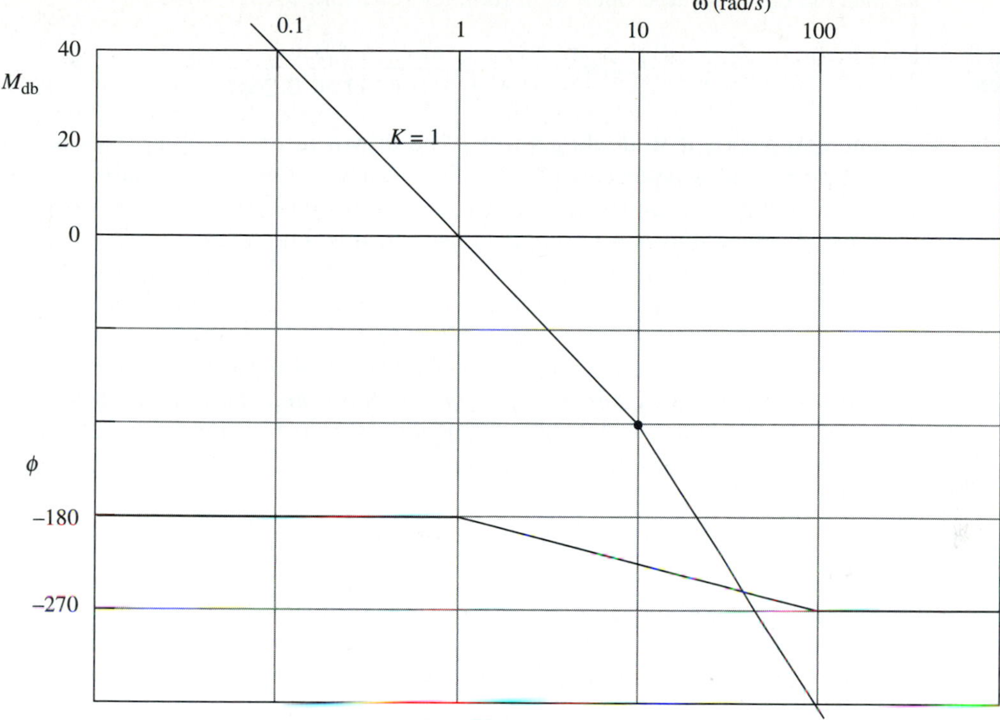

Fig. SP19.3.2

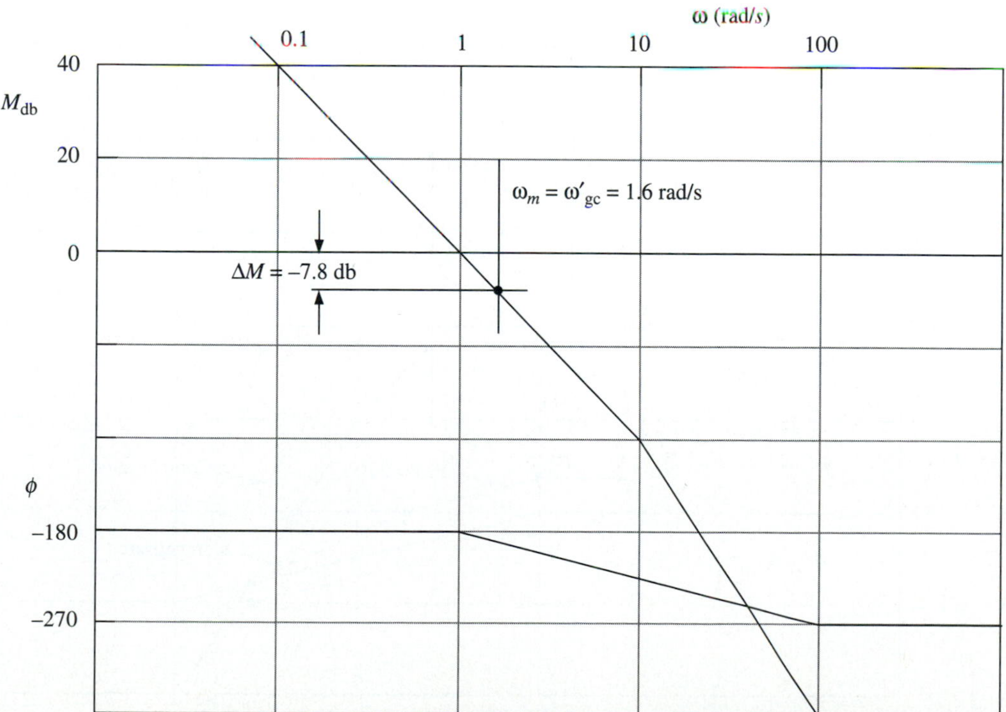

Fig. SP19.3.3

and the compensated open loop transfer function, for $K = 1$, is

$$G_cGH(s) = \frac{1 + 1.53s}{s^2(1 + 0.1s)(1 + 0.26s)}$$

The compensated Bode diagram may be drawn using the compensator zero and pole break frequencies of $\omega_z = 0.65$ rad/s and $\omega_p = 3.85$ rad/s, respectively. Using the procedures outlined in the previous two examples, the compensated Bode diagram will appear as shown in Fig. SP19.3.4. The diagram indicates a phase margin of around 27°, which is less than expected, and Fig. SP19.3.5, the unit step response of the system, confirms this. Since the 45% overshoot is too big, another stage of filter design will be performed. This time we will attempt to get, say 60° of phase lead, making the required $\alpha = 15$. Figure SP19.3.6 indicates that $\Delta M = -11.8$ db, and the new crossover frequency becomes $\omega_m = 2.0$ rad/s. This yields $T = 0.129$ and $\alpha T = 1.94$, making the new compensator of the form

$$G_c(s) = \frac{1 + 1.94s}{1 + 0.129s}$$

Figure SP19.3.7 shows an accurate plot of the compensated Bode diagram, suggesting a more acceptable phase margin of 50°, and this is confirmed in the unit step response shown in Fig. SP19.3.8. In this example, it has been shown that

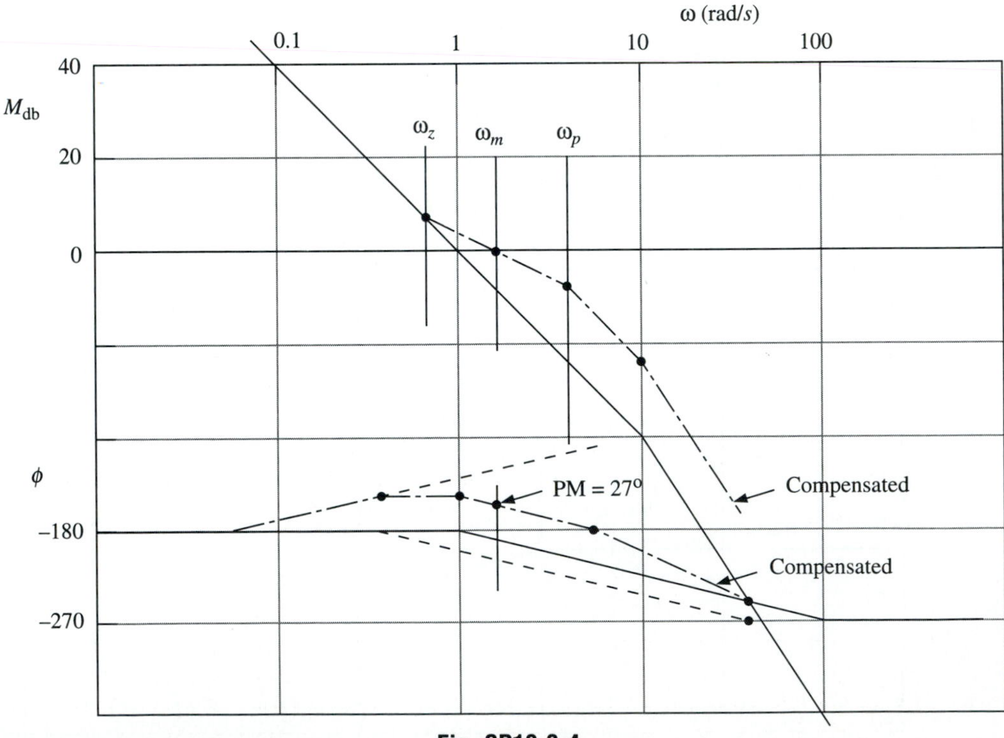

Fig. SP19.3.4

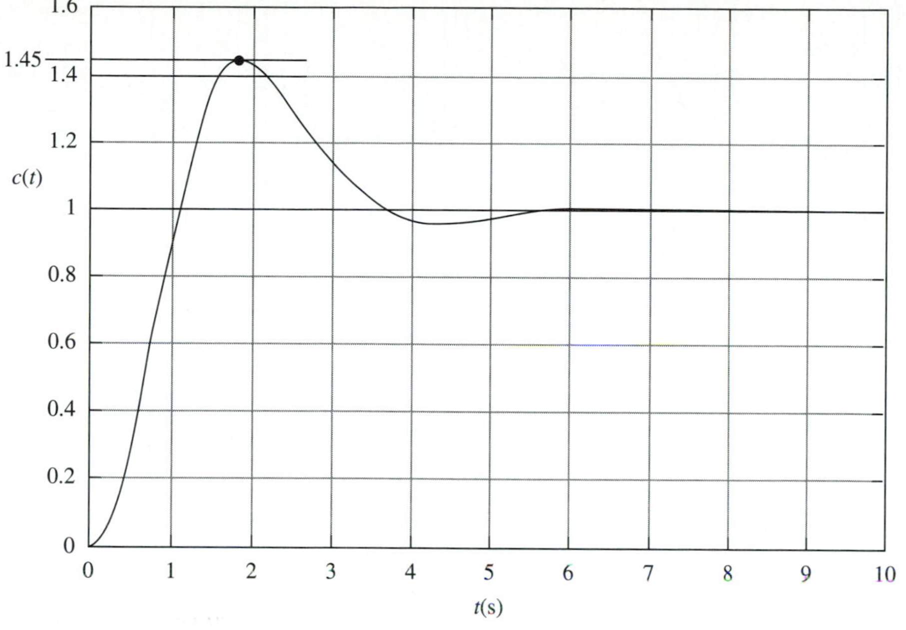

Fig. SP19.3.5

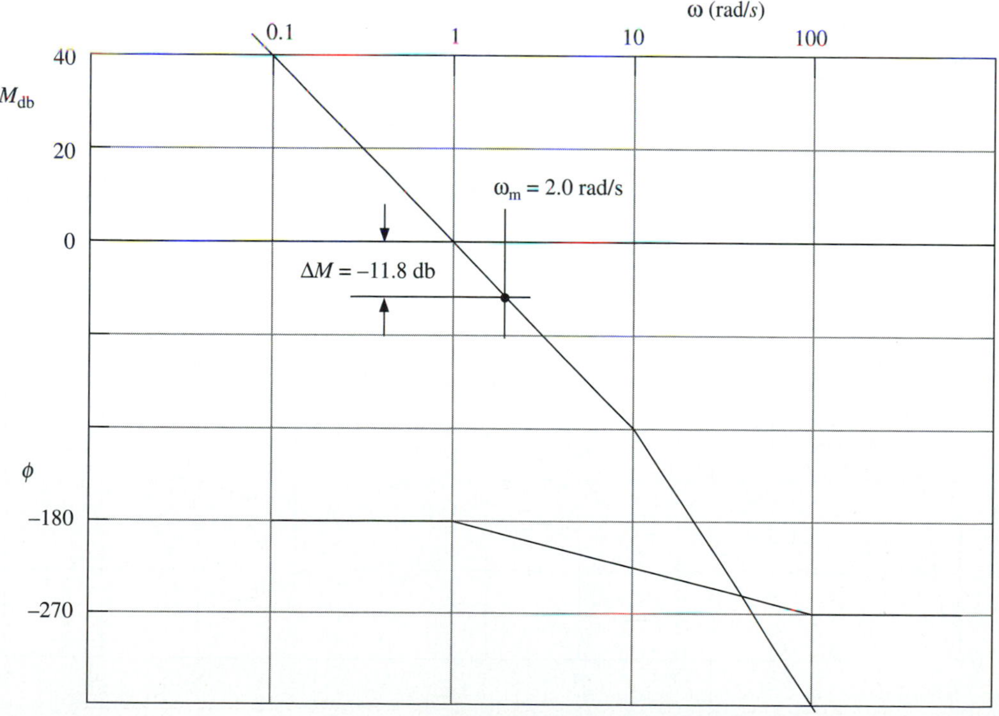

Fig. SP19.3.6

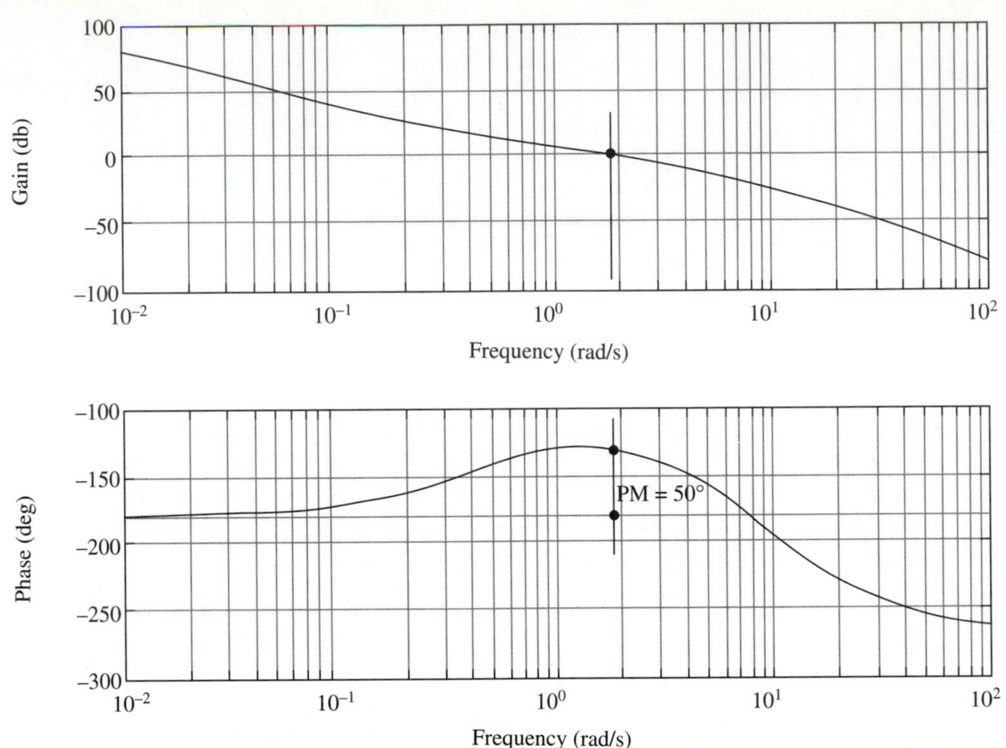

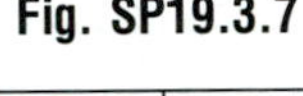

Fig. SP19.3.7

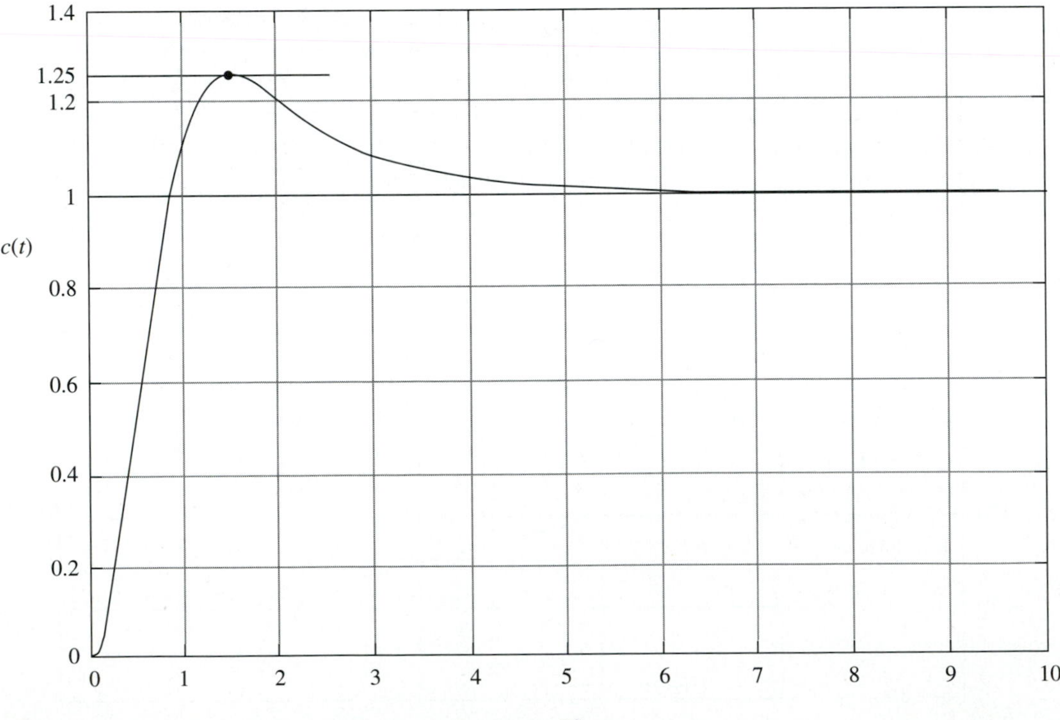

Fig. SP19.3.8

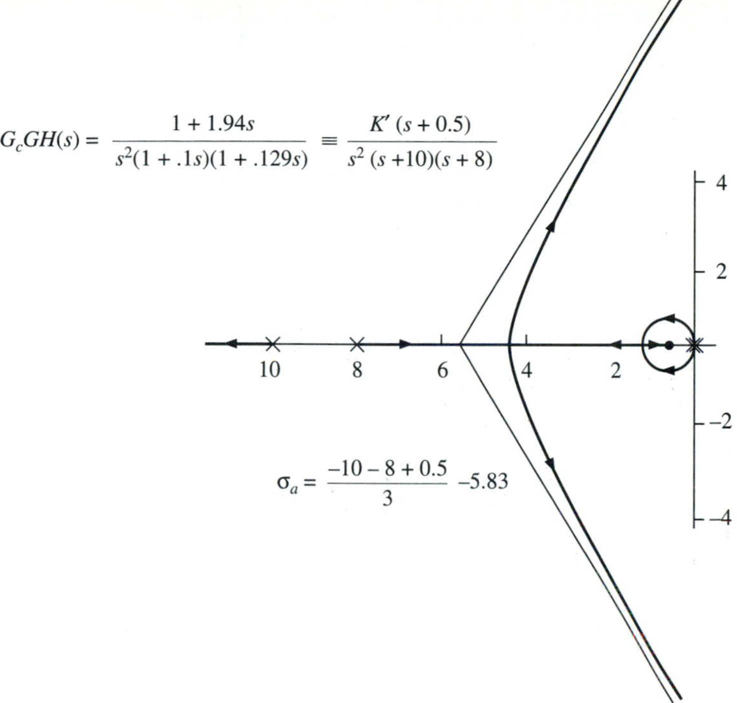

Fig. SP19.3.9

the original system has been stabilized as well as meets certain performance criteria. The effect of the stabilization process may be seen directly from the root locus of the compensated system, which is sketched in Fig. SP19.3.9. The order of the system is still 3 and. Therefore, given sufficient increase in gain, the system may become unstable. However, now there is a range of gain for which the system will be stable, and the gain $K \times \alpha = 15$ is clearly within this range.

◆ PROBLEMS

19.1 Show that the mechanical system illustrated in Fig. P19.1 may be represented as a phase lead element when the input is considered as R and the output as C. Determine the filter parameters α and T in terms of the various stiffness and damping constants.

19.2 Consider the unity-feedback system in Fig. P19.2. Design a phase lead compensation network that gives the closed-loop system:

a. A phase margin of 50°.

b. A velocity error of less than 10%.

19.3 Design a phase lead cascade compensation filter that provides the unity-feedback system

$$GH(s) = \frac{20}{(1 + 10s)^2}$$

with a percentage overshoot to a unit step of less than 15%.

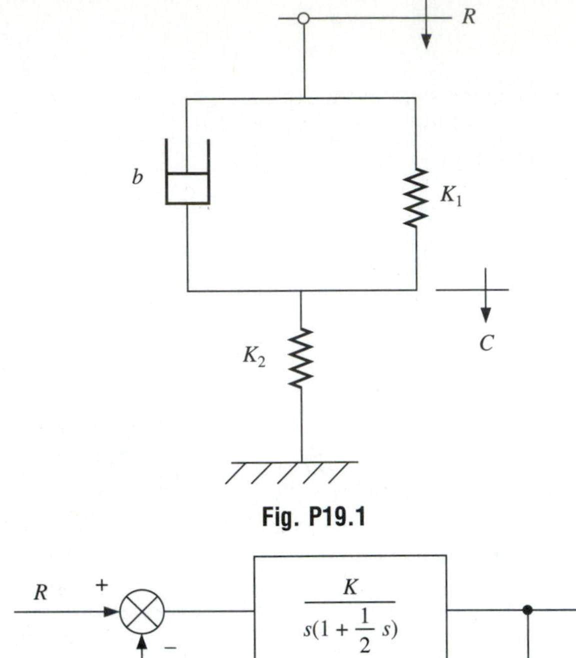

Fig. P19.1

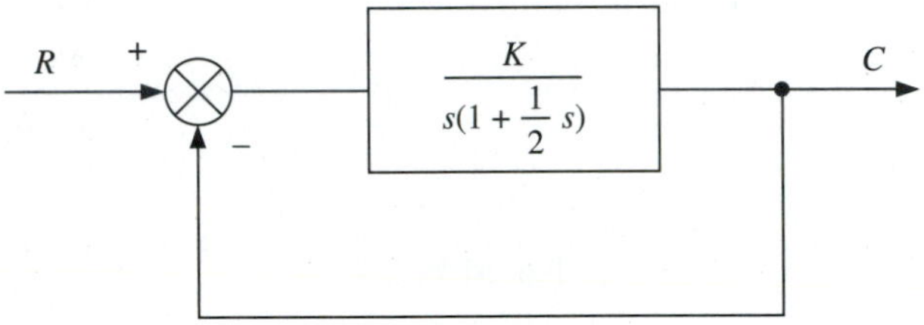

Fig. P19.2

19.4 Recall the system described in Sample Problem 19.3, where

$$G(s) = \frac{K}{s^2(1 + 0.1s)} \qquad H(s) = 1$$

Construct a phase lead compensator that meets the original 25% overshoot constraint but also provides the system with an acceleration constant of $K_a = 3$.

19.5 For the system shown in Fig. P19.5, design a phase lead compensator that gives the system a 45° phase margin. What is the open-loop bandwidth of the uncompensated and compensated systems?

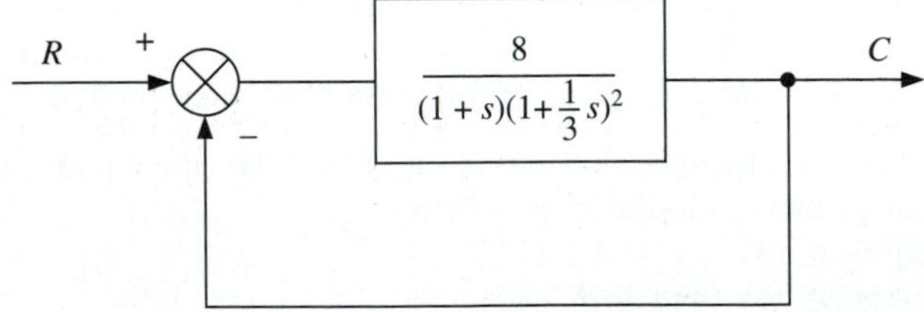

Fig. P19.5

19.6 Determine the stability of the closed-loop system shown in Fig. P19.6. Design a cascade phase lead compensator that provides the system with:

a. Open-loop bandwidth of at least 10 rad/s.

b. A 2% settling time of no more than 1 s.

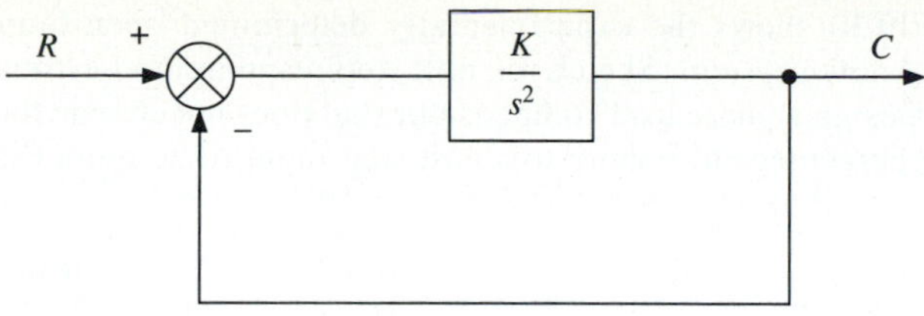

Fig. P19.6

Sketch the compensated-system unit step response and determine the percentage overshoot and peak time.

19.7 Design a phase lead filter that stabilizes the system shown in Fig. P19.7. What is the estimated settling time of the compensated system and the resulting gain margin?

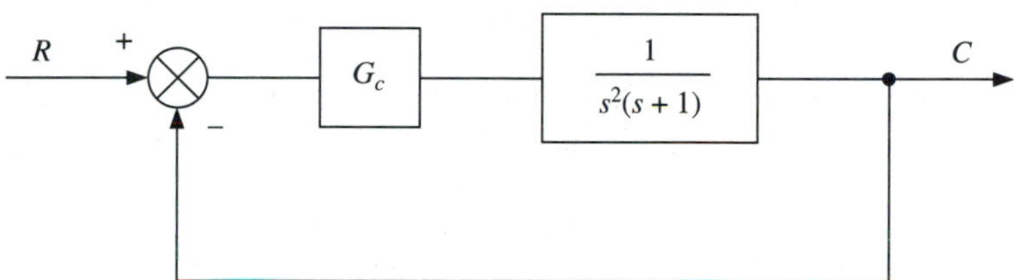

Fig. P19.7

19.8 Compute a suitable value of the gain K_a and the phase lead element G_c that provides an open-loop bandwidth of at least 25 Hz and a phase margin of 50° for the system shown in Fig. P19.8.

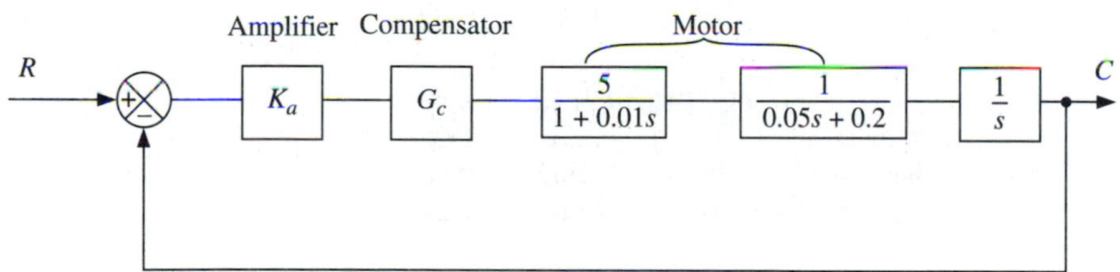

Fig. P19.8

19.9 Consider the system described by the transfer function

$$GH(s) = \frac{K}{s(1 + 0.25s)^2}$$

Sketch the root locus of the system and discuss its stability. Design the phase lead compensator G_c that meets the following performance requirements:

a. An error constant of 10.
b. A 20% maximum overshoot to a unit step.

For the resulting system determine:

c. The open-loop bandwidth.
d. The phase and gain margins.
e. The peak time and 2% settling time.

19.10 Figure P19.10 shows the experimentally determined open-loop frequency response of an unknown system. Sketch the unit step response and estimate the percentage overshoot. Design a phase lead compensator that does not change the error constant but reduces the percentage overshoot to a unit step input to no more than 15%.

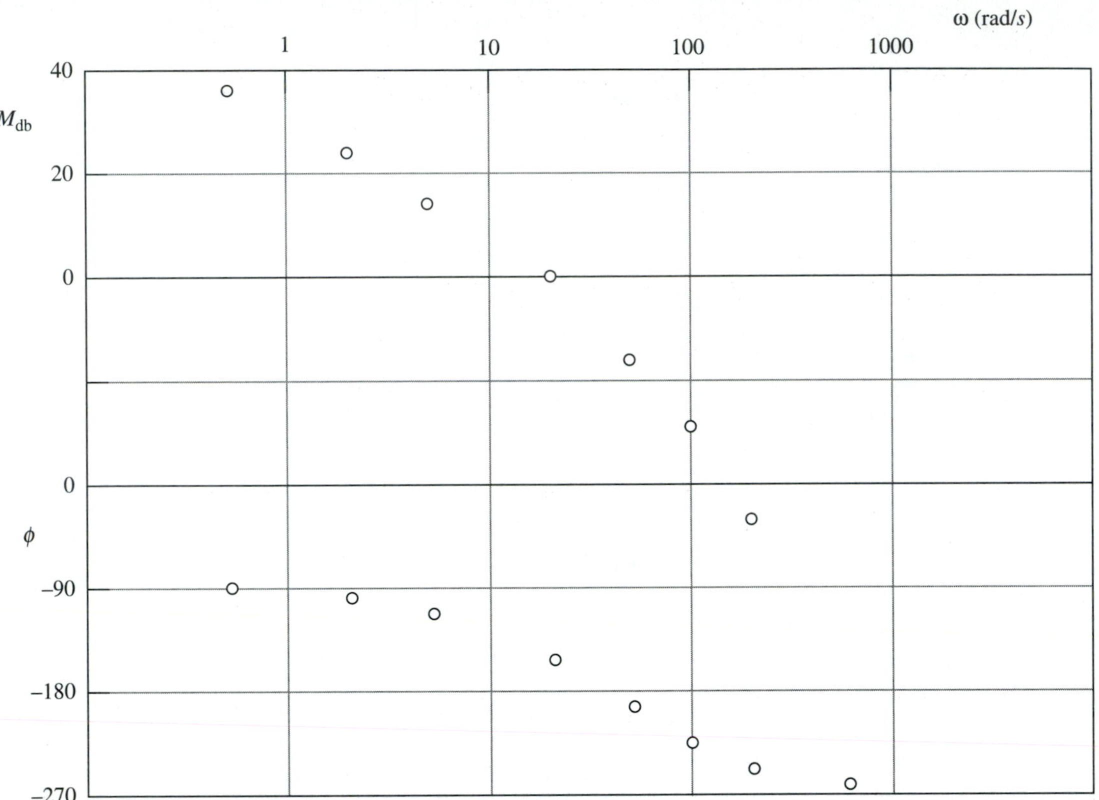

Fig. P19.10

MODULE 20

Phase Lag and Lead-Lag Compensation

As discussed in the previous module, it is possible to approach compensation from a slightly different perspective than that used in phase lead compensation, where we saw how the higher-frequency part of the phase angle plot was adjusted in order to meet a transient response requirement. In phase lag compensation, the magnitude part of the uncompensated Bode diagram is attenuated in order to reduce the gain crossover frequency, thereby allowing the uncompensated phase plot to produce the necessary phase margin. Phase lag elements and their transfer functions will be discussed, together with an outline of a process by which phase lag compensators may be designed. In the final section of the module, combined lead-lag compensation systems will be discussed, and it is shown to be a means of satisfying many performance constraints simultaneously.

◆ TRANSFER FUNCTION OF PHASE LAG ELEMENT

As with phase lead compensators, phase lag networks may be constructed from simple passive electrical components. Figure 20.1 shows the design of one such network. Analysis of the system reveals the following transfer function:

$$\frac{v_o}{v_i} = G_c(s) = \frac{1 + \tau s}{1 + \beta\tau s} \tag{20.1}$$

where the constants τ and β are given by

$$\tau = R_2C \qquad \beta = \frac{R_1 + R_2}{R_2} \tag{20.2}$$

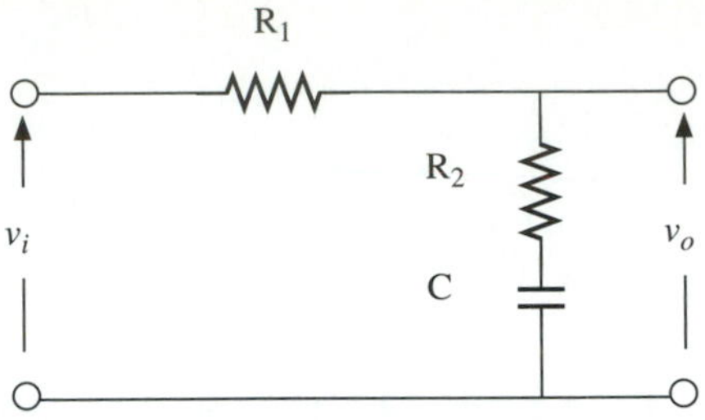

Fig. 20.1 Phase lag network.

Clearly $\beta > 1$, and once again the task of designing the phase lag compensator is the determination of τ and β. The Bode diagram of the element is shown in Fig. 20.2. There are many similarities between Figs. 20.2 and 19.6 showing the Bode diagram of the phase lead network, and many of the same features are common. For example, the maximum phase lag introduced is a function of β; its value occurs at the geometric mean of the break points, and this frequency is obtainable from

$$\omega_m = \frac{1}{\tau\sqrt{\beta}} \tag{20.3}$$

It is important to recognize, however, *that details of the phase part of the Bode diagram are of no consequence at all in the design of the compensator.* All that is important is the amount of attenuation provided, which as before is given by

$$\Delta M = -20 \log_{10} \beta \tag{20.4}$$

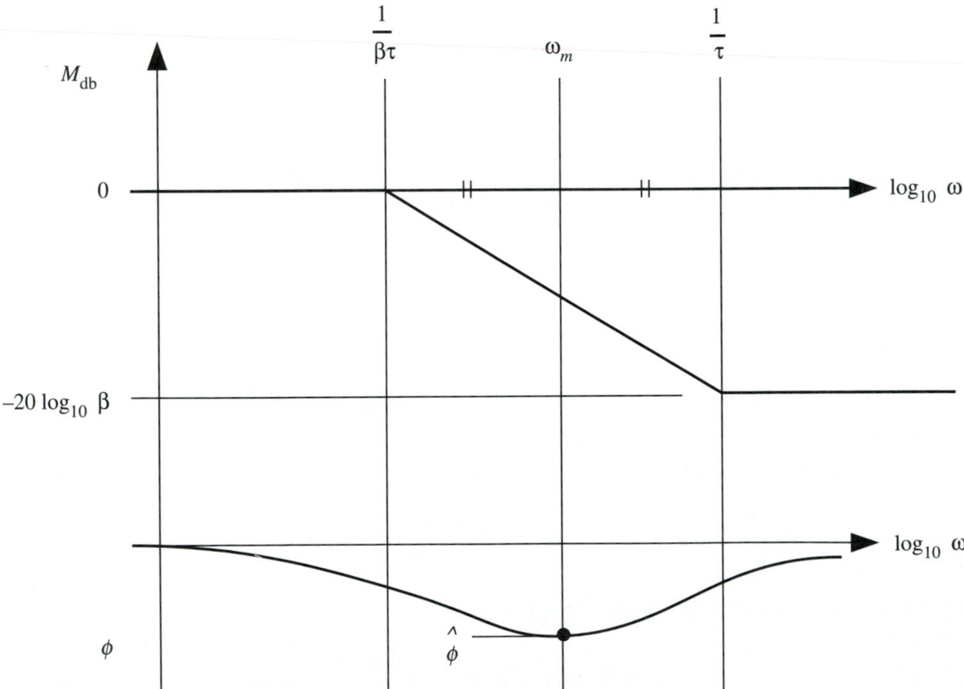

Fig. 20.2 Bode diagram of phase lag network.

Note that this attenuation is the high-frequency value, i.e., the attenuation above both break frequencies on the Bode diagram. We shall see that only equation 20.4 will be used in the compensator design.

◆ PHASE LAG COMPENSATION PROCESS

To illustrate the design process, we will investigate the same example as in the previous module, for which a phase lead compensator was designed. Recall the uncompensated system

$$GH(s) = \frac{K}{s(1 + 0.2s)} \qquad (20.5)$$

was subject to the following performance constraints:

1. no more than a 0.32% error to a unit ramp input, resulting in $K \geq 316$, and
2. maximum percentage overshoot of 20%, implying $K \leq 6.3$.

The design process is started as before, by plotting the Bode diagram of the system for a value of gain needed to satisfy the steady-state error requirements. This Bode diagram is shown in Fig. 20.3. We note from the diagram that the required phase

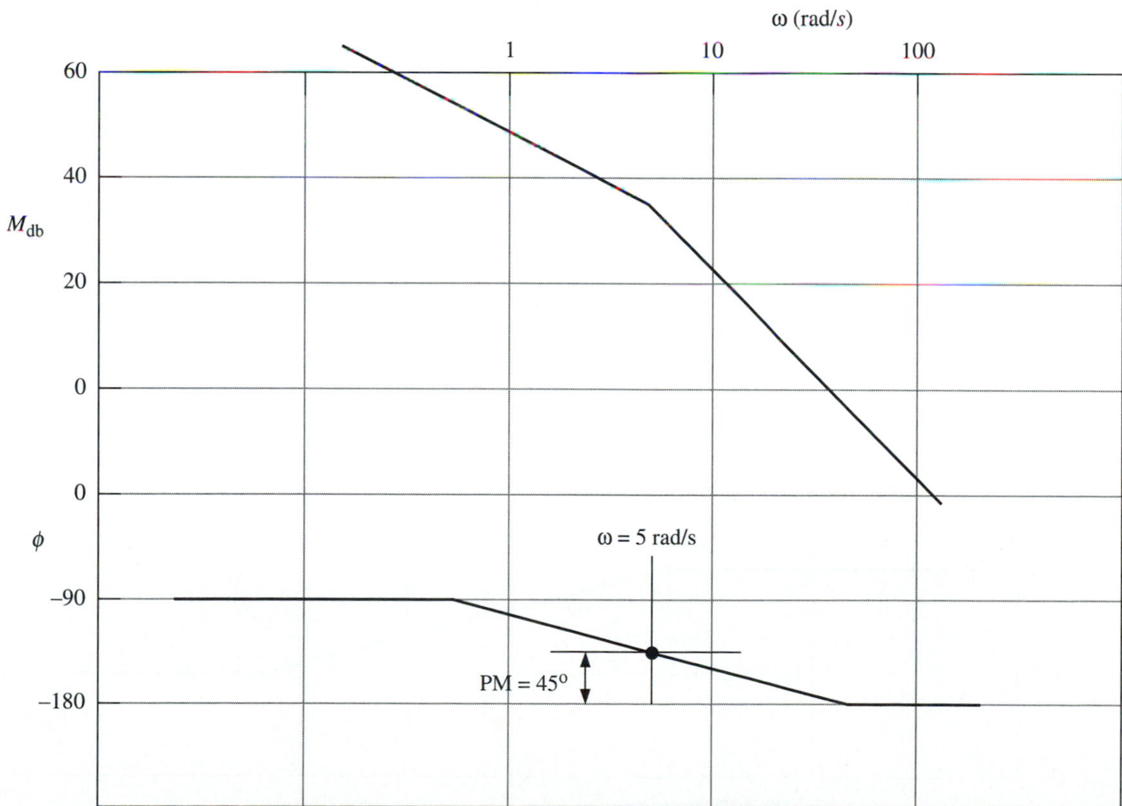

Fig. 20.3 Uncompensated system with $K = 316$.

margin (45°) occurs at a frequency of 5 rad/s. The general design approach is to use a phase lag element to reduce the uncompensated system gain from its current value, at this frequency, to 0 db, thereby meeting the transient part of the performance requirements. Because the low-frequency magnitude of the compensator is unchanged at unity, the steady-state part of the requirement will still be met.

A small modification to the above approach is applied, since it is assumed that in introducing the compensation network, no modification of the composite phase plot occurs. However, this will not be true. For this reason, a small safety margin of 10% is added to the phase margin requirement in order to allow for this. In the case of the example system, this means that the required phase is now 45 + 10% = 50°, which occurs at 3 rad/s, and this frequency now becomes the new gain crossover frequency ω'_{gc}. Figure 20.4 shows the Bode diagram with the relevant data marked and indicates that the uncompensated magnitude has to be reduced from 40 db at this frequency to zero. Since this attenuation has to be provided by the compensation network, equation 20.4 may be used to determine β:

$$20 \log_{10} \beta = 40 \qquad (20.6)$$

resulting in $\beta = 100$. All that remains now is to determine τ. This is a noncritical part of the design, since all we require is that the two break frequencies be at

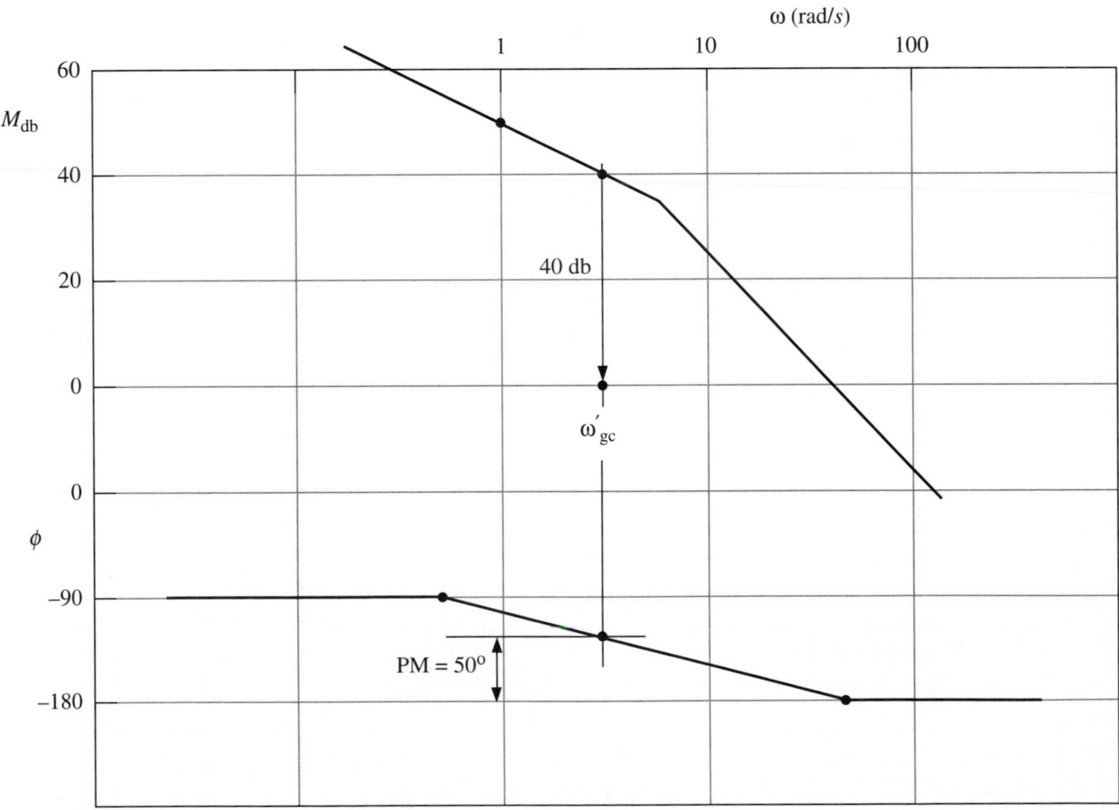

Fig. 20.4 Determining the new gain crossover frequency.

a lower frequency than ω'_{gc}. The only question is, how much lower? This is determined by two factors:

1. The higher break frequency $(1/\tau)$ should be at a frequency such that the residual phase lag due to the compensator is not more than the 10% allowed for in the design process.
2. The lower break frequency should not be too small, since it is clear that the bandwidth decreases, and a very low bandwidth may produce a sluggish response.

A common design guideline is to place the upper break frequency one decade below the new gain crossover frequency ω'_{gc}, which allows τ to be determined. Then the lower break frequency can be calculated from the known values of τ and β. In our example,

$$\frac{1}{\tau} = 0.1\omega'_{gc} = 0.3 \text{ rad/s} \tag{20.7}$$

resulting in $\tau = 3.34$ and $\beta\tau = 334$. The compensation network transfer function may be written directly as

$$G_c(s) = \frac{1 + 3.34s}{1 + 334s} \tag{20.8}$$

Figure 20.5 shows the compensated-system Bode diagram and confirms the phase margin is 45°. Note how the phase contribution from the compensation element has all but disappeared from the composite Bode diagram near the gain crossover frequency. If the phase was not close enough to the desired value, more iterations of the process may be undertaken, though, as in the case of the phase lead element, this is rarely necessary. The compensated open-loop transfer function becomes

$$\frac{316(1 + 3.34s)}{s(1 + 0.2s)(1 + 334s)} \tag{20.9}$$

Using a computer simulation, the closed-loop step response for the compensated and uncompensated system is shown in Fig. 20.6.

◆ COMMENTS ON PHASE LAG COMPENSATION

1. Phase lag compensation provides a form of integral control.
2. The phase lag method provides the necessary damping ratio in order to limit the overshoot to the required value.
3. The filter design process is somewhat simpler than the phase lead compensation system in that the selection of the pole and zero break frequencies is not too critical.
4. As can be seen from the Bode diagram for the compensated system, the phase lag technique reduces the open- and hence the closed-loop bandwidths. The open-loop system bandwidth is reduced from about 50 rad/s to about 1.5 rad/s.

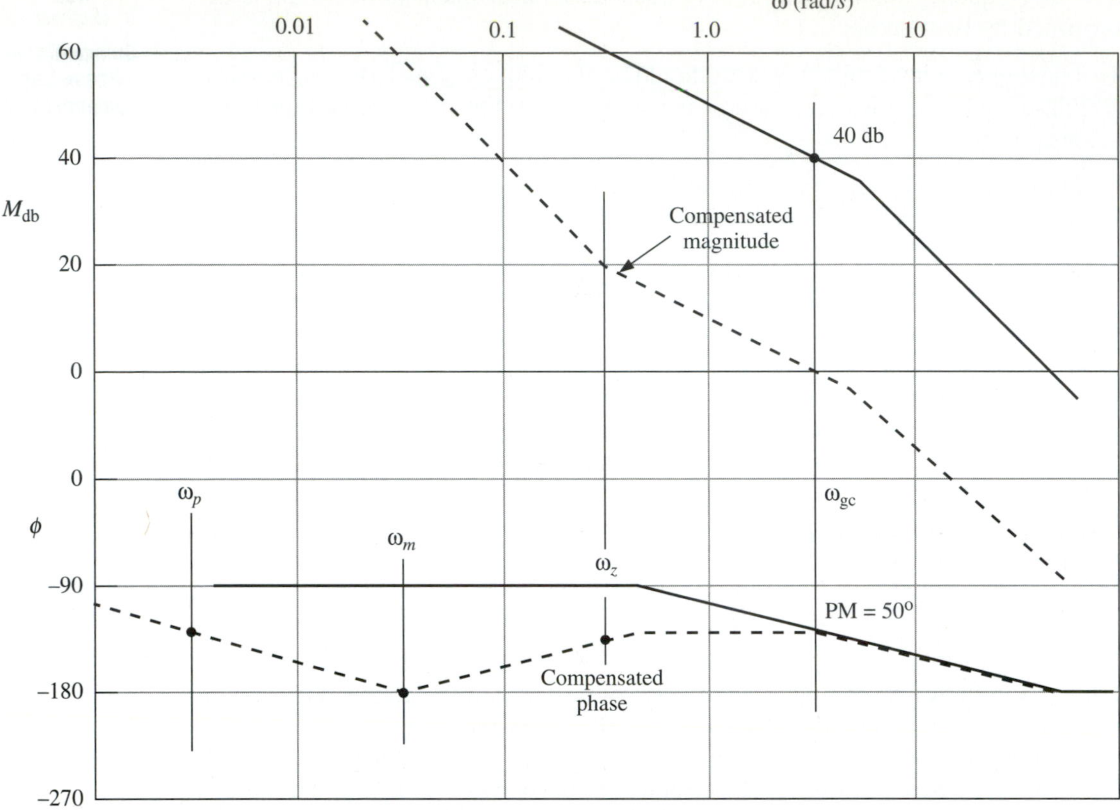

Fig. 20.5 Bode diagram of compensated system.

This is reflected in the closed-loop step response, which should be compared to Fig. 19.15b for the phase lead system. Using the phase lead, the open-loop bandwidth is about 100 rad/s, hence the faster response.

5. The settling time is also increased considerably, again due to the reduced bandwidth of the system.
6. Unlike phase lead compensation, phase lag compensation may change the phase margin by more than 90° (see Fig. P20.3).

In some cases, the designer has the option of using either the phase lead or the phase lag technique. However, the inclusion of further performance measures (such as the settling time) would dictate the use of one or the other. In the case of the phase lag approach, being able to physically realize the values of τ and β that the design process produces may also limit its use in practice.

◆ LEAD-LAG COMPENSATION

So far in this module and in the previous module, we have learned two ways to compensate systems so as to achieve multiple performance requirements. In

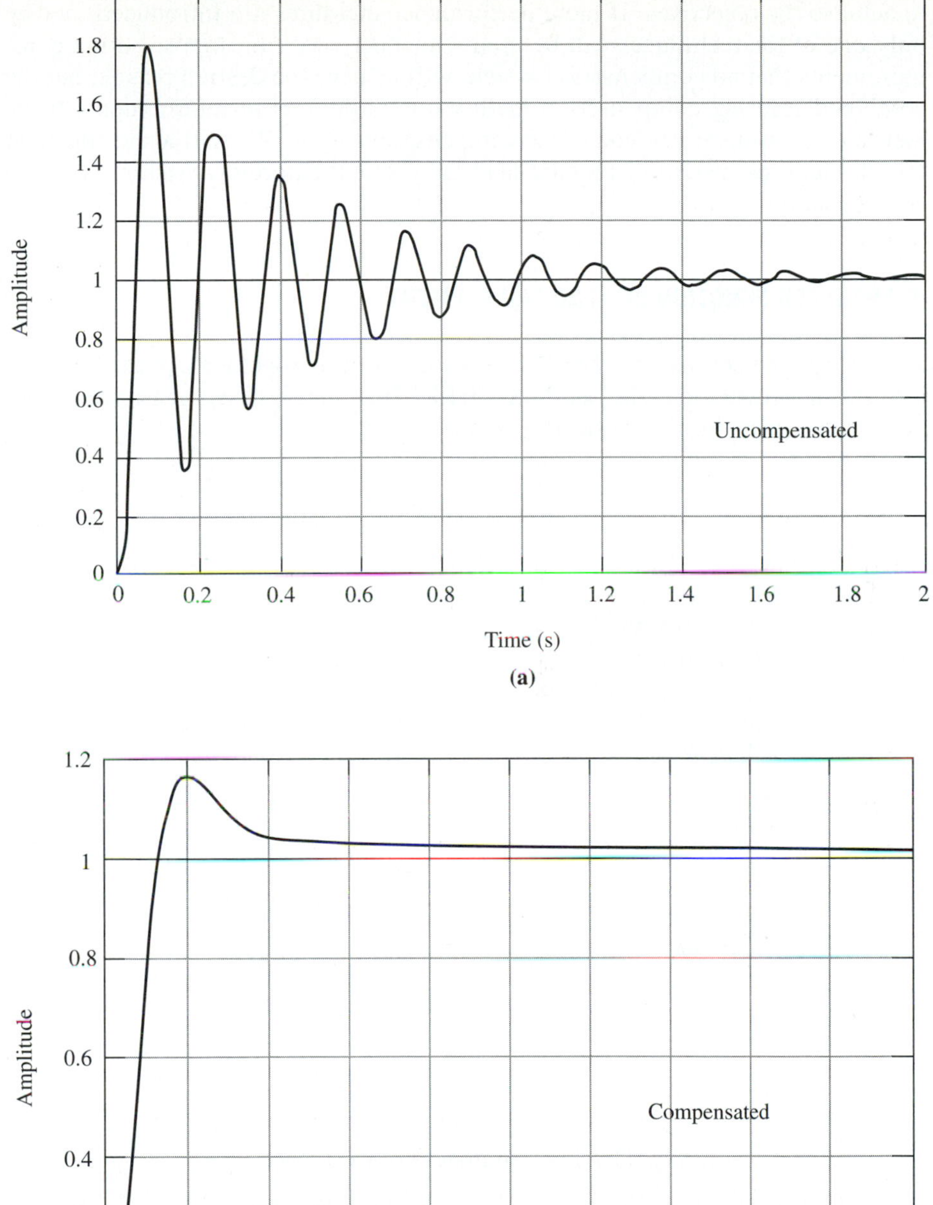

Fig. 20.6 Closed-loop step response of compensated system.

some cases, as in the example system just used, either method may be employed to achieve the objectives. If more performance measures are introduced, maybe only one of the techniques can be used. Obviously, we can impose so many requirements that no compensation system will achieve the desired results, but the combined lead-lag compensation method may allow more requirements to be met than by using either lead or lag compensation alone. We will again illustrate the physical realization of a phase lead-lag element before discussing its use in compensation.

◆ TRANSFER FUNCTION OF LEAD-LAG ELEMENT

A lead-lag element may be constructed by connecting together a phase lead and a phase lag network in series, as shown in Fig. 20.7. In this case, the transfer function of the compensation network becomes

$$G_c(s) = \frac{v_o}{v_i} = \frac{(1 + \alpha Ts)(1 + \tau s)}{(1 + Ts)(1 + \beta\tau s)} \tag{20.10}$$

where the values of α, β, τ, and T are independent and are functions of the network resistances and capacitances. Alternatively, it may be shown that the simpler network shown in Fig. 20.8 will have the same compensation characteristics. For this simpler case, the values of α, β, τ, and T will not be independent, but related. In order to have the maximum flexibility possible in the compensation provided by the lead-lag filter, we will assume that the filter takes the form shown in Fig. 20.7. The role of the buffer amplifier is twofold: (a) to decouple the two sections of the

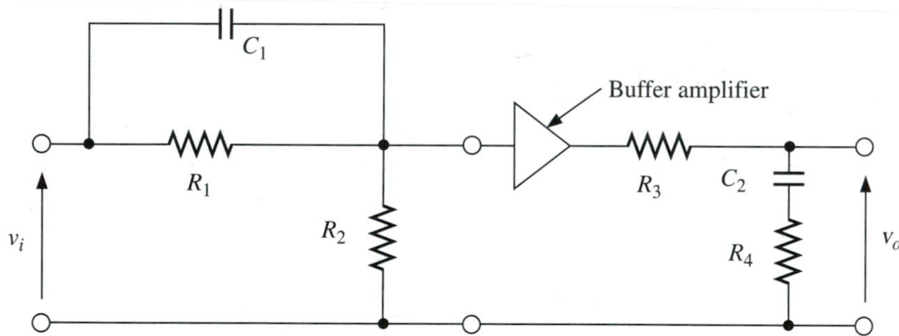

Fig. 20.7 Lead-lag from separate elements.

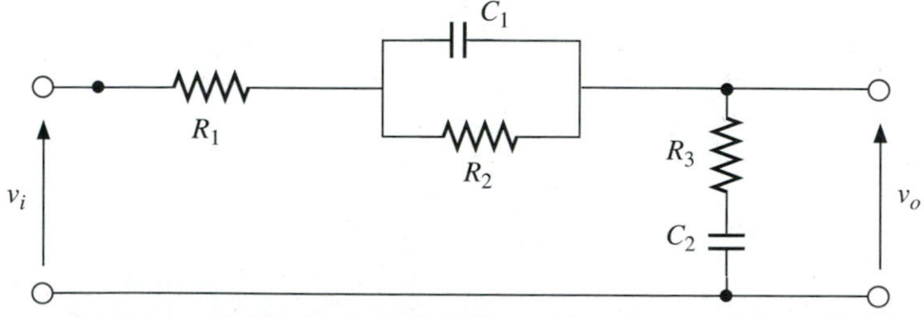

Fig. 20.8 Composite lead-lag filter.

filter allowing the composite transfer function to be written simply as the product of the individual elements and (b) to provide a fixed gain so that the lead part has unity gain at low frequencies, as discussed in the previous module. Details of the relationship between the poles and zeros of the filter and the component values will be the same as previously stated for the individual filter components. We may write the transfer function of the lead-lag filter as

$$G_c(s) = \frac{(1 + \tau s)(1 + \alpha T s)}{(1 + \beta \tau s)(1 + T s)} \tag{20.11}$$

In this equation it is assumed, as before, that the low-frequency gain of the lead section is unity, and suitable adjustment of the forward-path gain will be needed to achieve this. In the filter design it is usual to further assume that the two break frequencies of the lag portion are lower than the two break frequencies of the lead portion and that $\alpha < \beta$. This being the case, the Bode diagram of the filter will be as shown in Fig. 20.9. Note from the figure how the two elements of the filter are separated. Further features of the Bode diagram include the following:

1. The filter provides attenuation only and no gain.
2. The low-frequency gain is unity while the high-frequency gain is

$$\Delta M = -20 \log_{10}\left(\frac{\beta}{\alpha}\right) \tag{20.12}$$

3. The phase angle first lags, then leads, but the high- and low-frequency phases are both zero.
4. The maximum phase lag and the maximum phase lead occur between their respective filter break frequencies.

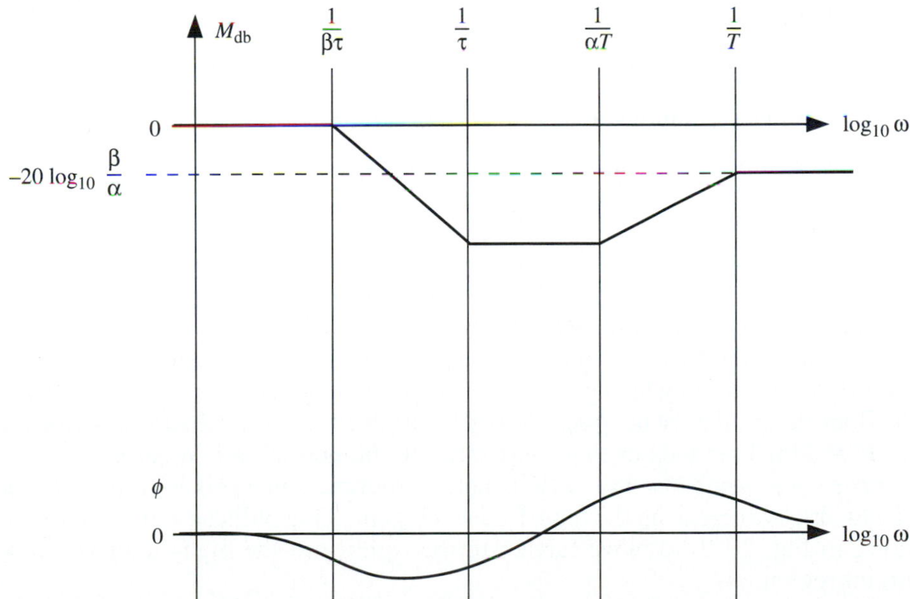

Fig. 20.9 Bode diagram of lead-lag filter.

The lead-lag network utilizes the best features of the individual lead and lag elements, usually without their disadvantages. For example, the lead-lag system allows the introduction of phase lead to stabilize a system, while providing attenuation at higher frequencies, thereby filtering out noise.

◆ LEAD-LAG COMPENSATION PROCESS

Unlike the previous filter design processes, the lead-lag design will be based on intelligent trial and error. Analytical procedures have been developed, but their complexity is rarely justified and often obscures what is actually being done to the uncompensated system. We will investigate the same system as before,

$$GH(s) = \frac{K}{s(1 + 0.2s)} \tag{20.13}$$

subject to the same performance constraints. It has been shown that either a lead or a lag compensation network will meet these two constraints, so further restrictions will now be applied. The existing constraints address mainly transient and time-domain performance, so the additional constraints will be concerned with the frequency response of the system, namely:

1. The open-loop gain crossover frequency has to be greater than 10 rad/s.
2. Frequencies above 100 rad/s must be attenuated by at least 30 db.

It is not known at this stage whether all of the performance requirements may be met, i.e., whether the system is *overconstrained*. In order to employ a methodical design procedure, we will allocate priorities to the stated requirements and ensure that our design meets the most important ones. Suppose that an order of priority is as follows:

1. Steady-state velocity error is less than 0.32%.
2. High-frequency attenuation is more than 30 db.
3. Gain crossover frequency is more than 10 rad/s.
4. Percentage overshoot is less than 20%.

Stating the constraints in this form also defines the filter design process. Figure 20.10 shows the uncompensated-system Bode diagram for $K = 316$, which is the gain needed to satisfy the steady-state error requirements, the highest priority. Also shown in the figure are the two frequency-domain constraints, the next most important, which may be represented by regions in which the magnitude part of the Bode diagram must not pass. Clearly the high-frequency constraint is violated for $K = 316$. This tells us that some form of compensation is required.

Phase lead compensation alone is not appropriate since this will increase the bandwidth by increasing the high-frequency gain. This will cause the magnitude curve in Fig. 20.10 to move further to the right, into the high-frequency constraint region.

Phase lag compensation is possible. Using the techniques described earlier in the module, it may be seen that the best performance would be obtained if the high-

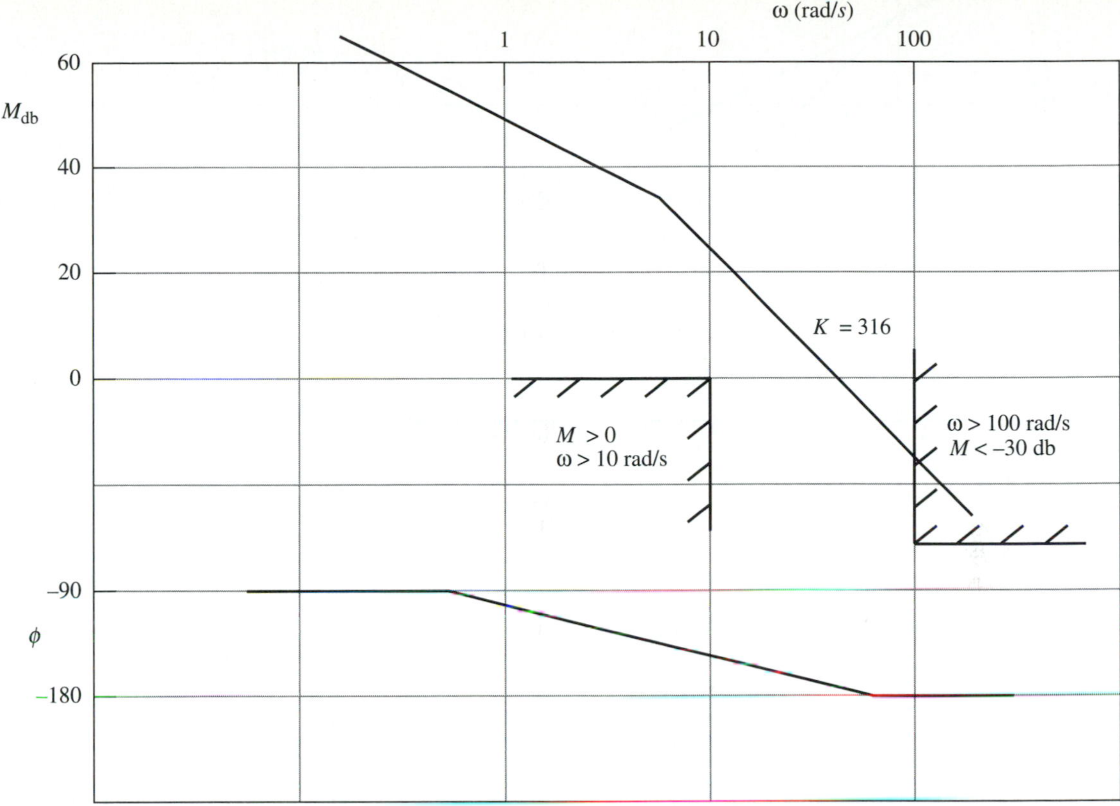

Fig. 20.10 Example system with frequency-domain constraints.

frequency part of the magnitude plot just touches the corner C_1 of the middle-frequency constraint region, as shown in Fig. 20.11. The remaining parts of the phase lag filter may be designed graphically using the fact that all slopes are integer multiples of ±20 db/decade, and the highest break frequency is one decade below C_1. For this filter design, the high-frequency attenuation is 24 db, yielding $\beta = 15.8$. The remaining portions of the filter may be designed in the usual way.

Unfortunately, a lag compensator does not meet the final performance criterion. Figure 20.11 shows that the phase margin is about 30°,[1] which would give about 35% overshoot. Our approach, therefore, is to design a lead-lag filter, where the lag portion is as specified above. In addition, we will introduce a lead component that will increase the phase margin in the region of the gain crossover frequency. The question immediately arises as to how much phase margin may be introduced, since phase lead will increase the high-frequency gain (above the lag magnitude, not the uncompensated magnitude), and the high-frequency constraint region must not be violated. This question is answered by realizing that the lead-lag filter does not modify the slope of the uncompensated magnitude at high frequency. Therefore a line of slope −40 db/decade may be drawn, passing through the corner C_2 of the high-frequency constraint region, as shown in

[1]Ignoring the phase contribution from the compensator at this frequency.

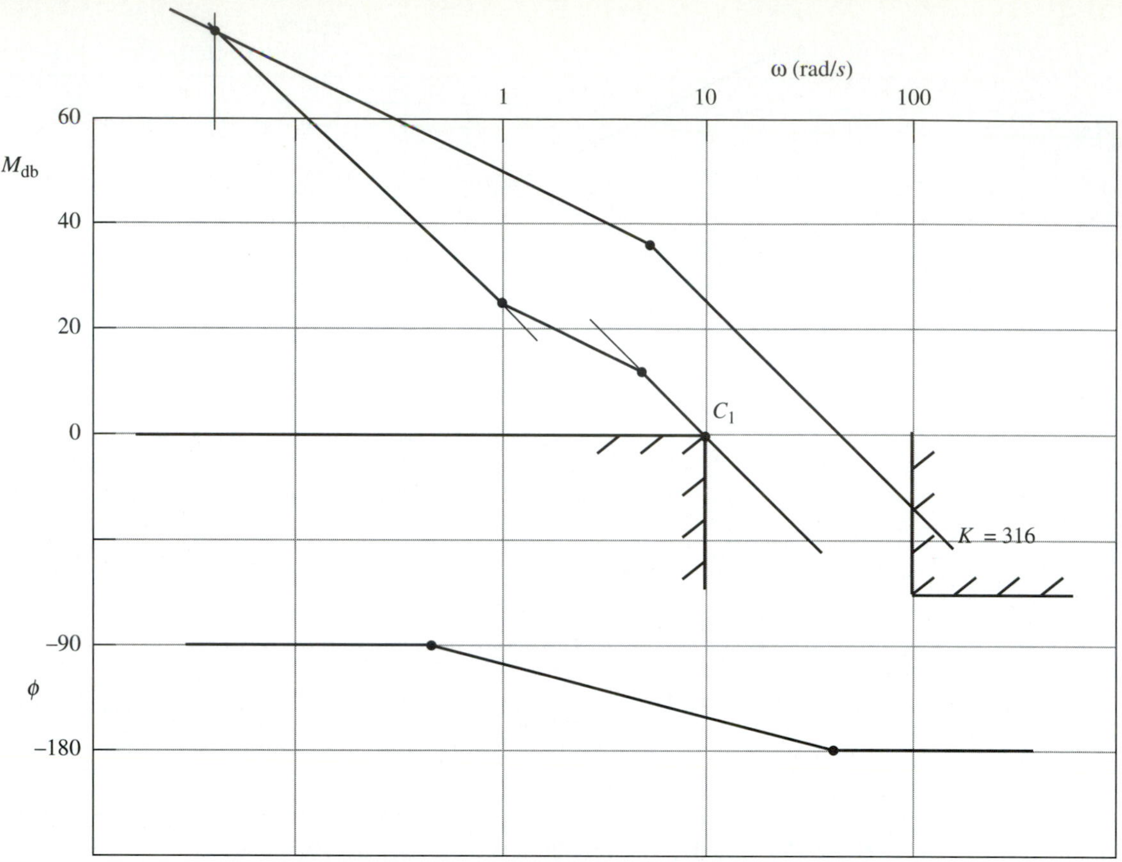

Fig. 20.11 Compensation using phase lag only.

Fig. 20.12. Although this does not allow us to obtain more information on break frequencies, the ratio of α and β is obtained from the high-frequency attenuation, as shown in Fig. 20.9. In this example,

$$\frac{\beta}{\alpha} = 5 \tag{20.14}$$

Figure 20.13 shows what information we have so far on the design of the filter, which consists of the lag portion and the high-frequency portion. Since we know $\beta/\alpha = 5$ and β from the lag portion design is 15.8, this determines $\alpha = 3.2$. Figure 13.17 indicates that the lead portion of the filter is capable of supplying a maximum additional phase lead of 30°, which, together with the existing phase of around 25°,[2] will provide sufficient limitation on the overshoot. In fact, we will design the lead portion to give us maximum phase angle, and therefore maximum damping ratio, and minimum percentage overshoot.

The design proceeds by placing the frequency of maximum phase lead ϕ_m at the current gain crossover frequency $\omega = 10$ rad/s. Since we know $\alpha = 3.2$, the

[2]Remember the crossover frequency will move to the right in Fig. 20.11.

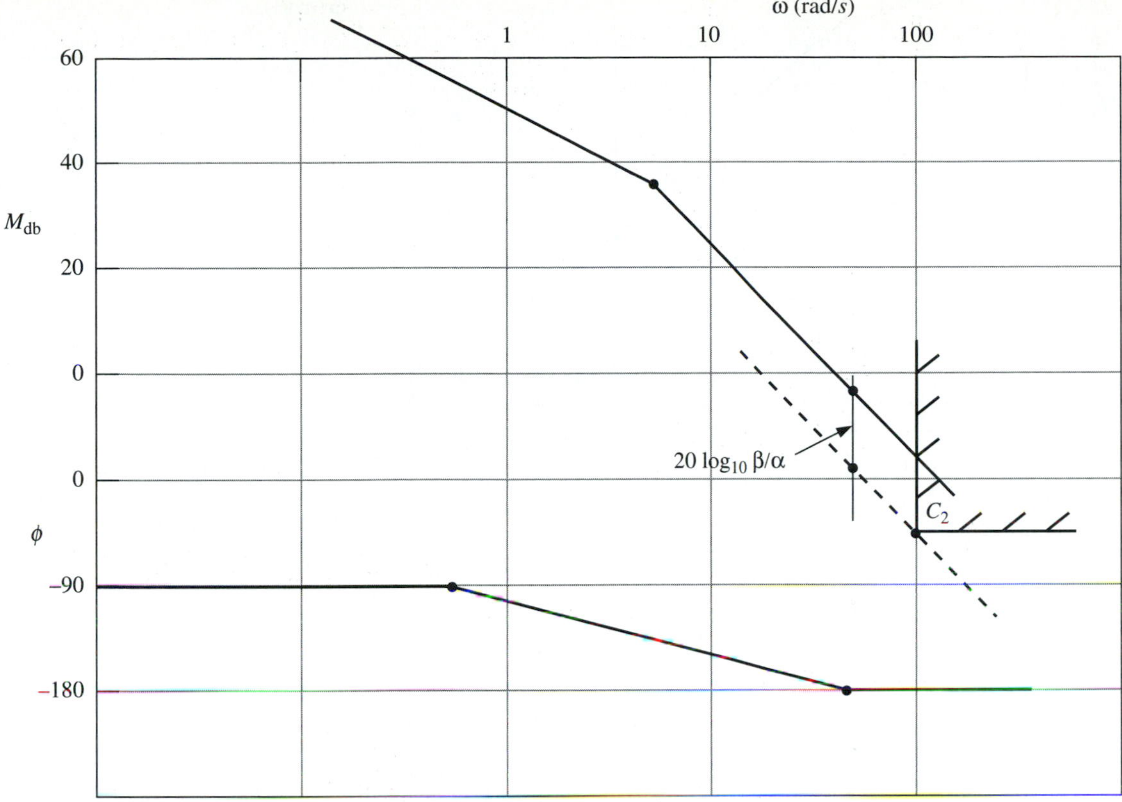

Fig. 20.12 High-frequency filter segment design.

break frequencies of the lead portion are defined:

$$T = \frac{1}{\omega_m \sqrt{\alpha}} = 0.056 \tag{20.15}$$

leading to $\alpha T = 0.179$. The break frequencies for the lead element become

$$\omega_z = 5.6 \text{ rad/s} \qquad \omega_p = 17.9 \text{ rad/s} \tag{20.16}$$

The only other frequency needed is the lowest break point of the lag element, which turns out to be

$$\frac{1}{\beta\tau} = \frac{1}{15.8} = 0.063 \tag{20.17}$$

The compensated Bode diagram is shown in Fig. 20.14. The compensation filter takes the form

$$G_c(s) = \frac{1 + s}{1 + 0.179s} \frac{1 + 15.8s}{1 + 0.056s} \tag{20.18}$$

Figure 20.14 indicates a phase margin of 45°, which suggests a percentage overshoot of around 20%. Figure 20.15 shows a computer-generated, accurate system

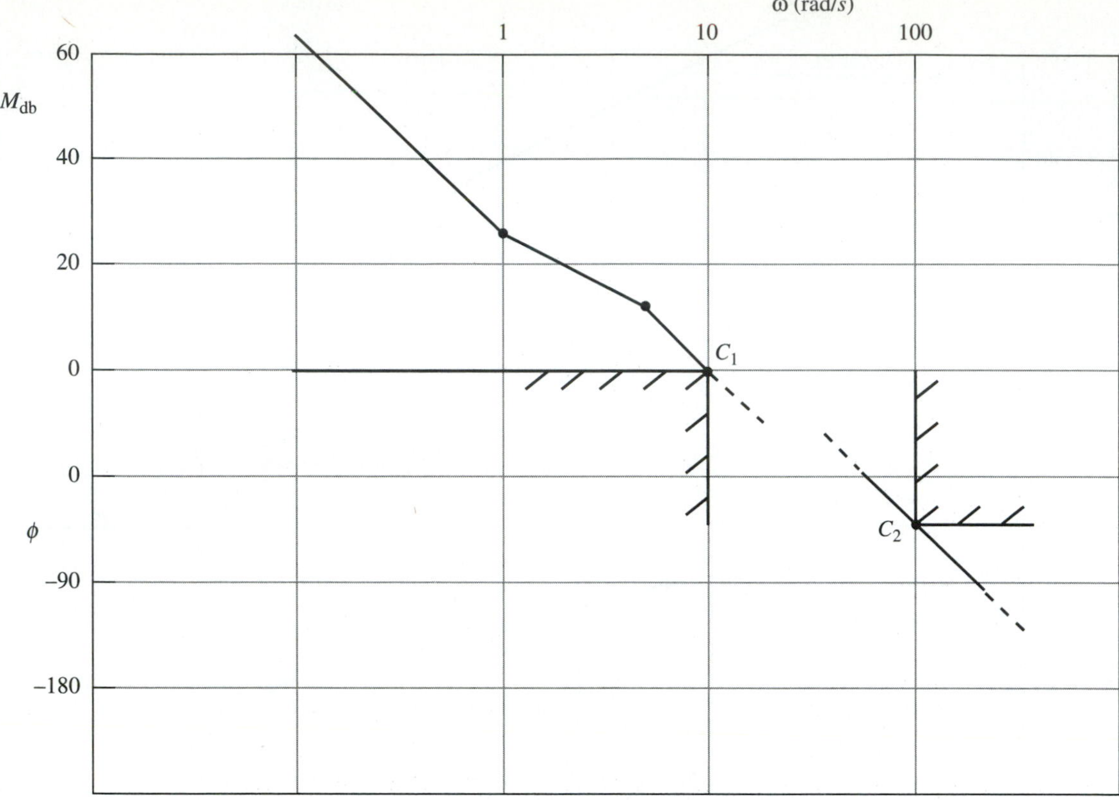

Fig. 20.13 Partial filter design.

Bode diagram, while Figure 20.16 shows the unit step response of the system. The overshoot is about 28%, a little larger than expected, but recall that this is a higher-order system, not a simple one of order 2, and the compensator introduces open- and closed-loop zeros. Further experiments with the filter parameters may produce better results, but for our purposes, the result is acceptable. The final open-loop transfer function with compensator becomes

$$G_cGH(s) = \frac{316(1 + s)(1 + 0.179s)}{s(1 + 15.8s)(1 + 0.056s)(1 + 0.2s)} \tag{20.19}$$

The system now meets all four performance constraints required of the open- and closed-loop response.

Sample Problem 20.1

A system has the following open-loop transfer function:

$$GH(s) = \frac{K}{s(1 + s)^2}$$

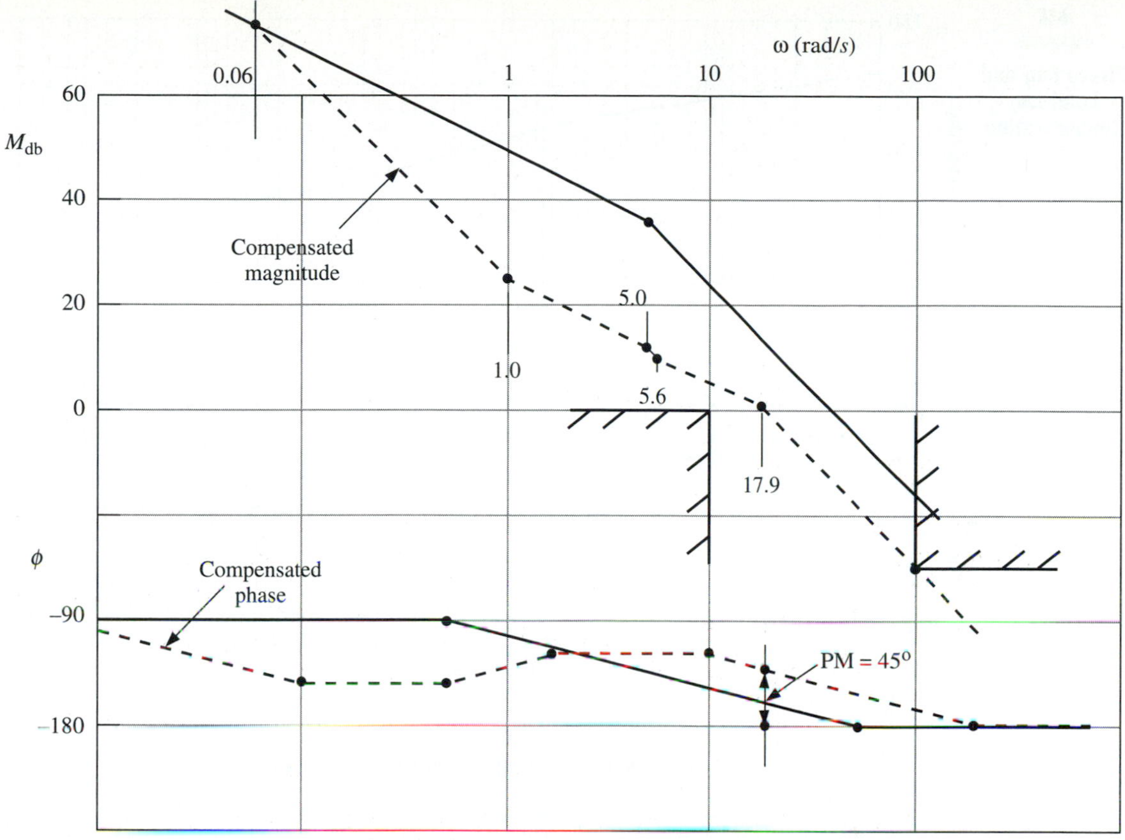

Fig. 20.14 Compensated Bode diagram.

Design a phase lag compensation element that enforces the following performance specifications on the closed loop system:

1. a damping ratio of $\zeta = 0.5$ and
2. a velocity error of no more than 10%.

Solution

The process begins by plotting the Bode diagram for the uncompensated system for the value of gain that satisfies the steady-state error requirement. For the system under study, the 10% velocity error translates into

$$K = K_v = \frac{1}{e_{ss}} = \frac{1}{0.1} = 10$$

Figure SP20.1.1 shows the Bode diagram for $K = 10$ and indicates that for this value of gain, the system is unstable. The compensator will both stabilize the system and meet the transient performance requirement. In order to do the latter, a phase margin requirement of

$$\phi = 50^\circ + 10\% = 55^\circ$$

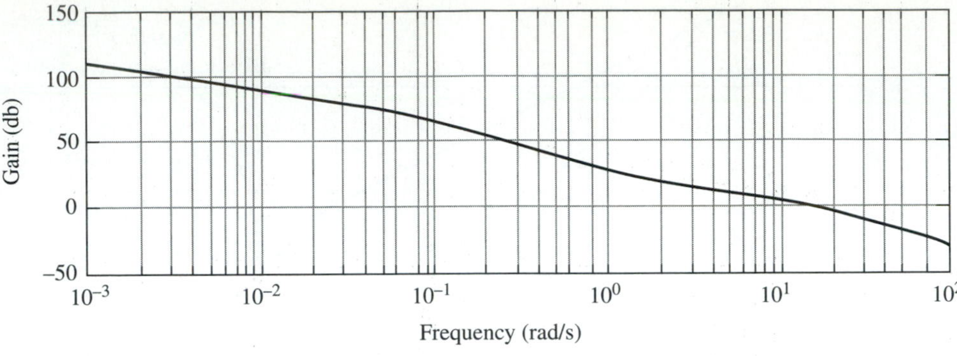

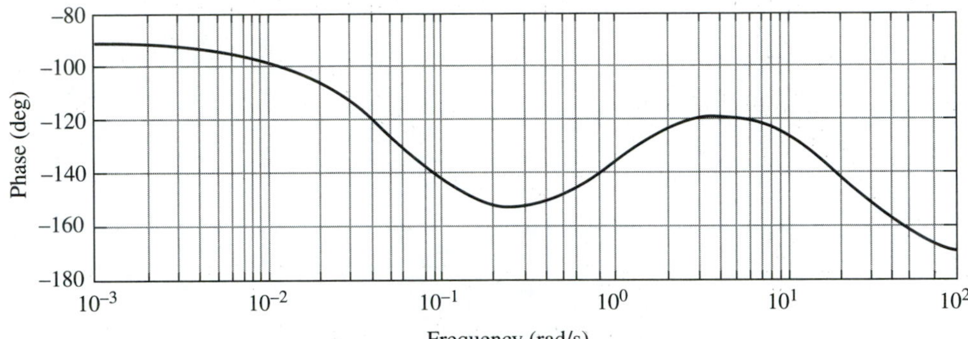

Fig. 20.15 Accurate system Bode diagram.

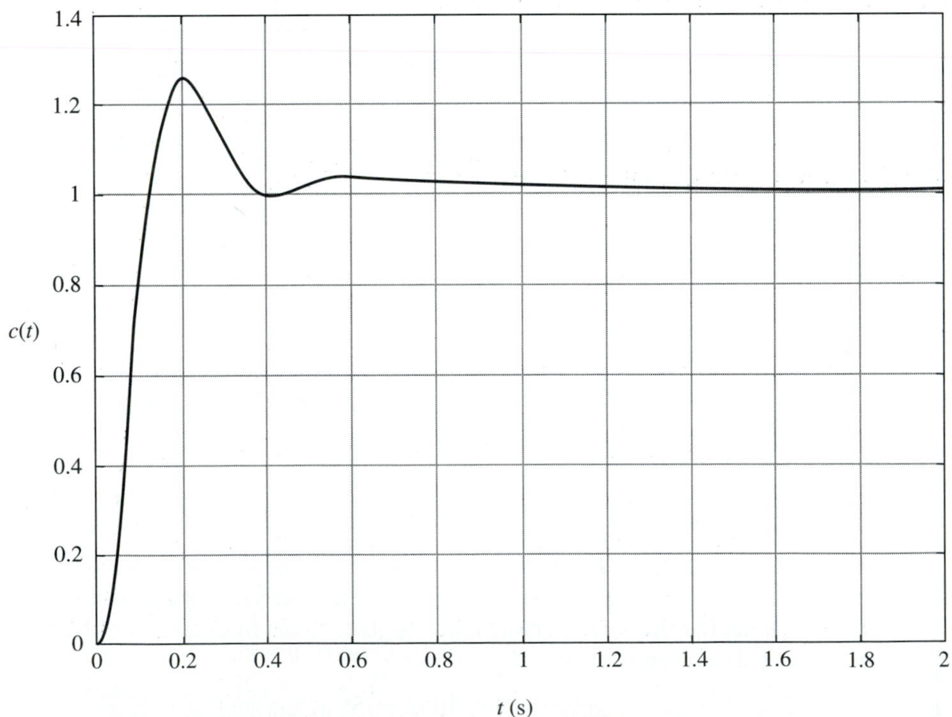

Fig. 20.16 Step response of compensated system.

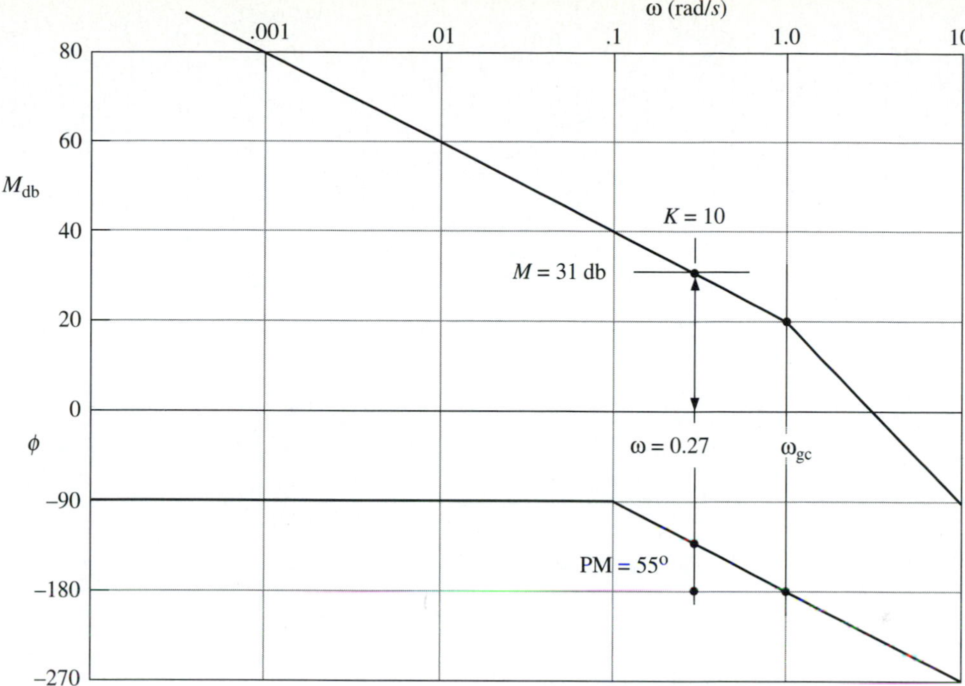

Fig. SP20.1.1

is needed. Inspecting Fig. SP20.1.1, this indicates that the phase achieves this value at $\omega = 0.27$ rad/s. At this frequency, the magnitude is 31 db. Substituting into

$$\Delta M = 20 \log_{10} \beta$$

yields

$$\beta = 35.5$$

The zero break frequency is set one decade below the new gain crossover frequency, i.e.,

$$\omega_z = \frac{1}{\tau} = 0.027 \text{ rad/s}$$

which gives $\tau = 37$ and leads to $\beta\tau = 1296$, giving the pole break frequency at

$$\omega_p = \frac{1}{\beta\tau} = 0.00077 \text{ rad/s}$$

The compensator network has the transfer function

$$G_c(s) = \frac{1 + \tau s}{1 + \beta\tau s} = \frac{1 + 37s}{1 + 1296s}$$

and leads to the compensated open-loop transfer function

$$G_c GH(s) \frac{10(1 + 37s)}{s(1 + s)^2(1 + 1296s)}$$

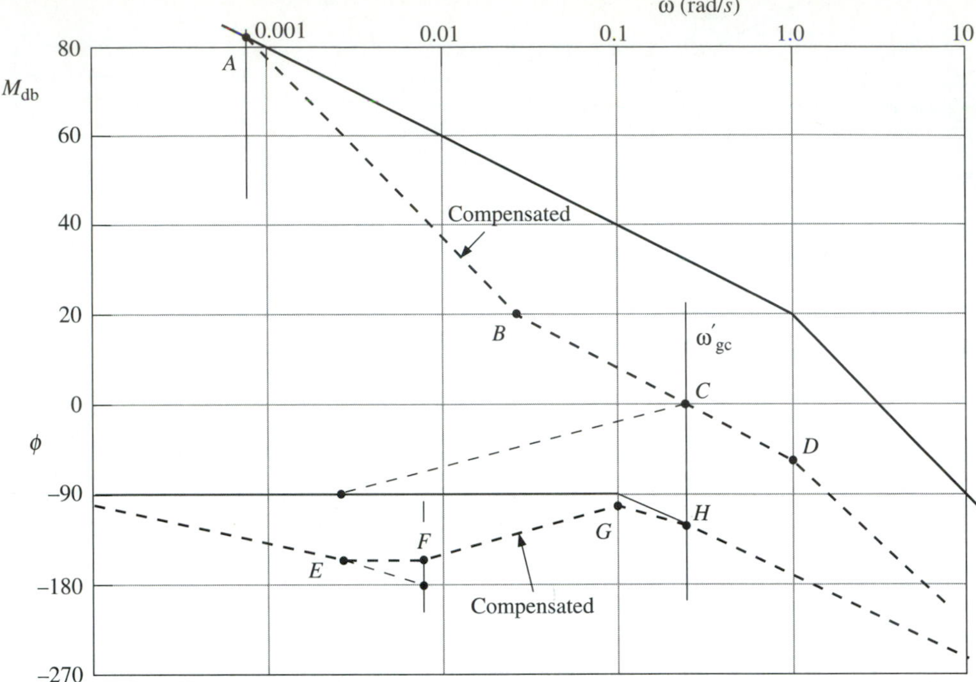

Fig. SP20.1.2

The compensated Bode diagram is shown in Figure SP20.1.2. For the magnitude part, a line is drawn from the uncompensated magnitude plot at the pole break frequency (0.00077 rad/s) (point A) at a slope of -40 db/decade. This line continues to B, which is the break point of the filter zero, when the slope reverts to the uncompensated value of -20 db/decade, until the double-system pole at $\omega = 1$ is reached. After this, the slope of the magnitude plot is -60 db/decade.

The phase plot is obtained by superimposing the numerator and denominator components of the filter onto the uncompensated phase and adding the two together. At low frequency, the compensator pole is dominant until the numerator component begins at point E. When the denominator component is finished, at point F, the phase increases until the system phase at G changes the slope by two units in the negative sense. At H, the effect of the compensator is finished, and the phase follows the uncompensated plot. The phase margin is obtained from point H in Fig. SP20.1.2 and is seen to be about 55°, which should give the response desired; however, a computer simulation of the step response shown in Fig. SP20.1.3 indicates an 18% overshoot. This illustrates the necessity of actually simulating the system response, since the presence of a closed-loop zero (from the open-loop filter zero) produces a less damped response than expected. In order to achieve the required 10% overshoot, one more iteration of the filter design will be undertaken. This time, we will aim for a phase margin of 65°. Figure 20.1.4 indicates that the new crossover frequency becomes $\omega'_{gc} =$

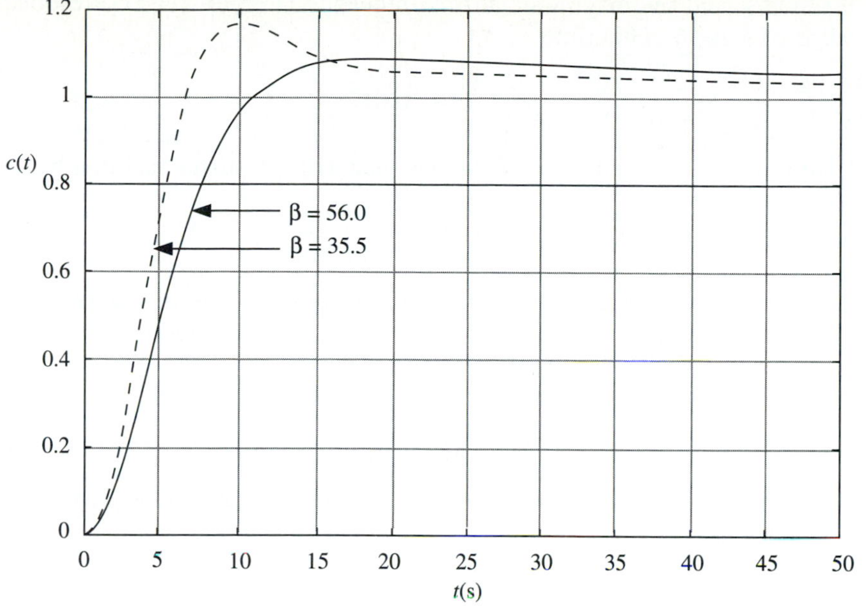

Fig. SP20.1.3

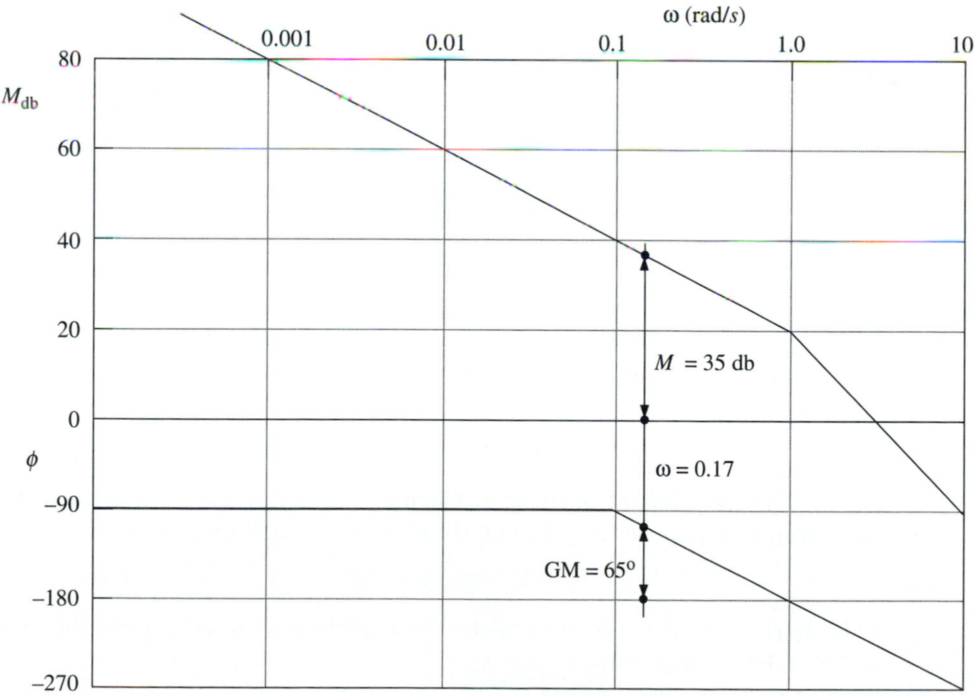

Fig. SP20.1.4

0.17 rad/s, and the magnitude at this frequency is 35 db. The corresponding value of β is 56. This time

$$\frac{1}{\tau} = 0.017$$

leading to $\tau = 59$ and $\beta\tau = 3304$. The new filter transfer function becomes

$$G_c(s) = \frac{1 + 59s}{1 + 3304s}$$

Simulating the compensated-system step response leads to the second plot shown in Fig. 20.1.3. The response now meets the required transient performance criterion. The final open-loop transfer function is

$$G_cGH(s) = \frac{10(1 + 59s)}{s(1 + s)^2(1 + 3304s)}$$

Notice how the reduced open-loop bandwidth for $\beta = 56$ produces a more sluggish response.

Sample Problem 20.2

Solve Sample Problem 19.2 using a phase lag compensation network, and compare the step response to that with phase lead compensation.

Solution

Recall that the open-loop transfer function is of the form

$$GH(s) = \frac{K}{s(1 + 0.5s)(1 + 0.1s)}$$

subject to the following requirements:

1. damping ratio of $\zeta = 0.7$ and
2. steady-state velocity error of less than 25%.

Figure SP20.2.1 shows the Bode diagram of the uncompensated system, plotted for the gain K that meets the steady-state error requirement, namely $K = 4$. We wish to find the new gain crossover frequency when the phase margin is given by

$$\phi = 70 + 10\% = 77°$$

Figure SP20.2.1 shows this to occur when $\omega = 0.4$ rad/s, at which frequency the magnitude is 20 db. The required value of β is obtained from

$$20 \log_{10} \beta = 20$$

yielding $\beta = 10$. This gives the zero break point one decade below the new gain crossover frequency,

$$\frac{1}{\tau} = 0.04 \text{ rad/s}$$

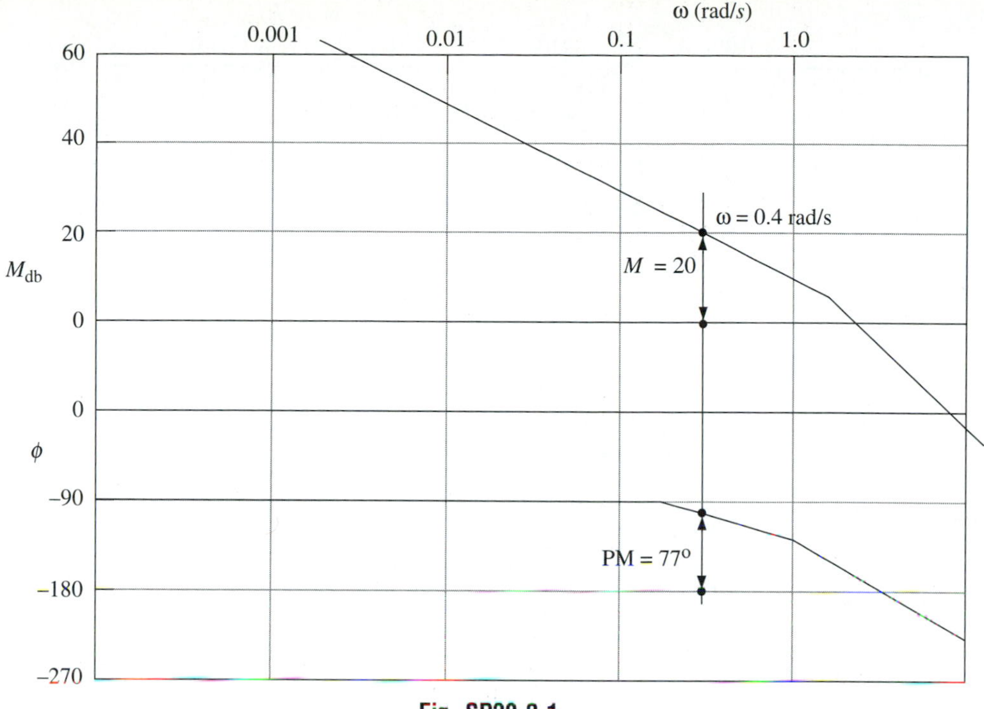

Fig. SP20.2.1

resulting in $\tau = 25$ and $\beta\tau = 250$. The compensation network may be written as

$$G_c(s) = \frac{1 + \tau s}{1 + \beta\tau s} = \frac{1 + 25s}{1 + 250s}$$

The break frequencies of the zero and pole become $\omega_z = 0.04$ rad/s and $\omega_p = 0.004$ rad/s, respectively, and the compensated Bode diagram is as shown in Fig. SP20.2.2. The magnitude part is plotted using the same technique described for the previous problem, starting at point *A*, changing the slope by −20 db/decade, moving to *B*, and changing the slope back again. The magnitude must pass through the design point *C* and continue until the system break point *D* is encountered. Most of the effect of the phase lag network is in the lower frequencies, so the high-frequency end of the Bode diagram beyond the gain crossover frequency need not be plotted accurately.

The phase is also plotted as before, where the components of the filter are added graphically to that of the uncompensated system, producing the compensated phase *EFGHIJ* in Fig. SP20.2.2. The new phase margin is obtained from point *I* and is estimated to be 70°. Figure SP20.2.3 shows an accurate Bode diagram confirming the phase margin. Finally, Fig. SP20.2.4 shows the unit step response of the system, indicating that the transient performance criterion has been met. However, observe the lengthy time scale in Fig. SP20.2.4, suggesting a sluggish response. Compare this with Fig. SP19.2.6, which achieves its peak

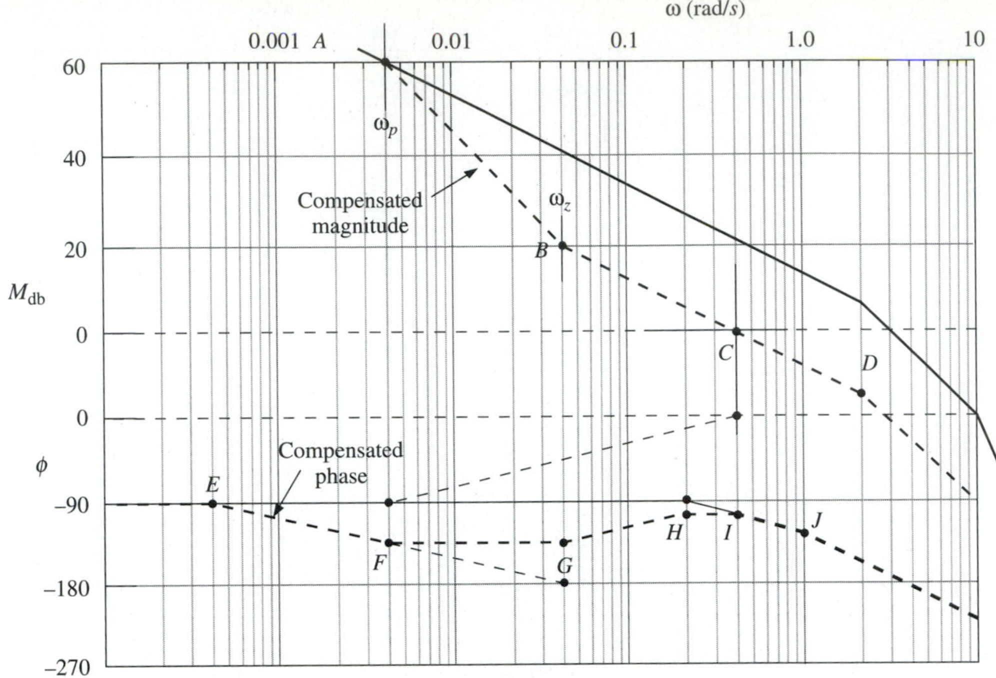

Fig. SP20.2.2

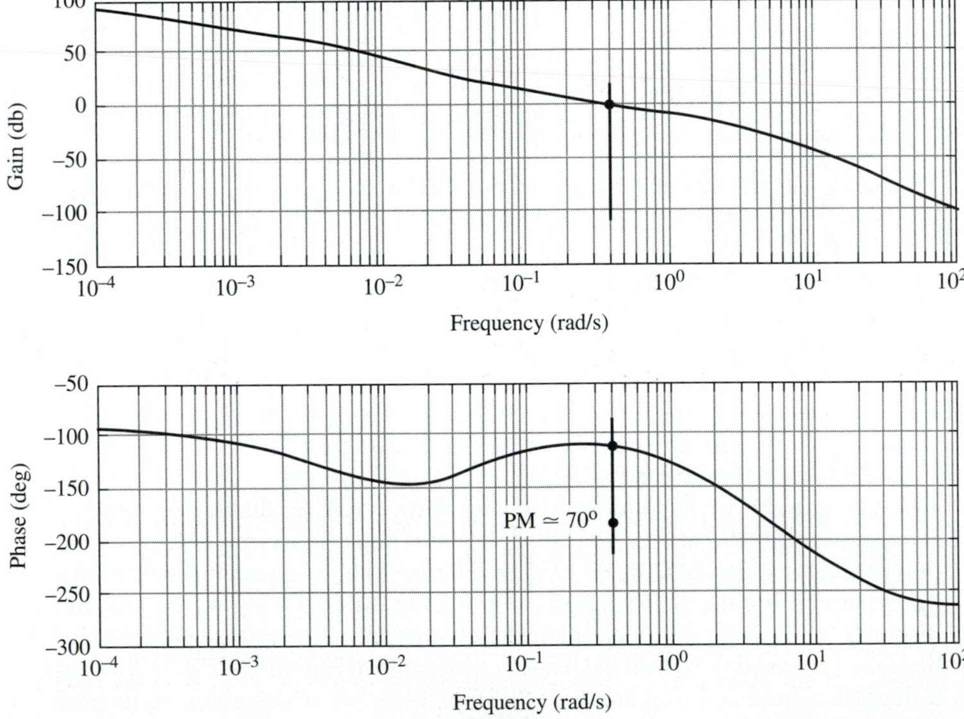

Fig. SP20.2.3

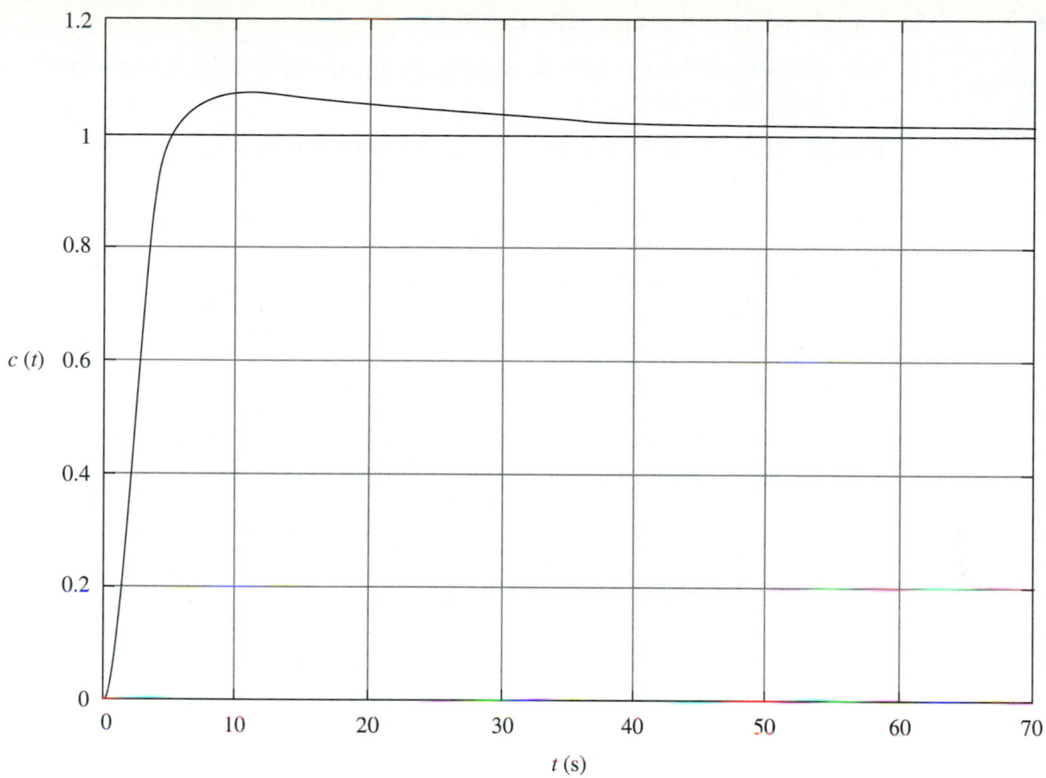

Fig. SP20.2.4

value at about $T_p = 0.6$ s compared with $T_p = 10.0$ s for the lag-compensated system. This is a characteristic feature of lag compensation, but if speed of response is not important, then the performance is acceptable.

Note that phase lag elements are simple to construct since they consist of purely passive components. Phase lead, on the other hand, requires an increase in gain to cancel the $1/\alpha$ intrinsic attenuation. If this can be provided by simply adjusting the forward-path gain of the plant, then the phase lead is also comprised of purely passive components. If, however, the forward-path gain is not adjustable, a separate amplifier will be needed, thereby making the phase lag method the preferred solution based solely on the grounds of simplicity and cost.

Sample Problem 20.3

A closed-loop control system has an open-loop transfer function given by

$$GH(s) = \frac{K}{s(1 + 0.2s)(1 + 0.1s)}$$

Design any kind of cascade compensation filter that ensures that the closed-loop system meets the following performance measures, listed in order of priority:

1. The velocity error constant $K_v = 40$.

2. The closed-loop damping ratio $\zeta \geq 0.35$.
3. The compensated open-loop bandwidth is not to exceed the uncompensated open-loop bandwidth.
4. The 2% settling time for a step input is to be less than 2 s.

Solution

We begin by first satisfying the steady-state error requirement by setting $K = 40$ and plotting the uncompensated Bode diagram. This is shown in Fig. SP20.3.1. From the figure it is seen that the system is unstable. Considering the third performance requirement, it appears that phase lead compensation alone will not be acceptable, since this will increase the open-loop bandwidth above its present value of about 15 rad/s. The next option is to consider phase lag compensation. Adding 10% to the phase margin requirement implies a need for about 40°, which, as shown in Fig. SP20.3.2, occurs at $\omega = 2.5$ rad/s. The magnitude at this frequency is 24 db, implying a value of $\beta = 15.8$. The break frequency of the pole is one decade below the new gain crossover frequency

$$\tau = \frac{1}{0.25} = 4.0$$

giving $\beta\tau = 63.4$. The compensation network transfer function becomes

$$G_c(s) = \frac{1 + 4s}{1 + 63.4s}$$

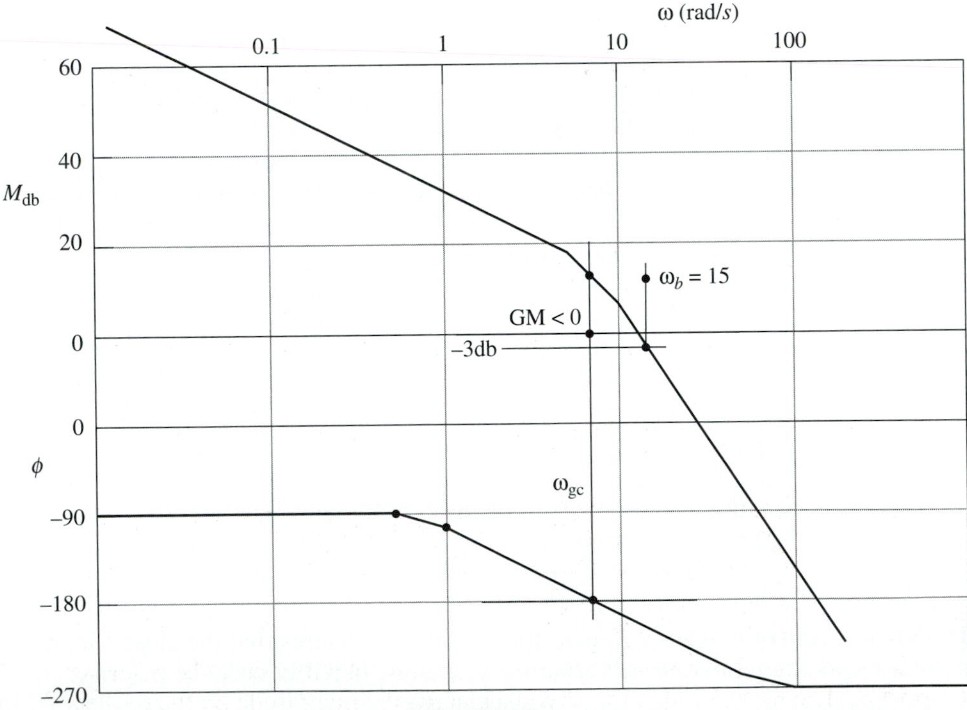

Fig. SP20.3.1

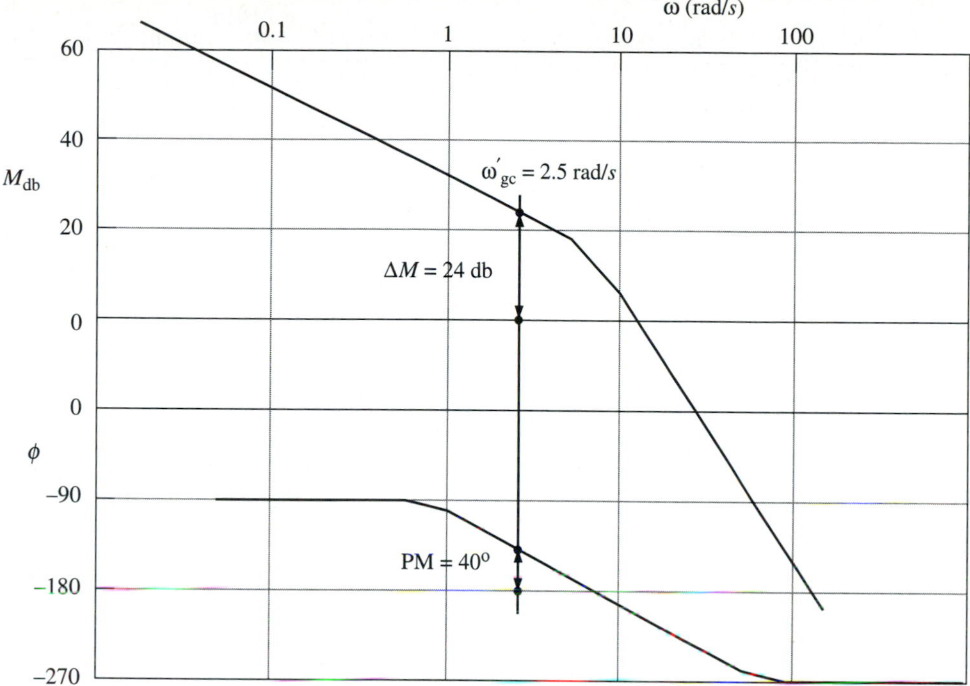

Fig. SP20.3.2

The compensator pole and zero break frequencies are $\omega_p = 0.015$ rad/s and $\omega_z = 0.25$ rad/s. The magnitude part of the compensated Bode diagram using phase lag alone is as shown in Fig. SP20.3.3 and clearly satisfies the third design requirement. The question remains as to whether the last performance constraint regarding the settling time has been met. Since

$$T_s = \frac{4}{\zeta\omega_n}$$

and the settling time and damping ratio are fixed,

$$\omega_n = \frac{4}{T_s\zeta} = 5.71$$

Recalling from Module 17 that an estimate of the natural frequency of a higher-order system may be obtained from the gain crossover frequency of the open-loop system, Fig. SP20.3.3 indicates that the natural frequency is 2.5 rad/s, less than half that required to satisfy the settling time constraint.

We will attempt to design a system with $\omega_n = 6$ rad/s using a lead-lag compensation filter. In fact, this constraint on ω_n and the bandwidth constraint may be represented as prohibited regions on the Bode diagram, as indicated in Fig. SP20.3.4.

The first step is to assume that a phase lag element will reduce the magnitude to zero when $\omega = 6$ rad/s, thereby just placing the magnitude on the corner of the low-frequency constraint (Fig. SP20.3.5).

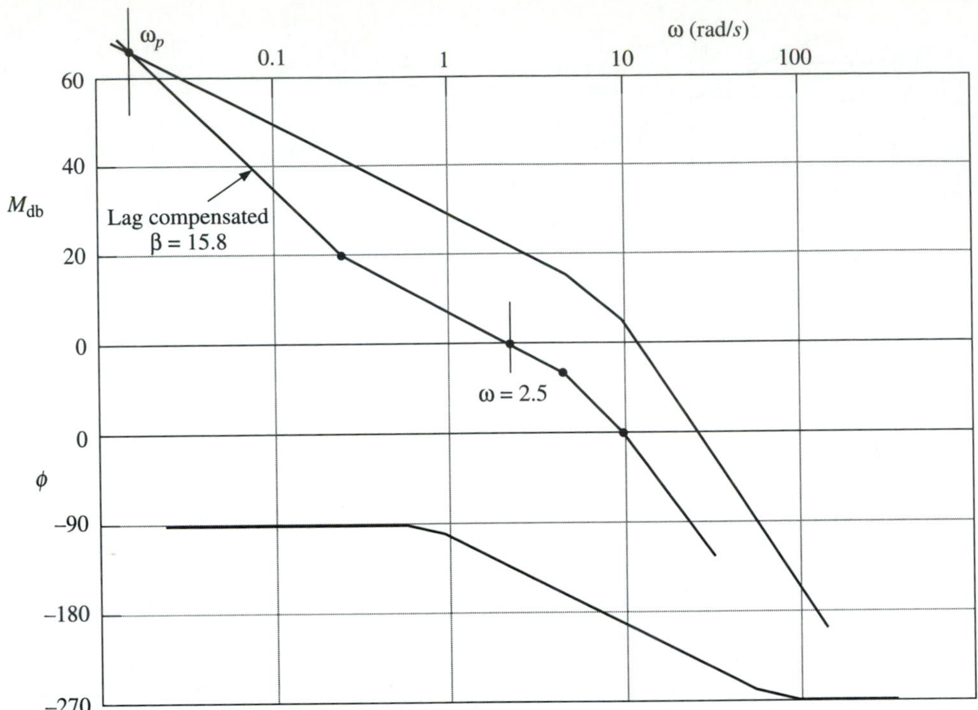

Fig. SP20.3.3

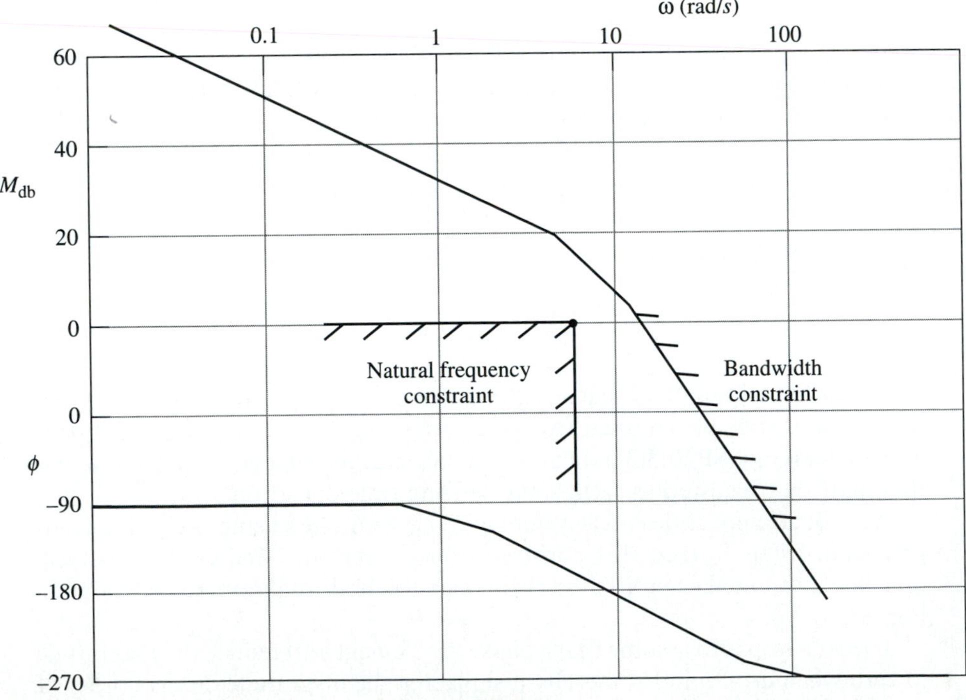

Fig. SP20.3.4

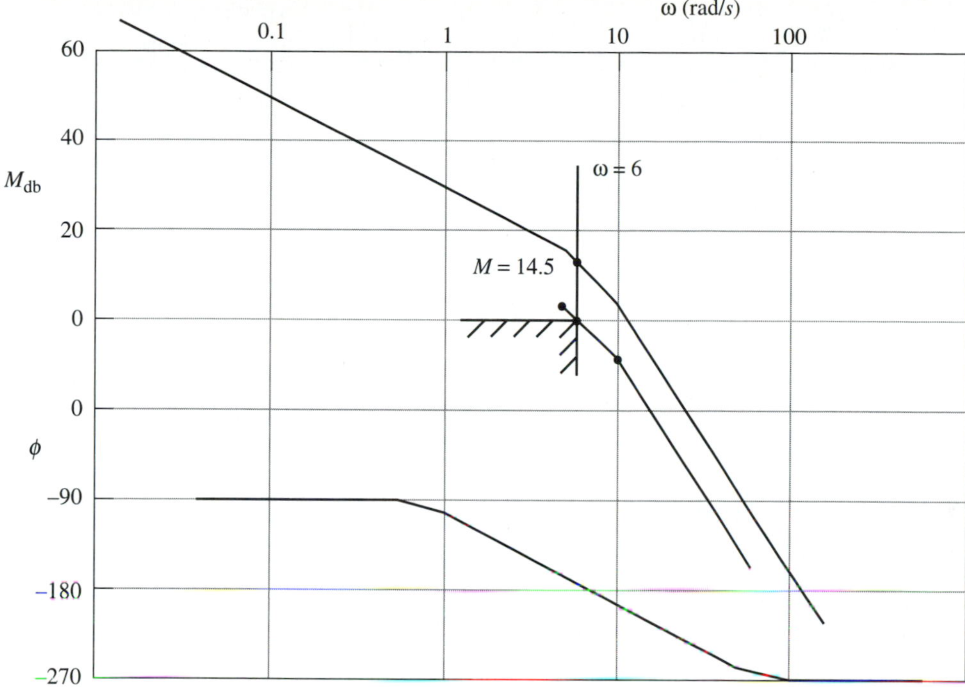

Fig. SP20.3.5

The magnitude of 14.5 db at this frequency translates into $\beta = 5.3$. It follows that

$$\tau = \frac{1}{0.6} = 1.67$$

and $\beta\tau = 8.83$. The lag part of the filter becomes

$$G_{c1}(s) = \frac{1 + 1.67s}{1 + 8.83s}$$

with pole and zero break frequencies of 0.6 and 0.11 rad/s, respectively. The effect of lag compensation only is shown in Fig. SP20.3.6. We now consider the lead part of the compensation. From Fig. SP20.3.6, it is clear that the lag-compensated system is barely stable, so all of the required phase lead will have to be supplied by the lead part of the filter. This implies a required phase margin of 40°, which in turn defines $\alpha = 4.5$. Note that since the design is bandwidth-limited, we must have zero attenuation at high frequencies, implying $\alpha = \beta = 5.3$. Since this value is close to the desired value of $\alpha = 4.5$, we will design for maximum bandwidth by setting $\alpha = 5.3$. This yields

$$\Delta M = -10 \log_{10} \alpha = -7.3 \text{ db}$$

leading to the new gain crossover frequency $\omega_m = 9$ rad/s. Then we obtain

$$T = \frac{1}{\omega_m \sqrt{\alpha}} = 0.048$$

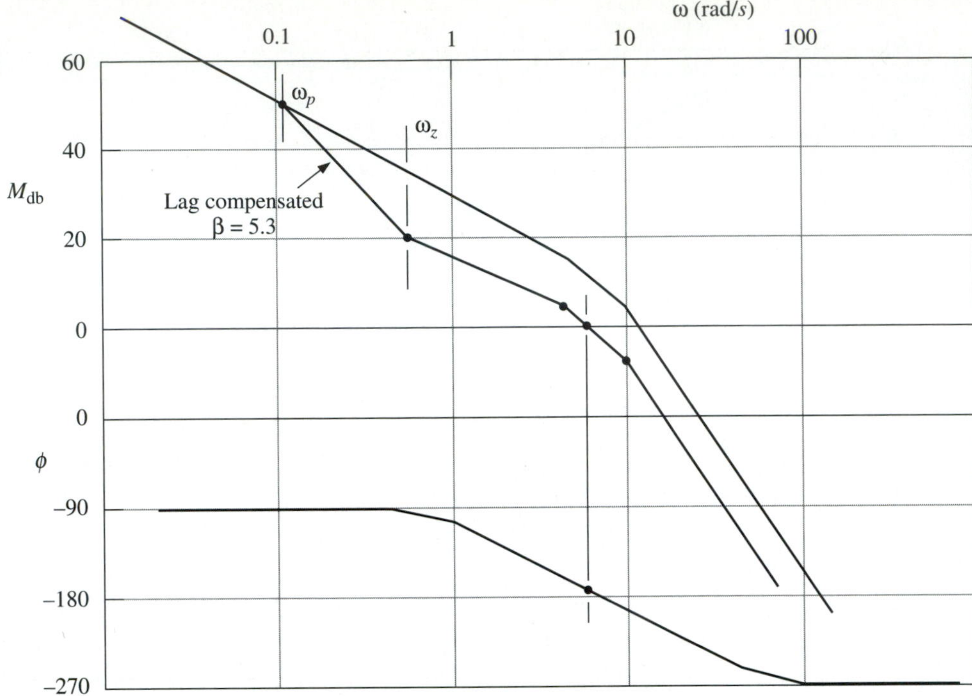

Fig. SP20.3.6

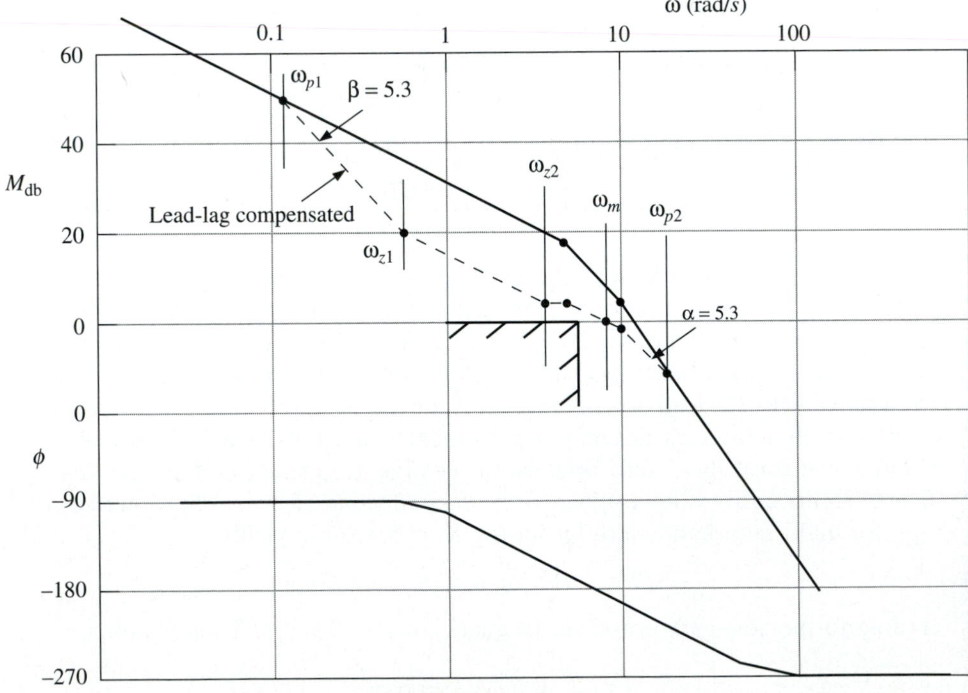

Fig. SP20.3.7

and $\alpha T = 0.256$. The lead part of the compensation filter may be written as

$$G_{c2}(s) = \frac{1 + 0.256s}{1 + 0.048s}$$

Pole and zero break frequencies for the lead component are $\omega_p = 20.8$ and $\omega_z = 3.9$ rad/s. The combined lead-lag compensator is

$$G_c(s) = G_{c1}(s)G_{c2}(s)$$

resulting in the compensated-system open-loop transfer function

$$G_cGH(s) = \frac{10(1 + 1.67s)(1 + 0.256s)}{s(1 + 0.2s)(1 + 0.1s)(1 + 8.83s)(1 + 0.048s)}$$

The approximate-magnitude part of the Bode diagram may be plotted in the usual way, and this is shown in Fig. SP20.3.7 while an accurate plot is shown in Fig. SP20.3.8. In Fig. SP20.3.8, the asymptotic uncompensated phase is also shown to illustrate first the phase lag, then the phase lead introduced by the compensator. The phase margin is seen to be about 35°. Finally, Fig. SP20.3.9 shows the unit step response of the system and indicates that all performance specifications have been achieved.

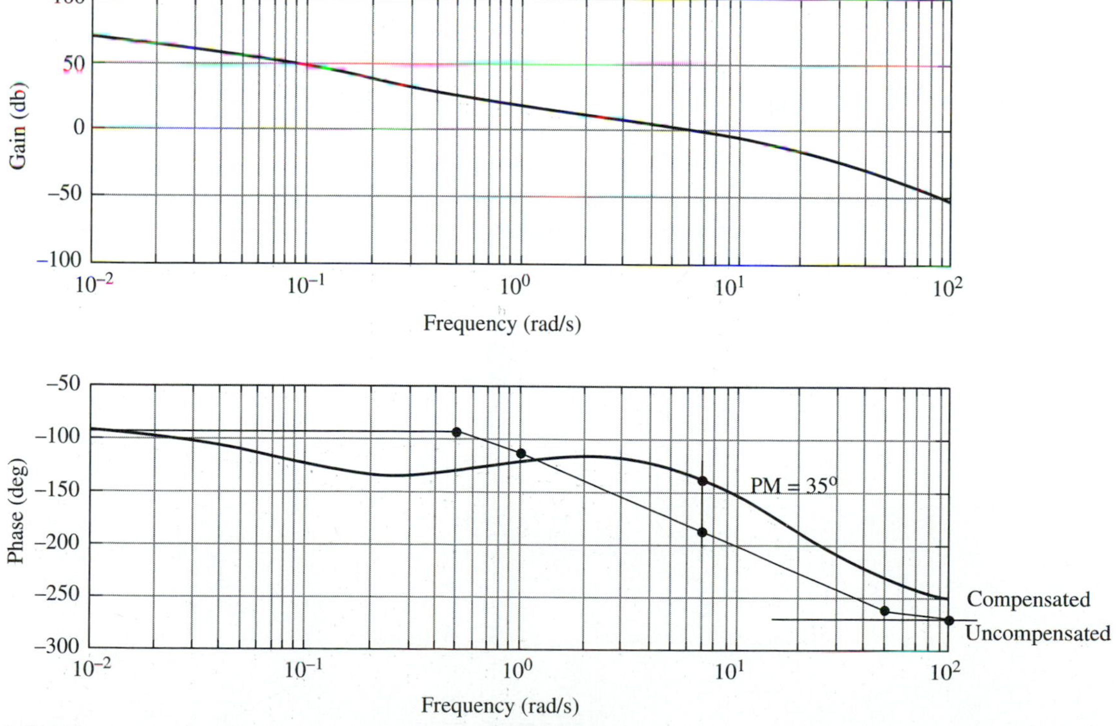

Fig. SP20.3.8

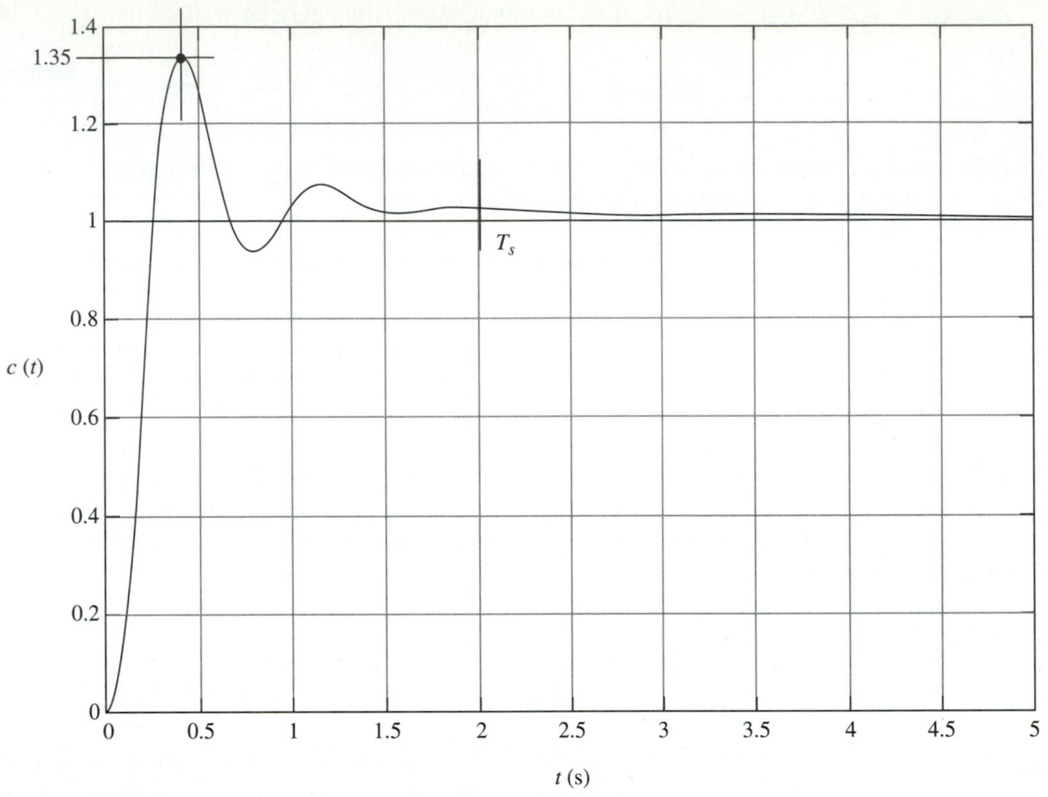

Fig. SP20.3.9

◆ PROBLEMS

20.1 Show that the mechanical system illustrated in Fig. P20.1 may be considered as a phase lag element, where the input is defined to be displacement R and the output is displacement C. Determine the filter parameters τ and β in terms of the various damping and stiffness constants.

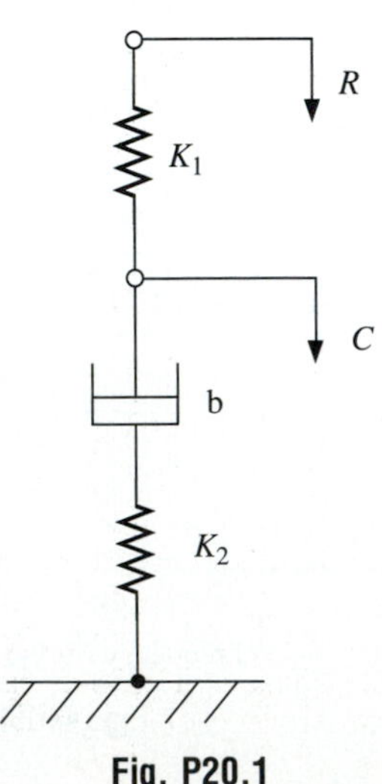

Fig. P20.1

20.2 A unity-feedback control system has the open-loop transfer function

$$GH(s) = \frac{K}{s(1 + 0.67s)}$$

Design a suitable phase lag compensation filter that ensures that the closed-loop system meets the following performance requirements:
a. A velocity error no more than 20%.
b. A maximum percentage overshoot to a step of 10%.

20.3 Compute the phase lag compensation element that gives the system shown in Fig. P20.3 a phase margin of around 135°. Write down the compensated, open-loop transfer function.

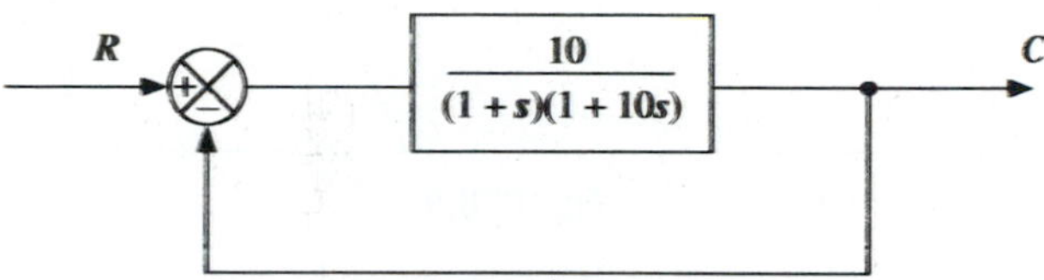

Fig. P20.3

20.4 Design a phase lag cascade compensation filter G_c for the open-loop system

$$GH(s) = \frac{K}{s(1 + 0.2s)(1 + 0.4s)}$$

such that the closed-loop system has a damping ratio $\zeta = 0.7$ and a velocity error $K_v = 100$. What is the estimated settling time of the compensated-system output when the input is a unit step? Sketch the root locus of the uncompensated and compensated systems.

20.5 Consider the system shown in Fig. P20.5. Design a compensator G_c so that the error constant remains unchanged but the compensated system has a phase margin of 35°. For the compensated system, estimate and sketch the unit step response.

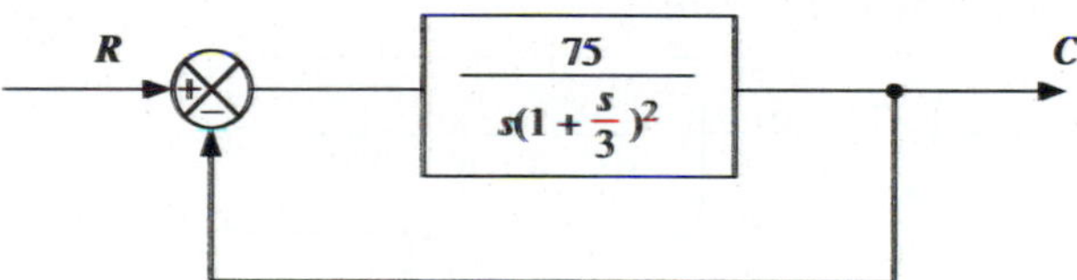

Fig. P20.5

20.6 For the system described by the open-loop transfer function

$$G(s) = \frac{K}{(1 + 0.1s)^2(1 + 0.2s)^2}$$

find the phase lag compensator that satisfies the following conditions:
a. A position error constant of 20.
b. A phase margin of not more than 30°.

20.7 For the system shown in Fig. P20.7, compute a phase lag compensator that gives the closed-loop system a phase margin of around 35°. Sketch the root locus of the uncompensated and compensated systems.

20.8 In Fig. P20.8, the error coefficient is to remain unaltered. Select an appropriate G_c that gives the system at least a 45° phase margin and maximizes the open-loop bandwidth.

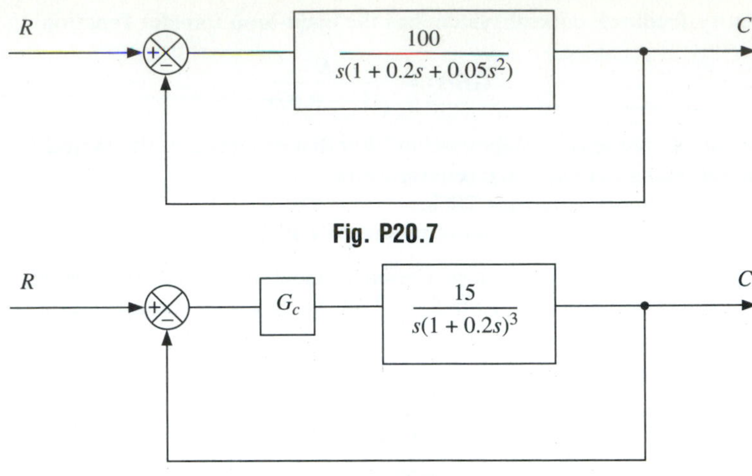

Fig. P20.7

Fig. P20.8

20.9 Consider the system that has unity feedback and forward-path transfer function

$$G(s) = \frac{12}{s^2}$$

The closed-loop system has to meet the following performance criteria:

a. Phase margin of 30°.

b. No increase in open-loop bandwidth.

c. No change in error constants.

Construct a lead-lag compensation network that achieves these objectives.

20.10 For the system shown in Fig. P20.10, design a lead-lag compensation network that meets the following performance requirements:

a. A 2% settling time for a step input not greater than 1 s.

b. Percentage overshoot for a step input not to exceed 20%.

c. Velocity error constant $K_v = 20$.

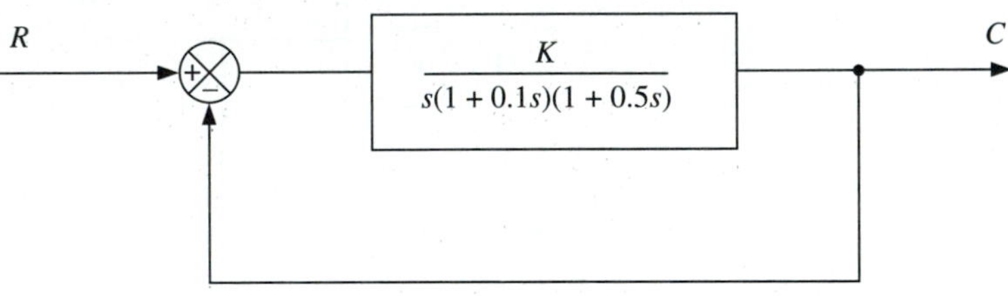

Fig. P20.10

MODULE 21

Multimode Controllers

Multimode controller design is a system compensation process in which the *type* of the system is changed so as to meet desired performance requirements. In the previous module, the student may have noticed that the steady-state error requirements were chosen to be consistent with the type of uncompensated system. For example, if the system were of type 1, a requirement on the velocity error was stipulated; if the system were type 0, a position error constant was given; and so on. The student may wonder how a type 1 system may be compensated so as to make its velocity error vanish. Clearly, the compensated system can no longer be of type 1. This form of compensation usually involves designing the controller in the loop (Fig. 21.1) to take a particular form, and the overall process is known as mode control. To begin with, simple proportional control will be reviewed, in which the forward-path gain is adjusted to obtain "good" response. First- and second-order systems subjected to external disturbances will be considered; however, the results obtained may be extended to higher-order systems without loss of generality.

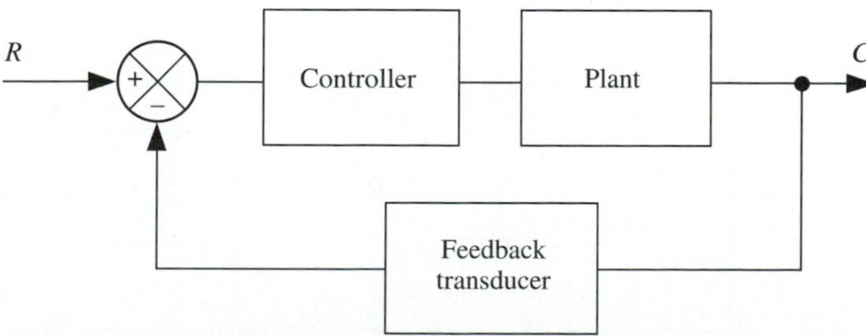

Fig. 21.1 Definition of controller.

◆ PROPORTIONAL CONTROL

A proportional controller is defined to be one in which the output is simply proportional to the input, where the constant of proportionality is the gain of the controller (Fig. 21.2). Proportional control is sometimes referred to as single-mode control. Consider initially the first-order system with a disturbance input, shown in figure 21.3. The transfer functions relating the output C to the two inputs R and T are

$$\frac{C}{R} = \frac{K/(1 + K)}{1 + \tau s/(1 + K)} \tag{21.1}$$

and

$$\frac{C}{T} = \frac{1/(1 + K)}{1 + \tau s/(1 + K)} \tag{21.2}$$

If the input R is a unit step, the steady-state output becomes

$$c_{ss}(t) = \frac{K}{1 + K} \tag{21.3}$$

If the disturbance T is a unit step, the corresponding steady-state output becomes

$$c_{ss}(t) = \frac{1}{1 + K} \tag{21.4}$$

These results may have been expected since both open-loop transfer functions are of zero order. In the case of a first-order system of type zero, good performance means making the steady-state error as small as possible, ideally making it zero. If K is made large, the output for unit R approaches unity, while the output for unit T approaches zero, which is a good result, but note that the error cannot be made equal to zero, or the disturbance rejection complete, since there will be prac-

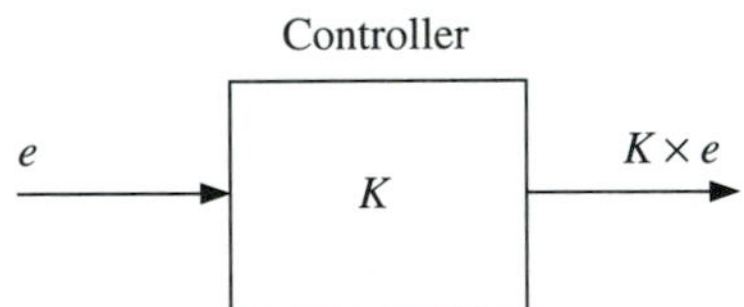

Fig. 21.2 Proportional controller.

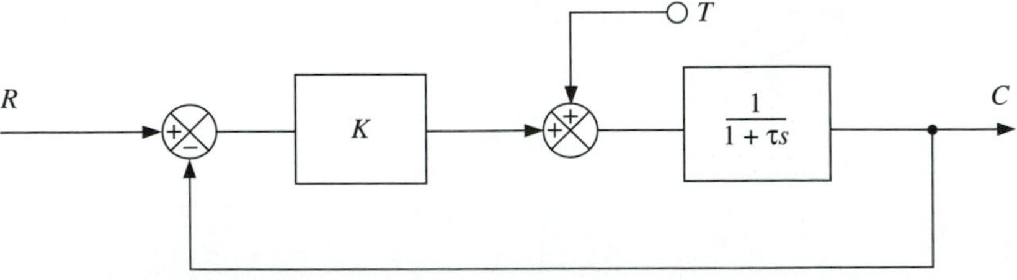

Fig. 21.3 First-order type 1 system with disturbance.

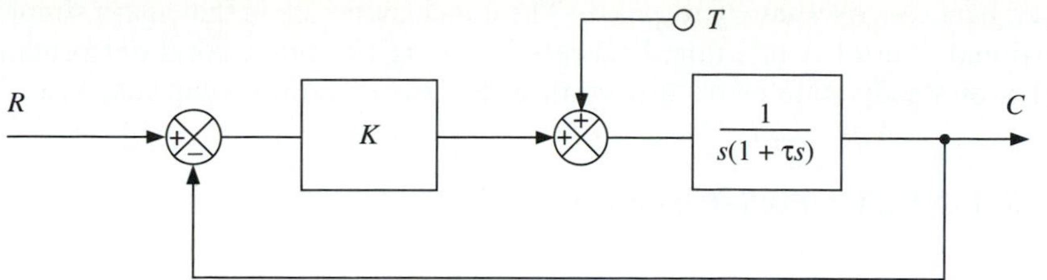

Fig. 21.4 Second-order type 1 system with disturbance.

tical limits on the magnitude of K. We conclude that simple proportional control (adjusting K) of a first-order system is of limited success.

Now consider a second-order system of type 1, also subjected to external disturbances, as shown in Fig. 21.4. Again we assume proportional control and determine the transfer functions from the two inputs R and T to the output C as

$$\frac{C}{R} = \frac{K}{\tau s^2 + s + K} \tag{21.5}$$

and

$$\frac{C}{T} = \frac{1}{\tau s^2 + s + K} \tag{21.6}$$

The steady-state output for a unit step input R is

$$c_{ss}(t) = 1 \tag{21.7}$$

and for a unit step disturbance T it is

$$c_{ss}(t) = \frac{1}{K} \tag{21.8}$$

Since the system is of type 1, the steady-state error for the regular input is zero, as expected, but the disturbance does produce a finite error. Although this may be made small by letting K be large (subject to practical limitations), the root locus of the system shown in Fig. 21.5 indicates that large K produces a small damping

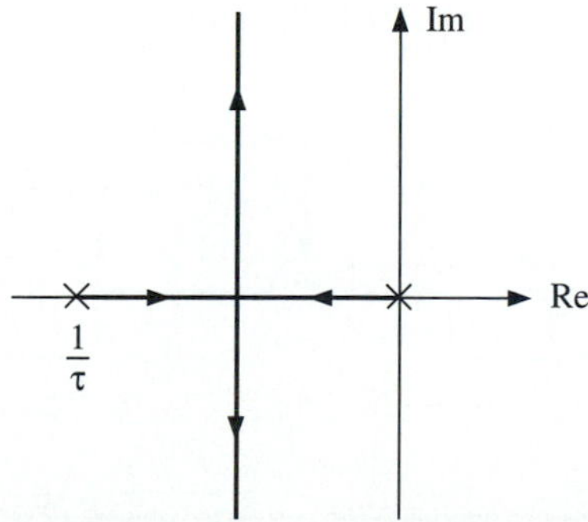

Fig. 21.5 Root locus of second-order system.

ratio, hence an oscillatory response. The conclusion here is that again simple proportional control is of limited success in trying to obtain good performance in terms of steady-state error, disturbance rejection, and transient response.

◆ PROPORTIONAL-PLUS-INTEGRAL CONTROL

Consider now a proportional-plus-integral (PI) controller, where the output is proportional to the input plus the integral of the input. A controller of this type is easily implemented using operational amplifiers in summing, gain, and integrating configurations. This type of control is called two-mode control. The formulation of the controller is shown in Fig. 21.6, while its transfer function is seen to be

$$\theta_o = \left[K_p + \frac{K_i}{s} \right] \theta_i \tag{21.9}$$

or

$$\frac{\theta_o}{\theta_i} = \frac{K_p s + K_i}{s} \tag{21.10}$$

Suppose this PI controller replaces the simple proportional controller in the first-order system just considered. The new system looks like that shown in Fig. 21.7. The new transfer functions relating the output to the two inputs become

$$\frac{C}{R} = \frac{K_p s + K_i}{\tau s^2 + s(1 + K_p) + K_i} \tag{21.11}$$

and

$$\frac{C}{T} = \frac{s}{\tau s^2 + s(1 + K_p) + K_i} \tag{21.12}$$

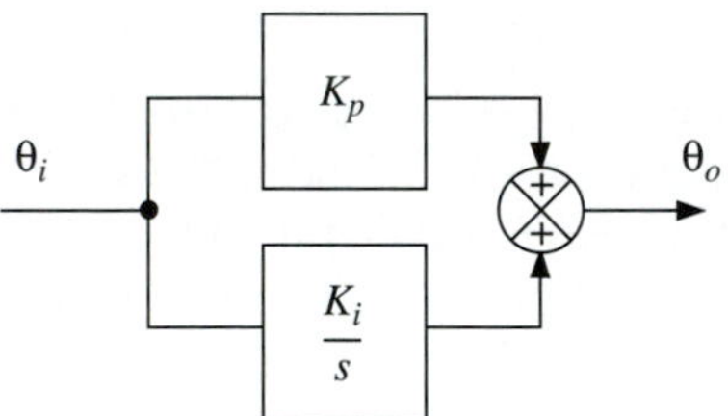

Fig. 21.6 PI controller.

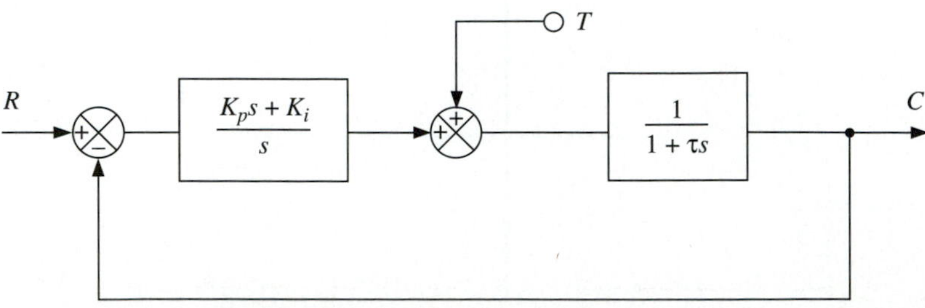

Fig. 21.7 PI control of first-order system.

This time, when the input R is a unit step, the steady-state output C becomes

$$c_{ss}(t) = 1 \tag{21.13}$$

and the steady-state output for unit T is

$$c_{ss}(t) = 0 \tag{21.14}$$

Proportional-plus-integral control has reduced the steady-state error to zero and completely rejected the external disturbance. Note that this has been achieved because the system type has been changed from 0 to 1, due to the controller.

We now turn to the second-order system shown in Fig. 21.8. In this system the transfer functions become

$$\frac{C}{R} = \frac{K_p s + K_i}{\tau s^3 + s^2 + K_p s + K_i} \tag{21.15}$$

and

$$\frac{C}{T} = \frac{s}{\tau s^3 + s^2 + K_p s + K_i} \tag{21.16}$$

Letting R be a unit step produces

$$c_{ss}(t) = 1 \tag{21.17}$$

and for the unit disturbance

$$c_{ss}(t) = 0 \tag{21.18}$$

It is seen that the disturbance is completely rejected without resorting to high forward-path gains. There is a small problem, however. Consider drawing the root locus of the system (ignoring the disturbance) using the open-loop transfer function

$$GH(s) = \frac{K_p s + K_i}{s^2(1 + \tau s)} \tag{21.19}$$

The locus, shown in Fig. 21.9, consists of a double pole at the origin, another pole at $s = 1/\tau$, and a zero at $s = -K_i/K_p$. In order for the system to be stable, the zero must be closer to the imaginary axis than the system pole, i.e.,

$$\left|\frac{1}{\tau}\right| > \left|\frac{K_i}{K_p}\right| \tag{21.20}$$

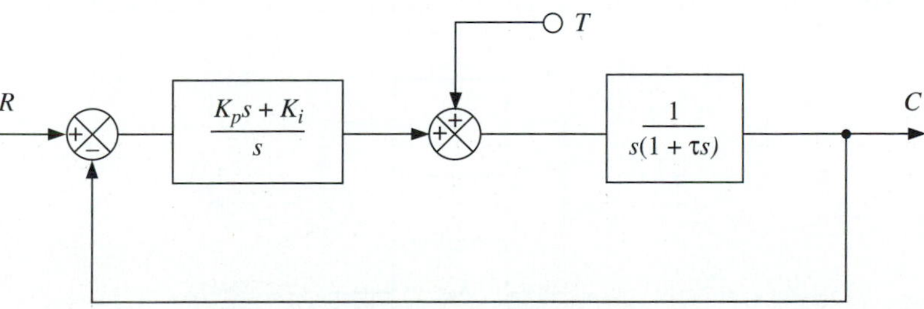

Fig. 21.8 PI control of second-order system.

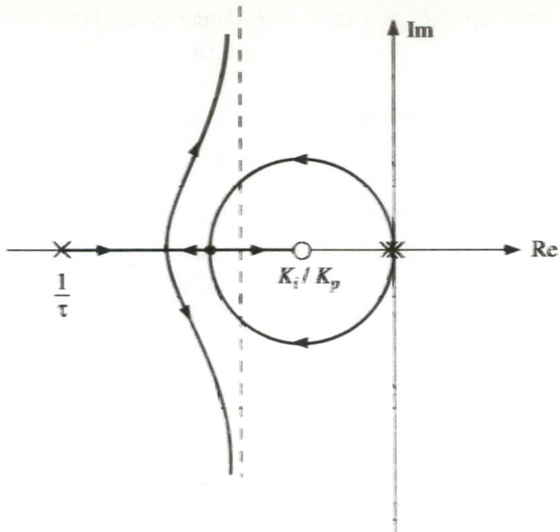

Fig. 21.9 Root locus of PI-controlled second-order system.

This places some restrictions on the selection of the proportional and integral gains, and as can be seen from the root locus, the system may suffer from small amounts of damping if the system pole ($s = 1/\tau$) is close to the imaginary axis.

It may be questioned, why not construct just an integral controller, which surely must be simpler to physically realize than a PI controller? Inspection of equation 21.15 shows that if $K_p = 0$, the term in s will disappear from the denominator of the closed-loop transfer function. Since this is the characteristic equation (when the denominator is set equal to zero), Routh's stability criterion stipulates that the system will be unstable due to the nonexistence of one of the coefficients. For this reason, integral control is not used alone, but with a proportional term added.

◆ PROPORTIONAL-PLUS-DERIVATIVE CONTROL

Proportional-plus-integral control is able to eliminate steady-state errors but has little effect on transient response. Proportional-plus-derivative (PD) control is able to influence the transient response by changing the damping ratio directly. We define a PD controller as one in which the controller output is the sum of two signals, one proportional to the input plus another proportional to the derivative of the input, as shown in Fig. 21.10. This type of control is also called two-mode control.

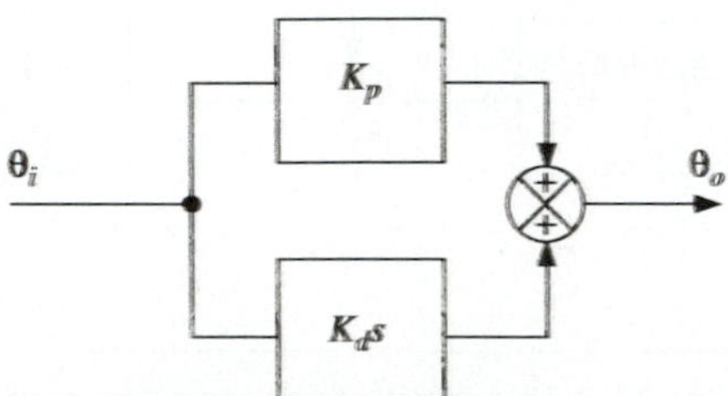

Fig. 21.10 PD controller.

and it may be appreciated that a PD controller may be constructed at low cost from operational amplifier building blocks. The corresponding transfer function of the controller is

$$\frac{\theta_o}{\theta_i} = K_p + K_d s \tag{21.21}$$

We will begin by investigating only second-order systems, since the concept of damping ratio is meaningless for first-order systems. In addition, we will write a slightly different form for the plant, since we will focus on a particular system that has been discussed before, namely an actuated inertial load with damping. This system, together with a PD controller, is shown in Fig. 21.11. The transfer function relating C and R is

$$\frac{C}{R} = \frac{K_p + K_d s}{Js^2 + s(c + K_d) + K_p} \tag{21.22}$$

while that relating C to T is

$$\frac{C}{T} = \frac{1}{Js^2 + s(c + K_d) + K_p} \tag{21.23}$$

Again, it is observed that the steady-state error due to unit step input R is zero, while the steady-state error due to unit disturbance T is

$$c_{ss}(t) = \frac{1}{K_p} \tag{21.24}$$

This may be compared to the case of simple proportional control by setting $K_d = 0$, yielding

$$\frac{C}{R} = \frac{K_p}{Js^2 + cs + K_p} \tag{21.25}$$

and

$$\frac{C}{T} = \frac{1}{Js^2 + cs + K_p} \tag{21.26}$$

which gives the same steady-state error for each step input. Notice that for proportional control, the disturbance rejection is enhanced by making K_p large, but if the system is inherently lightly damped, no control over the transient response may be effected. In the case of PD control, however, K_p may still be made as large as necessary to reject any disturbance, but the new damping coefficient is $c + K_d$,

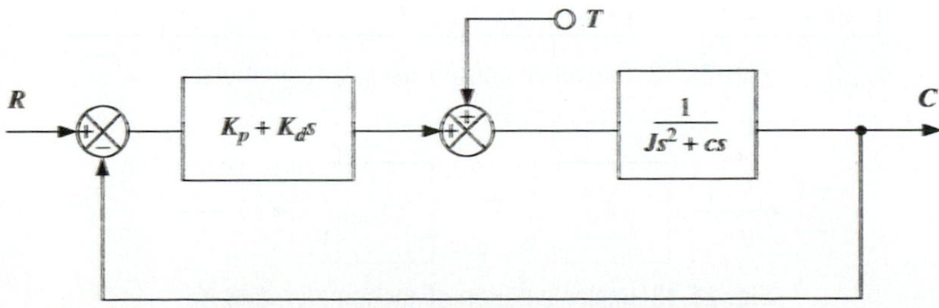

Fig. 21.11 Second-order PD control system.

allowing control over the damping ratio through K_d. In fact, if we inspect the characteristic equation from the denominator of the closed-loop transfer function and compare it to the standard form for a second-order system

$$s^2 + 2\zeta\omega_n s + \omega_n^2 = Js^2 + (c + K_d)s + K_p \tag{21.27}$$

we see that

$$2\zeta\omega_n = \frac{c + K_d}{J} \tag{21.28}$$

and

$$\omega_n^2 = K_p/J \tag{21.29}$$

Given a value of K_p from the last equation that will satisfy a disturbance rejection criterion, K_d may be obtained from equation 21.28 to satisfy a required damping ratio. Note, however, that the closed-loop system has a zero, and this must be taken into account in determining the transient response.

Another form of derivative control may occur in systems such as the one described, where the plant is an inertial load. The velocity of the load may be sensed directly with a tachometer, and this signal fed back to the controller, which is now simply a proportional controller, as shown in Fig. 21.12. In practice, the tachometer element K_T would consist of two individual elements, as shown in Fig. 21.13. In this case, the actual tachometer would have a fixed transfer function K_{tach} having units of volts per rpm, followed by an adjustable gain K_d, allowing the designer to add a particular contribution of velocity signal in the control loop. For our purposes

$$K_T = K_d \times K_{\text{tach}} \tag{21.30}$$

Calculating the relevant transfer functions reveals

$$\frac{C}{R} = \frac{K_p}{Js^2 + s(c + K_T) + K_p} \tag{21.31}$$

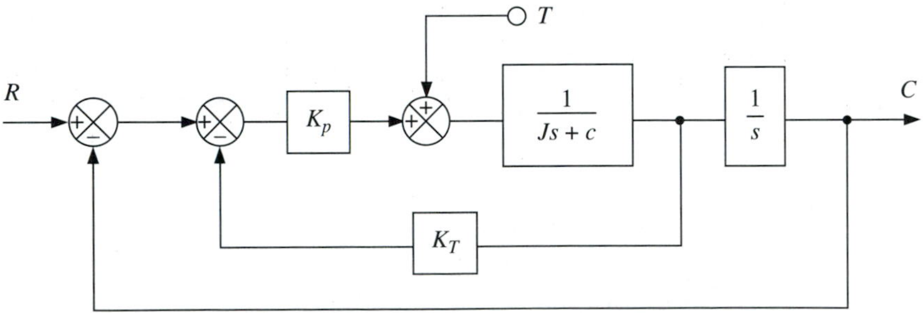

Fig. 21.12 Derivative control using a tachometer.

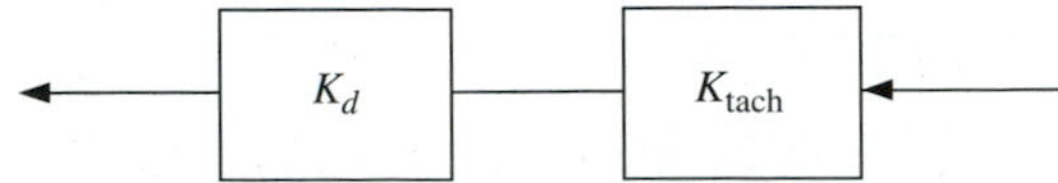

Fig. 21.13 Implementation of tachometer gain K_T.

and

$$\frac{C}{T} = \frac{1}{Js^2 + s(c + K_T) + K_p} \tag{21.32}$$

It is seen that the effect of the tachometer is the same as PD control, at least as far as the characteristic equation is concerned, in that additional damping is introduced through the gain K_T. Note that in this case there is no closed-loop zero, thereby making the prediction of the system response somewhat easier. Again, derivative control is never used alone but is always used with a proportional component. The reason this time is purely practical—noise. Consider the signal shown in Fig. 21.14a, which indicates a linearly increasing voltage with a small noise component added to it. Figure 21.14b shows the derivative of the signal, which in the absence of noise would be a horizontal line. The high-frequency noise is magnified considerably by the differentiator, since it is calculating the slope of the signal. For this reason, pure differentiation is always avoided. Tachometer feedback tends to filter out most of the noise due to the inertia of the tachometer rotor.

◆ PROPORTIONAL-PLUS-INTEGRAL-PLUS-DERIVATIVE CONTROL

Proportional-plus-integral-plus-derivative (PID) control, sometimes referred to as three-mode control, is a combination of the three modes previously discussed. The controller block diagram is as shown in Fig. 21.15, while its transfer function may be written as

$$\frac{\theta_o}{\theta_i} = K_p + \frac{K_i}{s} + K_d s \tag{21.33}$$

Such devices are very effective in adjusting the performance of control systems, and they are available as commercial components intended to be added to the forward path of a control system in order to improve its performance. The individual settings for K_p, K_i, and K_d are made using potentiometers on the device itself. The obvious question, of course, is what settings should be chosen. This problem is known as *tuning* the controller, and it may be accomplished in many ways. Before

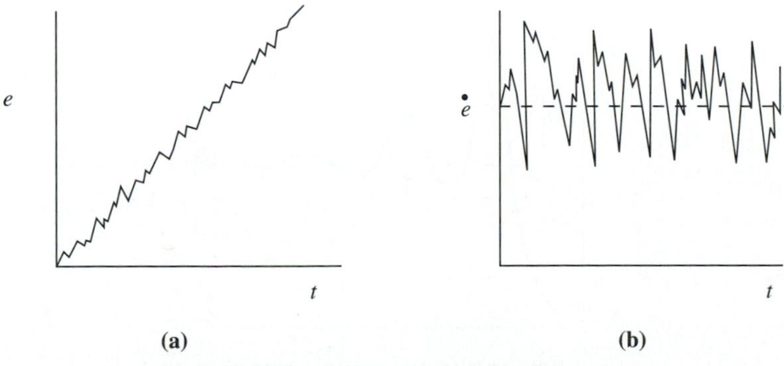

Fig. 21.14 (a) Noisy signal. (b) Its derivative.

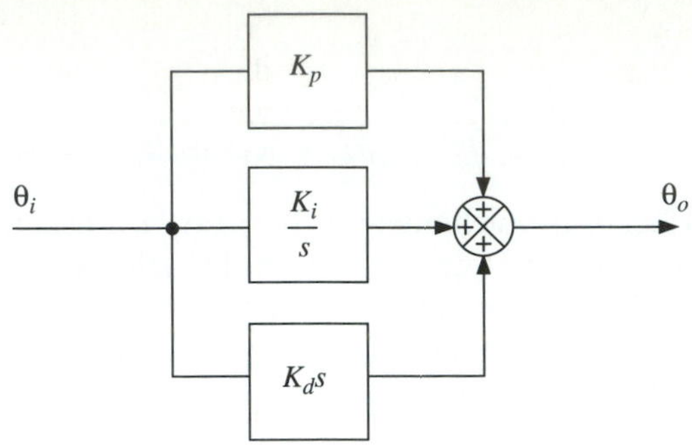

Fig. 21.15 PID controller.

discussing these, however, we rewrite the transfer function in a different form:

$$\frac{\theta_o}{\theta_i} = K_p\left(1 + \frac{1}{T_i s} + T_d s\right) \tag{21.34}$$

Tuning the controller will involve the selection of K_p, T_i, and T_d. Three methods of tuning the controller will be considered:

Trial-and-Error Method

In this approach, the value of K_p is selected first in order to achieve the transient response desired while setting T_i and T_d to zero. Then T_i is adjusted in order to satisfy any steady-state error requirements. This will probably degrade the transient response due to the inclusion of a closed-loop zero. The transient response may be restored by selecting a suitable T_d. Several iterations of this process may be required before the system behaves as desired.

For certain modes of control, Ziegler and Nichols have developed empirical settings for the three parameters in a PID controller. These settings were derived from the observation of many real systems in which PID controllers were adjusted in order to obtain optimum performance. In this context, optimum performance is defined as the "quarter-decay" response, where the ratio of successive overshoots decay to 25% of the previous value. This concept is illustrated in Fig. 21.16. Rather

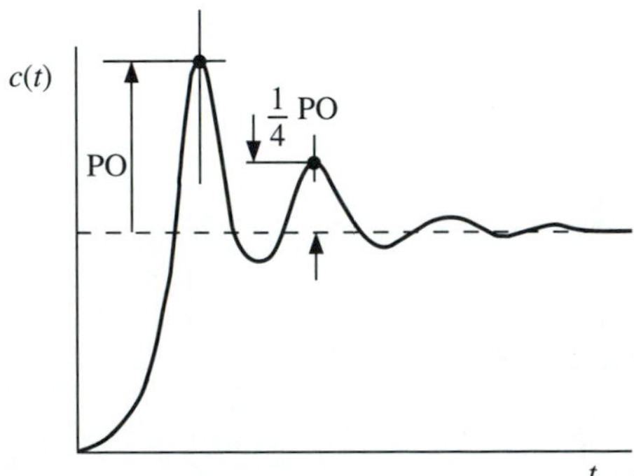

Fig. 21.16 Quarter-decay response.

surprisingly, they found that the settings were much the same for many types of control systems, and these settings have found considerable application in compensating diverse forms of control systems. They also have the advantages that a transfer function of the plant is not required and the compensation process may be performed rapidly, on-site.

Ziegler and Nichols First Method

This method only applies if the open-loop step response of the plant has no overshoot. If this is the case, the unit step response is first obtained (Fig. 21.17) and takes the form shown in Fig. 21.18. From the step response, we obtain the maximum slope P, which will usually be at the point of inflexion, and the time L, which is where the tangent of maximum slope intersects the time axis. The settings for the parameters in the controller are obtained from P and L, depending on the type of mode control required, and are shown in Table 21.1.

Ziegler and Nichols Second Method

This method is based on the closed-loop marginal-stability condition. Here, the plant is assumed to be in the closed-loop mode with only proportional gain control, as shown in Fig. 21.19. It is assumed that K_p is increased to the point where the system exhibits oscillatory output of constant magnitude for any disturbing input, as shown in Fig. 21.20. This defines the critical gain $K_p = K'_p$. The parameters required in the controller in order to obtain optimum response are calculated from

Fig. 21.17 Performing an open-loop step response of the plant.

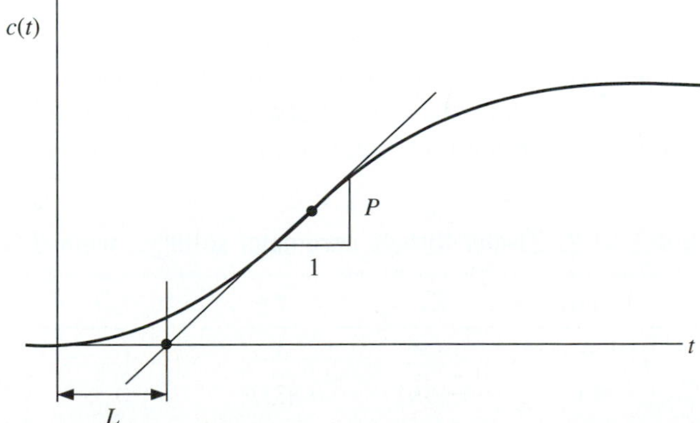

Fig. 21.18 Step response.

TABLE 21.1. Ziegler-Nichols parameter settings, method 1

Control	K_p	T_i	T_d
P	$\frac{1}{PL}$	∞	0
PI	$\frac{0.9}{PL}$	$\frac{10L}{3}$	0
PID	$\frac{1.2}{PL}$	$2L$	$0.5L$

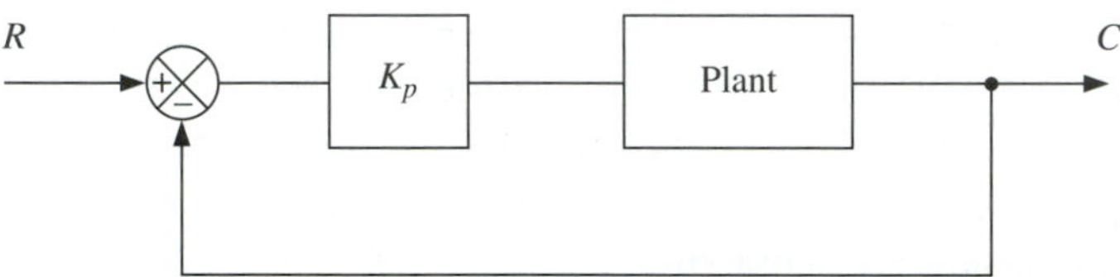

Fig. 21.19 Ziegler-Nichols system configuration, method 2.

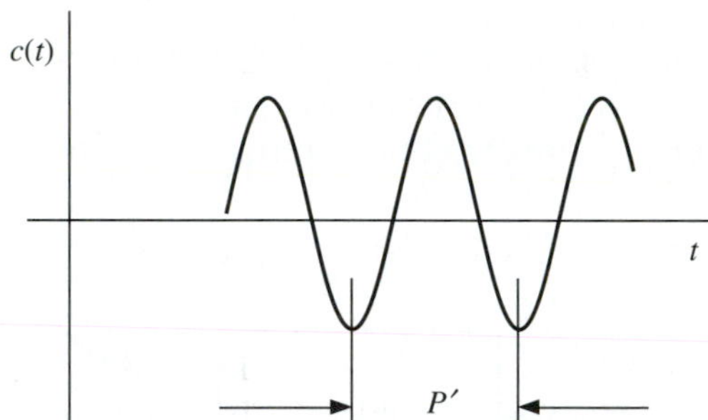

Fig. 21.20 System output for critical gain K'_p.

the critical gain K'_p needed to just make the system oscillate with constant amplitude output and the period of the oscillations P'. Again, the parameters depend upon the type of control law required and are shown in Table 21.2. Keep in mind the restrictions applying to the Ziegler-Nichols methods, since many common systems do not comply. For example, in order to apply the second method, the open-loop plant must be of at least third order; otherwise no limiting stability condition will be encountered. These methods may be applied to systems whose transfer

TABLE 21.2. Ziegler-Nichols parameter settings, method 2

Control	K_p	T_i	T_d
P	$0.5K'_p$	∞	0
PI	$0.45K'_p$	$0.833P'$	0
PID	$0.6K'_p$	$0.5P'$	$0.125P'$

function is known, since the step response and frequency response may usually be predicted from root locus or Bode diagram analysis.

Sample Problem 21.1

Consider the electromechanical servomechanism shown in Fig. SP21.1.1, where $J = 0.005$ kg/m^2 and $c = 0.02$ N-s/m. Design a controller that gives the system zero steady-state error for a velocity input and dominant closed-loop poles with a damping ratio $\zeta = 0.707$. Estimate the unit step response of the compensated system.

Solution

The simplest form of control would be just proportional control, in which case the open-loop transfer function of the system becomes

$$GH(s) = \frac{K_p/J}{s(s + c/J)} = \frac{200K_p}{s(s + 4)}$$

Since this system is of type 1, it will have a finite steady-state error for a velocity input, and PI control will be needed to eliminate this. The controller transfer function becomes

$$G_c = \frac{K_p s + K_i}{s}$$

while the new open-loop transfer function becomes

$$GH(s) = \frac{200K_p(s + K_i/K_p)}{s^2(s + 4)}$$

The root locus of the system will have two poles at the origin and the system pole at $s = -4$. We are able to place the zero anywhere we choose, but it has to be between the system pole and the origin; otherwise the resulting closed-loop system will be unstable. Suppose we choose to place this zero at $s = -0.25$. This produces

$$K_p = 4K_i$$

Although an accurate root locus may be drawn, a rough sketch will be adequate for our purposes. This is shown in Fig. SP21.1.2. The locus begins at the origin, loops around the zero, and then reenters the real axis around $s = -0.5$. One branch then moves to the zero, while the other meets the other closed-loop

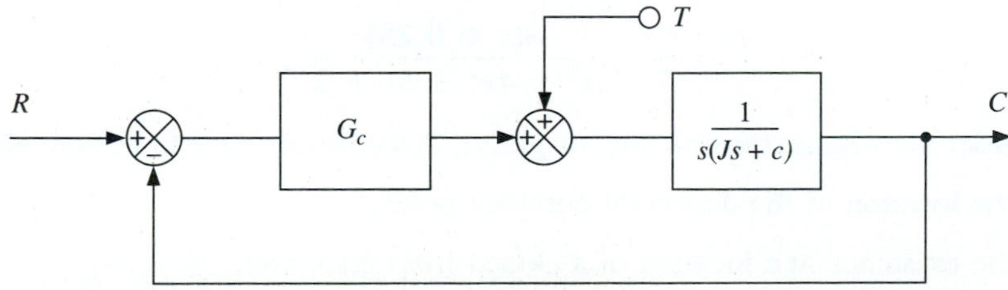

Fig. SP21.1.1

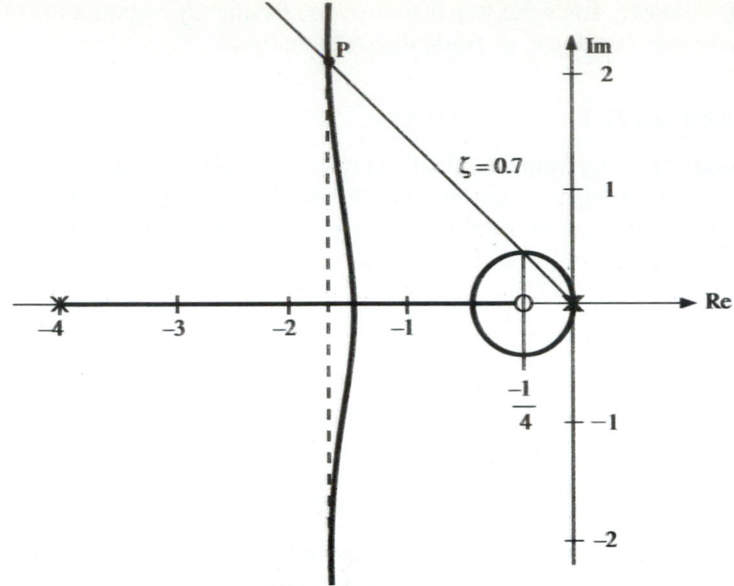

Fig. SP21.1.2

pole moving to the right from the open-loop pole at $s = -4$. After meeting, they branch away from the real axis and asymptotically approach the line passing through the real axis at

$$\sigma_a = \tfrac{1}{2}(-0 - 0 - 4 + 0.25) = -1.875$$

A line drawn at 45° to the imaginary axis defines where the closed-loop poles should be to give the required damping ratio, and this may be calculated directly from the root locus at point P as $K \approx 8$. Equating this to the Evans gain of the open-loop transfer function determines K_p, and hence K_i. The result is

$$K_p = \frac{8}{200} = 0.04, \; K_i = 0.01$$

The resulting open-loop transfer function becomes

$$GH(s) = \frac{8(s + 0.25)}{s^2(s + 4)}$$

and the closed-loop transfer function

$$\frac{C}{R} = \frac{8(s + 0.25)}{s^3 + 4s^2 + 8s + 2}$$

In order to estimate the unit step response, three factors have to be considered:

1. the location of the dominant complex poles,
2. the existence and location of a closed-loop zero, and
3. the location of any dominant real-axis closed-loop poles.

The dominant closed-loop poles are located approximately at $s = -2 \pm 2j$, which suggests about a 5% overshoot and a peak time of about

$$T_p = \frac{\pi}{\omega_d} = 1.4$$

The closed-loop system, in addition to the complex poles, has a zero at -0.25 and another pole close to -0.5. It might be argued that because these are close to each other, their effects on the expected response should cancel out, although it might be less damped due to the zero being closer to the imaginary axis than the pole. Fig. SP21.1.3 shows this to be the case, having about 15% overshoot and a peak time of around 1.6 s. The peak time is quite close to that expected from the equivalent second-order system, although the PO is somewhat larger. This is no doubt explained by the closer proximity of the closed-loop zero to the imaginary axis than the third closed-loop pole.

Sample Problem 21.2

The system shown in Fig. SP21.2.1 is a mechanical positioning system that possesses negligible natural damping. (a) Design a PD controller and (b) use tachome-

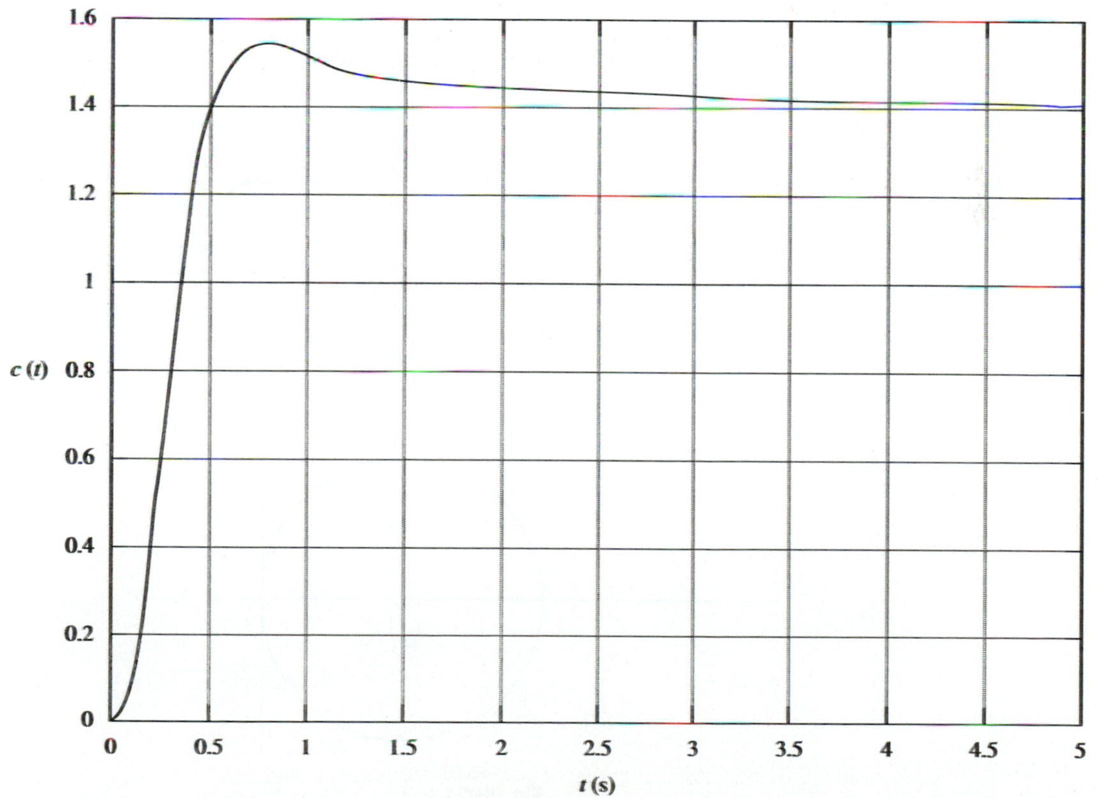

Fig. SP21.1.3

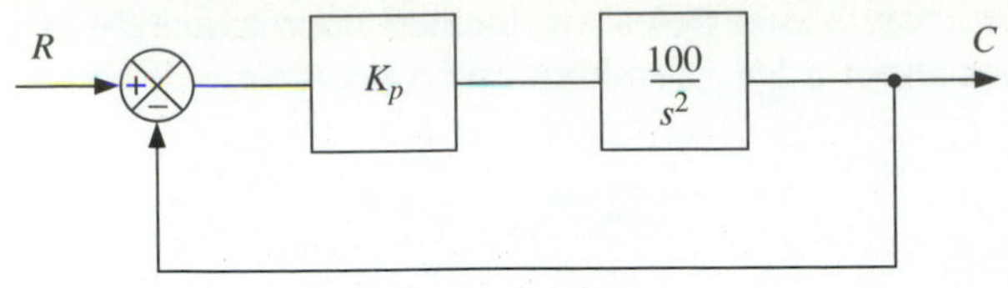

Fig. SP21.2.1

ter feedback to make the closed-loop system have an undamped natural frequency $\omega_n = 1.0$ rad/s and the unit step response of the closed-loop system have an overshoot of no more than 10%.

Solution

(A) PD CONTROLLER DESIGN. The closed-loop system for this approach is shown in Fig. SP21.2.2, where the problem is to select K_p and K_d to meet the performance requirements. The open-loop transfer function may be written as

$$GH(s) = \frac{100K_d(s + K_p/K_d)}{s^2}$$

The root locus is as shown in Fig. SP21.2.3 although the position of the zero has not yet been determined. The closed-loop transfer function may be written as

$$\frac{C}{R} = \frac{100K_d(s + K_p/K_d)}{s^2 + 100K_d s + 100K_p}$$

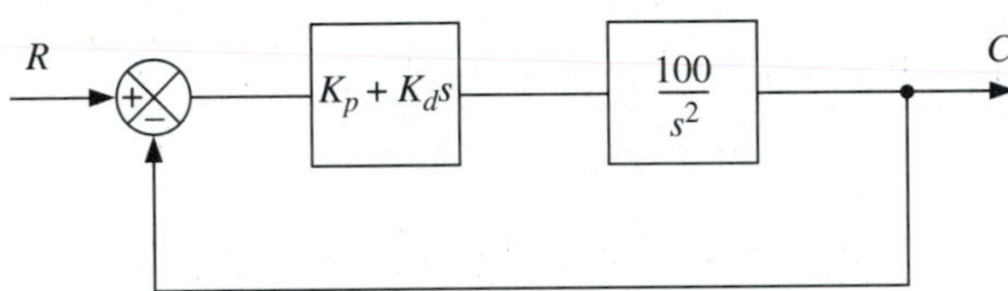

Fig. SP21.2.2

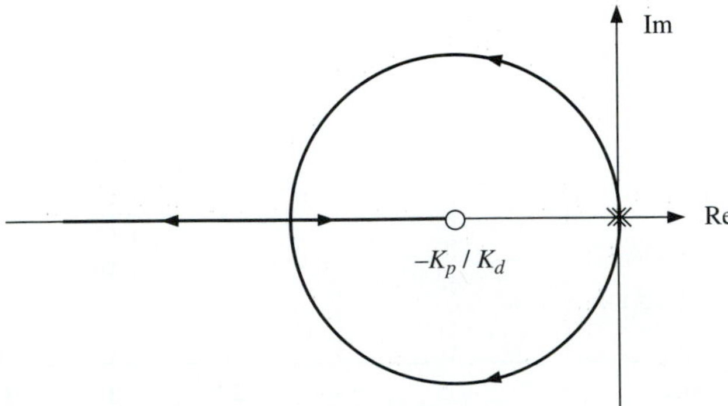

Fig. SP21.2.3

Inspecting the characteristic equation yields

$$2\zeta\omega_n = 100K_d$$

and

$$\omega_n^2 = 100K_p$$

From the problem statement, $\omega_n = 1.0$ rad/s; hence $K_p = 0.01$, but ζ is not known. If there were no closed-loop zero present, the percentage overshoot requirement could be met with $\zeta = 0.6$ from Fig. 4.17, but the closed-loop zero will require a higher damping ratio in order to obtain the necessary 10% overshoot. Let us guess that $\zeta = 0.707$. This gives the closed-loop transfer function

$$\frac{C}{R} = \frac{1.414(s + 1)}{s^2 + 1.414s + 1}$$

Using a MATLAB program, the unit step response is obtained as shown in Fig. SP21.2.4, which indicates a 20% overshoot. Next, increase the damping ratio to $\zeta = 1.0$, producing the closed-loop transfer function

$$\frac{C}{R} = \frac{2(s + 1)}{s^2 + 2s + 1}$$

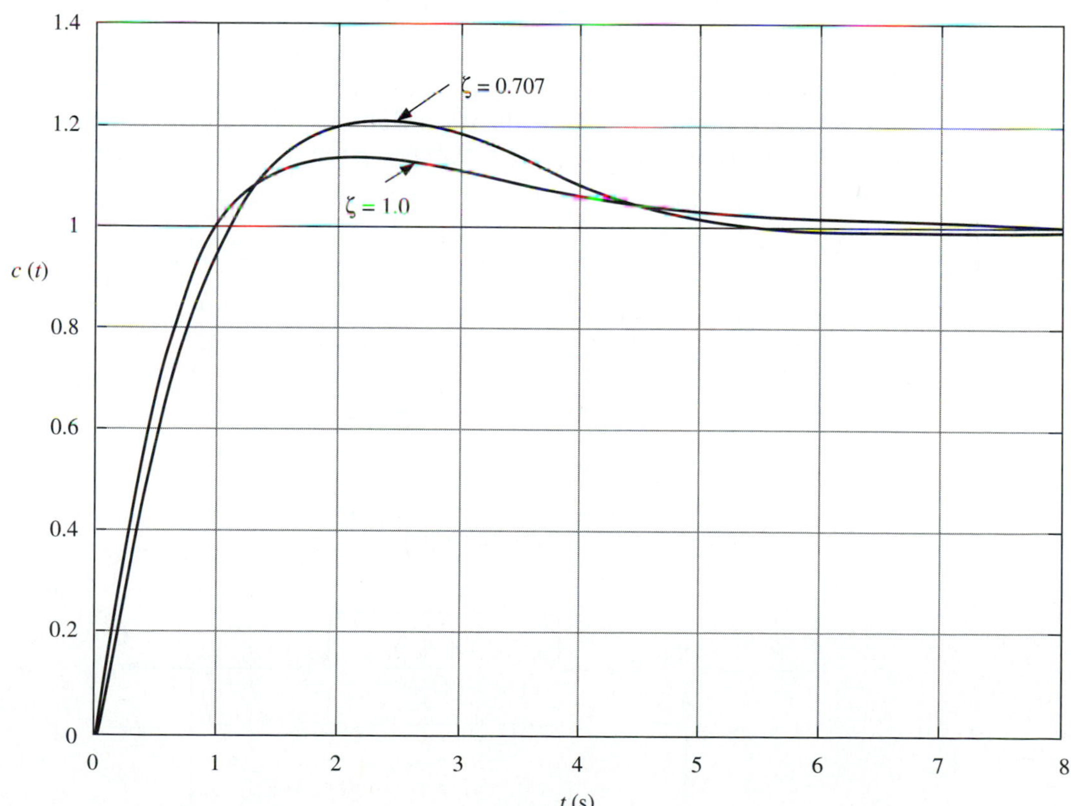

Fig. SP21.2.4

The unit step response of this system is also shown in Fig. SP21.2.4, indicating about 12% overshoot. The corresponding gain values are $K_p = 0.01$ and $K_d = 0.02$. Clearly, another iteration would produce closer results.

(B) TACHOMETER FEEDBACK. Using tachometer feedback, the block diagram will be as shown in Fig. SP21.2.5. Note, in this figure, the velocity of the output is obtained by differentiating the output. Closing the inner loop produces first the block diagram shown in Fig. SP21.2.6 and then the resulting closed-loop transfer function

$$\frac{C}{R} = \frac{100K_p}{s^2 + 100K_pK_Ts + 100K_p}$$

Observe the absence of the closed-loop zero, making the calculation of the control parameters exact. This time we know $\zeta = 0.6$. Therefore

$$100K_p = \omega_n^2 = 1$$

leading, as before, to $K_p = 0.01$. In addition

$$100K_pK_T = 2\zeta\omega_n = 1.2$$

yielding $K_T = 1.2$. The closed-loop transfer function then becomes

$$\frac{C}{R} = \frac{1}{s^2 + 1.2s + 1}$$

Again using MATLAB, the unit step response is computed, as shown in Fig. SP21.2.7. In this latter case, the step response could have been estimated fairly accurately from the peak time, damped natural frequency ω_d, and percentage overshoot.

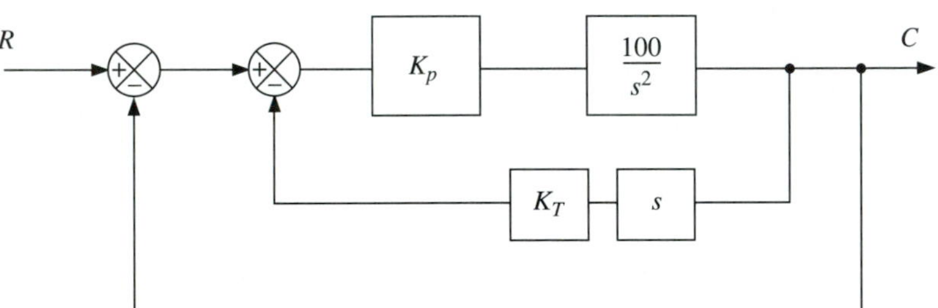

Fig. SP21.2.5

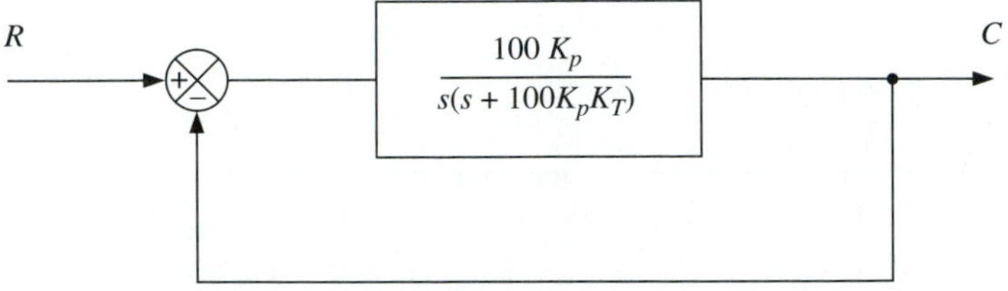

Fig. SP21.2.6

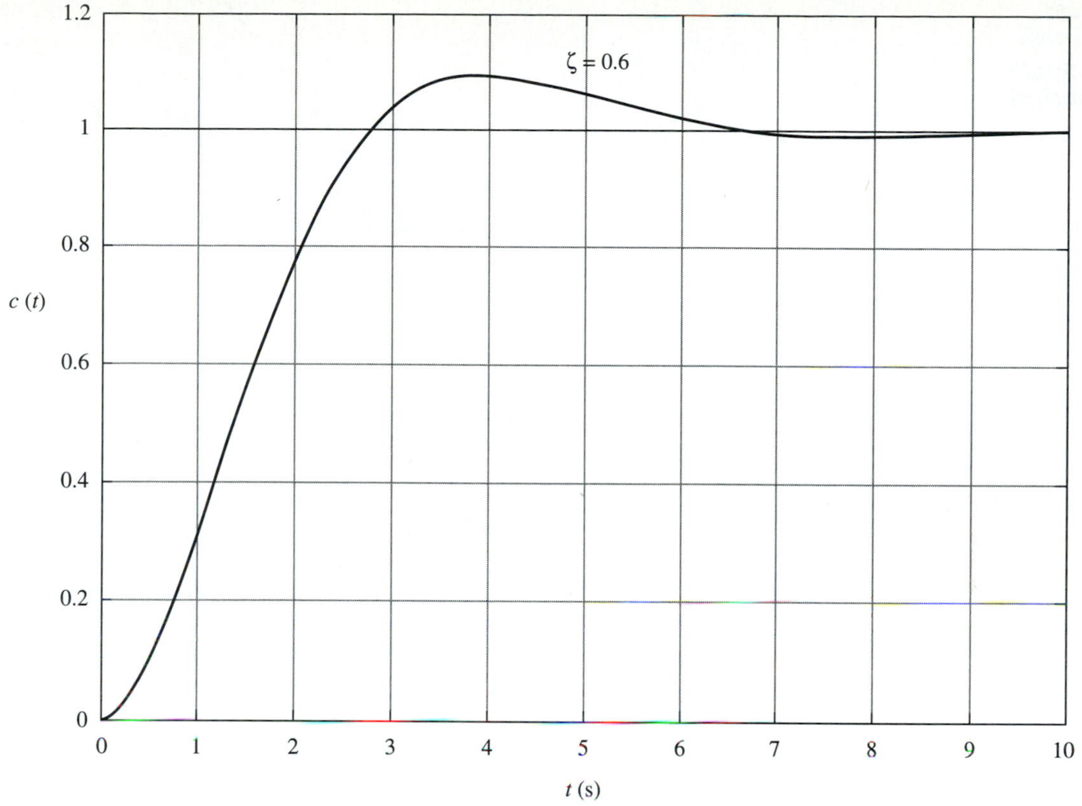

Fig. SP21.2.7

Sample Problem 21.3

For the system shown in Fig. SP21.3.1, design a PID controller using a Ziegler-Nichols method for tuning the coefficients, and compute the resulting step response of the closed-loop system.

Solution

Since we know the transfer function of the plant, we will use the second Ziegler-Nichols method, in which we drive the system to the point of instability using proportional gain only. Figure SP21.3.2 shows the root locus of the system, confirming that the system can indeed be made marginally stable and the

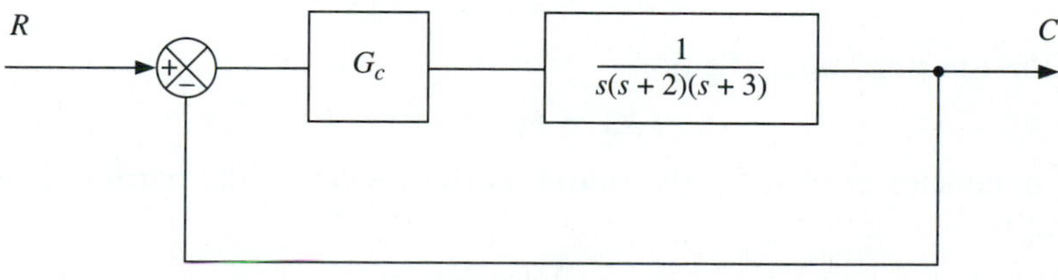

Fig. SP21.3.1

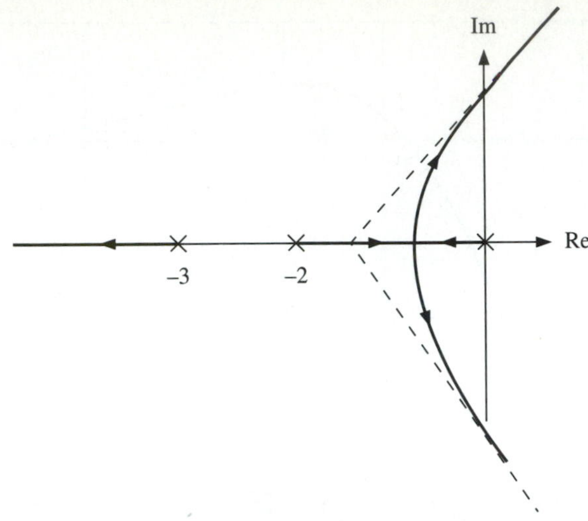

Fig. SP21.3.2

method is therefore applicable to this case. The open-loop transfer function becomes

$$GH(s) = \frac{K_p}{s(s + 2)(s + 3)}$$

and the closed-loop transfer function is

$$\frac{C}{R} = \frac{K_p}{s^3 + 5s^2 + 6s + K_p}$$

The characteristic equation is

$$s^3 + 5s^2 + 6s + K_p = 0$$

Considering the system on the point of instability, we can set $s = j\omega$, since the closed-loop poles will be where the locus passes through the imaginary axis in Fig. SP21.3.2. Substituting into the characteristic equation gives

$$-j\omega^3 - 5\omega^2 + 6j\omega + K_p = 0$$

Collecting terms yields

$$(K_p - 5\omega^2) + j\omega(6 - \omega^2) = 0$$

Equating both real and imaginary components to zero produces the frequency at marginal stability

$$\omega = \sqrt{6} = 2.45 \text{ rad/s}$$

and the limiting value of gain

$$K_p' = 5\omega^2 = 30$$

The parameter needed for the tuning is the period of the harmonic output, which is

$$P' = \frac{2\pi}{\omega} = 2.57 \text{ s}$$

Using the parameters in Table 21.2, the controller transfer function may be written directly as

$$G_c = 18\left(1 + \frac{1}{1.28s} + 0.321s\right)$$

which may be written as

$$G_c = \frac{18}{s}(0.321s^2 + s + 0.78)$$

This makes the open-loop transfer function

$$G_cGH(s) = \frac{18(0.321s^2 + s + 0.78)}{s^2(s + 2)(s + 3)}$$

After some manipulation, the closed-loop transfer function becomes

$$\frac{C}{R} = \frac{5.78s^2 + 18s + 14}{s^4 + 5s^3 + 11.78s^2 + 18s + 14}$$

Using MATLAB, the unit step response is shown in Fig. SP21.3.3, which indicates successive overshoots decreasing by about one-quarter.

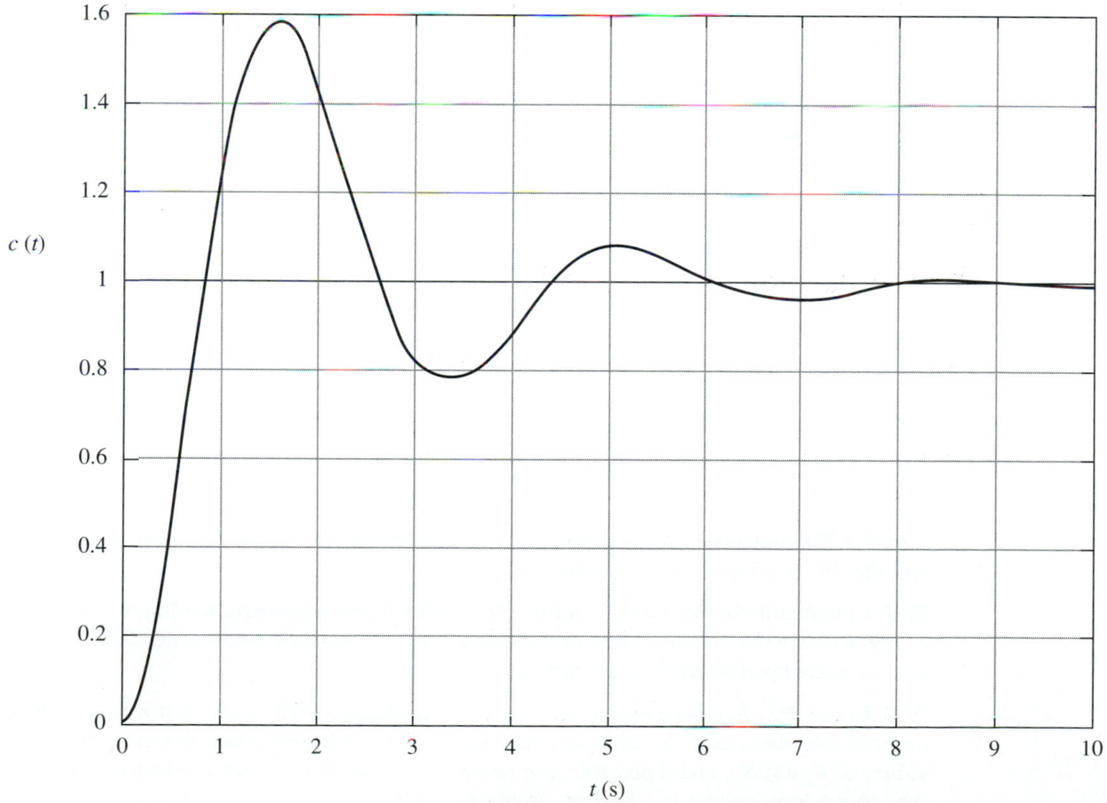

Fig. SP21.3.3

21.1 Show that the mechanical system illustrated in Fig. P21.1 may be considered as a PI controller, where the input is defined to be x and the output is y.

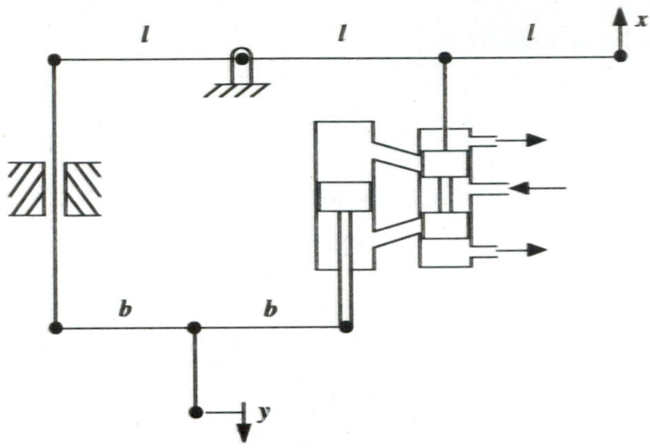

Fig. P21.1

21.2 A unity-feedback control system has the open-loop transfer function

$$GH(s) = \frac{40}{(s + 4)(s + 2)}$$

Use the root locus method to design a suitable PI controller such that the system has zero steady-state position error and the dominant closed-loop poles have a damping ratio of 0.5. Estimate the percentage overshoot, peak time, and 2% settling time of the closed-loop, unit step response.

21.3 Figure P21.3 shows the initial part of the unit step response of an unknown system. Use the Ziegler-Nichols tuning rules to design a PI controller for the system.

21.4 Because the system shown in Fig. P21.4 has to have zero velocity error, it has been decided to control it using a PI control law. Use the Ziegler-Nichols method to determine the controller parameters and sketch the closed-loop step response.

21.5 Show that the system in Fig. P21.5 may represent a PD controller when the time constant $\tau = C/K$ is small.

21.6 For the unity-feedback system described by the open-loop transfer function

$$G(s) = \frac{5}{(s + 10)^2}$$

design a PD controller such that the closed-loop system has a damping ratio of $\zeta = 0.7$ and the 2% settling time is less than 0.1 s.

21.7 Consider the system shown in Fig. P21.7, which has tachometer feedback. Calculate K_p and K_T such that the dominant closed-loop system poles have a damping ratio $\zeta = 0.4$ and an undamped natural frequency $\omega_n = 2.5$ rad/s.

21.8 In Fig. P21.8, it is required to make the output reject 90% of the disturbance while maintaining dominant closed-loop poles with a 0.707 damping ratio. Select appropriate values of K_p and K_d, and if possible, use computer methods to predict the step response of the closed-loop system to unit step inputs R and T.

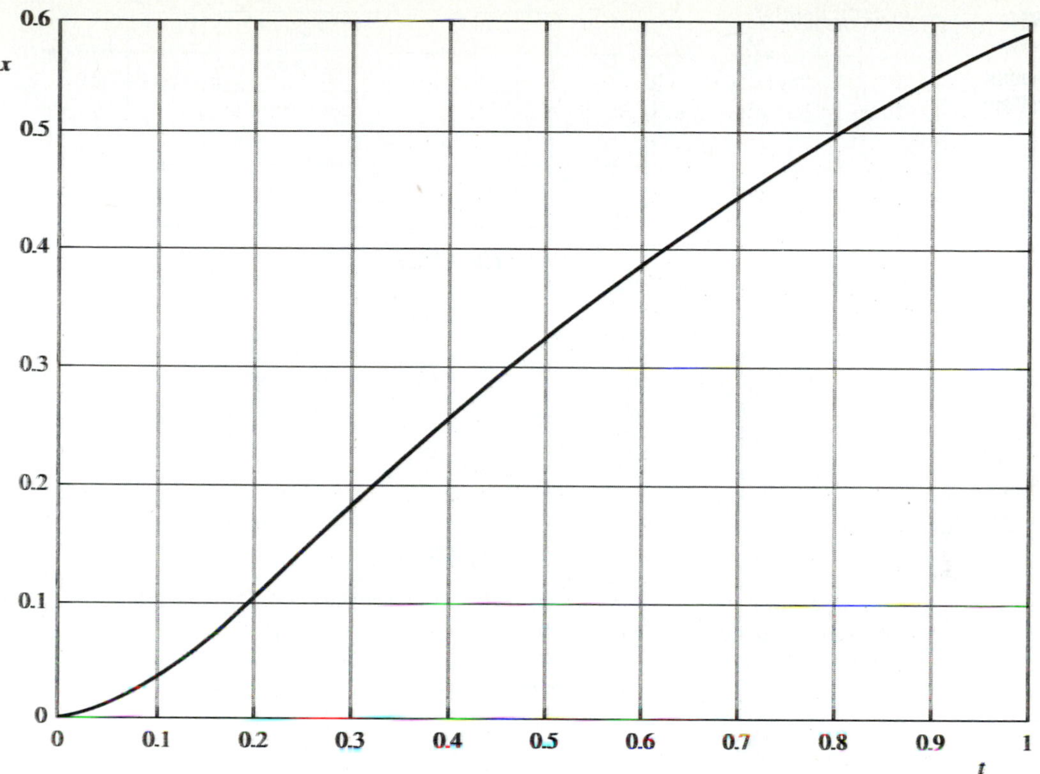

Fig. P21.3

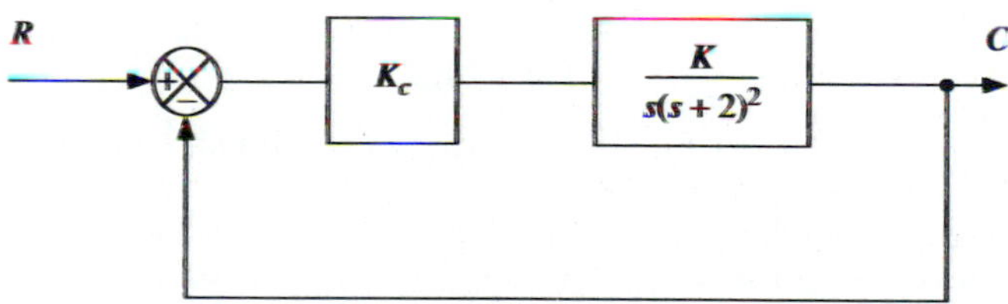

Fig. P21.4

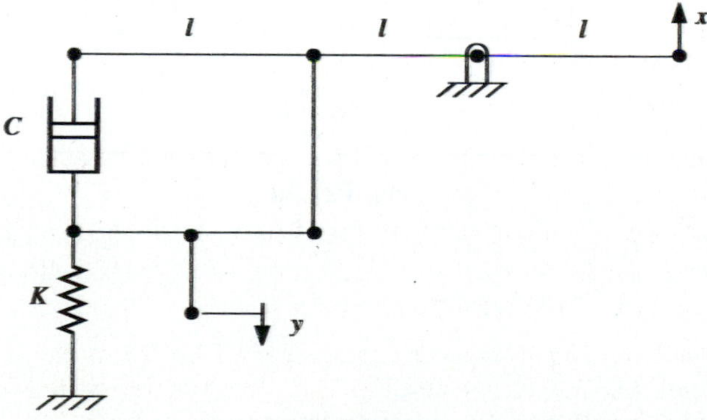

Fig. P21.5

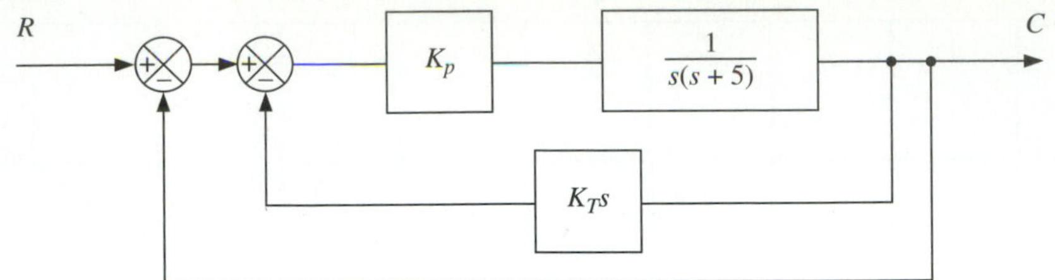

Fig. P21.7

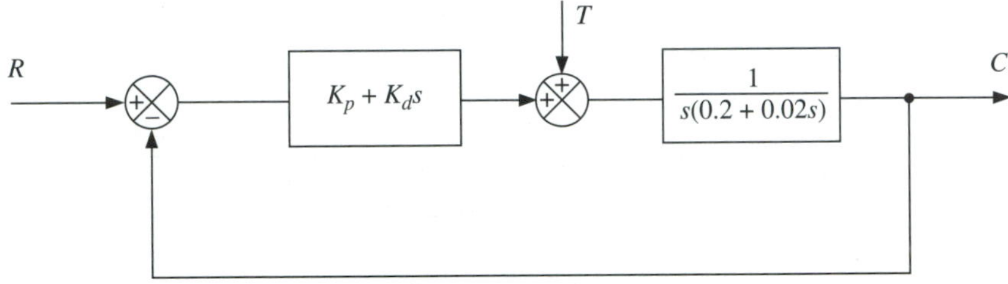

Fig. P21.8

21.9 Use the Ziegler-Nichols tuning rules to determine the PID controller coefficients for the system shown in Fig. P21.9.

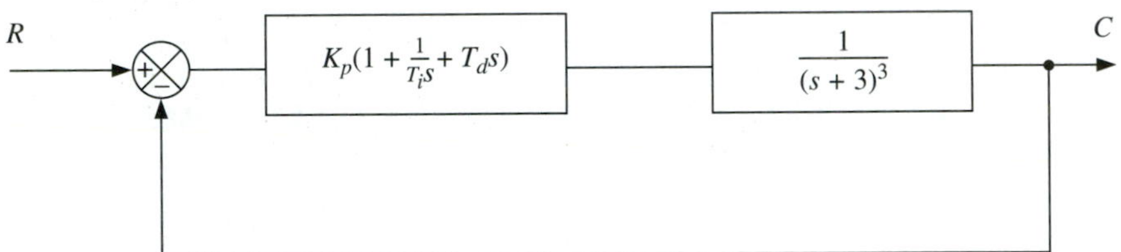

Fig. P21.9

21.10 For the system shown in Fig. P21.10, design a multimode controller that meets the following performance requirements:

a. Percentage overshoot not to exceed 10%.

b. Velocity error constant $K_v = 20$.

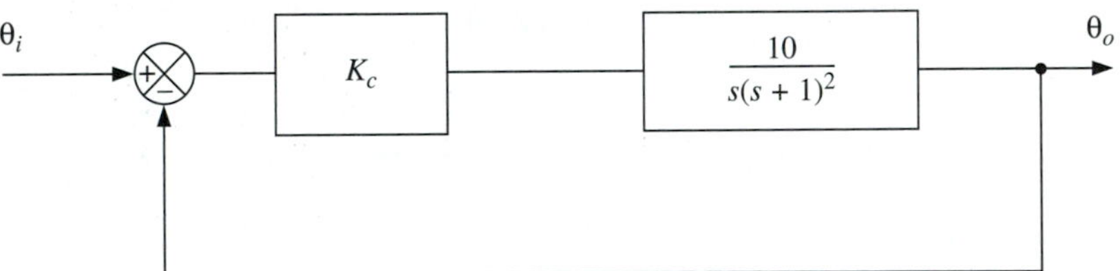

Fig. P21.10

MODULE 22

State-Space System Descriptions

So far, all of the analysis methods studied—root locus, Nyquist analysis, Bode diagrams, etc.—are commonly called *classical* methods. Such methods are useful for investigating single-input, single-output (SISO) systems. More complex systems may have multiple inputs and multiple outputs (MIMO), which complicates the situation considerably. The system shown in Fig. 22.1, for example, has

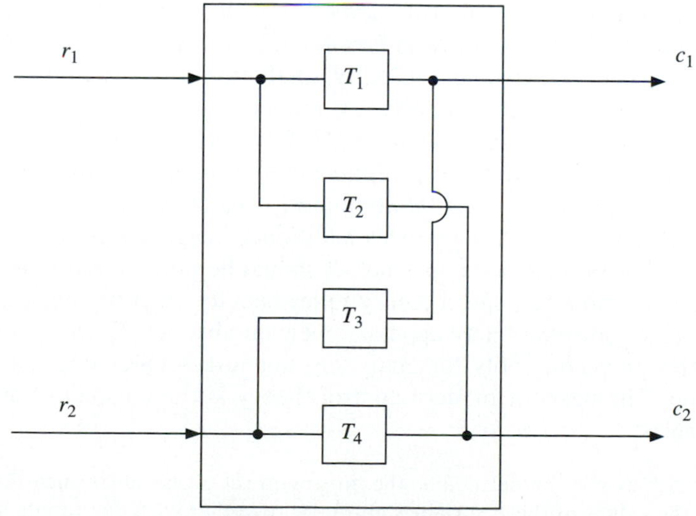

Fig. 22.1 MIMO system.

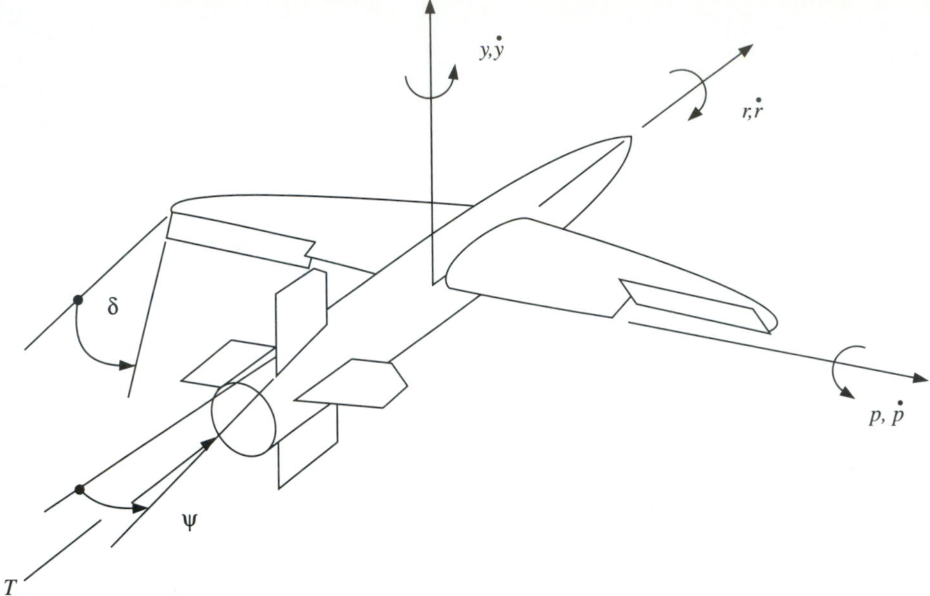

Fig. 22.2 Missile control system.

two inputs and two outputs. For this system, we need to analyze not one but four transfer functions, namely,

$$\frac{c_1}{r_1}, \quad \frac{c_2}{r_1}, \quad \frac{c_1}{r_2}, \quad \frac{c_2}{r_2} \tag{22.1}$$

Designing, stabilizing, and compensating such a system is obviously more complex than a SISO system, and the problem becomes more difficult as the number of inputs and outputs increase. Real systems, such as the missile shown in Fig. 22.2, may have their flight dynamics considered as outputs, namely, roll angle, roll rate, and pitch and yaw angles and rates, while inputs might be thrust, elevator angle, and rudder angle. Clearly, analyzing this system using classical methods would be very tedious and error prone.

An alternative approach for handling MIMO systems is in common use and is usually referred to as *modern* control theory. Its principle advantages are that it is easily implemented on a digital computer, it handles nonlinear systems and time-varying systems, and it provides additional insights into system behavior that transfer function analysis does not. It should be pointed out that, except for simple systems, modern control theory relies heavily on computers, and the design of system controllers may appear somewhat abstract. Even so, these methods do offer powerful tools for analyzing not just MIMO systems, but SISO systems too. The basis of modern control theory is the concept of state and of state variables:

> State variables may be defined as the minimum set of variables such that a knowledge of the values of these variables at time t_0, together with any inputs u to the system for $t \geq t_0$, will allow the behavior of the system to be determined for $t \geq t_0$.

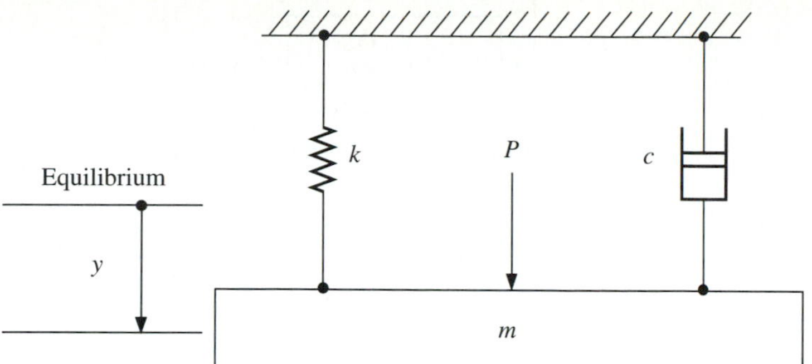

Fig. 22.3 Mechanical system.

The system state is known when all of the n state variables are known, and the state variables are usually collected into an $n \times 1$ state vector $\mathbf{x}$. For a particular system, the number of state variables is fixed and is equal to the system order,[1] but the choice of state variables is not unique. The only restriction on the choice of state variables is that they are independent. In order to develop state-space concepts, we will begin with a simple example that is a SISO system. Consider the dynamic system shown in Fig. 22.3. It is known that the equation of motion of this system is given by

$$m\ddot{y} + c\dot{y} + ky = P \tag{22.2}$$

We will write a description of the system using state variables and compare it to the differential equation above. First, we determine the number of state variables in this system, i.e., how many variables are needed to completely specify the state of the system. Clearly, the position y is needed, but is this all? Further thought reveals that, not just position, but velocity is also needed to completely specify the system state, making two state variables. This is the correct answer, since the system is of second order. One might wonder, from a purely physical point of view, if the acceleration is also required. However, the equation of motion reveals

$$\ddot{y} = P - \frac{c}{m}\dot{y} - \frac{k}{m}y \tag{22.3}$$

indicating that the acceleration may be derived from a linear combination of the other state variables (assuming the input P is known), position and velocity, and is not, therefore, independent of them. This excludes the acceleration from being a state variable. Suppose we consider the system in Fig. 22.3 to be represented by the block diagram in Fig. 22.4, in which the input is considered the force P while the output is the displacement of the mass y. The state variables position and velocity may be defined as

$$x_1 = y \tag{22.4}$$

$$x_2 = \dot{y} = \dot{x}_1 \tag{22.5}$$

[1]The number of state variables may also be determined as the number of initial conditions needed to solve the system differential equations or the number of *first-order* differential equations needed to define the system.

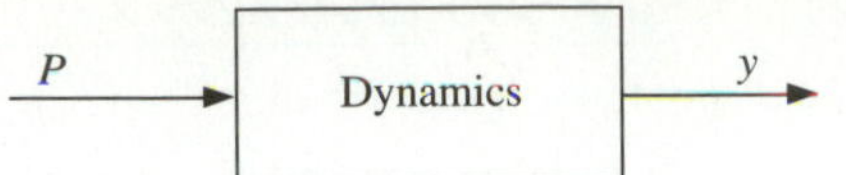

Fig. 22.4 Block diagram of mechanical system.

The equation of motion may be written in the form

$$\ddot{y} = -\frac{k}{m}y - \frac{c}{m}\dot{y} + \frac{P}{m} \tag{22.6}$$

or in terms of the state variables

$$\dot{x}_2 = -\frac{k}{m}x_1 - \frac{c}{m}x_2 + \frac{P}{m} \tag{22.7}$$

Equations 22.5 and 22.7 may be written in the compact matrix form

$$\begin{bmatrix} \dot{x}_1 \\ \dot{x}_2 \end{bmatrix} = \begin{bmatrix} 0 & 1 \\ -\dfrac{k}{m} & -\dfrac{c}{m} \end{bmatrix} \begin{bmatrix} x_1 \\ x_2 \end{bmatrix} + \begin{bmatrix} 0 & 0 \\ 0 & \dfrac{1}{m} \end{bmatrix} \begin{bmatrix} 0 \\ P \end{bmatrix} \tag{22.8}$$

which may be written as

$$\dot{\mathbf{x}} = \mathbf{Ax} + \mathbf{Bu} \tag{22.9}$$

This is the standard form of the system state equation. In this text, bold uppercase characters refer to matrices, while bold lowercase letters refer to vectors. Note that the state-space model is that of an open-loop system since no feedback is present. We will extend these ideas to closed-loop systems later. It may be seen that the same method of writing the system description in state variable form may be applied to the general nth order system

$$\frac{d^n y}{dt^n} + a_{n-1}\frac{d^{n-1}y}{dt^{n-1}} + a_{n-2}\frac{d^{n-2}y}{dt^{n-2}} + \cdots + a_1\frac{dy}{dt} + a_0 y = Ku(t) \tag{22.10}$$

by defining the state variables

$$x_1 = y \tag{22.11}$$

$$x_2 = \dot{x}_1 = \dot{y} \tag{22.12}$$

$$x_3 = \dot{x}_2 = \ddot{y} \tag{22.13}$$

$$\vdots \tag{22.14}$$

$$x_n = \dot{x}_{n-1} = \frac{d^{n-1}y}{dt^{n-1}} \tag{22.15}$$

From the system differential equation, we have

$$\dot{x}_n = -a_{n-1}x_n - a_{n-2}x_{n-1} - \cdots - a_1 x_2 - a_0 x_1 + Ku(t) \tag{22.16}$$

which allows the system to be written in the form

$$\begin{bmatrix} \dot{x}_1 \\ \dot{x}_2 \\ \vdots \\ \dot{x}_{n-1} \\ \dot{x}_n \end{bmatrix} = \begin{bmatrix} 1 & 0 & 0 & \cdots & 0 \\ 0 & 1 & 1 & \cdots & 0 \\ \vdots & \vdots & \vdots & \ddots & \vdots \\ 0 & 0 & 0 & \cdots & 1 \\ -a_0 & -a_1 & -a_2 & \cdots & -a_{n-1} \end{bmatrix} \begin{bmatrix} x_1 \\ x_2 \\ \vdots \\ x_{n-1} \\ x_n \end{bmatrix} + \begin{bmatrix} 0 & \cdots & 0 \\ 0 & \cdots & 0 \\ \vdots & \ddots & \vdots \\ 0 & \cdots & 0 \\ 0 & \cdots & K \end{bmatrix} \begin{bmatrix} 0 \\ 0 \\ \vdots \\ 0 \\ u \end{bmatrix} \tag{22.17}$$

which may also be written as

$$\dot{\mathbf{x}} = \mathbf{Ax} + \mathbf{Bu} \tag{22.18}$$

Equation 22.17 is a particular state representation of a system that is sometimes referred to as a *first companion form,* and in it the **A** matrix has the coefficients of the differential equation arranged on the bottom row, with all other elements zero except for the superdiagonal, which is unity. Equation 22.18 is a fundamental expression in state-space analysis, from which the following points may be noted:

1. The $\dot{\mathbf{x}}$ vector has dimensions $n \times 1$.
2. The **A** matrix has dimensions $n \times n$ and is known as the system, or plant, matrix.
3. The vector **x** is the $n \times 1$ state vector.
4. The control distribution matrix **B** is an $n \times n$ matrix, although not all elements may be present.
5. Vector **u** is an $n \times 1$ column vector of inputs, and again some elements may be zero.

Although the output in the mechanical system example corresponded to one of the system states ($y = x_1$), this is not always the case, and the system output may have to be constructed from the state vector in the form

$$\mathbf{y} = \mathbf{Cx} \tag{22.19}$$

where **y** is taken as a general system output vector comprising the m outputs. In some cases, there may be a direct connection between the system input and output, avoiding the plant altogether, in which case

$$\mathbf{y} = \mathbf{Cx} + \mathbf{Du} \tag{22.20}$$

The following points are noted regarding the dimensions of the matrices and vectors in the above equation:

1. The **y** vector has dimensions $m \times 1$.
2. The **C** matrix has dimensions $n \times m$.
3. Vector **x** is the $n \times 1$ state vector, as before.
4. The matrix **D** has dimensions $m \times n$.
5. Vector **u** is an $n \times 1$ column vector of control inputs.

We have also seen that some systems have, not just an intentional input **u**, but an unintentional input, or disturbance, **d**. The most general state equations may be written as

$$\dot{\mathbf{x}} = \mathbf{Ax} + \mathbf{Bu} + \mathbf{Wd} \tag{22.21}$$

$$\mathbf{y} = \mathbf{Cx} + \mathbf{Du} \tag{22.22}$$

The first of these equations is usually referred to as the system state equation, while the second is the output equation. In many practical examples $\mathbf{D} = 0$, and

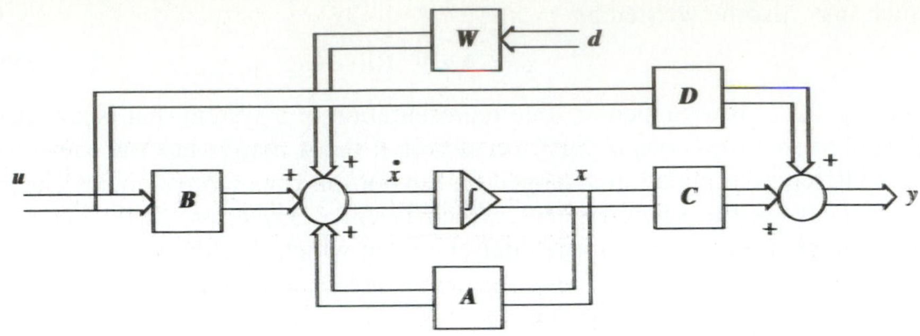

Fig. 22.5 Block diagram of general state-space representation.

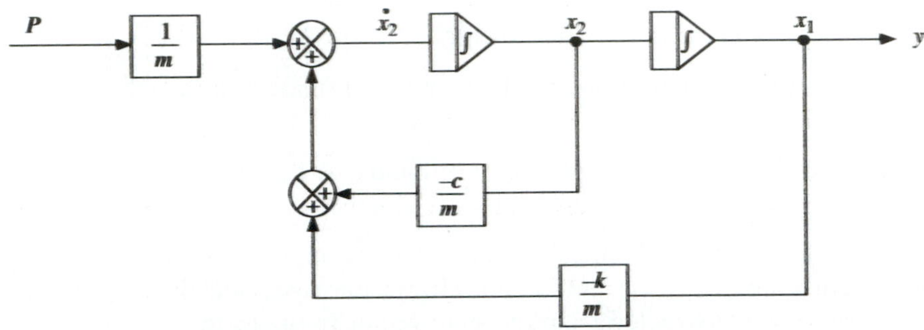

Fig. 22.6 Block diagram for mechanical system.

d may be neglected. A block diagram such as the one shown in Fig. 22.5 helps to visualize the relationship between the various elements and signals. The double lines represent vectors, rather than single variables, and the integrator symbolizes *n* integrators, one for each successive state variable. Although a feedback path is evident (through **A**), the representation is still of an open-loop system. When applied to a particular system, such as the mechanical system shown in Fig. 22.3 and represented by equations 22.5 and 22.7, the diagram takes the form shown in Fig. 22.6. Such diagrams not only provide insight into the structure of the system but could form the basis for solving the equations (calculating *y* for any given *u*) using an analog computer or digital simulation of an analog computer.

◆ STATE-SPACE FORM FROM TRANSFER FUNCTIONS

So far, we have developed state-space representations from the system differential equations. It is possible to develop the same result from the transfer function. In this section, we will consider only SISO systems, so **y** and **u** will be scalars. Consider the transfer function

$$\frac{y}{u} = \frac{K}{s^n + a_{n-1}s^{n-1} + \cdots + a_1 s + a_0} \tag{22.23}$$

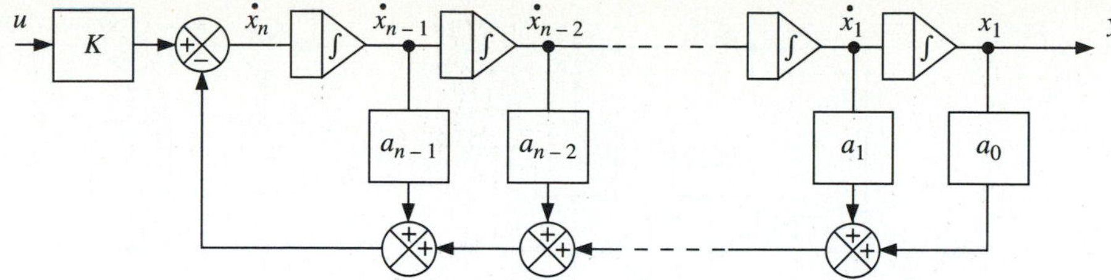

Fig. 22.7 Block diagram of simple transfer function.

Multiplying out and considering the s operator to be equivalent to differentiation, we obtain

$$\frac{d^ny}{dt^n} + a_{n-1}\frac{d^{n-1}y}{dt^{n-1}} + a_{n-2}\frac{d^{n-2}y}{dt^{n-2}} + \cdots + a_1\frac{dy}{dt} + a_0y = Ku(t) \qquad (22.24)$$

which is identical to equation 22.10. The state equations may then be written directly as in equations 22.11–22.16 and are represented in block diagram form in Fig. 22.7. Now consider a transfer function with numerator dynamics

$$\frac{y}{u} = \frac{K(s^n + b_{n-1}s^{n-1} + \cdots + b_1s + b_0)}{s^n + a_{n-1}s^{n-1} + \cdots + a_1s + a_0} \qquad (22.25)$$

This may be written in terms of a third variable z as

$$\frac{y}{u} = \frac{y}{z} \times \frac{z}{u} \qquad (22.26)$$

where

$$\frac{z}{u} = \frac{K}{s^n + a_{n-1}s^{n-1} + \cdots + a_1s + a_0} \qquad (22.27)$$

This is the same as the simple transfer function, so its state equations may be readily written, and its block diagram is as in Fig. 22.8. This is identical to Fig. 22.7 with z replacing x. The remaining part of the transfer function may be written as

$$\frac{y}{z} = s^n + b_{n-1}s^{n-1} + \cdots + b_1s + b_0 \qquad (22.28)$$

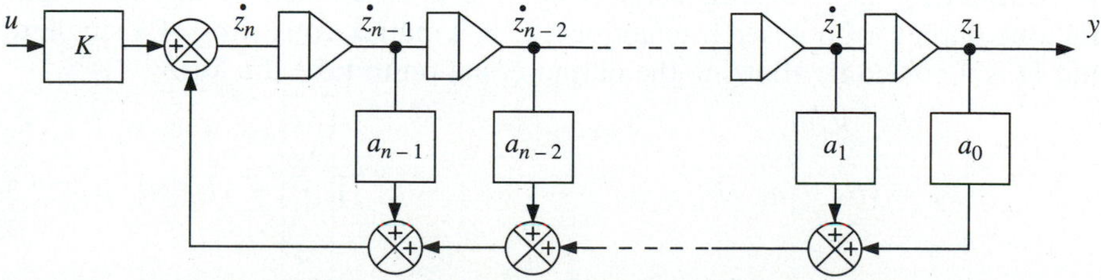

Fig. 22.8 Block diagram of part of complex transfer function.

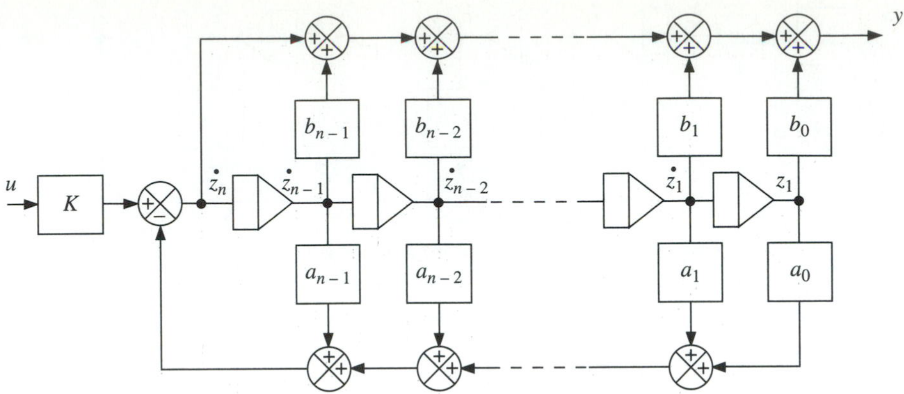

Fig. 22.9 Complete block diagram of complex transfer function.

which, again equating s to differentiation, becomes

$$y = \frac{d^n z}{dt^n} + b_{n-1}\frac{d^{n-1}z}{dt^{n-1}} + \cdots + b_1\frac{dz}{dt} + b_0 z \tag{22.29}$$

$$= \dot{z}_n + b_{n-1}\dot{z}_{n-1} + \cdots + b_1\dot{z}_1 + b_0 z \tag{22.30}$$

The output y may be constructed from the inputs to the n integrators by multiplying by the appropriate b coefficients and adding, as shown in Fig. 22.9. It has been assumed that the denominator is of the same order as the numerator and that all s terms are present.

Now that the block diagram of the system has been obtained, we will derive the state and output equations from it. It is known that this will result in

$$\dot{\mathbf{x}} = \mathbf{A}\mathbf{x} + \mathbf{B}u \tag{22.31}$$

and

$$y = \mathbf{C}\mathbf{x} + \mathbf{D}u \tag{22.32}$$

so the problem reduces to that of finding the matrices **A**, **B**, **C**, and **D**. Since the state equations are the same as 22.11–22.16, **A** and **B** will be the same as those in 22.17. The matrices **C** and **D** are found by considering the contribution to the output from the ith state. Reverting to our original state variable x in Fig. 22.9, we may write each contribution as

$$y_i = b_{i-1}x_i - a_{i-1}x_i \tag{22.33}$$

by considering the two paths connecting x_i to y. Since there is only one input u and one output y, both scalar quantities, **C** is a matrix consisting of a single row and **D** is a constant, allowing the output equation to take the form

$$y = [b_0 - a_0 \quad b_1 - a_1 \quad \cdots \quad b_{n-1} - a_{n-1}]\begin{bmatrix} x_1 \\ x_2 \\ \vdots \\ x_n \end{bmatrix} + [K][u] \tag{22.34}$$

The contribution of u directly to the output will be noted. If the numerator is of lower order than the denominator, which is usually the case, then the direct path from u to y in Fig. 22.9 will not be present. This also corresponds to the case where $\mathbf{D} = 0$, and none of the a_i will contribute to the output in equation 22.34, leaving

$$y = [b_0 \quad b_1 \quad \cdots \quad b_{n-1}] \begin{bmatrix} x_1 \\ x_2 \\ \vdots \\ x_n \end{bmatrix} \tag{22.35}$$

It may be seen that the **A** matrix is obtained from the denominator of the transfer function, while the numerator determines the **C** matrix.

◆ TRANSFER FUNCTION FROM STATE-SPACE FORM

In some cases, the transfer function matrix will need to be evaluated from the state-space form and the output equation. The problem may be stated in the form: given **A**, **B**, **C**, and **D**, find **y**/**u**. We begin with the system equation

$$\dot{\mathbf{x}} = \mathbf{Ax} + \mathbf{Bu} \tag{22.36}$$

and take Laplace transforms of both sides of this equation, assuming initial conditions are zero, to obtain

$$s\mathbf{x} = \mathbf{Ax} + \mathbf{Bu} \tag{22.37}$$

It is assumed that the variable **x** now refers to the s domain, and where

$$s\mathbf{x} = \begin{bmatrix} sx_1 \\ sx_2 \\ \vdots \\ sx_n \end{bmatrix} \tag{22.38}$$

this becomes

$$(s\mathbf{I} - \mathbf{A})\mathbf{x} = \mathbf{Bu} \tag{22.39}$$

or

$$\mathbf{x} = (s\mathbf{I} - \mathbf{A})^{-1}\mathbf{Bu} \tag{22.40}$$

Substituting for **x** into the output equation yields

$$\mathbf{y} = \mathbf{C}(s\mathbf{I} - \mathbf{A})^{-1}\mathbf{Bu} + \mathbf{Du} \tag{22.41}$$

resulting in the transfer function

$$\frac{\mathbf{y}}{\mathbf{u}} = \mathbf{C}(s\mathbf{I} - \mathbf{A})^{-1}\mathbf{B} + \mathbf{D} \tag{22.42}$$

Expanding the inverse into the adjoint and determinant

$$\frac{\mathbf{y}}{\mathbf{u}} = \frac{\mathbf{C}\,\mathrm{adj}(s\mathbf{I} - \mathbf{A})\mathbf{B} + \mathbf{D}\det(s\mathbf{I} - \mathbf{A})}{\det(s\mathbf{I} - \mathbf{A})} \tag{22.43}$$

From this last equation, it may be seen that the system poles, usually referred to in state-space analysis as the *eigenvalues*, may be found by solving the determinant, which will be an nth-order polynomial in s:

$$\det(s\mathbf{I} - \mathbf{A}) = 0 \tag{22.44}$$

The transfer function matrix G will be a matrix of q columns and p rows relating the p outputs and q inputs, as in

$$\begin{bmatrix} y_1 \\ \vdots \\ y_p \end{bmatrix} = \begin{bmatrix} G_{11} & \cdots & G_{1q} \\ \vdots & \ddots & \vdots \\ G_{p1} & \cdots & G_{pq} \end{bmatrix} \begin{bmatrix} u_1 \\ \vdots \\ u_q \end{bmatrix} \tag{22.45}$$

For SISO systems, this will reduce to a single, scalar equation.

◆ TRANSFORMATION OF STATE VARIABLE AND INVARIABILITY OF SYSTEM EIGENVECTORS

The selection of state variables for a particular system is not unique, although it may be shown that any two sets of state variables are linearly related. It follows that new state variables may be defined as a simple linear transformation of the old state variables. For example, we may construct a new set of state variables $\mathbf{z}$ from the old ones $\mathbf{x}$ by

$$\mathbf{z} = \mathbf{T}\mathbf{x} \tag{22.46}$$

Clearly $\mathbf{z}$ must be the same dimension as $\mathbf{x}$, and $\mathbf{T}$ must be a square matrix of full rank. It follows that the inverse of $\mathbf{T}$ exists, and therefore,

$$\mathbf{T}^{-1}\mathbf{z} = \mathbf{x} \qquad \mathbf{T}^{-1}\dot{\mathbf{z}} = \dot{\mathbf{x}} \tag{22.47}$$

Substituting into the state equation

$$\mathbf{T}^{-1}\dot{\mathbf{z}} = \mathbf{A}\mathbf{T}^{-1}\mathbf{z} + \mathbf{B}\mathbf{u} \tag{22.48}$$

resulting in

$$\dot{\mathbf{z}} = \mathbf{T}\mathbf{A}\mathbf{T}^{-1}\mathbf{z} + \mathbf{T}\mathbf{B}\mathbf{u} \tag{22.49}$$

Similarly, substituting into the output equation produces

$$\mathbf{y} = \mathbf{C}\mathbf{T}^{-1}\mathbf{z} + \mathbf{D}\mathbf{u} \tag{22.50}$$

giving the state and output equations in terms of the new state vector $\mathbf{z}$ as

$$\dot{\mathbf{z}} = \overline{\mathbf{A}}\mathbf{z} + \overline{\mathbf{B}}\mathbf{u} \tag{22.51}$$

and

$$\mathbf{y} = \overline{\mathbf{C}}\mathbf{z} + \mathbf{D}\mathbf{u} \tag{22.52}$$

where

$$\overline{\mathbf{A}} = \mathbf{T}\mathbf{A}\mathbf{T}^{-1} \qquad \overline{\mathbf{B}} = \mathbf{T}\mathbf{B} \qquad \overline{\mathbf{C}} = \mathbf{C}\mathbf{T}^{-1} \tag{22.53}$$

We will now show that the eigenvalues of the $\mathbf{x}$ system are unchanged by the transformation to the new state vector $\mathbf{z}$. The eigenvalues of the new system are found from

$$
\begin{aligned}
\det(s\mathbf{I} - \overline{\mathbf{A}}) &= \det(s\mathbf{I} - \mathbf{TAT}^{-1}) \\
&= \det(s\mathbf{TT}^{-1} - \mathbf{TAT}^{-1}) \\
&= \det[(s\mathbf{T} - \mathbf{TA})\mathbf{T}^{-1}] \\
&= \det[(\mathbf{T}s - \mathbf{TA})\mathbf{T}^{-1}] \\
&= \det[\mathbf{T}(s\mathbf{I} - \mathbf{A})\mathbf{T}^{-1}] \\
&= \det(\mathbf{T})\det(s\mathbf{I} - \mathbf{A})\det(\mathbf{T}^{-1}) \\
&= \det(\mathbf{T})\det(\mathbf{T}^{-1})\det(s\mathbf{I} - \mathbf{A}) \\
&= \det(\mathbf{TT}^{-1})\det(s\mathbf{I} - \mathbf{A})
\end{aligned}
$$

producing the final result

$$\det(s\mathbf{I} - \overline{\mathbf{A}}) = \det(s\mathbf{I} - \mathbf{A}) \tag{22.54}$$

This confirms that the eigenvalues, which play an important part in the system response, are independent of the state variables chosen to represent the system.

◆ CANONICAL FORMS AND DECOUPLED SYSTEMS

It is particularly useful if our standard state-space equations

$$\dot{\mathbf{x}} = \mathbf{Ax} + \mathbf{Bu} \tag{22.55}$$

may be written such that the system matrix **A** takes the form

$$\mathbf{A} = \begin{bmatrix} \lambda_1 & 0 & 0 & \cdots & 0 \\ 0 & \lambda_2 & 0 & \cdots & 0 \\ \vdots & \vdots & \vdots & \ddots & \vdots \\ 0 & 0 & \cdots & 0 & \lambda_n \end{bmatrix} \tag{22.56}$$

This form is commonly known as the *Jordan canonical form*. In this array, only the leading diagonal elements are nonzero, and these elements will later be shown to be equal to the system poles, or *eigenvalues,* as they are sometimes known. The advantages of writing the system in this form will only become apparent in the next few modules, but they may be stated here as follows:

1. The eigenvalues provide information on the time response of the system.
2. This form indicates a system structure in which each state variable is decoupled from the other state variables.
3. This form is required for controllability studies of the system.

If the system is written in Jordan form and without loss of generality, we may assume that the system has a single input u and write the ith row of the state equation as

$$\dot{x}_i = \lambda_i x_i + u \tag{22.57}$$

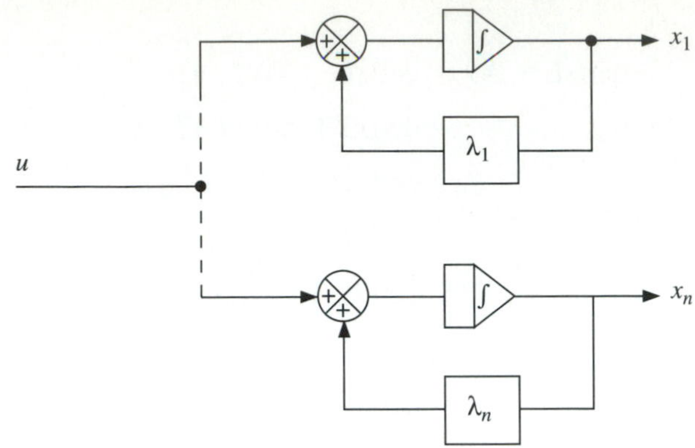

Fig. 22.10 Block diagram for Jordan canonical form.

leading to a block diagram relating u to the state variables $\mathbf{x}$ as shown in Fig. 22.10. The output $\mathbf{y}$ may be formed from the state variables in the usual manner. As was shown in the previous section, it is possible to transform a system's state variables into another set of state variables using a simple linear transformation. This transformation changes the $\mathbf{A}$, $\mathbf{B}$, and $\mathbf{C}$ matrices. The question arises, therefore, of how we can transform any system in such a manner that its system matrix takes the Jordan canonical form shown in equation 22.56. Consider the problem posed as

$$\dot{\mathbf{x}} = \mathbf{Ax} + \mathbf{Bu} \tag{22.58}$$

Let

$$\mathbf{x} = \mathbf{Mz} \tag{22.59}$$

or

$$\mathbf{z} = \mathbf{M}^{-1}\mathbf{x} \tag{22.60}$$

Substituting into the state equation yields

$$\mathbf{M}\dot{\mathbf{z}} = \mathbf{AMx} + \mathbf{Bu} \tag{22.61}$$

Hence

$$\dot{\mathbf{z}} = \mathbf{M}^{-1}\mathbf{AMz} + \mathbf{M}^{-1}\mathbf{Bu} \tag{22.62}$$

which may be written as

$$\dot{\mathbf{z}} = \mathbf{Jz} + \overline{\mathbf{B}}\mathbf{u} \tag{22.63}$$

where

$$\mathbf{J} = \mathbf{M}^{-1}\mathbf{AM} \tag{22.64}$$

The problem now[2] is to find $\mathbf{M}$ such that

$$\mathbf{J} = \begin{bmatrix} \lambda_1 & 0 & 0 & \cdots & 0 \\ 0 & \lambda_2 & 0 & \cdots & 0 \\ \vdots & \vdots & \vdots & \ddots & \vdots \\ 0 & 0 & \cdots & 0 & \lambda_n \end{bmatrix} \tag{22.65}$$

[2] The method used to find the eigenvalues is applicable only if they are real and unrepeated. Other methods, beyond the scope of this book, treat cases of repeated and complex eigenvalues.

From equation 22.64 we have

$$\mathbf{MJ} = \mathbf{AM} \tag{22.66}$$

where the $n \times n$ matrix $\mathbf{M}$ is considered to be written in terms of its column vectors $\mathbf{m}_1, \mathbf{m}_2, \ldots, \mathbf{m}_n$:

$$[\mathbf{m}_1\mathbf{m}_2\cdots\mathbf{m}_n]\begin{bmatrix} \lambda_1 & 0 & 0 & \cdots & 0 \\ 0 & \lambda_2 & 0 & \cdots & 0 \\ \vdots & \vdots & \vdots & \ddots & \vdots \\ 0 & 0 & \cdots & 0 & \lambda_n \end{bmatrix} = \mathbf{A}[\mathbf{m}_1\mathbf{m}_2\cdots\mathbf{m}_n] \tag{22.67}$$

This becomes

$$[\lambda_1\mathbf{m}_1 \quad \lambda_2\mathbf{m}_2 \quad \cdots \quad \lambda_n\mathbf{m}_n] = [\mathbf{Am}_1 \quad \mathbf{Am}_2 \quad \cdots \quad \mathbf{Am}_n] \tag{22.68}$$

Considering individual column vectors

$$\lambda_i\mathbf{m}_i = \mathbf{Am}_i \qquad i = 1, n \tag{22.69}$$

Hence

$$(\lambda_i\mathbf{I} - \mathbf{A})\mathbf{m}_i = 0 \tag{22.70}$$

However, since

$$\mathbf{m}_i \neq \mathbf{0} \tag{22.71}$$

the matrix $(\lambda_i\mathbf{I} - \mathbf{A})$ must be singular, which implies

$$\det(\lambda_i\mathbf{I} - \mathbf{A}) = 0 \tag{22.72}$$

We can solve equation 22.72 to find λ_i, the eigenvalues, and then solve

$$(\lambda_i\mathbf{I} - \mathbf{A})\mathbf{m}_i = 0 \tag{22.73}$$

for each column vector $\mathbf{m}_i$ corresponding to the eigenvalue λ_i. These column vectors are known as the *eigenvectors* of the system. The matrix $\mathbf{M}$ is then constructed from its n column vectors $\mathbf{m}_1\mathbf{m}_2\ldots\mathbf{m}_n$. Note that the modified control distribution matrix becomes

$$\overline{\mathbf{B}} = \mathbf{M}^{-1}\mathbf{B} \tag{22.74}$$

and that the output equation may be written in terms of the new state variable vector

$$\mathbf{y} = \mathbf{Cx} \tag{22.75}$$

$$= \mathbf{CMz} \tag{22.76}$$

$$= \overline{\mathbf{C}}\mathbf{z} \tag{22.77}$$

The Jordan canonical form[3] may be summarized, therefore, as

$$\dot{\mathbf{z}} = \mathbf{Jz} + \overline{\mathbf{B}}\mathbf{u} \tag{22.78}$$

$$\mathbf{y} = \overline{\mathbf{C}}\mathbf{z} \tag{22.79}$$

[3] The MATLAB command `[m,j]=eig(a)` find the matrices **M** and **J** directly from **A**.

This relationship is easily shown by considering the ith block in Fig. 22.10:

$$\dot{x}_i = u + \lambda_i x_i \tag{22.80}$$

Writing this in the form

$$\dot{x}_i - \lambda_i x_i = u \tag{22.81}$$

and taking Laplace transforms yield

$$(s - \lambda_i)x_i = u \tag{22.82}$$

Hence

$$x_i = \frac{u}{s - \lambda_i} \tag{22.83}$$

Since the output in this example is simply the sum of the state variables (Fig. 22.10),

$$y = \sum_{i=1}^{n} x_i = \frac{u}{s - \lambda_i} + \frac{u}{s - \lambda_2} + \cdots + \frac{u}{s - \lambda_n} \tag{22.84}$$

The system transfer function may then be written as

$$\frac{y}{u} = \frac{K}{(s - \lambda_1)(s - \lambda_2)\cdots(s - \lambda_n)} \tag{22.85}$$

We shall see that for a stable system the eigenvalues are negative; hence it is clear that the system poles and the eigenvalues are the same, implying that the eigenvalues may be derived directly from the transfer function if it is available in the above factored form.

In this section we have learned how to write SISO and MIMO systems in state-space form and developed techniques for transforming system descriptions between state-space and transfer function forms. In particular, we have studied a particular description, the Jordan canonical form, which gives direct insight into the response of the system through the eigenvalues, or poles, which were shown to be independent of the choice of eigenvectors used to describe the system.

Sample Problem 22.1

Consider the two-degree-of-freedom system shown in Fig. SP22.1.1. Calculate the state equation and the output equation and sketch a block diagram representation of the system showing the states and how the output y is generated.

Solution

We note the system has two inputs, F_1 and F_2, and two outputs, y_1 and y_2. Applying Newton's second law produces the following equations of motion in terms of z_1 and z_2, the displacements of the masses from their equilibrium position:

$$F_1 + K_3(z_2 - z_1) - K_1 z_1 - c_1 \dot{z}_1 = m_1 \ddot{z}_1$$

$$F_2 - K_3(z_2 - z_1) - K_2 z_2 - c_2 \dot{z}_2 = m_2 \ddot{z}_1$$

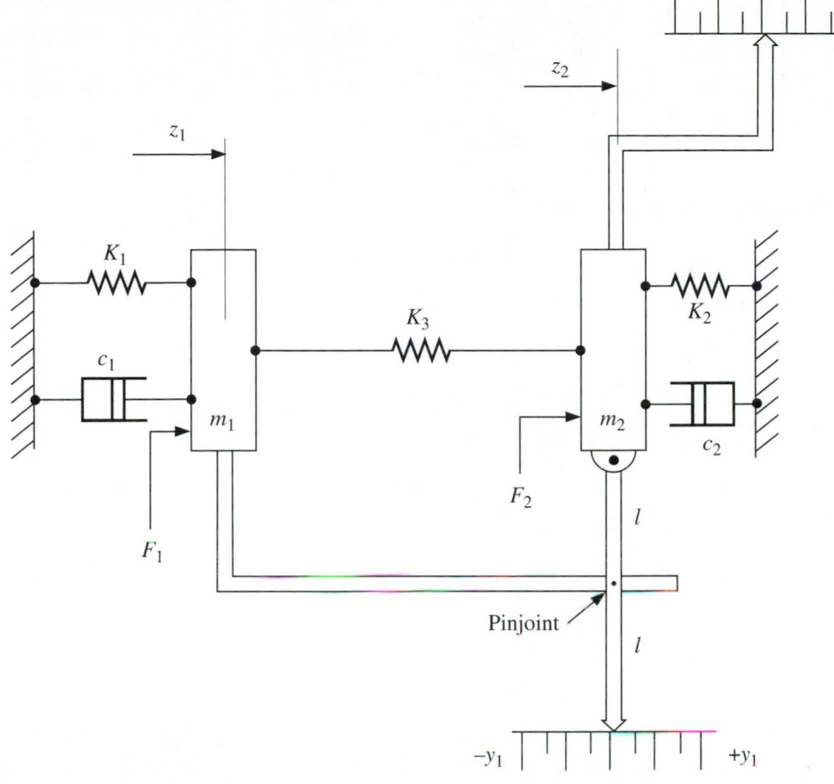

Fig. SP22.1.1

The next problem is to determine the number of state variables. By realizing that there are two second-order differential equations describing the system, or by observing that each mass has to be specified by a position and velocity, it is seen that four state variables will be sufficient. A suitable choice would be

$$x_1 = z_1$$

$$x_2 = \dot{z}_1 = \dot{x}_1$$

$$x_3 = z_2$$

$$x_4 = \dot{z}_2 = \dot{x}_3$$

The state equations are formulated by writing the definitions of the state variables and the equations of motion in the form

$$\dot{x}_1 = x_2$$

$$\dot{x}_2 = \ddot{z}_1 = \frac{F_1}{m_1} + \frac{K_3}{m_1}x_3 - x_1\left(\frac{K_3}{m_1} + \frac{K_1}{m_1}\right) - \frac{c_1}{m_1}x_2$$

$$\dot{x}_3 = x_4$$

$$\dot{x}_4 = \ddot{z}_2 = \frac{F_2}{m_2} - x_3\left(\frac{K_3}{m_2} + \frac{K_2}{m_2}\right) + \frac{K_3}{m_2}x_1 - \frac{c_2}{m_2}x_4$$

This enables the state equations to be written as

$$\begin{bmatrix} \dot{x}_1 \\ \dot{x}_2 \\ \dot{x}_3 \\ \dot{x}_4 \end{bmatrix} = \begin{bmatrix} 0 & 1 & 0 & 0 \\ -\dfrac{K_1 + K_3}{m_1} & -\dfrac{c_1}{m_1} & \dfrac{K_3}{m_1} & 0 \\ 0 & 0 & 1 & 0 \\ \dfrac{K_3}{m_2} & 0 & -\dfrac{K_2 + K_3}{m_2} & -\dfrac{c_2}{m_2} \end{bmatrix} \begin{bmatrix} x_1 \\ x_2 \\ x_3 \\ x_4 \end{bmatrix} + \begin{bmatrix} 0 & 0 & 0 & 0 \\ \dfrac{1}{m_1} & 0 & 0 & 0 \\ 0 & 0 & 0 & 0 \\ 0 & \dfrac{1}{m_2} & 0 & 0 \end{bmatrix} \begin{bmatrix} F_1 \\ F_2 \\ 0 \\ 0 \end{bmatrix}$$

which is in the required form

$$\dot{\mathbf{x}} = \mathbf{Ax} + \mathbf{Bu}$$

There are two outputs y_1 and y_2, where

$$y_1 = z_2 + 2z_1 = 2x_1 - x_3$$

$$y_2 = z_2 = x_3$$

The output equation takes the form

$$\mathbf{y} = \mathbf{Cx}$$

since $\mathbf{D} = 0$, yielding

$$\begin{bmatrix} y_1 \\ y_2 \end{bmatrix} \begin{bmatrix} 2 & 0 & -1 & 0 \\ 0 & 0 & 1 & 0 \end{bmatrix} \begin{bmatrix} x_1 \\ x_2 \\ x_3 \\ x_4 \end{bmatrix}$$

In order to draw a block diagram representation of the system, we begin by drawing the two inputs F_1 and F_2. Inspection of the equations of motion indicates that $\dot{x}_2$ and $\dot{x}_4$ are the sum of a number of signals. Assume that these signals are available, and we are able to construct $\dot{x}_2$ and $\dot{x}_4$. Successive integrations give us x_1, x_2, x_3, and x_4. The diagram so far is as shown in Fig. SP22.1.2. From the equations of motion, the missing signals may now be generated from the state variables and incorporated into the diagram, as shown in Fig. SP22.1.3. Finally, the output vector $\mathbf{y}$ may be constructed from the appropriate state variables and included in the block diagram, the final form of which is shown in Fig. SP22.1.4.

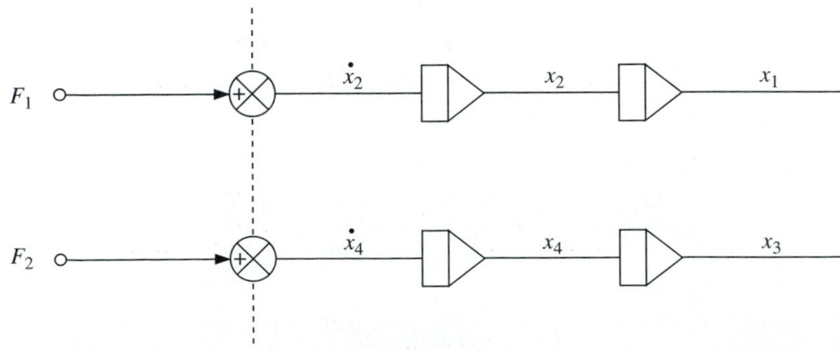

Fig. SP22.1.2

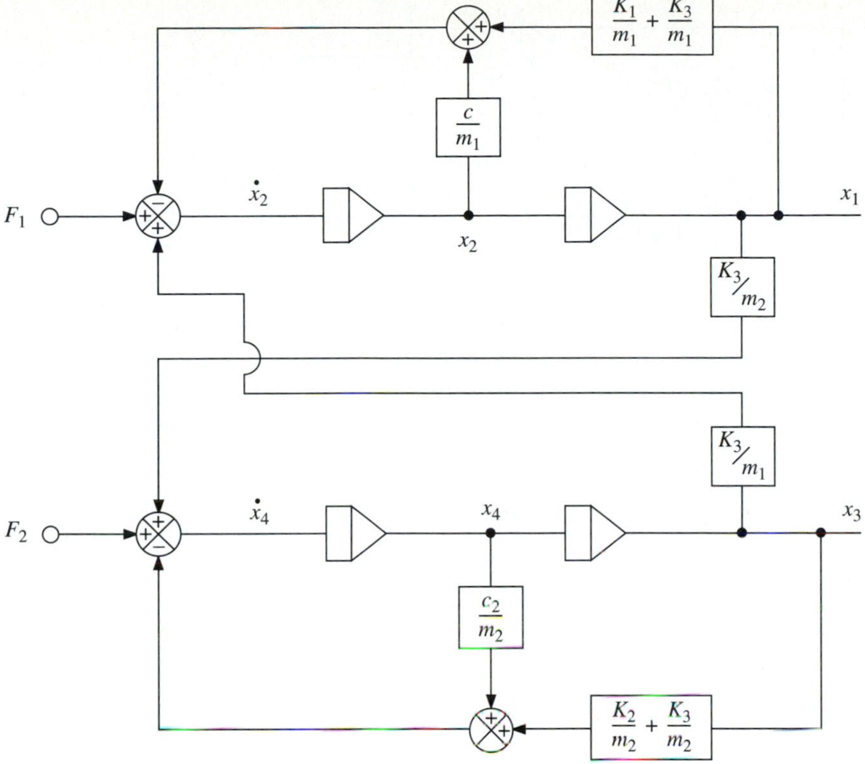

Fig. SP22.1.3

SAMPLE PROBLEM 22.2

Develop the state-space representation of the electromechanical system shown in Fig. SP22.2.1, where the input is the field voltage v_i and the output is (a) angular velocity ω and (b) angular position θ. For case (a), develop the system transfer function from the state-space formulation. Assume that the motor is field controlled and the armature current is therefore constant and also that there is frictional resistance to the rotation of the load, proportional to the speed of rotation.

Solution

Recalling the analysis of the motor-controlled inertial load in Module 5, this open-loop system may be described by Kirchhoff's voltage law for the field winding:

$$v_i - \iota R - L\dot{\iota} = 0$$

Also, the interaction of the field and armature magnetic fields produces a torque proportional to the field current,

$$\tau_m = K_m \iota$$

and finally, Newton's second law for the rotational system gives

$$\tau_m - c\omega = J\dot{\omega}$$

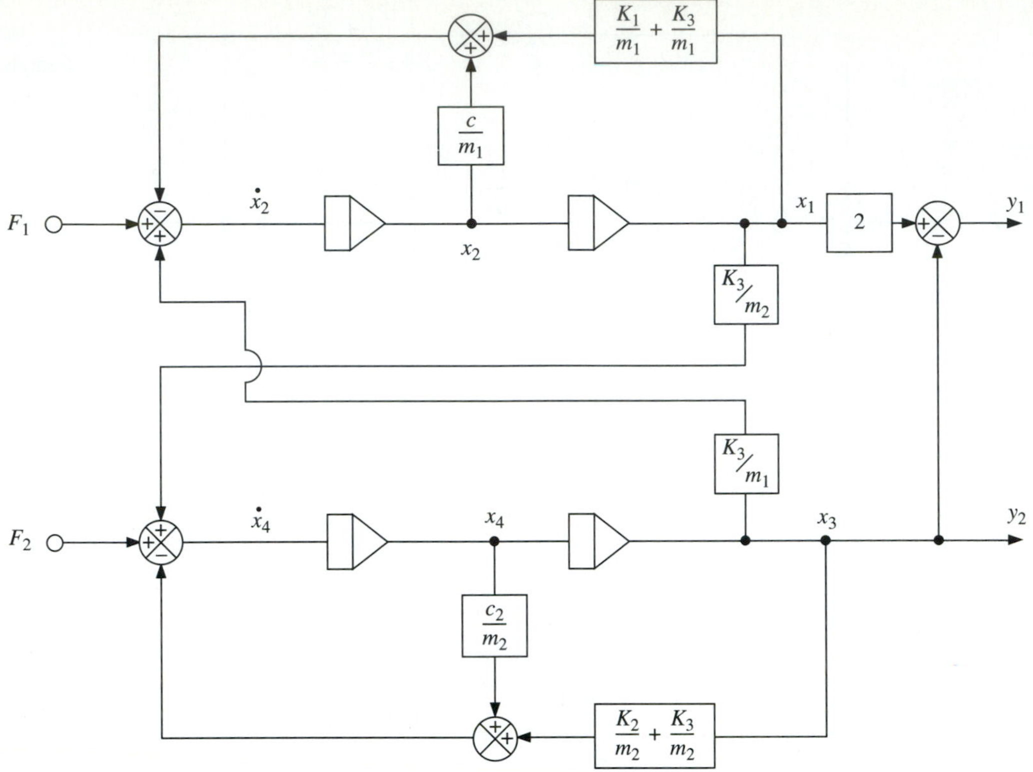

Fig. SP22.1.4

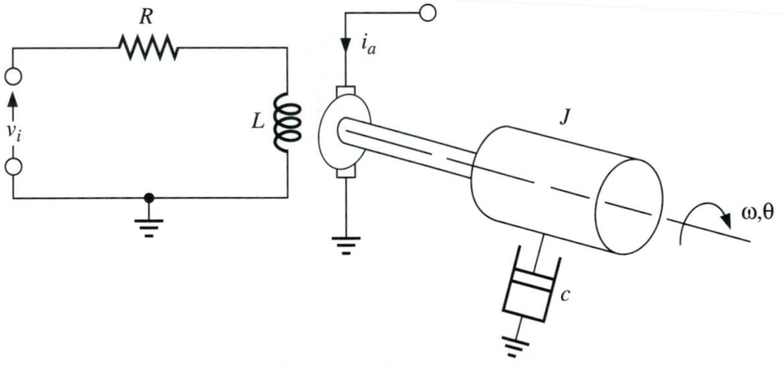

Fig. SP22.2.1

Since the system is described by two first-order differential equations, the system may be described by two state variables. A suitable choice might be the field current ι and the rotational speed ω for case (a) in the problem statement. The state-space representation will take the form

$$\begin{bmatrix} \dot{i} \\ \dot{\omega} \end{bmatrix} = \begin{bmatrix} a_{11} & a_{12} \\ a_{21} & a_{22} \end{bmatrix} \begin{bmatrix} \iota \\ \omega \end{bmatrix} + \begin{bmatrix} b_{11} & b_{12} \\ b_{21} & b_{22} \end{bmatrix} \begin{bmatrix} v_i \\ 0 \end{bmatrix}$$

The elements of **A** and **B** are identified from the system differential equations, aided by writing them in the form

$$\dot{\iota} = -\frac{R}{L}\iota + \frac{v_i}{L}$$

$$\dot{\omega} = \frac{K_m}{J}\iota - \frac{c}{J}\omega$$

resulting in

$$\begin{bmatrix} \dot{\iota} \\ \dot{\omega} \end{bmatrix} = \begin{bmatrix} -\dfrac{R}{L} & 0 \\ \dfrac{K_m}{J} & -\dfrac{c}{J} \end{bmatrix} \begin{bmatrix} \iota \\ \omega \end{bmatrix} + \begin{bmatrix} \dfrac{1}{L} & 0 \\ 0 & 0 \end{bmatrix} \begin{bmatrix} v_i \\ 0 \end{bmatrix}$$

which is of the required form

$$\dot{\mathbf{x}} = \mathbf{A}\mathbf{x} + \mathbf{B}\mathbf{u}$$

For the case where the system output is to be the angular position θ, the set of equations is supplemented by

$$\omega = \dot{\theta}$$

This is also a first-order differential equation, bringing to three the total for the system. This system is represented by three state variables, a suitable choice being

$$x_1 = \iota \qquad x_2 = \omega \qquad x_3 = \theta$$

Writing the system equations as

$$\dot{\iota} = -\frac{R}{L}\iota + v_i$$

$$\dot{\omega} = K_m\iota - \frac{c}{J}\omega$$

$$\dot{\theta} = \omega$$

the state equations may be written as

$$\begin{bmatrix} \dot{\iota} \\ \dot{\omega} \\ \dot{\theta} \end{bmatrix} = \begin{bmatrix} -\dfrac{R}{L} & 0 & 0 \\ \dfrac{K_m}{J} & -\dfrac{c}{J} & 0 \\ 0 & 1 & 0 \end{bmatrix} \begin{bmatrix} \iota \\ \omega \\ \theta \end{bmatrix} + \begin{bmatrix} \dfrac{1}{L} & 0 & 0 \\ 0 & 0 & 0 \\ 0 & 0 & 0 \end{bmatrix} \begin{bmatrix} v_i \\ 0 \\ 0 \end{bmatrix}$$

which is also of the required form

$$\dot{\mathbf{x}} = \mathbf{A}\mathbf{x} + \mathbf{B}\mathbf{u}$$

The output equation for system (a) may be determined by noting that the output is defined to be the angular speed ω, i.e.,

$$y = \omega$$

This may be expressed in the form

$$\mathbf{y} = \mathbf{Cx} + \mathbf{Du}$$

as

$$\begin{bmatrix} y_1 \\ y_2 \end{bmatrix} = \begin{bmatrix} 0 & 1 \\ 0 & 0 \end{bmatrix}\begin{bmatrix} \iota \\ \omega \end{bmatrix} + \begin{bmatrix} 0 & 0 \\ 0 & 0 \end{bmatrix}\begin{bmatrix} v_i \\ 0 \end{bmatrix}$$

whereas, for the second system, in which the output is defined to be

$$\mathrm{y} = \theta$$

the output equations in matrix form become

$$\begin{bmatrix} y_1 \\ y_2 \\ y_3 \end{bmatrix} = \begin{bmatrix} 0 & 0 & 1 \\ 0 & 0 & 0 \\ 0 & 0 & 0 \end{bmatrix}\begin{bmatrix} \iota \\ \omega \\ \theta \end{bmatrix} + \begin{bmatrix} 0 & 0 & 0 \\ 0 & 0 & 0 \\ 0 & 0 & 0 \end{bmatrix}\begin{bmatrix} v_i \\ 0 \\ 0 \end{bmatrix}$$

For the velocity system under consideration

$$\mathbf{A} = \begin{bmatrix} -\dfrac{R}{L} & 0 \\ \dfrac{K_m}{J} & -\dfrac{c}{J} \end{bmatrix}$$

$$\mathbf{B} = \begin{bmatrix} \dfrac{1}{L} \\ 0 \end{bmatrix}$$

$$\mathbf{C} = \begin{bmatrix} 0 & 1 \\ 0 & 0 \end{bmatrix}$$

$$\mathbf{D} = \begin{bmatrix} 0 & 0 \\ 0 & 0 \end{bmatrix}$$

Since the transfer function is given by

$$\frac{\mathbf{y}}{\mathbf{u}} = \frac{\mathbf{C}\ \mathrm{adj}(s\mathbf{I} - \mathbf{A})\mathbf{B} + \mathbf{D}\ \det(s\mathbf{I} - \mathbf{A})}{\det(s\mathbf{I} - \mathbf{A})}$$

and $\mathbf{D} = 0$, this reduces to

$$\frac{\mathbf{y}}{\mathbf{u}} = \frac{\mathbf{C}\ \mathrm{adj}(s\mathbf{I} - \mathbf{A})\mathbf{B}}{\det(s\mathbf{I} - \mathbf{A})}$$

We obtain

$$(\mathbf{sI} - \mathbf{A}) = \begin{bmatrix} s + \dfrac{R}{L} & 0 \\ -\dfrac{K_m}{J} & s + \dfrac{c}{J} \end{bmatrix}$$

hence

$$\det(\mathbf{sI} - \mathbf{a}) = \left(s + \frac{R}{L}\right)\left(s + \frac{c}{J}\right)$$

Recalling the adjoint of a matrix is the transpose of the cofactors,

$$\operatorname{adj}(s\mathbf{I} - \mathbf{A}) = \begin{bmatrix} s + \dfrac{c}{J} & \dfrac{K_m}{J} \\ 0 & s + \dfrac{R}{L} \end{bmatrix}^T$$

$$= \begin{bmatrix} s + \dfrac{c}{J} & 0 \\ \dfrac{K_m}{J} & s + \dfrac{R}{L} \end{bmatrix}^T$$

The transfer function becomes

$$\frac{y}{u} = \frac{\omega}{v_i} = \frac{\begin{bmatrix} 0 & 1 \\ 0 & 0 \end{bmatrix}\begin{bmatrix} s + \dfrac{R}{L} & 0 \\ \dfrac{K_m}{J} & s + \dfrac{c}{J} \end{bmatrix}\begin{bmatrix} \dfrac{1}{L} \\ 0 \end{bmatrix}}{\left(s + \dfrac{R}{L}\right)\left(s + \dfrac{c}{J}\right)}$$

which becomes

$$\frac{\omega}{v_i} = \frac{\begin{bmatrix} \dfrac{K_m}{J} & s + \dfrac{R}{L} \\ 0 & 0 \end{bmatrix}\begin{bmatrix} \dfrac{1}{L} \\ 0 \end{bmatrix}}{\left(s + \dfrac{R}{L}\right)\left(s + \dfrac{c}{J}\right)}$$

resulting in

$$\frac{\omega}{J} = \frac{K_m/JL}{\left(s + \dfrac{R}{L}\right)\left(s + \dfrac{c}{J}\right)} = \frac{K_m}{(Js + c)(Ls + R)}$$

This result could have also been obtained directly from the system differential equations.

Sample Problem 22.3

An open-loop system is described by the following set of equations:

$$\dot{x}_1 = -4x_1 + 3x_2 + u$$

$$\dot{x}_2 = u$$

$$y = x_1$$

1. Find the eigenvalues of the system.
2. Transform the system to Jordan canonical form, and write down the new state and output equations.

3. Calculate the system transfer function and confirm the result from the equations above.
4. Draw the block diagrams of the original and diagonalized systems.

Solution

(1) EIGENVALUES. The original system may be written in the form

$$\dot{\mathbf{x}} = \mathbf{A}\mathbf{x} + \mathbf{B}\mathbf{u}$$

as

$$\begin{bmatrix} \dot{x}_1 \\ \dot{x}_2 \end{bmatrix} = \begin{bmatrix} -4 & 3 \\ 0 & 0 \end{bmatrix} \begin{bmatrix} x_1 \\ x_2 \end{bmatrix} + \begin{bmatrix} 1 & 0 \\ 1 & 0 \end{bmatrix} \begin{bmatrix} u \\ 0 \end{bmatrix}$$

The output equation is of the form

$$\mathbf{y} = \mathbf{C}\mathbf{x}$$

where

$$y = [1 \quad 0] \begin{bmatrix} u \\ 0 \end{bmatrix}$$

The eigenvalues are found from

$$\det(s\mathbf{I} - \mathbf{A}) = 0$$

where

$$(s\boldsymbol{I} - \mathbf{A}) = \begin{bmatrix} s + 4 & -3 \\ 0 & s \end{bmatrix}$$

This yields

$$s(s + 4) = 0$$

giving two eigenvalues of $s = 0$ and $s = -4$.

(2) JORDAN FORM. Since we already have the eigenvalues, the eigenvectors are obtained from equation 22.73 for each eigenvalue in turn.

(i) $\lambda_1 = 0$. Equation 22.73 becomes

$$(\lambda_1 \mathbf{I} - \mathbf{A})\mathbf{m}_1 = 0$$

which, for our system, becomes

$$\begin{bmatrix} 4 & -3 \\ 0 & 0 \end{bmatrix} \begin{bmatrix} m_1 \\ m_2 \end{bmatrix} = 0$$

This yields

$$4m_1 - 3m_2 = 0$$

Hence the eigenvector may be written as

$$\mathbf{m}_1 = \begin{bmatrix} m_1 \\ \frac{4}{3}m_1 \end{bmatrix} = \begin{bmatrix} 1 \\ \frac{4}{3} \end{bmatrix}$$

since an eigenvector may be scaled by an arbitrary constant, $\mathbf{m}_1$ in this case.

(ii) $\lambda_2 = -4$. This time we obtain from

$$(\lambda_2\mathbf{I} - \mathbf{A})\mathbf{m}_2 = 0$$

the matrix equation

$$\begin{bmatrix} 0 & -3 \\ 0 & -4 \end{bmatrix}\begin{bmatrix} m_1 \\ m_2 \end{bmatrix} = 0$$

which yields

$$-3m_2 = 0 \qquad -4m_2 = 0$$

i.e., $m_2 = 0$. The corresponding eigenvector is

$$\mathbf{m}_2 = \begin{bmatrix} m_1 \\ 0 \end{bmatrix} = \begin{bmatrix} 1 \\ 0 \end{bmatrix}$$

The matrix $\mathbf{M}$ is composed of the two eigenvectors as

$$\mathbf{M} = [\mathbf{m}_1 \quad \mathbf{m}_2] = \begin{bmatrix} 1 & 1 \\ \frac{4}{3} & 0 \end{bmatrix}$$

The Jordan form will be

$$\dot{\mathbf{z}} = \mathbf{Jz} + \overline{\mathbf{B}}\mathbf{u}$$

where

$$\mathbf{J} = \mathbf{M}^{-1}\mathbf{AM}$$

The inverse of $\mathbf{M}$ is calculated as

$$\mathbf{M}^{-1} = \frac{\text{adj } \mathbf{M}}{\det \mathbf{M}} = \frac{\begin{bmatrix} 0 & -\frac{4}{3} \\ -1 & 1 \end{bmatrix}^T}{-\frac{4}{3}} = -\frac{3}{4}\begin{bmatrix} 0 & -1 \\ -\frac{4}{3} & 0 \end{bmatrix}$$

Hence

$$\begin{aligned} \mathbf{J} = \mathbf{M}^{-1}\mathbf{AM} &= \mathbf{M}^{-1}\begin{bmatrix} -4 & 3 \\ 0 & 0 \end{bmatrix}\begin{bmatrix} 1 & 1 \\ \frac{4}{3} & 0 \end{bmatrix} \\ &= -\frac{3}{4}\begin{bmatrix} 0 & -1 \\ -\frac{4}{3} & 1 \end{bmatrix}\begin{bmatrix} 0 & -4 \\ 0 & 0 \end{bmatrix} \\ &= -\frac{3}{4}\begin{bmatrix} 0 & 0 \\ 0 & \frac{4}{3} \end{bmatrix} \\ &= -\frac{3}{4}\begin{bmatrix} 0 & 0 \\ 0 & \frac{16}{3} \end{bmatrix} \\ &= \begin{bmatrix} 0 & 0 \\ 0 & -4 \end{bmatrix} \end{aligned}$$

Inspecting this expression for **J** reveals that it is a matrix of zeros with the system eigenvalues along the leading diagonal, as expected. The output equation for the new state variables **z** is obtained from

$$\overline{\mathbf{C}} = \mathbf{CM}$$

giving

$$\mathbf{C} = [1 \quad 0]\begin{bmatrix} 1 & 1 \\ \frac{4}{3} & 0 \end{bmatrix}$$

$$= [1 \quad 1]$$

The new control distribution matrix $\overline{\mathbf{B}}$ becomes

$$\overline{\mathbf{B}} = \mathbf{M}^{-1}\mathbf{B}$$

which yields

$$\overline{\mathbf{B}} = -\frac{3}{4}\begin{bmatrix} -1 & 0 \\ -\frac{1}{3} & 0 \end{bmatrix}$$

The new state equation becomes

$$\begin{bmatrix} \dot{z}_1 \\ \dot{z}_2 \end{bmatrix} = \begin{bmatrix} 0 & 0 \\ 0 & -4 \end{bmatrix}\begin{bmatrix} z_1 \\ z_2 \end{bmatrix} + \begin{bmatrix} \frac{3}{4} & 0 \\ \frac{1}{4} & 0 \end{bmatrix}\begin{bmatrix} u \\ 0 \end{bmatrix}$$

The new output equation becomes

$$y = [1 \quad 1]\begin{bmatrix} z_1 \\ z_2 \end{bmatrix}$$

(3) Transfer Function. The transfer function is obtained from equation 22.43,

$$\frac{\mathbf{y}}{\mathbf{u}} = \frac{\mathbf{C}\ \mathrm{adj}(s\mathbf{I} - \mathbf{A})\mathbf{B}}{\det(s\mathbf{I} - \mathbf{A})}$$

since $\mathbf{D} = 0$. The denominator yields

$$\det(s\mathbf{I} - \mathbf{A}) = s(s + 4)$$

The numerator becomes

$$\mathbf{C}\ \mathrm{adj}(s\mathbf{I} - \mathbf{A})\mathbf{B} = [1 \quad 0]\begin{bmatrix} s & 0 \\ 3 & s+4 \end{bmatrix}^T\begin{bmatrix} 1 & 0 \\ 1 & 0 \end{bmatrix}$$

$$= [s \quad 3]\begin{bmatrix} 1 & 0 \\ 1 & 0 \end{bmatrix}$$

$$= s + 3$$

The transfer function becomes

$$\frac{y}{u} = \frac{s + 3}{s(s + 4)}$$

This result may be verified by taking the Laplace transform of the system equations to obtain

$$sx_1 = -4x_1 + 3x_2 + u$$
$$sx_2 = u$$
$$y = x_1$$

Eliminating x_2 gives

$$sx_1 = -4x_1 + \frac{3u}{s} + u$$
$$x_1(s + 4) = \frac{3u}{s} + u$$
$$y(s + 4) = \frac{(3u + us)}{s}$$

resulting in

$$\frac{y}{u} = \frac{s + 3}{s(s + 4)}$$

(4) Block Diagrams. The original system block diagram is obtained from the equations of motion and takes the form shown in Fig. SP22.3.1. For the Jordan form, the state and output equations in the transformed state variable **z** are expanded,

$$\dot{z}_1 = \tfrac{3}{4}u$$
$$\dot{z}_2 = -4z_2 + \tfrac{1}{4}u$$
$$y = z_1 + z_2$$

The block diagram is drawn as before, resulting in Fig. SP22.3.2. Note how the state variables are decoupled in the Jordan form but that the system eigenvalues are unchanged by the linear transformation

$$\mathbf{z} = \mathbf{M}^{-1}\mathbf{x}$$

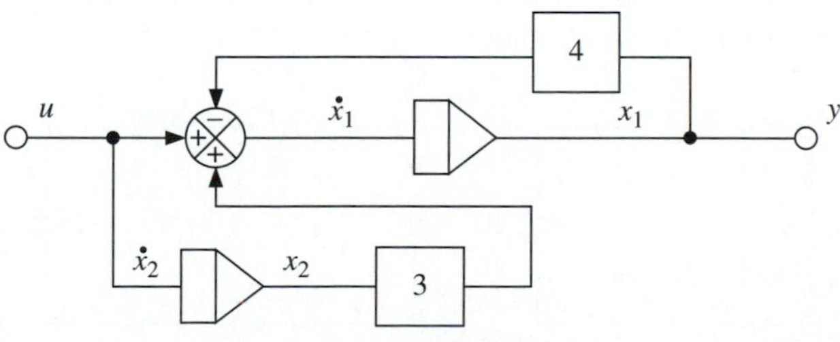

Fig. SP22.3.1

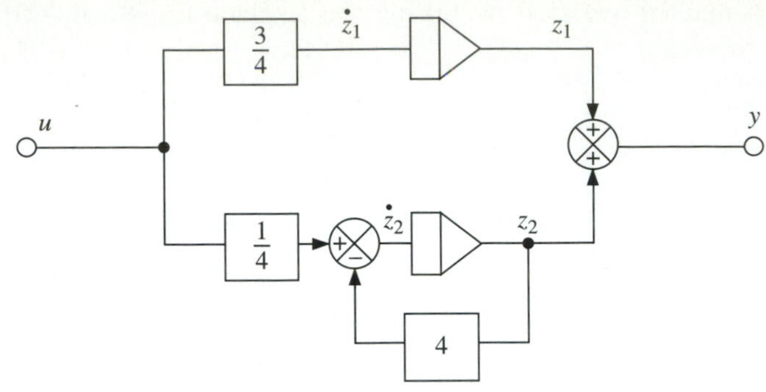

Fig. SP22.3.2

◆ PROBLEMS

22.1 Determine the state-space form for the mechanical system shown in Fig. P22.1.

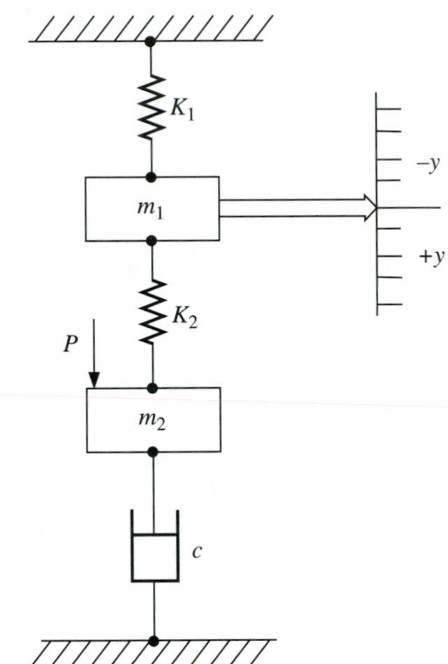

Fig. P22.1

22.2 For the electrical systems shown in Fig. P22.2, calculate the state-space form of the system equations. Write down the output equation in each case when the output variable is defined to be the voltage across the capacitor.

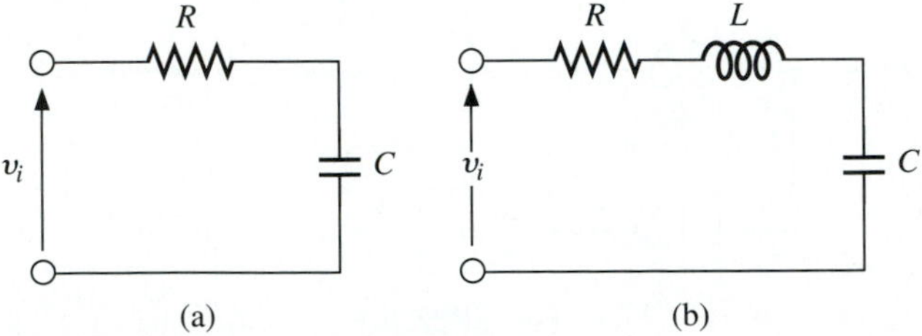

Fig. P22.2

22.3 For the rotational system shown in Fig. P22.3, determine the state-space representation, considering the output of the system to be the speed of the inertial load J_2. Neglect the inertia of the gears. Draw a block diagram representation of the state-space form of the system.

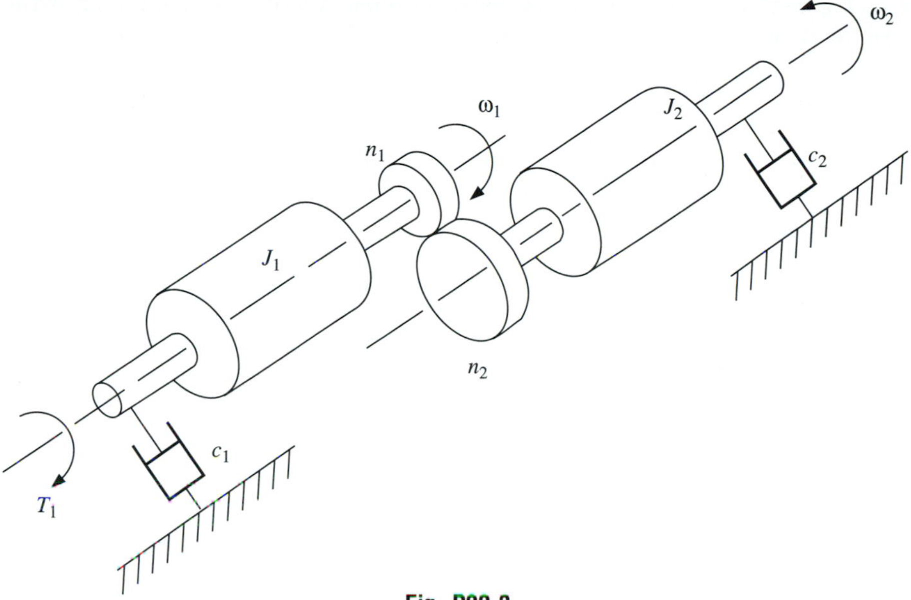

Fig. P22.3

22.4 For the system shown in Fig. P22.4, determine the state-space representation and draw a block diagram representing the system structure.

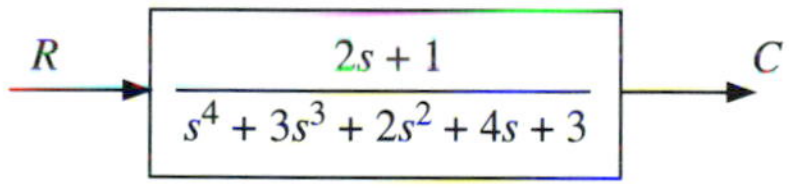

$$R \rightarrow \boxed{\frac{2s+1}{s^4+3s^3+2s^2+4s+3}} \rightarrow C$$

Fig. P22.4

22.5 Find the transfer functions from input **u** to output y of the following system represented in state-space form:

$$\begin{bmatrix} \dot{x}_1 \\ \dot{x}_2 \end{bmatrix} = \begin{bmatrix} -1 & 0 \\ 2 & -4 \end{bmatrix} \begin{bmatrix} x_1 \\ x_2 \end{bmatrix} + \begin{bmatrix} 1 & 1 \\ 0 & 1 \end{bmatrix} \begin{bmatrix} u_1 \\ u_2 \end{bmatrix}$$

$$y = [1 \quad 1]\mathbf{x}$$

Sketch the block diagram of the system in state-space form.

22.6 Consider the system described by the state variable $\mathbf{x}$ and the following matrices:

$$\mathbf{A} = \begin{bmatrix} 1 & 2 \\ 0 & 3 \end{bmatrix}$$

$$\mathbf{B} = \begin{bmatrix} 1 & 0 \\ 1 & 0 \end{bmatrix}$$

$$\mathbf{C} = \begin{bmatrix} 0 & 0 \\ 1 & 2 \end{bmatrix}$$

$$\mathbf{D} = 0$$

Write the equations in the new state vector **z** where

$$\begin{bmatrix} z_1 \\ z_2 \end{bmatrix} = \begin{bmatrix} 0 & 1 \\ 1 & 2 \end{bmatrix} \begin{bmatrix} x_1 \\ x_2 \end{bmatrix}$$

22.7 Figure P22.7 shows a block diagram of a system in Jordan canonical form. Write down the state-space equations and find the transfer function of the system.

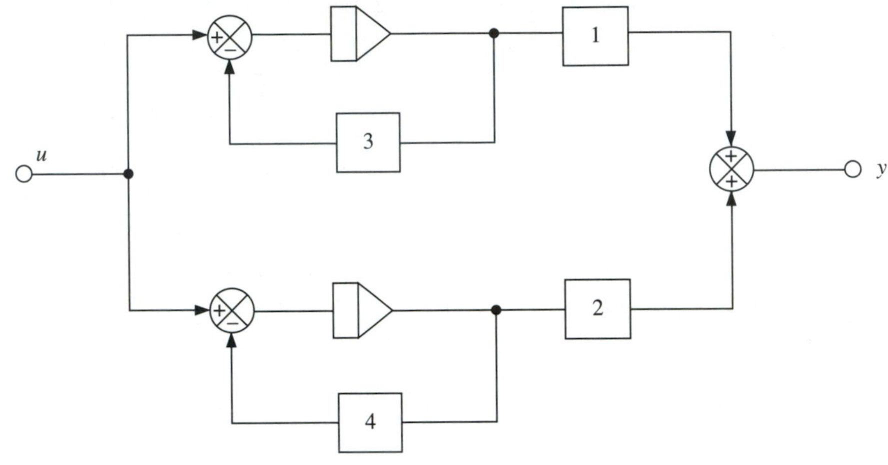

Fig. P22.7

22.8 A system is described by the transfer function

$$GH(s) = \frac{15(s + 2)}{(s + 1)(s^2 + 6s + 8)}$$

Write the Jordan form of the system state equations, and sketch a block diagram of its structure.

22.9 A system is written in the form

$$\dot{\mathbf{x}} = \begin{bmatrix} 1 & 3 \\ 2 & -3 \end{bmatrix} \mathbf{x} + \begin{bmatrix} 1 \\ 4 \end{bmatrix} u$$

Find the system eigenvalues and rewrite it in Jordan canonical form.

22.10 Use a computer program (such as MATLAB) to perform the necessary matrix operations needed to diagonalize the system

$$\dot{\mathbf{x}} = \begin{bmatrix} 2.15 & 0 & 3.87 \\ 0 & -1.08 & 9.34 \\ 0.35 & 7.75 & 2.66 \end{bmatrix} \mathbf{x} + \begin{bmatrix} 8.77 \\ 3.21 \\ 12.80 \end{bmatrix} \mathbf{u}$$

and

$$y = [2.22 \quad -0.76 \quad 7.29]\, u$$

Write down the system eigenvalues.

State-Space System Response, Controllability, and Observability

Now that techniques have been developed to represent more complex systems in the state-space form

$$\dot{\mathbf{x}} = \mathbf{Ax} + \mathbf{Bu} \tag{23.1}$$

$$\mathbf{y} = \mathbf{Cx} + \mathbf{Du} \tag{23.2}$$

methods will now be developed for obtaining the time response. This means calculating $\mathbf{x}(t)$ for a given $\mathbf{u}$ and initial conditions $\mathbf{x}_0$ for $t > 0$. Recognizing that the ouput equation is of algebraic form, we assume $\mathbf{y}$ will be easy to compute if we know $\mathbf{x}$. Attention is therefore focused on the state equation.

◆ DIRECT NUMERICAL SOLUTION OF THE STATE EQUATION

One of the principal advantages of the state-space approach is the ease with which the resulting equations are solved numerically by computer. We begin by approximating the derivative of $\mathbf{x}$ by

$$\dot{\mathbf{x}} = \frac{\mathbf{x}(t + \Delta t) - \mathbf{x}(t)}{\Delta t} \tag{23.3}$$

which may be written

$$\mathbf{x}(t + \Delta t) = \mathbf{x}(t) + \dot{\mathbf{x}}\,\Delta t \tag{23.4}$$

This may be considered a recursive expression to compute the value of $\mathbf{x}$ at time $t + \Delta t$, knowing the value of $\mathbf{x}$, and $\dot{\mathbf{x}}$ at time t. Substituting for $\dot{\mathbf{x}}$ from the state equation yields

$$\mathbf{x}(t + \Delta t) = \mathbf{x}(t) + [\mathbf{Ax}(t) + \mathbf{Bu}]\,\Delta t \tag{23.5}$$

So far we have considered the elements of **A** and **B** to be linear and time invariant, but the above expression is still applicable if the elements (hence the system) are time dependent and/or nonlinear. All that is required is a knowledge of the elements of the state and control distribution matrices at each instant in time.

Beginning at time $t = 0$ and knowing the initial conditions for all state variables, the above equation may be used to generate the values of the state variable at incremental time steps. Although not developed here, it may be appreciated that coding this equation should not be too difficult.

◆ SOLUTION USING STATE TRANSITION MATRIX

The problem may be stated as: Find $\mathbf{x}(t)$ from

$$\dot{\mathbf{x}} = \mathbf{A}\mathbf{x} + \mathbf{B}\mathbf{u} \tag{23.6}$$

given $\mathbf{u}(t)$ and a known $\mathbf{x}$ at some time t_0. This time t_0 may be zero; hence we know the initial state of $\mathbf{x}$, but what follows is not limited to $\mathbf{t}_0 = 0$. The approach we use is similar to the complementary function and particular integral methods used for linear differential equations in a single variable. In fact, if we review the solution method for the scalar case, we see the complementary function would be the solution to

$$\dot{x} = ax \tag{23.7}$$

It is assumed that we know all initial conditions $x_0, \dot{x}_0, \ddot{x}_0, \cdots$. Further assume a solution to the scalar differential equation in the form of a Taylor series

$$x(t) = x_0 + \dot{x}_0 t + \frac{\ddot{x}_0 t^2}{2!} + \cdots \tag{23.8}$$

From equation 23.7, however, when $t = 0$, we may write

$$\dot{x}_0 = a x_0 \tag{23.9}$$

$$\ddot{x}_0 = a\dot{x}_0 = a^2 x_0 \cdots \tag{23.10}$$

Substituting yields

$$x(t) = x_0 + a x_0 t + \frac{a^2 x_0 t^2}{2!} \cdots \tag{23.11}$$

$$= \left(1 + at + \frac{at^2}{2!} + \cdots\right) x_0 \tag{23.12}$$

$$= e^{at} x_0 \tag{23.13}$$

which is usually written as

$$x = x_0 e^{at} \tag{23.14}$$

Applying the same approach to the matrix equation, we arrive at

$$\mathbf{x} = \left(\mathbf{I} + \mathbf{A}t + \frac{\mathbf{A}^2 t^2}{2!} + \cdots\right) \mathbf{x}_0 \tag{23.15}$$

Each term in the parentheses is an $n \times n$ matrix. This results in

$$\mathbf{x} = e^{\mathbf{A}t}\mathbf{x}_0 \tag{23.16}$$

$$= \mathbf{\Phi}(t)\mathbf{x}_0 \tag{23.17}$$

where the $n \times n$ matrix $\mathbf{\Phi}(t)$ is known as the state transition matrix, since it transforms the initial state $\mathbf{x}_0$ into the current state $\mathbf{x}$. Note that although it was assumed that the initial state was known at $t = 0$, this does not have to be the case, and $\mathbf{x}_0$ may define the state at some general time $t = t_0$. We may then write

$$\mathbf{x}(t) = e^{\mathbf{A}(t-t_0)}\mathbf{x}_0 = \mathbf{\Phi}(t - t_0) \tag{23.18}$$

Turning now to the particular integral part of the solution, we may write

$$\dot{\mathbf{x}} - \mathbf{A}\mathbf{x} = \mathbf{B}\mathbf{u} \tag{23.19}$$

or, multiplying each side by the state transition matrix,

$$e^{-\mathbf{A}t}[\dot{\mathbf{x}} - \mathbf{A}\mathbf{x}] = e^{-\mathbf{A}t}\mathbf{B}\mathbf{u} \tag{23.20}$$

this may be expressed as

$$\frac{d}{dt}[e^{-\mathbf{A}t}\mathbf{x}] = e^{-\mathbf{A}t}\mathbf{B}\mathbf{u} \tag{23.21}$$

We integrate this equation from t_0 to a *fixed* time t; hence the variable of integration needs to be something other than t. Let this variable be τ. This produces

$$e^{-\mathbf{A}t}\mathbf{x} = \int_{t_0}^{t} e^{-\mathbf{A}\tau}\mathbf{B}\mathbf{u}(\tau)\, d\tau \tag{23.22}$$

which produces the result

$$\mathbf{x} = \int_{t_0}^{t} e^{\mathbf{A}(t-\tau)}\mathbf{B}\mathbf{u}(\tau)\, d\tau \tag{23.23}$$

The complete solution is the sum of the complementary function and particular integral

$$\mathbf{x} = e^{\mathbf{A}(t-t_0)}\mathbf{x}_0 + \int_{t_0}^{t} e^{\mathbf{A}(t-\tau)}\mathbf{B}\mathbf{u}(\tau)\, d\tau \qquad (t > t_0) \tag{23.24}$$

This may also be written in terms of the state transition matrix as

$$\mathbf{x} = \mathbf{\Phi}(t - t_0)\mathbf{x}_0 + \int_{t_0}^{t} \mathbf{\Phi}(t - \tau)\mathbf{B}\mathbf{u}(\tau)\, d\tau \tag{23.25}$$

If t_0 is chosen as zero, this equation reduces to

$$\mathbf{x} = e^{\mathbf{A}t}\mathbf{x}_0 + \int_{0}^{t} e^{\mathbf{A}(t-\tau)}\mathbf{B}\mathbf{u}(\tau)\, d\tau \qquad (t > 0) \tag{23.26}$$

It will be shown later, in Sample Problem 23.1, that obtaining the time response of a system in this manner is time consuming due to the complexity in calculating $\mathbf{\Phi}$. Another method for calculating $\mathbf{\Phi}$ utilizing Laplace transforms provides a faster alternative.

◆ SOLUTION USING LAPLACE TRANSFORMS

Beginning with the state equation

$$\dot{\mathbf{x}} = \mathbf{A}\mathbf{x} + \mathbf{B}\mathbf{u} \tag{23.27}$$

taking Laplace transforms gives

$$s\mathbf{x} - \mathbf{x}_0 = \mathbf{A}\mathbf{x} + \mathbf{B}\mathbf{u} \tag{23.28}$$

or

$$(s\mathbf{I} - \mathbf{A})\mathbf{x} = \mathbf{x}_0 + \mathbf{B}\mathbf{u} \tag{23.29}$$

Hence

$$\mathbf{x} = \frac{\mathbf{x}_0}{s\mathbf{I} - \mathbf{A}} + \frac{\mathbf{B}\mathbf{u}}{s\mathbf{I} - \mathbf{A}} \tag{23.30}$$

Taking the inverse transform[1] produces

$$\mathbf{x} = e^{\mathbf{A}t}\mathbf{x}_0 + \int_0^t e^{\mathbf{A}(t-\tau)}\mathbf{B}\mathbf{u}(\tau)\, d\tau \tag{23.31}$$

as before. The Laplace transform of the state transition matrix is known as the *resolvent* matrix and may be written as

$$\mathbf{\Phi}(s) = \mathscr{L}[\mathbf{\Phi}(t)] = (s\mathbf{I} - \mathbf{A})^{-1} \tag{23.32}$$

This enables us to find the state transition matrix quickly by calculating $s\mathbf{I} - \mathbf{A}$, inverting it to find the resolvent, and taking the inverse transform, element by element, to find $\mathbf{\Phi}$.

◆ SYSTEM STABILITY

The determination of system stability is based on the same principle as that for SISO systems: by investigating the location of the closed-loop poles on the complex plane. As was seen in Module 22, the closed-loop poles, or eigenvalues, are determined from

$$\det(s\mathbf{I} - \mathbf{A}) = 0 \tag{23.33}$$

Note that this determinant has to be evaluated anyway in calculating the resolvent. As before, if any of the eigenvalues are positive or have a positive real part, the system is unstable.

[1] The first term is obtained by comparing with the inverse scalar transform

$$\mathscr{L}^{-1}\left[\frac{1}{s + a}\right] = e^{-at}$$

while the generalized forcing term is obtained from standard tables of transform pairs.

In order to demonstrate the concepts of controllability and observability, consider the SISO system described by the equations

$$\dot{\mathrm{x}} = \begin{bmatrix} -5.5 & -1.1 & -1.8 & -2.6 \\ 1.0 & -3.6 & 0.7 & 0.9 \\ 0.5 & 0.9 & -2.8 & 0.4 \\ 0.0 & 0.4 & 0.7 & -2.1 \end{bmatrix} \mathrm{x} + \begin{bmatrix} -0.2 \\ 0.3 \\ -0.2 \\ 0.3 \end{bmatrix} u \tag{23.34}$$

and

$$y = [3 \quad 5 \quad 7 \quad 5]\mathbf{x} \tag{23.35}$$

This appears to be a fourth-order system with four state variables x_1, x_2, x_3, and x_4. Let us evaluate the transfer function. From equation 22.43, this is given by

$$\frac{y}{u} = \mathbf{C}(s\mathbf{I} - \mathbf{A})^{-1}\mathbf{B} \tag{23.36}$$

In this case

$$s\mathbf{I} - \mathbf{A} = \begin{bmatrix} s + 5.5 & 1.1 & 1.8 & 2.6 \\ -1.0 & s + 3.6 & -0.7 & -0.9 \\ -0.5 & -0.9 & s + 2.8 & -0.4 \\ 0.0 & -0.4 & -0.7 & s + 2.1 \end{bmatrix} \tag{23.37}$$

Inverting this array[2] and multiplying by **C** and **B** yield

$$\frac{y}{u} = \frac{60 + 47s + 12s^2 + s^3}{120 + 154s + 71s^2 + 14s^3 + s^4} \tag{23.38}$$

$$= \frac{(s + 3)(s + 4)(s + 5)}{(s + 2)(s + 3)(s + 4)(s + 5)} \tag{23.39}$$

$$= \frac{1}{s + 2} \tag{23.40}$$

This leads to the conclusion that because of three pole-zero cancellations, the system is actually *first* order, not fourth order, as suggested by the state equations. How is this paradox resolved, and what is the system order? To understand this better, we write the system in Jordan canonical form. The appropriate transformation is

$$\mathbf{z} = \mathbf{M}^{-1}\mathbf{x} \tag{23.41}$$

where **M** is obtained using the techniques outlined in Module 22, and turns out to be

$$\mathbf{M} = \begin{bmatrix} 1 & 2 & 3 & 4 \\ 2 & 3 & 4 & 1 \\ 3 & 4 & 1 & 2 \\ 4 & 1 & 2 & 3 \end{bmatrix} \tag{23.42}$$

[2] Arrays greater than 2 × 2, such as this one, may require a symbolic manipulation program such as Mathematica.

Transforming **B** and **C** leads to the new state equation

$$\dot{\mathbf{z}} = \begin{bmatrix} -2 & 0 & 0 & 0 \\ 0 & -3 & 0 & 0 \\ 0 & 0 & -4 & 0 \\ 0 & 0 & 0 & -5 \end{bmatrix} \mathbf{z} + \begin{bmatrix} 1 \\ 0 \\ 1 \\ 0 \end{bmatrix} u \tag{23.43}$$

and the transformed output equation

$$y = [1 \quad 1 \quad 0 \quad 0]\mathbf{z} \tag{23.44}$$

The equations describing the system become

$$\dot{z}_1 = -2z_1 + u \tag{23.45}$$

$$\dot{z}_2 = -3z_2 \tag{23.46}$$

$$\dot{z}_3 = -4z_3 + u \tag{23.47}$$

$$\dot{z}_4 = -5z_4 \tag{23.48}$$

and

$$y = z_1 + z_2 \tag{23.49}$$

Clearly the eigenvalues of the **z** system are the same as those of the **x** system, as expected. The state and output equations for the diagonalized **z** system may be represented in block diagram form as in Fig. 23.1. Since the eigenvectors of the two systems are known to be the same, considerable information about the **x** system is available from Fig. 23.1. The following important points should be noted:

1. State z_1 is affected by the input u and this state also influences the output y.
2. State z_2 is unaffected by the input but does influence the output.
3. State z_3 is affected by the input but does not influence the output.
4. State z_4 is unaffected by the input and does not influence the output.

If a *state* is affected by the input, it is said to be controllable; otherwise it is uncontrollable. Similarly, if a state influences the output, it is said to be observable; otherwise it is unobservable. A *system* is uncontrollable if *any* z_i is unaffected by u.[3] Similarly, a system is unobservable if *any state* z_i does not appear in the output equation. The example system described above is therefore both uncontrollable and unobservable since z_2 and z_4 are uncontrollabe and z_3 and z_4 are unobservable.

It should further be noted that *only the controllable and observable modes appear in the transfer function.* In this case, only the single mode with eigenvalue $s = -2$ is both controllable and observable, which is why the transfer function reduces to

$$\frac{y}{u} = \frac{1}{s + 2} \tag{23.50}$$

[3] A more formal definition is that a system is uncontrollable if any z_i cannot be driven from some initial state $z_i(0)$ using unbounded $u(t)$ in finite time t.

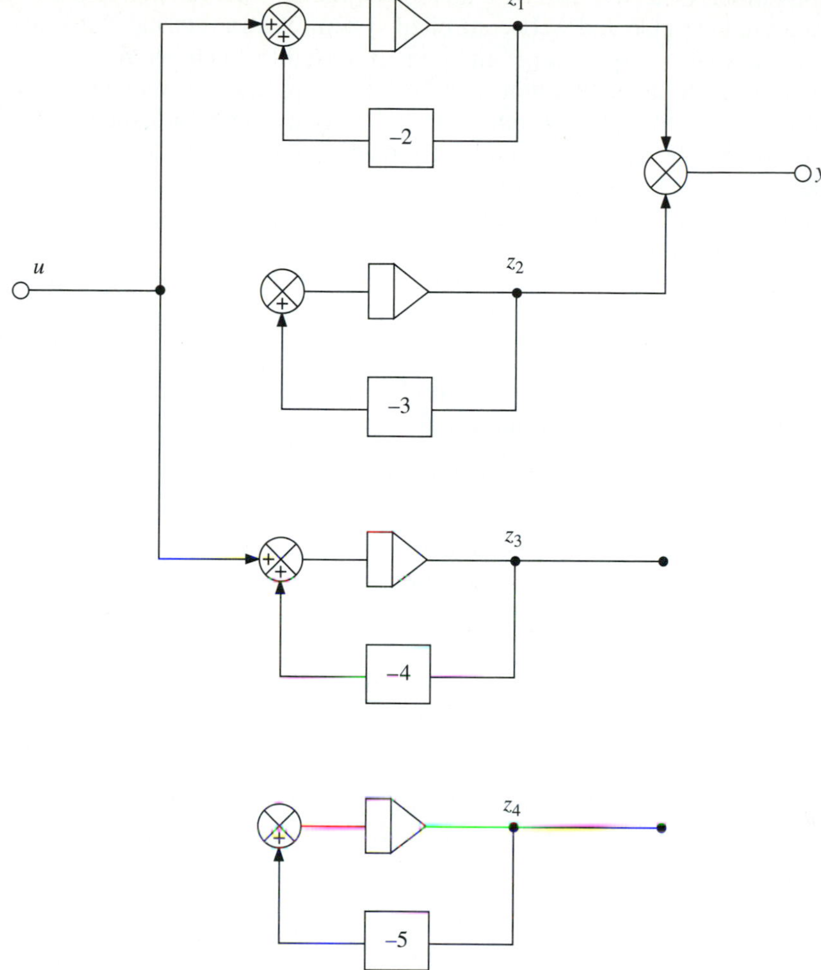

Fig. 23.1 Block diagram of diagonalized system.

One might argue whether we should concern ourselves with the other modes, since these appear to be stable and would disappear with time. This is fortunate in this example, but another system may have uncontrollable and unobservable states that are unstable. The transfer function would still predict a stable system; yet some of the internal states, which may correspond to physical parameters, would be unstable. An example might be of the motor shown in Fig. 23.2.[4] The transfer function relates output speed ω to input voltage v_i, but since the system

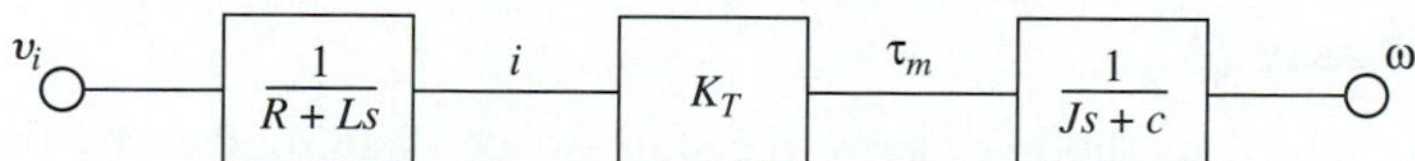

Fig. 23.2 System with internal state variables.

[4] This system is in fact both controllable and observable but is given as an example of how an internal state variable corresponding to a physical quantity may behave unexpectedly.

is second order, there will be two states, say i and ω. The transfer function indicates a perfectly stable and well-controlled system, but an internal state variable, such as i, may become very large and perhaps unstable. Such behavior is not predicted from the transfer function. For this reason pole-zero cancellation, as observed in the fourth-order system, is to be avoided, since the cancellation may remove potentially dangerous modes from the response and lead the designer into believing that all is well with the system. It is for this reason that such cancellation methods were not treated as a serious compensation technique in earlier modules of the book.

The next question to answer is how to recognize uncontrollable and unobservable systems. In the example given, the **x** system was first diagonalized. Then, from the transformed **B** and **C** matrices the following may be observed:

1. In diagonal form, a system is uncontrollable if there is a zero *row* in the transformed control distribution matrix $\overline{\mathbf{B}}$.

2. In diagonal form, a system is unobservable if there is a zero *column* in the transformed output matrix $\overline{\mathbf{C}}$.

It is also possible to determine whether a system is controllable and observable directly from the original system equations

$$\dot{\mathbf{x}} = \mathbf{A}\mathbf{x} + \mathbf{B}\mathbf{u} \tag{23.51}$$

$$\mathbf{y} = \mathbf{C}\mathbf{x} + \mathbf{D}\mathbf{u} \tag{23.52}$$

To determine controllability, we form the *controllability matrix* $\mathbf{C}_T$ from

$$\mathbf{C}_T = [\mathbf{B} \quad \mathbf{AB} \quad \mathbf{A}^2\mathbf{B} \quad \mathbf{A}^3\mathbf{B} \quad \cdots \quad \mathbf{A}^{n-1}\mathbf{B}] \tag{23.53}$$

The system is controllable if the rank of $\mathbf{C}_T$ is n, the order of the system. Similarly, from the original system equations we may form the *observability matrix* $\mathbf{O}_T$ from

$$\mathbf{O}_T = \begin{bmatrix} \mathbf{C} \\ \mathbf{CA} \\ \mathbf{CA}^2 \\ \vdots \\ \mathbf{CA}^{n-1} \end{bmatrix} \tag{23.54}$$

The test for observability is that the rank of $\mathbf{O}_T$ must also be n, the system order.

Although the concepts of controllability and observability have important implications in their own right, they become particularly significant once closed-loop MIMO system performance is considered. This is addressed in the next module.

Sample Problem 23.1

Use numerical methods to solve the state-space system given by the equation

$$\begin{bmatrix} \dot{x}_1 \\ \dot{x}_2 \end{bmatrix} = \begin{bmatrix} -1 & 1 \\ 0 & -2 \end{bmatrix} \begin{bmatrix} x_1 \\ x_2 \end{bmatrix} + \begin{bmatrix} 0 & 0 \\ 1 & 0 \end{bmatrix} \begin{bmatrix} u \\ 0 \end{bmatrix}$$

for the initial conditions $x_1(0) = -1$, $x_2(0) = 0$, when the input $u(t)$ is a unit step applied at $t = 0$.

Solution

This will be done by writing a computer program that codes the recursive equation

$$\mathbf{x}(t + \Delta t) = \mathbf{x}(t) + [\mathbf{Ax}(t) + \mathbf{Bu}]\,\Delta t$$

Almost any computer language may be used to do this, but MATLAB is particularly useful since it handles the matrix operations easily. The following code achieves the desired results:

```
dt=0.01;
x=[-1 0]';
u=[1 0]';
a=[-1 1; 0 -2];
b=[ 0 0; 1 0];
for i=1:500
m1=a*x;
m2=b*u;
m3=(m1+m2)*dt;
xdash=x+m3;
x=xdash;
x1(i)=x(1);
x2(i)=x(2);
end
```

The results may then be plotted using the commands

```
t=[0:1:499];
plot(t,x1,t,x2)
```

The results are as shown in Fig. SP23.1.1. It is seen that the solution is obtained in a straightforward manner using this programming technique and is easily extended to higher order systems and systems with nonlinear coefficients. In addition, the elements of **A** or **B** may be written as functions of time (the index i) to represent systems with time-varying coefficients.

Sample Problem 23.2

For the same system as that given in Sample Problem 23.1, determine the time response of the system using (a) the series expansion and (b) the resolvent for the state transition matrix.

Solution

We first use the series expansion to evaluate $\mathbf{\Phi}$:

$$\mathbf{\Phi} = e^{\mathbf{A}t} = \mathbf{I} + \mathbf{A}t + \frac{\mathbf{A}^2t^2}{2!} + \cdots$$

$$= \mathbf{I} + \begin{bmatrix} -1 & 1 \\ 0 & -2 \end{bmatrix} t + \begin{bmatrix} -1 & 1 \\ 0 & -2 \end{bmatrix}\begin{bmatrix} -1 & 1 \\ 0 & -2 \end{bmatrix}\frac{t^2}{2} + \cdots$$

$$= \begin{bmatrix} 1 & 0 \\ 0 & 1 \end{bmatrix} + \begin{bmatrix} -1 & 1 \\ 0 & -2 \end{bmatrix} t + \begin{bmatrix} 1 & -3 \\ 0 & -2 \end{bmatrix}\frac{t^2}{2} + \cdots$$

$$= \begin{bmatrix} 1 - t + \frac{1}{2}t^2 \cdots & t - \frac{3}{2}t^2 + \cdots \\ 0 & 1 - 2t + \frac{4}{2}t^2 - \cdots \end{bmatrix}$$

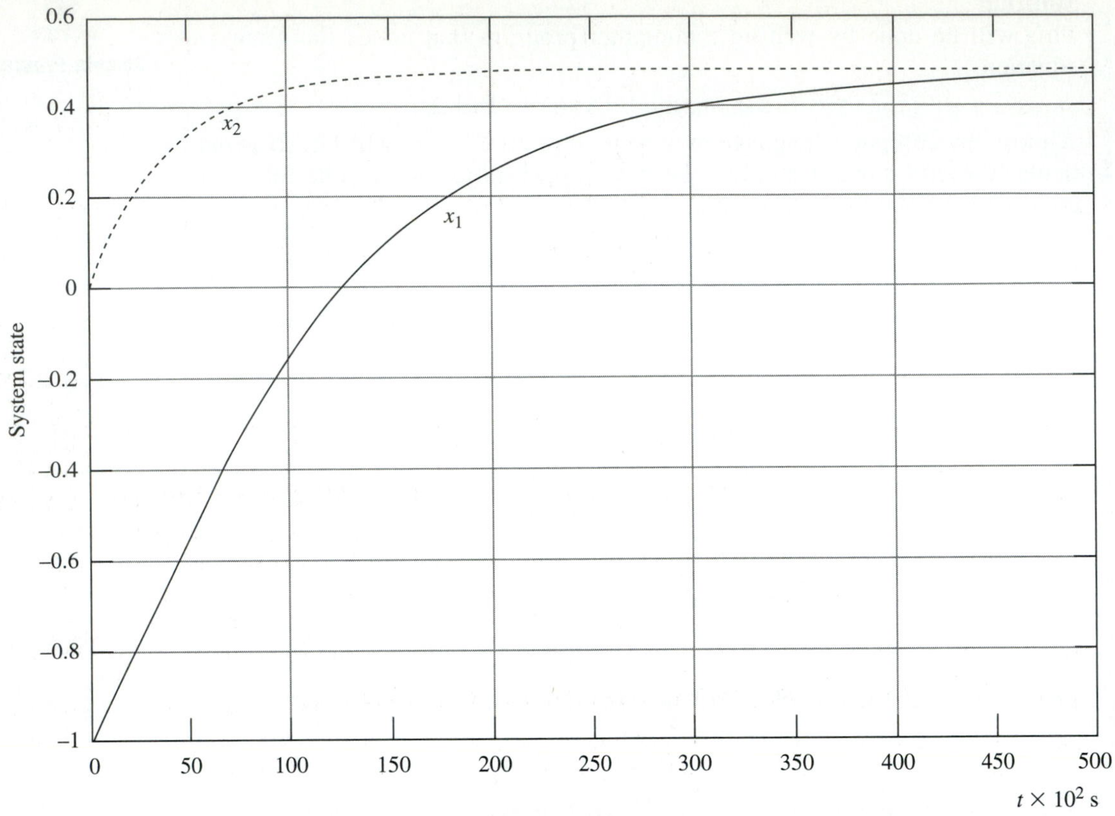

Fig. SP23.1.1

Inspecting the individual elements of $\mathbf{\Phi}$, it may be apparent that

$$1 - t + \tfrac{1}{2}t^2 - \cdots = e^{-t}$$

and perhaps that

$$1 - 2t + \tfrac{4}{2}t^2 - \cdots = e^{-2t}$$

but it requires some insight to see that

$$t - \tfrac{3}{2}t^2 + \cdots = e^{-t} - e^{-2t}$$

However, this makes the state transition matrix

$$\mathbf{\Phi} = \begin{bmatrix} e^{-t} & e^{-t} - e^{-2t} \\ 0 & e^{-2t} \end{bmatrix}$$

If we calculate $\mathbf{\Phi}$ using the resolvent,

$$\mathbf{\Phi} = (s\mathbf{I} - \mathbf{A})^{-1}$$

In this example

$$s\mathbf{I} - \mathbf{A} = \begin{bmatrix} s & 0 \\ 0 & s \end{bmatrix} - \begin{bmatrix} -1 & 1 \\ 0 & -2 \end{bmatrix}$$

$$= \begin{bmatrix} s + 1 & -1 \\ 0 & s + 2 \end{bmatrix}$$

The inverse becomes[5]

$$(s\mathbf{I} - \mathbf{A})^{-1} = \frac{\begin{bmatrix} s+2 & 1 \\ 0 & s+1 \end{bmatrix}}{(s+1)(s+2)}$$

$$= \begin{bmatrix} \dfrac{1}{s+1} & \dfrac{1}{(s+1)(s+2)} \\ 0 & \dfrac{1}{s+2} \end{bmatrix}$$

$$= \begin{bmatrix} \dfrac{1}{s+1} & \dfrac{1}{s+1} - \dfrac{1}{s+2} \\ 0 & \dfrac{1}{s+2} \end{bmatrix}$$

Thus the state transition matrix is

$$\mathbf{\Phi} = \mathscr{L}^{-1}[(s\mathbf{I} - \mathbf{A})^{-1}] = \begin{bmatrix} e^{-t} & e^{-t} - e^{-2t} \\ 0 & e^{-2t} \end{bmatrix}$$

as before. Now that $\mathbf{\Phi}$ has been calculated, the time response may be derived. Since

$$\mathbf{x} = e^{\mathbf{A}t}\mathbf{x}_0 + \int_0^t e^{\mathbf{A}(t-\tau)}\mathbf{Bu}(\tau)\, d\tau$$

the solution will be calculated from the two terms in this equation, corresponding to the transient and steady-state solutions. The transient part is

$$\mathbf{x}_t = e^{\mathbf{A}t}\mathbf{x}_0$$

$$= \begin{bmatrix} e^{-t} & e^{-t} - e^{-2t} \\ 0 & e^{-2t} \end{bmatrix} \begin{bmatrix} -1 \\ 0 \end{bmatrix}$$

$$= \begin{bmatrix} -e^{-t} \\ 0 \end{bmatrix}$$

The forced part of the response is

$$\mathbf{x}_f = \int_0^t e^{\mathbf{A}(t-\tau)}\mathbf{Bu}(\tau)\, d\tau$$

From the state equation for a unit step input, we may write

$$\mathbf{Bu}(\tau) = \begin{bmatrix} 0 \\ 1 \end{bmatrix}$$

[5] Again, the symbolic inverse becomes quite complex for arrays larger than 2 × 2.

Hence

$$\mathbf{x}_f = \int_0^t \begin{bmatrix} e^{-(t-\tau)} & e^{-(t-\tau)} - e^{-2(t-\tau)} \\ 0 & e^{-2(t-\tau)} \end{bmatrix} \begin{bmatrix} 0 \\ 1 \end{bmatrix} d\tau$$

$$= \int_0^t \begin{bmatrix} e^{-(t-\tau)} - e^{-2(t-\tau)} \\ e^{-2(t-\tau)} \end{bmatrix} d\tau$$

$$= \begin{bmatrix} e^{-t}[e^{\tau})]_0^t - e^{-2t}[\frac{1}{2}e^{2\tau}]_0^t \\ e^{-2t}[\frac{1}{2}e^{-2\tau}]_0^t \end{bmatrix}$$

$$= \begin{bmatrix} e^{-t}[e^{t} - 1] - e^{-2t}[\frac{1}{2}e^{2t} - \frac{1}{2}] \\ e^{-2t}[\frac{1}{2}e^{2t} - \frac{1}{2}] \end{bmatrix}$$

Adding the transient and forced response yields

$$\mathbf{x}(t) = \begin{bmatrix} \frac{1}{2} - 2e^{-t} + \frac{1}{2}e^{-2t} \\ \frac{1}{2} - \frac{1}{2}e^{-2t} \end{bmatrix}$$

The student may verify that the time response for x_1 and x_2 is the same as that shown in Fig. SP23.1.1 of the previous example.

SAMPLE PROBLEM 23.3

A system is given in the form

$$\dot{\mathbf{x}} = \begin{bmatrix} -2 & -2 \\ 0 & -1 \end{bmatrix}\mathbf{x} + \begin{bmatrix} 0 \\ 1 \end{bmatrix}u$$

$$y = [1 \quad 2]\mathbf{x}$$

1. Diagonalize the system and determine whether it is controllable and observable.
2. Confirm the result using the controllability and observability matrices.
3. Calculate the system transfer function.

Solution

To diagonalize the system, we first calculate the eigenvalues from

$$\det(s\mathbf{I} - \mathbf{A}) = 0$$

where

$$s\mathbf{I} - \mathbf{A} = \begin{bmatrix} s & 0 \\ 0 & s \end{bmatrix} - \begin{bmatrix} -2 & -2 \\ 0 & -1 \end{bmatrix} = \begin{bmatrix} s+2 & 2 \\ 0 & s+1 \end{bmatrix}$$

Thus

$$\det(s\mathbf{I} - \mathbf{A}) = (s+1)(s+2) = 0$$

yielding the eigenvalues $s = \lambda_1 = -1$ and $s = \lambda_2 = -2$. Following the procedure in Module 22, the transformation matrix **M** is found from the eigenvectors

$$\mathbf{M} = \begin{bmatrix} -2 & 1 \\ 1 & 0 \end{bmatrix}$$

This enables the system to be written in diagonalized form,

$$\dot{\mathbf{z}} = \mathbf{J}\mathbf{z} + \overline{\mathbf{B}}\mathbf{u}$$

and

$$\mathbf{y} = \overline{\mathbf{C}}\mathbf{z}$$

where

$$\mathbf{J} = \mathbf{M}^{-1}\mathbf{A}\mathbf{M} = \begin{bmatrix} -1 & 0 \\ 0 & -2 \end{bmatrix}$$

$$\overline{\mathbf{B}} = \mathbf{M}^{-1}\mathbf{B} = \begin{bmatrix} 1 \\ 2 \end{bmatrix}$$

$$\overline{\mathbf{C}} = \mathbf{C}\mathbf{M} = [0 \quad 1]$$

Inspecting the diagonalized control distribution matrix $\overline{\mathbf{B}}$, the system is seen to be controllable; however, the transformed output matrix $\overline{\mathbf{C}}$ has the first element zero, indicating that z_1 does not contribute to the output. The system is therefore unobservable.

Confirming the result from the original system involves calculating the controllability and observability matrices. For the system under study, the controllability matrix is

$$\mathbf{C}_T = [\mathbf{B} \quad \mathbf{AB}]$$

Since

$$\mathbf{AB} = \begin{bmatrix} -2 \\ -1 \end{bmatrix}$$

we obtain

$$\mathbf{C}_T = \begin{bmatrix} 0 & -2 \\ 1 & -1 \end{bmatrix}$$

The rank of a matrix is the number of independent rows. Since the second row is *not* a linear multiple of the first row, the rank of $\mathbf{C}_T$ is 2, which is the order of the system. The system is therefore controllable. The observability matrix is found from

$$\mathbf{O}_T = \begin{bmatrix} \mathbf{C} \\ \mathbf{CA} \end{bmatrix}$$

Since

$$\mathbf{CA} = [-2 \quad -4]$$

we obtain

$$\mathbf{O}_T = \begin{bmatrix} 1 & 2 \\ -2 & -4 \end{bmatrix}$$

Because the second row is -2 times the first row, the rank is 1. Since there is only one independent row in the observability matrix and the system order is 2, the system is unobservable.[6]

[6] The `rank(a)` command in MATLAB may be used for higher order matrices. In addition, the commands `ctrb(a,b)` and `obsv(c,a)` calculate $\mathbf{C}_T$ and $\mathbf{O}_T$ directly.

The transfer function is obtained from equation 22.43, which requires the calculation of the resolvent

$$(s\mathbf{I} - \mathbf{A})^{-1} = \frac{\begin{bmatrix} s+1 & -2 \\ 0 & s+2 \end{bmatrix}^T}{(s+2)(s+1)}$$

$$= \begin{bmatrix} \dfrac{1}{s+2} & \dfrac{-2}{(s+1)(s+2)} \\ 0 & \dfrac{1}{s+1} \end{bmatrix}$$

The transfer function is then given by

$$\frac{y}{u} = [1 \quad 2] \begin{bmatrix} \dfrac{1}{s+2} & \dfrac{-2}{(s+1)(s+2)} \\ 0 & \dfrac{1}{s+1} \end{bmatrix} \begin{bmatrix} 0 \\ 1 \end{bmatrix}$$

$$= \frac{2(s+1)}{(s+1)(s+2)}$$

Canceling the zero and pole at $s = -1$ produces the final result

$$\frac{y}{u} = \frac{2}{s+2}$$

Note how the system appears to be first order when it is known to be second order. The cancellation of a pole and zero appears to reduce the system from second to first order, and this ability to cancel such terms is typical of a system that is either uncontrollable or unobservable. In fact, the mode that is canceled is not eliminated from the system; it is only "hidden" from the input in the case of an uncontrollable system or from the output in the case of an unobservable system.

◆ PROBLEMS

23.1 The mechanical system shown in Fig. P23.1 is pulled down a distance of 1.0 cm and released. Derive a state-space model of the system and determine the displacement of the block using the state transition matrix.

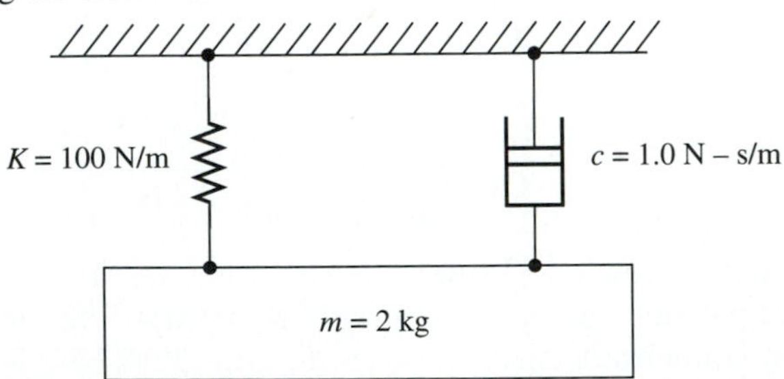

Fig. P23.1

23.2 For the same system shown in Fig. P23.1, write a computer program to predict the response and compare the results to the previous problem.

23.3 A system is defined by

$$\mathbf{A} = \begin{bmatrix} -2 & 0 \\ 1 & -1 \end{bmatrix}$$

$$\mathbf{B} = \begin{bmatrix} 0 \\ 1 \end{bmatrix}$$

$$\mathbf{C} = [3 \quad 2]$$

$$\mathbf{D} = 0$$

If the system has zero initial conditions, determine the output if the input is a unit impulse.

23.4 For the state-space system

$$\mathbf{A} = \begin{bmatrix} -1 & 1 \\ 0 & -2 \end{bmatrix}$$

$$\mathbf{B} = \begin{bmatrix} 2 \\ 0 \end{bmatrix}$$

the initial conditions are $x_1(0) = 0.1$ and $x_2(0) = -0.5$. Estimate the time-domain response of each state variable when the input is a unit step.

23.5 For the system given in the previous problem, confirm the results by programing the recursive state-space equation

$$\mathbf{x}(t + \Delta t) = \mathbf{x}(t) + [\mathbf{A}\mathbf{x}(t) + \mathbf{B}\mathbf{u}]\,\Delta t$$

If the output equation is defined by

$$\mathbf{C} = [1 \quad 2]$$

$$\mathbf{D} = \begin{bmatrix} 2 \\ 0 \end{bmatrix}$$

generate a plot of time response $y(t)$.

23.6 Determine whether the following system is controllable and observable:

$$\mathbf{A} = \begin{bmatrix} -3 & 2 \\ 1 & 0 \end{bmatrix}$$

$$\mathbf{B} = \begin{bmatrix} 4 \\ 4 \end{bmatrix}$$

$$\mathbf{C} = [3 \quad 1]$$

$$\mathbf{D} = 0$$

23.7 Assess the controllability and observability of the system described by the following transfer function:

$$\frac{y}{u} = \frac{5(s + 2)}{(s + 4)(s + 10)(s + 20)}$$

Draw a block diagram of the system states.

23.8 Diagonalize and thereby investigate the controllability and observability of the following system:

$$\dot{\mathbf{x}} = \begin{bmatrix} -1 & 0 & 1 \\ 1 & -3 & 0 \\ 0 & 0 & -4 \end{bmatrix} \mathbf{x} + \begin{bmatrix} 1 \\ 4 \\ 0 \end{bmatrix} u$$

$$\mathbf{y} = \begin{bmatrix} 0 & 1 & -1 \\ 1 & 0 & 2 \end{bmatrix} \mathbf{x}$$

23.9 In the following system, determine the relationship between b_1, b_2, c_1, and c_2 such that the system is both controllable and observable:

$$\mathbf{A} = \begin{bmatrix} -1 & 1 \\ 2 & 0 \end{bmatrix}$$

$$\mathbf{B} = \begin{bmatrix} b_1 \\ b_2 \end{bmatrix}$$

$$\mathbf{C} = [c_1 \quad c_2]$$

$$\mathbf{D} = 0$$

23.10 By calculating the transfer function for the following system, discuss its controllability and observability:

$$\dot{\mathbf{x}} = \begin{bmatrix} 0 & 1 & 0 \\ 1 & -2 & 0 \\ -2 & 0 & -1 \end{bmatrix} \mathbf{x} + \begin{bmatrix} 1 \\ 0 \\ -1 \end{bmatrix} u$$

$$\mathbf{y} = [2 \quad 0 \quad -1]\,\mathbf{x}$$

Check the answer by writing the state equations in canonical form, and sketch a block diagram of the diagonalized system states. Calculate $\mathbf{C}_T$ and $\mathbf{O}_T$ to confirm the result.

MODULE 24

State-Space Controller Design

So far, our studies of state-space methods have involved only open-loop systems in the form shown in Fig. 24.1, where it is assumed that **D** is zero. We are now in a position to investigate closed-loop systems and in particular determine the selection of a matrix **G** that feeds back the system states to provide a control vector **u** so that a "desirable" transient response is obtained. The structure of this feedback system is shown in Fig. 24.2. This system is a little different to that used

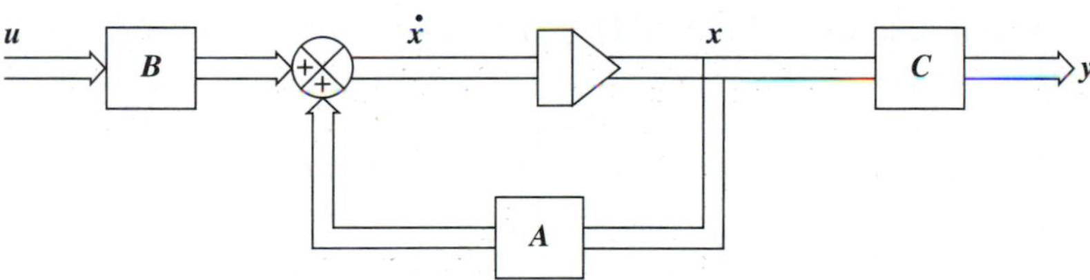

Fig. 24.1 State-space representation of open-loop system.

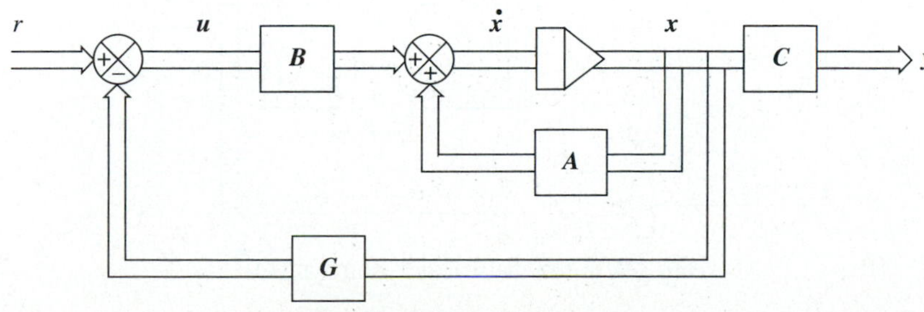

Fig. 24.2 Closed-loop system.

previously, where the output is fed back. Here, the state variables are fed back and subtracted from the input signal(s) **r**. In order to simplify matters, two initial assumptions will be made. First, the control matrix is assumed to be a vector $\mathbf{G} = \mathbf{g}$ where

$$\mathbf{g} = \begin{Bmatrix} g_1 \\ g_2 \\ g_3 \\ \cdot \\ \cdot \\ \cdot \\ g_n \end{Bmatrix} \tag{24.1}$$

so that the output of the control matrix is a scalar

$$r_m = \mathbf{g}'\mathbf{x} = [g_1 \quad \cdots \quad g_n] \begin{bmatrix} x_1 \\ \cdot \\ \cdot \\ \cdot \\ x_2 \end{bmatrix} \tag{24.2}$$

The second assumption is that the input and output are scalars; i.e., we are dealing only with a SISO system. This results in the modifications to the system block diagram shown in Fig. 24.3. It is further noted that in designing a control vector **g**, we are principally interested in placing the closed-loop poles on the complex plane so as to ensure good transient response. As we have seen, this may be accomplished independently of the input $r(t)$. Thus, it is customary to ignore completely the input and focus on the location of the system closed-loop poles obtained from the characteristic equation. Under these circumstances we may write

$$u = -r_m = -\mathbf{g}'\mathbf{x} \tag{24.3}$$

◆ DIRECT CALCULATION OF GAINS BY COMPARISON WITH CHARACTERISTIC EQUATION

Substituting equaton 24.3 into the state equation produces[1]

$$\dot{\mathbf{x}} = (\mathbf{A} - \mathbf{b}\mathbf{g}')\mathbf{x} \tag{24.4}$$

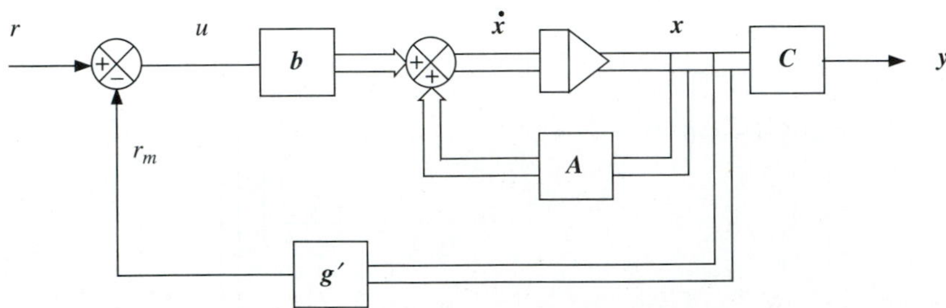

Fig. 24.3 Modified closed-loop system.

[1] Since u is a scalar, the control distribution matrix **B** must now be the vector **b**.

The eigenvectors of the closed-loop system will be given by

$$|s\mathbf{I} - (\mathbf{A} - \mathbf{bg}')| = 0 \tag{24.5}$$

Given that the desired location of the closed-loop poles are $-\lambda_1, -\lambda_2, \ldots, -\lambda_n$, the required characteristic equation may be constructed as

$$(s + \lambda_1)(s + \lambda_2)\cdots(s + \lambda_n) = 0 \tag{24.6}$$

Since both equations 24.5 and 24.6 will give an nth-order polynomial in s, the values of g_i may be obtained by equating the coefficients of s and solving the resulting set of n simultaneous equations. Consider the example of the system defined by

$$\mathbf{A} = \begin{bmatrix} -1 & 0 \\ 1 & -2 \end{bmatrix} \tag{24.7}$$

$$\mathbf{b} = \begin{bmatrix} 1 \\ 2 \end{bmatrix} \tag{24.8}$$

We will find the gain vector $\mathbf{g}$ such that the closed-loop system poles are located at $s = -1 \pm j$. Since

$$\mathbf{bg}' = \begin{bmatrix} 1 \\ 2 \end{bmatrix} [g_1 \quad g_2] = \begin{bmatrix} g_1 & g_2 \\ 2g_1 & 2g_2 \end{bmatrix} \tag{24.9}$$

we obtain

$$s\mathbf{I} - \mathbf{A} + \mathbf{bg}' = \begin{bmatrix} s & 0 \\ 0 & s \end{bmatrix} - \begin{bmatrix} -1 & 0 \\ 1 & -2 \end{bmatrix} + \begin{bmatrix} g_1 & g_2 \\ 2g_1 & 2g_2 \end{bmatrix} \tag{24.10}$$

$$= \begin{bmatrix} s + 1 + g_1 & g_2 \\ -1 + 2g_1 & s + 2 + 2g_2 \end{bmatrix} \tag{24.11}$$

Taking the determinant of the matrix, this leads to the characteristic equation

$$(s + 1 + g_1)(s + 2) + g_2 - g_2(2g_1 - 1) = 0 \tag{24.12}$$

or

$$s^2 + s(3 + g_1 + 2g_2) + (2 + 2g_1 + 3g_2) = 0 \tag{24.13}$$

Note that in expanding the characteristic equation, all terms involving the products $g_i g_j$ always cancel out, leaving a linear expression. The characteristic equation based on the desired pole placement is

$$(s + 1 + j)(s + 1 - j) = s^2 + 2s + 2 = 0 \tag{24.14}$$

Equating the coefficients of equations 24.13 and 24.14 yields

$$g_1 + 2g_2 = -1 \tag{24.15}$$

$$2g_1 + 3g_2 = 0 \tag{24.16}$$

This results in $g_1 = 3$ and $g_2 = -2$; hence

$$\mathbf{g} = \begin{bmatrix} 3 \\ -2 \end{bmatrix} \tag{24.17}$$

The control input may be constructed as

$$u = -\mathbf{g}'\mathbf{x} = -3x_1 + 2x_2 \tag{24.18}$$

This method, though quite simple for lower order systems, may become cumbersome for systems of more than third order. For such systems, alternative methods are better suited.

◆ POLE PLACEMENT VIA CONTROL CANONICAL FORM OF STATE EQUATIONS

The control canonical form of the state equations is similar to the Jordan canonical form, in that the system **A** matrix has information about the eigenvalues represented in a compact format. The form is

$$\dot{\mathbf{z}} = \mathbf{A}\mathbf{z} + \mathbf{b}u \tag{24.19}$$

and the corresponding output equation is

$$y = \mathbf{C}\mathbf{z} \tag{24.20}$$

where

$$\mathbf{A} = \begin{bmatrix} -a_1 & -a_2 & \cdots & -a_{n-1} & -a_n \\ 1 & 0 & \cdots & 0 & 0 \\ 0 & 1 & \cdots & 0 & 0 \\ \vdots & \vdots & \ddots & \vdots & \vdots \\ 0 & 0 & \cdots & 1 & 0 \end{bmatrix} \tag{24.21}$$

and

$$\mathbf{b} = \begin{bmatrix} 1 \\ 0 \\ 0 \\ \vdots \\ 0 \end{bmatrix} \tag{24.22}$$

The a's are the coefficients of the characteristic equation of the open-loop system

$$s^n + a_1 s^{n-1} + a_2 s^{n-2} + \cdots + a_{n-1}s + a_n = 0 \tag{24.23}$$

Note that if the system is specified in terms of a transfer function,

$$\frac{y}{u} = \frac{s^2 + c_1 s + c_0}{s^n + a_1 s^{n-1} + \cdots + a_{n-1}s + a_n} \tag{24.24}$$

the control canonical form may be written directly from it. Again, writing the closed-loop system in the form

$$\dot{\mathbf{z}} = (\mathbf{A} - \mathbf{b}\mathbf{g}')\mathbf{z} \tag{24.25}$$

and since

$$\mathbf{bg}' = \begin{bmatrix} g_1 & g_2 & \cdots & g_n \\ 0 & 0 & \cdots & 0 \\ \vdots & \vdots & \ddots & \vdots \\ 0 & 0 & \cdots & 0 \end{bmatrix} \tag{24.26}$$

we have

$$\mathbf{A} - \mathbf{bg}' = \begin{bmatrix} -(a_1 + g_1) & -(a_2 + g_2) & \cdots & -(a_{n-1} + g_{n-1}) & -(a_n + g_n) \\ 1 & 0 & \cdots & 0 & 0 \\ 0 & 1 & \cdots & 0 & 0 \\ \vdots & \vdots & \ddots & \vdots & \vdots \\ 0 & 0 & \cdots & 1 & 0 \end{bmatrix} \tag{24.27}$$

Given the desired closed-loop poles $\lambda_1, \lambda_2, \ldots, \lambda_n$, we may generate the corresponding closed-loop characteristic equation

$$s^n + a_1^* s^{n-1} + \cdots + a_{n-1}^* s + a_n^* = 0 \tag{24.28}$$

This closed-loop system, written in control canonical form, becomes

$$\mathbf{A}^* = \begin{bmatrix} -a_1^* & -a_2^* & \cdots & -a_{n-1}^* & -a_n^* \\ 1 & 0 & \cdots & 0 & 0 \\ 0 & 1 & \cdots & 0 & 0 \\ \vdots & \vdots & \ddots & \vdots & \vdots \\ 0 & 0 & \cdots & 1 & 0 \end{bmatrix} \tag{24.29}$$

By comparing equatons 24.27 and 24.29, it follows directly that

$$a_i^* = a_1 + g_i \tag{24.30}$$

Hence

$$g_i = a_i^* - a_i \tag{24.31}$$

Recall that g_i is the feedback gain associated with state variable z_i, a_i^* is a coefficient in the *desired* characteristic equation, and a_i is a coefficient in the open-loop system characteristic equation in control canonical form. It may be appreciated that if the system is expressible in control canonical form, the feedback vector **g** is easily evaluated. It seems, therefore, that our system should always be written in control canonical form in order to expedite the pole placement process.

There is one problem with this conclusion. Remember that in order to control the system, we feed back the product of the gain vector and the system states, i.e.,

$$u = -\mathbf{g}'\mathbf{x} \tag{24.32}$$

It follows that the state variables must be physically measurable.[2] It would be a remarkable coincidence that when our system is written in control canonical form, it just happened to produce a state vector in which all elements correspond to measurable quantities. The problem, therefore, it that we have a state vector **x** corresponding to states that may be physically measured, while there exists another state vector **z** corresponding to the same system but, when written in control canonical form, produces a set of state variables that may not be measurable. The gain vector evaluated from equation 24.31 would give us $\mathbf{g}_z$, the gain vector with which to multiply **z** in order to construct the control u, but unfortunately we cannot measure **z**, only **x**. The question is, what is $\mathbf{g}_x$?[3] In order to determine $\mathbf{g}_x$, the following strategy will be employed:

1. The system, in terms of the measurable state vector **x**, will be transformed into the control canonical form by means of the matrix **T**, where

$$\mathbf{z} = \mathbf{T}\mathbf{x} \tag{24.33}$$

   We state, without proof, that the transformation **T** that will convert the measurable system

$$\dot{\mathbf{x}} = \mathbf{A}\mathbf{x} + \mathbf{b}u \tag{24.34}$$

   into control canonical form is

$$\mathbf{T} = [\mathbf{C}_T\mathbf{W}]^{-1} \tag{24.35}$$

   where $\mathbf{C}_T$ is the previously encountered controllability matrix and **W** may be constructed from the characteristic equation (equation 24.23), obtained from the measurable system **A** matrix as

$$\mathbf{W} = \begin{bmatrix} 1 & a_1 & a_2 & \cdots & a_{n-1} \\ 0 & 1 & a_1 & \cdots & a_{n-2} \\ \vdots & \vdots & \vdots & \ddots & \vdots \\ 0 & \cdots & \cdots & \cdots & 1 \end{bmatrix} \tag{24.36}$$

2. From a knowledge of the desired closed-loop pole positions, the corresponding characteristic equation will be used to determine $\mathbf{g}_z$.

3. The control gain vector back in the **x** domain will be evaluated from **T**. Although we will not prove the result, it may be shown that

$$\mathbf{g}_x = \mathbf{T}'\mathbf{g}_z \tag{24.37}$$

Consider the previous example, where

$$\mathbf{A} = \begin{bmatrix} -1 & 0 \\ 1 & -2 \end{bmatrix} \tag{24.38}$$

$$\mathbf{b} = \begin{bmatrix} 1 \\ 2 \end{bmatrix} \tag{24.39}$$

[2] Situations where some or all of the system states cannot be measured will be addressed in Module 25.

[3] Note that the eigenvalues are the same for both systems; hence a_i^* will be identical.

We will find the gain vector $\mathbf{g}_x$ such that the closed-loop system poles are located at $s = -1 \pm j$. The controllability matrix becomes

$$\mathbf{C}_T = [\mathbf{B} \quad \mathbf{AB}] = \begin{bmatrix} 1 & -1 \\ 2 & -3 \end{bmatrix} \tag{24.40}$$

The characteristic equation of the open-loop system is

$$|s\mathbf{I} - \mathbf{A}| = 0 \tag{24.41}$$

where

$$s\mathbf{I} - \mathbf{A} = \begin{bmatrix} s+1 & 0 \\ -1 & s+2 \end{bmatrix} \tag{24.42}$$

This yields

$$s^2 + 3s + 2 = 0 \tag{24.43}$$

The matrix $\mathbf{W}$ defined in equation 24.36 may be written from the characteristic equation as

$$\mathbf{W} = \begin{bmatrix} 1 & 3 \\ 0 & 1 \end{bmatrix} \tag{24.44}$$

This results in

$$\mathbf{C}_T\mathbf{W} = \begin{bmatrix} 1 & 2 \\ 2 & 3 \end{bmatrix} \tag{24.45}$$

and

$$\mathbf{T} = [\mathbf{C}_T\mathbf{W}]^{-1} = \begin{bmatrix} -3 & 2 \\ 2 & -1 \end{bmatrix} \tag{24.46}$$

Since

$$\mathbf{z} = \mathbf{Tx} \tag{24.47}$$

we may write

$$\mathbf{x} = \mathbf{T}^{-1}\mathbf{z} \tag{24.48}$$

so that the original system

$$\dot{\mathbf{x}} = \mathbf{Ax} + \mathbf{b}u \tag{24.49}$$

becomes

$$\mathbf{T}^{-1}\dot{\mathbf{z}} = \mathbf{AT}^{-1}\mathbf{z} + \mathbf{b}u \tag{24.50}$$

or

$$\dot{\mathbf{z}} = \mathbf{TAT}^{-1}\mathbf{z} + \mathbf{Tb}u \tag{24.51}$$

Now that $\mathbf{T}$ is known, the transformed system becomes

$$\dot{\mathbf{z}} = \begin{bmatrix} -3 & -2 \\ 1 & 0 \end{bmatrix}\mathbf{z} + \begin{bmatrix} 1 \\ 0 \end{bmatrix}u \tag{24.52}$$

The system is now in control canonical form. The desired characteristic equation is, as before, obtained from the desired closed-loop pole locations

$$s^2 + 2s + 2 = 0 \tag{24.53}$$

while the characteristic equation of the system in control canonical form may be written directly from the system matrix as

$$s^2 + 3s + 2 = 0 \tag{24.54}$$

We may now identify the following terms:

$$a_1^* = 2 \tag{24.55}$$

$$a_2^* = 2 \tag{24.56}$$

$$a_1 = 3 \tag{24.57}$$

$$a_2 = 2 \tag{24.58}$$

Hence, from equation 24.31, $g_1 = -1$ and $g_2 = 0$. Since this is the control vector for the $\mathbf{z}$ system, we may write

$$\mathbf{g}_z = \begin{bmatrix} -1 \\ 0 \end{bmatrix} \tag{24.59}$$

Transforming back to the $\mathbf{x}$ system yields

$$\mathbf{g}_x = \mathbf{T}'\mathbf{g}_z = \begin{bmatrix} 3 \\ -2 \end{bmatrix} \tag{24.60}$$

which is the same as the previously obtained result. An important point to note is that in converting to control canonical form, the inverse of the matrix $\mathbf{C}_T\mathbf{W}$ is calculated. Clearly, if $\mathbf{C}_T$ is less than the full rank n, the inverse is not possible, and conversion to control canonical form cannot be made. It follows, therefore, that the closed-loop poles may only be placed in an arbitrary manner if and only if the system is controllable. This condition should always be evaluated even if poles are placed using direct comparison of the characteristic equations, since this method may appear to work even if the system is uncontrollable.

It may also be appreciated that it is not necessary to actually convert the $\mathbf{x}$ system into control canonical form, but only to obtain the matrix $\mathbf{T}$. Although in equation 24.31 the a_i are the coefficients of the system characteristic equation when written in control canonical form, the fact that the eigenvalues are invariant under a linear transformation of the state variables implies that the characteristic equation should be the same for $\mathbf{x}$ and the $\mathbf{z}$ systems. We may then write

$$\mathbf{g}_x = \mathbf{T}'\mathbf{g}_z = \mathbf{T}'(\mathbf{a}^* - \mathbf{a}) \tag{24.61}$$

where the $\mathbf{a}^*$ and $\mathbf{a}$ are column vectors of the desired and open-loop characteristic equations, respectively. Since

$$\mathbf{T} = [\mathbf{C}_T\mathbf{W}]^{-1} \tag{24.62}$$

then

$$\mathbf{g}_x = [(\mathbf{C}_T\mathbf{W})']^{-1}(\mathbf{a}^* - \mathbf{a}) \tag{24.63}$$

In the case of the previous example

$$\mathbf{g}_x = \begin{bmatrix} 1 & 2 \\ 2 & 3 \end{bmatrix}^{-1} \times \left\{ \begin{bmatrix} 2 \\ 2 \end{bmatrix} - \begin{bmatrix} 3 \\ 2 \end{bmatrix} \right\} \tag{24.64}$$

$$= \begin{bmatrix} 3 \\ -2 \end{bmatrix} \tag{24.65}$$

The control input may be written as

$$u = -\mathbf{g}_x'\mathbf{x} = -3x_1 + 2x_2 \tag{24.66}$$

◆ POLE PLACEMENT VIA ACKERMANN'S FORMULA

Another method for calculating $\mathbf{g}_x$ without determining the control canonical form for the system is by using Ackermann's formula. This expression relates the original system matrix $\mathbf{A}$, the desired closed-loop poles, and the gain vector needed to obtain them. The formula is

$$\mathbf{g}_x' = [0 \quad \cdots \quad 0 \quad 1]\mathbf{C}_T^{-1}\mathbf{\Lambda} \tag{24.67}$$

where $\mathbf{C}_T$ is the controllability matrix and

$$\mathbf{\Lambda} = \mathbf{A}^n + a_1^* \mathbf{A}^{n-1} + a_2^* \mathbf{A}^{n-2} + \cdots + a_{n-1}^* \mathbf{A} + a_n^* \mathbf{I} \tag{24.68}$$

Here $\mathbf{A}$ is the $\mathbf{x}$ system matrix and the a_i^* are the coefficients of the *desired* characteristic equation

$$s^n + a_1^* s^{n-1} + a_2 s^{n-2} + \cdots + a_{n-1}^* s + a_n^* = 0 \tag{24.69}$$

In terms of the same example that has been discussed, we have

$$\mathbf{A} = \begin{bmatrix} -1 & 0 \\ 1 & -2 \end{bmatrix} \tag{24.70}$$

$$a_1^* = 2 \tag{24.71}$$

$$a_2^* = 2 \tag{24.72}$$

Inverting the previously calculated controllability matrix,

$$\mathbf{C}_T^{-1} = \begin{bmatrix} 3 & -1 \\ 2 & -1 \end{bmatrix} \tag{24.73}$$

Hence, for $n = 2$ we obtain

$$\mathbf{\Lambda} = \mathbf{A}^2 + a_1^* \mathbf{A} + a_2^* \mathbf{I} \tag{24.74}$$

$$= \begin{bmatrix} 1 & 0 \\ -3 & 4 \end{bmatrix} + \begin{bmatrix} -2 & 0 \\ 2 & -4 \end{bmatrix} + \begin{bmatrix} 2 & 0 \\ 0 & 2 \end{bmatrix} \tag{24.75}$$

$$= \begin{bmatrix} 1 & 0 \\ -1 & 2 \end{bmatrix} \tag{24.76}$$

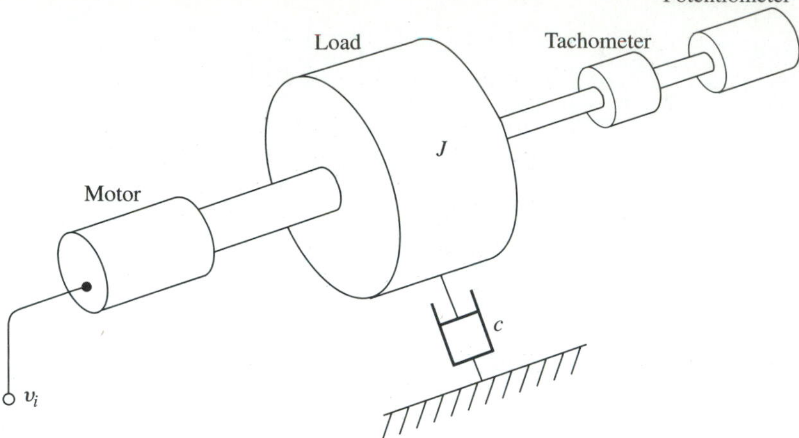

Fig. SP24.1.1

From 2.67 the control vector becomes

$$\mathbf{g}_x' = [0 \quad 1]\begin{bmatrix} 3 & -1 \\ 2 & -1 \end{bmatrix}\begin{bmatrix} 1 & 0 \\ -1 & 2 \end{bmatrix} \tag{24.77}$$

$$= [3 \quad -2] \tag{24.78}$$

This is the same result obtained previously.

Sample Problem 24.1

Consider the motor-positioned load shown in Fig. SP24.1.1, where both angular position and speed may be measured. Assume $J = 1.0$ kg-m^2, $c = 0.02$ N-m/s, and the motor torque constant $K_m = 8$ N-m/V. Assuming that the motor electrical dynamics may be neglected, calculate the feedback gain vector **g** that would give the closed-loop system critical damping and a time constant of 1 rad/s.

Solution

First, the open-loop state-space system equations will be derived. Since the electrical dynamics may be ignored, the motor simply takes a voltage as input and produces a torque as output. This torque then overcomes viscous friction and accelerates the load. Newton's equation for the motor and load becomes

$$K_m v_i - c\omega = J\dot{\omega}$$

The block diagram of the open-loop system is shown in Fig. SP24.1.2. We will select the state variables to be ω and θ. If we write the equation for the

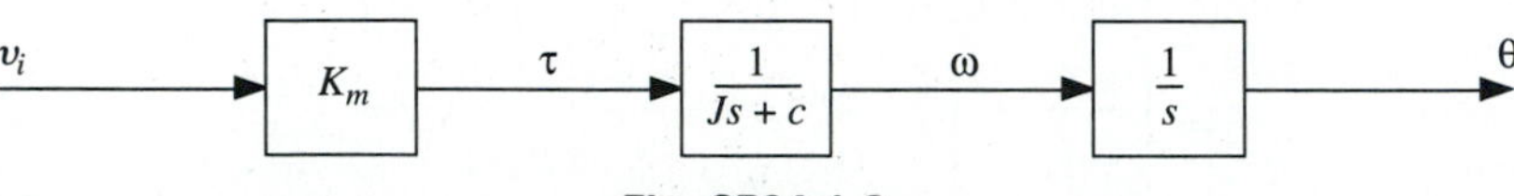

Fig. SP24.1.2

motor and the relationship between the state variables as

$$\dot{\omega} = -\frac{c}{J}\omega + \frac{K_m}{J}v_i$$

$$\dot{\theta} = \omega$$

the state equations may be readily written as

$$\begin{bmatrix} \dot{\omega} \\ \dot{\theta} \end{bmatrix} = \begin{bmatrix} -\frac{c}{J} & 0 \\ 1 & 0 \end{bmatrix} \begin{bmatrix} \omega \\ \theta \end{bmatrix} + \begin{bmatrix} \frac{K_m}{J} \\ 0 \end{bmatrix} u$$

which is in the form

$$\dot{\mathbf{x}} = \mathbf{Ax} + \mathbf{Bu}$$

This problem will be solved by direct comparison of the system characteristic equation with that formed from the desired response. First, the closed-loop configuration is drawn, as shown in Fig. SP24.1.3, which indicates that each state variable is multiplied by the gain g_i and then added to form the control input u. The problem is to determine $\mathbf{g}$. Next it must be determined whether the system is controllable. Substituting the known values, we find

$$\mathbf{A} = \begin{bmatrix} -0.02 & 0 \\ 1 & 0 \end{bmatrix}$$

$$\mathbf{b} = \begin{bmatrix} 8 \\ 0 \end{bmatrix}$$

The controllability matrix is then calculated as

$$\mathbf{C}_T = \begin{bmatrix} 8 & -0.16 \\ 0 & 8 \end{bmatrix}$$

which has the rank of 2. Since the system is controllable, we may proceed with the controller design. The closed-loop poles are given as the solution of the characteristic equation, which is obtained from equation 24.5:

$$|s\mathbf{I} - (\mathbf{A} - \mathbf{bg}')| = 0$$

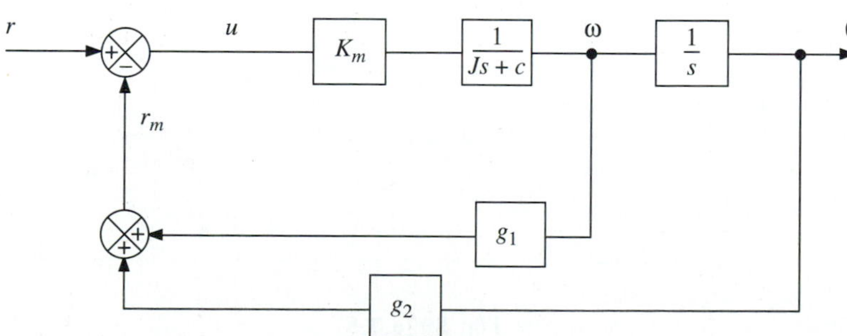

Fig. SP24.1.3

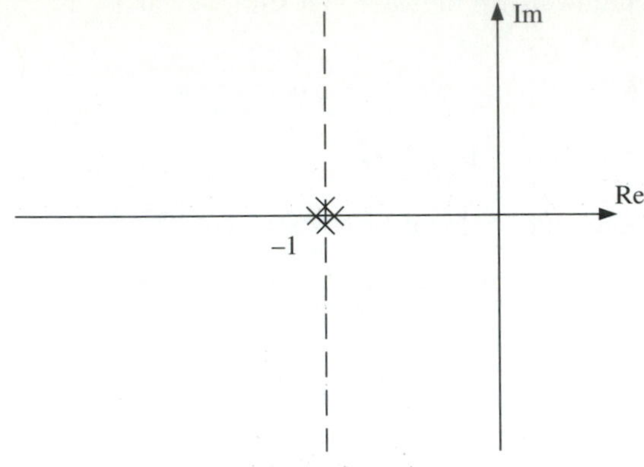

Fig. SP24.1.4

In this case

$$s\mathbf{I} - (\mathbf{A} - \mathbf{b}\mathbf{g}') = \begin{bmatrix} s & 0 \\ 0 & s \end{bmatrix} - \begin{bmatrix} -0.02 & 0 \\ 1 & 0 \end{bmatrix} + \begin{bmatrix} 4 \\ 0 \end{bmatrix} [g_1 \quad g_2]$$
$$= \begin{bmatrix} s + 0.02 + 8g_1 & 8g_2 \\ -1 & s \end{bmatrix}$$

The characteristic equation of the closed-loop system becomes

$$s^2 + s(0.02 + 8g_1) + 8g_2 = 0$$

The desired closed-loop characteristic equation may be derived from the closed-loop pole map, shown in Fig. SP24.1.4.[4] The characteristic equation becomes

$$(s + 1)^2 = s^2 + 2s + 1 = 0$$

Comparing the actual and desired characteristic equations and equating corresponding coefficients of s yield

$$0.02 + 8g_1 = 2$$
$$8g_2 = 1$$

which gives the result $g_1 = 0.2475$ and $g_2 = 0.1250$. This result may be compared to that obtained by successive block diagram reduction. If the inner velocity loop is closed first, the system becomes as shown in Fig. SP24.1.5. The

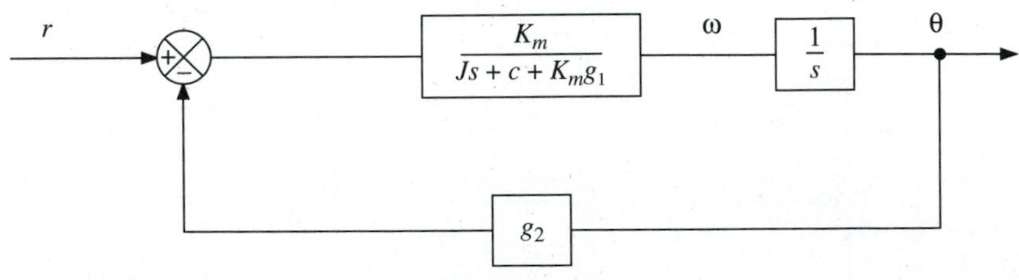

Fig. SP24.1.5

[4] Since the system is second order and critical damping is required, the closed-loop poles must be coincident at $s = -1$, twice.

closed-loop transfer function is then

$$\frac{\theta}{v_i} = \frac{K_m}{Js^2 + s(c + K_m g_1) + K_m g_2}$$

Substituting data into the above equation yields

$$\frac{\theta}{v_i} = \frac{8}{s^2 + s(0.02 + 8g_1) + 8g_2}$$

which is the same as obtained from the state-space approach. The control signal may then be constructed as

$$u = -0.2475\omega - 0.125\theta$$

and the system will have the required response. Note that it is possible to achieve partial selection of the system performance parameters if only the output, or θ, is fed back. This corresponds to there being no velocity feedback in the system for which the open-loop transfer function becomes

$$\frac{\theta}{u} = \frac{K_m g_2}{s(Js + c)}$$

The root locus for this system is shown in Fig. SP24.1.6 as a function of the variable gain g_2. Clearly, the damping may be set to any desired value, but the time constant cannot. This example distinguishes between *output feedback* and *state feedback* control. In the former, only some of the design requirements may be met, while in the latter case, all performance requirements may be met if full state variable feedback is achieved.

Sample Problem 24.2

An open-loop, third-order system is described by the equation

$$\dot{\mathbf{x}} = \begin{bmatrix} -1 & 0 & 2 \\ 0 & -2 & 1 \\ 1 & 0 & 3 \end{bmatrix} \mathbf{x} + \begin{bmatrix} 2 \\ 0 \\ 1 \end{bmatrix} u$$

Find a suitable gain vector **g** such that the closed-loop system is dominated by a pair of closed-loop poles having a damping ratio $\zeta = 0.707$ and a time constant of 0.5 rad/s.

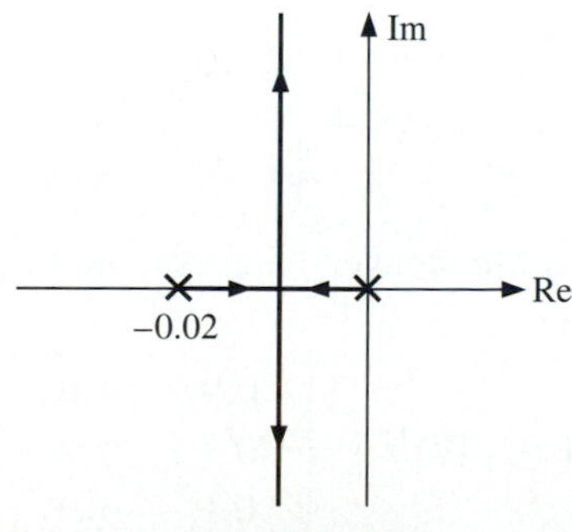

Fig. SP24.1.6

Solution

This problem will be solved by finding the transform **T** that would convert the system into control canonical form. First, however, the controllability of the system must be determined. From the given values of **A** and **b**, it may be shown that

$$\mathbf{C}_T = \begin{bmatrix} 2 & 0 & 10 \\ 0 & 1 & 3 \\ 1 & 5 & 15 \end{bmatrix}$$

which has the rank of 3. The system is therefore controllable. Next, the characteristic equation is derived. Recall that this equation will be the same for the given system and for the system transformed into control canonical form, since the eigenvalues are identical. We form

$$s\mathbf{I} - \mathbf{A} = \begin{bmatrix} s+1 & 0 & -2 \\ 0 & s+2 & -1 \\ -1 & 0 & s-3 \end{bmatrix}$$

The characteristic equation is derived from the determinant, which in turn may be easily calculated:

$$|s\mathbf{I} - \mathbf{A}| = s^3 - 9s - 10 = 0$$

It is important to note that Routh's criterion indicates that this is an unstable open-loop system that, because it is controllable, may be stabilized by means of suitable feedback gains. For a third-order system such as this, the matrix **W** may be constructed from equation 24.36 as

$$\mathbf{W} = \begin{bmatrix} 1 & a_1 & a_2 \\ 0 & 1 & a_1 \\ 0 & 0 & 1 \end{bmatrix}$$

where the a_i are obtained from the open-loop characteristic equation, as indicated in equation 24.23. Hence

$$\mathbf{W} = \begin{bmatrix} 1 & 0 & -9 \\ 0 & 1 & 0 \\ 0 & 0 & 1 \end{bmatrix}$$

We now form

$$\mathbf{C}_T\mathbf{W} = \begin{bmatrix} 2 & 0 & -8 \\ 0 & 1 & 3 \\ 1 & 5 & 6 \end{bmatrix}$$

The transform required to convert the system to control canonical form becomes

$$\mathbf{T} = (\mathbf{C}_T\mathbf{W})^{-1} = \begin{bmatrix} 0.9 & 4.0 & -0.8 \\ -0.3 & -2.0 & 0.6 \\ 0.1 & 1.0 & -0.2 \end{bmatrix}$$

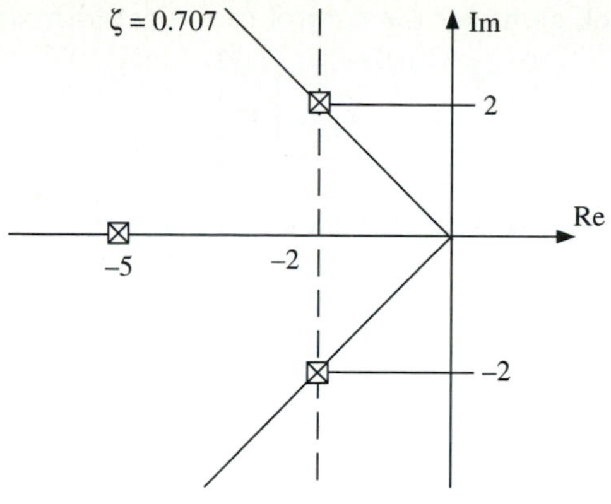

Fig. SP24.2.1

At this point, we could convert the system to control canonical form, but this is not necessary. All that is required is **T**. Next, we determine the form of the required characteristic equation. Since the system is third order and only the location of the dominant poles are specified, the remaining pole may be placed in an arbitrary position as long as the complex pair dominate the response. Figure SP24.2.1 shows the location of the system poles that satisfy the design requirements, with the real pole placed at $s = -5$. The desired characteristic equation becomes

$$(s + 5)(s + 2 + 2j)(s + 2 - 2j) = 0$$

or

$$s^3 + 9s^2 + 28s + 40 = 0$$

From the open-loop characteristic equation we have

$$a_1 = 0$$

$$a_2 = -9$$

$$a_3 = -10$$

while from the desired characteristic equation we have

$$a_1^* = 9$$

$$a_2^* = 28$$

$$a_3^* = 40$$

Since

$$g_i = a_i^* - a_i$$

then

$$g_1 = 9$$

$$g_2 = 37$$

$$g_3 = 50$$

Thus, the feedback gains for the control canonical form are

$$\mathbf{g}_z = \begin{bmatrix} 9 \\ 37 \\ 50 \end{bmatrix}$$

In terms of the original state variables **x**, the feedback gains are

$$\mathbf{g}_x = \mathbf{T}'\mathbf{g}_z = \begin{bmatrix} 2 \\ 12 \\ 5 \end{bmatrix}$$

The control signal then becomes

$$u = -2x_1 - 12x_2 - 5x_3$$

which is the required result.

Sample Problem 24.3

Consider the system described by the state-space equation

$$\dot{\mathbf{x}} = \begin{bmatrix} -1 & 0 & 2 & -1 \\ 0 & -2 & 3 & 0 \\ 1 & 0 & 1 & 2 \\ 0 & 2 & 3 & -1 \end{bmatrix}\mathbf{x} + \begin{bmatrix} 1 \\ 3 \\ 2 \\ 0 \end{bmatrix}u$$

Use Ackermann's formula to design a feedback gain vector such that the resulting closed-loop system is dominated by the three closed-loop poles shown in Fig. SP24.3.1.

Solution

Given **A** and **b**, the controllability matrix may be calculated as

$$\mathbf{C}_T = \begin{bmatrix} 1 & 3 & -9 & 72 \\ 3 & 0 & 9 & 72 \\ 2 & 3 & 30 & 15 \\ 0 & 12 & -3 & 111 \end{bmatrix}$$

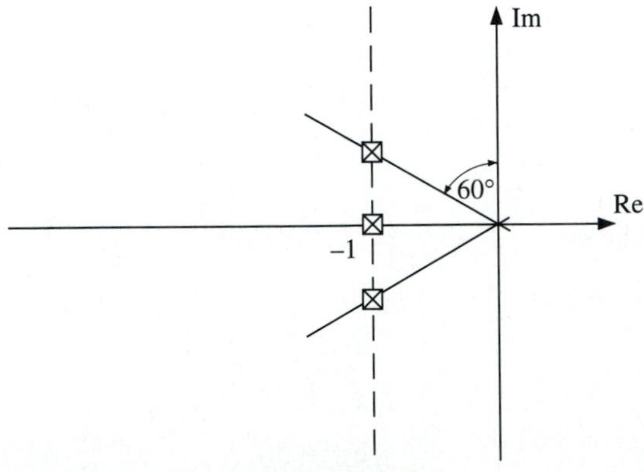

Fig. SP24.3.1

The rank of $\mathbf{C}_T$ is found to be 4, confirming that the system is controllable. Since the inverse of the controllability matrix is needed for Ackermann's formula, it will be calculated[5] at this time:

$$\mathbf{C}_T^{-1} = \begin{bmatrix} 4.2593 & -2.2963 & 1.8148 & -1.5185 \\ 1.1728 & -0.7531 & 0.5432 & -0.3457 \\ -0.3333 & 0.1852 & -0.1111 & 0.1111 \\ -0.1358 & 0.0864 & -0.0617 & 0.0494 \end{bmatrix}$$

Next, the desired characteristic equation is formed. Since the system is fourth order but only three pole positions are specified, the remaining pole will be placed arbitrarily at $s = -5$. This gives the required characteristic equation as

$$(s + 1)(s + 5)(s + 1 + 0.577j)(s + 1 - 0.577j) = 0$$

resulting in

$$s^4 + 8s^3 + 18.34s^2 + 18s + 6.667 = 0$$

Now, for $n = 4$, the array $\mathbf{\Lambda}$ from equation 24.68 becomes

$$\begin{aligned} \mathbf{\Lambda} &= \mathbf{A}^4 + 8\mathbf{A}^3 + 18.34\mathbf{A}^2 + 18\mathbf{A} + 6.667\mathbf{I} \\ &= \begin{bmatrix} 16.67 & 29.32 & 106.98 & 23.04 \\ 46.02 & 38.03 & 250.98 & 77.04 \\ 99.00 & 51.36 & 396.73 & 177.66 \\ 91.02 & 85.96 & 389.04 & 158.05 \end{bmatrix} \end{aligned}$$

The required feedback gain vector is then obtained from

$$\begin{aligned} \mathbf{g}_x' &= [0 \quad 0 \quad 0 \quad 1]\mathbf{C}_T^{-1}\mathbf{\Lambda} \\ &= [0.0942 \quad 0.3794 \quad 1.8838 \quad 0.3672] \end{aligned}$$

Note that MATLAB has the command $\mathbf{g}_x$= acker(A, b, dp), which calculates $\mathbf{g}_x$ directly from **A**, **b**, and the vector of desired closed-loop poles **dp**. Executing this command produces the same result as above.

◆ PROBLEMS

24.1 Consider the open-loop system described by the state equation

$$\dot{\mathbf{x}} = \begin{bmatrix} 0.5 & 0 \\ 1 & 2 \end{bmatrix}\mathbf{x} + \begin{bmatrix} 1 \\ 3 \end{bmatrix}u$$

By comparing the actual and desired characteristic equations, find the feedback control vector **g** such that the closed-loop poles are coincident at $s = -2$.

24.2 A system is described in control canonical form as

$$\dot{\mathbf{x}} = \begin{bmatrix} -1 & 2 \\ 1 & 0 \end{bmatrix}\mathbf{x} + \begin{bmatrix} 1 \\ 0 \end{bmatrix}u$$

Design a feedback gain vector that places the closed-loop poles at $s = -1$ and $s = -2$.

[5] MATLAB is used here.

24.3 The following system is to have less than a 5% overshoot to a step, while the time constant is to be no greater than 1.5 s. Design a feedback control gain vector to accomplish the task:

$$\dot{\mathbf{x}} = \begin{bmatrix} -3 & 1 & 2 \\ 1 & 0 & 0 \\ 0 & 1 & 0 \end{bmatrix} \mathbf{x} + \begin{bmatrix} 1 \\ 0 \\ 0 \end{bmatrix} u$$

24.4 An open-loop SISO control system is represented by the transfer function

$$\frac{y}{u} = \frac{8}{s^2 - 9}$$

Comment on the stability and controllability of the open-loop system. Find suitable feedback of the state variables such that the closed-loop system is critically damped.

24.5 Write the following system in control canonical form:

$$\dot{\mathbf{x}} = \begin{bmatrix} 1 & 0 \\ 2 & 3 \end{bmatrix} \mathbf{x} + \begin{bmatrix} 2 \\ 1 \end{bmatrix} u$$

Find a feedback vector such that the closed-loop poles satisfy

$$s^2 + 3s + 5 = 0$$

24.6 Given the open-loop system

$$G(s) = \frac{20}{s(s + 3)(s + 5)}$$

design a feedback controller so that the closed-loop response has 15% overshoot and a 2% settling time of 1 s for a step input, ensuring that any higher order poles are placed at least five times farther away from the imaginary axis than the domaint system pole(s). What difference will there be if there is a zero at $s = -2$ so that

$$G(s) = \frac{10(s + 2)}{s(s + 3)(s + 5)}$$

24.7 Consider the system described by

$$\mathbf{A} = \begin{bmatrix} -1 & 0 & 1 \\ 2 & -2 & 0 \\ 0 & 1 & -4 \end{bmatrix}$$

$$\mathbf{b} = \begin{bmatrix} 0 \\ 1 \\ 0 \end{bmatrix}$$

Determine the system characteristic equation, and by comparing it with that placing the system closed-loop poles at $s = -1, -2, -3$, find the corresponding gain vector $\mathbf{g}'$.

24.8 By identifying the transform that would convert the following system into control canonical form,

$$\dot{\mathbf{x}} = \begin{bmatrix} -5 & 1 & 3 \\ 0 & 0 & 1 \\ 1 & 2 & -3 \end{bmatrix} \mathbf{x} + \begin{bmatrix} 2 \\ 4 \\ 0 \end{bmatrix} u$$

find the control vector that makes the closed-loop system have the following characteristic equation:

$$(s + 1)(s^2 + 3s + 16) = 0$$

24.9 Use Ackermann's formula to design a controller that gives the system

$$\dot{\mathbf{x}} = \begin{bmatrix} 2 & -1 \\ 1 & -3 \end{bmatrix} \mathbf{x} + \begin{bmatrix} 2 \\ 0 \end{bmatrix} u$$

a damping ratio $\zeta = 0.4$ and a peak time of 2.5 s.

24.10 Use computer methods (such as MATLAB) to determine the feedback control vector for the system

$$\dot{\mathbf{x}} = \begin{bmatrix} -3 & -1 & 2 & 0 \\ 1 & -2 & 4 & 1 \\ 2 & 2 & 3 & -2 \\ -1 & 0 & 0 & 2 \end{bmatrix} \mathbf{x} + \begin{bmatrix} 1 \\ 2 \\ 4 \\ 0 \end{bmatrix} u$$

so that the closed-loop poles comprise two complex pairs each with damped natural frequency of 2 rad/s, the dominant pair having a time constant of 4 s, and the other pair having a time constant half of this.

MODULE 25

State-Space Observer Design

Module 24 demonstrated how the dynamics of a system may be altered by feeding back the state variables through a suitable gain vector **g** and subtracting it from the input r to form the control signal u. A question arises regarding situations where not all of the state variables are available from which the feedback control signal may be derived. This may happen because some of the state variables do not correspond to physical variables and therefore cannot be measured or because of the prohibitive cost in providing a sensor for each state variable in the system. Whatever the reason, the unavailability of all state variables prevents controller design as previously outlined, and alternative methods are needed.

◆ OBSERVER SYNTHESIS

The approach taken to solve this problem is to construct a model of the system under study, obtain all the state variables from the model, and then assume (subject to certain restrictions) that the computed state variables are good approximations to the true state variables. From these computed state variables, a suitable controller for the actual system may be constructed using the techniques described in Module 24.

Suppose we have the SISO system described in the familiar state-space form

$$\dot{\mathbf{x}} = \mathbf{A}\mathbf{x} + \mathbf{b}u \tag{25.1}$$

and

$$y = \mathbf{C}\mathbf{x} \tag{25.2}$$

The system may be represented in block diagram form as in Fig. 25.1. Assume that we have approximate values of **A**, **b**, and **C**, which may be denoted as $\overline{\mathbf{A}}$, $\overline{\mathbf{b}}$,

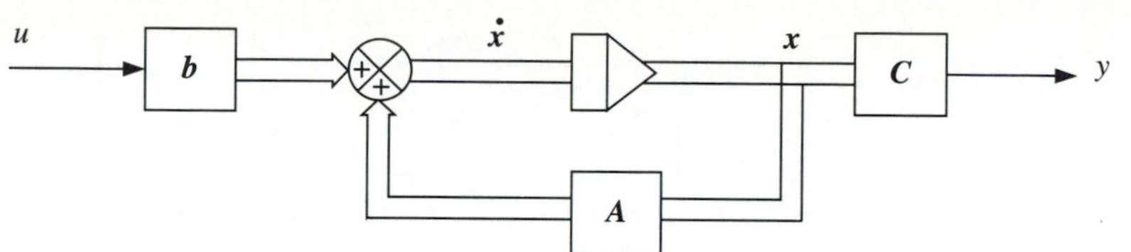

Fig. 25.1 Block diagram of actual system.

and $\overline{\mathbf{C}}$. In some cases we will know the system matrices exactly, in which case $\overline{\mathbf{A}} = \mathbf{A}$, $\overline{\mathbf{b}} = \mathbf{b}$, and $\overline{\mathbf{C}} = \mathbf{C}$. We could then construct a model[1] known as an *observer,* of a system we think is very close to the actual system, described by the equation

$$\dot{\overline{\mathbf{x}}} = \overline{\mathbf{A}}\overline{\mathbf{x}} + \overline{\mathbf{b}}u \tag{25.3}$$

and

$$\overline{y} = \overline{\mathbf{C}}\overline{\mathbf{x}} \tag{25.4}$$

The input u is not approximated, since a unit step, for example, is the same for the actual system and the observer. The block diagram of the observer is shown in Fig. 25.2. If the observer is accurate, then the state variables are simply obtained from it, as shown in Fig. 25.2, and may be used as input to a controller. The question arises, of course, as to what happens if the observer and actual systems are different to a substantial degree. Can we adjust the model to compensate for such discrepancies? Recognizing that although we cannot compare the state variables, we can compare the outputs $\overline{y}$ and y and derive a measure of the discrepancy of the two systems by subtracting them, as shown in Fig. 25.3. This error signal, known as the residual r, is then fed back to the observer summing junction. Consider the case where, due to the disparity between the systems, $y > \overline{y}$. The residual will be positive, increasing $\overline{\mathbf{x}}$, hence $\mathbf{x}$, and subsequently y, thereby reducing the residual r. This feedback forces the observer to behave more like the actual system in the presence of errors.

The structure shown in Fig. 25.3 has a problem, however. Suppose a difference between the output of the system and the observer exists and, therefore, the residual is finite and nonzero. An accurate value of the state variables $\overline{\mathbf{x}}$ will only be obtained after the transient response of the observer dynamics has

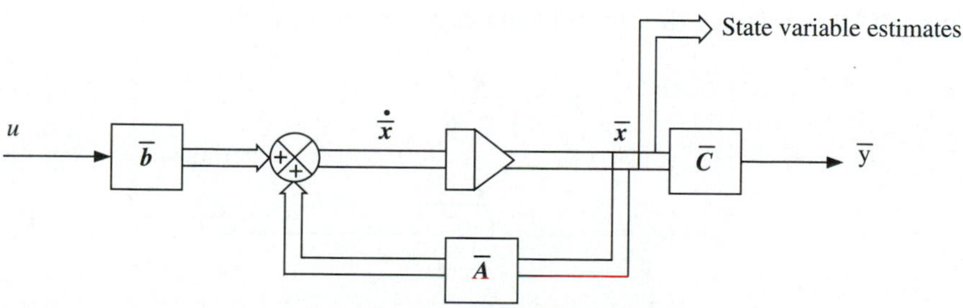

Fig. 25.2 Block diagram of observer.

[1] Normally a computer model.

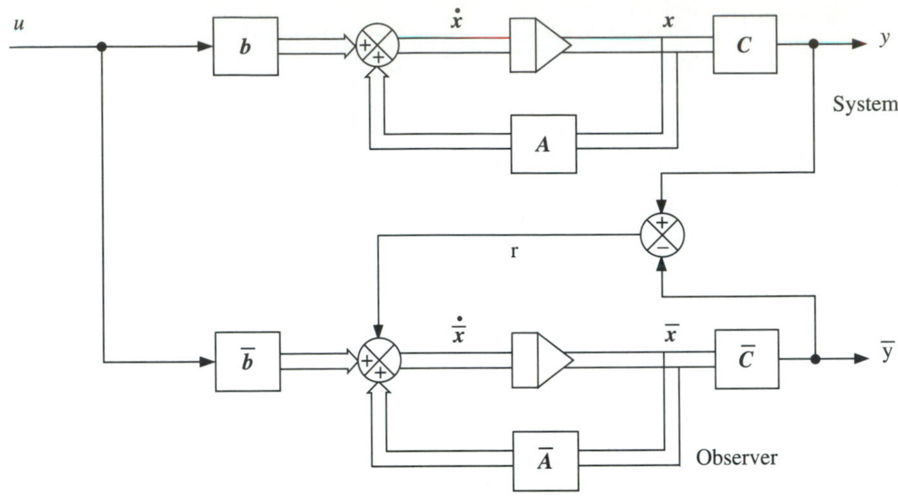

Fig. 25.3 Output error correction.

disappeared. If the observer is a reasonably good represenation of the actual system, it is apparent that the dynamics of the system and observer will also be very similar. It follows that if the state variables **x** are changing, then the observer state variables will also change, but subject to a transient response caused by the additional input due to the residual. The conclusion is that, under these circumstances, the observer state variables will never "track" the actual state variables except in the steady state. What is needed is to make the dynamics of the observer much faster than the dynamics of the actual system, so that the dynamics of the former disappear very quickly, allowing $\bar{\mathbf{x}}$ to follow **x** accurately. How can we change the dynamics of the observer?

Consider the simple feedback control system shown in Fig. 25.4 where there is a feedback gain K. The plant is a first-order lag with a time constant of 1 s. The closed-loop transfer function is

$$\frac{C}{R} = \frac{1/(1 + K)}{1 + s\left(\dfrac{1}{1 + K}\right)} \tag{25.5}$$

The effect of the feedback gain is to reduce the DC gain of the system, but more importantly, it reduces the system time constant from 1 s to

$$\tau = \frac{1}{1 + K} \tag{25.6}$$

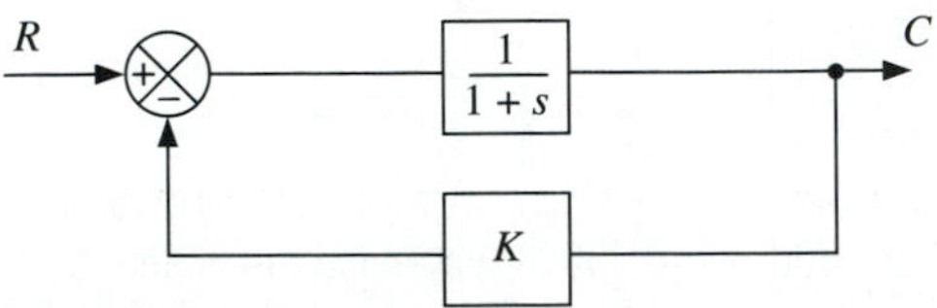

Fig. 25.4 Simple feedback system with feedback gain.

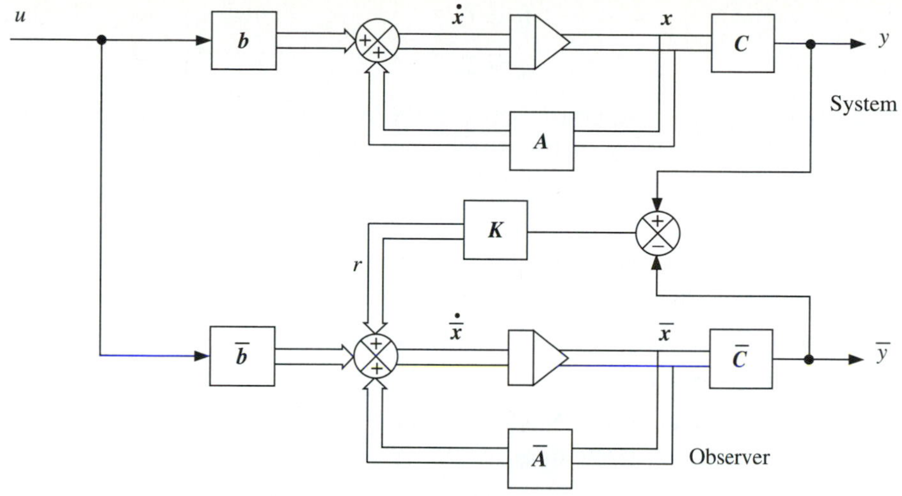

Fig. 25.5 Adjustable observer dynamics.

With suitable selection of K, the system time constant, and hence the system dynamics, can be fixed at any desired value. Applying this approach to our system-observer arrangement yields the construction shown in Fig. 25.5. In this figure, **K** is a column vector that produces a column vector for the residual **r**. The problem now is to determine **K** such that the dynamics of the observer (a) are faster than those of the system and (b) are well behaved, in that they are stable and appropriately damped. This is a pole placement exercise in selecting the eigenvalues of the observer and may be achieved using any of the techniques described in Module 24. The observer equation may be written in the form

$$\dot{\bar{\mathbf{x}}} = \bar{\mathbf{A}}\bar{\mathbf{x}} + \bar{\mathbf{b}}u + \mathbf{K}(y - \bar{y}) \tag{25.7}$$

$$= \bar{\mathbf{A}}\bar{\mathbf{x}} + \bar{\mathbf{b}}u + \mathbf{K}(y - \bar{\mathbf{C}}\bar{\mathbf{x}}) \tag{25.8}$$

$$= (\bar{\mathbf{A}} - \mathbf{K}\bar{\mathbf{C}})\bar{\mathbf{x}} + \bar{\mathbf{b}}u + \mathbf{K}y \tag{25.9}$$

The last two terms in the above equation may be considered as forcing terms and will determine the steady-state value of the estimated state vector. The first term is the one that determines the transient response and may be seen to be the solution of the unforced equation

$$\dot{\bar{\mathbf{x}}} = (\bar{\mathbf{A}} - \mathbf{K}\bar{\mathbf{C}})\bar{\mathbf{x}} \tag{25.10}$$

We wish to allocate the poles of $\bar{\mathbf{A}} - \mathbf{K}\bar{\mathbf{C}}$. Noting that the eigenvalues of a matrix are the same as those of the transpose of the matrix, we may write

$$(\bar{\mathbf{A}} - \mathbf{K}\bar{\mathbf{C}})' = (\bar{\mathbf{A}}' - \bar{\mathbf{C}}'\mathbf{K}') \tag{25.11}$$

Comparing the right-hand side of this equation with equation 24.4,

$$\dot{\mathbf{x}} = (\tilde{\mathbf{A}} - \tilde{\mathbf{b}}\mathbf{g}')\mathbf{x} \tag{25.12}$$

we see that the pole placement method is directly applicable to the design of the observer, by considering the following matrices to be equivalent:

$$\tilde{\mathbf{A}} \equiv \bar{\mathbf{A}}' \tag{25.13}$$

$$\tilde{\mathbf{b}} \equiv \overline{\mathbf{C}}' \tag{25.14}$$

$$\mathbf{g} \equiv \mathbf{K} \tag{25.15}$$

In equation 25.12 we distinguish between the system matrices **A** and **b** and the equivalent observer system matrices (with residual feedback) $\tilde{\mathbf{A}}$ and $\tilde{\mathbf{b}}$.

Summarizing so far, the observer design may be achieved using the following steps:

1. Based on the expected system dynamics, determine a characteristic equation that represents the required observer dynamics.[2]
2. Knowing the estimated system matrices $\overline{\mathbf{A}}$ and $\overline{\mathbf{C}}$, use the above equivalences to calculate $\tilde{\mathbf{A}}$ and $\tilde{\mathbf{b}}$.
3. Use one of the methods in Module 24 (such as Ackermann's formula) to derive the feedback gain vector **g**.
4. Use the last equivalence above to determine the observer gain vector **K**.

One other point has to be considered. It was noted in the pole placement methods used previously that the designer has complete control over the eigenvalues of the closed-loop system only if the system is controllable. For our observer design problem, this means that the observer has to be controllable also. Recall that the controllability matrix

$$\mathbf{C}_T = [\tilde{\mathbf{b}} \quad \tilde{\mathbf{A}}\tilde{\mathbf{b}} \quad \tilde{\mathbf{A}}^2\tilde{\mathbf{b}} \quad \tilde{\mathbf{A}}^3\tilde{\mathbf{b}} \quad \cdots \quad \tilde{\mathbf{A}}^{n-1}\tilde{\mathbf{b}}] \tag{25.16}$$

has to have rank equal to n, the system order. In terms of the observer system equivalent matrices, the above expression may be written as

$$\mathbf{C}_T = [\overline{\mathbf{C}}' \quad \overline{\mathbf{A}}'\overline{\mathbf{C}}' \quad (\overline{\mathbf{A}}')^2\overline{\mathbf{C}}' \quad \cdots \quad (\overline{\mathbf{A}}')^{n-1}\overline{\mathbf{C}}'] \tag{25.17}$$

Taking the transpose of the above equation yields

$$\mathbf{C}_T' = \begin{bmatrix} \overline{\mathbf{C}} \\ \overline{\mathbf{C}}\,\overline{\mathbf{A}} \\ \overline{\mathbf{C}}\,\overline{\mathbf{A}}^2 \\ \vdots \\ \overline{\mathbf{C}}\,\overline{\mathbf{A}}^{n-1} \end{bmatrix} = \mathbf{O}_T \tag{25.18}$$

which is the previously defined observability matrix. In words, the above equation states that since the transpose of the controllability matrix is equal to the observability matrix[3] and the rank of the controllability matrix has to equal the order of the system, then the rank of the observability matrix also has to equal the system order. It follows, therefore, that the observer poles may be placed in specified locations only if the original system is observable. Just as we always check whether a system is controllable before designing the feedback gain vector, we must also check that the system is observable before designing a state observer.

[2] For example, make the observer poles twice or five times as fast as the system poles and have critical damping.

[3] This is not generally the case, only in linear observer design.

A compensator is simply an observer that generates estimates of the state vector of a system in order that the system closed-loop poles may be placed in selected locations. From Module 24, if the state is measurable, then the vector **g** may be determined given a desired closed-loop characteristic equation. This is shown in Fig. 25.6. The state feedback is shown dashed because we now assume that **x** is not available, and an observer must be designed to provide it. We begin by drawing the general observer shown in Fig. 25.2 but rearranged slightly and placed before the system, or plant, as shown in Fig. 25.7. Note the subtle change between Figs. 25.2 and 25.7 regarding the specification of the system matrices. We are assuming that the estimates of the system matrices are accurate so that

$$\overline{\mathbf{A}} = \mathbf{A} \tag{25.19}$$

$$\overline{\mathbf{b}} = \mathbf{b} \tag{25.20}$$

$$\overline{\mathbf{C}} = \mathbf{C} \tag{25.21}$$

Note also that the input u is temporarily missing from the observer in Fig. 25.7. Assuming the observer is capable of generating the state vector, we could use it to generate the control signal u that may then be used in both the system and observer, as shown in Fig. 25.8. For a system where not all of the state variables are available, the compensation process may then be summarized as follows:

1. Given a system, check its controllability to determine whether state feedback may be used to arbitrarily place the closed-loop poles.

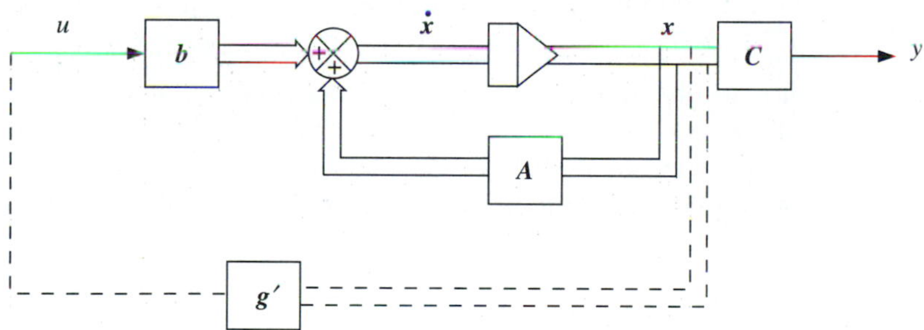

Fig. 25.6 Pole placement based on state availability.

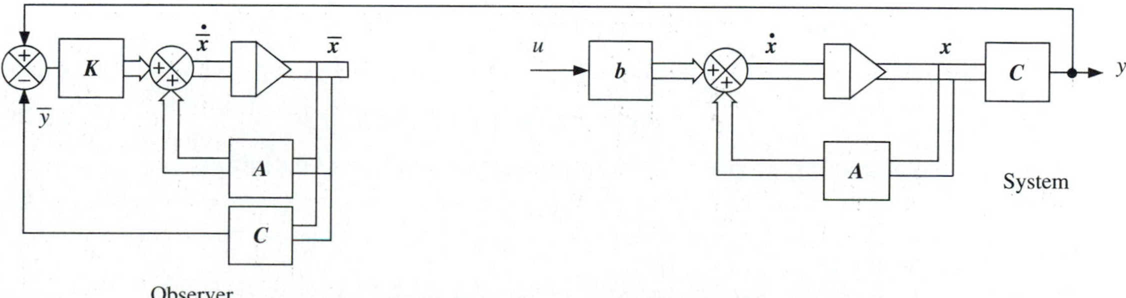

Fig. 25.7 Preliminary compensator design.

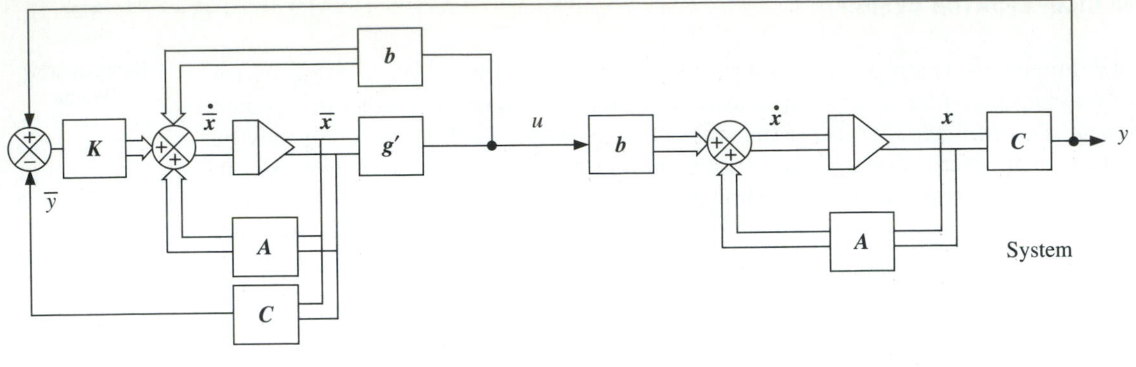

Fig. 25.8 Complete compensator design.

2. Check the system observability to determine whether an observer may be used to estimate the state.
3. Calculate the feedback gain vector, assuming that all state variables are available, using one of the methods outlined in Module 24, to place the closed-loop system poles at the desired locations.
4. Using the equivalent matrices, use pole placement techniques to locate the eigenvalues of the observer with respect to the desired poles of the system, and calculate the observer gain vector **K**.
5. Construct the resulting observer design in a computer model in order to generate the control system required by the plant.

Recall that in our initial discussion of controller design in Module 24 we disregarded the input disturbance r to the plant and developed the feedback vector **g** to provide "good" response. Figure 25.8 is now redrawn to include the input and is simply a rearrangement of the individual plant and observer diagrams. The complete system and compensator is shown in Fig. 25.9. The equation of the plant may be written as

$$\dot{\mathbf{x}} = \mathbf{A}\mathbf{x} + \mathbf{b}u \tag{25.22}$$

$$= \mathbf{A}\mathbf{x} + \mathbf{b}(r - \mathbf{g}'\bar{\mathbf{x}}) \tag{25.23}$$

$$= \mathbf{A}\mathbf{x} - \mathbf{b}\mathbf{g}'\bar{\mathbf{x}} + \mathbf{b}r \tag{25.24}$$

The equations for the observer are

$$\dot{\bar{\mathbf{x}}} = \mathbf{A}\bar{\mathbf{x}} - \mathbf{K}(\mathbf{y} - \bar{\mathbf{y}}) + \mathbf{b}u \tag{25.25}$$

$$= \mathbf{A}\bar{\mathbf{x}} - \mathbf{K}\mathbf{C}(\mathbf{x} - \bar{\mathbf{x}}) - \mathbf{b}\mathbf{g}'\bar{\mathbf{x}} + \mathbf{b}r \tag{25.26}$$

Subtracting equation 25.26 from equation 25.24 and defining

$$\mathbf{e} = \mathbf{x} - \bar{\mathbf{x}} \tag{25.27}$$

produces

$$\dot{\mathbf{e}} = (\mathbf{A} - \mathbf{K}\mathbf{C})\mathbf{e} \tag{25.28}$$

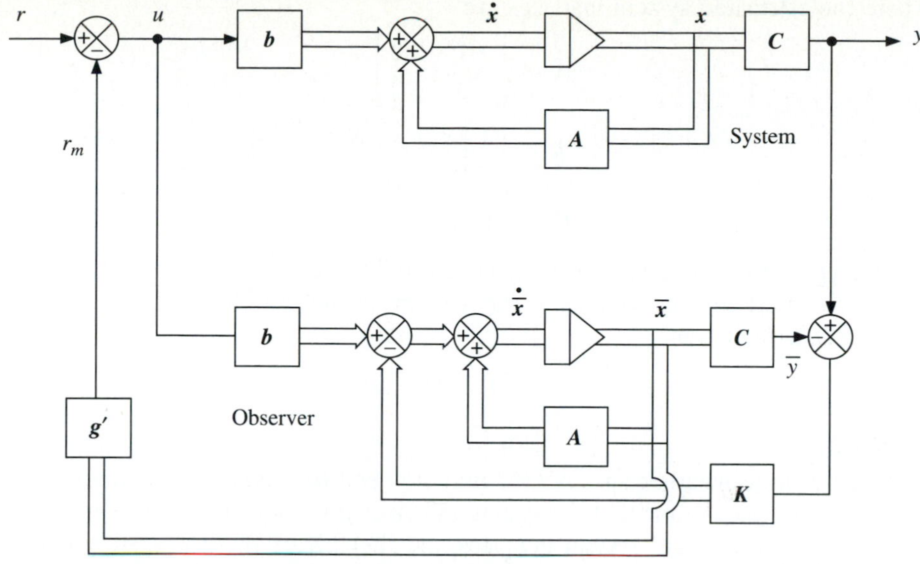

Fig. 25.9 Compensator with input.

We rearrange equation 25.24 as

$$\dot{\mathbf{x}} = \mathbf{A}\mathbf{x} + \mathbf{b}\mathbf{g}'\mathbf{x} - \mathbf{b}\mathbf{g}'\bar{\mathbf{x}} - \mathbf{b}\mathbf{g}'\mathbf{x} + \mathbf{b}r \tag{25.29}$$

$$= (\mathbf{A} - \mathbf{b}\mathbf{g}')\mathbf{x} + \mathbf{b}\mathbf{g}'\mathbf{e} + \mathbf{b}r \tag{25.30}$$

Now defining a system whose state is comprised of the actual system state, together with the error between the actual and estimated system states, we may write equations 25.27 and 25.30 as

$$\begin{bmatrix} \dot{\bar{\mathbf{x}}} \\ \dot{\mathbf{e}} \end{bmatrix} = \begin{bmatrix} \mathbf{A} - \mathbf{b}\mathbf{g}' & \mathbf{b}\mathbf{g}' \\ \mathbf{0} & \mathbf{A} - \mathbf{K}\mathbf{C} \end{bmatrix} \begin{bmatrix} \bar{\mathbf{x}} \\ \mathbf{e} \end{bmatrix} + \begin{bmatrix} \mathbf{b} \\ \mathbf{0} \end{bmatrix} r \tag{25.31}$$

and the output equation as

$$y = [\mathbf{C} \quad \mathbf{0}] \begin{bmatrix} \mathbf{x} \\ \mathbf{0} \end{bmatrix} \tag{25.32}$$

This formulation allows us to investigate the dynamics of the system and the observer and will be dealt with in detail in one of the sample problems that follow.

Sample Problem 25.1

A system is described by the state-space equations

$$\dot{\mathbf{x}} = \bar{\mathbf{A}}\mathbf{x} + \bar{\mathbf{b}}u$$
$$y = \bar{\mathbf{C}}\mathbf{x}$$

where the estimated system matrices are

$$\overline{\mathbf{A}} = \begin{bmatrix} -1 & 0 \\ 3 & -2 \end{bmatrix}$$

$$\overline{\mathbf{b}} = \begin{bmatrix} 1 \\ 2 \end{bmatrix}$$

$$\overline{\mathbf{C}} = [0 \quad 1]$$

Assuming that the state variables are not available for measurement and it is required to adjust the eigenvalues of the system, design an observer that has the characteristic equation

$$s^2 + 6s + 18 = 0$$

Solution

To determine whether an observer may be used to produce estimates of the states, we first assume that the system matrices given above are close approximations to the actual system matrices and then check that the system is observable. It is expedient to use the MATLAB command

```
r = rank(obsv(a,c))
```

which yields the rank of the observability matrix as 2. The system is thus observable, and an observer may be constructed that will yield estimates of the state variables. Out of interest, we check the dynamics of the system by forming the characteristic equation

$$\det(s\mathbf{I} - \mathbf{A}) = \begin{vmatrix} s + 1 & 0 \\ -3 & s + 2 \end{vmatrix} = 0$$

This yields the eigenvalues

$$s = -1.5 \pm 3.32j$$

The roots of the required characteristic equation of the observer give the eigenvalues

$$s = -3 \pm 3j$$

This indicates that the observer dynamics will have half the settling time of the system and have a damping ratio $\zeta = 0.707$, giving about a 5% overshoot. The next step is to use pole placement methods for the observer by calculating the equivalent matrices in equations 25.13 and 25.14. This gives

$$\tilde{\mathbf{A}} = \begin{bmatrix} -1 & 3 \\ 0 & -2 \end{bmatrix}$$

$$\tilde{\mathbf{b}} = \begin{bmatrix} 0 \\ 1 \end{bmatrix}$$

We will use Ackermann's formula to determine the required gain vector **g** since there is a useful MATLAB command

```
g = acker(a,b,despol)
```

where the vector despol contains the desired poles and may be specified, from above, with the command

```
despol = [-3 + 3*i, -3 -3*i]
```

MATLAB returns the value of **g**, and hence **K**, as

$$\mathbf{g} = \mathbf{K} = \begin{bmatrix} 4.333 \\ 3.000 \end{bmatrix}$$

If the observer is constructed as shown in Fig. 25.5 with **K** having the value above, its dynamics will be represented by the required characteristic equation.

Sample Problem 25.2

A system is described by the open-loop transfer function

$$G(s) = \frac{10(s + 2)}{(s + 4)(s + 1)}$$

If the system states cannot be measured, design an observer to estimate them such that the response time of the observer is at least five times faster than that of the plant and is critically damped.

Solution

First, we express the system transfer function in state-space form by expanding the numerator and denominator into polynomials:

$$G(s) = \frac{10(s + 2)}{s^2 + 5s + 4}$$

From this, the a_i and b_i in equation 22.25 may be identified. The $\overline{\mathbf{A}}$ matrix is obtained from the denominator of the transfer function.[4] Recalling equations 22.23 and 22.17, we may write directly

$$\overline{\mathbf{A}} = \begin{bmatrix} 0 & 1 \\ -4 & -5 \end{bmatrix}$$

$$\overline{\mathbf{b}} = \begin{bmatrix} 0 \\ 10 \end{bmatrix}$$

Since the numerator is of one order less than the denominator, equation 22.35 is used to obtain

$$\overline{\mathbf{C}} = [2 \quad 1]$$

Now that the system matrices have been identified,[5] the feasibility of designing a state observer may be determined by examining the rank of the observ-

[4] In this example, because the system transfer function is known, the system matrices are also known; hence $\overline{\mathbf{A}} = \mathbf{A}$, etc. We still require an observer, however, because the system state vector is unavailable for measurement.

[5] MATLAB has the command TF2SS, which will transform a transfer function into the equivalent state-space matrices. If the reader uses this command, a different formulation for $\overline{\mathbf{A}}$, $\overline{\mathbf{b}}$, and $\overline{\mathbf{C}}$ will result. Executing the reverse process SS2TF will, however, show that the same transfer function is generated. Recall that a set of state variables, and hence system matrices, is not unique.

ability matrix, which in this case takes the form

$$\mathbf{O}_T = \begin{bmatrix} 2 & 1 \\ -4 & -3 \end{bmatrix}$$

Since this has the rank of 2, an observer is feasible. The next step is to determine the observer dynamics. The system dynamics are obtained as before by calculating the characteristic equation from

$$\det(s\mathbf{I} - \mathbf{A}) = 0$$

which yields

$$s^2 + 5s + 4 = 0$$

which could have been calculated more simply from the denominator of the transfer function. This reveals the poles of the sysem are located at $s = -1$ and $s = -4$. The dominant pole thus has a time constant of 1 s. To provide observer dynamics five times faster than that of the system and to define the damping to be critical, we place both the observer poles at $s = -5$, as shown in Fig. SP25.2.1. This time, we will use the comparison of characteristic equations method to determine the required feedback gain vector. The matrices needed are

$$\tilde{\mathbf{A}} = \overline{\mathbf{A}}' = \begin{bmatrix} 0 & -4 \\ 1 & -5 \end{bmatrix}$$

$$\tilde{\mathbf{b}} = \overline{\mathbf{C}}' = \begin{bmatrix} 2 \\ 1 \end{bmatrix}$$

Referring to Module 24, the matrix

$$\tilde{\mathbf{b}}\mathbf{g}' = \begin{bmatrix} 2 \\ 1 \end{bmatrix} [g_1 \quad g_2] = \begin{bmatrix} 2g_1 & 2g_2 \\ g_1 & g_2 \end{bmatrix}$$

Hence the matrix

$$s\mathbf{I} - \tilde{\mathbf{A}} + \tilde{\mathbf{b}}\mathbf{g}' = \begin{bmatrix} s + 2g_1 & 4 + 2g_2 \\ -1 + g_1 & s + 5 + g_2 \end{bmatrix}$$

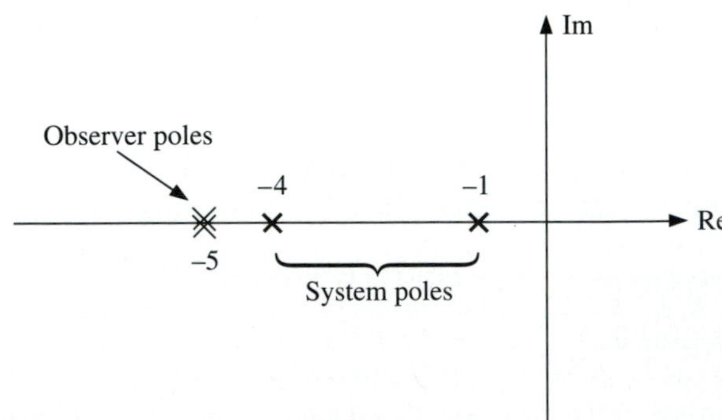

Fig. SP25.2.1

The characteristic equation of the observer is obtained from

$$\det(s\mathbf{I} - \tilde{\mathbf{A}} + \tilde{\mathbf{b}}\mathbf{g}') = 0$$

yielding

$$s^2 + s(5 + g_2 + 2g_1) + (4 + 6g_1 + 2g_2) = 0$$

Comparing the coefficients of this equation with the desired observer characteristic equation

$$(s + 5)^2 = s^2 + 10s + 25 = 0$$

gives two simultaneous linear equations in g_1 and g_2, which when solved yield

$$\mathbf{g} = \mathbf{K} = \begin{bmatrix} 5.5 \\ -6.0 \end{bmatrix}$$

This is the observer gain vector for the dynamics required.

Sample Problem 25.3

An open-loop system is described by the equations

$$\dot{\mathbf{x}} = \begin{bmatrix} -1 & 0 & 3 \\ 1 & -4 & 0 \\ 0 & 0 & -2 \end{bmatrix} \mathbf{x} + \begin{bmatrix} 1 \\ 2 \\ 5 \end{bmatrix} u$$

and

$$y = [4 \quad 2 \quad 3]\,\mathbf{x}$$

It is known that the system state variables cannot be measured directly.

1. Calculate the open-loop eigenvalues of the system.
2. Design an observer, the dynamics of which have a response time 10 times faster than that of the open-loop plant and the dominant roots have a damping ratio of 0.707.
3. Determine the feedback vector for both observer and plant such that the closed-loop system response time is half that of the open-loop system and is also critically damped.

Solution

Again, in this example, the system matrices are known. Since the system is third order, the manual calculation of the various matrices needed to solve this problem is quite laborious. Instead, we shall use this example to illustrate how commercial software can reduce this computational burden considerably. We will use MATLAB to solve this problem.[6]

[6] The commands that follow are available in the professional version of MATLAB and the control systems toolbox. Not all of the commands are available in the student version of MATLAB.

We begin by defining the open-loop system matrices as

```
a = [-1 0 3;1 -4 0;0 0 -2];
b = [1 2 5]';
c = [4 2 3];
d = 0;
```

To check whether an observer may be constructed, the rank of the observability matrix is calculated:

```
r = rank(obsv(a,c)) ⇒ r = 3
```

The notation above is that the command on the left produces the response on the right of the ⇒ symbol. To determine if the closed-loop poles may be placed in an arbitrary pattern, the system controllability is checked with

```
r = rank(ctrb(a,b)) ⇒ r = 3
```

Since the system is both controllable and observable, we may proceed with the design. The open-loop eigenvalues are found using

```
eol = eig(a) ⇒ eol = [-4 -1 -2]'
```

The dominant pole is located at $s = -1$. The observer poles will then be located on the s plane as shown in Fig. SP25.3.1. The desired observer poles are contained in the vector dpo where

```
dpo = [-10 + 10*i -10 -10*i -20];
```

Note that the third observer pole is located on the real axis at twice the distance from the imaginary axis as the dominant complex pair. The **K** vector for the observer feedback path may be found from equations 25.13 and 25.14 using Ackermann's formula by entering

```
k = acker(a',c',dpo) ⇒ k = [-123.4 78.7 123.0]
```

The closed-loop system poles will be located according to Fig. SP25.3.2 and are contained in the vector

```
dpp = [-2 -2 -10]
```

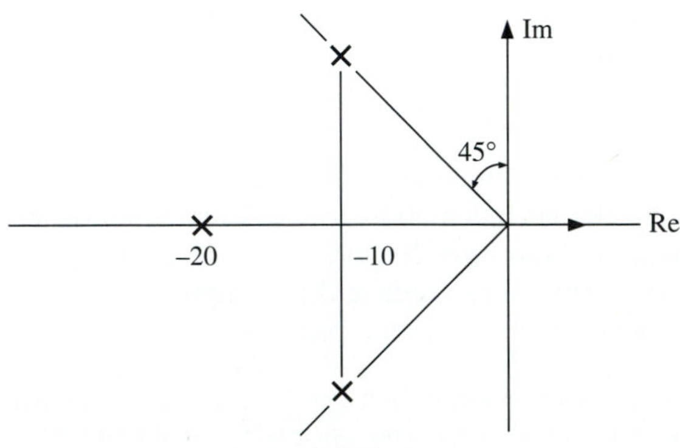

Fig. SP25.3.1

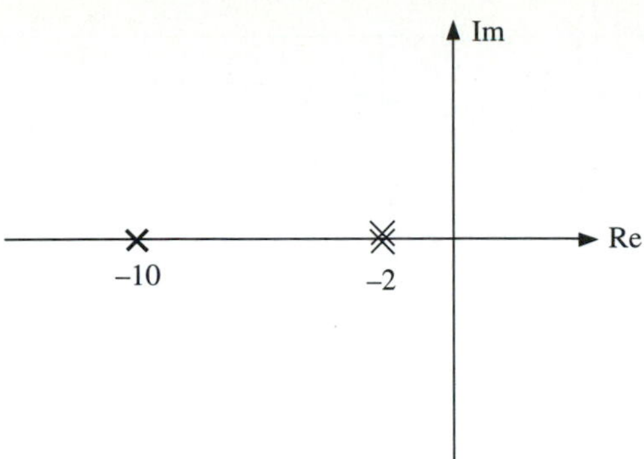

Fig. SP25.3.2

Again, Ackermann's formula can be used to calculate the required feedback gain vector **g** from

```
g = acker(a,b,dpp) ⇒ g = [-0.133 0.960 1.043]
```

This completes the design of both observer and feedback path, and the system will take the form shown in Fig. 25.9. It is instructive, however, to observe, through simulation, the behavior of the system to a step input. Beginning with the open-loop system, the step response may be obtained by first defining a suitable time vector over which the response is required, say $0 < t < 10$,

```
t = [0.0:0.05:10.0]
```

and then entering

```
yol = step(a,b,c,d,1,t)
```

The parameter 1 in the above command specified to which of the *n*th inputs the system is to respond. Issuing the command `plot(t,yol)` produces the open-loop system step response shown in Fig. SP25.3.3. We now return to the observer. The observer system matrix is obtained from equation 25.10 as

```
ao = a - k*c
```

which yields a 3×3 matrix. Typing

```
eob = eig(ao) ⇒ eob = [-20 -10+10i -10-10i]
```

confirms the observer dynamics. To obtain the closed-loop system dynamics, we use equation 25.31, which is obtained from

```
acl = [a-b*g'  b*g'; zeros(a)  a-k*c];
bcl = [b;0;0;0];
ccl = [c 0 0 0];
dcl = d
```

This generates a 6×6 array for `acl`, the eigenvalues of which are found from

```
ecl = eig(acl) ⇒ ecl = [-2 -2 -10 -20 10+10i -10-10i]
```

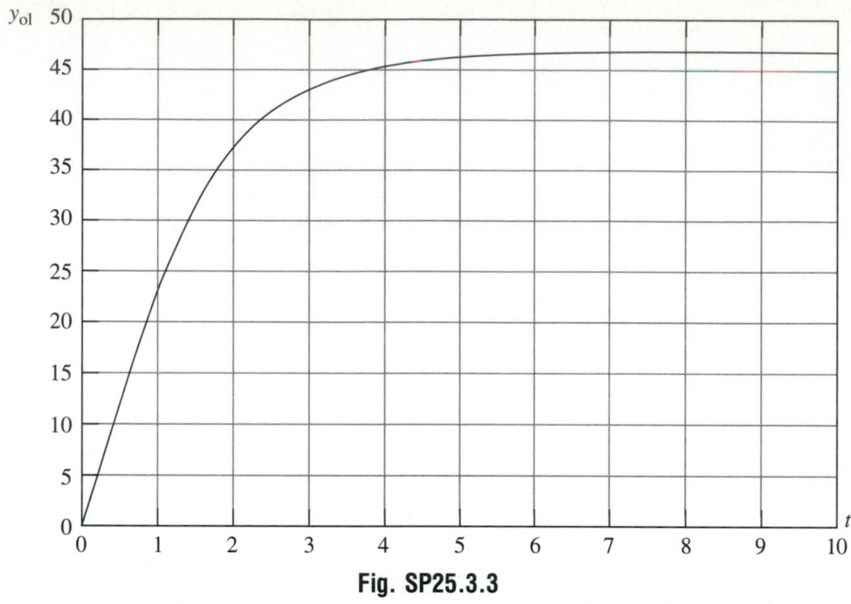

Fig. SP25.3.3

This indicates that the eigenvalues of the closed-loop system are simply the sum of those of the plant and the observer. This will always be the case if the system matrices are known exactly.[7] The step response of the closed-loop system may be generated by

```
ycl = step(acl,bcl,ccl,dcl,1,t); plot(t,ycl)
```

producing the result shown in Fig. SP25.3.4. Note that the steady-state output is smaller than in the case of the open-loop system, but according to

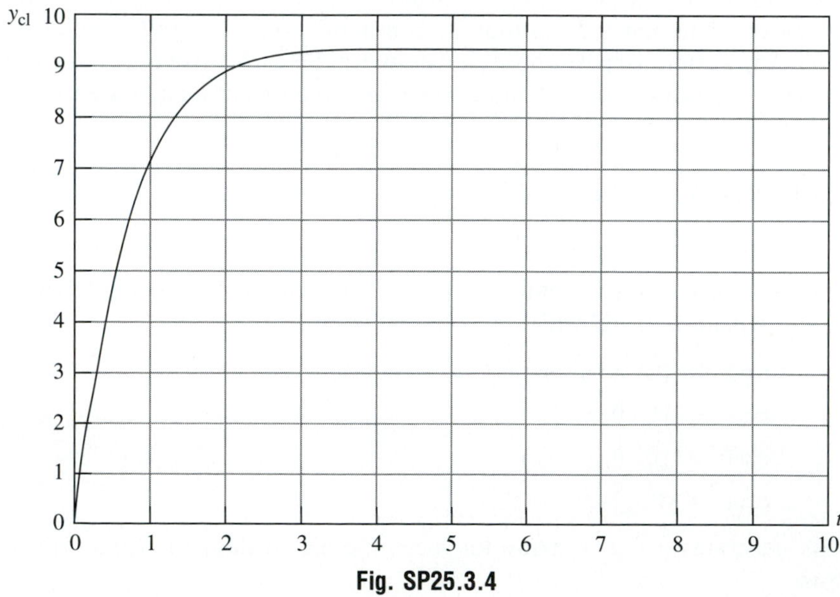

Fig. SP25.3.4

[7] This is called the separation principle, though not proven rigorously here.

equation 25.5, this would be expected. The dynamics of the system meet the specifications. Finally, the error between the actual system state and that estimated by the observer can be obtained from

```
[y,x] = step(acl,bcl,ccl,dcl,1,t);
```

since this will return the state vector time history in **x**. Examining the second set of three states, namely **e**, reveals that all elements are zero for all time! This suggests that the estimates are exactly the true states. This is only true if the initial conditions of the plant and observer are the same, which they are in this case (all zero).

We may simulate the system to show how quickly the estimates of the state variables converge to the true values. To do this, we look at the response of the system to zero input r, but with initial conditions present. We see that a measure of the error $\mathbf{e} = \mathbf{x} - \overline{\mathbf{x}}$ may be observed at the output, since

$$y_e = y - \overline{y} = \mathbf{C}(\mathbf{x} - \overline{\mathbf{x}}) = \mathbf{Ce}$$

If we assume initial conditions for $\mathbf{x}$ to be [1 1 1], while those for $\overline{\mathbf{x}}$ are zero, then e will indicate the difference between them. However, since $\mathbf{e}$ is a vector, we will plot the scalar output y_e from the above equation. This is shown in Fig. SP25.3.5. Also shown in the figure is the same output but for $\mathbf{K} = \mathbf{0}$. This is the case where the dynamics of the observer are equal to those of the open-loop plant and indicate the longer time it takes for the observer to track the plant. The ability to adjust the observer dynamics is apparent.

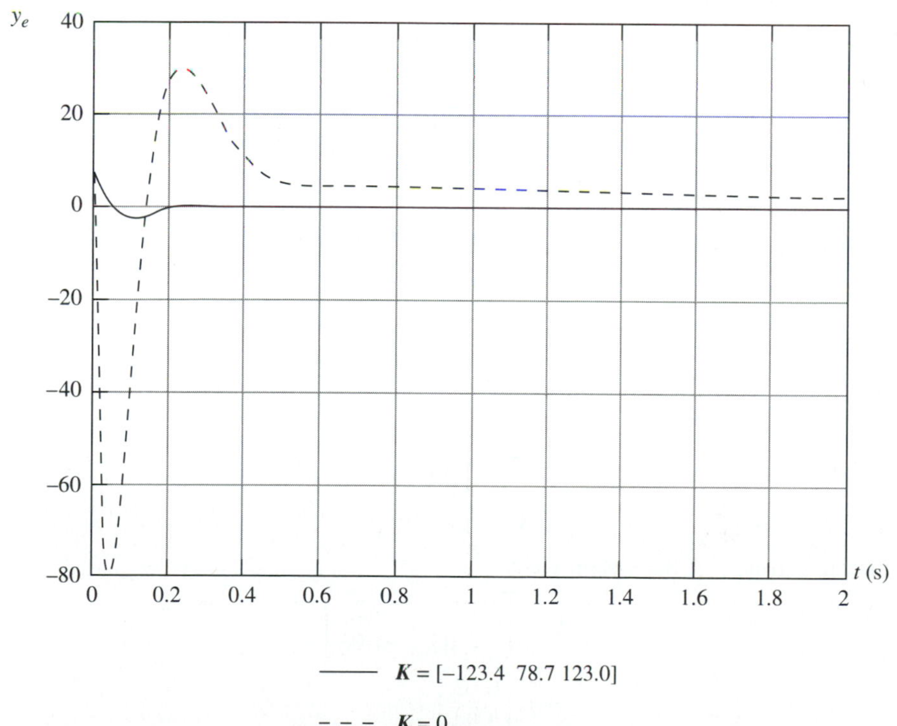

Fig. SP25.3.5

◆ PROBLEMS

25.1 For the system

$$\dot{\mathbf{x}} = \begin{bmatrix} 1 & 0 & 1 \\ 1 & -1 & 2 \\ 3 & 0 & -2 \end{bmatrix} \mathbf{x} + \begin{bmatrix} 0 \\ 0 \\ 3 \end{bmatrix} u$$

and

$$y = [3 \quad 4 \quad 3]\,\mathbf{x}$$

determine whether the closed-loop system poles may be placed in arbitrary locations and if an observer may be constructed that will predict the system states.

25.2 Determine which of the following systems may have observers designed for them:

$$\frac{y}{u} = \frac{2}{s(s+1)}$$

$$\frac{y}{u} = \frac{s+1}{(s+2)(s+3)}$$

$$\frac{y}{u} = \frac{10(s+2)}{s^2(s+1)}$$

$$\frac{y}{u} = \frac{12(s^2+6s+8)}{s^3+13s^2+54s+72}$$

25.3 Design an observer for the system described by the matrices

$$\mathbf{A} = \begin{bmatrix} 2 & -1 \\ 0 & 2 \end{bmatrix}$$

$$\mathbf{b} = \begin{bmatrix} 2 \\ 3 \end{bmatrix}$$

$$\mathbf{C} = [1 \quad 1]$$

such that the observer poles satisfy the characteristic equation

$$s^2 + 3s + 1 = 0$$

25.4 A system is actually described by

$$\overline{\mathbf{A}} = \begin{bmatrix} 1 & 2 \\ 0 & -1 \end{bmatrix}$$

$$\overline{\mathbf{b}} = \begin{bmatrix} 0 \\ 2 \end{bmatrix}$$

$$\overline{\mathbf{C}} = [2 \quad 2]$$

while an estimate of the system yields

$$\mathbf{A} = \begin{bmatrix} 1.1 & 2.2 \\ -0.05 & -1.02 \end{bmatrix}$$

$$\mathbf{b} = \begin{bmatrix} 0.03 \\ 1.94 \end{bmatrix}$$

$$\mathbf{C} = [1.94 \quad 1.97]$$

Design an observer so that the response is critically damped with a 2% settling time of 0.2 s. Calculate the eigenvalues of the observer and the actual system.

25.5 Consider the system described by

$$\dot{\mathbf{x}} = \begin{bmatrix} -4 & 2 \\ 0 & 3 \end{bmatrix} \mathbf{x} + \begin{bmatrix} 2 \\ 5 \end{bmatrix} u$$

and

$$y = [2 \quad 3]\,\mathbf{x}$$

Design an observer for the system such that the closed-loop poles of the observer have half the response time as the plant and at the same time have a damping ratio $\zeta = 0.5$.

25.6 The system states for the following system are not physically measurable:

$$\dot{\mathbf{x}} = \begin{bmatrix} -5 & 7 & 6 \\ 0 & -2 & 8 \\ 0 & 4 & 0 \end{bmatrix} \mathbf{x} + \begin{bmatrix} 2 \\ 2 \\ 3 \end{bmatrix} u$$

and

$$y = [1 \quad 2 \quad 4]\,\mathbf{x}$$

Design an observer that will provide the system states and in which the dominant poles have a time constant of 0.2 s and a damping ratio $\zeta = 0.6$.

25.7 An open-loop system is described in state-space form by

$$\dot{\mathbf{x}} = \begin{bmatrix} -1 & 0 \\ 2 & 3 \end{bmatrix} \mathbf{x} + \begin{bmatrix} 2 \\ 1 \end{bmatrix} u$$

and

$$y = [0 \quad 2]\,\mathbf{x}$$

Design an observer that has a dominant time constant that is 20% that of the plant. Then evaluate a feedback vector for the open-loop system such that the closed-loop poles satisfy the characteristic equation

$$s^2 + 4s + 10 = 0$$

Determine the eigenvalues of the closed-loop plant.

25.8 In the following system, the state variables are not measurable:

$$\dot{\mathbf{x}} = \begin{bmatrix} -2 & 1 \\ 0 & -3 \end{bmatrix} \mathbf{x} + \begin{bmatrix} -2 \\ 1 \end{bmatrix} u$$

and

$$y = [1 \quad -5]\,\mathbf{x}$$

Design a feedback gain vector that will give the closed-loop system a 2% settling time of 0.5 s and a damping ratio $\zeta = 0.8$. Design an observer to provide the state variables such that its dynamic response is 50% faster than the open-loop plant and is critically damped. Simulate the open- and closed-loop unit step response of the plant.

25.9 For the open-loop system described by the transfer function

$$\frac{y}{u} = \frac{10(s + 0.5)}{s(s + 2)(s + 5)}$$

design an observer such that all of its poles are located on the complex plane at $s = -1$. What feedback gain vector is required to place the dominant closed-loop eigenvalues such that the system has peak time of 0.1 s and $\zeta = 0.3$? Plot the time response of the error vector $(\mathbf{x} - \overline{\mathbf{x}})$ when the initial conditions on the system state are [1 1 1]′ and zero for the observer.

25.10 A system is described by the following state-space matrices:

$$\mathbf{A} = \begin{bmatrix} -1 & -1 & 0 & 2 \\ 1 & -2 & 0 & 0 \\ 0 & 0 & -6 & 2 \\ 0 & 7 & -1 & -1 \end{bmatrix}$$

$$\mathbf{b} = \begin{bmatrix} 2 \\ 3 \\ 0 \\ -1 \end{bmatrix}$$

$$\mathbf{C} = [0 \quad 1 \quad 0 \quad 5]$$

$$\mathbf{D} = 0$$

Examine the controllability and observability of the system. Design an observer to estimate the system state such that its dominant poles satisfy the characteristic equation

$$s^2 + s + 4 = 0$$

Calculate the feedback gain vector that places the dominant closed-loop plant poles so that they satisfy the equation

$$s^2 + 2s + 6 = 0$$

If the input r is a unit step, show how all plant state variables and the output y vary with time.

Wave Energy Absorbtion Device

Figure 26.1 shows the design of a system intended to extract energy from offshore ocean waves. It consists of a prismatic structure about 50 m in length that is free to rotate about a horizontal axis perpendicular to the wave direction. This axis may be considered fixed in space. As shown in Fig. 26.2, the structure, sometimes referred to as a "duck" because of its nodding motion, oscillates about the axis of rotation due to the incident-wave forces on its upstream side. Connected to the axis of the duck is a series of hydraulic pumps that produces a flow of hydraulic fluid proportional to the rotational speed of the duck. This periodic flow is rectified with hydraulic valves, in much the same way as AC current is rectified, and drives a hydraulic motor connected to a generator that produces electricity. Given the following data, we have to investigate the performance of the system and suggest

Fig. 26.1 Wave power device.

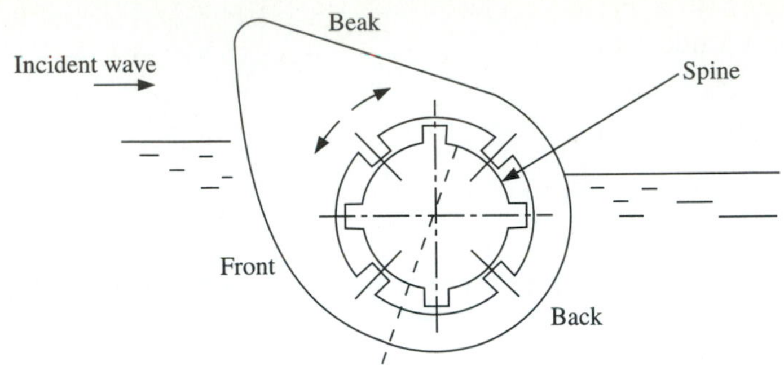

Fig. 26.2 End view of duck.

ways in which feedback control may be used to optimize the overall system efficiency:

1. The period of incident waves averages 8 s, with most waves in the ±20% range.
2. The radius of gyration $\bar{r}$ of the duck is 1.5 m and the mass per unit length is 4000 kg.
3. The system experiences resistive torques opposing rotation of c_n = 100 N-m-s/rad due to water resistance and c_p = 2000 N-m-s/rad due to the power take-off pumps.
4. In order to ensure that the duck faces into the waves and to resist rotation through excessive angles, a torsional spring of stiffness K_s = 1000 N-m/rad connects the duck body to the shaft.

The first job is to determine the response of the duck to the waves, which may be considered as a periodic forcing torque τ. Clearly, we are interested in the frequency response of the system to forcing torques having the same frequency as the waves. The equation of motion of the duck becomes

$$\tau - (c_p + c_n)\dot{\theta} - K_s\theta = I\ddot{\theta} \tag{26.1}$$

Since the value of the duck inertia becomes

$$J = m\bar{r}^2 = 4500 \text{ kg-m}^2 \tag{26.2}$$

the equation of motion becomes

$$4500\ddot{\theta} + 2100\dot{\theta} + 1000\theta = \tau \tag{26.3}$$

which may be written as

$$\ddot{\theta} + 0.47\dot{\theta} + 0.22\theta = \frac{\tau}{4500} \tag{26.4}$$

or in Laplace form

$$\frac{\theta}{\tau} = \frac{1/4500}{s^2 + 0.47s + 0.22} \tag{26.5}$$

This is a second-order system, and its frequency response, or Bode diagram, may be found from Module 15 (Fig. 15.17) by calculating the following parameters:

$$\omega_n = \sqrt{0.22} = 0.47 \text{ rad/s} \tag{26.6}$$

and

$$2\zeta\omega_n = 0.47 \rightarrow \zeta = 0.5 \tag{26.7}$$

The frequency response of the duck is shown in Fig. 26.3. Calculating the frequency of the incident waves, we find that

$$0.65 < \omega_{\text{wave}} < 0.98 \tag{26.8}$$

with the average wave frequency at $\overline{\omega} = 0.75$ rad/s. Since this is nearly twice the natural frequency of the duck structure, it is seen that the duck is not responsive to the wave frequencies encountered. In fact, at $\omega = 0.75$, the response is -8 db down, indicating that the response is only 40% of that due to lower frequencies.[1] This means that because the amplitude of rotation of the duck is small, its velocity will also be small, producing less flow of hydraulic fluid from the pumps coupled to the duck, hence producing less power. We will try to modify the duck dynamics using feedback in order to make it more responsive.

Basically, we want to increase the bandwidth of the system, enabling it to respond to the wave frequencies encountered, and this may be done by changing ω_n, which in turn means changing either the system inertia or the system stiffness. We will select the latter course of action. We will choose to place the new natural frequency at $\omega_n = 1.0$ rad/s in order to accommodate all wave frequencies of interest. Since

$$\omega_n = \sqrt{\frac{K}{J}} = 1.0 \text{ rad/s} \tag{26.9}$$

the spring stiffness has to be increased to

$$K = \omega_n^2 J = 4500 \text{ N-m/rad} \tag{26.10}$$

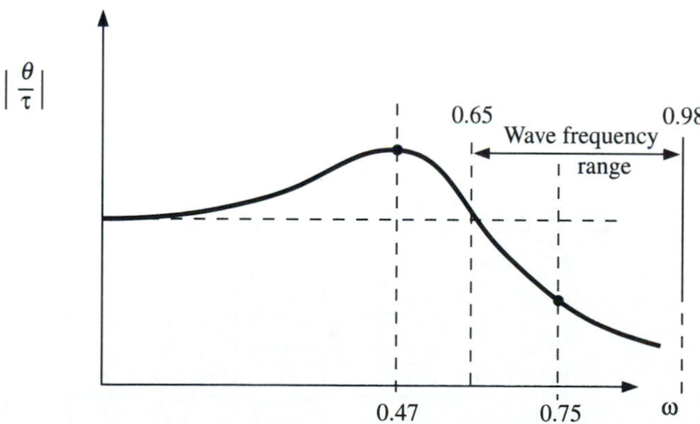

Fig. 26.3 Open-loop frequency response of system.

[1]Since $20 \log_{10} M = -8$, $M = 0.4$.

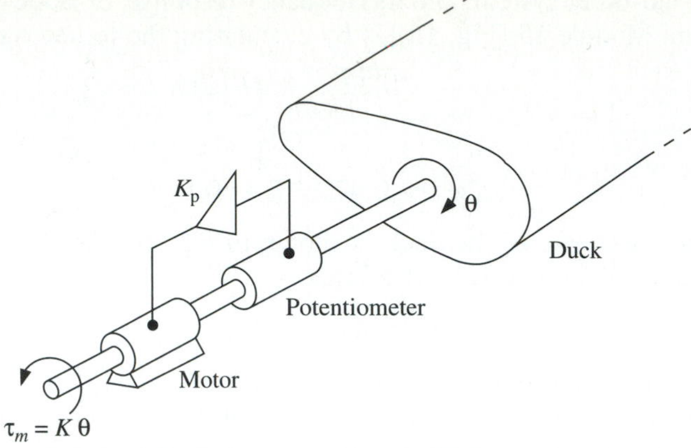

Fig. 26.4 Modifying system stiffness using feedback.

This may be done using feedback as shown in Fig. 26.4, where a potentiometer senses duck rotation. This voltage is amplified by a gain K_p, and then the signal is used to drive a motor of torque constant $K_T = 2000$ N-m/V in the *opposite* direction to that of duck rotation. The new equation of motion of the system with feedback becomes

$$\tau - (c_n + c_p)\dot{\theta} - K_s\theta - K_pK_T\theta = J\ddot{\theta} \tag{26.11}$$

or

$$J\ddot{\theta} + (c_n + c_p)\dot{\theta} + (K_s + K_pK_T)\theta = \tau \tag{26.12}$$

The corresponding block diagram of the system is shown in Fig. 26.5. Since we require the total spring stiffness to be

$$K_s + K_pK_T = 4500 \text{ N-m/rad} \tag{26.13}$$

we obtain the required feedback gain $K_p = 1.75$. Note that now the transfer function from torque to rotation is

$$\frac{\theta}{\tau} = \frac{1/4500}{s^2 + 0.47s + 1} \tag{26.14}$$

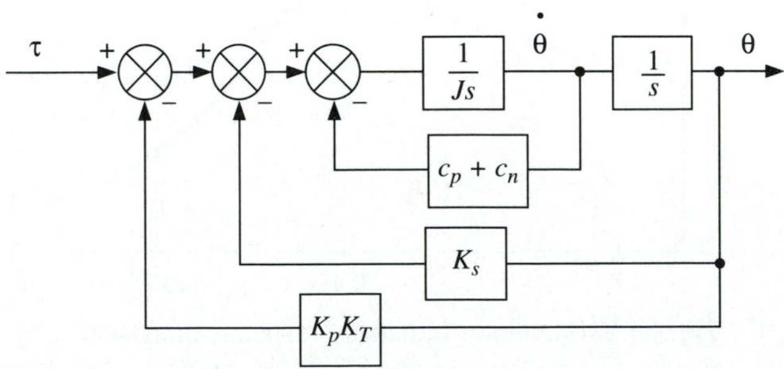

Fig. 26.5 Modified block diagram.

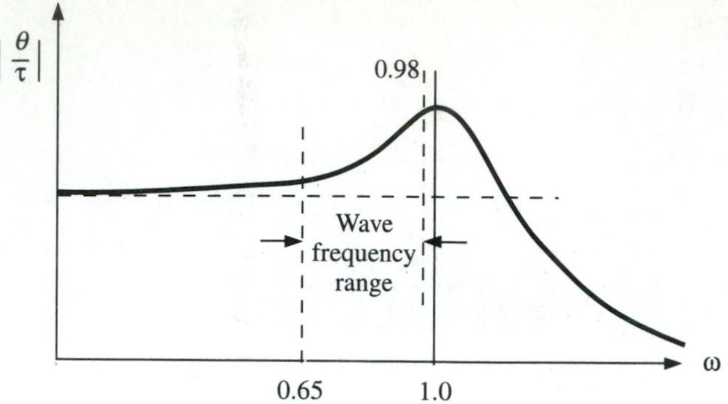

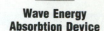

Fig. 26.6 Modified Bode diagram.

The undamped natural frequency is unity, as required, but the damping ratio has become $\zeta = 0.23$. The Bode diagram of the modified system is as shown in Fig. 26.6. If the resonance peak in Fig. 26.6 is objectionable (for example, it may cause unacceptably large oscillations of the duck), additional damping may be provided by using velocity feedback from a tachometer mounted in the rotational axis of the duck.

MODULE 27

Missile Attitude Controller

Suppose we are to design a control system that will ensure a missile achieves a particular angle with respect to a vertical line, as shown in Fig. 27.1. Such a system may be used to guide the missile on a particular trajectory. Adjustment of the attitude is achieved by means of small thruster jets that apply a force normal to the axis of the missile. The location of the normal force is fixed by the location of the thrusters, but the force exerted may be controlled. The main missile rocket exerts a constant thrust T along the missile axis. Assume the mass of the missile remains constant and the effects of gravitational and aerodynamic forces may be neglected.

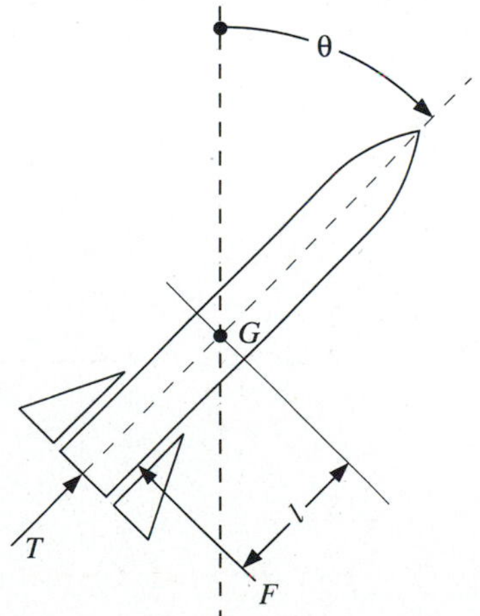

Fig. 27.1 Missile attitude control problem.

The longitudinal motion of the missile is determined by the components of T in the horizontal and vertical directions and is of no interest to us at the moment. The attitude of the missile is determined by the moments about the center of gravity G. Newton's second law about the mass center G yields

$$\sum M_G = Fl = J\ddot{\theta} \tag{27.1}$$

where J is the moment of inertia of the missile about G. Suppose that at $t = 0$, $\theta = \theta_0$, and we wish the attitude to be $\theta = \theta_d$. How are the various system parameters, θ, θ_0, θ_d, F, J, etc., related? Taking the Laplace transform of the equation of motion reveals

$$Fl = J(s^2\theta - s\theta_0) \tag{27.2}$$

This may be written as

$$Js^2\theta = Fl + sJ\theta_0 \tag{27.3}$$

resulting in

$$\theta = \frac{l}{Js^2}F + \frac{1}{s}\theta_0 \tag{27.4}$$

This concludes the modeling phase of the problem.

◆ BLOCK DIAGRAM REPRESENTATION

The relationship between the missile attitude and the two parameters F and θ_0 may be represented by the partial block diagram shown in Fig. 27.2. This diagram shows that the attitude angle depends on two inputs: (1) the initial angle θ_0 and (2) the normal force F. It is now necessary to decide upon an appropriate control strategy to define how the thruster force F is to be adjusted. Based upon our knowledge of the way feedback works, a suitable approach would be to make F proportional to the error between the desired and actual missile angle. The control equation may be written as

$$F = K(\theta_d - \theta) \tag{27.5}$$

In order to check that we have the correct expression, suppose that $\theta_0 = 30°$ and $\theta_d = 0$, indicating that it is necessary to return the missile to the vertical position.

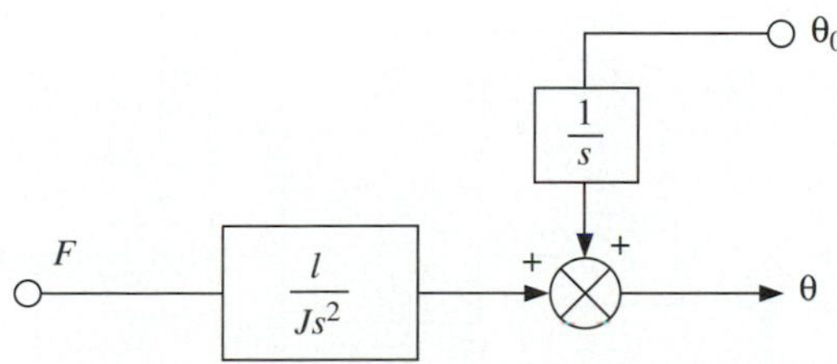

Fig. 27.2 Missile dynamics block diagram.

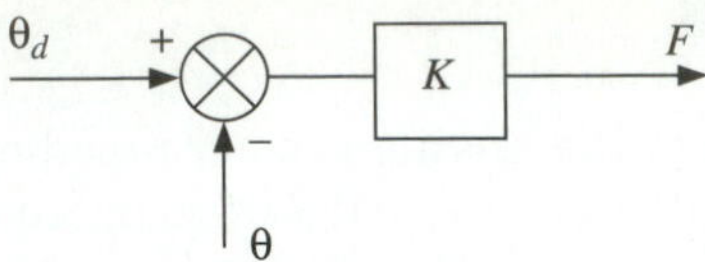

Fig. 27.3 Controller block diagram.

This equation yields $F = -30K$, which, from Fig. 27.1, suggests a moment about G that would indeed produce the motion required. The control equation is, therefore, correct. The control equation may also be represented in block diagram form, as shown in Fig. 27.3. Note that simple proportional control is tried first, since it is the least expensive and complicated controller to build. If it works, it will be an acceptable solution. Putting the two block diagrams together produces the complete block diagram of the system, as shown in Fig. 27.4. The closed-loop transfer function may be written as

$$\theta = \frac{\theta_0}{s} + \frac{l}{Js^2}[K(\theta_d - \theta)] \tag{27.6}$$

$$\theta\left(1 + \frac{Kl}{Js_2}\right) = \frac{\theta}{s} + \frac{Kl}{Js^2}\theta_d \tag{27.7}$$

$$\theta\left(s^2 + \frac{Kl}{J}\right) = s\theta_0 + \frac{Kl}{J}\theta_d \tag{27.8}$$

resulting in

$$\theta = \frac{s}{s^2 + Kl/J}\theta_0 + \frac{Kl/J}{s^2 + Kl/J}\theta_d \tag{27.9}$$

The closed-loop response is clearly dependent upon both the initial condition θ_0 and the demanded attitude angle θ_d. The response to either input is determined from the characteristic equation

$$1 + GH(s) = 1 + \frac{Kl}{Js^2} = 0 \tag{27.10}$$

which results in

$$s^2 + \frac{Kl}{J} = 0 \tag{27.11}$$

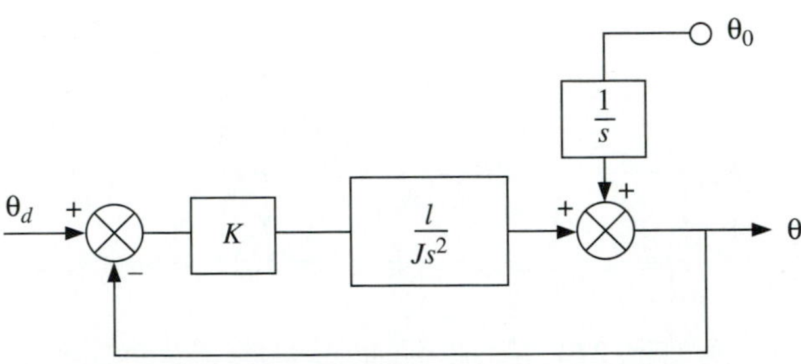

Fig. 27.4 Attitude control system block diagram.

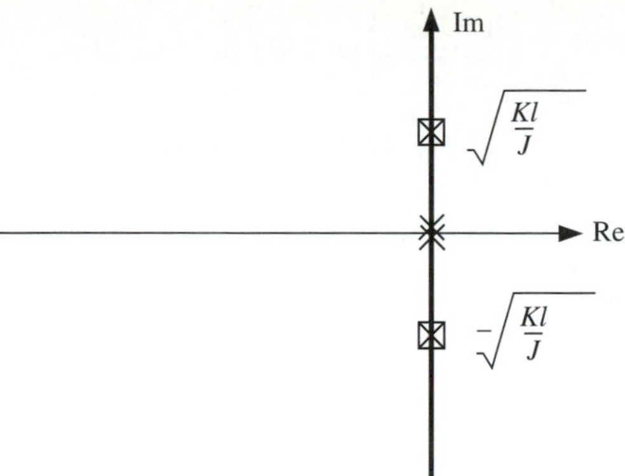

Fig. 27.5 Root locus of attitude control system.

This equation gives two closed-loop poles at $s = \pm\sqrt{Kl/J}$, which, due to their location on the imaginary axis, would give an oscillatory response with zero damping. This is also apparent from the root locus shown in Fig. 27.5. We will consider a particular case, known as pitch-over. This occurs as soon as the missile is launched from a vertical tube and is required to rotate to some finite angle θ_d. This simplifies the response somewhat since $\theta_0 = 0$. The expected response of a proportionally controlled system, for $\theta_d = 10°$ say, would be as shown in Fig. 27.6. Such response is obviously unacceptable, and improvements to the control system are required. An approach will be taken toward the design of an appropriate controller using a multimode controller.

◆ MULTIMODE CONTROLLER DESIGN

It is apparent that what the simple proportional controller lacks is damping, and a PD controller should remedy this. The form of the controller is shown in Fig. 27.7. With this new controller, the closed-loop system becomes as shown in Fig. 27.8. The open-loop transfer function becomes

$$GH(s) = \frac{l(K_p + K_d s)}{Js^2} = \frac{K_d l(K_p/K_d + s)}{Js^2} \tag{27.12}$$

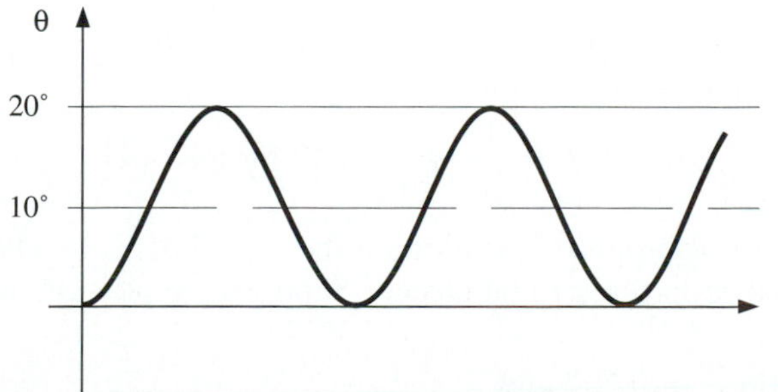

Fig. 27.6 Pitch-over step response of proportionally controlled system.

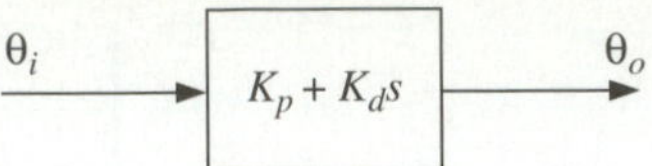

Fig. 27.7 PD controller.

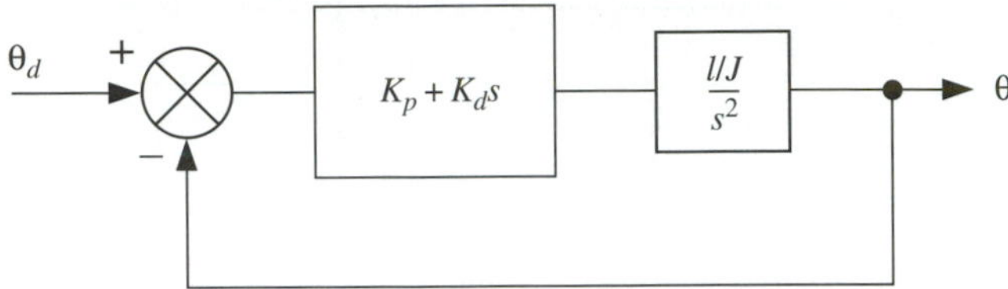

Fig. 27.8 Compensated closed-loop system.

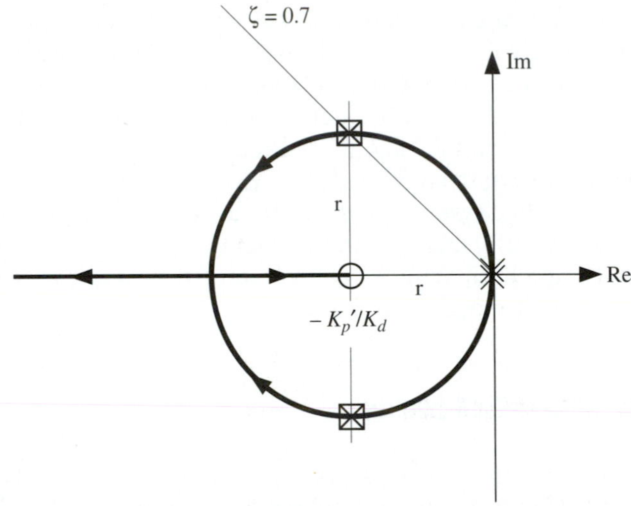

Fig. 27.9 Compensated-system root locus.

The root locus of this system is shown in Fig. 27.9, where the location of the zero is determined by the ratio of the controller gains. MATLAB confirms that the off-axis part of the root locus is circular, making the following design method particularly easy. First we need some data regarding the missile, namely its moment of inertia and length. Suppose the missile is 2 m in overall length and has a mass of 600 kg. Assuming the missile may be represented by a slender cylinder, the value of J may be calculated from

$$J = \frac{1}{12} m L^2 = 200 \text{ kg-m}^2 \tag{27.13}$$

We next define an appropriate location on the complex plane where we wish the dominant closed-loop poles to be placed. Suppose we pick the following design objectives:

1. damping ratio $\zeta = 0.707$ and
2. a time constant of 0.5 s.

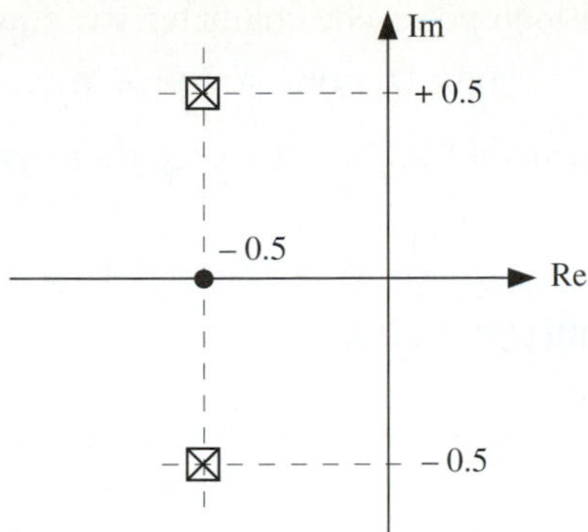

Fig. 27.10 Location of desired closed-loop poles.

This places the closed-loop poles in the locations shown in Fig. 27.10. From the time constant requirement we obtain

$$\tau = 0.5 = \frac{K_p}{K_d} \tag{27.14}$$

The gain at the closed-loop poles may be obtained graphically and equated to the gain of the open-loop transfer function, in Evans form, from equation 27.12:

$$\frac{K_d l}{J} = \frac{(r\sqrt{2})^2}{r} = \frac{2K_p}{K_d} \tag{27.15}$$

resulting in

$$K_d^2 l = 2K_p J \tag{27.16}$$

Solving equations 27.14 and 27.16 for the given data yields

$$K_p = 100 \qquad K_d = 200 \tag{27.17}$$

These are the required controller gains that place the closed-loop poles in the required locations. If the root locus had not been circular, the design would have proceeded in an iterative manner:

1. Estimate K_p and K_d based on a "circular" assumption.
2. Plot the root locus accurately.
3. Find locations on the root locus that satisfy the two design requirements. If none can be found, return to step 1 and change the controller gains.
4. Calculate the gain at the closed-loop poles and, knowing the location of the zero, solve for the two controller gains.

For this case, the closed-loop pole locations may also be determined directly from the characteristic equation

$$1 + GH = 1 + \frac{(K_p + K_d s)l}{Js^2} = 0 \tag{27.18}$$

leading to

$$Js^2 + K_d ls + K_p l = 0 \tag{27.19}$$

For the required closed-loop poles, the characteristic equation must be

$$200s^2 + 200s + 100 = 0 \tag{27.20}$$

Comparing the coefficients of the above two equations yields the same values of controller gains.

◆ STATE SPACE CONTROLLER DESIGN

To begin with, we write the equations of the open-loop system in state space form. The open loop system may be represented as shown in figure 27.11. Since the system is second order, there will be two-state variables, x_1, x_2, corresponding to the physical variables θ, $\dot{\theta}$. We may write

$$\dot{x}_1 = x_2 \tag{27.21}$$

$$\dot{x}_2 = \frac{l}{J}u \tag{27.22}$$

This enables the state equations to be written in the form

$$\begin{bmatrix} \dot{x}_1 \\ \dot{x}_2 \end{bmatrix} = \begin{bmatrix} 0 & 1 \\ 0 & 0 \end{bmatrix} \begin{bmatrix} x_1 \\ x_2 \end{bmatrix} + \begin{bmatrix} 0 \\ l/J \end{bmatrix} u \tag{27.23}$$

The output equation may be written as

$$y = \mathrm{Cx} = [1 \quad 0] \begin{bmatrix} x_1 \\ x_2 \end{bmatrix} \tag{27.24}$$

from which **A**, **B**, and **C** may be identified. MATLAB may be used to check that the system is controllable, using the `ctrb(a, b)` command. This reveals the controllability matrix to be

$$\mathbf{C}_{\mathrm{T}} = \begin{bmatrix} 0 & 0.005 \\ 0.005 & 0 \end{bmatrix} \tag{27.25}$$

the rank of which is 2. The closed-loop poles may be specified using the command

```
dp = [-0.5 + 0.5*i  -0.5 - 0.5*i];        (27.26)
```

and the required gain vector calculated using Ackermann's method from

```
k = acker(a, b, dp)        (27.27)
```

This yields the feedback gain vector as

$$\mathbf{k} = [100 \quad 200] \tag{27.28}$$

Is it coincidence that these feedback gains are the same as the proportional and derivative gains calculated for the multimode controller? Considering the

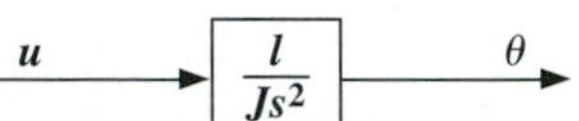

Fig. 27.11 Open-Loop System

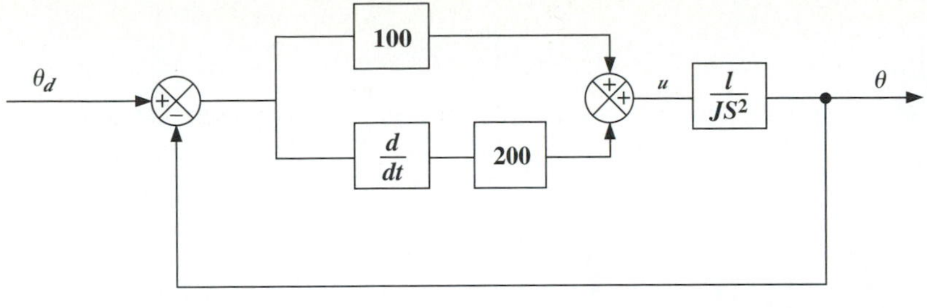

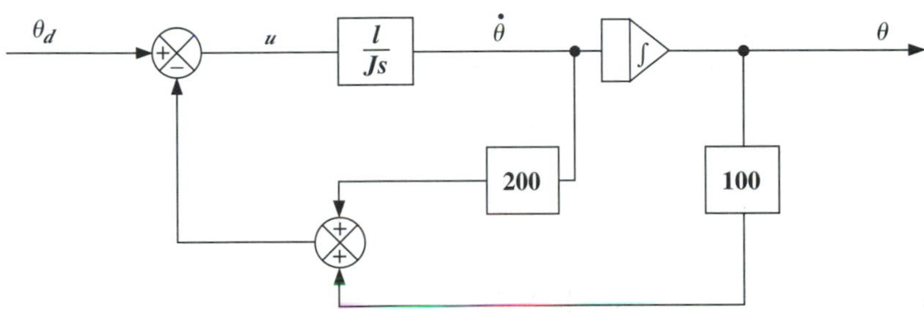

Fig. 27.12 Comparison of Two-Design Approaches

representations of the two approaches shown in figure 27.12, it may be appreciated that they are completely equivalent, and it is not, therefore, surprising that they produce the same results. Note however, that as the order of the open-loop system increases beyond 2, the multimode controller approach will always involve a compromise in performance, while state-space designs are extendable (at least theoretically) to the case where *all* system poles are located as desired, as long as all state variables are measurable or may be estimated.

MODULE 28

Robotic Hand Design

In this problem, we look at the design of a hybrid control system, one that controls both position and force. The application area is focused on a robotic hand, such as that shown in Fig. 28.1. We will study in detail the design of a multijointed finger module, several of which make up a complete hand. In this design, the position of

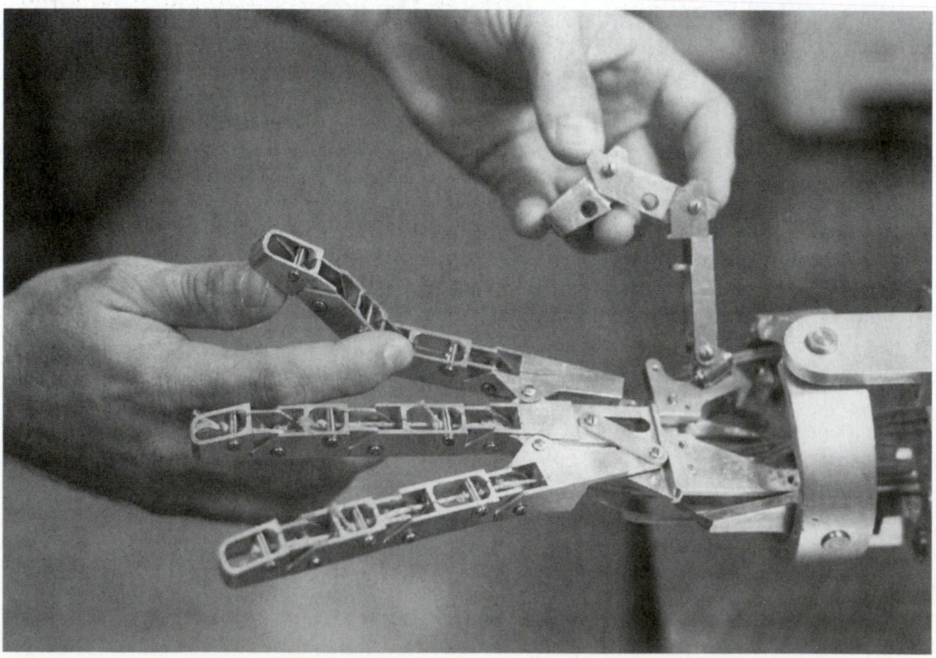

Fig. 28.1 Humanlike robotic hand.

each finger joint is determined either by a computer when the grasping task is automated or by a glove worn by a human that senses the finger joint angles as the human performs the task the robot is to accomplish. Either way, we assume that finger joint angles are available for all values of time. The finger design is known as direct drive in the sense that small gear motors are located in each finger link driving the next joint "downstream" of the joint. The last link has no motor in it, as shown in Fig. 28.2. In particular, we will study the last two links of the finger, where the last link (link 3) is driven by the last motor (motor 3). Positioning of this link by the motor is accomplished by means of simple feedback from a potentiometer mounted on the motor shaft. The motor drives the finger shaft through bevel gears. A compact actuator/sensor unit is illustrated in Fig. 28.3. The control loop is shown in Fig. 28.4, where an additional velocity feedback loop has been added to adjust the damping of the response. It is particularly important in this case to design for not less than critical damping so that no overshoot occurs in the position loop. Consider the task of grasping a light bulb to see why this is necessary! This system is similar to several we have seen already and will be characterized by a second-order system where the dynamics may be set by suitable choice of the forward-path and velocity gains. For the particular hardware used in the finger design, the following data apply:

$$K_p = 5.7 \text{ V/rad} \qquad \text{(potentiometer constant)}$$

$$K_T = 0.2 \text{ N-m/V} \qquad \text{(motor torque constant)}$$

$$J = 1.75 \times 10^{-5} \text{ N-m-s}^2\text{/rad} \qquad \text{(motor inertia)}$$

$$c = 2.0 \times 10^{-5} \text{ N-m-s/rad} \qquad \text{(damping constant)}$$

$$N = 1000 \qquad \text{(reduction gear ratio)}$$

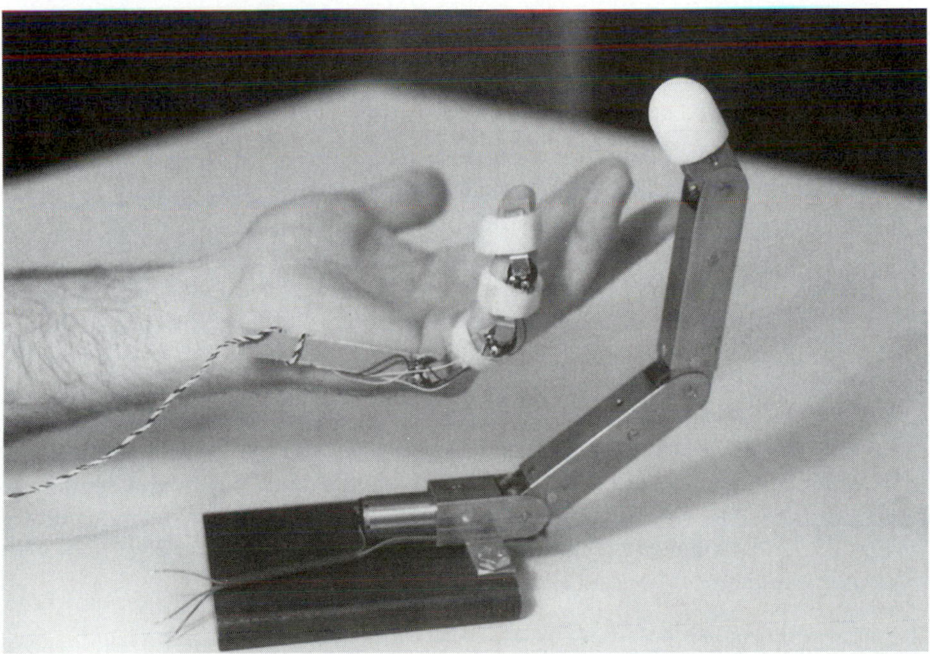

Fig. 28.2 Direct-drive finger design.

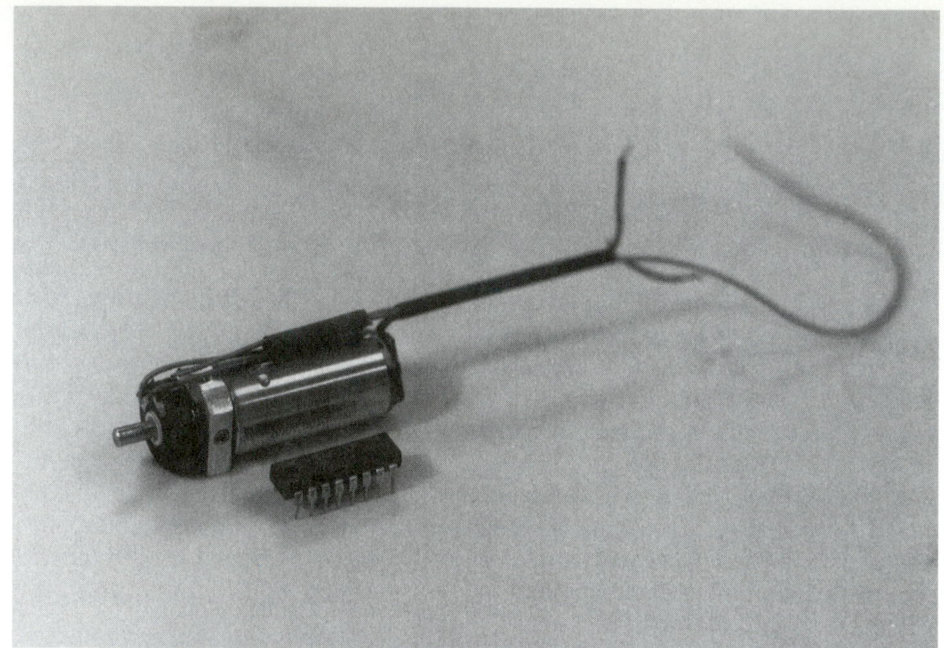

Fig. 28.3 Motor and potentiometer unit.

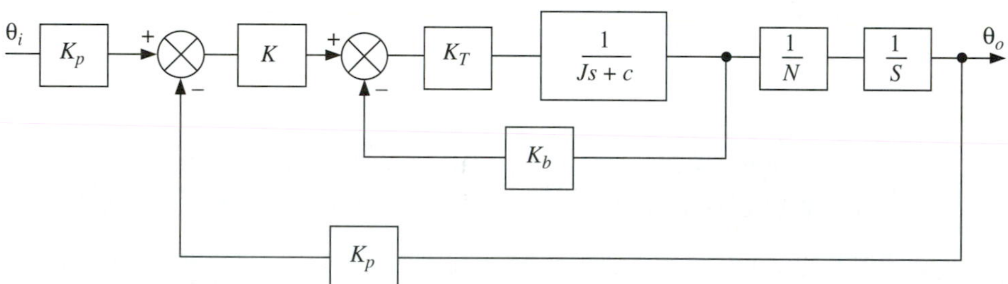

Fig. 28.4 Position control system for finger joint.

In this case, the total inertia seen by the motor is actually

$$J = J_m + \frac{J_l}{N^2} \tag{28.1}$$

where J_m is the motor armature inertia and J_l is the inertia of the link attached to the motor through the gear reduction. Considering the value of the gear ratio, it is reasonable to ignore J_l and set J equal to J_m. The values of K_v and K remain to be determined. The velocity loop may be reduced to the transfer function shown in Fig. 28.5. The reduced block diagram takes the form shown in Fig. 28.6. The closed-loop transfer function becomes

$$\frac{\theta_o}{\theta_i} = \frac{KK_pK_T}{Ns(Js + c + K_TK_b) + KK_pK_T} \tag{28.2}$$

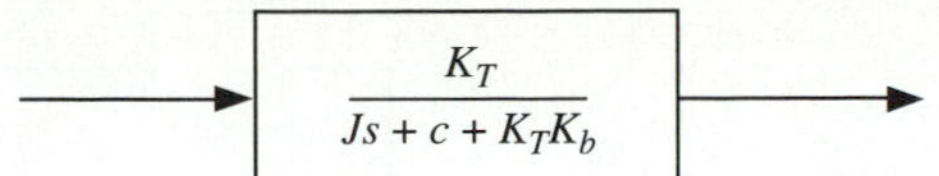

Fig. 28.5 Closure of velocity loop.

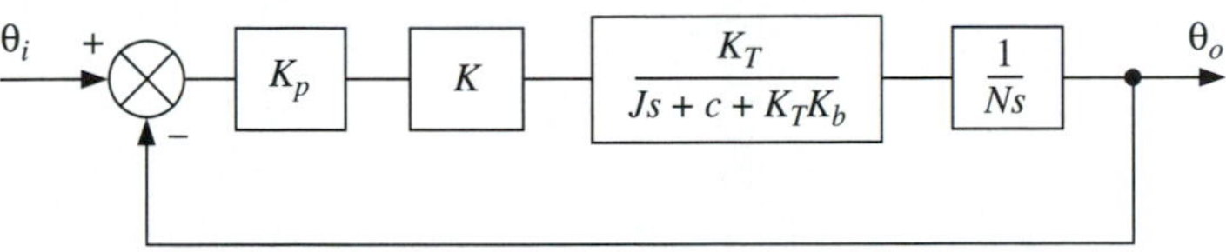

Fig. 28.6 Reduced block diagram.

Equating the characteristic equation to that of a standard second-order system yields

$$JNs^2 + sN(c + K_T K_b) + KK_T K_p = s^2 + 2\zeta\omega_n s + \omega_n^2 \tag{28.3}$$

Substituting in the values of the known parameters and setting $\zeta = 1$ give

$$\omega_n^2 = 65.14K \tag{28.4}$$

$$2\omega_n = 11.4K_b + 1.14 \tag{28.5}$$

Using the equivalent time constant concept for second-order systems from Module 5, we might select

$$\frac{1}{\zeta\omega_n} = 0.5 \tag{28.6}$$

to achieve a finger movement of about one-half second, comparable with a fast human finger movement. This results in $\omega_n = 2$ rad/s, producing $K = 0.06$ and $K_b = 0.25$.

Now consider the additional problem of grasping objects with controlled fingertip forces. When an object is grasped, it exerts a force on the fingertip, as shown in Fig. 28.7. Since this force is equivalent to a moment τ_d about the finger axis, it may be regarded as a disturbing torque, as shown in Fig. 28.8. Note that since τ_d is applied to the finger shaft, i.e., on the load side of the gearbox, the torque ex-

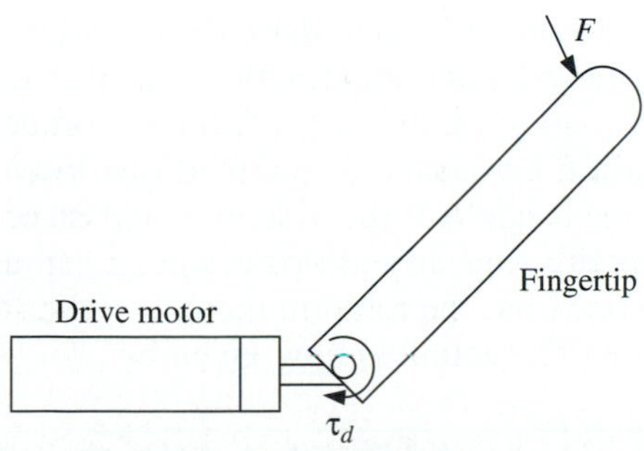

Fig. 28.7 Fingertip force.

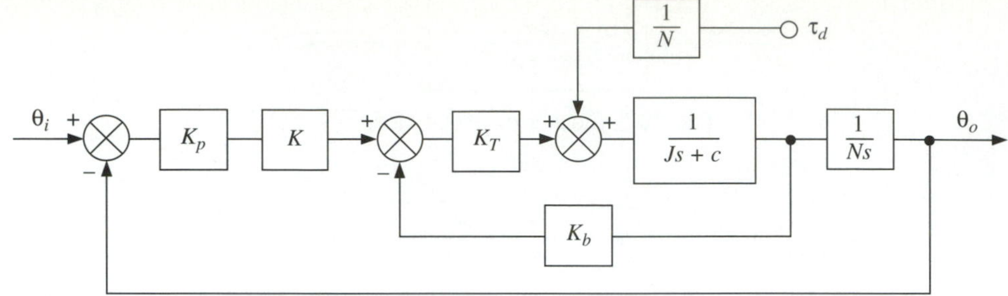

Fig. 28.8 Position and control system with disturbance.

erted on the motor shaft is τ_d/N. The output as a function of the two inputs is determined as

$$\theta_o[Ns(Js + c) + K_TK_bNs + KK_TK_p] = \frac{\tau_d}{N} + KK_TK_p\theta_i \tag{28.7}$$

Considering $\theta_i = 0$ and τ_d to be a unit step, the steady-state angular rotation of the link will be

$$\theta_o = \frac{1}{KK_TK_pN} = 0.0146 \text{ rad} \tag{28.8}$$

This is the natural stiffness of the servo system and is the stiffness that would be felt if the output shaft were grasped and turned. The potentiometer would detect the rotation and generate a torque to restore the position of the shaft to that defined by θ_i.

The question now is how to change this stiffness so that if the finger closes a small amount, large interaction forces are not generated. The stiffness may be defined by

$$S = \frac{\tau_d}{\theta_o} = NKK_TK_p \tag{28.9}$$

This is a function of only one variable, K, which has to be set to 0.06 in order to get the required response time. It seems that with the existing system configuration it will not be possible to design for both position and force control simultaneously. The problem is solved by realizing that if the torque τ_d can be measured in some way, this signal could be used to drive the output θ_d to a larger value than would otherwise be achieved. Functionally, this is achieved in the block diagram shown in Fig. 28.9. In this figure, the signal from the torque-measuring device is multiplied by some gain E and added to the feedback loop at the first summing junction. If a unit torque is applied, the system would respond with a rotation of 0.833° as before, except this time an additional voltage causes a larger rotation θ_o. Since the input is the same but the rotation increases, the stiffness will have decreased. For this system, the output is now given by

$$\theta_o = \frac{1}{Ns(Js + c)}\left(\frac{\tau_d}{N} - K_T[K_bNs\theta_o + K(E\tau_d + K_p\theta_i - K_p\theta_o)]\right) \tag{28.10}$$

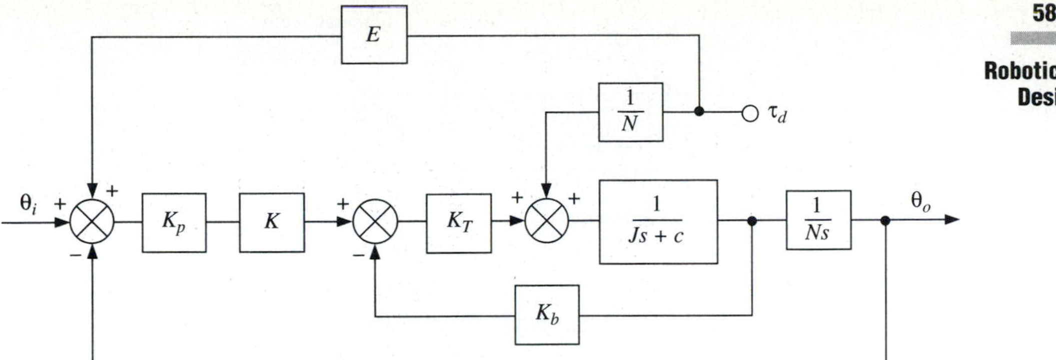

Fig. 28.9 Force feedback for variable stiffness.

Grouping terms produces

$$\theta_o[Ns(Js + c) + NsK_tK_b + KK_TK_p] = \tau_d\left(KK_TE + \frac{1}{N}\right) + KK_TK_p\theta_i \tag{28.11}$$

Again considering $\theta_i = 0$ and applying the final-value theorem for a unit step torque input, we obtain

$$\theta_o = \frac{\tau_d(NKK_TE + 1)}{NKK_TK_p} \tag{28.12}$$

Finally the stiffness is calculated as

$$S = \frac{\tau_d}{\theta_o} = \frac{NKK_TK_p}{NKK_TE + 1} \tag{28.13}$$

For the data given, this reduces to

$$S = \frac{68.4}{1 + 12E} \tag{28.14}$$

For $E = 0$, this reduces to the previous result, equation 28.9. Figure 28.10 shows how the stiffness varies with the gain E. In the final design, the fingertip is con-

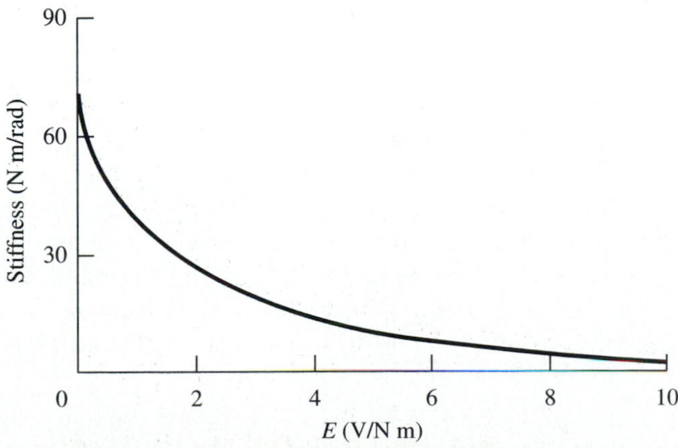

Fig. 28.10 Variation in joint stiffness with gain *E*.

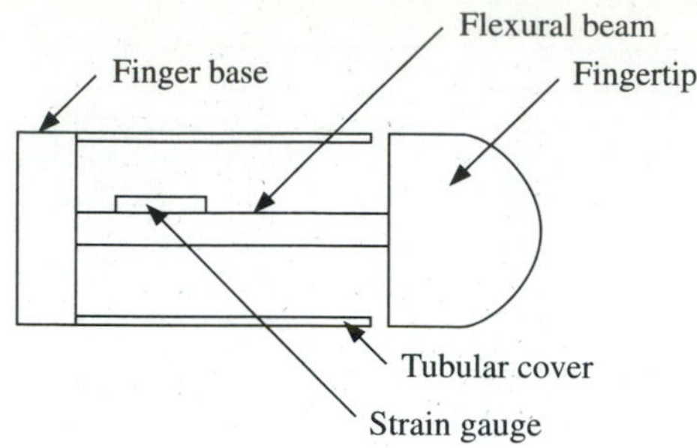

Fig. 28.11 Torque sensor.

nected to the last finger link by means of a cantilever beam to which is attached a strain gauge, as indicated in Fig. 28.11. The applied torque is computed from the strain gauge signal. For each task that the hand performs, the appropriate finger stiffness has to be known. For example, picking up an egg would require a low stiffness compared to grasping a hand tool such as a wrench. The corresponding value of E is selected from Fig. 28.10 and is then set in the control loop. This is an example of *adaptive* control, where system parameters may be adjusted in real time as the control system performs various functions. With the arrangement shown in Fig. 28.9, the hand may perform force- and position-controlled tasks independently.

Pumped Storage Flow Control System

A pumped storage system is a method in which a pump and some storage tanks are used to "even out" the fluctuating demands for water flow by periodically pumping water from a secondary source into the storage tank. Consider such a system shown in Fig. 29.1. In this system, water is drawn off to supply a turbine to produce electrical power. The amount of water flow, q_o, depends upon the fluctuating electrical power requirements of the turbine and generator. A small pump supplies the storage tanks from a secondary water supply, perhaps a small reservoir, when the demand for power, and hence output flow, is small. The two storage tanks are connected together through a shut-off valve, which may be modeled as a linear resistance R_1, while the resistance of the turbine supply pipe may be modeled also as a linear resistance R_2.

In the open-loop mode, the pump could be made to switch on when the water level in each tank is low and switch off when it is nearly full, but this type of control will not give uniform output flow q_o, since this will be governed by the head h_2. We are to design a closed-loop system that will maintain the output flow q_o at some desired value q_d by adjusting the speed of the pump, hence the input flow q_i.

The first stage is to develop an open-loop model of the plant, comprising the two connected tanks, with inflow q_i and outflow q_o. Imagining a control volume

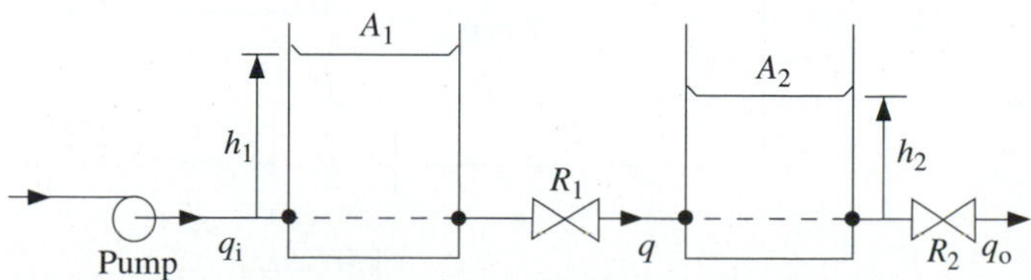

Fig. 29.1 Pump storage system.

drawn around the first tank, the principle of conservation of mass for an incompressible fluid yields

$$q_i - q = A_1\dot{h}_1 \tag{29.1}$$

where q is the flow from tank 1 to tank 2 and A_1 is the cross-sectional area of tank 1. Similarly for tank 2 we obtain

$$q - q_o = A_2\dot{h}_2 \tag{29.2}$$

The equation relating flow and head drop for the linear resistance R_1 is

$$h_1 - h_2 = qR_1 \tag{29.3}$$

and for the other resistance

$$h_2 = q_oR_2 \tag{29.4}$$

assuming that the turbine inlet head is taken as zero. The above equations represent a set of four equations in the five unknowns h_1, h_2, q_i, q, and q_o. It is possible to obtain an expression relating the input and output flows q_i and q_o. The student may verify that after some manipulation we obtain

$$q_i = \ddot{q}_o[A_1A_2R_1R_2] + \dot{q}_o[A_1R_1 + A_2R_2 + A_1R_2] + q_o \tag{29.5}$$

Clearly the open-loop system may be represented as a second-order transfer function, as shown in Fig. 29.2. To construct a closed-loop flow control system using the simplest approach, we might measure the flow q_o and compare it to the required flow q_d and vary the pump speed with this flow error. This proportional system is shown in schematic form in Fig. 29.3. It is assumed that the flow meter dynamics may be ignored; hence its transfer function is unity, but with units of volts per cubic meter per second. Similarly, the demanded flow would be set by a potentiometer also having unity gain and the same units. A variable gain K allows some

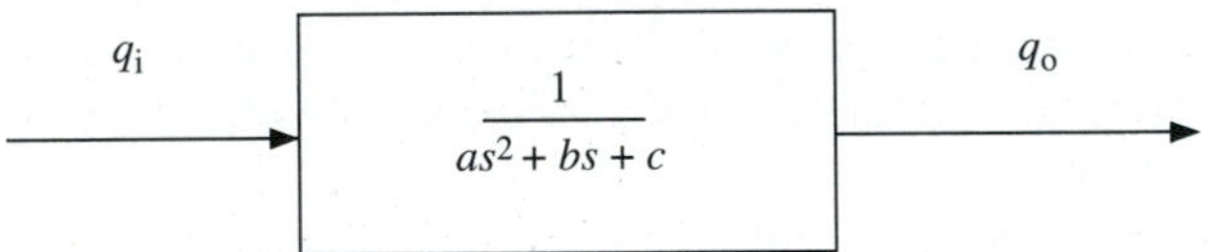

Fig. 29.2 Open-loop transfer function.

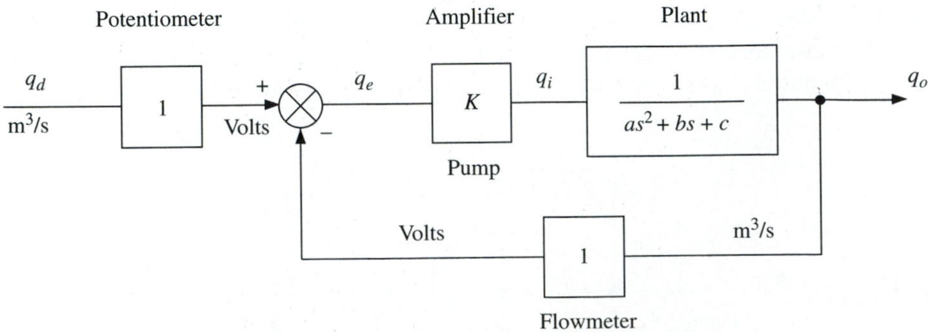

Fig. 29.3 Proportional closed-loop flow control.

measure of adjusting the system performance as well as amplifying the voltage to drive the pump. The gain block K, therefore, represents both the amplifier and pump and will have units of cubic meters per second per volt. The system open-loop transfer becomes

$$GH(s) = \frac{K}{as^2 + bs + 1} \tag{29.6}$$

Suppose the following data apply to the system:

Tank 1 is 2 m in diameter, tank 2 is 3 m in diameter, and the linear resistances have the values $R_1 = 10\ \text{s/m}^2$ and $R_2 = 20\ \text{s/m}^2$.

The characteristic equation of the system becomes

$$71096s^2 + 942.7s + 1 = 0 \tag{29.7}$$

which has the roots $s = -0.0012, -0.0121$. The dominant time constant is 833 s, or about 14 min. From the open-loop transfer function, the system is of type 0 and will be subject to a steady-state error for step inputs. The system response is second order, and the amount of damping may be adjusted through K. The root locus of the proportionally controlled system is shown in Fig. 29.4. In order to eliminate the steady-state error to step changes in demanded flow, a PI controller will be used, as shown in Fig. 29.5. The transfer function of the controller may be written as

$$G_c(s) = K_p\left[\frac{s + K_i/K_p}{s}\right] \tag{29.8}$$

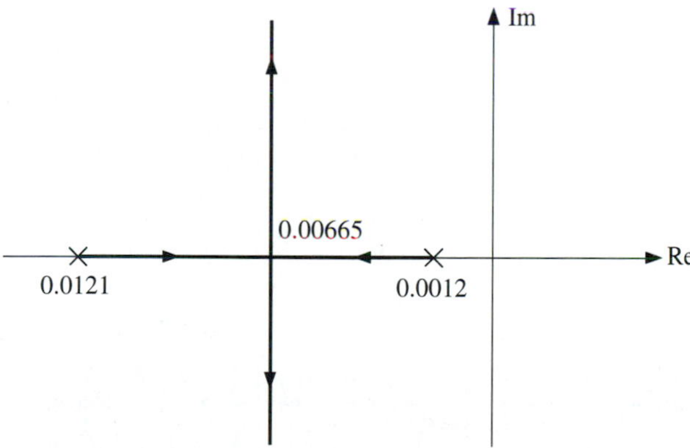

Fig. 29.4 Root locus of proportional-control system.

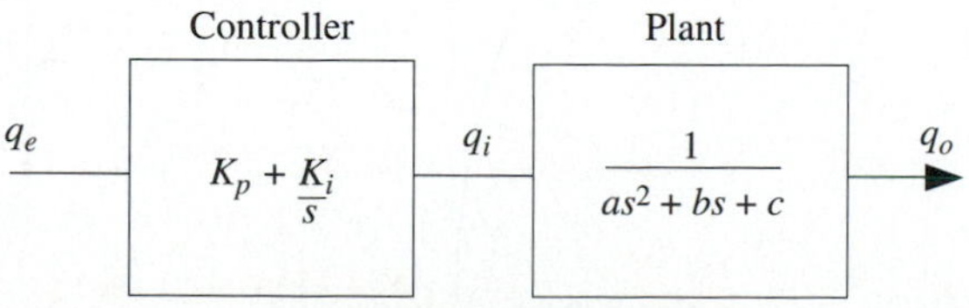

Fig. 29.5 PI controller.

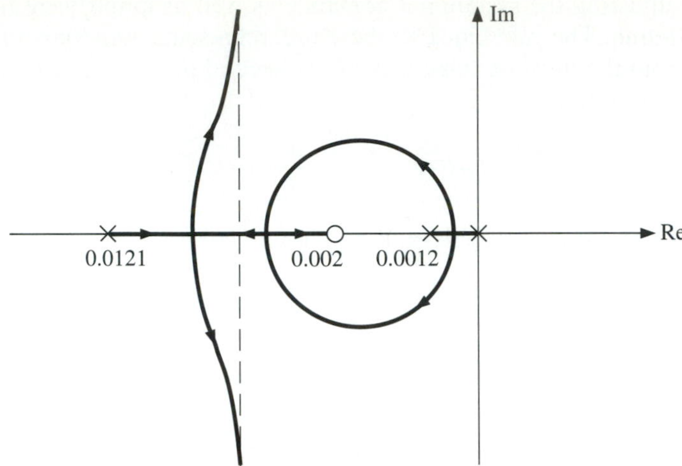

Fig. 29.6 Compensated root locus.

Our compensation strategy will be in two parts. First the zero is placed at $s = -0.002$ so as to make the locus near the origin loop around the zero before branching away from the real axis. The anticipated root locus is shown in approximate form in Fig. 29.6. The second part of the strategy is to make the complex closed-loop poles have a damping ratio $\zeta = 0.707$, i.e., to lie on a line inclined at 45° to the real axis. This will determine the value of the Evans gain. To estimate this more precisely, a semiaccurate root locus will be sketched. The asymptotes intersect the real axis at $s = -0.00565$, producing the plot shown in Fig. 29.7. Assuming

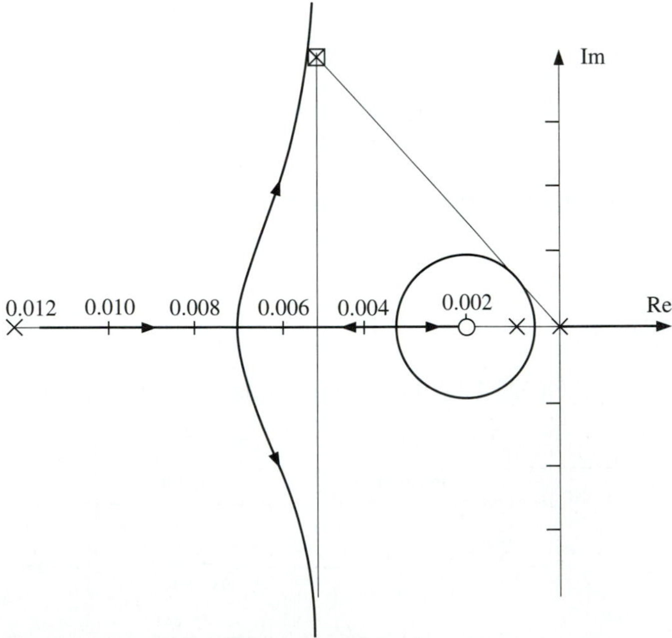

Fig. 29.7 Semiaccurate root locus.

the locus is close to the asymptotes, the gain required to locate the closed-loop poles with the required damping ratio may be calculated graphically as $K_E = 0.0000731$. From the open-loop transfer function, the Evans gain and K_p are related by

$$K_E = \frac{K_p}{71{,}096} \tag{29.9}$$

resulting in $K_p = 5.196$. Since the open-loop zero is placed at $s = -0.002$, we obtain

$$\frac{K_i}{K_p} = 0.002 \tag{29.10}$$

resulting in $K_i = 0.01$. Since the system is third order, there will be three closed-loop poles located approximately at $s = -0.005$ and $s = -0.005 \pm 0.005j$. We would expect the closed-loop system response to a unit step input to have a dominant time constant of around 200 s and also to exhibit some oscillatory component with a frequency of around 0.005 rad/s. Figure 29.8 shows an accurate root locus plotted using MATLAB and indicates that the loop is not circular and does not

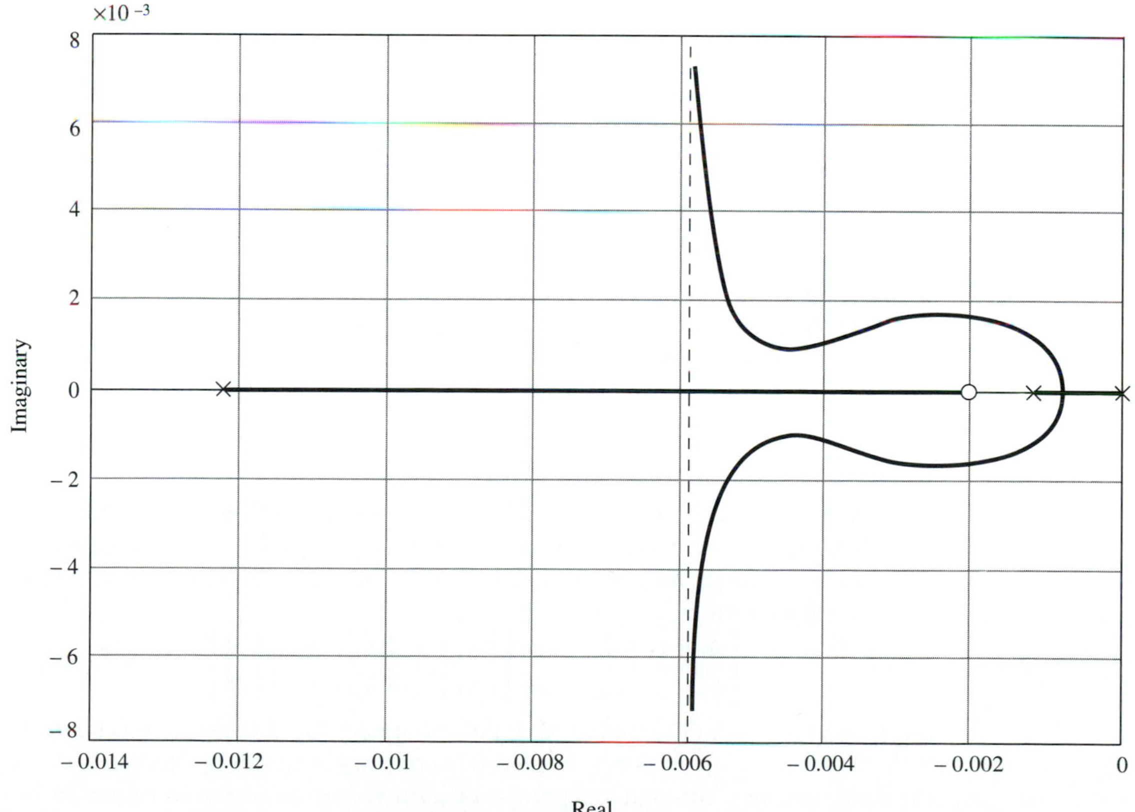

Fig. 29.8 Accurate root locus.

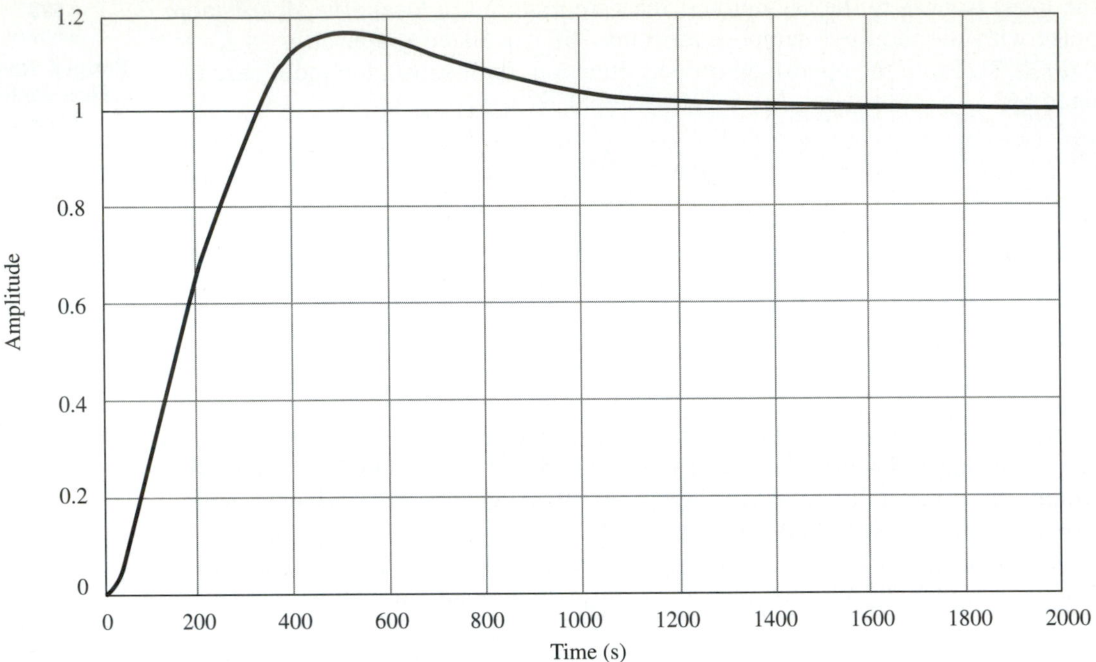

Fig. 29.9 Closed-loop step response.

actually meet back on the real axis but branches off to infinity. Our esimates for the closed-loop poles still look reasonable. Using the MATLAB command `cloop` enables the closed-loop transfer function to be obtained from the open loop transfer function, and the command `step` produces the closed-loop unit step response shown in Fig. 29.9. It is confirmed that the response has zero steady-state error with the expected rise time, primary time constant, and oscillatory components.

Although the P+I controller meets all of the required performance specifications, we have not looked at the effect of the control system on the height of water in the two tanks. We might want to check to ensure that the tanks do not overflow or run dry, and to perhaps address the more general question of whether we can control not just q_o but h_1 and h_2 as well. We found earlier that

$$h_2 = q_o R_2 \tag{29.11}$$

so we may conclude that h_2 and q_o cannot be controlled independently. However, there is no reason we cannot control h_1 and q_o independently. More insight into the system dynamics may be obtained by recasting the equations in the following state-space form.

$$\begin{bmatrix} \dot{q}_o \\ h_1 \end{bmatrix} = \begin{bmatrix} a_{11} & a_{1,2} \\ a_{2,1} & a_{22} \end{bmatrix} \begin{bmatrix} q_o \\ h_1 \end{bmatrix} + \begin{bmatrix} b_{11} & b_{12} \\ b_{21} & b_{22} \end{bmatrix} \begin{bmatrix} q_i \\ 0 \end{bmatrix} \tag{29.12}$$

Expressing the system in this form, it may be stated that if the system is controllable, the closed-loop poles may be placed in arbitrary locations by feeding back all the state variables[1] through an appropriate gain vector, as shown in figure 29.10.

[1]Assuming they may be easily measured.

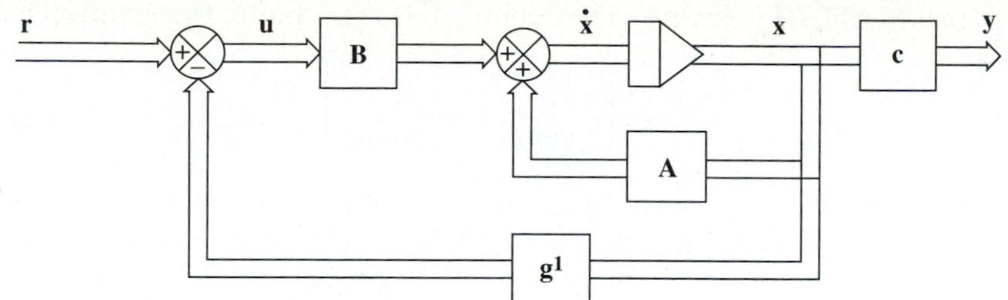

Fig. 29.10 Full State Vector Feedback

The output equation may be directly written as

$$y = [1 \quad 0]\begin{bmatrix} q_o \\ h_1 \end{bmatrix} \tag{29.13}$$

The process of determining the elements of **A** and **b** begins by observing that in equations 29.1–29.4 we need to eliminate first h_2, then q. The student may confirm the result of eliminating h_2 to produce

$$q_i - q = A_1 h_1 \tag{29.14}$$

$$q - q_0 = A_2 R_2 \dot{q}_o \tag{29.15}$$

$$h_1 - q_o = qR_1 \tag{29.16}$$

Now eliminating q produces

$$\dot{q}_0 = h_1\left[\frac{1}{A_2 R_1 R_2}\right] - q_o\left[\frac{1}{A_2 R_2} + \frac{1}{A_1 R_1}\right] \tag{29.17}$$

$$\dot{h}_1 = q_o\left[\frac{R_2}{R_1 A_1}\right] - h_1\left[\frac{1}{R_1 A_1}\right] + q_i\left[\frac{1}{A_1}\right] \tag{29.18}$$

Hence, the elements of matrices in the state equation become

$$a_{11} = -\left[\frac{1}{A_1 R_1} + \frac{1}{A_2 R_2}\right] \tag{29.19}$$

$$a_{12} = \frac{1}{A_2 R_1 R_2} \tag{29.20}$$

$$a_{21} = \frac{R_2}{R_1 A_1} \tag{29.21}$$

$$a_{22} = -\frac{1}{A_1 R_1} \tag{29.22}$$

$$b_{11} = b_{12} = b_{21} = 0 \tag{29.23}$$

$$b_{22} = \frac{1}{A_1} \tag{29.24}$$

Substituting numerical data and letting

$$\mathbf{x} = \begin{bmatrix} q_o \\ h_1 \end{bmatrix}, \qquad \mathbf{u} = \begin{bmatrix} q_i \\ 0 \end{bmatrix} \tag{29.25}$$

we obtain

$$\dot{\mathbf{x}} = \begin{bmatrix} -0.0097 & 0.000177 \\ 0.159 & -0.00796 \end{bmatrix}\mathbf{x} + \begin{bmatrix} 0 & 0 \\ 0.0796 & 0 \end{bmatrix}\mathbf{u} \tag{29.26}$$

Using the MATLAB `ctrb(a,b)` command, we obtain the controllability matrix

$$C_T = \begin{bmatrix} 0 & 0 \\ 0.0796 & -0.0006 \end{bmatrix} \tag{29.27}$$

which has the rank of 2. Since the system is controllable, we may use Ackerman's method for pole placement. Based on the root locus of the open-loop system, a reasonable location for the two-closed loop poles might be to place them coincident at $s = -0.005$. Putting these in a vector of desired closed-loop poles p, the MATLAB command `g=acker(a,b,p)` produces the required feedback vector

$$\acute{\mathbf{g}} = [3.565 \quad -0.0962] \tag{29.28}$$

The control input becomes

$$\mathbf{u} = \acute{\mathbf{g}}\mathbf{x} = [3.565 \quad -0.0962]\begin{bmatrix} q_o \\ h_1 \end{bmatrix} \tag{29.29}$$

Note, finally, that the selection of only two-state variables, and not h_2, is as expected from this second order system.

Ship Steering Control System

We now look at the problem of designing a ship steering control system, basically an autopilot but for two-dimensional motion only. Consider the effect of a rudder deflection δ on the angular position ψ relative to an arbitrary, but fixed, reference direction (such as due north), as shown in Fig. 30.1. The relationship between δ and ψ is given by what is known as Nomoto's equation and may be expressed as

$$\frac{\dot{\psi}}{\delta} = \frac{-K(1 + T_3 s)}{(1 + T_1 s)(1 + T_2 s)} \tag{30.1}$$

It should be noted that (1) a fixed δ results in a fixed rotation *rate* $\dot{\psi}$ and (2) a negative sign occurs in the transfer function because, in ships with conventional rudder geometry, a clockwise rotation of the rudder produces a counterclockwise rotation of the ship. The dynamics of rotation indicate that the rotation rate asymptotically

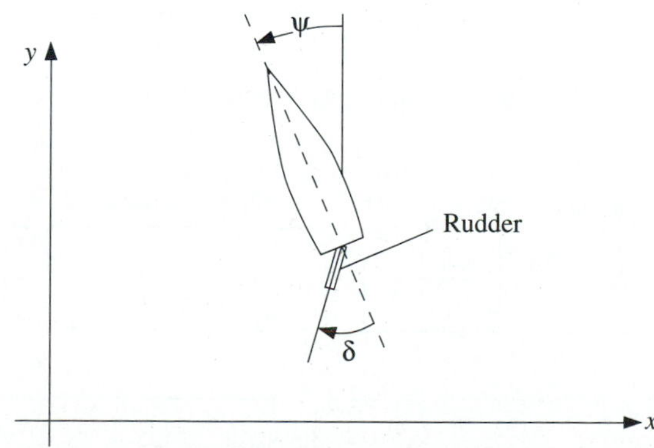

Fig. 30.1 Ship kinematics.

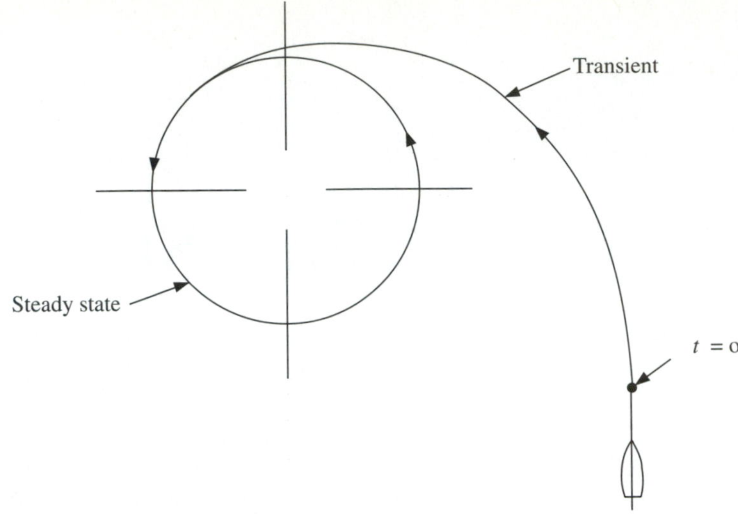

Fig. 30.2 Expected path due to fixed-rudder deflection.

approaches a constant value, so if the ship were moving in a straight line and a constant rudder rotation occurred, the ship would spiral in toward a circular path, which then gives the vessel a constant rate of rotation, as shown in Fig. 30.2. We may also model the steering gear as a simple lag, as shown in Fig. 30.3, where θ is the wheel rotation causing the rudder rotation δ. A simple control system may now be proposed, involving either a computer or a human on the bridge, in which the achieved ship direction obtained from a compass is compared to that desired, and corrective action results in a rotation of the wheel θ. This system is shown in Fig. 30.4. Considering typical data for a medium-size oil tanker 950 ft long weighing 150,000 dead weight tons traveling at 10.8 knots (17.28 ft/s), we have

$$T_1 = 23.71 \text{ s} \tag{30.2}$$

$$T_2 = -2446.4 \text{ s} \tag{30.3}$$

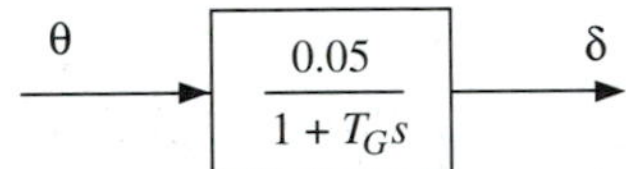

Fig. 30.3 Steering gear transfer function.

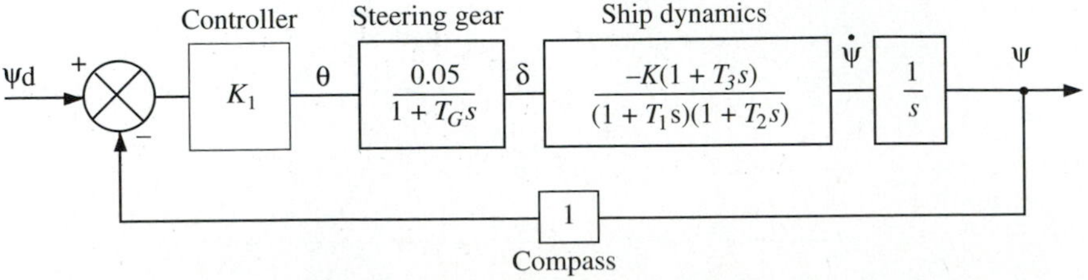

Fig. 30.4 Proposed steering system.

$$T_3 = 35.69 \text{ s} \tag{30.4}$$

$$K = 0.474 \text{ s}^{-1} \tag{30.5}$$

$$T_G = 11 \text{ s} \tag{30.6}$$

Substituting into the transfer functions reveals the open-loop transfer function to be

$$GH(s) = \frac{1.325 \times 10^{-6} K_1 (s + 0.028)}{s(s + 0.091)(s + 0.042)(s - 0.00041)} \tag{30.7}$$

Figure 30.5 shows the root locus for the system and indicates, rather surprisingly, that the system is unstable! It is a little known fact that *most* large ships are course unstable. What this means is that if the ship were traveling in a straight line and the wheel fixed in the corresponding position, the ship would eventually veer off course. Because the time constant associated with the instability is very large (tens of minutes), an attentive pilot will correct for the small course deviations as they occur. However, the pilot is attempting to control an unstable system. We will now attempt to design a better controller that not only stabilizes the steering system but also meets the following performance parameters:

1. The ship should not have more than a 5% overshoot to a step change in rudder angle.
2. The 2% settling time for the same input should occur in a time equivalent to five ship lengths at the design speed.

The damping ratio corresponding to condition 1 is obtained from Module 4 as $\zeta = 0.707$. The settling time may be obtained as

$$t_s = \frac{d}{v} = \frac{5 \times 950}{17.28} = 275 \text{ s} \tag{30.8}$$

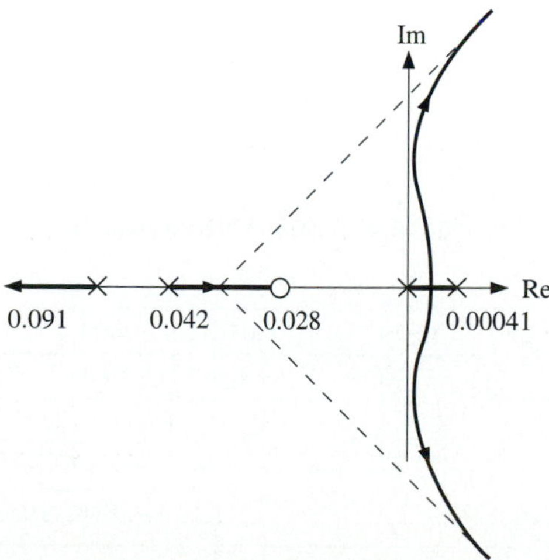

Fig. 30.5 Root locus of steering system.

Since the settling time is given by

$$t_s = \frac{4}{\zeta\omega_n} = \frac{4}{\sigma} \tag{30.9}$$

we obtain

$$\sigma = 0.014 \tag{30.10}$$

These two performance measures may be represented in the complex plane by the boundaries shown in Fig. 30.6. The problem with the root locus in Fig. 30.5 is that it has three asymptotes going to infinity and will always be unstable for large gains. In order to make the ship unconditionally stable, we need to introduce an extra open-loop zero. This could be done by (a) using a PD controller or (b) using velocity feedback. We will use method (b) assuming that $\dot{\psi}$ is available from a rate gyroscope. The proposed system is now shown in Fig. 30.7, where the design objectives will be met by correctly selecting K_1 and K_2. We begin with the closed-loop transfer function of the velocity loop, which we will denote as G^*, where

$$G^*(s) = \frac{-0.005K(1 + T_3)}{(1 + T_G s)(1 + T_1 s)(1 + T_2 s) - 0.005KK_2(1 + T_3 s)} \tag{30.11}$$

The open-loop transfer function for the complete system now becomes

$$GH(s) = \frac{K_1 G^*}{s} = \frac{-0.005KK_1(1 + T_G)}{s[(1 + T_G s)(1 + T_1 s)(1 + T_2 s) - 0.005KK_2(1 + T_3 s)]} \tag{30.12}$$

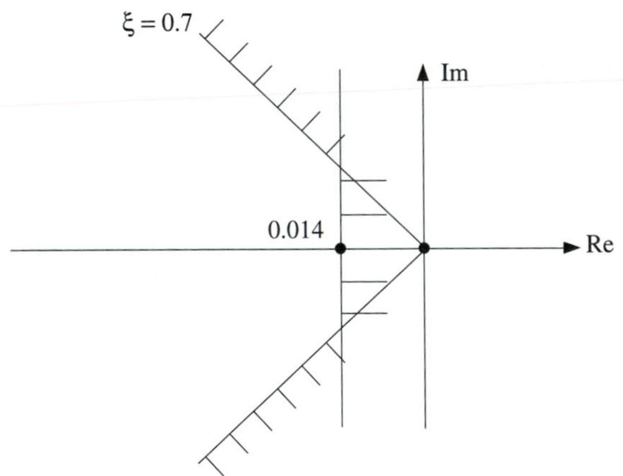

Fig. 30.6 Performance boundaries.

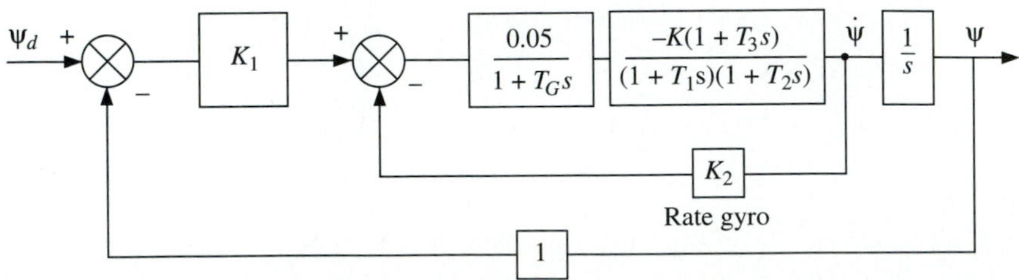

Fig. 30.7 Proposed control system.

Substituting the numerical values and expanding result in

$$GH(s) = \frac{0.0237K_1(1 + 35.69s)}{638046s^4 + 84653s^3 + s^2(2411 + 0.846K_2) + s(0.024K_2 - 1)} \tag{30.13}$$

We will use the root locus to design the system, but we cannot place the open-loop poles until a value of K_2 is estimated. Further, since the denominator of the above expression must be factored into the form $(s + \sigma_1)(s + \sigma_2)\cdots$, the multiplier for the term s^4 must be unity. We divide the above transfer function by the coefficient of s^4 to obtain the Evans form

$$GH(s) = \frac{1.325 \times 10^{-5}K_1(s + 0.0280)}{s^4 + 0.1327s^3 + s^2[(2411 + 0.846K_2)/638046] + s[(0.024K_2 - 1)/638046]} \tag{30.14}$$

We now guess values of K_2 and investigate the resulting root locus for the complete open-loop system.

$K_2 = 500$. For this case the denominator factors into the following open-loop poles:[1]

$$s = 0 \tag{30.15}$$

$$s = -0.0045 \tag{30.16}$$

$$s = -0.048 \tag{30.17}$$

$$s = -0.08 \tag{30.18}$$

These poles, together with the open-loop zero, are plotted in Fig. 30.8, which also shows the performance boundaries and the resultant root locus. Clearly, this root

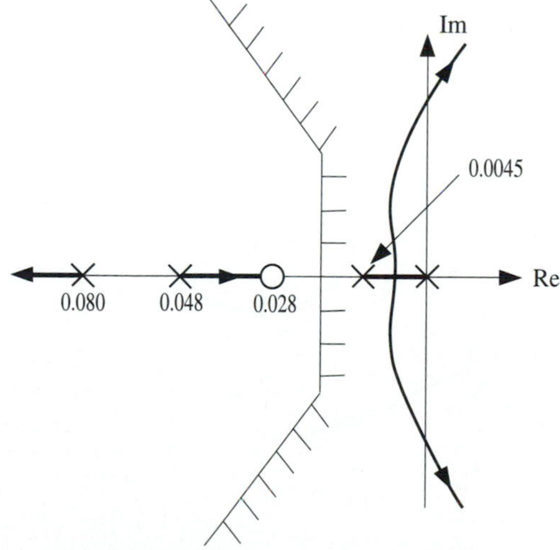

Fig. 30.8 Root locus for $K_2 = 500$.

[1]Using the MATLAB command `roots`.

locus is unsatisfactory since the two closed-loop poles near the imaginary axis will always violate the settling time performance boundary. We try increasing the value of K_2.

$K_2 = 1000$. Again solving the polynomial in the denominator to determine the open-loop poles results in

$$s = 0 \tag{30.19}$$

$$s = -0.0091 \tag{30.20}$$

$$s = -0.062 \pm 0.012j \tag{30.21}$$

Plotting the open-loop poles and sketching the root locus produces the result shown in Fig. 30.9. This is better, since the branch from the two poles near the origin loop around the zero and will eventually move to the left of the performance boundary. The only drawback is that it might take a relatively high gain to achieve this, by which time the branches from the complex open-loop poles may be underdamped or even unstable. We will try again to increase K_2 further.

$K_2 = 5000$. This value of velocity feedback gain produces the open-loop poles

$$s = 0 \tag{30.22}$$

$$s = -0.024 \tag{30.23}$$

$$s = -0.054 \pm 0.070j \tag{30.24}$$

The corresponding poles and root locus are shown in Fig. 30.10. This gives the most promising result so far since the entire root locus appears to be to the left of the settling time constraint. An accurate root locus of this system[2] is shown in

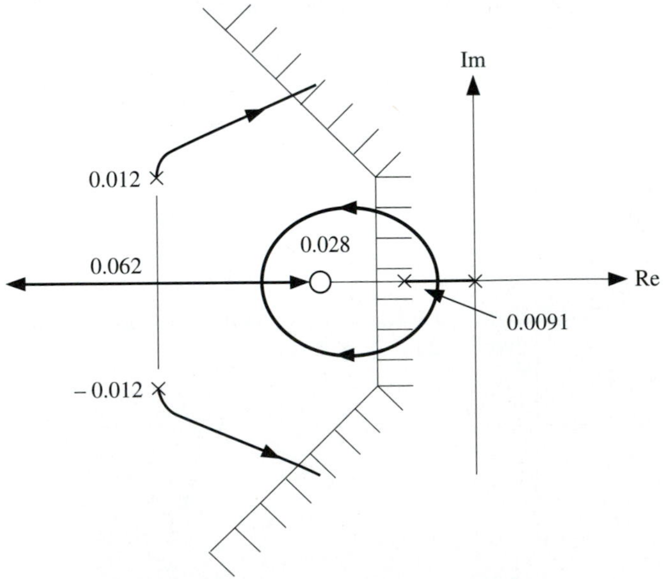

Fig. 30.9 Root locus for $K_2 = 1000$.

[2]Using MATLAB's `rlocus` command.

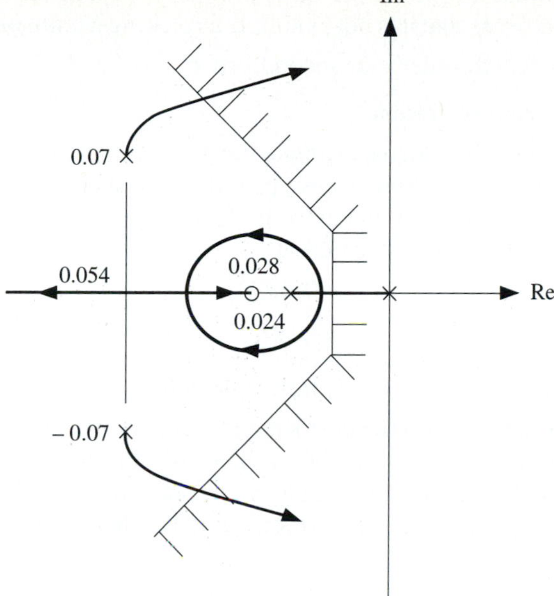

Fig. 30.10 Root locus for $K_2 = 5000$.

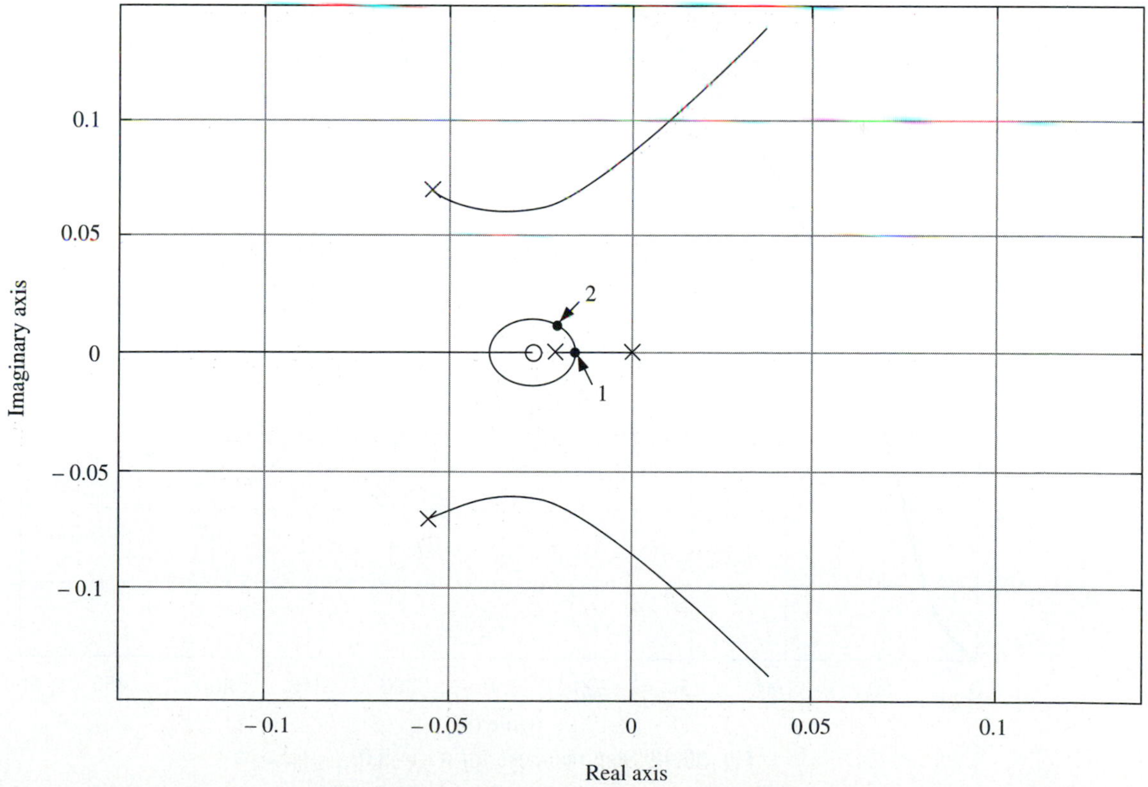

Fig. 30.11 Accurate root locus for $K_2 = 5000$.

Fig. 30.11. We now have to select the value of K_1 that we think satisfies the performance constraints. Note that it is not as simple as picking a point on the locus since:

1. The system is fourth order, not second order.
2. A closed-loop zero is present.

We will pick a point on the locus, evaluate the gain, and then simulate the closed-loop response to a unit step. Suppose we pick the gain such that the two dominant poles are coincidently located at 1 on Fig. 30.11. This corresponds to a gain of $K_1 = 5$ and closed-loop poles located at

$$s = -0.0165 \tag{30.25}$$

$$s = -0.0165 \tag{30.26}$$

$$s = -0.05 \pm 0.066 \tag{30.27}$$

Superficially, these lie within the performance constraints, but when we simulate the closed-loop step response,[3] we obtain the result shown in Fig. 30.12. Observe the dominant influence of the double pole on the real axis. This does not quite meet the settling time requirement, so we attempt to make the response faster by select-

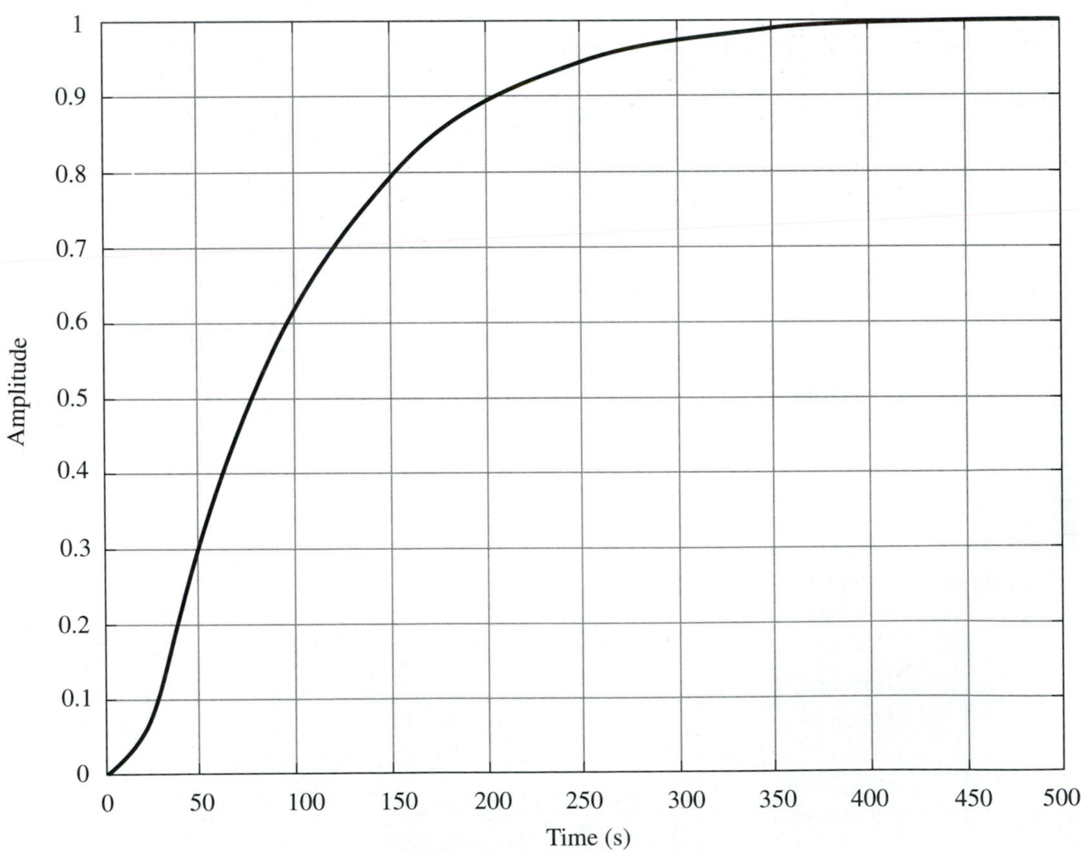

Fig. 30.12 Step response for $K_1 = 5.0$.

[3]Using MATLAB commands `cloop` and `step`.

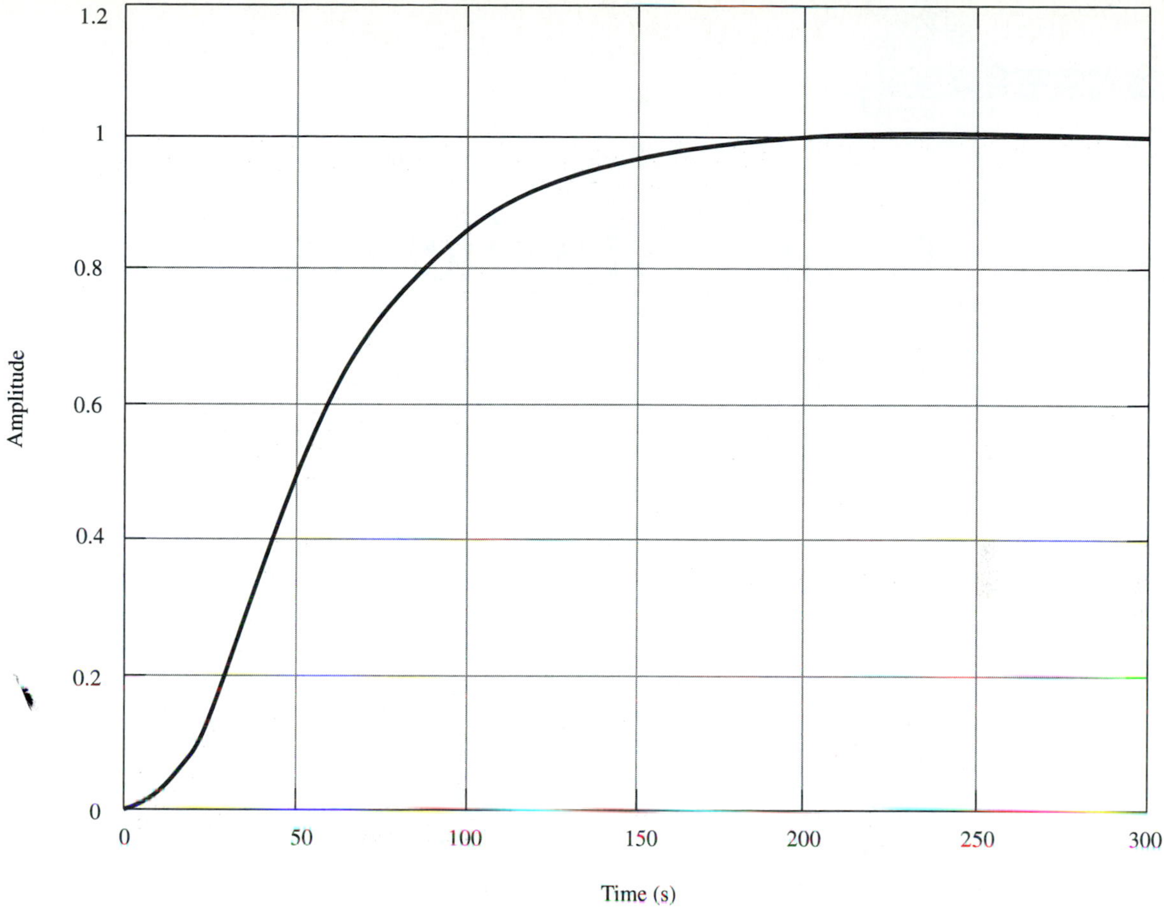

Fig. 30.13 Step response for $K_1 = 8.45$.

ing the gain at point 2 in Fig. 30.11. This corresponds to a gain of $K_1 = 8.45$ and closed loop poles at

$$s = -0.02 \pm 0.010j \tag{30.28}$$

$$s = -0.06 \pm 0.064j \tag{30.29}$$

Again simulating the closed-loop step response, the result is shown in Fig. 30.13. This response does satisfy both performance measures, and we may now conclude that the system will behave as desired for $K_1 = 8.45$ and $K_2 = 5000$. The following will be noted about the design process:

1. The analytical techniques used (e.g., root locus) serve as a useful guide, but they do not always provide a completely analytical approach to solving the problem.
2. The final design should always be simulated (or better still experimentally operated) so that higher-order effects may be observed.
3. It is rarely possible to obtain all possible values of design parameters satisfying the performance criteria. Finding a single set of parameters (K_1, K_2) may be difficult enough.

MODULE 31

Cruise Missile Altitude Control System

We will consider the design of a cruise missile altitude control system in which we desire the missile to fly at a constant height above the ground and then navigate its way to the target. In this problem, we will be concerned only with ensuring that the missile flies at a demanded altitude. This problem was covered briefly in Module 18, but here we will analyze the system in detail.

As the missile flies along, it receives signals from a downward-looking radar. In general, this signal will give information about the height of the missile above the ground, but objects such as trees and buildings will also influence the signal. This situation is shown in Figs. 31.1 and 31.2. We need to ensure that the missile responds to undulations in terrain but not to the spikes produced by smaller objects, so that we obtain the smoother flight path shown in Fig. 31.3. This will reduce fuel consumption and extend the missile range. Note that if a frequency analysis of the radar signal is performed, it exhibits a bimodal behavior, as shown in Fig. 31.4. Low-frequency components are due to the slowly changing terrain, while higher frequencies are generated by smaller objects. Our control system must reject the high-frequency noise, yet respond to signals with a low-frequency content.

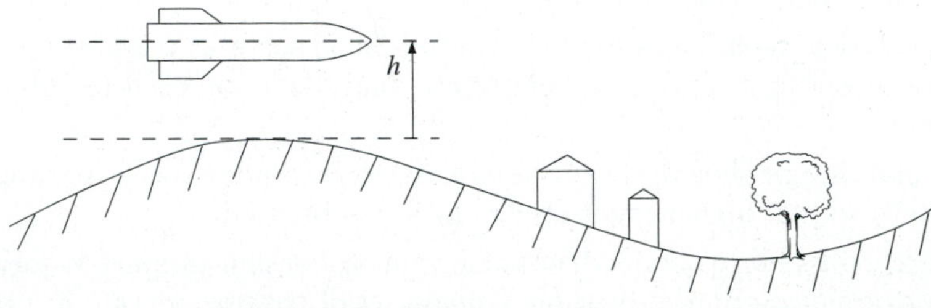

Fig. 31.1 Ground-skimming missile.

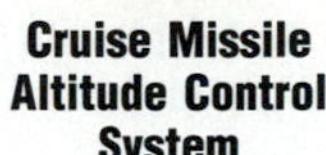

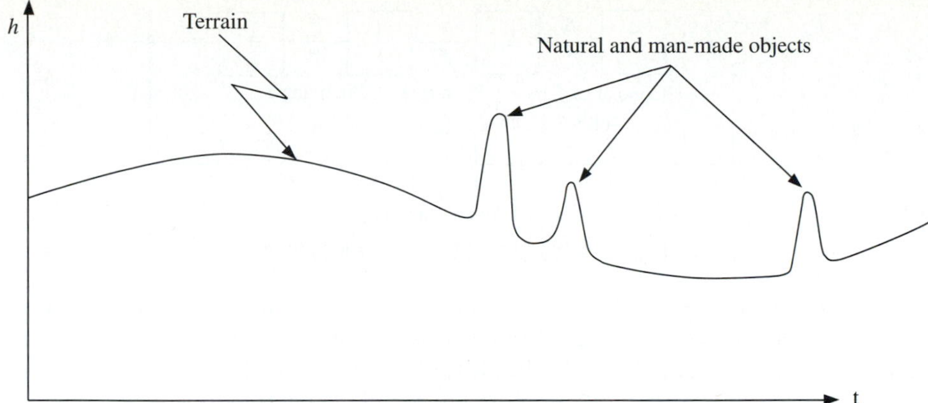

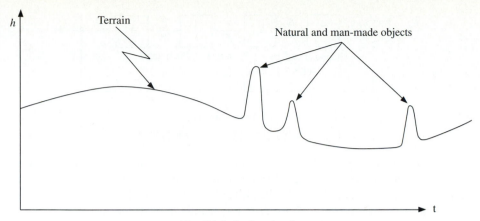

Fig. 31.2 Radar signal.

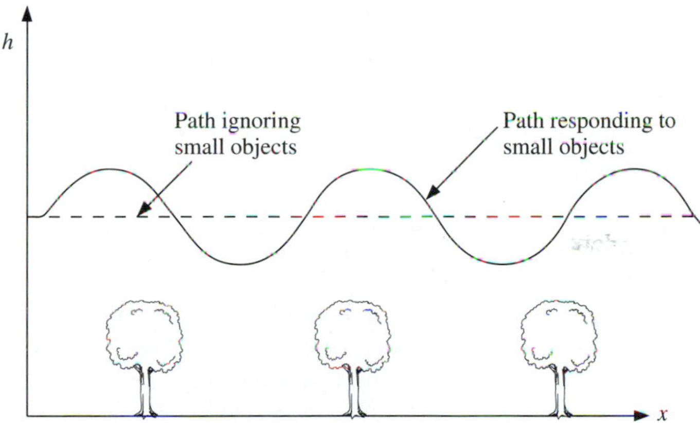

Fig. 31.3 Possible flight paths.

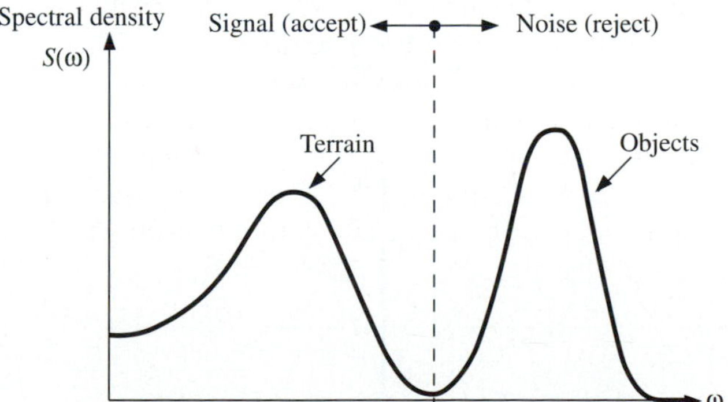

Fig. 31.4 Frequency analysis of radar signal.

The first attempt to design an altitude control system is to compare the radar signal to the demanded height. This error is multiplied by a proportional gain K_a and fed to an actuator connected to a control surface (see Fig. 31.5). If the resulting control surface rotation is δ, we might expect it to produce a pitch rate of the

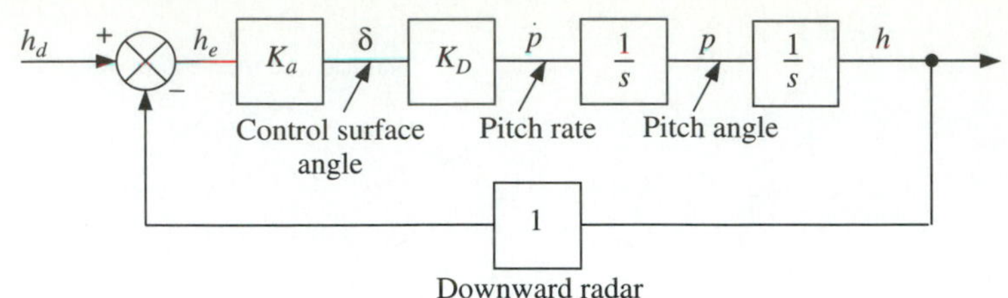

Fig. 31.5 Simple altitude control system.

missile $\dot{p}$, where K_D is proportional to the missile inertia and therefore a measure of the missile dynamics. This rate may be integrated to give the pitch angle p, which in turn is integrated to give the altitude h. Assuming the downward-looking radar can measure the altitude exactly, the simple control loop is complete. It is seen that the open-loop transfer function is given by

$$GH(s) = \frac{K_a K_D}{s^2} \tag{31.1}$$

The root locus of the system is sketched in Fig. 31.6 and the Bode diagram in Fig. 31.7. Clearly the system has no damping, and would not perform in a stable manner. The design requirements may be expressed as follows:

1. Stabilize the system.
2. Design the system so that signal is accepted but noise is rejected.
3. Minimize the overshoot to a step input.

In order to proceed with the design, which will be performed in the frequency domain, some quantitative data are needed. Suppose that the noise is caused by objects whose sizes range from 10 to 50 m. Can we determine the resultant frequencies these objects produce in the altimeter signal? Consider a succession of these generalized objects as shown in Fig. 31.8. If the objects are separated by a space equal to the object size and the missile is traveling at 400 km/h, then the time taken to fly over one cycle of object and intermediate space is

$$t = \frac{2 \times l \times 60}{400 \times 1000} \tag{31.2}$$

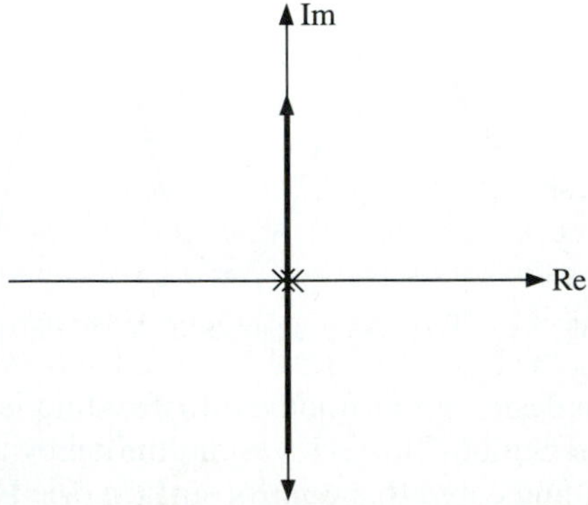

Fig. 31.6 Root locus of system with proportional control.

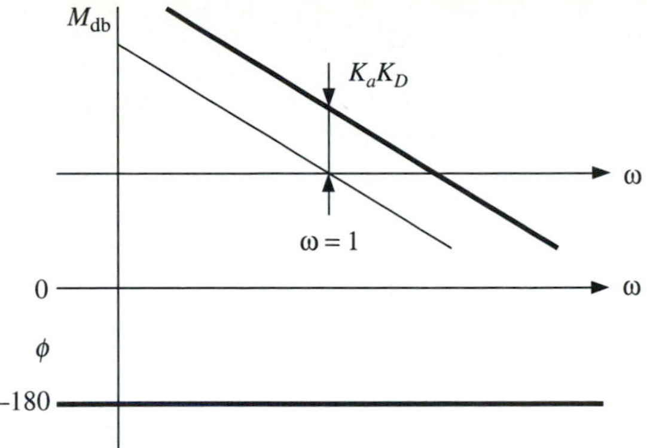

Fig. 31.7 Bode diagram of system.

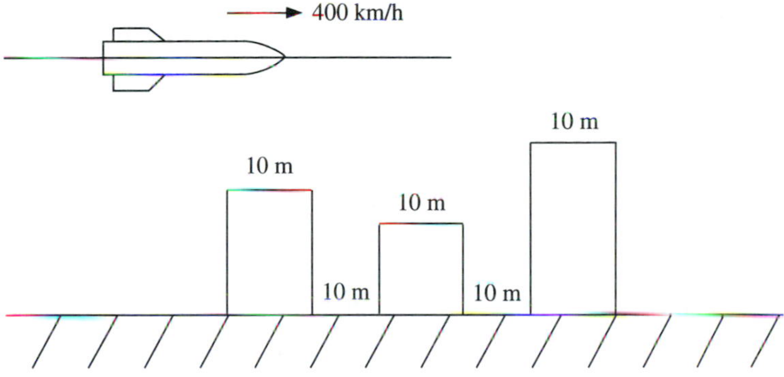

Fig. 31.8 Multiple "noise" objects.

This time is the period of the noise signal, from which the frequency may be obtained. Substituting $l = 10$ and $l = 50$ we obtain the frequency range of the noise to be

$$419 < \omega_{\text{noise}} < 2095 \tag{31.3}$$

We will therefore define noise to be any frequency component of the altimeter signal above 400 rad/s. In order for the system to reject noise, we might specify that the resultant Bode diagram must have at least 20 db of attenuation at this frequency. This imposes a constraint boundary in the system Bode diagram, as shown in Fig. 31.9. It should be recognized, however, that this boundary in the frequency domain applies to the closed-loop system. By considering the proportional-control system shown in Fig. 31.5, it will be the transfer function h/h_d that has to reject noise above 400 rad/s, not the open-loop transfer function. To achieve the desired result, we plot the Bode diagram of a closed-loop transfer function that satisfies the performance constraint. Such a transfer function is shown in Fig. 31.10 and is obtained by placing a line of -40 db/decade against the boundary corner. In addition, we select a horizontal boundary line at 0 db to avoid any resonance effects. Then, rather arbitrarily, we synthesize a simple transfer function that fits

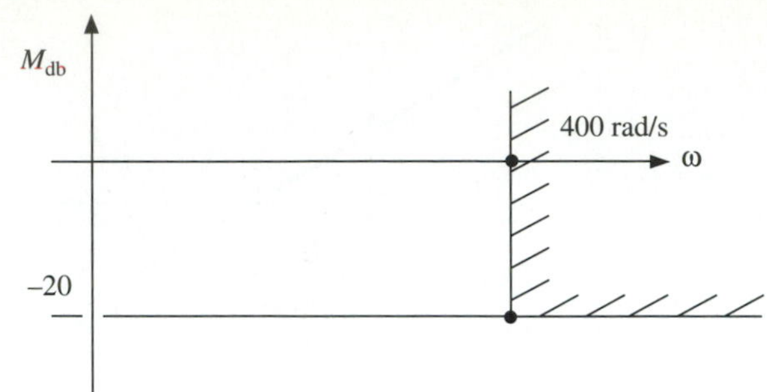

Fig. 31.9 Constraint boundary in the frequency domain.

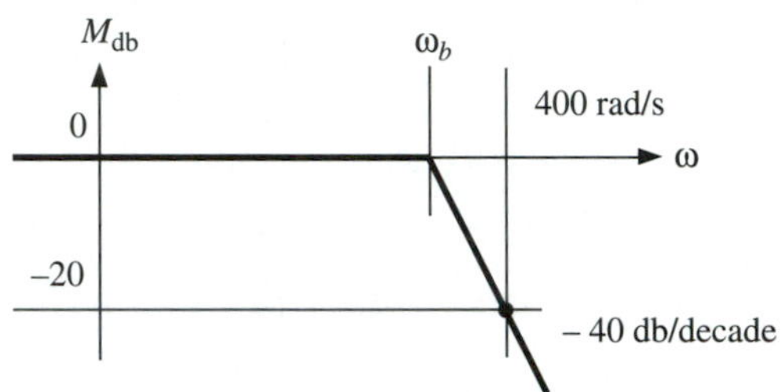

Fig. 31.10 Bode diagram satisfying performance boundaries.

within the boundaries. In this case we choose

$$\frac{h}{h_d} = \frac{1}{(1 + sT)^2} \tag{31.4}$$

The numerator is set to unity for obvious reasons, and a second-order denominator is selected to enssure rapid attenuation with increasing frequency. This feature is commonly known as rolloff. Extrapolating the high-frequency asymptote back to the 0-db line reveals $\omega_b = 103$ rad/s, yielding the result $T = 0.0077$. The question now is, what open-loop transfer function will give the closed-loop frequency response shown in Fig. 31.10? We could plot the Nichols chart and pick off the Bode diagram of the open-loop transfer function and then estimate its analytical form, but in this case a direct approach will be used. We know that if G^* is the closed-loop transfer function and G the open-loop transfer function, then assuming unity feedback,

$$\frac{h}{h_d} = \frac{G}{1 + G} = G^* \tag{31.5}$$

Solving for the open-loop transfer function results in

$$G = \frac{G^*}{1 - G^*} \tag{31.6}$$

Substituting the desired closed-loop transfer function for G^* produces

$$G = \frac{1}{s(sT^2 + 2T)} = \frac{64.9}{s(1 + 0.00385s)} \tag{31.7}$$

We have an estimated open-loop transfer function

$$G' = \frac{K_a K_D}{s^2} \tag{31.8}$$

The question now is how to make the known open-loop transfer function G' equal, or perhaps approximate, the desired open-loop transfer function G. What is needed is to eliminate one of the poles at the origin and add a lag term with a time constant of 0.00385. It will be unwise to add pure differentiation in the controller, but we can approximate the desired result by using a lead-lag controller of the form

$$G_c = \frac{1 + sT_1}{1 + sT_2} \tag{31.9}$$

We will set $T_2 = 0.004$ to approximate the lag, while the zero must dominate the other poles. A suitable choice will be $T_1 = 1$. The compensated open-loop transfer function then becomes

$$GG_c(s) = \frac{64.9(1 + s)}{s^2(1 + 0.004s)} \tag{31.10}$$

where $K_a K_D = 64.9$. The closed-loop transfer function is

$$\frac{h}{h_d} = \frac{64.9(1 + s)}{0.004s^3 + s^2 + 64.9s + 64.9} \tag{31.11}$$

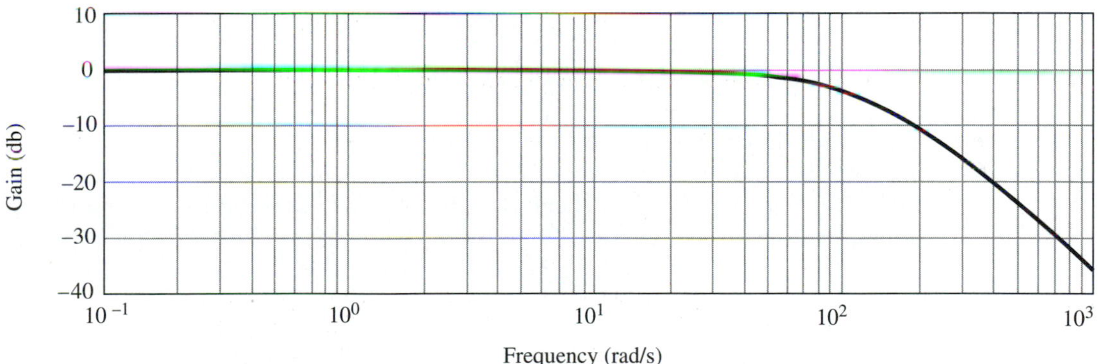

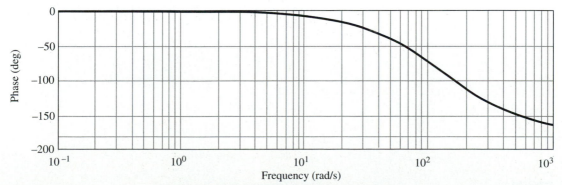

Fig. 31.11 Final closed-loop frequency response.

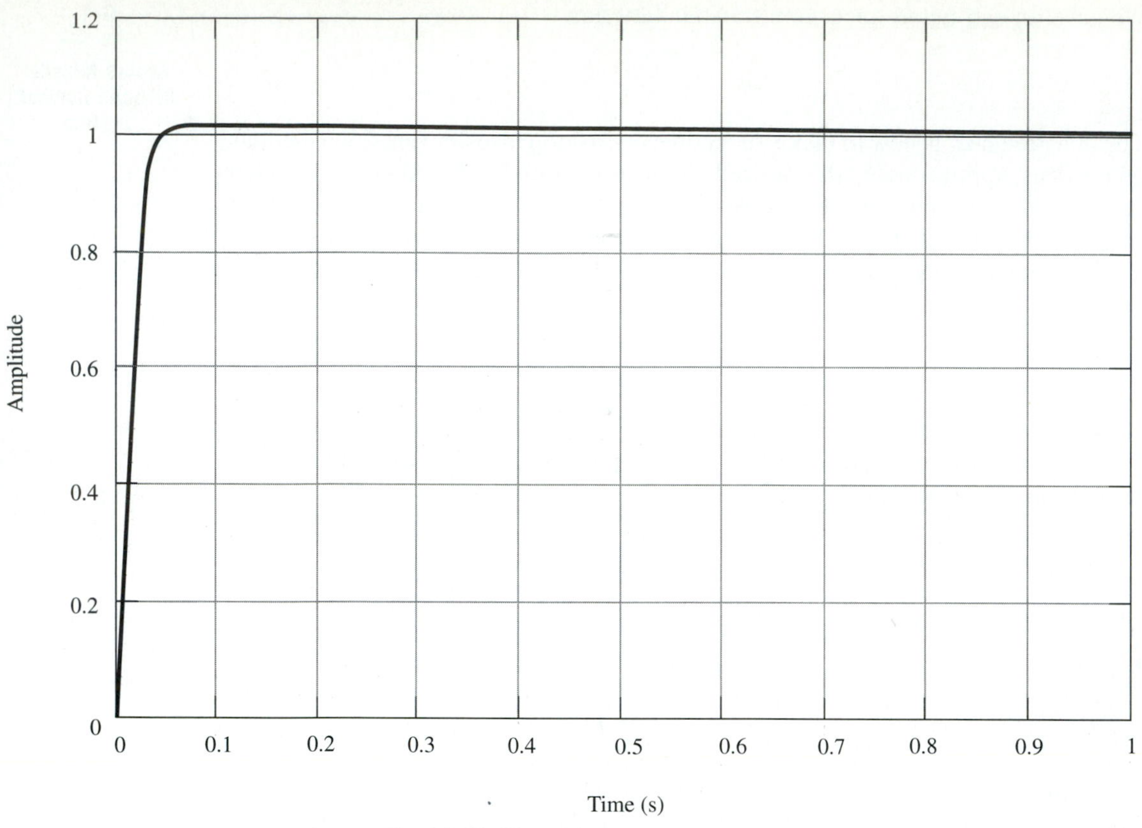

Fig. 31.12 Closed-loop step response.

The closed-loop frequency response and unit step response may be computed and plotted. These are shown in Figs. 31.11 and 31.12, respectively. Compare Fig. 31.11 with Fig. 31.10 to check that the system comes within the frequency-domain constraint boundaries. From these figures it may be seen that all of the performance requirements have been met.

Machine Tool Power Drive System with Flexibility

Figure 32.1 shows a single axis of a heavy machine tool in which a motor drives a lead screw through a reduction gear. The lead screw moves the workpiece under the cutter to produce the finished item. A computer sends desired positions of the table to a control system that attempts to place the table in the correct position as rapidly as possible. Because of the length of the lead screw and flexibility of the reducing gears, the angular twist along its length cannot be neglected, implying that its rotation at the gear box will not be the same as its rotation at the table, leading to some unpredictability in the actual displacement of the workpiece. Our job is to investigate the effect that a flexible lead screw has on the performance of the system and to design a controller that takes this into account and places the workpiece at a desired location with no overshoot. This requirement is of critical importance if movement of the workpiece occurs while cutting is in progress; otherwise uncertain profiles will be cut into the material being machined.

We begin the controller design by first obtaining the transfer function of the plant. In this case, this may be thought of as the transfer function between the application of motor torque τ_m and rotation of the end of the leadscrew, θ_l. A model of the system is shown in Fig. 32.2, where the compliance of the drive system is assumed constant and modeled with the torsional spring of stiffness k. Assuming the damping torque is proportional to the angular velocity, the equations of motion of the motor and load shaft can be written as

$$\tau_m - c_m\omega_m - T_1 = J_m\ddot{\theta}_m \tag{32.1}$$

$$k\left(\frac{\theta_m}{N} - \theta_t\right) - c_l\omega_l = J_l\ddot{\theta}_l \tag{32.2}$$

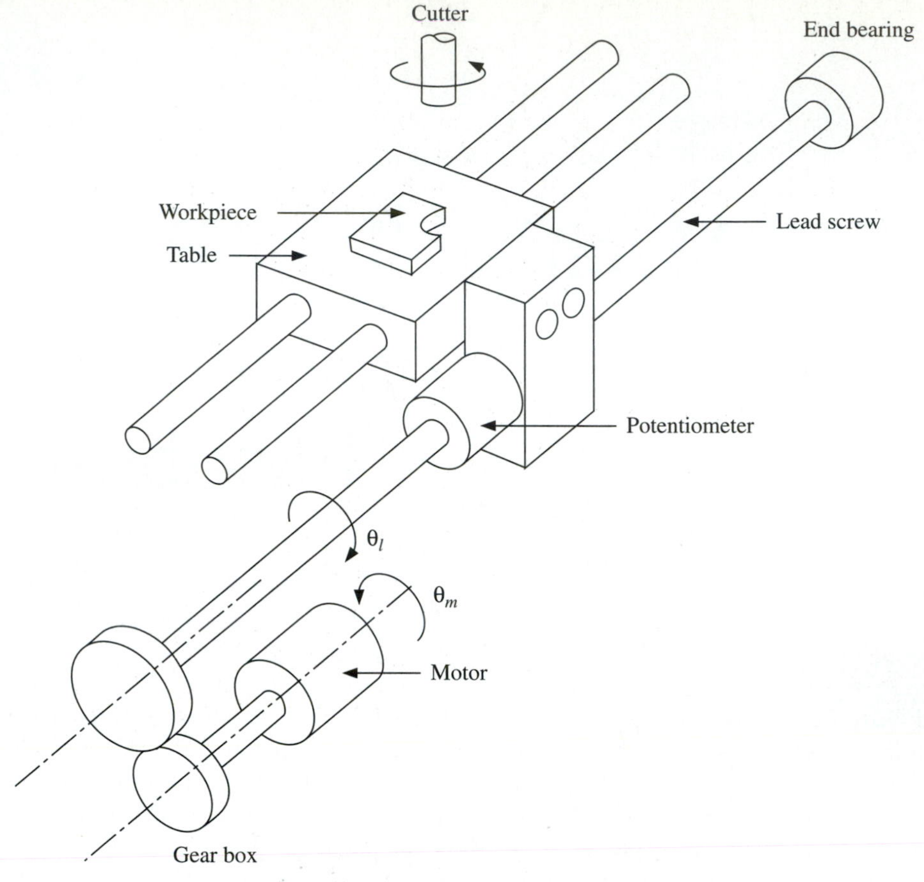

Fig. 32.1 Machine tool drive system.

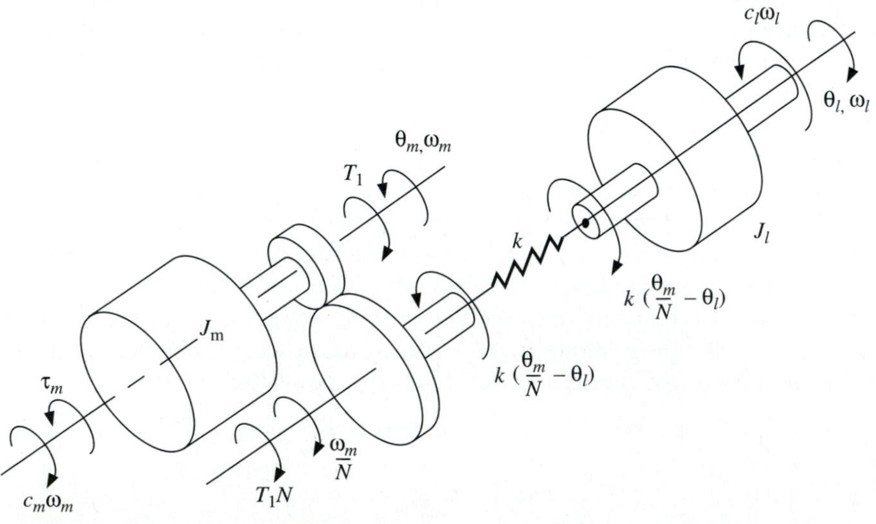

Fig. 32.2 Model of drive system.

Since the middle shaft with the larger gear on it has zero inertia, we may write

$$T_1 N = k\left(\frac{\theta_m}{N} - \theta_l\right) \tag{32.3}$$

These are three equations in the four unknowns T_1, τ_m, θ_m, and θ_l. Eliminating T_1 and θ_m, we arrive at the transfer function

$$\frac{\theta_l}{\tau_m} = \frac{k/N}{(J_l s^2 + c_l s + k)(J_m s^2 + c_m s + k/N^2) - k^2/N^2} \tag{32.4}$$

The simplest control system we could build would measure θ_l with a potentiometer and compare it to a demanded value, driving the motor with the resultant error. Such a proportional system is shown in Fig. 32.3. To obtain a qualitative feel for system response, the root locus will be sketched by ignoring all damping in the system. The open-loop transfer function reduces to

$$G(s) = \frac{k/N}{J_l J_m s^4 + s^2(J_l + J_m)} \tag{32.5}$$

This results in two poles at the origin and two others on the imaginary axis, leading to a root locus of the form shown in Fig. 32.4. If damping is present, the terms k^2/N^2 will still cancel out, leaving the one pole at the origin. The two

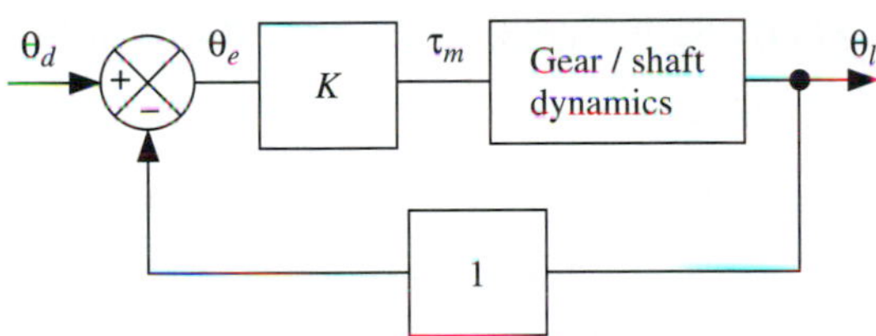

Fig. 32.3 Proportional control system.

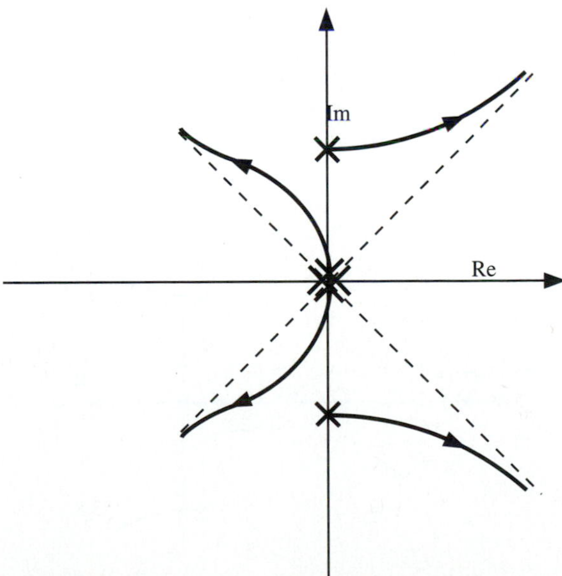

Fig. 32.4 Root locus of system with no damping.

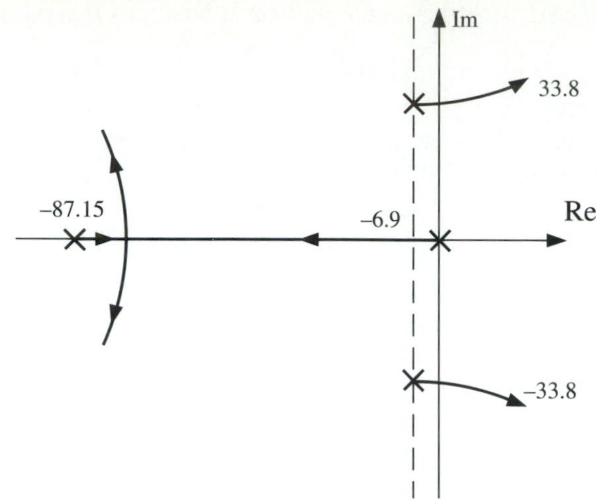

Fig. 32.5 Root locus of system with damping.

other poles probably move to the left of the imaginary axis. Consider the following data:

$$J_m = 0.0001 \text{ N-m-s}^2/\text{rad}, \qquad J_l = 0.001 \text{ N-m-s}^2/\text{rad},$$
$$c_m = c_l = 0.01 \text{ N-m-s/rad}, \qquad k = 1 \text{ N-m/rad}, \qquad N = 5$$

The characteristic equation reduces to

$$s^4 + 101s^3 + 2400s^2 + 104000s = 0 \tag{32.6}$$

which has the roots $s = 0$, -87.15, and $s = -6.9 \pm 33.8j$. This leads to the approximate root locus shown in Fig. 32.5. Although the system may be stable for small gains, its response will be very lightly damped and therefore unacceptable. If a PD controller is used with a zero at $s = -20$, as indicated in Fig. 32.6, the situation is improved but very little. A sample step response of the system with PD control is shown in Fig. 32.7. Although the response achieves a steady state in about 1 s, this is quite long for high-speed machining. In addition, the oscillatory nature of the response would produce undulating edges on the ma-

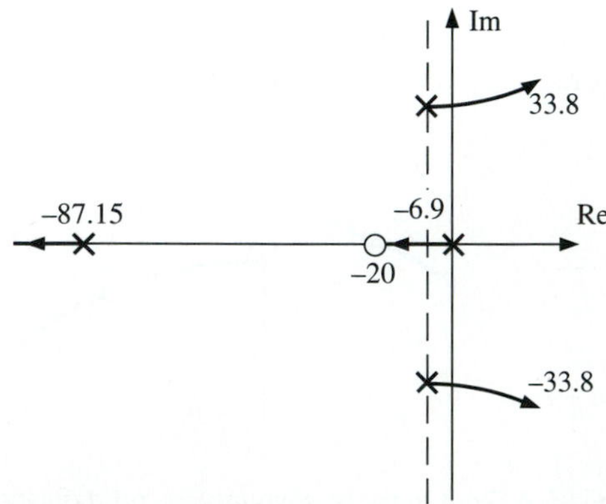

Fig. 32.6 Root locus of system with PD control.

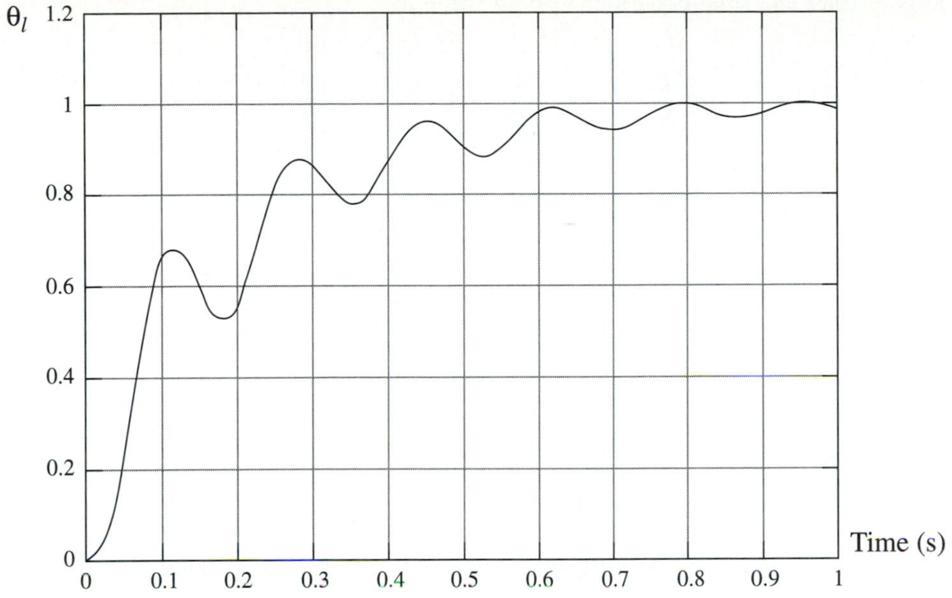

Fig. 32.7 Step response of system with PD control.

chined part. In order to design a better control system, we will use state-space techniques. Suppose we define the state variable vector **x** such that

$$x_1 = \theta_m \tag{32.7}$$

$$x_2 = \dot{\theta}_m \tag{32.8}$$

$$x_3 = \theta_l \tag{32.9}$$

$$x_4 = \dot{\theta}_l \tag{32.10}$$

We may now write

$$\dot{x}_1 = x_2 \tag{32.11}$$

$$\dot{x}_2 = \ddot{\theta}_m \tag{32.12}$$

$$\dot{x}_3 = x_4 \tag{32.13}$$

$$\dot{x}_4 = \ddot{\theta}_l \tag{32.14}$$

In the above equations, we have to obtain $\ddot{\theta}_m$ and $\ddot{\theta}_l$ in terms of the other state variables. From equations 32.1, 32.2, and 32.3, we have

$$\ddot{\theta}_m = -\frac{c_m}{J_m}\dot{\theta}_m - \frac{k}{J_m N^2}\theta_m + \frac{k}{J_m N}\theta_l + \frac{\tau_m}{J_m} \tag{32.15}$$

$$= -\frac{c_m}{J_m}x_2 - \frac{k}{J_m N^2}x_1 + \frac{k}{J_m N}x_3 + \frac{\tau_m}{J_m} \tag{32.16}$$

and

$$\ddot{\theta}_m = \frac{k}{J_l N}\theta_m - \frac{k}{J_l}\theta_l - \frac{c_l}{J_l}\dot{\theta}_l \tag{32.17}$$

$$= \frac{k}{J_l N}x_1 - \frac{k}{J_l}x_3 - \frac{c_l}{J_l}x_4 \tag{32.18}$$

This enables the state equations to be written as

$$\begin{bmatrix} \dot{x}_1 \\ \dot{x}_2 \\ \dot{x}_3 \\ \dot{x}_4 \end{bmatrix} = \begin{bmatrix} 0 & 1 & 0 & 0 \\ \dfrac{-k}{J_m N^2} & \dfrac{-c_m}{J_m} & \dfrac{k}{J_m N} & 0 \\ 0 & 0 & 0 & 1 \\ \dfrac{k}{J_l N} & 0 & \dfrac{-k}{J_l} & \dfrac{-c_l}{J_l} \end{bmatrix} \begin{bmatrix} x_1 \\ x_2 \\ x_3 \\ x_4 \end{bmatrix} + \begin{bmatrix} 0 \\ \dfrac{1}{J_m} \\ 0 \\ 0 \end{bmatrix} \tau_m \tag{32.19}$$

Writing in the form

$$\dot{\mathbf{x}} = \mathbf{A}\mathbf{x} + \mathbf{b}u \tag{32.20}$$

and substituting the data for this problem, the system matrices reduce to

$$\mathbf{A} = \begin{bmatrix} 0 & 1 & 0 & 0 \\ -400 & -100 & 2000 & 0 \\ 0 & 0 & 0 & 1 \\ 200 & 0 & -1000 & -10 \end{bmatrix} \qquad \mathbf{b} = \begin{bmatrix} 0 \\ 10000 \\ 0 \\ 0 \end{bmatrix} \tag{32.21}$$

The output is θ_l; hence the state-space output equation may be written as

$$y = [0 \quad 0 \quad 1 \quad 0] \begin{bmatrix} x_1 \\ x_2 \\ x_3 \\ x_4 \end{bmatrix} \tag{32.22}$$

In order to determine the controllability of the system, we calculate the matrix $\mathbf{C}_T$ for this sytem as[1]

$$\mathbf{C}_T = [B \quad AB \quad A^2B \quad A^3B] \tag{32.23}$$

The rank of $\mathbf{C}_T$ turns out to be 4, the order of the system, indicating that it is controllable. We will now use pole placement methods to locate the closed-loop poles at locations of our choice. Since critical damping is required, we will locate all four system poles at $s = -20$. MATLAB's `acker` command will be used to determine the appropriate feedback gain vector. We first set the vector of desired closed-loop poles:

```
dp= [-20 -20 -20 -20]
```

Then, after defining the system matrices, the command

```
k=acker(a,b,dp)
```

yields the result

$$k = [0.03 \quad -0.003 \quad -0.007 \quad -0.0225] \tag{32.24}$$

The closed-loop step response may be checked using the following commands:

```
acl=a-b*k
bcl=b
ccl=c
dcl=0
eig(acl)
step(acl,bcl,ccl,dcl,1)
```

[1] Or use the MATLAB command `ct=ctrb(a,b)`.

Fig. 32.8 Step response for full state feedback.

The intermediate calculation of the eigenvalues of the closed-loop system matrix ensures all poles are placed at $s = -20$. The resulting step response is shown in Fig. 32.8. Although this method seems to produce the desired result, some potential drawbacks should be recognized:

1. The arbitrary placement of the poles does not take into account the level of control effort required and may exceed that of the motors moving the workpiece.
2. Since all state variables have to be fed back, θ_m, $\dot{\theta}_m$, θ_l, and $\dot{\theta}_l$ all have to be measured. This will require two encoders, or potentiometers, and two tachometers. This will produce a more complex and expensive system.

Assuming these problems may be overcome, the power of the pole placement method may be readily observed. Even faster response may be obtained by specifying alternative closed-loop pole locations.

APPENDIX

Review of Laplace Transforms and Their Use in Solving Differential Equations

The definition of the Laplace transform $F(s)$ of a time-dependent function $f(t)$ is

$$\mathscr{L}[f(t)] = F(s) = \int_0 f(t)e^{-st}\,dt \tag{1}$$

Fortunately, the Laplace transform does not need to be evaluated using integration each time it is needed. In almost all cases, this has been done before for the functions of interest, and the results tabulated, as shown in Table 1.

◆ LINEAR PROPERTIES

If a and b are constants, then

$$\mathscr{L}[af_1(t) + bf_2(t)] = a\mathscr{L}f_1(t) + b\mathscr{L}f_2(t) = aF_1(s) + bF_2(s) \tag{2}$$

◆ SHIFTING THEOREM

If $F(s)$ is the Laplace transform of $f(t)$, then

$$\mathscr{L}[e^{-at}f(t)] = F(s + a) \tag{3}$$

This means that the Laplace transform of a time function multiplied by e^{-at} may be obtained by substituting $s + a$ for s in the Laplace transform of the time function alone.

$f(t)$	$F(s)$
Unit impulse $\delta(t)$	1
1	$\dfrac{1}{s}$
t^n	$\dfrac{n!}{s^{n+1}}$
e^{-at}	$\dfrac{1}{s+a}$
$t^n e^{-at}$	$\dfrac{n!}{(s+a)^{n+1}}$
$\sin \omega t$	$\dfrac{\omega}{s^2+\omega^2}$
$\cos \omega t$	$\dfrac{s}{s^2+\omega^2}$
$e^{-at} \sin \omega t$	$\dfrac{\omega}{(s+a)^2+\omega^2}$
$e^{-at} \cos \omega t$	$\dfrac{s+a}{(s+a)^2+\omega^2}$
$\sin(\omega t \pm \phi)$	$\dfrac{\omega \cos \phi \pm s \sin \phi}{s^2+\omega^2}$
$\sinh \omega t$	$\dfrac{\omega}{s^2-\omega^2}$
$\cosh \omega t$	$\dfrac{s}{s^2-\omega^2}$
$\dfrac{d^n}{dt^n} f(t)$	$s^n F(s) - s^{n-1} f(0) - \cdots - f^{n-1}(0)$

◆ TIME DIFFERENTIALS

Again if $F(s)$ is the Laplace transform of $f(t)$, then the Laplace transform of the time derivative of $f(t)$ is given by

$$\mathcal{L}[f'(t)] = sF(s) - f(0-) \tag{4}$$

where $f(0-)$ is the value of $f(t)$ just before $t = 0$. This result may be extended to the second derivative

$$\mathcal{L}[f''(t)] = s^2F(s) - sf(0-) - f'(0-) \tag{5}$$

where $f'(0-)$ is the value of the time derivative of $f(t)$ evaluated just before $t = 0$. The Laplace transform of higher-order differentials may be found from the formula given in Table 1.

◆ FINAL-VALUE THEOREM

If we are interested in the value that $f(t)$ takes as $t \to \infty$ and only the Laplace transform $F(s)$ is available, the result may be obtained from the final-value theorem as

$$\lim_{t\to\infty} f(t) = \lim_{s\to 0} sF(s) \tag{6}$$

This states that the final value of $f(t)$ is given by the limit as s approaches zero, of s multiplied by the Laplace transform of $f(t)$.

◆ INVERSE TRANSFORMS

The process of changing from the s domain to the time domain is known as inverse Laplace transformation and may be stated mathematically as

$$f(t) = \mathcal{L}^{-1}[F(s)] \tag{7}$$

In simple cases the inverse transform may be found directly from Table 1. However, in most practical cases $F(s)$ requires some manipulation before this can be done.

◆ SOLVING LINEAR DIFFERENTIAL EQUATIONS

Suppose the solution to the differential equation

$$\ddot{y} + 2y = t \tag{8}$$

is required, subject to the initial conditions

$$y(0-) = -1 \qquad \dot{y}(0-) = 1 \tag{9}$$

The solution proceeds by first taking the Laplace transform of both sides:

$$\mathcal{L}[\ddot{y} + 2y] = \mathcal{L}[t] \tag{10}$$

$$s^2Y - sy(0) - \dot{y}(0) + 2Y = \frac{1}{s^2} \tag{11}$$

$$s^2Y + s - 1 + 2Y = \frac{1}{s^2} \tag{12}$$

$$Y(s^2 + 2) = \frac{1}{s^2} + 1 - s \tag{13}$$

$$Y = \frac{1}{s^2(s^2+2)} + \frac{1}{s^2+2} - \frac{s}{s^2+2} \tag{14}$$

The solution of the differential equation is $y(t)$, where

$$y(t) = \mathcal{L}^{-1}[Y(s)] = \mathcal{L}^{-1}\left[\frac{1}{s^2(s^2+2)} + \frac{1}{s^2+2} - \frac{s}{s^2+2}\right] \tag{15}$$

$$= \mathcal{L}^{-1}\left[\frac{c_1}{s^2} + \frac{c_2}{s} + \frac{c_3 s + c_4}{s^2+2}\right] + \mathcal{L}^{-1}\left[\frac{1}{s^2+2} - \frac{s}{s^2+2}\right] \tag{16}$$

Evaluating the partial-fraction coefficients and combining with the other terms produce

$$y(t) = \mathcal{L}^{-1}\left(\frac{0.5}{s^2} + \frac{0.5}{s^2 + 2} - \frac{s}{s^2 + 2}\right) \tag{17}$$

$$= 0.5t + 0.35 \sin t\sqrt{2} - \cos t\sqrt{2} \tag{18}$$

which is the required solution.

Index

Acceleration error, 130–132
Accelerometer, 22
Ackermann's formula, 539
Actuating signal, 125
Angle (phase) condition, in root locus, 149
Asymptotes, determination of in root locus, 158
Asymptotic frequency response, 227, 231, 232
Automatic control system, 4
Automatically guided vehicle, 20

Back-EMF, 30
Bandwidth, 55, 83, 364
Block diagram representation, 1
Block diagram simplification, 23
Bode diagrams, 289
  compound, 293
  deviation between approximate and accurate, 291–292, 300
  elemental, 297
  from root locus, 341
  plotting procedure, 303
  second-order element, 299–302
Breakaway point, on root locus, 159
Break frequency, on Bode diagram, 291

Canonical forms, 497
Cauchy's theorem, 241, 244
Characteristic equation, in root locus, 148
Closed loop, 5
Closed-loop frequency response, 361, 365–374
  from Bode diagram, 371
  from Nyquist diagram, 365
Closed-loop pole:
  off-axis location on root locus, 155
  sum of real parts on root locus, 179
Closed-loop time response from open-loop phase margin, 344
Closed-loop zero:
  effect of, 114
  occurrence of, 117
Compensation:
  phase lag, 431
  phase lead, 396
  lead-lag, 440
Compensator design, state-space systems, 555
Complex block diagrams, simplification, 4
Complex plane:
  desirable areas of, 205–207
  performance constraints, 203
Conditional stability, 272
  in Bode diagram, 319
Conformal transformations, 243
  computation of, 258
Constant magnitude (M-circle) locus, 367–369
Constant phase (N-circle) locus, 370–371
Contours, s-plane, 242
Control surfaces:
  aircraft, 16
  submarine, 20
Control systems, 1

Controllability, 519
  effect on compensator design, 554
  matrix, 522
Controller, 3
  state-space controller design, 531
Critical damping, 65, 76
Critical point, encirclement of, 247
Cut-off frequency, 363

Damped natural frequency, 64
Damping ratio, 6
  from poles, 79
  specification of, 206
DC (direct current) motor, 11
Decade, 290
Decibels, 83
Decoupled systems, 497
Delay time, 92
Design, in the complex plane, 199
Design constraints, multiple, 396
Differencing junction, 4
Differential equations, 19
  solving using Laplace transforms, 622
Disturbance:
  block diagram representation, 94
  effect on speed, 98
  external, 5
Disturbance rejection, open- and closed-loop, 96
Disturbance rejection ratio, 97
Dive-planes, submarine, 23
Dominant poles, 47

Eigenvalues and eigenvectors, 496
Electro-hydraulic servo valve, 16
Encirclement, of a pole or zero, 252
Engine speed control, 7–8, 134
Error, 3
Error constant, position, 128
  velocity, 129
  acceleration, 130
Errors, steady state, 125
  as a function of system type, 132
External, unwanted disturbances, 5

Feedback control, 1
Feedback path, 3, 7
Feedback transfer function, 3
Final value theorem, 622
First-order system:
  electrical, 38
  hydraulic, 38
  mechanical, 37
  with feedback, 43
Flapper-nozzle valve, 50
Flexible drive, in machine tool, 613

Forward path transfer function, 3
Frequency-domain specifications, 361
Frequency response, 223–224
  closed-loop, 363
  correlation with step response, 344–346
  test, 84

Gain, calculated from root locus, 152
Gain and phase margins, on Bode diagram, 326
Gain crossover frequency, 320
Gain margin:
  from Bode diagram, 321
  from Nichols chart, 375
  from Nyquist diagram, 275
Gearbox, 58–60
Governor, 8

Harmonic response:
  first-order system, 41
  second-order system, 76
Higher-order systems, reduction of, 111
  estimated step response, 112–114
Hydraulic actuator, 16
Hydraulic control system, 16–17

Impulse response:
  first-order system, 39
  steady-state error due to, 127
Instability, in ship steering systems, 514
Inverse Laplace transforms, 622

Jet pipe servo valve, 17

Kirchhoff's laws, 38

Lag compensation, 431–436
Laplace transforms, 620–623
  linear properties of, 620
Lead compensation, 396–412
Lead-lag compensation, 436
  process, 440
Lead-lag element, transfer function, 438
Liquid-level control systems, first order, 38–39
  flow control, 589–596
Load, reflected through gearbox, 60, 94

Magnitude equation, for root locus, 149–150
Mapping, conformal, 241
Maximum overshoot, 82
M-Circles, 368
Minimum phase systems, 248

Missile attitude control, 574–581, 606–612
Modified Nyquist contour, 248
Motor torque constant, 59
Motors:
  as second-order systems, 57
  field and armature control, 9
Multi-mode controller, 463
  design example, 577
Multiple inputs, 5

N-Circles, 368
Nichols chart, 372
Non-unity feedback systems, Nichols chart, 377
Nyquist contour, 246,
  modified, 248
Nyquist diagram:
  asymptotic values, table of, 231
  compound elements, 232
  computing magnitude and phase, 235
  elemental components, 226
  from transfer functions, 225
  second-order system, 228–230
Nyquist stability criterion, 241–252
  alternative approach to, 254
  comments on, 252

Observability, 519
  effect on compensator design, 554
  observability matrix, 522
Observer design, 550–568
Observer dynamics, 552
Open-loop control, 5
Open-loop transfer function, defined, 25
Operational amplifier, 13
Output equation, state-space form, 491
Overshoot:
  function of damping ratio, 66
  second-order system, 65
  third-order system, 115

Partial fraction coefficients, 8
PD (proportional plus derivative) control, 468
Peak time, 82
Percentage overshoot, 82–83
  effect of zero, 117
  third-order system, 115
Performance specification, in the time domain, 81–83
Phase crossover frequency, 320
Phase lag compensation, 431–440
  element transfer function, 431
  element Bode diagram, 432
  limits of applicability, 435
  process, 433
Phase lead compensation, 396
Phase lead compensation: *(cont.)*
  element transfer function, 399
  Bode diagram of, 401
  limits of applicability, 409
  process, 402
Phase margin:
  from Bode diagram, 321
  from Nichols chart, 375
  from Nyquist diagram, 277
  related to damping ratio, 346
PI (proportional plus integral) control, 466
Pick-off point, 4
PID (proportional plus integral plus derivative) controller tuning:
  trial and error method, 472
  Ziegler and Nichols first method, 473
  Ziegler and Nichols second method, 473, 481–483
Plant, 3
Pneumatic control system, 50–52
Poles:
  and eigenvalues, 500
  and zeros, 45
Pole placement:
  from canonical forms, 534
  using Ackermann's formula, 539
Position control, with velocity feedback, 101
Position error, 128
Position and stiffness control, 585
Potentiometer, 12
Proportional control, 464
Proportional plus derivative (PD) control, 468
Proportional plus integral (PI) control, 466
Proportional plus integral plus derivative (PID) control, 471
Pumped storage flow control system:
  multi-mode control, 589–593
  state-space controller design, 593–596

Ramp response:
  first-order system, 41
  second-order system, 75
Rate feedback, 99–102
Real axis pole, effect of, 112
Resistance in hydraulic circuits, 39
Resonance peak:
  function of damping ratio, 366
  gain selection from Nichols chart, 377
  gain selection from Nyquist diagram, 374
Resonance, 77
  frequency of, 78
Rise time, 81
Robotic hand design, 582–588
Root locus:
  angle of emergence, 175–176
  gain from, 178
  imaginary axis intersection, 160, 176

Root locus: *(cont.)*
- magnitude and phase equations, 148
- multi-loop systems, 199
- number of segments, 173
- real-axis breakaway point, 159, 164
- real-axis segments, 173
- rules for plotting, 173

Routh stability criterion, 145
Routh's array, 146

Second-order system:
- disturbance rejection, 93
- electrical, 63
- electro-mechanical, 57
- harmonic response, 76
- ramp response, 75
- step response, 64
- time constant, 82

Serial elements, reduction of, 3
Servomechanisms, 6
Settling time, 82
Shifting theorem, of Laplace transforms, 620
Ship steering control, 597–605
Signal-noise, 362
Spool valve, 21
Stability:
- of state-space system, 518
- Routh's criterion, 145

State equation, solution of, 515–518
- numerical solution, 515
- using state transition matrix, 516
- using Laplace transforms, 518

State-space observer design, 550
State-space, system description, 487
- from transfer functions, 492

State-space compensator design, 555
State variable feedback gains, 531
- from canonical form, 534
- from Ackermann's formula, 539
- selecting by comparing with characteristic equation, 532

State variables, invariability of eigenvalues, 496
Steering control, of ships, 512
Steady-state error, 76, 83, 125–132
- from Bode diagram, 323
- from root locus, 204

Step input, steady-state error due to, 128
Step response:
- first-order system, 40
- second-order system, 64

Step response: *(cont.)*
- third-order system, 116

Stiffness modification, 489, 503
Summing (differencing) junction, 2–3
System type, defined, 126
Systems, feedback control, 1

Tachometer feedback, 470
Tachometer, 20
- feedback, 99–102

Third-order systems, 112
- step response, 114

Time constant:
- first-order system, 37–38
- second-order systems, 82

Time differential, Laplace transform of, 621
Time-domain performance specifications, 81
Time response from Bode diagram, higher order system, 346
Torque sensor, 588
Transfer function:
- definition, 2
- dimensions of, 11
- first-order system, 39
- motor, 61
- second-order system, 62
- from state-space form, 495

Transient response, and system poles, 78, 80
- specification of, 81

Turbine, 8
Type, system, 126

Undamped natural frequency, 62
- from closed loop poles, 79

Unity feedback, 3
- equivalent of non-unity feedback system, 132

Unstable system, stabilization, 210

Velocity error, 129
Velocity feedback, 99, 470

Wave energy absorber, 569–573

Ziegler and Nichols PID controller tuning methods, 473